THE OPTICAL PRINCIPLES OF THE DIFFRACTION OF X-RAYS

THE OPTICAL PRINCIPLES OF THE DIFFRACTION OF X-RAYS

By

R. W. JAMES

D.Sc., F.R.S.

LATE PROFESSOR OF PHYSICS IN THE UNIVERSITY OF CAPE TOWN

OX BOW PRESS

Woodbridge, Connecticut

1982

Originally published as The Crystalline State — Vol. II in 1948
Editor: Sir Lawrence Bragg

This reprint published in agreement with Bell & Hyman Ltd, England

ISBN: 0-918024-23-4
Library of Congress Card Number: 82-80706
Printed in the United States of America

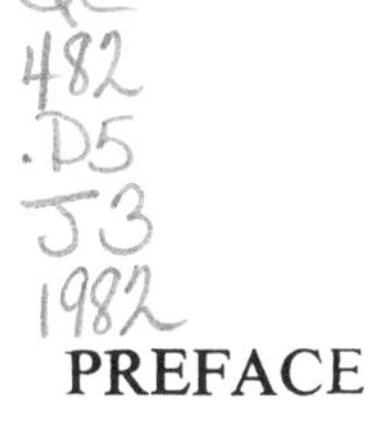

PREFACE

THE second volume of this work, the first of the more detailed volumes planned to follow the general account of the subject given in Vol. I, is intended to provide an outline of the general optical principles underlying the diffraction of X-rays by matter, which may serve as a foundation on which to base subsequent discussions of actual methods and results. Accordingly, reference has been made to experiments only where these illustrate principles, and all details of actual techniques, and of their application to specific problems, have been considered as lying beyond the scope of the volume.

Difficulties connected with the departure of the author in charge of the preparation of the volume to take up new work in South Africa, and, still more, the war, have greatly delayed its completion. Fortunately, the topics dealt with in the volume are such as do not change greatly with time ; but the work has become during the writing somewhat longer than was intended in the original plan, since it has become necessary to include much material not published when the volume was first contemplated. It is hoped that the work has on the whole benefited by this. For example, Chapter V, which deals with thermal motion in crystals, contained in the original draft, completed in 1936, a fairly detailed account of the Faxén-Waller theory, almost in the form in which it appears in the present volume; but the recent experimental work of Lonsdale and others on diffuse scattering has given point and completeness to the whole treatment.

The writer wishes particularly to express his indebtedness to Prof. H. Bethe, and Prof. R. Peierls, F.R.S., both of whom gave invaluable help during the writing of the more theoretical parts of Chapters III, IV, and V. Prof. Peierls in particular read and criticised the whole of Chapter V in its original form. Prof. D. R. Hartree, F.R.S., also gave great help in the course of many discussions during the writing of Chapter III, and read it and criticised it in its completed form. Thanks are due to many workers who have allowed the reproduction of diagrams and photographs from their works, and due acknowledgement is made in each case in the text.

THE UNIVERSITY OF CAPE TOWN
November, 1947

R. W. JAMES

NOTE TO REPRINT OF 1962

When this book was reprinted in 1954 a fifth appendix, on the use of Fourier transforms in diffraction problems, was included. In the further reprint of 1962 a sixth appendix has been added, giving a brief account of the dynamical theories of absorption and energy propagation in perfect crystals. This appendix follows the treatment of the dynamical theory given in Chapter VIII, in which absorption was not considered, and is to be read in conjunction with it.

R. W. J.

CONTENTS

CHAPTER I

THE GEOMETRICAL THEORY OF DIFFRACTION BY SPACE-LATTICES

PAGE

1. DIFFRACTION BY A SIMPLE SPACE-LATTICE
   - (*a*) The Laue theory 1
   - (*b*) The interference conditions in terms of the reciprocal lattice 6
   - (*c*) The width of a diffraction maximum and its dependence on crystal size. The interference function 8

2. SOME APPLICATIONS OF THE PRINCIPLES TO PARTICULAR PROBLEMS
   - (*a*) Introductory 14
   - (*b*) The method of the rotating or oscillating crystal . . 15
   - (*c*) The method of the powdered crystal 18
   - (*d*) The Laue method 19

3. THE EFFECT OF RANDOM IRREGULARITIES OF THE LATTICE ON THE DIFFRACTION PHENOMENA. THE TEMPERATURE EFFECT
   - (*a*) Introductory 20
   - (*b*) Diffraction by a lattice whose points are subject to random displacements 21
   - (*c*) The general scattering when the lattice-points vibrate independently 24

4. SOME EFFECTS DUE TO MULTIPLE REFLECTION 25

CHAPTER II

THE INTENSITY OF REFLECTION OF X-RAYS BY CRYSTALS

1. DIFFRACTION BY A COMPOSITE LATTICE
   - (*a*) The structure amplitude 27
   - (*b*) The atomic scattering factor 29
   - (*c*) The structure factor 31
   - (*d*) Phase changes on scattering and their effect on the structure factor 32
   - (*e*) The temperature effect and the structure factor . . . 34

2. REFLECTION FROM CRYSTALS WHEN DYNAMICAL INTERACTION MAY BE NEGLECTED
   - (*a*) Introductory 34
   - (*b*) The amplitude reflected by a plane sheet of atoms . . 35
   - (*c*) Reflection by a crystal slab of infinite lateral extent, composed of a number of planes 36
   - (*d*) Reflection by a small crystal of any form 39
   - (*e*) The integrated reflection 41
   - (*f*) The integrated reflection derived from an integration in the reciprocal-lattice space 41

PAGE

(*g*) The crystal mosaic 43
(*h*) Intensity formulae for mosaic crystals 44
(i) Reflection from a face 44
(ii) Reflection through a crystal plate 45
(*i*) Reflection from powdered crystals 46
(*j*) Secondary extinction 49
(*k*) Summary of formulae for intensity of reflection by mosaic crystals 50

3. Reflection from Perfect Crystals
(*a*) Introductory 52
(*b*) The refractive index of the crystal 53
(*c*) Deviations from Bragg's law 54
(*d*) Reflection from a set of crystal planes, taking into account multiple reflections 55
(*e*) Crystal with negligible absorption 57
(*f*) The integrated reflection from the face of a perfect crystal . 59
(*g*) Primary extinction 60
(*h*) The perfect crystal with absorption—Prins's method . . 62
(*i*) The composite lattice in the perfect crystal 65

4. Ewald's Dynamical Theory
(*a*) Introductory 66
(*b*) The electromagnetic field due to a dipole-wave in a crystal lattice 67
(*c*) The alternative methods of description of a dipole-wave in a lattice 70
(*d*) The relationship of the interference wave-field to the dipole-waves 72
(*e*) The dynamical problem 73
(*f*) The dispersion equation and the dispersion surface . . 75
(*g*) The dispersion surface for the case of two waves . . 76
(*h*) The transition from the case of a single wave to that of two waves 79
(*i*) The field due to a dipole-wave in a semi-infinite crystal The boundary waves 80
(*j*) The relation between the external and internal waves considered in terms of the boundary conditions . . . 83
(*k*) One primary and one secondary wave in the finite crystal . 84
(*l*) Unsymmetrical reflection 88
(*m*) Conclusion 90

CHAPTER III

THE ATOMIC SCATTERING FACTOR

1. Introductory 93

2. The Classical Calculation of Scattering Factors
(*a*) General considerations 94
(*b*) Scattering by atoms arranged in a crystal lattice . . . 95
(*c*) Scattering by atoms distributed at random 98
(*d*) Coherent and incoherent radiation 100
(*e*) Numerical calculation of scattering factors on classical basis 101

PAGE

3. THE TREATMENT OF SCATTERING BY THE METHODS OF WAVE-MECHANICS
(*a*) Introductory 101
(*b*) The wave-equation 103
(*c*) The calculation of the perturbed wave-function . . . 105
(*d*) The calculation of the current-density and of the corresponding scattered radiation 107
(*e*) The scattering of radiation of frequency large compared with the atomic absorption frequencies 108
(i) Coherent scattering 108
(ii) Incoherent scattering 110
(*f*) The dispersion terms 112
(*g*) The general form of the scattering factor for the many-electron atom 113
(*h*) Approximate forms of the expressions for coherent and total scattering for a many-electron atom. The exclusion principle 114

4. NUMERICAL METHODS
(*a*) The method of the self-consistent field 117
(*b*) The self-consistent field, including exchange—Fock's equation 121
(*c*) The Pauling-Sherman method 123
(*d*) The Thomas-Fermi method 123
(*e*) The calculation of *f*-curves from the charge distributions . 125
(*f*) Comparison of *f*-curves based on different atomic models. A discussion of the available tables of *f*-values . . . 129
(*g*) Limitations of applicability of the tabulated *f*-curves . . 132

CHAPTER IV

THE ANOMALOUS SCATTERING AND DISPERSION OF X-RAYS

1. THEORETICAL
(*a*) Introductory 135
(*b*) Scattering by a classical dipole-oscillator 135
(*c*) The relation between the scattering and the absorption and refractive index in a medium containing dipoles . . 137
(*d*) The breadth of the absorption lines 138
(*e*) Comparison between the classical and quantum-theory expressions for the scattering factor. The oscillator strength 140
(*f*) Extension to the many-electron atom 143
(*g*) The oscillator-density for the continuum of positive energy states 144
(*h*) The determination of the oscillator-density from the photo-electric absorption 146
(*i*) The imaginary part of the scattering factor 148
(*j*) The total scattering factor *f* 149
(*k*) The variation of the scattering factor with frequency . . 149
(*l*) The effect of the imaginary component of *f* on the amplitude and phase of the scattered wave 151
(*m*) The influence of damping of the oscillators on *f* . . 154

PAGE

(n) The calculation of oscillator strengths from the atomic wave-functions . . . 157
(o) The quadrupole terms in the scattering factor . . . 161
(p) The total scattering factor, including quadrupole terms . 164
(q) The dependence of δf on the angle of scattering . . 166

2. EXPERIMENTAL VERIFICATION OF THE DISPERSION FORMULAE
(a) Introductory . . . 167
(b) Measurement of the refractive index from the deviations from Bragg's law . . . 168
(c) Comparison with theory . . . 171
(d) Refractive indices by total reflection . . . 171
(e) Refractive indices from the deviations produced by prisms . 177
(f) Tests of the dispersion formulae from the measurement of atomic scattering factors . . . 180
(g) General conclusions . . . 188

CHAPTER V

THE INFLUENCE OF TEMPERATURE ON THE DIFFRACTION OF X-RAYS BY CRYSTALS

1. THEORETICAL
(a) Introductory . . . 193
(b) The normal co-ordinates of the lattice vibrations . . 195
(c) Plane waves in a lattice . . . 196
(d) Waves in a finite lattice . . . 197
(e) The calculation of $\overline{\{\kappa \mathbf{S} \cdot (\mathbf{u}_n - \mathbf{u}_{n'})\}^2}$ in terms of the normal co-ordinates . . . 198
(f) The relation between the mean-square amplitude and the temperature . . . 199
(g) The calculation of the average scattering from a lattice in thermal movement . . . 201
(h) The scattering function considered with the aid of the reciprocal lattice . . . 202
(i) The diffuse maxima considered as optical 'ghosts' . . 205
(j) The distribution of intensity in the diffuse maxima . . 207
(k) The surfaces of equal diffusion for a cubic crystal . . 210
(l) The shapes of the surfaces of equal diffusion in some particular cases . . . 213
(m) The numerical evaluation of the temperature factor e^{-2M} for a cubic crystal by Debye's method . . . 215
(n) Determination of the characteristic temperature . . 220
(o) The contribution of the different frequency ranges of the elastic spectrum to M . . . 223
(p) The limitations of the Debye-Waller formula for M . . 225
(q) Formal representation of the temperature factor for a complex crystal . . . 226
(r) The temperature factor for crystals not having cubic symmetry . . . 227
(s) Rotations of molecules and groups of atoms in crystals . 229

PAGE

2. EXPERIMENTAL TESTS OF THE FORMULAE FOR THE DEBYE TEMPERATURE FACTOR
   (a) Introductory 231
   (b) The temperature factor for sylvine (KCl) 233
   (c) The mean-square amplitude of the atomic vibrations . . 236
   (d) The relation between β and the elastic constants . . 238

3. THE EXPERIMENTAL STUDY OF DIFFUSE SCATTERING
   (a) Introductory 239
   (b) The study of diffuse patterns on Laue photographs . . 241
   (c) The radial streaks 244
   (d) The relation between the characteristic, Laue, and diffuse reflections 246
   (e) The qualitative effect of temperature on the diffuse maxima 247
   (f) The shape of the diffuse maxima and its dependence on crystal structure 248
   (g) The shapes of the diffuse maxima for cubic crystals . . 250
   (h) Primary and secondary extra reflections 252
   (i) Laval's investigation of diffuse scattering 253
   (j) The quantitative dependence of the intensity of the diffuse scattering on temperature 257
   (k) Jauncey's scattering formula and its investigation . . 258

CHAPTER VI

EXPERIMENTAL TESTS OF THE INTENSITY FORMULAE

1. PRIMARY AND SECONDARY EXTINCTION
   (a) Introductory 268
   (b) The correction for primary extinction 270
   (c) Secondary extinction 274
   (d) Correction for secondary extinction in a crystal composed of small independent blocks 276
   (e) Unsymmetrical reflection from a crystal mosaic . . 278
   (f) The integrated reflection from a mosaic of small blocks . 281
   (g) Reflection curves from actual crystals 282
   (h) Evidence of crystal imperfection from the absolute intensities of X-ray reflections 285
   (i) Experimental determination of secondary extinction . . 287
   (j) Other methods of estimating secondary extinction . . 292
   (k) Effects of simultaneous existence of primary and secondary extinction 294
   (l) Artificial alteration of the state of perfection of crystals . 295

2. QUANTITATIVE TEST OF THE FORMULA FOR THE MOSAIC CRYSTAL
   (a) Tests from comparison of observed and calculated structure factors 299

3. EXPERIMENTAL TESTS OF THE REFLECTION FORMULAE FOR PERFECT CRYSTALS
   (a) Introductory 304
   (b) The principle of the double-crystal spectrometer . . 306
   (c) The parallel arrangement for the double-crystal spectrometer 308

PAGE
(*d*) Absence of dispersion in the parallel arrangement . . . 312
(*e*) The reflection curve and the dispersion in the (1, 1) arrangement of the double-crystal spectrometer 314
(*f*) The effect of polarisation of the incident radiation on the double-reflection curve 315
(*g*) Integrated reflections from the double-crystal spectrometer . 316
(*h*) The relation between the widths of the double-reflection and single-reflection curves 317
(*i*) The diamond as a perfect crystal 318
(*j*) Tests of the Darwin-Prins formula 322

4. Intensity Measurements from Powdered Crystals
(*a*) Introductory 332
(*b*) The transmission method 333
(*c*) The reflection method and the focusing condition . . 334
(*d*) The mixed-powder method, and the substitution method . 337

CHAPTER VII

THE USE OF FOURIER SERIES IN CRYSTAL ANALYSIS

1. Triple, Double, and Single Series and their Application
(*a*) Introductory 342
(*b*) The derivation of the triple series 343
(*c*) The calculation of the coefficients of the series . . . 345
(*d*) The physical interpretation of the terms of the series . . 345
(*e*) Reality conditions for the density distribution . . . 346
(*f*) Case of crystal with symmetry centre 347
(*g*) The Fourier series and the reciprocal lattice 348
(*h*) Examples of the use of triple series 349
(*i*) The double series 351
(*j*) The projection of a slice of the crystal cell 355
(*k*) Methods of handling the double series 356
(*l*) Practical application of the double series when the projection is centrosymmetrical 360
(*m*) Some examples of the use of the double series when the projection has a centre of symmetry 361
(*n*) The use of the double series when the projection has no centre of symmetry 365
(*o*) One-dimensional series 369
(*p*) Fourier methods requiring no knowledge of the phases of the spectra: Patterson's series 371
(*q*) The properties of Patterson's series 373
(*r*) Methods of increasing the resolution of the Patterson distribution 376
(*s*) Harker's application of the Patterson method . . . 377
(*t*) An example of the use of the Patterson-Harker method . 380
(*u*) Summary 383

2. The Fourier Projection considered as an Optical Image
(*a*) Introductory 385
(*b*) The reciprocal relation between the net of a two-dimensional grating and the array of spectra produced by it . . . 386

PAGE

(*c*) The relationship between the two-dimensional grating and the projection of a crystal structure on a plane 388
(*d*) Abbe's theory of image formation 390
(*e*) The relationship between the spectra and the Fourier components of the image 392
(*f*) Diffraction effects in the Fourier projection . . . 396
(*g*) Some examples of false detail due to diffraction . . 401

3. Application of the Methods of Fourier Analysis to the Determination of the Electron Distribution in Atoms
(*a*) Fourier integrals 403
(*b*) The determination of the radial electron density, U(*r*), from the atomic scattering factor, *f* 404
(*c*) Series for the radial electron distribution 408

CHAPTER VIII

LAUE'S DEVELOPMENT OF THE DYNAMICAL THEORY—KOSSEL LINES

1. The Dynamical Theory in Terms of a Continuous Charge Distribution
(*a*) Introductory 413
(*b*) The fictitious dielectric constant and polarisation expressed as a Fourier series 414
(*c*) The fundamental equations of the wave-field . . . 416
(*d*) The case in which two waves only are appreciable . . 419
(*e*) Determination of the wave-points when the crystal is bounded by a plane surface and a primary wave enters it from outside 422
(*f*) The calculation of the ratio $\mathbf{D}_m/\mathbf{D}_o$ 425
(*g*) Total reflection at a crystal surface (Case II) . . . 425
(*h*) The application of the boundary conditions in Case II (reflection) 427
(*i*) The reflection coefficient in Case II 428
(*j*) Detailed discussion of the values of *x* for a non-absorbing crystal in Case II 429
(*k*) Extinction within the range of total reflection . . . 430
(*l*) The intensity of the wave-field in the crystal in the reflection problem (Case II), as a function of the angle of incidence . 431
(*m*) The transmission coefficient in Case I 435
(*n*) The intensity of the wave-field in the crystal in Case I . 436

2. Diffraction Phenomena when the Source of Radiation lies within the Crystal
(*a*) Introductory 438
(*b*) The reciprocity theorem in optics 439
(*c*) The application of the reciprocity theorem to the problem of the diffraction of radiation excited within the crystal . . 440
(*d*) Detailed consideration of the diffraction cones corresponding to the reflection problem (Case II) 441
(*e*) The total excess or defect of intensity in the Kossel lines . 442

PAGE

(*f*) The nature of the diffraction cones corresponding to Case I (transmission) in the dynamical problem 443
(*g*) Composite cones 445
(*h*) Experimental details 446
(*i*) The geometry of the cones 448
(*j*) Seemann's wide-angle diagrams 452
(*k*) Divergent-beam photography 452
(*l*) Laue's explanation of Kikuchi lines 455

CHAPTER IX

THE SCATTERING OF X-RAYS BY GASES, LIQUIDS AND AMORPHOUS SOLIDS

1. Introductory
(*a*) General survey of the subject 458
(*b*) Formulae for incoherent scattering 461
(*c*) General formulae for the diffraction of coherent radiation by assemblages of atoms 463

2. The Scattering of Coherent Radiation by Assemblages of Monatomic Molecules
(*a*) Scattering by a gas consisting of point atoms 465
(*b*) Scattering by a gas consisting of monatomic molecules of finite size 469
(*c*) Experimental tests of the scattering formula for monatomic gases 472
(*d*) The probability function $W(r)$ and the density function $\rho(r)$. 474
(*e*) Formulae for scattering by monatomic liquids . . . 475
(*f*) Examples of scattering by monatomic liquids . . . 478

3. The Scattering of Coherent Radiation by Assemblages of Polyatomic Molecules
(*a*) Introductory 480
(*b*) Scattering by gases consisting of polyatomic molecules . 482
(*c*) Scattering by gases consisting of diatomic molecules . . 483
(*d*) Scattering by molecules with more than two atoms . . 486
(*e*) Experimental methods 487
(*f*) The effect of temperature on scattering by gases . . 488
(*g*) Scattering of electrons by gas molecules 493
(*h*) Scattering by liquids with complex molecules . . . 494
(*i*) The structure of water 497

4. Analysis of Diffraction Patterns due to Powdered Crystals and Vitreous Solids
(*a*) Introductory 501
(*b*) Powdered crystals containing only one kind of atom . . 502
(*c*) Crystal powder or solid containing more than one kind of atom 506
(*d*) Some examples 504
(*e*) The analysis of vitreous solids 508

CHAPTER X

DIFFRACTION BY SMALL CRYSTALS AND ITS RELATIONSHIP TO DIFFRACTION BY AMORPHOUS MATERIAL

PAGE

1. A Comparison between Diffraction by Crystal Powders and by Polyatomic Molecules
   - (a) Introductory 513
   - (b) The interference function and the structure factor in terms of the reciprocal lattice 513
   - (c) Scattering by irregular assemblages of crystallites . . 517
   - (d) The linear crystallite 518
   - (e) The transition from the case of the linear crystallite to that of the diatomic molecule 520
   - (f) The interference function $I_0(\xi, \eta, \zeta)$ expressed as a Fourier series 522
   - (g) Scattering from crystallites in random orientation . . 525

2. The Effect of Crystal Size on the Widths of the Diffraction Spectra
   - (a) Introductory 528
   - (b) The breadths of Debye-Scherrer rings 529
   - (c) The generality of the expression for the integral line-breadth 535
   - (d) Calculation of the integral line-breadth in some special cases: the Scherrer constant 536
   - (e) The use of approximation functions: Laue's formula . 537
   - (f) Experimental application of the formulae for line-breadth . 540
   - (g) Some examples of the diffraction of electrons by very small single crystals 545
   - (h) The effect of the external form of the crystal on the interference function 548
   - (i) The crystal-form factor expressed as a surface integral . 550
   - (j) The crystal-form factor expressed as the sum of a set of integrals along the crystal edges 552
   - (k) The optical analogy to the interference function . . 555

3. The Effect of certain Types of Fault and Imperfection of the Lattice on the Diffraction Spectra
   - (a) Scattering by crystals with non-identical unit cells based on a regular lattice 555
   - (b) Diffraction by a distorted lattice 560
   - (c) Diffraction by lattices with periodic distortions . . . 563
   - (d) Experimental evidence of structures with periodic faults . 568

4. Diffraction by Fibrous Materials
   - (a) Introductory 571
   - (b) Diffraction by elongated crystallites with parallel arrangement and random orientation about the direction of their lengths 572
   - (c) Effects of lack of parallelism of the fibres 577
   - (d) The elongated two-dimensional structure with a number of rows 578
   - (e) The elongated three-dimensional crystallite . . . 580
   - (f) The external interference terms 582
   - (g) The structure factor and the atomic scattering factor . 585
   - (h) Summary of conclusions 588

APPENDICES

PAGE

I. SUMMARY OF VECTOR FORMULAE 592

II. THE RECIPROCAL LATTICE 598

III. TABLES FOR ESTIMATING THE CORRECTION TO BE APPLIED TO THE SCATTERING FACTOR ON ACCOUNT OF DISPERSION BY THE K ELECTRONS 608

IV. DERIVATION OF THE FOURIER INTEGRAL 611

V. THE APPLICATION OF THE FOURIER TRANSFORM TO DIFFRACTION PROBLEMS 613

VI. THE DYNAMICAL THEORY OF ABSORPTION AND PROPAGATION OF ENERGY IN PERFECT CRYSTALS 632

INDEX OF SUBJECTS 654

INDEX OF AUTHORS 662

CHAPTER I

THE GEOMETRICAL THEORY OF DIFFRACTION BY SPACE-LATTICES

1. Diffraction by a simple Space-lattice

(*a*) *The Laue theory:* In this section we shall consider the elementary theory of the diffraction of X-rays by a space-lattice consisting of identical scattering points. The method to be followed is in principle that used by Laue [1] in his original treatment of the subject, and the mathematical methods employed are of general application to a variety of diffraction problems.

Let us consider a small parallelepiped of the crystal lattice through which there sweeps a train of plane waves. Each lattice-point becomes the centre of a spherical scattered wavelet, and our problem is to find the combined effect of these wavelets at a point outside the crystal, and at a distance from it large in comparison with its linear dimensions. In the elementary discussion of the problem certain important simplifying assumptions are introduced. We suppose first of all that the primary X-ray beam travels through the crystal with the velocity of light. This is equivalent to neglecting the interaction between the incident wave-train and the scattered wavelets, an effect that would cause the wave-train to travel through the crystal with a modified phase velocity. We are in fact taking the refractive index of the crystal for X-rays to be unity. In so far as the positions of the diffraction maxima are concerned the errors introduced by doing so are small, and will be considered in due course. Secondly, we shall assume that each scattered wavelet travels through the crystal without being rescattered by other lattice-points. Such rescattering must occur, and its effects may become very important. For small crystal fragments, however, we make little error by neglecting it, and the conditions under which it is permissible to do so will be discussed in detail in a later section. Finally, we shall suppose that no absorption either of the incident or of the scattered radiation takes place in the piece of crystal considered, an assumption that again limits the validity of the calculation to small crystals, although it should be pointed out that a 'small' crystal in this connection may yet contain many thousands of millions of lattice-points.

We shall consider a small parallelepiped of a crystal lattice, having a unit cell defined by the translation vectors **a**, **b**, and **c**. The rows parallel to **a**, **b**, and **c** contain respectively N_1, N_2, and N_3 points, so that N, the total number of lattice-points in the block, is equal to $N_1N_2N_3$.

As a first step, we have to calculate the phase difference, at a point at which the intensity of the scattered radiation is to be found, between the wavelets scattered from any two of the lattice-points. The formulae are much simplified, and much labour is saved, by the use of vector notation.* Let A_1, A_2, fig. 1, be the two lattice-points, and let

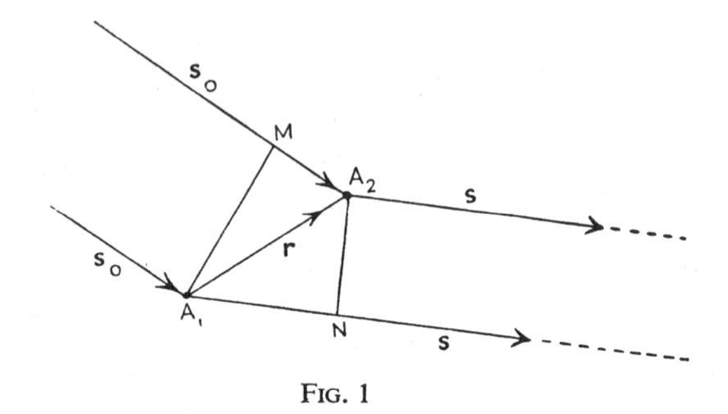

FIG. 1

their distance apart be defined both in magnitude and direction by the vector $\mathbf{r}$. Let the direction of the incident wave-normal be defined by the unit vector $\mathbf{s}_0$. A_1 and A_2 each scatter the incident wave, and we consider the phase difference between the two scattered wavelets arriving at a point Q, which we suppose to be at a distance R very large compared with the length of A_1A_2. The directions A_1Q, A_2Q may then be taken as parallel, and each may be defined by a unit vector $\mathbf{s}$. The path difference d between the two scattered wave-trains travelling in the direction $\mathbf{s}$ is $A_1N - A_2M$, the difference between the projections of A_1A_2 on the directions of scattering and incidence. In vector notation, this may be expressed by saying that d is the difference between the scalar products of $\mathbf{r}$ and $\mathbf{s}$ and of $\mathbf{r}$ and $\mathbf{s}_0$, or

$$d = \mathbf{r}\cdot\mathbf{s} - \mathbf{r}\cdot\mathbf{s}_0$$

$$= \mathbf{r}\cdot\mathbf{s} - \mathbf{s}_0. \qquad (1.1)$$

The vector $\mathbf{s} - \mathbf{s}_0$ has a simple geometrical interpretation. Let OA, OB, fig. 2, represent the unit vectors $\mathbf{s}_0$ and $\mathbf{s}$. Then AB is the vector $\mathbf{s} - \mathbf{s}_0$, and it is evident that it is in the direction of the normal to a plane that would reflect the direction of incidence $\mathbf{s}_0$ into the direction of scattering $\mathbf{s}$. It will be convenient to speak of this plane as the *reflecting plane*, even when no actual physical reflection is implied by the use of the term.

If θ is the glancing angle of incidence on, or of reflection from, this plane, the angle of scattering is 2θ, and we see from fig. 2, since $\mathbf{s}_0$ and $\mathbf{s}$ are of magnitude unity, that

$$|\mathbf{s} - \mathbf{s}_0| = 2 \sin \theta. \qquad (1.2)$$

* A summary of the principal vector formulae used in this volume will be found in Appendix I.

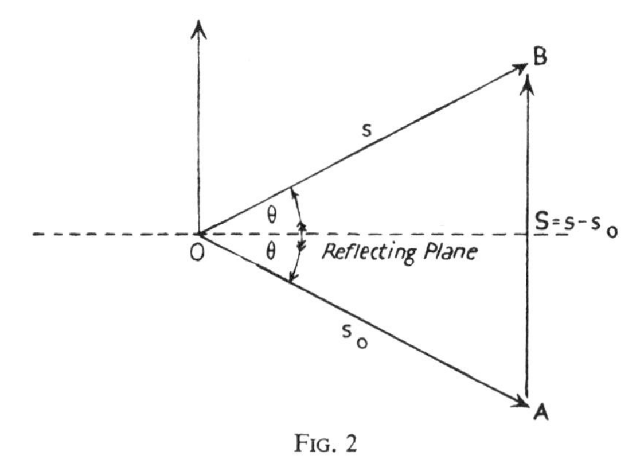

FIG. 2

The phase difference is obtained from the path difference by multiplication by $2\pi/\lambda$, λ being the wave-length of the radiation. Writing for brevity

$$2\pi/\lambda=\kappa, \tag{1.3}$$

and using (*1*.1), we obtain for the phase difference ϕ at the point Q, between the waves scattered by A_1 and A_2,

$$\phi=\kappa\mathbf{r}\cdot\mathbf{s}-\mathbf{s}_0, \tag{1.4}$$

which is equal to $4\pi(\sin\theta)/\lambda$ times the projection of A_1A_2 on the normal to the reflecting plane. In what follows we shall put

$$\mathbf{s}-\mathbf{s}_0=\mathbf{S}, \tag{1.5}$$

and equation (*1*.4) then becomes

$$\phi=\kappa\mathbf{r}\cdot\mathbf{S}. \tag{1.6}$$

Suppose now that the displacement due to the wave which is scattered from a point at the origin of the lattice and reaches Q at time t is $\Phi_0 e^{i\omega t}/R$. The quantity Φ_0 represents the amplitude of the scattered wave at unit distance from the scattering point in the direction **s**. It will in general be a function of the angle of scattering, depending on the nature of the scattering unit, and on the wave-length and state of polarisation of the incident radiation.

By equation (*1*.6), the displacement y at Q at time t due to the wave from the lattice-point at a vector distance **r** from the origin is given by

$$y=\frac{\Phi_0}{R}e^{i\omega t+i\kappa\mathbf{r}\cdot\mathbf{S}}, \tag{1.7}$$

since Q is supposed to be at a distance from the crystal so large in comparison with its linear dimensions that the distances of all the lattice-points from Q may be taken as constant, and equal to R, in so far as the effect of distance on amplitude is concerned. The resultant disturbance at Q due to the whole crystal may be obtained by summing (*1*.7) over all the lattice-points.

For a lattice-point,

$$\mathbf{r} = u\mathbf{a} + v\mathbf{b} + w\mathbf{c}, \tag{1.8}$$

u, v, w being integers whose values range from zero to $N_1 - 1$, $N_2 - 1$, $N_3 - 1$ respectively. If we put this value for $\mathbf{r}$ into (*1*.7), and sum over all the lattice-points, that is to say, over all values of u, v, w, we obtain for the total displacement Y due to the whole crystal

$$Y = \frac{\Phi_0}{R} e^{i\omega t} \sum_{u=0}^{u=N_1-1} e^{i\kappa u\mathbf{a}\cdot\mathbf{S}} \sum_{v=0}^{v=N_2-1} e^{i\kappa v\mathbf{b}\cdot\mathbf{S}} \sum_{w=0}^{w=N_3-1} \tag{1.9}$$

To calculate the intensity of the scattered radiation, we must multiply Y by its conjugate complex quantity Y*, obtained by changing i to $-i$ in (*1*.9). The three factors under the summation signs in (*1*.9) may be considered separately. Each is a geometrical progression, and the sum of the first is

$$\sum_a = \frac{1 - e^{i\kappa N_1 \mathbf{S}\cdot\mathbf{a}}}{1 - e^{i\kappa \mathbf{S}\cdot\mathbf{a}}}. \tag{1.10}$$

Multiplying (*1*.10) by its conjugate expression, we obtain

$$\left|\sum_a\right|^2 = \frac{1 - \cos(\kappa N_1 \mathbf{S}\cdot\mathbf{a})}{1 - \cos(\kappa \mathbf{S}\cdot\mathbf{a})} = \frac{\sin^2 N_1 \Psi_1}{\sin^2 \Psi_1}, \tag{1.11}$$

where

$$\Psi_1 = \tfrac{1}{2}\kappa \mathbf{S}\cdot\mathbf{a}. \tag{1.12}$$

Dealing with each of the factors in (*1*.9) in the same way, we have finally for the intensity at the point Q

$$|Y|^2 = J = \frac{|\Phi_0|^2}{R^2} \frac{\sin^2 N_1\Psi_1}{\sin^2\Psi_1} \frac{\sin^2 N_2\Psi_2}{\sin^2\Psi_2} \frac{\sin^2 N_3\Psi_3}{\sin^2\Psi_3}, \tag{1.13}$$

where

$$\begin{aligned} \Psi_1 &= \tfrac{1}{2}\kappa\mathbf{S}\cdot\mathbf{a} = \kappa a \sin\theta \cos\alpha \\ \Psi_2 &= \tfrac{1}{2}\kappa\mathbf{S}\cdot\mathbf{b} = \kappa b \sin\theta \cos\beta \\ \Psi_3 &= \tfrac{1}{2}\kappa\mathbf{S}\cdot\mathbf{c} = \kappa c \sin\theta \cos\gamma, \end{aligned} \tag{1.14}$$

α, β, γ being the angles which the vector **S** makes with the directions of **a**, **b**, and **c** respectively.

Each of the last three factors in (*1*.13) is of the type met with in the theory of the ordinary plane diffraction grating. Consider, for example, the first factor. It has maxima equal in value to N_1^2 when $\Psi_1 = h\pi$, h being integral or zero.

The value of the fraction changes from N_1^2 to zero when Ψ_1 changes from $h\pi$ to $(h \pm 1/N_1)\pi$. Between each pair of main maxima corresponding to successive integral values of h are $N_1 - 2$ subsidiary maxima. The two lateral subsidiary maxima on either side of a main maximum have values about five per cent. of that of the main maximum, but the successive maxima fall off very rapidly in intensity. If N_1 is large, and in practical cases it is usually at least several thousand, it is evident that the fraction has appreciable values only for values of Ψ_1 that are very nearly integral multiples of π. Since each angular factor in (*1*.13) has a negligible value unless the condition for it to be a maximum is very nearly fulfilled, it follows that if the total intensity J is to have an appreciable value all three factors must have a maximum value. This leads to the value

$$|Y|^2_{max} = J_{max} = \frac{|\Phi_0|^2}{R^2} N_1^2 N_2^2 N_3^2 = \frac{|\Phi_0|^2}{R^2} N^2, \qquad (1.15)$$

for the maxima of (*1*.13), when the conditions

$$\begin{aligned} \mathbf{S}\cdot\mathbf{a} &= 2a \sin\theta \cos\alpha = h\lambda \\ \mathbf{S}\cdot\mathbf{b} &= 2b \sin\theta \cos\beta = k\lambda \\ \mathbf{S}\cdot\mathbf{c} &= 2c \sin\theta \cos\gamma = l\lambda \end{aligned} \qquad (1.16)$$

are satisfied, h, k, and l being integers.

The conditions (*1*.16) have a very simple geometrical interpretation. $\cos\alpha$, $\cos\beta$ and $\cos\gamma$ are the direction cosines of the vector S, or, in other words, of the normal to the reflecting plane, referred to the three crystal axes a, b, and c. Equations (*1*.16) state that these direction cosines are proportional to h/a, k/b, and l/c respectively. The successive lattice-planes of the crystal having indices (h, k, l) intersect the axes a, b, c at intervals a/h, b/k, c/l respectively, and the direction cosines of the normal to these planes are therefore also proportional to the same quantities. The planes (h, k, l) must thus be parallel to the reflecting plane, and it will be seen that conditions (*1*.16) imply that no appreciable scattered beam will be produced by the crystal lattice unless the directions of incidence and scattering are so related that one may be considered as derived from the other by reflection in one of the lattice-planes of the crystal. The conditions imposed by (*1*.16) are, however, more stringent than this; for, if $d(hkl)$ is the spacing of the lattice-planes (h, k, l),

$$d(hkl) = \frac{a}{h}\cos\alpha = \frac{b}{k}\cos\beta = \frac{c}{l}\cos\gamma,$$

since S is in the direction of the normal to the plane (h, k, l). Inserting this in equations (*1*.16), we obtain the condition

$$2d(hkl)\sin\theta = \lambda. \qquad (1.17)$$

We have thus deduced in a formal way the well-known 'reflection conditions' for the diffraction of X-rays by a crystal lattice. The diffracted beam corresponding to an interference maximum must always be in such a direction that it can be considered as produced by the reflection of the incident beam in one of the sets of lattice-planes, while, at the same time, the relation (*1*.17) must hold between the spacing of the lattice-planes in question and the wave-length of the radiation. The term 'spacing' here has to be interpreted in the following extended sense. If h, k, l have a common factor n, the corresponding maximum may be considered either as an nth order reflection from lattice-planes with the true lattice-spacing appropriate to these planes, or as a first order reflection from a set of planes parallel to the true lattice-planes, but with a spacing $d(hkl)$ that is $1/n$ of the true spacing. The simple reflection form of the interference conditions was first pointed out by W. L. Bragg[2], although it is, of course, implicit in the more complicated expression derived by Laue, and is a consequence of the three interference conditions that must be fulfilled simultaneously if a maximum is to be produced with a three-dimensional diffraction grating.

(*b*) *The interference conditions in terms of the reciprocal lattice:* The interference conditions take a very elegant geometrical form when expressed in terms not of the space-lattice itself but of a lattice that is reciprocal to it in the sense now to be defined. Let **a**, **b**, **c**, be the primitive translations of the crystal lattice, and **a***, **b***, **c***, those of the reciprocal lattice. Then **a*** is perpendicular to both **b** and **c**, **b*** to both **c** and **a**, and **c*** to both **a** and **b**. In vector notation, these relations are expressed by the equations

$$\mathbf{a}^*\cdot\mathbf{b} = \mathbf{a}^*\cdot\mathbf{c} = \mathbf{b}^*\cdot\mathbf{c} = \mathbf{b}^*\cdot\mathbf{a} = \mathbf{c}^*\cdot\mathbf{a} = \mathbf{c}^*\cdot\mathbf{b} = 0, \qquad (1.18)$$

for the vanishing of the scalar product of two vectors implies that they are mutually perpendicular.

The magnitudes of the reciprocal vectors are fixed by the further equations

$$\mathbf{a}^*\cdot\mathbf{a} = \mathbf{b}^*\cdot\mathbf{b} = \mathbf{c}^*\cdot\mathbf{c} = 1, \qquad (1.19)$$

which, when taken in conjunction with (*1*.18), mean that $|\mathbf{a}^*|$ is the reciprocal of the spacing of the a planes of the crystal lattice, $|\mathbf{b}^*|$ the reciprocal of the spacing of the b planes, and $|\mathbf{c}^*|$ the reciprocal of the spacing of the c planes. The idea of the reciprocal lattice is used throughout this volume, and a simple mathematical treatment of its more important properties will be found in Appendix II. We shall here state without proof two properties of the lattice that make it of great value in treating problems of diffraction by space-lattices.

(i) The vector $\mathbf{r}^*(hkl)$ to the point (h, k, l) of the reciprocal lattice is normal to the planes (hkl) of the crystal lattice.

(ii) The magnitude of the vector $\mathbf{r}^*(hkl)$ is the reciprocal of the spacing of the planes (hkl) of the crystal lattice.

We must here interpret the term 'spacing' in the general sense considered in the last paragraph, corresponding to the reflection condition (*1*.17). If h, k, and l have no common factor, the reciprocal-lattice point (h, k, l) is that point on the row through the origin of which it is a member that lies nearest to the origin, and $|\mathbf{r}^*(hkl)|$ is the reciprocal of the true spacing of the planes (hkl). The next point on the same row is the point $(2h, 2k, 2l)$, and the magnitude of the corresponding vector is the reciprocal of half the true spacing, and so of the value of d in (*1*.17) corresponding to the second order spectrum from the planes (hkl) ; and in general, the nth point from the origin in a given row corresponds to the nth-order reflection from the crystal-planes to which the direction of the row is perpendicular. Every reciprocal-lattice point corresponds in this sense to a possible spectrum from the crystal lattice, and the whole array of spectra that could in any circumstances be given by the crystal is represented by the points of the reciprocal lattice.

It is now possible to express the reflection condition (*1*.17) in terms of the reciprocal lattice. In fig. 3, let $\overrightarrow{PO}$ be the vector $\mathbf{s}_0/\lambda$, that is to say, a vector of length $1/\lambda$ in the direction of the wave-normal of the incident radiation. Similarly, let $\overrightarrow{PQ}$ be $\mathbf{s}/\lambda$, the corresponding vector in the direction in which the scattering is considered, and suppose that

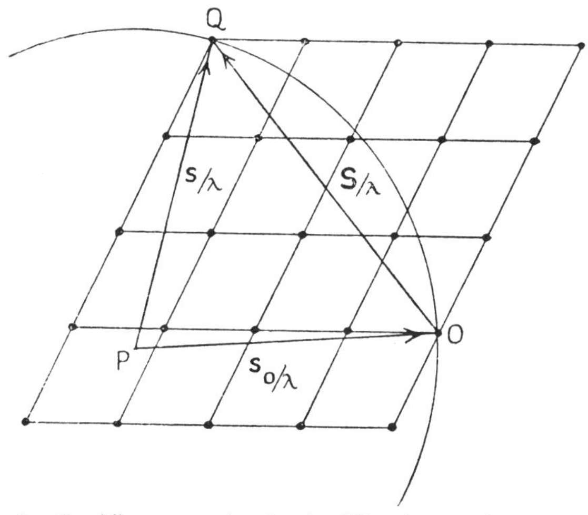

FIG. 3. Ewald's construction for the diffraction maxima, using the reciprocal lattice and the sphere of reflection

$\overrightarrow{PQ}$ is in the direction of a diffraction maximum. Then we have seen that $\overrightarrow{OQ}$, which is the direction of the vector $\mathbf{S}$, must be normal to one of the lattice-planes, say to (hkl). The length of OQ is $(2/\lambda)\sin\theta$, which, when the diffraction conditions are fulfilled, is equal to $1/d(hkl)$, by (*1*.17). $\overrightarrow{OQ}$ is therefore in the direction of a vector in the reciprocal lattice, and is equal to it in magnitude. Thus, if the vector $\overrightarrow{PO}$ is drawn so that O lies at the origin of the reciprocal lattice, Q must lie at another point (h, k, l) of this lattice. We have therefore the following construction. Let a vector $\overrightarrow{PO}$ of length $1/\lambda$ be drawn in the direction of the incident wave-normal, O being the origin of the reciprocal lattice. With P as centre and radius $1/\lambda$ describe a sphere, which, of course, passes through O. Then the interference conditions (*1*.16) are equivalent to the statement that no diffracted beam can arise unless another of the reciprocal-lattice points, say Q, lies on the sphere. If it does so, then PQ is the direction of the wave-normal of a possible diffracted beam. The sphere used in this construction is called the 'sphere of reflection' (Ewald's *Ausbreitungskugel*). The use of the reciprocal-lattice construction, in which the consideration of an array of planes is replaced by that of an array of points, is a great simplification in discussing the geometrical problems met with in the application of X-ray diffraction; indeed, many which are easily treated in this way are most intractable by the ordinary methods, and it has become an essential part of the crystallographer's mathematical equipment.

The idea of a reciprocal vector system was introduced by Willard Gibbs [3] about 1884. Ewald [4], in 1913, used the reciprocal lattice to represent interference by a space-lattice for the special case of orthogonal crystals. Later in the same year, M. v. Laue [5], independently of the work of Gibbs, used reciprocal vectors in discussing interference by space-lattices of any form, and laid the foundation of the general treatment of interference phenomena by this method, while in an important paper published in 1921, Ewald [6] stressed the importance of the reciprocal lattice as an alternative method of considering the geometry of space-lattices, apart from any question of interference.*

(c) *The width of a diffraction maximum and its dependence on crystal size. The interference function :* For a crystal of finite size, appreciable scattering will occur over a certain range of angles in the neighbourhood of the exact direction corresponding to a diffraction maximum, and this range will be greater the smaller the crystal fragment. Equation (*1*.13) gives the intensity diffracted by a small block of crystal as a function of S, the vector that is the difference of the unit vectors $\mathbf{s}$ and $\mathbf{s}_0$ in the directions of diffraction and incidence, which is therefore uniquely defined by the conditions of diffraction.

In order to deal with the problem of the relation between the width of a diffraction maximum and the size of the crystal, and for many other purposes, it is convenient to express **S** in terms of the reciprocal-lattice vectors. Let us write

$$\mathbf{S}/\lambda = \xi\mathbf{a}^* + \eta\mathbf{b}^* + \zeta\mathbf{c}^*, \tag{1.20}$$

so that ξ, η, ζ are the components of $\mathbf{S}/\lambda$ parallel to the three reciprocal axes, expressed as multiples of the reciprocal translations $\mathbf{a}^*$, $\mathbf{b}^*$, and $\mathbf{c}^*$.

The quantity Ψ_1 of equation (*1*.13) is then given by

$$\Psi_1 = \pi\mathbf{a}\cdot(\xi\mathbf{a}^* + \eta\mathbf{b}^* + \zeta\mathbf{c}^*) = \pi\xi,$$

* See also Ewald, *Zeit. f. Krist*, **93**, 396 (1936), for some account of the history of the use of the reciprocal lattice.

since $$\mathbf{a}\cdot\mathbf{a}^* = 1,\ \mathbf{a}\cdot\mathbf{b}^* = \mathbf{a}\cdot\mathbf{c}^* = 0,$$

by the properties of the reciprocal vectors; and similarly $\Psi_2 = \pi\eta$, $\Psi_3 = \pi\zeta$, so that equation (*1*.13) becomes

$$J = \frac{|\Phi_0|^2}{R^2} I_0(\xi, \eta, \zeta), \qquad (1.21)$$

where

$$I_0(\xi, \eta, \zeta) = \frac{\sin^2(\pi N_1 \xi)}{\sin^2(\pi\xi)} \cdot \frac{\sin^2(\pi N_2 \eta)}{\sin^2(\pi\eta)} \cdot \frac{\sin^2(\pi N_3 \zeta)}{\sin^2(\pi\zeta)}. \qquad (1.21a)$$

Equation (*1*.21) is the product of two factors, the first of which depends on the scattering power of an individual lattice-unit under the conditions considered, and the second of which, $I_0(\xi, \eta, \zeta)$, depends upon the regular arrangement of these units to form the lattice. The function I_0 is called by Laue the *interference function.* It may be thought of as representing a periodic distribution in the reciprocal-lattice space having the following significance. For given directions of incidence and scattering, the vector S/λ is fixed, and the value of $I_0(\xi, \eta, \zeta)$ at the point in the reciprocal-lattice space lying at the extremity of the vector S/λ, drawn from the origin, gives the scattered intensity under the conditions considered, expressed in terms of the intensity scattered under the same conditions by a single unit of the structure. The method of considering diffraction problems in terms of the distribution of the appropriate interference function in the reciprocal-lattice space is one of great power, and we shall make frequent use of it in this book.

The principal maxima of $I_0(\xi, \eta, \zeta)$ evidently occur when

$$\xi = h,\ \eta = k,\ \zeta = l, \qquad (1.22)$$

h, k, and l being any integers, and these conditions are, of course, equivalent to the conditions (*1*.16) for the formation of the spectrum *hkl*. By (*1*.20), we see that they are also equivalent to the statement that S/λ is a vector from the origin to the reciprocal-lattice point (h, k, l), which again we saw in the last paragraph to be the condition for the formation of the spectrum *hkl*.

The maxima of the three factors of (*1*.21a) are not indefinitely sharp, and appreciable diffraction will occur when the extremity of S/λ lies within a small region surrounding the point (h, k, l). We may easily see what are the boundaries of this region. If the diffracted intensity is to be appreciable ξ, η, and ζ must lie respectively within the limits $h \pm 1/N_1$, $k \pm 1/N_2$, and $l \pm 1/N_3$. Suppose a small parallelepiped to be described having the reciprocal-lattice point $Q\ (h, k, l)$ at its centre, and its edges parallel to the reciprocal axes, with lengths $2a^*/N_1$, $2b^*/N_2$, and $2c^*/N_3$. The diffracted intensity will be appreciable if the conditions of incidence and diffraction are such that the extremity of the vector S/λ lies within this parallelepiped. In considering diffraction from a crystal of finite size by Ewald's construction, we must therefore replace each point of the reciprocal lattice by a small volume—in the case we are considering, the small parallelepiped just discussed. If the sphere of reflection passes through this volume there will be a scattered beam, whose intensity will fall to zero for the directions corresponding to the intersections of the sphere with the boundaries of the parallelepiped, and will be greatest when the sphere passes near its mid-point.

In fig. 4, the points represent the $a^* b^*$ plane of a reciprocal net, and the parallelograms surrounding the points are the sections by this plane

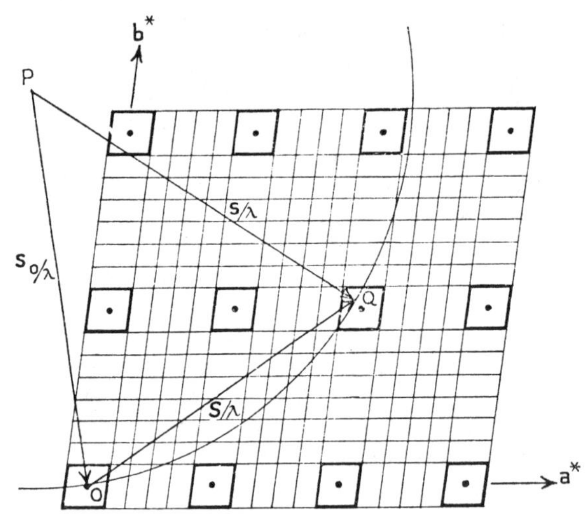

FIG. 4. Principal and subsidiary maxima in the interference function

of the parallelepiped described above, drawn for the case of a crystal block having 6 and 8 lattice-points parallel to the a and b edges respectively. To avoid confusion, no attempt has been made to represent the lattice in three dimensions. The trace of the sphere of reflection is also shown in the diagram for the direction of incidence determined by the unit vector $\mathbf{s}_0$, and the vector $\mathbf{S}/\lambda$ corresponding to the direction of diffraction defined by the unit vector $\mathbf{s}$. The sphere of reflection does not pass exactly through the lattice-point Q, but through the parallelepiped surrounding it, and there would be appreciable diffracted intensity for the direction of $\mathbf{s}$ shown in the diagram. The parallelepipeds are here relatively large, owing to the very small size assumed for the crystal block.

It is easy to extend this scheme to include the subsidiary maxima. The value of the interference function is zero anywhere on the planes $\xi = p_1/N_1$, $\eta = p_2/N_2$, $\zeta = p_3/N_3$, p_1, p_2, p_3 being any integers positive or negative except the integers hN_1, kN_2, lN_3 respectively. The traces of these planes of zero value are shown in the reciprocal-lattice plane $a^* b^*$ in fig. 4. They divide the whole reciprocal-lattice space into small cells of four types. The cells of the first type, those for which $p_1 = hN_1 \pm 1$, $p_2 = kN_2 \pm 1$, $p_3 = lN_3 \pm 1$, we have already considered. They are those at the mid-points of which lie the main intensity maxima, for which all three factors of the interference function have principal maxima. Cells with a volume half that of those containing the main maxima lie with their mid-points along the lines that join the reciprocal-lattice points and are parallel to $\mathbf{a}^*$, $\mathbf{b}^*$, and $\mathbf{c}^*$. On these lines, and nearly but not exactly at the mid-points of the small cells, lie the strongest subsidiary maxima, for which two of the factors of the interference function have principal maxima and the third a subsidiary maximum. Cells having a volume a quarter that of the cells containing the main maxima have their mid-points on the three sets of planes $\xi = h$, $\eta = k$, $\zeta = l$, that is to say, on the faces of the reciprocal-lattice cells. Subsidiary maxima for which only one of the three factors has a principal maximum lie nearly at the mid-points of these cells. The remaining small cells into which the planes of zero intensity divide the reciprocal-lattice space have one eighth of the volume of the cells containing the main maxima, and the corresponding subsidiary maxima, which lie near their mid-points, are such that all three of the factors of the interference function themselves have subsidiary maxima. If the crystal lattice contains more than a very small number of cells, the intensities of the subsidiary maxima of the interference function will be entirely negligible in comparison with those of the main maxima, except for those immediately surrounding the latter. As we have seen, the strongest subsidiary maxima lie along the edges of the reciprocal-lattice cells, where two of the factors of the interference function have their maximum value. The maximum regions of the interference function thus extend in directions perpendicular to the faces of the parallelepipedal crystal block. We shall see in Chapter X, § 2(*i*), that this is a particular case of a general result for any crystal bounded by plane faces.

The distribution of the interference function in the space of the reciprocal lattice is the three dimensional analogue of the diffraction pattern produced by monochromatic light by a two-dimensional net. Let a beam of monochromatic light fall normally on a plane two-dimensional grating, and let the radiation diffracted by the grating then fall on a lens. A diffraction pattern is formed in the focal plane of the lens, and it is easy to show that the centres of the diffraction spots of this pattern lie at the points of a net which is the reciprocal net of the original grating.* The breadth of a diffraction spot depends on the number of effective elements in the original grating, and we have here an actual distribution of intensity about the points of the reciprocal net which depends on the number of the grating elements, and which

* The actual diffraction pattern is a slightly deformed reciprocal net. If r is the distance of any spot from the centre of the diffraction pattern, and R is its distance from the optical centre of the lens, which is, of course, equal to f the focal length of the lens for the central spot, r must be reduced in the ratio f/R in order to give the true reciprocal net. The distortion is very small for small angles of diffraction.

corresponds in two dimensions to the density distribution $I_0(\xi, \eta, \zeta)$ in the space of the three-dimensional reciprocal lattice. In the three-dimensional case, however, all the maxima are not produced at once, as they are with a two-dimensional net, but only those are possible through which the sphere of reflection passes corresponding to the particular wave-length and angle of incidence used. We shall discuss the relationship between the crystal and the optical grating at greater length in Chapter VII.

Using the geometrical methods described in this section we may easily see in a general way how the intensity of scattering will vary with the angle of scattering for a given angle of incidence; or how the total scattering, or the scattering in a given direction, or through a given angle, will depend on the direction of incidence of the radiation relative to the crystal axes. For actual crystal fragments of irregular shape the problem cannot be solved in the same detail as for the simple shape we have considered; but the same general principles will apply, and the reciprocal-lattice point is to be replaced by a small volume, of the order of the reciprocal of the volume of the crystal fragment. Since the linear dimensions of this volume are of the order of the reciprocal of the linear dimensions of the crystal, while the radius of the sphere of reflection is the reciprocal of the wave-length, it is evident from fig. 4 that the range of angles in the region of a maximum over which appreciable reflection will occur is of the order of the wave-length of the radiation used, divided by the linear dimensions of the crystal fragment, an angle which is already extremely small if the crystal has as many as a thousand points in each row. A more detailed discussion of the case in which the crystal fragment is bounded by plane faces but is not parallelepipedal will be found in Chapter X, § 2(*i*).

We shall conclude this discussion by considering an actual case, in order to give some idea of the numerical magnitudes involved in a typical problem. Potassium chloride, owing to the similarity in scattering power of potassium and chlorine, behaves very nearly as a simple cubic lattice, and we shall here assume it to be such a lattice with a cell-edge of 3·14A. For a cubic cell, the magnitudes of the reciprocal axes are simply the reciprocals of those of the crystal axes, and are in the same directions. For potassium chloride, we have, therefore, $a^* = b^* = c^* = 0{\cdot}318\text{A}^{-1}$. If we use molybdenum K$\alpha$ radiation, $\lambda = 0{\cdot}71\text{A}$, the radius of the sphere of reflection, $1/\lambda$, is $1{\cdot}40\text{A}^{1}$.

In fig. 5, a single net of the reciprocal lattice, that for which $h = 0$, is shown, and the sphere of reflection for the spectrum 001* referred to the simple cubic lattice is drawn in to scale.

The radius of this sphere is a small multiple of the spacing of the reciprocal lattice, and this is typical for X-rays. For electron waves, such as those used in experiments in electron diffraction, the wave-

* This spectrum would be 002 referred to the true unit cell of the KCl lattice, which is not simple cubic as we have supposed here.

length is about ten times smaller, and the radius of the sphere of reflection ten times greater, than in the case of X-rays. On the other hand, with visible light, λ is about ten thousand times greater than for X-rays; the corresponding sphere of reflection would have a radius far

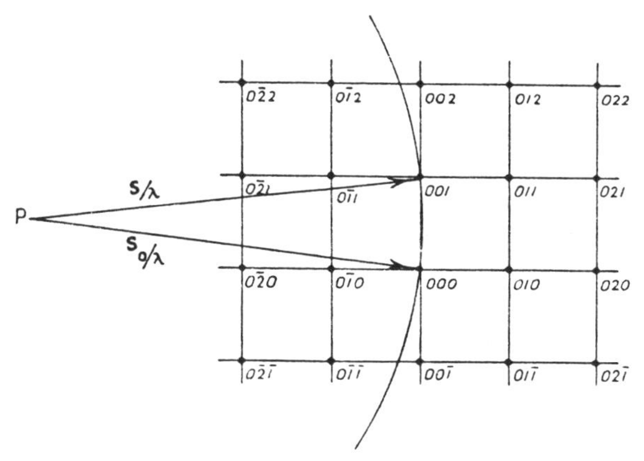

FIG. 5. Reciprocal net and sphere of reflection for KCl considered as a simple cubic lattice. Drawn to scale for $\lambda = 0{\cdot}71$A

smaller than the spacing of the reciprocal lattice, and the Ewald construction for a diffracted wave would be impossible.

If we were dealing with a small cube of the crystal, each edge of which had a length 1000 times the lattice-spacing, the small cube around the reciprocal-lattice point, which defines the breadth of the spectra, would have edges 1/500 of those of the reciprocal cell, and would be too small to draw to scale in the diagram. The angular range on either side of the maximum within which all the scattering would be confined would be of the order of $\frac{0{\cdot}318}{1000} \div 1{\cdot}40 = 2{\cdot}3 \times 10^{-4}$ radians or 46 seconds of arc. It is evident, therefore, that unless the crystal fragment is exceedingly small, we can for nearly all purposes neglect the angular width of the diffracted beam, which will generally be less than the unavoidable divergence of the incident beam, and in calculating the geometrical relations of X-ray diffraction patterns treat the reciprocal-lattice points as mathematical points. In dealing with diffraction by colloidal particles, and by the very smallest particles used in the powder method, we may not make this assumption, and we shall return to the detailed discussion of this matter in Chapter X.

Before leaving the problem of diffraction by a finite lattice, we should remind ourselves that the simplifying assumptions of unit refractive index, and absence of re-scattering, which we made at the outset, underlie the whole of the discussion. If these restrictions are removed, it is found that some latitude in the angle of scattering is allowed, even for a large crystal. The lattice-points must lie near, but not necessarily exactly on, the sphere of reflection for diffraction to occur. We shall return to this when discussing Ewald's dynamical theory of crystal optics, in Chapters II and VIII.

2. Some Applications of the Principles to Particular Problems

(*a*) *Introductory:* We shall now consider briefly the application of the principles we have deduced to a few of the chief methods used to observe the diffraction of X-rays by crystals.

It is evident from the construction of fig. 3 that, for any arbitrary angle of incidence and wave-length, that is to say, for any arbitrary direction and length of the vector $\overrightarrow{\mathrm{PO}}$ with respect to the crystal lattice, no point of the reciprocal lattice will in general lie on the sphere of reflection, and no diffracted beam can be produced. By rotating the crystal, and therefore the reciprocal lattice, or by suitable adjustment of the direction of incidence of the radiation on the crystal, points may be brought on to the sphere, and diffraction can occur. This principle underlies all those methods in which monochromatic radiation is used; viz.: the X-ray-spectrometer method, in which the spectra are examined one by one, by suitably adjusting the angle of incidence; the rotating- and oscillating-crystal methods, in which as the crystal rotates the reciprocal-lattice points pass in turn through the sphere of reflection, the spectra which flash out as they do so being received on a photographic plate; and the powdered-crystal method, in which a random orientation of crystal fragments produces in effect a rotation of the average crystal in all directions. In another method, which was actually that first used by Friedrich, Knipping and Laue in their pioneer experiments, the direction of incidence is kept fixed, and the fulfilment of the diffraction conditions is attained by using 'white' radiation giving a continuous range of wave-lengths. We shall consider these two types of method in turn.

In the methods using monochromatic radiation we may use directly the Ewald construction described in I, § 1 (*b*), in which a definite wave-length was assumed throughout. In fig. 6 let O be the origin of the reciprocal lattice of the crystal, and let the sphere of reflection be drawn corresponding to some direction of incidence PO. Let the crystal perform a complete rotation about an axis AB which passes through O. Then certain points of the reciprocal lattice pass through the sphere of reflection, and would give rise to spectra during the rotation. These points are evidently those which lie inside the tore which would be obtained by allowing the sphere of reflection to make a complete rotation about the same axis AB with respect to the lattice. It is evident that all possible tores of this kind, corresponding to all possible axes of rotation and angles of incidence, are contained within a sphere of centre O and radius $2/\lambda$, the diameter of the sphere of reflection. This sphere may be called the *limiting sphere*. It contains all the reciprocal-lattice points corresponding to spectra which can, in any

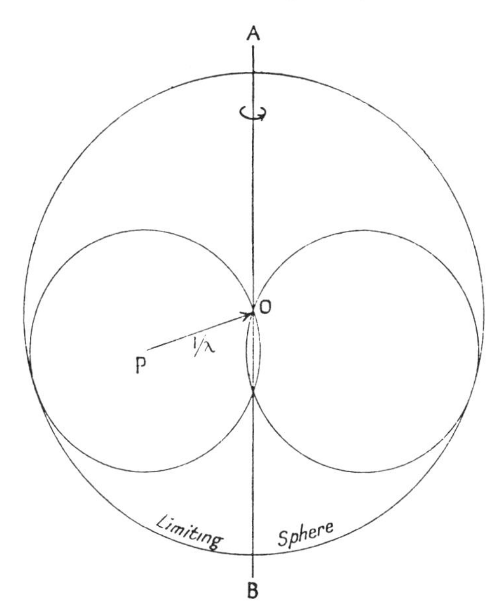

FIG. 6. The tore of reflection and the limiting sphere

circumstances, be obtained from the crystal, using radiation of wave-length λ.

For the values of λ used in most experiments the limiting sphere will contain a considerable number of reciprocal-lattice points, which we may take as being proportional to its volume. The number of diffraction spectra obtainable from a given crystal with wave-length λ is therefore proportional to $1/\lambda^3$.

(*b*) *The method of the rotating or oscillating crystal:* In this method the crystal is rotated, either completely or through a certain angular range, about an axis which is usually perpendicular to the direction of the incident X-ray beam, and parallel to an important zone-axis of the crystal lattice, say, for example, the c axis. The a^* and b^* translations of the reciprocal lattice, and the net-planes defined by them, are then perpendicular to the axis of rotation. These net-planes, whose distance apart is $1/c$, cut the sphere of reflection in a series of circles of latitude, the corresponding polar axis of the sphere being parallel to the axis of rotation. As the crystal rotates, any reciprocal-lattice point that passes through the sphere must do so somewhere on one of these circles of latitude. All points having the same value of the l-index pass through the same circle; for the reciprocal-lattice planes we are here considering are those for each of which l=constant. The points for which $l=0$ pass through the equator, those for which $l=\pm 1$ through the first circles above and below the equator, and so on. The direction of any diffracted beam which arises during the rotation is thus always parallel to one of the generating lines of the set of cones which have the centre of the sphere as apex and pass through the circles of latitude. A part of the reciprocal lattice and the sphere of reflection are shown in fig. 7.

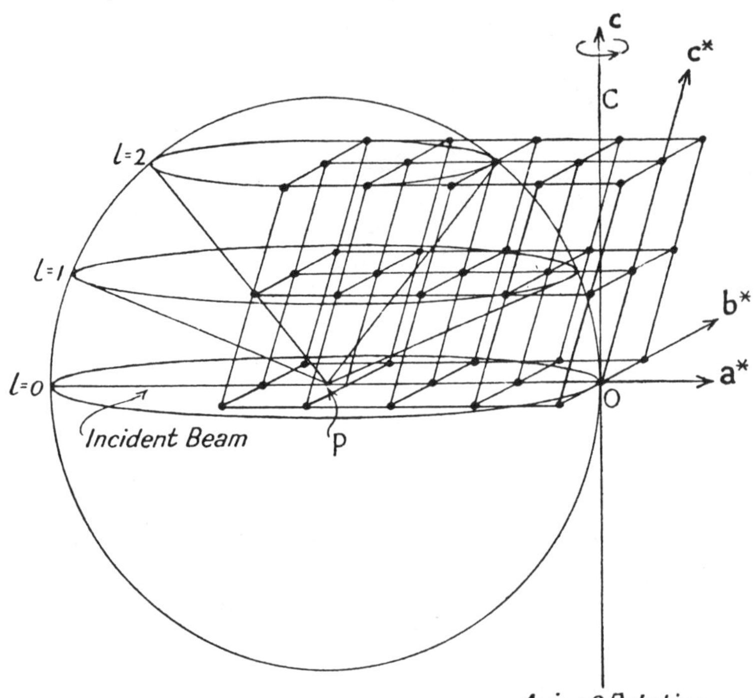

FIG. 7. Illustrating the formation of layer lines

The rotation is about OC, the c axis, and PO is the direction of incidence. The circles of latitude in which the reciprocal net-planes for which $l=0$, 1, and 2 intersect the sphere are shown, together with the generating lines of the corresponding cones.

If a very small crystal lying on the axis of rotation is used in the experiment, the diffracted beams will all lie on a set of cones whose angles are given by the construction just described, whose apex lies in the crystal, and whose common axis is the axis of rotation. In one method of recording the spectra, which is often used, the photographic film is in the form of a cylinder whose axis is the axis of rotation of the crystal. The conical surfaces containing the diffracted beams intersect this cylinder in a series of parallel lines, which are known as the layer lines, along which all the spectra recorded on the film must be distributed. In the case we are considering, where the c axis is the axis of rotation, the spots on any one layer line have the same l-index. All those lying on the line formed by the intersection of the equatorial plane with the film have $l=0$, those on the lines next above and below the zero line $l=\pm 1$, and so on. It is evident that the spacing of the layer lines is a measure of the reciprocal of the primitive translation which is parallel to the axis of rotation.

Suppose now that the axis of rotation is parallel to one of the axes of the *reciprocal* lattice, say to **c***, and therefore perpendicular to the net-planes defined by the crystal-lattice translations **a** and **b**. Rows of reciprocal-lattice points along each of which h and k remain constant are now parallel to the axis of rotation, and any such row describes a cylinder as the crystal rotates, which will intersect the sphere of reflection in a certain curve. The directions of the possible diffracted beams corresponding to points in the row must be parallel to the generating lines of the cone, no longer circular, which has the centre of the sphere as apex and passes through the curve of intersection of sphere and cylinder. This is illustrated by fig. 8. The locus of the corresponding spectra on the photographic film is then the curve of intersection of this cone with the cylindrical film, whose axis is a line

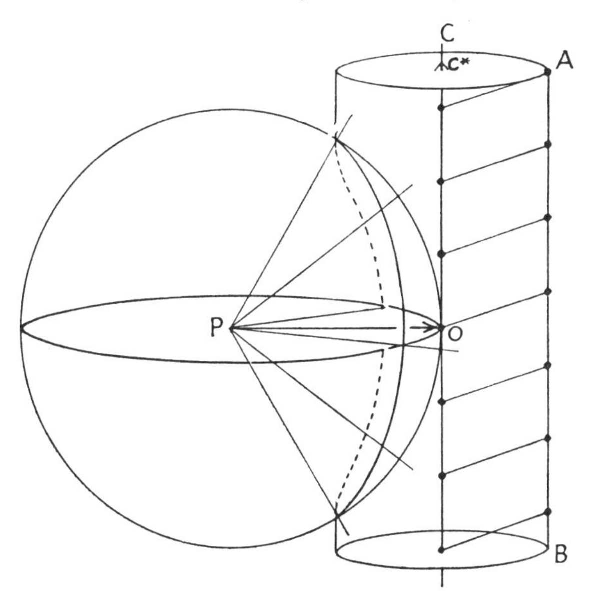

FIG. 8. Illustrating the formation of row lines

through P, the centre of the sphere of reflection, and parallel to OC. This curve is not a simple figure, and its mathematical form cannot be considered in detail in this volume, but it will be seen that to each row of points in the reciprocal lattice there corresponds a row of spectra, or for a complete rotation a pair of rows, called 'row lines', which are not straight, but in the equatorial region of the film are nearly so, and are approximately perpendicular to the equatorial line.

If the lattice is orthogonal, the crystal axes and the reciprocal axes coincide in direction, and the spectra must lie both on layer lines and on row lines, and are determined by their intersection. With the c axis, which is now also the c^* axis, as rotation axis all the spectra on any given row line have the same h and k indices, and all those on any layer line the same l-index. If the crystal is not orthogonal, the reciprocal axes and the crystal axes are not in general coincident. The row lines are then no longer symmetrical about, and approximately perpendicular to, the equatorial layer line but slope across it at an angle.

This simple discussion of rotation photographs has been introduced at this stage as an example of the use of the reciprocal lattice. The question of the assignment of indices to the spectra in such photographs lies beyond the scope of this volume.

(*c*) *The method of the powdered crystal:* Suppose that the crystal specimen is in the form of a fine powder whose particles have entirely random orientation. If the number of particles is large enough, we may suppose small crystals of all possible orientations to be present, and all orientations to be equally probable. To represent this case by means

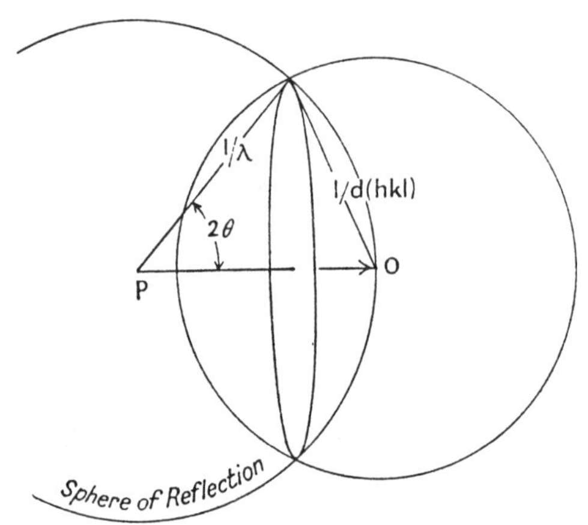

FIG. 9. Reciprocal-lattice construction for powder haloes

of the reciprocal-lattice construction, we may suppose the reciprocal lattice to be rotated about the origin in all possible directions. The locus of any one lattice-point during this rotation is a sphere. If the point in question lies within the limiting sphere of I, § 2(*a*), the sphere that is its locus during the rotation will intersect the sphere of reflection in a small circle whose axis is the direction of the incident beam. Any line joining the centre of the sphere of reflection to a point on this small circle is a possible direction for a diffraction maximum. Each crystal plane (h, k, l) whose corresponding reciprocal-lattice point lies within the limiting sphere will thus give rise to a diffraction halo having the direction of incidence as axis.

The formation of these haloes is illustrated in fig. 9. The radius of the sphere described by the reciprocal-lattice point (h, k, l) is $1/d(hkl)$, and 2θ, the semi-angle of the halo, and the angle of deviation of the diffracted beam is, from the figure, evidently given by the equation

$$(2/\lambda)\sin\theta = 1/d(hkl),$$

which is the Bragg condition.

(*d*) *The Laue method:* In the Laue method, the direction of incidence relative to the crystal axes is kept constant, and the interference conditions are fulfilled by using radiation varying continuously in wave-length over a certain range. Spectra are formed in directions corresponding to the reflection of the incident radiation by the lattice-planes, but each is formed by a different wave-length, the appropriate one being selected by the crystal spacing from the continuous radiation. The reciprocal-lattice construction for the diffracted beams is easily extended to include this case.

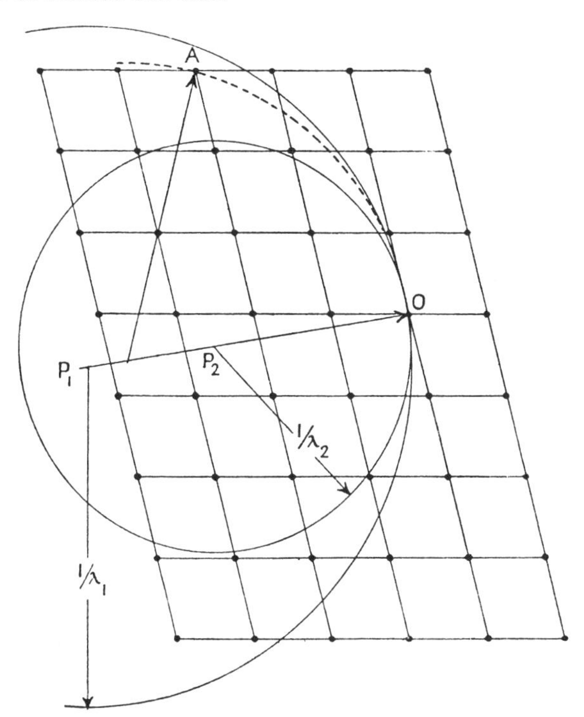

FIG. 10. Reciprocal-lattice construction for the Laue method

In fig. 10, P_1P_2O represents as before the direction of the incident radiation, O being a point in the reciprocal lattice which is chosen as origin. Let λ_1 and λ_2 be the wave-lengths of the limits of the continuous spectrum used in the experiment. The length P_1O is made equal to $1/\lambda_1$ and P_2O to $1/\lambda_2$, and spheres are drawn with centres P_1 and P_2, and passing through O. Each point of the reciprocal lattice which lies between these two spheres corresponds to a possible diffracted beam; for any such point lies on a sphere passing through O and having its centre between P_1 and P_2. This sphere is the sphere of reflection for some wave-length lying between the limits λ_1 and λ_2. Its centre can readily be found, and the line joining it to the lattice-point gives at once the direction of the corresponding diffracted beam. The sphere, and the direction of the diffracted beam, are shown in fig. 10 for the point A. Another method of dealing with the production of Laue spots in terms of the reciprocal lattice is considered in Chapter V, § 3 (*c*).

3. The Effect of Random Irregularities of the Lattice on the Diffraction Phenomena. The Temperature Effect

(*a*) *Introductory:* The theory of diffraction by a crystal lattice developed in Sections 1 and 2 of this chapter shows that, for a crystal of measurable size, the diffraction conditions must be very exactly fulfilled if there is to be any appreciable scattered intensity. According to equation (*1*.13), an incident beam whose direction is not such as to satisfy conditions (*1*.16) should pass through the crystal without being appreciably scattered. In all directions except those corresponding to the lattice spectra, mutual interference of the scattered wavelets produces a resultant whose intensity approaches zero. When a beam of X-rays passes through a real crystal, a measurable amount of general scattering does occur, even when the interference conditions are not fulfilled, and although it is true that the intensity of this scattering is small in comparison with that of the interference maxima, it is much larger than would be predicted by equation (*1*.13).

We have so far taken into account only the coherent scattering, which is accompanied by no change of wave-length. But in a real crystal, where the scattering points are atoms, Compton scattering will also occur, and a certain fraction of the radiation, which increases with the angle of scattering, will be scattered with a small increase of wave-length. Such radiation is 'incoherent', and does not produce interference maxima, nor, of course, will it interfere destructively in other directions, so that the general scattering from a crystal will consist in part of Compton radiation. Making due allowance for this, however, we find a general scattering of radiation of unchanged wave-length which is greater than that allowed by the simple theory. This is due to the fact that in no actual crystal is the lattice so regular as we have assumed in our treatment of the problem; and any departure from strict regularity in the arrangement of the scattering particles will tend to weaken the interference maxima, and to increase the radiation scattered in other directions.

On very general grounds we should expect an effect of this kind. Suppose any scattering point to be displaced from its correct position in the lattice. The exact co-operation in the directions of the interference maxima of the waves scattered by this point with those scattered by the remaining points of the lattice is now destroyed, and the intensities of the diffraction maxima are thereby reduced ; but there can be no loss of scattered energy on the whole, and there must therefore be an increase in the scattered intensity in directions other than those of the maxima. Indeed, the practically complete destruction by mutual interference of the scattered waves in directions other than those of the maxima depends on the regularity of the lattice, and if this regularity is destroyed the general scattering must increase. The same is true if one of the points is replaced by another which scatters differently from the rest.

The type of irregularity we are now considering is not to be confused with that usually spoken of as a *mosaic* structure, in which such dislocations occur in the lattice that an apparently single crystal may in effect consist of a number of optically independent fragments. We are here dealing with a single lattice, into which certain random irregularities are introduced, either in the exact position, or in the nature, of the scattering points.

The most obvious cause of such irregular displacements of the atoms in a real crystal from the ideal lattice positions is the thermal vibration. The effect of this on the scattering was pointed out by Debye as early as 1913, and was treated mathematically by him in a series of well-known papers.[7] Laue[8] has treated the case of mixed crystals, in which a lattice is supposed to contain points having more than one kind of scattering power, which are distributed at random amongst the available lattice sites. The difference between this case and that in which the lattice consists of identical points whose scattering power is the average of those of the individual points lies in the general scattering. A permanent distortion of the lattice, resulting in a change of the general scattering analogous to that characteristic of a temperature effect, may be produced in some cases by cold working a metal.

(*b*) *Diffraction by a lattice whose points are subject to random displacements :* We shall now show how the expression for the intensity of the scattered radiation has to be modified to take into account random departures of the lattice-points from their ideal positions. Suppose the points, when in their ideal positions, to be fixed by the vectors $\mathbf{r}_1$, $\mathbf{r}_2, \dots \mathbf{r}_n, \mathbf{r}_m, \dots$. By (*1.7*), the resultant disturbance due to the whole lattice is

$$Y = \frac{\Phi_0}{R} e^{i\omega t} \sum_n e^{i\kappa \mathbf{r}_n \cdot \mathbf{S}}, \qquad (1.23)$$

the sum being taken over all the lattice-points. To get the intensity, we have to take the square of the modulus of Y, which we do by multiplying the series on the right-hand side of (*1*.23) by the corresponding series in which i is replaced by $-i$. This gives for the intensity J_0 due to the undisturbed lattice the double summation

$$J_0 = \frac{|\Phi_0|^2}{R^2} \sum_n \sum_m e^{i\kappa(\mathbf{r}_n - \mathbf{r}_m \cdot \mathbf{S})}. \qquad (1.24)$$

Each summation has to be extended over all values of n and m, which range from 1 to N, N being the total number of lattice-points. Equation (*1*.24) is, of course, quite general for any array of scattering points, and does not depend on their being arranged in a lattice. For the parallelepipedal space-lattice it readily reduces to the form (*1*.13).

Let us now assume small vector displacements $\mathbf{u}_1, \mathbf{u}_2, \ldots \mathbf{u}_n, \mathbf{u}_m, \ldots$ of the lattice points from the ideal positions. The vectors $\mathbf{r}_n$ and $\mathbf{r}_m$ in (*1*.24) then have to be replaced by $\mathbf{r}_n + \mathbf{u}_n$ and $\mathbf{r}_m + \mathbf{u}_m$, and the expression for the intensity becomes

$$J = \frac{|\Phi_0|^2}{R^2} \sum_n \sum_m e^{i\kappa(\mathbf{r}_n - \mathbf{r}_m \cdot \mathbf{S})} e^{i\kappa(\mathbf{u}_n - \mathbf{u}_m \cdot \mathbf{S})}. \qquad (1.25)$$

If we are dealing with oscillations of the points, such as occur in thermal movement, it will be necessary to take the mean of (*1*.25), which represents the intensity due to some instantaneous configuration of points, over a period long compared with the period of vibration of any point. The first factor under the summation signs does not vary with time, and it is the second that has to be averaged. Expression (*1*.25) can only be used if the frequency of oscillation of the points is small compared with that of the incident radiation, so that any given configuration can be considered as persisting over a time long compared with the period of the wave, an assumption which is fully justified in the cases we shall consider.

To discuss the mean of the factor involving the displacements of the atoms n and m, we put, for brevity,

$$\kappa(\mathbf{u}_n - \mathbf{u}_m \cdot \mathbf{S}) = p_{n,m}. \qquad (1.26)$$

If we write, for the moment, p for any of the quantities $p_{n,m}$,

$$\overline{e^{ip}} = 1 + \overline{ip} - \frac{\overline{p^2}}{2!} - \frac{\overline{ip^3}}{3!} + \ldots = 1 - \frac{\overline{p^2}}{2} + \frac{\overline{p^4}}{24} - \ldots; \qquad (1.27)$$

for the mean value of the terms involving odd powers of p will be zero, since positive and negative values of the difference of the displacement of the two points parallel to any given direction will be equally likely.

Equation (*1*.27) may, to a close approximation, be written in the form

$$\overline{e^{ip}} = e^{-\frac{1}{2}\overline{p^2}}, \qquad (1.28)$$

a result given by Debye and by Waller, but first rigorously derived by Ott [9] in 1935.

The mean value of (*1*.25) therefore becomes

$$\bar{J} = \frac{|\Phi_0|^2}{R^2} \sum_n \sum_m e^{i\kappa(\mathbf{r}_n - \mathbf{r}_m \cdot \mathbf{S})} \, e^{-\frac{1}{2}\overline{p_{n,m}^2}}. \tag{1.29}$$

Now

$$p_{n,m} = \frac{4\pi \sin\theta}{\lambda} (u_{nS} - u_{mS}), \tag{1.30}$$

where u_{nS} and u_{mS} are the components of the displacements of the two lattice-points n and m parallel to the direction of the vector S defining the reflecting plane. We have therefore to calculate the mean value of $(u_{nS} - u_{mS})^2$. We may write

$$\overline{(u_{nS} - u_{mS})^2} = \overline{u_{nS}^2} + \overline{u_{mS}^2} - 2\overline{u_{nS}u_{mS}}. \tag{1.31}$$

The assumptions made by Debye in his first papers on the temperature effect were equivalent to the supposition that the oscillations of the different lattice-points were independent, and that all possessed the same mean energy. If this were so, we could put

$$\overline{u_{nS}u_{mS}} = 0; \quad \overline{u_{nS}^2} = \overline{u_{mS}^2} = \overline{u_S^2}, \tag{1.32}$$

where $\overline{u_S^2}$ is the mean-square elongation of any point of the lattice from its mean position in a direction parallel to the vector S. Now, as Debye himself pointed out in a later paper, the assumption of the independence of the lattice-points is not justifiable in considering the thermal vibrations of a real lattice. The atoms of the lattice are coupled together by the lattice forces, and the direction of vibration of one point must influence those of its neighbours. We cannot therefore put $\overline{u_{nS}u_{mS}} = 0$, and the value of $\overline{p_{nm}^2}$ will depend on the pair of lattice-points that are being considered. This leads to important consequences, which will be discussed in detail in Chapter V, but to complete our preliminary account of the geometrical theory of diffraction by a simple lattice we shall consider the case of a set of points all of which do vibrate independently, and in the same way, so that equations (*1*.32) may be taken as applicable.

We return now to equation (*1*.29). The double summation contains N^2 terms, N being the total number of lattice-points. In N of these terms $n = m$, and for each of these the exponential factor is equal to unity. For all terms involving a pair of different lattice-points $\frac{1}{2}\overline{p_{nm}^2}$ is the same, and by (*1*.30), (*1*.31), and (*1*.32) is equal to 2M, if

$$M = 8\pi^2 \overline{u_S^2} (\sin^2\theta)/\lambda^2. \tag{1.33}$$

Equation (*1*.29) may thus be written

$$\bar{J} = \frac{|\Phi_0|^2}{R^2} \left\{ N + e^{-2M} \sum_n{}' \sum_m{}' e^{i\kappa(\mathbf{r}_n - \mathbf{r}_m \cdot \mathbf{S})} \right\}, \tag{1.34}$$

where the dashes denote that terms for which $n = m$ are not to be included in the summations. If this restriction were removed, the double summation in (*1*.34) would be equal to J_0, the intensity scattered by the undisplaced lattice. The second term of (*1*.34) is thus equal to $e^{-2M}(J_0 - |\Phi_0|^2 N/R^2)$, since each of the N terms missing from the double summation is equal to unity. The expression for the scattered intensity thus becomes finally

$$\bar{J} = \frac{|\Phi_0|^2}{R^2} N(1 - e^{-2M}) + J_0 e^{-2M}. \qquad (1.35)$$

The second term in (*1*.35) represents the lattice spectra. It has appreciable values only when the interference conditions (*1*.16) are fulfilled, and we see that with the simple assumptions we have made as to the nature of the lattice vibrations the interference maxima are just as sharp for the vibrating lattice as they are when the lattice-points are undisplaced. Their intensities are, however, all reduced by the factor e^{-2M}, which is often known as the Debye factor. Its exponent is proportional to the mean square of the displacements of the lattice-points perpendicular to the reflecting plane, and also to $(\sin^2\theta)/\lambda^2$; so that, for a given set of displacements, the diminution of the intensities of the spectra becomes more important for the larger angles of scattering.

Formally, since J_0 is proportional to $|\Phi_0|^2$, we may represent the effect of the lattice displacements on the interference maxima by treating the vibrating lattice as a lattice at rest, composed of points having a scattering power Φ given by

$$\Phi = \Phi_0 e^{-M}, \qquad (1.36)$$

instead of the original scattering power Φ_0.

(*c*) *The general scattering when the lattice-points vibrate independently:* The first term in (*1*.35) has no sharp maxima. It contains two factors that depend on the angle of scattering, $|\Phi_0|^2$, which as we shall see later decreases with increasing θ, and $1 - e^{-2M}$, which is zero for zero angle of scattering, rises steadily as e^{-2M} decreases with increasing θ, and has its greatest value for scattering through 180°. The product of the two factors corresponds to a very broad maximum of general scattering at moderate angles. The intensity distribution in the reciprocal lattice representing this scattering is spherically symmetrical about the origin.

The intensity scattered in any direction is proportional to N, the number of points in the crystal fragment, whereas the *maximum* intensity in the exact direction of the diffraction spectra is proportional to N^2. As we shall see in the next chapter, however, the methods of measuring the intensities of the spectra in actual practice involve an integration over the whole diffraction maximum, the breadth of which is proportional to 1/N. The integrated intensity over a range of angles for both spectra and general scattering is thus proportional to N, that is to say, to the volume of the crystal, and the disparity between them is not so great as the formulae would at first sight suggest. We shall return to this question in Chapter V.

The more accurate treatment, in which the coupling of the oscillations of the lattice-points is taken into account, gives the same result as that deduced above for the interference maxima. The term giving the general scattering is, however, modified, and shows weak maxima, coinciding in position with the interference maxima, but far more diffuse. This effect, the existence of which was predicted by Faxén and Waller, has only recently been properly investigated experimentally. Its study has led to results of the greatest interest and importance, which will be discussed in detail in Chapter V. The important result that the interference maxima are reduced by the factor e^{-2M} remains true. In Chapter V also, we shall take up the question of the calculation of the value of M for actual crystals.

4. Some Effects due to Multiple Reflection

Let O be the origin of the reciprocal lattice, and Q_1 the point (h_1, k_1, l_1), corresponding to the spectrum m_1. Any point L in the plane P_1 that bisects the line OQ_1 at right-angles is a possible centre of a sphere of reflection corresponding to the production of m_1 in wave-length 1/LO by reflection of the incident ray LO into the direction LQ_1. Let Q_2 be a second reciprocal-lattice point (h_2, k_2, l_2), corresponding to a spectrum m_2. Then, if L lies at any point in the line of intersection of the plane P_1 with the plane P_2 that bisects the line OQ_2 at right-angles, the spectra m_1 and m_2 are produced simultaneously by reflection of LO into LQ_1 and LQ_2, and each must be weaker than it would have been had it alone been produced. This effect is actually observed.

All incident rays that can give rise to simultaneous reflection in m_1 and m_2 form a flat pencil, containing the line of intersection of P_1 and P_2, and converging on O. The corresponding rays reflected in the spectrum m_1 form a flat pencil containing the same line and converging on Q_1. The wave-length varies across the pencil, being always equal to 1/LO. A *continuous* spectrum m_1, formed by varying the angle of incidence and the wave-length, will be crossed by a line of lower intensity, representing the intersection of the pencil reflected through Q_1 with the photographic plate. Narrow bright lines due to this cause, in general cutting the spectrum obliquely, have been observed by Berg,[10] and the phenomenon has been called *Aufhellung*, or brightening up, which refers, of course, to the appearance of the photographic record. The brighter lines actually represent a *reduction* of intensity.

Further, if Q_1 and Q_2 both lie on the sphere of reflection, the reflected ray LQ_2 is correctly set to be reflected again along LQ_1, the reciprocal-lattice vector belonging to this second reflection being Q_2Q_1 : that is to say, the second reflection takes place from $(h_1\text{-}h_2, k_1\text{-}k_2, l_1\text{-}l_2)$, although it appears in the direction LQ_1, as if it had been the direct reflection of LO by the planes $(h_1k_1l_1)$. This effect will tend to increase the observed intensity of m_1, but will not in general outweigh the effect of simultaneous reflection, discussed above, in decreasing it, since the double reflection is in general weak. An important case arises, however, when the spectrum m_1 is forbidden by space-group or structure-factor requirements, while both m_2 and m_1-m_2 occur as fairly strong spectra. It may then be possible at appropriate settings of the crystal to observe an apparent occurrence of the forbidden spectrum as a result of this double reflection. Or, if a spectrum is very weak, its intensity as observed may sometimes be greatly affected in this way, as Renninger [11] has shown for the 222 spectrum of diamond (see Chap. VI, p. 339), for which the double reflection may be many times stronger than the true spectrum. Renninger has used the name *Umweganregung* for this effect, i.e. excitation by a roundabout path. The term ' indirect reflection ' is used in English works, and it is sometimes also called the Renninger effect. Its possibility seems to have been pointed out by Darbyshire and Cooper,[12] for electron diffraction, and also by Kossel,[13] but its systematic investigation is due to Renninger.

The existence of the effects discussed in this paragraph shows that the simple assumption made in this chapter, that there is no rescattering of the primary scattered wavelets, is inadequate. The fuller theory, in which rescattering is taken into account, is given in Chapter II, § 3.

REFERENCES

1. M. v. Laue, *Sitz. math. phys. Klasse Bayer. Akad. Wiss.* p. 303, (1912); *Ann. d. Physik*, **41**, 971 (1913); *Enzyklopädie der math. Wissen.*, **V** 24, 359 (1915).
2. W. L. Bragg, *Proc. Camb. Phil. Soc.*, **17**, 43 (1913).
3. Willard Gibbs, *Vector Analysis*, reprinted in his collected works, 1st Ed., 1906, 2nd Ed., 1928.
4. P. P. Ewald, *Physikal. Zeit.*, **14**, 465, 1038 (1913).
5. M. v. Laue, *Communication to the Solvay Congress* 1913, published 1921; *Jahrbuch der Radioakt. u. Elektronik*, **11**, 308 (1917).
6. P. P. Ewald, *Zeit. f. Krist.*, **56**, 129 (1921).
7. P. Debye, *Verh. der deutsch. Phys. Ges.*, **15**, 678, 738, 857 (1913); *Ann. d. Physik*, **43**, 49 (1914).
8. M. v. Laue, *Ann. d. Physik*, **56**, 497 (1918); **78**, 167 (1925).
9. H. Ott, *Ann. d. Physik*, **23**, 169 (1935).
10. O. Berg, *Veröffentlichung a.d. Siemens-Konzern*, **5**, 89 (1926).
11. M. Renninger, *Zeit. f. Physik*, **106**, 141 (1937).
12. J. A. Darbyshire and E. R. Cooper, *Proc. Roy. Soc.*, **A 152**, 104 (1935).
13. W. Kossel, *Ann. d. Physik*, **25**, 512 (1936).

CHAPTER II

THE INTENSITY OF REFLECTION OF X-RAYS BY CRYSTALS

1. Diffraction by a Composite Lattice

(*a*) *The structure amplitude:* In Chapter I, we considered diffraction by a simple space-lattice. We have now to extend our treatment to the type of structure that is produced by applying the lattice translations to a group of points, not necessarily all identical. Each point of the group will build up a space-lattice like the one discussed in Chapter I, and the whole structure will consist of s parallel and interpenetrating lattices, all alike, one for each of the s points of the unit group, or 'basis', of the structure. This is the kind of geometrical scaffolding upon which any real crystal may be considered as built; for the crystal pattern is produced by the continued repetition in space of a group of atoms, and is only in exceptional cases based on a simple lattice.

We shall now consider what modifications to the formulae for the resultant scattered amplitude and intensity have to be introduced if each point of the lattice considered in Chapter I is replaced by such a group of points. Suppose that associated with the lattice-point chosen as origin are s points, whose positions are fixed by the vectors $\boldsymbol{\rho}_1, \boldsymbol{\rho}_2, \ldots \boldsymbol{\rho}_j, \ldots \boldsymbol{\rho}_s$. Then associated with the lattice-point defined by the vector $\mathbf{r}_n$ is a similar group fixed by vectors $\mathbf{r}_n + \boldsymbol{\rho}_1, \ldots \mathbf{r}_n + \boldsymbol{\rho}_j, \ldots \mathbf{r}_n + \boldsymbol{\rho}_s$. Suppose that the scattering powers of the points, as introduced in Chapter I, § (1)*a*, are $\phi_1, \phi_2, \ldots \phi_s$. In place of the expression (*1*.21) for the resultant wave scattered by the whole structure, we must now write

$$\mathrm{Y} = \frac{e^{i\omega t}}{\mathrm{R}} \left\{ \phi_1 e^{i\kappa \mathbf{S}\cdot\boldsymbol{\rho}_1} + \ldots + \phi_s e^{i\kappa \mathbf{S}\cdot\boldsymbol{\rho}_s} \right\} \sum_n e^{i\kappa \mathbf{S}\cdot\mathbf{r}_n} \tag{2.1}$$

$$= \frac{e^{i\omega t}}{\mathrm{R}} \mathrm{A} \sum_n e^{i\kappa \mathbf{S}\cdot\mathbf{r}_n}, \tag{2.2}$$

where the summation is over the N points of the fundamental lattice. On comparing the last expression with (*1*.21), we see that the effect of replacing the single lattice-point of scattering power Φ_0 by the group of points is merely to replace Φ_0 in the expression for the scattered wave by the more complicated quantity

$$\mathrm{A} = \phi_1 e^{i\kappa \mathbf{S}\cdot\boldsymbol{\rho}_1} + \ldots + \phi_s e^{i\kappa \mathbf{S}\cdot\boldsymbol{\rho}_s} \tag{2.3}$$

$$= \sum_j \phi_j e^{i\kappa \mathbf{S}\cdot\boldsymbol{\rho}_j}. \tag{2.3a}$$

We shall call A the '*structure amplitude*', a name introduced by Ewald, to denote the fact that its value depends essentially on the structure of the group associated with each lattice-point. It is the amplitude, at unit distance, of the wave scattered by the unit group of s points.

In practice we are as a rule interested in the values of A corresponding to the main diffraction maxima, which occur in the same positions and have the same breadth for the composite as for the simple lattice. The diffraction maxima are characterised by the condition that S/λ is a vector in the reciprocal lattice, say $\mathbf{r}^*$, where

$$\mathbf{r}^* = h\mathbf{a}^* + k\mathbf{b}^* + l\mathbf{c}^*. \tag{2.4}$$

This gives for the structure amplitude for the spectrum hkl, in terms of the reciprocal vector,

$$\mathrm{A} = \sum_j \phi_j \, e^{2\pi i \mathbf{r}^* \cdot \boldsymbol{\rho}_j}, \tag{2.5}$$

in which the summation is supposed to extend over all the s points in the group. Now let us put

$$\boldsymbol{\rho}_j = u_j\mathbf{a} + v_j\mathbf{b} + w_j\mathbf{c} \tag{2.6}$$

so that u_j, v_j, and w_j are the co-ordinates of the point fixed by the vector $\boldsymbol{\rho}_j$, expressed as fractions of the sides of the unit lattice-cell.

Then, by (2.4) and (2.6),

$$\mathbf{r}^* \cdot \boldsymbol{\rho}_j = hu_j + kv_j + lw_j, \tag{2.6a}$$

since $\mathbf{a}\cdot\mathbf{a}^* = \mathbf{b}\cdot\mathbf{b}^* = \mathbf{c}\cdot\mathbf{c}^* = 1$; $\mathbf{b}\cdot\mathbf{a}^* = \mathbf{c}\cdot\mathbf{a}^* = \ldots = 0$, by the properties of the reciprocal vectors.

Thus

$$\mathrm{A}(hkl) = \sum_j \phi_j \, e^{2\pi i (hu_j + kv_j + lw_j)}. \tag{2.7}$$

Equation (2.7) was derived in a simple geometrical way in Vol. 1, p. 96 ; we have here shown its relationship to the more general theory of lattices that we have been discussing. The relation of (2.5) to the more strictly geometrical way of considering the matter is easy to see. The hkl spectrum may be considered as produced by the reflection of the incident beam at the lattice-planes (hkl). When the unit of structure is a group of s points, there are s parallel space-lattices, and each single (hkl) plane becomes a group of s parallel planes, this group repeating itself in the distance $d(hkl)$. The phase of a wave reflected from a plane of the group distant d from a plane chosen as standard differs from that of a wave reflected from the latter by $2\pi d/d(hkl)$; for the phase of the waves reflected by two planes at a distance $d(hkl)$ apart differs by 2π. Now $\mathbf{r}^*\cdot\boldsymbol{\rho}_j$ is equal to $d/d(hkl)$ if the plane through the origin is the standard plane. For $\mathbf{r}^* d(hkl)$ is the unit vector in the direction of $\mathbf{r}^*$, which is normal to the plane (hkl), so that $\boldsymbol{\rho}_j\cdot\mathbf{r}^* d(hkl)$ is the projection of $\boldsymbol{\rho}_j$ on the normal to the planes (hkl), that is to say, the distance d between the planes of the set passing through the origin and through the point defined by the vector $\boldsymbol{\rho}_j$.

(*b*) *The atomic scattering factor:* At this stage it is necessary to define rather more precisely the quantity ϕ, the scattering factor, which we have used in the preceding sections. In an actual crystal, groups of atoms or molecules are associated with each lattice-point, and it is the radiation scattered by these which builds up the interference maxima. The scattering will be done almost entirely by the electrons, and we can think of each unit cell of the structure as occupied by a certain average distribution of scattering matter, which, in a sense to be discussed later on, may be considered as an average electron density. The electrons are, however, concentrated round the atomic nuclei, and to a fairly close approximation we can think of the scattering by the unit cell as the resultant of the scattering by the individual atoms which it contains, taking each atom as spherically symmetrical, and ascribing to it a characteristic scattering power. It is clear that this is an approximation. When the atoms are combined to form molecules, there must be sharing of the electrons between atoms, and a distortion of the average charge-distribution from true spherical symmetry around the atomic nuclei; but, for reasons that will become clear later, the error we make by assuming such a spherical distribution is small, and in what follows we shall do so.

We are interested only in the *coherent* scattering, which has the same wave-length as the incident radiation, since only this is effective in forming interference maxima. It is convenient to take as the unit of scattered amplitude that due to a free classical electron. For our present purpose, which is merely to obtain a convenient unit of scattering, the fact that actual electrons do not behave as classical electrons is of no consequence; it is the natural unit in terms of which the quantum-mechanical scattering by electrons, in such cases as interest us, is also expressed. The problem of the scattering of radiation by a free electric charge was first solved by J. J. Thomson [1] in terms of the classical electrodynamics, and we shall quote his results.

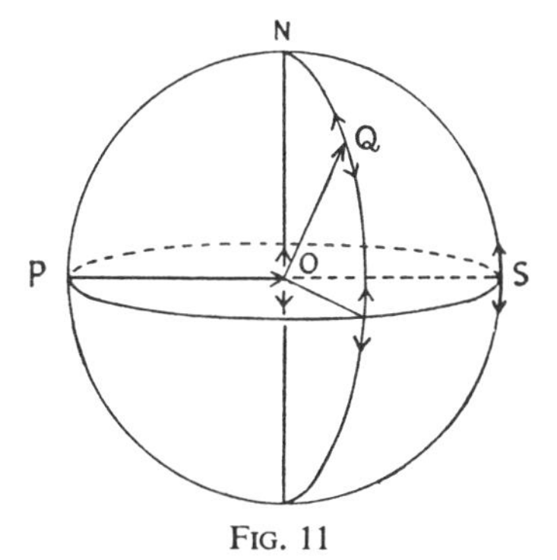

FIG. 11

Let the electron, when at rest, lie at O, fig. 11, the centre of a sphere of radius R, and let PO, an equatorial radius of this sphere, be the direction of propagation of the incident wave, which is plane-polarised, with the electric vector in the direction of the polar axis ON. Let $\mathbf{E}_0 e^{i\omega t}$ be the value of the electric vector in the incident wave at the point O at any time t. Then, at the same instant, the value of the electric vector E_s in the scattered wave at a point Q on the surface of the sphere, whose polar distance is ϕ, is given by

$$E_s = -E_0 \frac{e^2}{mc^2} \frac{\sin \phi}{|R|} e^{i\omega(t-|R|/c)}. \qquad (2.8)$$

In this formula, which is valid only when the radius R is large in comparison with the wave-length of the radiation, e and m are the charge and mass of the electron, and c the velocity of light in free space. The electric vector in the scattered wave is along the meridian QN, and its value is independent of the longitude of Q, being axially symmetrical about ON, the direction of the incident electric vector. The negative sign in (2.8) indicates that the phase of the scattered wave is opposed to that of the incident wave in the sense that the electric field in the scattered wave at a point such as S, fig. 11, in the line of incidence, is opposed to that in the direct wave at the same point (cf. Chapter IV, § 1(b), p. 137).

For our present purpose it is convenient to consider the polarisation of the incident wave relative to the reflecting plane corresponding to the directions of incidence and scattering, PO and OQ.

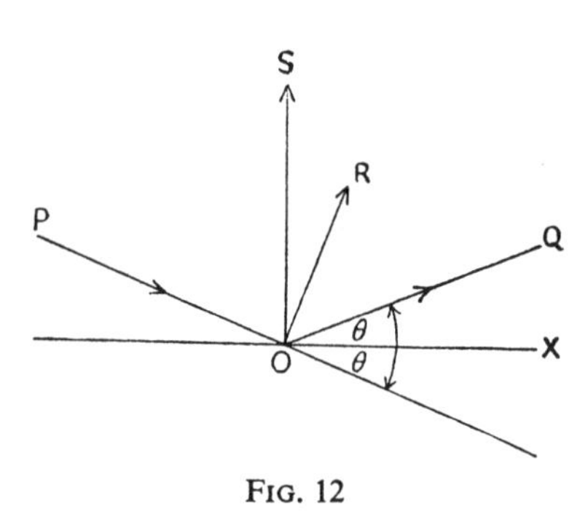

FIG. 12

In fig. 12, let PO, OQ lie in the plane of the paper, and let OX be the trace of the reflecting plane, and OS the direction of the vector S, which is normal to it. We consider two cases of polarisation in perpendicular directions.

(i) Let the direction of the incident electric vector be parallel to the reflecting plane and normal to the plane POQ. In fig. 11, Q will then lie on the equator of the sphere, and the scattered amplitude at a distance R is $-(e^2/mc^2)/R$ times the incident amplitude.

(ii) Let the incident electric vector lie in the plane POQ, and so along OR, which is perpendicular to PO, and makes an angle $\pi/2-2\theta$ with the direction of scattering OQ. Then the scattered amplitude will be $-\cos 2\theta(e^2/mc^2)/R$ times the incident amplitude.

We shall now assume the scattered amplitude due to a given atom to be f times that due to a single classical electron under identical conditions. The number f is called the *atomic scattering factor*, or often simply the *f-factor* of the atom. The corresponding German term is *Atomformfaktor*, the not altogether satisfactory translation of which, atomic-form factor, is sometimes used. For radiation of frequency large compared with any of the atomic absorption frequencies, the value of f approaches Z, the total number of electrons in the atom, for small angles of scattering ; but it decreases rapidly with increasing angle of scattering in a manner depending on the distribution of electrons in the atom, which will be discussed in detail in Chapter III.

When f as here defined is positive, the electric vector in the scattered wave in the direction of incidence is opposed to that in the direct wave, corresponding to a phase change of π on scattering. It should be noted, however, that f is in general a complex number, corresponding to a phase change on scattering neither 0 nor π.

Consider now a crystal composed of atoms all of one kind, with their centres at the points of a space-lattice. With the radiation polarised as in case (i) above (electric vector perpendicular to the plane containing the directions of incidence and scattering), we should have to write in the equations of II, §1(a) for the resultant scattering

$$\phi = -f(e^2/mc^2). \qquad (2.9)$$

On the other hand, if the polarisation corresponds to case (ii) (electric vector in the plane containing the directions of incidence and scattering), we must write

$$\phi = -f(e^2/mc^2)\cos 2\theta. \qquad (2.10)$$

If the incident radiation is unpolarised, the average value of the electric vector in each of the two perpendicular directions in the primary wavefront is the same, and to get the scattered intensity we have to take the average of the *intensities* corresponding to the two directions of polarisation ; for the phases of the vibrations in the two directions in unpolarised light are independent. In the expression for the intensity that we have derived in the preceding sections we have to put, for a lattice consisting of one kind of atom only,

$$|\phi|^2 = \left(\frac{e^2}{mc^2}\right)^2 |f|^2 \frac{1+\cos^2 2\theta}{2}. \qquad (2.11)$$

(c) *The structure factor :* If each unit of the structure is composed of a number of atoms, with scattering factors $f_1, f_2, \dots f_s$, the structure amplitude A, equation (2.3), becomes

$$A = -F(e^2/mc^2) \quad \text{or} \quad A = -F(e^2/mc^2)\cos 2\theta, \qquad (2.12)$$

according to whether the state of polarisation is (i) or (ii). Here

$$F = \sum_j f_j \, e^{i\kappa \mathbf{S}\cdot\boldsymbol{\rho}_j}. \qquad (2.13)$$

$\boldsymbol{\rho}_j$ is now the vector fixing the centre of an atom, and the summation is to be taken over all the atoms in the unit cell. F we shall call the '*structure factor*'. Like f, it is dimensionally a pure number, and its value depends essentially on the arrangement of the atoms in the unit group.

For the direction of the spectrum hkl, the structure factor F(hkl) is given by

$$F(hkl) = \sum_j f_j \, e^{2\pi i (hu_j + kv_j + lw_j)}, \qquad (2.14)$$

u_j, v_j, w_j, being the fractional co-ordinates of the centre of the atom whose scattering factor is f_j, and the summation again being over all the atoms in the unit cell.

(*d*) *Phase changes on scattering and their effect on the structure factor:* So long as we are dealing with a lattice consisting of identical atoms, the question of a possible phase change on scattering is of no real importance ; for it cannot affect the intensity of the resultant scattered wave, but only its phase. This is no longer true when the unit cell contains more than one kind of atom, for the phase change on scattering may differ for the different atoms, thus introducing into the structure factor phase differences other than those due to the positions of the atoms in the unit cell.

Such phase changes may be included in equation (*2*.13) for F without changing its form, if we consider the atomic scattering factors $f_1, f_2, \ldots$ to be complex numbers. A complex amplitude always includes a phase factor ; for let us suppose that $f = f' + if''$.

Then we can put

$$f' + if'' = |f|(\cos\delta + i\sin\delta) = |f|\,e^{i\delta},$$

where

$$\tan\delta = f''/f', \text{ and } |f| = \sqrt{f'^2 + f''^2},$$

corresponding to a real amplitude factor equal to the modulus of f, and a phase factor $e^{i\delta}$.

Expression (*2*.13) for F is therefore general, and includes any phase change on scattering, so long as the f-factors are understood to be complex. It is perhaps well to point out at this stage that since we can never measure the amplitudes and phases of X-ray spectra, but only their intensities, which are proportional to $|F(hkl)|^2$, it is always the modulus of F, and not F itself that is determined experimentally.

Phase changes on scattering become important only when the frequency of the incident radiation is very nearly equal to the critical absorption frequency of one of the atoms in the unit cell. In the majority of cases this will not be so, and we can treat any phase change that may occur on scattering as being the same for all the atoms of the unit. Under these conditions, we obtain from (*2*.14)

$$|F(hkl)|^2 = \left\{\sum |f_j| \cos 2\pi(hu_j + kv_j + lw_j)\right\}^2 + \left\{\sum_j |f_j| \sin 2\pi(hu_j + kv_j + lw_j)\right\}^2. \quad (2.15)$$

$|F(hkl)|$ may in this case be called the '*geometrical structure factor*', since it depends only on the atomic positions. It is this factor which is used in calculating the space-group conditions, but it must be remembered that equality of phase change on scattering for all the constituent atoms of the structure is, in principle, a necessary condition for its employment.

Upon this condition too rests the assumption, usually made in crystal analysis, that $|F(hkl)| = |F(\bar{h}\bar{k}\bar{l})|$, a result which follows at once from (*2*.15), and which means that the intensity of reflection from opposite sides of the same set of crystal planes is the same. This is sometimes known as Friedel's law.[2] A consequence of its truth is that the actual symmetry of a crystal may always be lower by the absence of a centre of symmetry than the apparent symmetry as observed by X-rays; or, in other words, that X-rays cannot show the polarity of crystal structures. In principle, however, this statement is untrue if the different atoms in the polar structure scatter X-rays with different phase changes, and this has been confirmed in some beautiful experiments with zinc-blende by Coster, Knol and Prins.[3] In the ZnS crystal (Vol. I, p. 53), planes of zinc and sulphur alternate parallel to the face (111), as shown in fig. 13, the distance between the close pairs of Zn and S planes being one quarter of the whole (111) spacing. The structure is polar, for the aspect of each of the close pairs of planes depends upon whether it is being viewed from one side or the other.

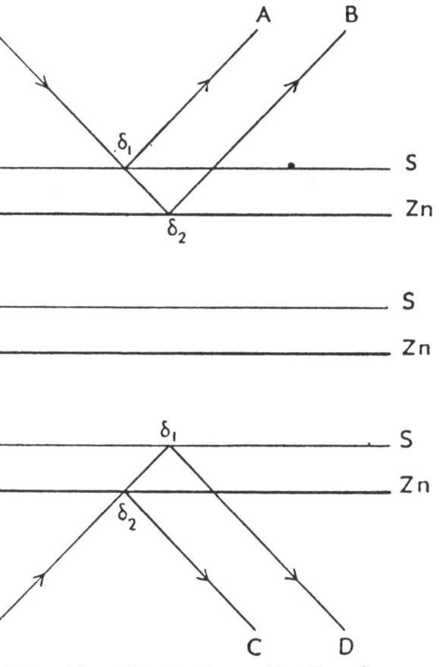

FIG. 13. Illustrating reflection from opposite faces of a polar crystal

Let us suppose now that phase changes δ_1 and δ_2 are produced respectively when atoms of sulphur and zinc scatter. The phase difference between the rays B and A when the reflection occurs from one side of the planes is $\delta_0 + \delta_2 - \delta_1$, δ_0 being the phase difference corresponding to the separation of the Zn and S planes. On the other hand, the phase difference between the rays D and C when the reflection takes place from the other side of the planes is $\delta_0 + \delta_1 - \delta_2$, and these are not equal unless $\delta_1 = \delta_2$, when each becomes equal to δ_0.

Coster, Knol, and Prins used the $L\alpha_1$ and $L\alpha_2$ radiations from gold, which have wave-lengths 1273·77 X units and 1285·02 X units respectively. Now the K absorption edge of zinc lies at 1280·5 X units, between the $L\alpha_1$ and $L\alpha_2$ lines, and we should expect, from the theory of scattering, that the phase change on scattering from Zn would differ considerably for the two radiations. On the other hand, since the absorption edge for S lies nowhere near the L lines of gold, we should expect the phase change on scattering from S to be the same for each kind of radiation. One might therefore expect the relative intensities of reflection of the $L\alpha_1$ and $L\alpha_2$ radiations of gold to differ appreciably according to whether the reflection takes place from (111) or from ($\bar{1}\bar{1}\bar{1}$). The photometer curves obtained by Coster, Knol, and Prins are shown in fig. 14, (Pl. I), and exactly confirm this prediction.

(*e*) *The temperature effect and the structure factor:* Expressions (*2*.13) and (*2*.14) for the structure factor are deduced on the assumption that the atoms are at rest. In real crystals they will be in vibration, owing to the thermal agitation and the zero-point energy of the lattice. We can easily allow for this formally by writing for f in the expressions for the structure factor

$$f = f_0 \, e^{-M}, \tag{2.16}$$

where f is the effective scattering factor under the actual conditions and f_0 is the scattering factor for the atom at rest. M is given by equation (*1*.33), $\overline{u_s^2}$ being now the mean-square displacement of the atom in a direction normal to the reflecting planes. In a composite unit cell the different types of atom will not as a rule have the same mean-square displacements, so that M will in general be different for each atom in the unit cell. The temperature effect is discussed in detail in Chapter V.

2. Reflection from Crystals when Dynamical Interaction may be Neglected

(*a*) *Introductory:* In this section we shall deduce formulae for the intensities of the spectra produced when X-rays are diffracted by a crystal whose volume, although containing a large number of unit cells, is yet so small that the assumption that there is no appreciable re-scattering or absorption of the radiation is valid. The problem is still essentially that dealt with in Chapter I, but we shall here be concerned with the calculation of the intensities in a form suitable for direct comparison with experiment.

We have seen that one way of considering the formation of diffraction spectra is to think of them as produced by the agreement in phase of those fractions of the incident radiation that are reflected at the successive atomic planes as the X-ray beam sweeps through the crystal. The crystal units lying on any net-plane of the crystal give rise to spherical scattered wavelets, and these combine according to Huyghens' principle to form what is effectively a reflected wave. We may consider the amplitude of the beam scattered by the crystal as a whole to be the resultant of all such reflected waves. This method of using the reflection from an atomic plane as the unit of amplitude was introduced by W. L. Bragg, and intensity formulae were deduced in terms of it by C. G. Darwin,[4] and later by A. H. Compton,[5] and by W. L. Bragg, R. W. James and C. H. Bosanquet,[6] on whose treatment the method given here is based.

Plate I — See other side

PLATE I

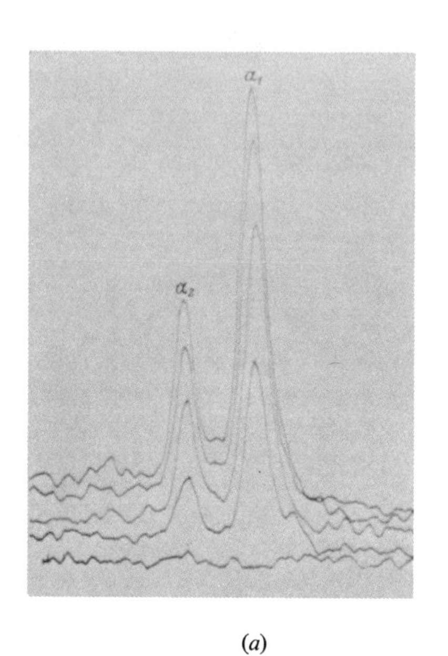

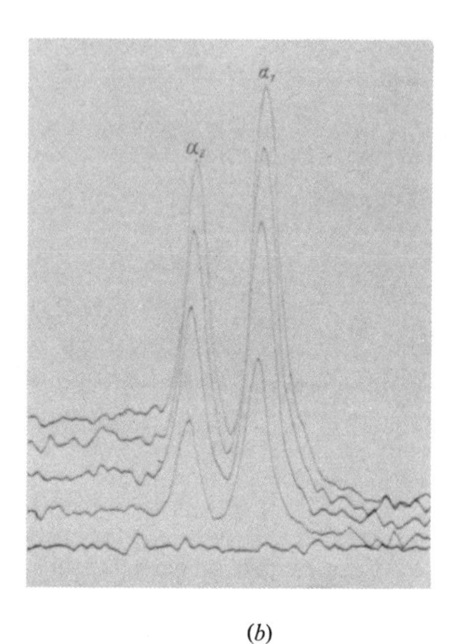

(*a*) (*b*)

FIG. 14. The gold lines $L\alpha_1$ and $L\alpha_2$ reflected in the first order from the octahedral faces of zincblende, (*a*) from the front face, (*b*) from the rear face (Koster, Knol, and Prins)

(*Zeit. f. Physik*, **63**, 345 (1930))

(*b*) *The amplitude reflected by a plane sheet of atoms:* We shall first consider the amplitude of the wave reflected by an infinite plane sheet of atoms, each of which scatters the incident X-rays.

Suppose A, fig. 15, is the source of the radiation, and let the amplitude of the reflected wave be required at B. Let the plane APB be normal to the plane of atoms, and let AP, PB make equal angles θ with this plane. Then P is such that the distance APB is the shortest distance from A to B *via* the plane. Let M be a point of the plane such that the distance AMB is greater by $\lambda/2$ than the distance APB. Then M is on the edge of the first Fresnel zone corresponding to the points A and B, and the whole zone boundary, which is the locus of points such as M, is an ellipse having P as centre and the trace of the plane APB on the reflecting plane as major axis. Proceeding in this way, we may divide the surface of the plane up into the successive Fresnel zones, all of which are elliptical. In the usual Fresnel construction, we are dealing with a wave-front which is supposed to contain the sources of an indefinitely great number of Huyghens wavelets. Here we have a set of actual sources, and, if we are to apply the Fresnel construction, we must show that the scattering points are closely enough set on the planes, in comparison with the areas of the zones, for them to be considered effectively as having a continuous distribution. The area of the first zone is easily shown to be

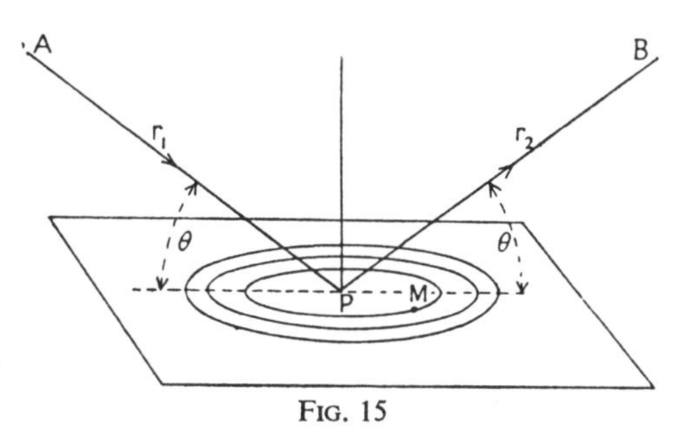

FIG. 15

$$\frac{\pi r_1 r_2}{r_1 + r_2} \frac{\lambda}{\sin\theta}, \tag{2.17}$$

where r_1 and r_2 are the distances AP and BP respectively. Now let us suppose that r_1 is large, corresponding to a nearly plane incident wave, and that r_2 is one centimetre. The area of the first zone is then $\pi\lambda/\sin\theta$. For the least favourable case we take $\sin\theta = 1$, and since $\lambda \sim 10^{-8}$ cm., the area is then about 3×10^{-8} sq. cm. This seems very small, but we must remember that in an average crystal net-plane there are about 10^{15} atoms per sq. cm., so that the number in the first zone in the case we have considered is about thirty million. We shall not, therefore, make any appreciable error in assuming a continuous distribution of scattering points.

The ordinary Fresnel construction shows that the resultant amplitude is equal to half that due to the scattering points lying within the first Fresnel zone; and the resultant amplitude due to the first zone is $2/\pi$

times the sum of the amplitudes due to the individual points. Further, the phase of the resultant wave at B is a quarter of a period behind that of a wavelet scattered by the point lying at the pole P. Let there be n atoms per unit area, and let A be the amplitude at unit distance of the wave scattered by a single atom when the incident amplitude is unity. All the atoms in the first zone may be considered as lying at the same distance r_2 from B, and each therefore contributes an amplitude A/r_2. Using (*2*.17), and the results of the Fresnel construction, we may therefore write

$$\frac{\text{Amplitude at B}}{\text{Incident Amplitude}} = \frac{1}{2} \frac{\pi r_1 r_2}{r_1 + r_2} \frac{\lambda}{\sin\theta} \frac{nA}{r_2} \frac{2}{\pi}.$$

If the incident wave is plane, or nearly so, we may neglect r_2 in comparison with r_1, and in this case q, the ratio of the reflected to the incident amplitude, is given by

$$q = \frac{n\lambda A}{\sin\theta}.$$

This expression takes no account of the phase-lag of a quarter of a period referred to above. We may allow for this by writing the reflection coefficient in the form $-iq$, since $e^{-i\pi/2} = -i$. If f is the atomic scattering factor of an atom, $A = -f(e^2/mc^2)$ when the incident wave is polarised with the electric vector perpendicular to the plane APB. In this case therefore

$$q = -\frac{n\lambda}{\sin\theta} f \frac{e^2}{mc^2}. \tag{2.18}$$

This is not the most convenient form of the expression when we are dealing with one of a number of parallel planes arranged at a distance a apart to form a crystal. If N is the number of atoms per unit volume, $n = Na$, so that (*2*.18) becomes

$$q = -\frac{Na\lambda}{\sin\theta} f \frac{e^2}{mc^2}. \tag{2.19}$$

It is easy to extend the proof we have given to the composite crystal unit, when each simple atomic plane becomes a group of parallel planes. The necessary modification of (*2*.19) is simply the replacement of f by F, the structure factor of the unit group of atoms, N being then the number of unit groups in unit volume. The complex reflection coefficient, which takes account of the phase as well as the amplitude of the scattered wave, may then be written

$$-iq = +i(Na\lambda/\sin\theta)F(e^2/mc^2), \tag{2.20}$$

a formula first given by C. G. Darwin.[4]

(c) *Reflection by a crystal slab of infinite lateral extent, composed of a number of planes:* We shall now consider the amplitude at a point

Q, fig. 16, when a train of plane waves polarised with the electric vector perpendicular to the plane of incidence falls at a glancing angle θ upon a series of parallel scattering planes, whose spacing is a. The radiation reaching Q from the first plane comes effectively from the area round P which contains the first few Fresnel zones. Let QP be produced backwards cutting the successive planes in P′, P″, ... Then, so far as the point Q is concerned, the effective region of the crystal is confined to a small cylinder, having QPP′P″ ... as axis and a diameter a few times that of the first Fresnel zone. The diameter of this region depends on the distance QP, which we shall here take as large in comparison with the total thickness of the set of planes considered, so that the Fresnel zones on all the planes may be taken as equal in area.

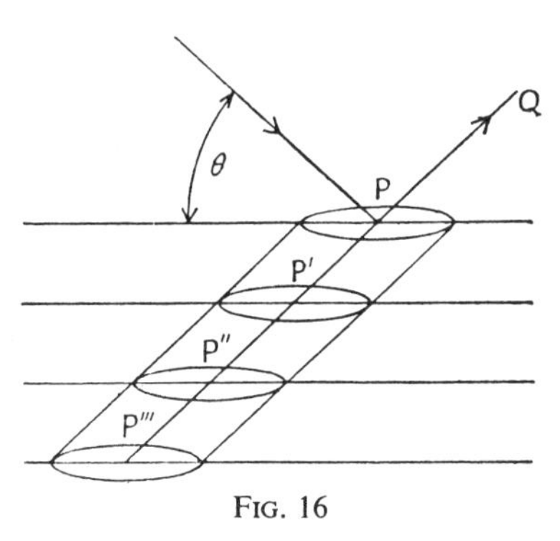

FIG. 16

Each plane scatters an amplitude q given by (2.20), but there is a phase difference between the amplitudes contributed by successive planes equal to $4\pi a(\sin\theta)/\lambda$. Let θ_0 be the angle corresponding to exact agreement in phase between the reflections from successive planes, that is to say, to the direction of a spectrum. We shall consider the reflection only in the near neighbourhood of θ_0, and write

$$\theta = \theta_0 + \epsilon,$$

ϵ being small. When $\theta = \theta_0$, the phase difference between the waves reflected from successive planes is an integral number of times 2π, and so effectively zero. If θ becomes $\theta_0 + \epsilon$, this effective phase difference becomes, since ϵ is small,

$$\frac{4\pi a}{\lambda}\{\sin(\theta_0+\epsilon)-\sin\theta_0\} = \frac{4\pi a\epsilon}{\lambda}\cos\theta_0 = \delta. \qquad (2.21)$$

We shall now calculate the resultant amplitude due to p successive planes, making the following assumptions :

1. That the crystal is so thin that there is no appreciable absorption of the radiation in the p sheets.

2. That the effect of the interaction of the scattered and incident radiation-fields can be neglected. This is equivalent to assuming a refractive index unity, as we have so far done throughout the work.

The amplitude of the incident radiation being A_0, A, that of the resultant reflected wave at Q, is given by

$$\frac{A}{A_0} = q\,\{1 + e^{-i\delta} + \ldots + e^{-(p-1)i\delta}\} = q\,\frac{1-e^{-ip\delta}}{1-e^{-i\delta}}\,. \qquad (2.22)$$

Let I_ϵ and I_0 be the corresponding intensities, or the energy passing per sq. cm. per sec., in the incident and reflected beams respectively ; and let us put

$$I_\epsilon/I_0 = R(\epsilon).$$

Then $R(\epsilon)$ is obtained by multiplying (*2*.22) by its conjugate complex quantity, and this gives, exactly as for equation (*1*.11),

$$R(\epsilon) = |q|^2 \frac{\sin^2(pB\epsilon)}{\sin^2(B\epsilon)}, \qquad (2.23)$$

where B has been written for $2\pi a(\cos\theta_0)/\lambda$.

The expression for the intensity due to p planes is thus of the same form as that due to a diffraction grating of p elements, so that the range of values of the angle of incidence over which the reflected intensity will be appreciable is very small if p is at all large. We have seen in Chapter I, §1(*c*) that the angular range of reflection is of the order of seconds of arc if p is of the order of a few thousand ; but nevertheless there is always a finite range of reflection, and it is convenient to take as a measure of the intensity of a spectrum the integrated intensity, $\int R(\epsilon)\,d\epsilon$, over the range of appreciable reflection in the region of the intensity maximum.

In the case we are now considering, the crystal planes and the wave-fronts of the incident beam are supposed to be of infinite lateral extent. The reflected wave-fronts are therefore also infinite, the intensity being everywhere constant, and we have to consider the energy reflected through some finite area.

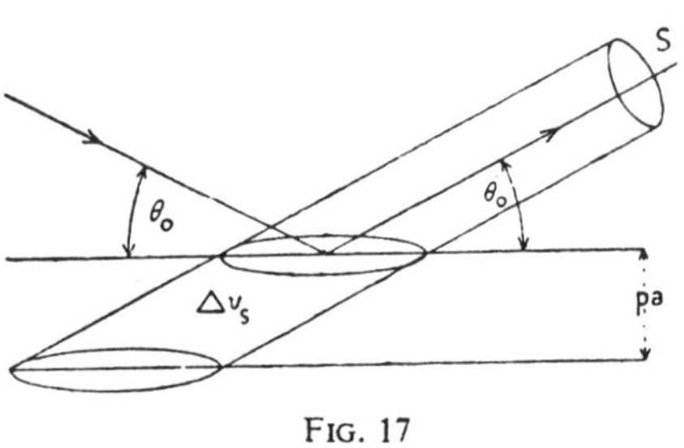
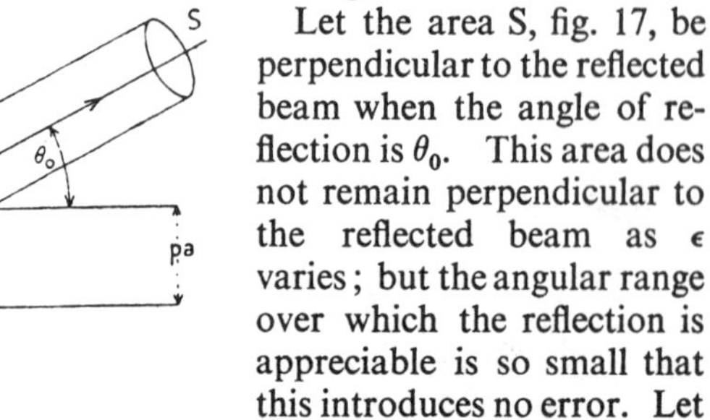

FIG. 17

Let the area S, fig. 17, be perpendicular to the reflected beam when the angle of reflection is θ_0. This area does not remain perpendicular to the reflected beam as ϵ varies; but the angular range over which the reflection is appreciable is so small that this introduces no error. Let us suppose the crystal to rotate with a uniform angular velocity ω about an axis parallel to the reflecting planes and perpendicular to the direction of incidence. Then it takes a time $d\epsilon/\omega$ to rotate through the very small angular range $d\epsilon$, and throughout this range the energy per sq. cm. per sec. in the reflected beam is $I_0R(\epsilon)$. If E_s is the total energy passing through S when the crystal rotates with angular velocity ω through the whole reflecting range, we have

$$E_s = I_0 S \int R(\epsilon)\frac{d\epsilon}{\omega}, \qquad (2.24)$$

or
$$\frac{E_s\omega}{I_0}=S\int R(\epsilon)\,d\epsilon, \tag{2.25}$$

which is independent of the rate of rotation, since E_s varies inversely as ω. Equation (2.25) thus furnishes us with a way of determining $\int R(\epsilon)\,d\epsilon$, and it will be convenient for the time being to base our argument on this method, although the results obtained are more general.

By (2.23), the last equation becomes

$$\frac{E_s\omega}{I_0}=\frac{S}{B}\int |q|^2\,\frac{\sin^2(pB\epsilon)}{\sin^2(B\epsilon)}\,d(B\epsilon). \tag{2.26}$$

The integrand in (2.26) has appreciable values only when ϵ is very small. We can thus assume $|q|^2$ to be constant during the integration, and equal to its value for $\theta=\theta_0$, or $\epsilon=0$, and may also take $\sin(B\epsilon)$ as equal to $B\epsilon$. If now we put $pB\epsilon=x$, (2.26) becomes

$$\frac{E_s\omega}{I_0}=\frac{S\,|q|^2 p}{B}\int_{-\infty}^{+\infty}\frac{\sin^2 x}{x^2}\,dx=\frac{S\,|q|^2\,\pi p}{B}.$$

We may take the limits of integration in the last equation as $\pm\infty$; for, since the integrand has values appreciably different from zero only near $x=0$, the precise limits are unimportant. Putting in the values of q and B from (2.20) and (2.23) we get

$$\frac{E_s\omega}{I_0}=\frac{N^2\lambda^3}{\sin 2\theta_0}\,|F|^2\left(\frac{e^2}{mc^2}\right)^2\frac{paS}{\sin\theta_0}. \tag{2.27}$$

Now $paS/\sin\theta_0$ is the volume of the crystal slice intercepted by the cylinder which has S for base, and whose axis is in the direction of reflection θ_0. Let us call this volume Δv_s. Then

$$\frac{E_s\omega}{I_0}=Q_1\Delta v_s,\ \text{where}\ Q_1=\frac{N^2\lambda^3}{\sin 2\theta_0}\,|F|^2\left(\frac{e^2}{mc^2}\right)^2. \tag{2.28}$$

We have so far supposed the incident radiation to be polarised with the electric vector perpendicular to the plane of incidence. Had the electric vector lain in the plane of incidence, the appropriate value of the structure factor would have been $F\cos 2\theta$, instead of F. In this case the last equation would have taken the form

$$\frac{E_s\omega}{I_0}=Q_2\Delta v_s,\ \text{where}\ Q_2=Q_1\cos^2 2\theta_0. \tag{2.29}$$

If the incident radiation is unpolarised, we must write

$$E_s\omega/I_0=Q\,\Delta v_s,\ \text{where}\ Q=\tfrac{1}{2}(Q_1+Q_2)=Q_1\,\frac{1+\cos^2 2\theta_0}{2}.$$

(*d*) *Reflection by a small crystal of any form:* We shall now suppose the crystal to be of finite lateral extent, and to have any form, and shall calculate the total energy reflected from such a crystal fragment as it is rotated through the reflecting position.

Consider, as before, the energy passing through a plane AA′ (fig. 18) perpendicular to the reflected beam, and let this plane be divided into a number of elements. As the crystal rotates through the reflecting position, the total energy which passes through any element of area such as *ds* in fig. 18 is Q times the volume of the crystal fragment intercepted by the cylinder having the element *ds* as base, and the direction θ_0 as axis. If this is so, the energy reflected by the whole crystal must evidently be Q times its total volume.

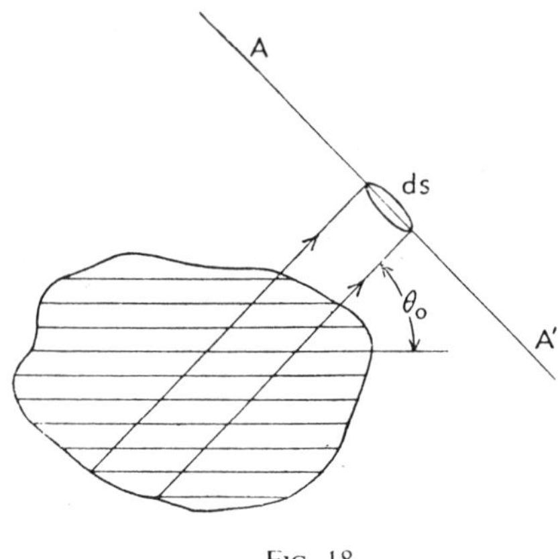

FIG. 18

This argument presupposes that the radiation reflected by the small cylindrical crystal element defined by *ds* has not spread appreciably by the time it arrives at the plane AA′. This, in its turn, requires that the areas of the crystal planes cut off by the cylinder should each contain a number of complete Fresnel zones. Now the size of the Fresnel zones depends on the distance of the plane AA′ from the crystal, and, by choosing this small enough, we can satisfy the required conditions for all the elements of volume. The plane must not, of course, be chosen too near the crystal, or the zones become so small that the distribution of scattering points in them cannot be taken as continuous. The discussion in II, §2(*b*) shows, however, that, with a wave-length of 10^{-8} cm. there will still be about a thousand points in the first zone, even if the distance of the plane from the crystal is reduced to 10^{-5} cm. At this distance, the area of the first zone will be about 10^{-12} cm.^{2}, and its linear dimensions, therefore, of the order of 10^{-6} cm. It is evident then that the argument may be applied to exceedingly minute crystals. Since we are interested only in the total energy reflected from the crystal, which will be the same for all planes such as AA′, and not in the distribution of energy over these planes, it is only necessary that it should be possible in principle to find a plane to which the argument we have used may be applied.

We have thus the following very important result :

The total energy, E, which is reflected by a piece of crystal so small that the absorption of radiation within it may be neglected, as it rotates with uniform angular velocity ω about an axis parallel to a set of crystal planes, and through a range of angles including a spectrum reflected from these planes, is proportional to the volume Δv of the piece of crystal. We have, for the case of unpolarised radiation,

$$\frac{E\omega}{I_0} = Q\,\Delta v, \tag{2.30}$$

I_0 being the energy incident per unit area in the beam, and Q being given by

$$Q = \frac{N^2\lambda^3}{\sin 2\theta_0} |F|^2 \left(\frac{e^2}{mc^2}\right)^2 \frac{1+\cos^2 2\theta_0}{2}. \tag{2.31}$$

(*e*) *The integrated reflection :* The quantity $E\omega/I_0$ is called the *integrated reflection* from the crystal element. Its definition in terms of ω, although convenient for the derivation of the formula, and for its application to the case of reflection from large crystals, is nevertheless rather artificial. The value of the integrated reflection does not depend on ω, and it must be possible to define it without bringing in the idea of angular velocity. It is easy to do this for the crystal element.

Let $P(\theta)\,I_0$ be the radiation reflected in the direction θ by the whole crystal fragment, so that $P(\theta)$ may be called the reflecting power of the whole element. Then the total energy E reflected during the rotation considered above is evidently given by

$$E = \int P(\theta)\,I_0 \frac{d\theta}{\omega}, \quad \text{or} \quad \frac{E\omega}{I_0} = \int P(\theta)\,d\theta. \tag{2.32}$$

We may thus define the integrated reflection from the crystal element by the integral in (2.32). We have finally

$$\int P(\theta)\,d\theta = Q\Delta v, \tag{2.33}$$

a formula that we shall use in considering reflection by powdered crystals. We have here expressed the integration in terms of θ instead of ϵ, as we can do, since $\theta = \theta_0 + \epsilon$.

(*f*) *The integrated reflection derived from an integration in the reciprocal-lattice space :* It is interesting to deduce the formula for the integrated reflection from a small crystal block by the methods of §1(*c*) of Chapter I, which involve the idea of the distribution of the interference function in the reciprocal-lattice space. If J is the intensity of the radiation scattered by a crystal block under the conditions defined by the point (ξ, η, ζ) of the reciprocal-lattice space,

$$J = |F|^2 \left(\frac{e^2}{mc^2}\right)^2 BI_0/R^2. \tag{2.34}$$

J is here the energy passing per unit area per unit time through a plane perpendicular to the direction of scattering, and at a distance R from the crystal which is large in comparison with its linear dimensions.

is a factor depending on the state of polarisation of the incident radiation, which is equal to $(1+\cos^2 2\theta)/2$ for the case of unpolarised radiation.

In fig. 19, let M be the reciprocal-lattice point corresponding to the spectrum considered, and let the shaded area around it denote the region within which the interference function I_0 has appreciable values. Let the sphere of reflection be drawn corresponding to the exact reflection of the spectrum M. The sphere then passes through M; LO is the direction of incidence, and LM the direction corresponding to the maximum intensity of the spectrum. For a given direction of incidence LO, diffraction can take place in a variety of directions corresponding to the intersection of the sphere of reflection with the intensity distribution around M.

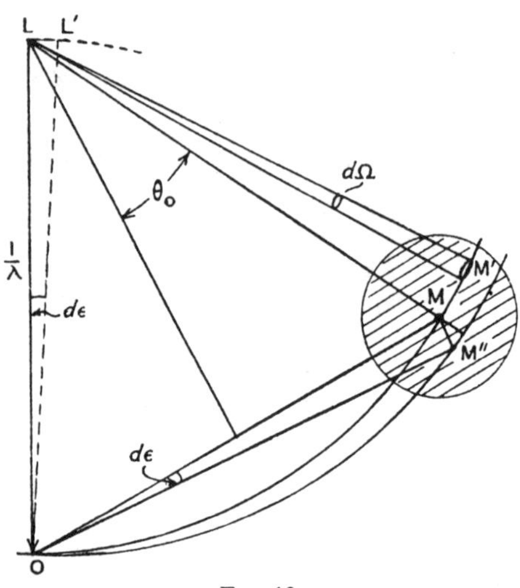

FIG. 19

If $d\Omega$ is a small solid angle containing such a direction of diffraction LM′, the energy diffracted within it for the given angle of incidence is $JR^2d\Omega$ since $R^2d\Omega$ is the area intercepted by the solid angle on the surface of a sphere of radius R. The total energy diffracted for this angle of incidence is therefore $\int JR^2d\Omega$.

We are concerned in practice with diffraction over only a very small total range of solid angles, so that if (x, y) are the rectangular coordinates of the intersection of LM′ with a plane through M and perpendicular to LM, $d\Omega = dx\,dy/LM^2 = \lambda^2\,dx\,dy$. Thus

$$\int JR^2d\Omega = \lambda^2R^2\int Jdx\,dy = J_1. \qquad (2.35)$$

Now suppose the angle of incidence to be varied in such a way that LO moves in a plane containing O, M, and the original direction of LO. This is equivalent to the motion of sweeping the crystal through the reflecting position on an X-ray spectrometer, for example. If LO moves through a small angle $d\epsilon$, a point M″ on the sphere, originally coinciding with M, moves perpendicular to OM through a small distance $|S|\,d\epsilon/\lambda$, and the sphere of reflection in the immediate neighbourhood of M moves nearly parallel to itself through a distance $|S|\cos\theta_0\,d\epsilon/\lambda = dz$ say, where z is perpendicular to x, and y. The total integrated reflection $\int P(\theta)d\theta$ will be the sum of the reflections for all the positions of the crystal relative to the incident beam, and so to $\int J_1 d\epsilon$. By (*2*.35) therefore, since $|S| = 2\sin\theta$,

$$\int P(\theta)d\theta = \lambda^2 R^2 \int J dx\, dy\, d\epsilon = \frac{\lambda^3 R^2}{\sin 2\theta_0}\int J dx\, dy\, dz. \qquad (2.36)$$

We now substitute the value of J given by (*2*.34) in the integral, and since the integrand has appreciable values only very near M, the reciprocal-lattice point, we may use for F its value at that point, and take it outside the integral sign. Thus

$$\int P(\theta)d\theta = \frac{\lambda^3 F_M^2}{\sin 2\theta_0}\left(\frac{e^2}{mc^2}\right)^2 B \int I_0 dx\, dy\, dz. \qquad (2.37)$$

The integral is to be taken through the relevant region of the reciprocal-lattice space, of which $dx\, dy\, dz$ is an element of volume. The integration may be carried out with respect to ξ, η, and ζ by substituting for $dx\, dy\, dz$ the element of volume $v^* d\xi\, d\eta\, d\zeta = d\xi\, d\eta\, d\zeta/v$ where v^* and v are respectively the volumes of the unit cell in the reciprocal lattice and the crystal lattice.

For a crystal having the form of a parallelepiped, (*1*.21a) shows that the required integral is the product of three integrals of the type

$$K = \frac{1}{\pi}\int \frac{\sin^2(N_1\pi\xi)}{\sin^2(\pi\xi)}\, d(\pi\xi).$$

If N_1 is large, $\pi\xi$ will always be small in the relevant region, and we may replace $\sin^2(\pi\xi)$ by $(\pi\xi)^2$ in the integrand, and may then extend the limits of integration to $\pm\infty$. Thus

$$K = \frac{N_1}{\pi}\int_{-\infty}^{+\infty} \frac{\sin^2 u}{u^2}\, du = N_1, \qquad (2.38)$$

and

$$\int I_0\, dx\, dy\, dz = N_1 N_2 N_3/v. \qquad (2.39)$$

If now n is the number of lattice-cells per unit volume, and V is the total volume of the crystal, $nV = N_1 N_2 N_3$, so that $N_1 N_2 N_3/v = n^2 V$, and by (*2*.37)

$$\int P(\theta)\, d\theta = \frac{n^2\lambda^3}{\sin 2\theta_0}\left(\frac{e^2}{mc^2}\right)^2 F_M^2 BV, \qquad (2.40)$$

which is equivalent to equations (*2*.30) and (*2*.31) if the incident radiation is unpolarised.

(*g*) *The crystal mosaic:* We have seen that the range of angles over which a piece of crystal will reflect an X-ray spectrum should be of the order of only a few seconds if the thickness of the crystal exceeds a few thousand atomic planes. Now many real crystals reflect over ranges much greater than this, in some cases over a degree or more, and such behaviour suggests that the apparently single crystal consists in reality of a series of independent crystal regions which are nearly but not quite parallel to one another. To treat such a case mathematically, we shall suppose that each such region consists of a small crystal block, like that considered in the last section, throughout which the planes are quite regular and parallel, and which is small enough for absorption in it to be neglected. The complete crystal is supposed to be composed of a very large number of such blocks, whose planes are nearly parallel, but distributed in orientation over a range of angles small, yet considerably greater than the angular range of reflection from a single block. Such a composite crystal we shall refer to as a *mosaic*, a term introduced by Ewald.

The reflections from the different blocks of a mosaic are optically independent, in the sense that no regular phase relationships exist between them, so that to get the total intensity reflected by the crystal we have simply to add the *intensities* reflected by the individual blocks.

In real crystals such sharply defined little blocks probably do not exist, and the dislocations may be due to warping of the planes, or to other causes. Their optical behaviour with respect to X-rays will, however, be similar to that of the ideal mosaic, if exact regularity does not extend over more than a few thousand atomic planes without some small dislocation. Not all real crystals are mosaics. As we shall see, perfect crystals, or very close approximations to them, do exist. The treatment of reflection from a perfect crystal is, however, rather complicated, and we have preferred to deal first with the easier case of the mosaic, which does, moreover, appear to correspond fairly closely to the state of a large number of real crystals.

(*h*) *Intensity formulae for mosaic crystals :*

(i) Reflection from a face: We now consider the case of a large piece of crystal with an extended face parallel to the mean direction of a set of planes in the mosaic blocks which would form a spectrum at a glancing angle θ_0. An incident X-ray beam of intensity I_0 and cross-section S_0 falls on the face, which is large enough to intercept the whole of it.

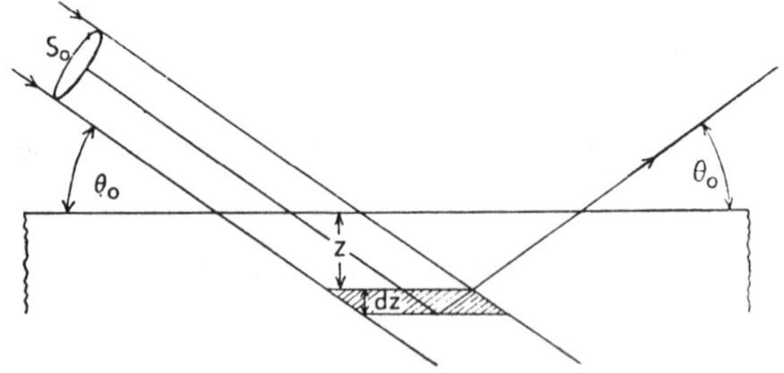

FIG. 20. Reflection from an extended face of a mosaic crystal

Let dE be the contribution to the total energy E reflected by the crystal during a rotation with angular velocity ω through the reflecting range from a block of volume dv lying in the path of the incident beam, and at a depth z below the surface of the crystal (fig. 20). The radiation reflected from this block has traversed in all a distance $2z \operatorname{cosec} \theta_0$ in the crystal before it leaves it, and has had its intensity reduced thereby by a factor $e^{-2\mu z \operatorname{cosec} \theta_0}$, where μ is the linear absorption coefficient of the radiation in the crystal. By (2.30) we have therefore for dE the equation

$$\frac{\omega\, dE}{I_0} = Qe^{-2\mu z \operatorname{cosec} \theta_0}\, dv. \tag{2.41}$$

The total energy reflected by the whole crystal during a rotation through the range over which the mosaic reflects is obtained by integrating (2.41) over the whole volume of the crystal which is irradiated. We may take as a convenient element of volume for the integration a slice of the irradiated region of thickness dz, parallel to the crystal surface and at a distance z below it; for the absorption factor will be the same for the whole of this element. Its volume is $S_0 \operatorname{cosec} \theta_0\, dz$, and if the thickness of the crystal is so great that no radiation gets through it, we have from (2.41)

$$\frac{E\omega}{I_0} = Q \int_0^\infty S_0\, e^{-2\mu z \operatorname{cosec} \theta_0} \operatorname{cosec} \theta_0\, dz = \frac{QS_0}{2\mu}. \tag{2.41a}$$

We have taken the angle of incidence as constant and equal to θ_0 in the integration, since appreciable reflection is confined to a very small range of angles on either side of this direction.

The energy current per unit area in the incident beam is I_0 so that $I_0S_0(=I)$ is the total energy falling per unit time on the crystal. We have therefore

$$\frac{E\omega}{I} = \frac{Q}{2\mu}. \tag{2.42}$$

Suppose now that $R(\theta)$ is the fraction of the total incident beam of radiation that is reflected at an angle θ. Then, just as in equation (2.24), we have

$$E = I \int R(\theta) \frac{d\theta}{\omega}, \quad \text{or} \quad E\omega/I = \int R(\theta)\, d\theta. \tag{2.42a}$$

The quantity $E\omega/I$, or $\int R(\theta)\, d\theta$, is known as the *integrated reflection* from a crystal face, and has the physical dimensions of a pure number. The formula we have deduced here is of great importance in connection with the measurement of intensities of reflection from large crystals by means of the X-ray spectrometer.

(ii) Reflection through a crystal plate: Instead of reflecting from a crystal face it is often more convenient to reflect through a plate of crystal from a set of planes nearly normal to its surface.

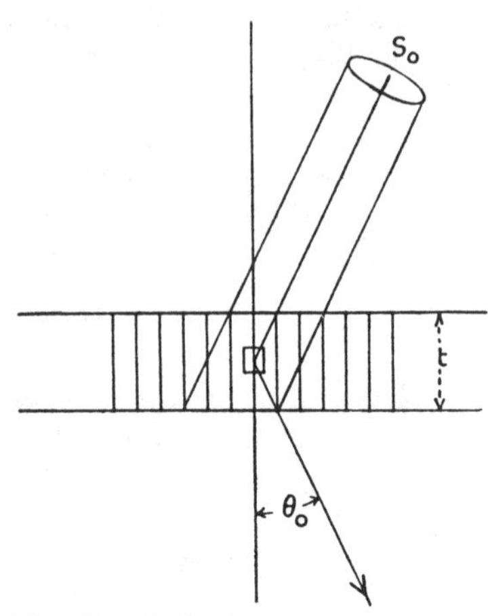

FIG. 21. Reflection through a crystal plate

Let us consider a spectrum formed by reflection at a glancing angle θ_0 from a set of planes which are normal to the surface of a parallel-sided plate of crystal of thickness t (fig. 21). The total distance travelled in the crystal by the radiation reflected from each element of volume is here the same, and is equal to $t \sec \theta_0$. If the crystal is a mosaic, we have at once from (*2.30*)

$$\frac{E\omega}{I_0} = Qe^{-\mu t \sec \theta_0} S_0 t \sec \theta_0, \quad (2.43)$$

for $S_0 t \sec \theta_0$ is the total volume of the crystal plate irradiated by the incident beam. If I is the total energy incident in unit time on the crystal, $I = I_0 S_0$, and

$$\int R(\theta)\, d\theta = E\omega/I = Qt' e^{-\mu t'}, \quad (2.44)$$

where $t' = t \sec \theta_0$, and is the distance travelled in the crystal by any ray. We notice here that the integrated reflection has a maximum value for a given kind of crystal and a given wave-length if

$$t' = 1/\mu. \quad (2.45)$$

This restricts the application of the method to crystals of moderate absorption coefficient for the radiation employed; but it is in many cases a very powerful method of comparing rapidly the intensities of a large number of spectra all belonging to the same zone.

The formula is rather more complicated if the reflecting planes are not perpendicular to the surface of the crystal plate. The general formula for a crystal plate of thickness t can be written in the form [8]

$$\int R(\theta)\, d\theta = \frac{Qq}{\mu} \frac{e^{-pt} - e^{-qt}}{q - p}, \quad (2.46)$$

where $p = \mu \sec \chi \sec (\theta - \eta)$, $q = \mu \sec \chi \sec (\theta + \eta)$. Here, χ is the angle between the normal to the surface of the plate and the plane of reflection, which contains the axis of rotation, and η is the angle between the axis of rotation and the projection of the normal to the plate on the plane of reflection.

(*i*) *Reflection from powdered crystals:* In the examples we have so far considered, we have given the crystal an opportunity to reflect at all settings relative to the incident beam by rotating it. Another way to achieve the same end is to powder the crystal. If the orientation of the fragments is random, we are again in effect giving the crystal an opportunity to reflect at all possible settings. This is the principle of the Debye-Scherrer method. (See Vol. I., p. 33). We shall now calculate the intensity reflected into a given halo.

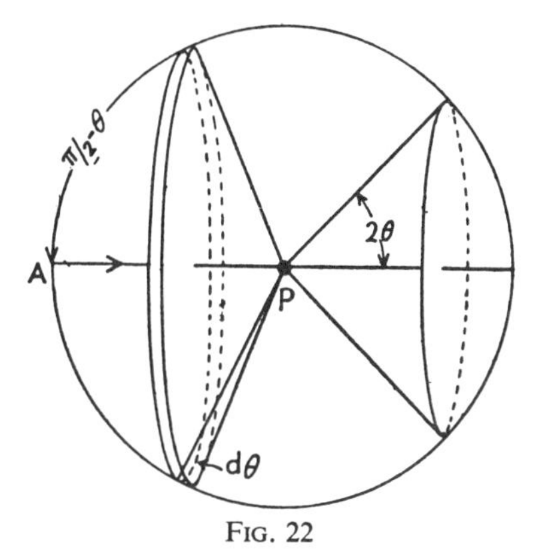

FIG. 22

Let P, (fig. 22), be a small mass of powdered crystal, whose volume V is so small that we can neglect absorption of the radiation in it, and is yet large enough to contain so many powder particles that their orientations may be considered as having a continuous random distribution. Let AP be the direction of the incident beam, whose intensity is I_0, and let us consider those spectra reflected from the crystal particles at a glancing angle θ. The diffracted rays all lie on a cone of semi-vertical angle 2θ, whose axis is the direction of the incident beam, and if the orientation of the particles is truly random, the intensity distribution about this axis for a given angle of diffraction will be uniform. Now the normals to the crystal planes which contribute to this halo all lie in the neighbourhood of a cone whose semi-vertical angle, reckoned from A, is $\frac{\pi}{2}-\theta$ (fig. 22). Let us consider all fragments for which the normals to the plane in question lie between the two cones whose semi-vertical angles are $\frac{\pi}{2}-\theta$ and $\frac{\pi}{2}-(\theta+d\theta)$. With a random distribution of orientations, the chance that a normal lies between these two cones is equal to the ratio of the area cut off between the two cones from the surface of any sphere with P as centre to the whole area of the surface, or to $\frac{1}{2}\cos\theta\, d\theta$. Suppose there are N particles in all; a number $\frac{1}{2}N\cos\theta\, d\theta$ will be set with their normals lying within the given range, and if $\overline{P}(\theta)$ is the average reflecting power of a particle, as defined on p. 41, the total energy reflected by them into the halo will be $\frac{1}{2}NI_0\overline{P}(\theta)\cos\theta\, d\theta$. P, the total energy reflected into the halo by the planes of the type considered, may be obtained by integrating this last expression over all values of θ. Again only a very small range of angles in the neighbourhood of θ_0 is effective, and we take $\cos\theta$ as constant, and equal to $\cos\theta_0$ in the integration. This gives, by (2.33)

$$P=\tfrac{1}{2}NI_0\cos\theta_0\int\overline{P}(\theta)\,d\theta=\tfrac{1}{2}NI_0\cos\theta_0\, Q\,\overline{dv},$$

or
$$\frac{P}{I_0}=\tfrac{1}{2}\cos\theta_0\,QV,$$
if $\overline{dv}$ is the average volume of a crystal particle, and $N\,\overline{dv}=V$, the total volume of the powder.

We have also to take into account the possibility of a number of sets of planes with different crystallographic indices having the same spacing, and so reflecting into the same halo. Let there be p such separate sets: then finally
$$\frac{P}{I_0}=\frac{pQ\cos\theta_0}{2}V. \tag{2.47}$$
The opposite sides of the same set of planes must here be taken as distinct, that is to say, we count (hkl) and ($\bar{h}\bar{k}\bar{l}$) as two different sets of planes.

Equation (2.47) gives the energy reflected in unit time into the whole halo. To calculate that reflected into a short length l of the halo, small compared with its radius, we must multiply (2.47) by $l/2\pi r\sin 2\theta_0$, r being here the distance of the halo from the crystal specimen. This gives
$$\frac{P}{I_0}=\frac{pl}{8\pi r\sin\theta_0}QV. \tag{2.48}$$
In the various methods which employ powdered crystals, absorption in the specimen has to be taken into account, and the allowance to be made depends on the precise geometrical arrangement (cf. for example, A. J. Bradley, *Proc. Phys. Soc.*, **47**, 879 (1935)). The necessary corrections will be discussed in the volume on methods. We shall here consider only one interesting example.

Suppose a beam of X-rays of intensity I_0 and cross-section S_0, whose direction AB lies in the plane of the paper in fig. 23, to fall on a parallel-sided slab of powdered crystal of thickness t, whose surface is perpendicular to the plane of the diagram. Consider the radiation reflected through the slab in a direction making an angle $2\theta_0$ with that of the incident beam. We shall calculate P_t, the energy that falls into the slit of an ionisation chamber whose length perpendicular to the plane of the diagram is l, and which is distant r from the crystal.

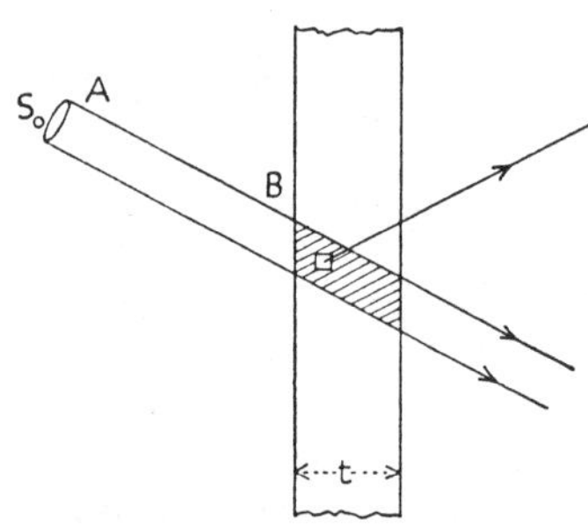

FIG. 23. Reflection by slab of powdered crystal

Both the directly transmitted beam and that reflected from any part of the slab have traversed the same distance, $t\sec\theta_0$, in the crystal, and the total volume irradiated is $(\rho'/\rho)\,S_0 t\sec\theta_0$, where ρ' and ρ are the densities of the powder slab and of the crystal in bulk, respectively. If then I_t is the total energy transmitted per unit time in the direction of the incident beam, the above considerations, together with (2.48), lead at once to the result

$$\frac{P_t}{I_t} = \frac{plt}{4\pi r \sin 2\theta_0} \frac{\rho'}{\rho} Q. \tag{2.49}$$

By moving the ionisation chamber to receive first the directly transmitted beam and then the reflected beam, the ratio P_t/I_t may be measured, and hence Q may be determined.

As in the case of reflection through a crystal slice, there is an optimum thickness for the slab of powder, P_t being a maximum when $t \sec \theta_0 = 1/\mu'$, where μ' is the linear absorption coefficient of the slab for the radiation. This method has been used by Bearden[9] to measure absolute intensities from powdered crystals.

Instead of the transmitted beam, the radiation diffracted from the surface of a slab of crystal powder so thick that it transmits none of the incident radiation is sometimes used.[10] Let α be the glancing angle made by the incident beam with the surface of the slab, and let β be the corresponding angle for the diffracted beam, so that $\alpha + \beta = 2\theta_0$. Consider a layer of the powder slab of thickness dz, parallel to the surface of the slab, and at a depth z below it. If the incident beam has a cross-section S_0, the volume of the crystal particles in this layer irradiated by it is $S_0(\rho'/\rho) \operatorname{cosec} \alpha \, dz$. The radiation in any diffracted beam coming from this layer has traversed a distance $z(\operatorname{cosec} \alpha + \operatorname{cosec} \beta)$ in the powder slab before reaching the surface, and the effective linear absorption coefficient is $\mu(\rho'/\rho)$, μ being the linear absorption coefficient of the crystal in bulk. Using these data, and equation (2.48), it is easy to show by an obvious modification of the method used in §2(*h*) (i) of this chapter, for the mosaic crystal, that

$$\frac{P}{I} = \frac{plQ}{8\pi\mu r \sin \theta_0 \left(1 + \dfrac{\sin \alpha}{\sin \beta}\right)}. \tag{2.50}$$

Here, I is the total power, or energy passing in unit time, in the incident beam, and P is the power in that part of the diffracted beam that enters an ionisation-chamber slit of height l at a distance r from the crystal specimen.

(*j*) *Secondary extinction:* In the formulae that we have deduced for the integrated reflection from mosaic crystals it has so far been assumed that the appropriate absorption coefficient is the ordinary linear coefficient for the crystal. This is, however, not quite correct. The radiation which reaches any block lying at a depth below the surface in a mosaic crystal has passed through other blocks, some of which will be set with their planes so nearly parallel to the block we are considering that they will reflect radiation at the same setting of the crystal. While any block is reflecting, therefore, the intensity of the radiation reaching it is reduced not only by the ordinary processes of absorption, but also by reflection from the nearly parallel blocks through which the incident beam has passed; and the intensity of the reflected beam also suffers diminution in a similar way on the outward path. The contribution from every crystal block to the integrated intensity has suffered this extra diminution, and formally this may be allowed for by using in the expressions for the integrated reflection an effective absorption coefficient, in excess of the ordinary coefficient by an amount which is greater the smaller the range in orientation of the mosaic blocks, and the greater the intensity of the spectrum concerned.

The existence of this effect was pointed out by W. H. Bragg,[11] and was measured for rock salt by W. L. Bragg, James and Bosanquet,[7] who concluded that the increase in the absorption coefficient appropriate to any spectrum was proportional to the intensity of that spectrum. The matter was considered theoretically by C. G. Darwin,[12] who showed that instead of the ordinary linear absorption coefficient μ_0 an effective coefficient μ_ϵ given by

$$\mu_\epsilon = \mu_0 + gQ \tag{2.51}$$

should be used, where g is a constant coefficient for a given crystal specimen, depending on the nature of the mosaic, and Q has the significance of (2.31). This effect was termed by Darwin *secondary extinction.* Its full discussion must be postponed until we have considered reflection by large perfect crystals; we have introduced it here to complete the discussion of the formulae for mosaic crystals. A detailed discussion of the effect will be found in Chapter VI, § 1.

(*k*) *Summary of formulae for intensity of reflection by mosaic crystals:* Before proceeding further we shall give a brief summary of the formulae for the intensity of reflection which we have so far obtained. Throughout, the assumption has been made that there is no interaction between incident and scattered radiation, and that each wavelet taking part in the interference has been scattered once only.

1. All the following formulae are based on that for reflection from a small crystal element of volume dV in which absorption may be neglected. This is

$$\frac{E\omega}{I_0} = \int P(\theta)\,d\theta = Q\,dV. \tag{2.52}$$

E is the energy reflected by the crystal when it rotates with uniform angular velocity ω about an axis parallel to a set of crystal planes which reflect a spectrum at a glancing angle θ_0. I_0 is the energy incident per unit area per unit time in the incident beam. $P(\theta)$ is the reflecting power, at an angle θ, of the crystal element per unit incident intensity.

2. For reflection from the face of a large crystal mosaic,

$$\frac{E\omega}{I}=\int R(\theta)d\theta=\frac{Q}{2\mu}. \tag{2.53}$$

3. For reflection through a plate of mosaic crystal,

$$\frac{E\omega}{I}=\int R(\theta)d\theta=Qt'e^{-\mu t'}, \tag{2.54}$$

where t' is the distance travelled by any ray in the crystal.

In (2) and (3), I is the total energy incident on the crystal in unit time, and $R(\theta)$ is the reflecting power per unit *total* incident energy: μ is the effective linear absorption coefficient, corrected for secondary extinction by (2.51).

4. For a small volume V of crystal powder, in which absorption can be neglected,

$$\frac{P}{I_0}=\tfrac{1}{2}p\cos\theta_0\,QV. \tag{2.55}$$

P is the total energy reflected per unit time into a halo corresponding to a glancing angle θ_0; p is the number of sets of co-operating planes, (hkl) and $(\bar{h}\bar{k}\bar{l})$ being counted as different sets.

5. For P_l the energy reflected into a length l of the halo at a distance r from the crystal,

$$\frac{P_l}{I_0}=\frac{pl}{8\pi r\sin\theta_0}QV. \tag{2.56}$$

6. For reflection *through* a slab of powdered crystal of thickness t,

$$\frac{P_t}{I_t}=\frac{plt}{4\pi r\sin 2\theta_0}\frac{\rho'}{\rho}Q. \tag{2.57}$$

P_t/I_t is the ratio of the energy reflected through the slab into a length l of a halo at a distance r, to the total directly transmitted energy, and ρ' and ρ are the densities of the slab of powder and of the crystal in bulk respectively.

7. For reflection from the surface of a thick slab of crystal powder

$$\frac{P}{I}=\frac{plQ}{8\pi\mu r\sin\theta_0(1+\sin\alpha/\sin\beta)}. \tag{2.58}$$

P/I is the ratio of the energy reflected into a length l of a halo at distance r to the total incident energy; μ is the linear absorption coefficient of the crystal in bulk; α is the glancing angle of incidence of the radiation on the slab, and β the corresponding angle for the diffracted beam. In all these formulae, if the incident radiation is supposed unpolarised,

$$Q=\frac{N^2\lambda^3}{\sin 2\theta_0}|F|^2\left(\frac{e^2}{mc^2}\right)^2\frac{1+\cos^2 2\theta_0}{2}.$$

In the next section we shall have to consider under what conditions the assumptions on which these formulae are based are valid, and how they have to be modified when these conditions are no longer fulfilled.

3. Reflection from Perfect Crystals

(*a*) *Introductory:* We have now to take into account certain factors which have been neglected in the preceding sections. It was there assumed that the primary beam passed through the crystal with the velocity of light in free space, c, and without loss of intensity. The latter assumption evidently cannot be true when a spectrum is produced. The very existence of a reflected beam shows that energy must have been abstracted from the primary beam, quite apart from any losses due to photoelectric absorption, or to incoherent scattering. With an infinitely thick crystal, the assumption would in fact lead to an infinite reflected beam—an evident absurdity. So long as we are dealing with small crystals the assumption is justified, since the losses by scattering at each plane are small, and are greater than the losses by ordinary absorption, but it ceases to be valid for a large crystal.

When the angle of incidence has an appropriate value, a reflected wave is produced whose intensity is of the same order of magnitude as that of the incident wave. This reflected wave will also meet the crystal planes at the correct angle for reflection, and so will be reflected a second time, back into the direction of the primary beam. This effect we have so far entirely neglected. If regularity of the crystal planes persists over a large volume of the crystal, the twice reflected beam that builds up will have definite phase-relationships with the primary beam, and must modify its amplitude considerably. A full treatment of diffraction by a crystal lattice must therefore include the effect of such multiple reflections.

Finally, the assumption that the radiation travels through the crystal with a velocity c cannot be justified. In addition to the reflected wave from each plane, a wave is scattered in the forward direction whose phase differs by about a quarter of a period from that of the incident wave. Scattered and incident waves together build up a wave differing slightly in phase from the incident wave, and this happens at each plane of atoms which the wave traverses. The resultant wave thus passes through the crystal with a modified phase-velocity, and it is, of course, the phase-velocity that interests us in diffraction problems. We shall therefore expect slight modifications of the Bragg law for reflection, which depends on the phase retardation between the different reflected waves in the crystal.

All the points just discussed were fully dealt with by Darwin [4] in two remarkable papers published within two years of the discovery of diffraction by crystals. The treatment we shall give here is modelled on that of Darwin, rather than on that of Ewald, who later, and independently, worked out the same ideas in his well-known papers on the dynamical theory of diffraction by a crystal. Ewald's treatment is rather more general, but Darwin's is probably easier to understand, and certainly follows more naturally on the work of the preceding sections. Ewald's theory will, however, be discussed later.

(*b*) *The refractive index of the crystal:* We shall first discuss the modification of the phase velocity in the crystal. In § 2(*b*) of this chapter we calculated the amplitude of the wave reflected from an infinite plane sheet of atoms. Now exactly the same construction applies to the wavelets scattered on the other side of the plane, which must build up to give a scattered wave travelling in the same direction as the incident wave. The amplitude of this wave is evidently to be obtained by replacing f in (2.18) and (2.19) by $f(0)$, the value of the atomic scattering factor for radiation scattered in the direction of the incident wave. We should perhaps write instead of f, $f(2\theta)$, to signify that it refers to radiation scattered through an angle 2θ.

By (2.20), we can therefore write for the amplitude of this forward scattered wave

$$-iq_0 = +i(n\lambda/\sin\theta)f(0)\,e^2/mc^2, \tag{2.59}$$

n being the number of atoms per unit area of the plane.*

The amplitude of the primary beam after transmission through a plane is the resultant of the incident amplitude A and the forward scattered amplitude $-iq_0$A, or A$(1-iq_0)$, which, if q_0 is small, is to a close approximation Ae^{-iq_0}. Thus the primary beam suffers a change in phase of $-q_0$ in passing through the plane.

Let the beam now pass in succession through p planes whose spacing is a. Suppose that the glancing angle of incidence θ is not that corresponding to reflection, so that no appreciable reflected beam builds up, and we have only to consider the primary beam. After it has passed through p planes, its phase has changed by $-pq_0$ owing to this cause. The wave has travelled a distance $x = pa/\sin\theta$ in passing through the p planes. This alone would correspond to a phase-lag of $(2\pi/\lambda)\,x$, so that the total phase-lag in the distance x is $(2\pi/\lambda)\,x + pq_0$.

Now, by (2.59), since $n = \mathrm{N}a$, N being the number of atoms per unit volume, we have

$$q_0 = -(\mathrm{N}a\lambda/\sin\theta)f(0)\,e^2/mc^2 = -(2\pi/\lambda)(a/\sin\theta)\,\delta,$$

* The assumption that both reflected and forward scattered waves lag in phase by $\pi/2$ behind the incident wave, which is symbolised by the factor $-i$, is not strictly accurate, for it will not lead to conservation of energy if q and q_0 are real. Let us suppose that the amplitudes of the forward and reflected waves are $\rho_1 e^{i\phi}$ and $\rho_2 e^{i\phi}$ times that of the incident wave respectively, ρ_1 and ρ_2 being real amplitude factors, and ϕ a phase factor. Assuming unit incident amplitude, the total transmitted amplitude is $1+\rho_1 e^{i\phi}$, and for conservation of energy we must have

$$|1+\rho_1 e^{i\phi}|^2 + |\rho_2 e^{i\phi}|^2 = 1,$$

which leads directly to $\cos\phi = -(\rho_1^2+\rho_2^2)/2\rho_1$. Now the coefficients ρ_1 and ρ_2 are very small compared with unity, of the order of 10^{-4} for strong reflections, so that ϕ is very nearly indeed equal to, but slightly greater than, $\pi/2$. This slight excess over $\pi/2$ produces a diminution in the transmitted amplitude which allows for the energy deflected into the reflected wave. (Ewald, *Phys. Zeit.*, **21**, 617, (1920).)

where
$$\delta = \frac{\lambda^2 e^2}{2\pi mc^2} \mathrm{N} f(0) = -\frac{q_0 \sin\theta}{\kappa a}. \tag{2.60}$$

Thus the total phase-lag in a distance x may be written $(2\pi/\lambda)\, x(1-\delta)$. A distance x in the crystal is thus equivalent, in so far as phase-lag is concerned, to a distance $(1-\delta)\, x$ in free space; or, in other words, the crystal has a refractive index of $1-\delta$ for the waves. If n is the refractive index,

$$n = 1 - \delta = 1 - \frac{\lambda^2 e^2}{2\pi mc^2} \mathrm{N} f(0). \tag{2.61}$$

If the crystal units are not single atoms, but groups, we have simply to substitute $\mathrm{F}(0) = \Sigma f(0)$ for $f(0)$.

The quantity δ is small compared with unity. Without going into details at this stage, we may calculate its order of magnitude. If the frequency of the incident radiation is considerably greater than the critical absorption frequencies of the scattering atoms, we may put $f(0) = \mathrm{Z}$, the total number of electrons in the atom. If ρ is the density of the crystal, A the atomic weight of the scattering atoms, or the sum of the atomic weights if the unit is composite, and m_H the mass of the hydrogen atom, we may put

$$\delta = \frac{e^2}{2\pi mc^2 m_H} \frac{\mathrm{Z}}{\mathrm{A}} \rho\lambda^2 = 2{\cdot}72 \times 10^{10} \frac{\mathrm{Z}}{\mathrm{A}} \rho\lambda^2.$$

For MoKα, $\lambda = 0{\cdot}71 \times 10^{-8}$ cm., and $\delta = 1{\cdot}36 \times 10^{-6} \rho \mathrm{Z}/\mathrm{A}$. Since Z/A is always about one half, we see that n is less than unity by a few parts in a million for most substances. We shall return to the discussion of the numerical values of refractive indices in Chapter IV.

Nothing in the argument we have used in deducing the refractive indices depends essentially on a crystalline arrangement of the points. So long as the reflection conditions are not fulfilled, and no diffracted beam is produced, the refractive index for a crystal will be equal to that for an amorphous substance of the same composition and density. Matters are more complicated if diffracted waves are produced, for these must interact again with the primary beam, and a full discussion requires the methods of the dynamical theory.

(c) *Deviations from Bragg's law:* Let us consider the case of a crystal whose face is parallel to the reflecting planes. Let P_1 and P_2 be two successive planes. If the reflection condition is to be fulfilled, the reflected beams from P_1 and P_2 should differ in phase by $2m\pi$, m being integral, and the phase of the wave being measured above the plane P_1. Both direct and reflected beams have their phases changed by $-q_0$ in passing through a sheet of atoms, and the wave from P_2 has passed through P_1 twice, once in the primary and once in the reflected beam. The path difference between the two beams is $2a \sin\theta$, and Bragg's law, which only takes this path difference into account, gives for the reflection condition

$$2a \sin \theta_0 = m\lambda. \tag{2.62}$$

When, however, we allow for the extra phase changes due to transmission, we get the condition

$$(2\pi/\lambda)\, 2a \sin \theta_1 + 2q_0 = 2m\pi, \tag{2.63}$$

or, by (2.60), $2a \sin \theta_1 = m\lambda + (2a/\sin \theta_1)\delta$, which, since $2a \sin \theta_1$ is very nearly equal to $m\lambda$, may be written

$$2a \sin \theta_1 = m\lambda \left(1 + \frac{\delta}{\sin^2 \theta_1}\right). \tag{2.64}$$

We see, therefore, that $(2a \sin \theta)/m$ is not constant, as the elementary law requires, but depends slightly on the order of the spectrum. We may calculate the deviation from the reflection law. Let θ_0 be the Bragg angle, and θ_1 the actual angle of a spectrum, and let $\Delta\theta_0 = \theta_1 - \theta_0$. Then, since $\Delta\theta_0$ is small,

$$2a \sin \theta_0 + 2a \cos \theta_0\, \Delta\theta_0 = m\lambda + 2a\delta/\sin \theta_1,$$

or, using (2.62),

$$\Delta\theta_0 = \delta/\sin \theta_0 \cos \theta_0 = 2\delta/\sin 2\theta_0, \tag{2.65}$$

nearly enough; which, since $\delta \sim 10^{-5}$ or 10^{-6}, is of the order of seconds of arc only. The existence of this deviation from the Bragg law was first pointed out by Darwin[4] in 1913. It appears, however, to have been forgotten, and small as it is, was rediscovered experimentally by Stenström[13] in Siegbahn's laboratory in 1919, and was independently explained by Ewald.[14] The deduction that we have given applies to small crystals and mosaic crystals, since we have neglected the effect of multiple reflections. In the next section we shall see how the result must be modified if these are taken into account.

(*d*) *Reflection from a set of crystal planes, taking into account multiple reflections:* We shall now proceed to calculate the intensity of reflection from the surface of a crystal composed of an ideally regular array of lattice-planes containing atoms all of one kind, and spaced at a distance a apart. In doing this, we shall follow closely the method of Darwin,[4] but shall introduce a few modifications due to Prins.[15]

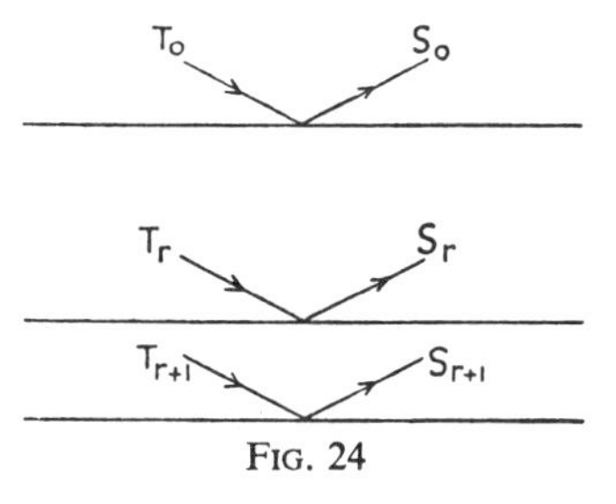

FIG. 24

Let any plane be denoted by a serial number r, which is zero for the surface plane, and let T_r and S_r, fig. 24, represent the displacements in the primary and reflected waves respectively at a point just above the rth plane. What we wish to determine is then

S_0/T_0, the ratio of the amplitude of the reflected beam leaving the surface to that of the incident beam on entering it. The whole idea underlying the calculation is that of the interchange of energy between the primary and the reflected beams. S_r, for example, must consist of that part of T_r which is reflected at the upper surface of the rth plane together with that part of S_{r+1} which is transmitted through the rth plane from below. Similarly, T_{r+1} consists of that part of T_r which is transmitted through the rth plane from above together with that part of S_{r+1} which is reflected from the lower side of the rth plane. We assume a reflection coefficient $-iq$ and a transmission coefficient $(1-iq_0)$, q and q_0 having the values given by (2.20) and (2.59).

Now S_{r+1} is the value of the secondary beam just above the $(r+1)$th plane. In calculating the fraction of it transmitted by the rth plane, what has to be multiplied by $1-iq_0$ is its value just below the rth plane. This is not S_{r+1} but $S_{r+1}e^{-i\phi}$, where

$$\phi=(2\pi/\lambda)\,a\sin\theta=\kappa a\sin\theta, \tag{2.66}$$

and is the phase difference corresponding to the path difference $a\sin\theta$ for two successive planes when the glancing angle of incidence is θ. The above considerations lead directly to the equations

$$S_r=-iqT_r+(1-iq_0)\,e^{-i\phi}\,S_{r+1}, \tag{2.67}$$

$$T_{r+1}=(1-iq_0)\,e^{-i\phi}\,T_r-i\bar{q}\,e^{-2i\phi}\,S_{r+1}, \tag{2.68}$$

where $\bar{q}$ is the reflection coefficient from the lower side of the plane, in a composite crystal, with absorption, not necessarily equal to q. In equation (2.67) we now substitute for S_{r+1} from (2.68), and for S_r from the corresponding equation with r diminished by one unit. This leads to

$$(1-iq_0)\,(T_{r-1}+T_{r+1})=\{q\bar{q}\,e^{-i\phi}+(1-iq_0)^2\,e^{-i\phi}+e^{i\phi}\}\,T_r, \tag{2.69}$$

a set of equations that are solved by the substitution

$$T_{r+1}=xT_r, \tag{2.70}$$

x being independent of r.

Apart from the phase factor, T_r varies slowly from plane to plane. The phase change from plane to plane is also very nearly equal to $m\pi$, since we suppose that the direction θ is very nearly that corresponding to a spectrum. We may therefore put

$$x=(1-\xi)\,e^{-im\pi}, \tag{2.71}$$

where ξ is a small quantity, which may be complex. We also define a quantity v by

$$\kappa a\sin\theta=\phi=m\pi+v. \tag{2.72}$$

v is zero when θ is equal to θ_0, corresponding to reflection according to Bragg's law.

Substituting from (*2*.70) and (*2*.71) in (*2*.69), we get

$$(1-iq_0)\left\{1-\xi+\frac{1}{1-\xi}\right\}=q\bar{q}e^{-iv}+(1-iq_0)^2e^{-iv}+e^{iv},$$

and this, if we expand, and retain only the squares of the small quantities ξ, q_0, q, $\bar{q}$ and v, gives quite simply

$$\xi^2=q\bar{q}-(q_0+v)^2. \tag{2.73}$$

Now $T_{r+1}=xT_r$, and this gives at once, from (*2*.67) and (*2*.68), $S_{r+1}=xS_r$. If we substitute this in (2.67), and put $r=0$, we get

$$\frac{S_0}{T_0}=\frac{-iq}{1-x(1-iq_0)e^{-i(m\pi+v)}}, \tag{2.74}$$

which, if the squares of the small quantities be neglected in its denominator, gives, using (*2*.71) and (*2*.73),

$$\frac{S_0}{T_0}=\frac{-q}{q_0+v\pm\sqrt{(q_0+v)^2-q\bar{q}}}. \tag{2.75}$$

The expression (*2*.75) gives the ratio of the amplitudes of the reflected and incident beams at the surface of the crystal. If there is absorption in the crystal q, $\bar{q}$, and q_0 may be complex, as we shall see below. The ratio of the intensity of the reflected beam to that of the incident beam is given by $|S_0/T_0|^2$, the square of the modulus of the right-hand side of the equation (*2*.75). Which sign is to be taken for the square-root may be determined by the condition necessary for the conservation of energy, that $|S_0/T_0|^2$ cannot exceed unity.

We have so far made no allowance for the absorption of the radiation on its passage through the atomic planes, owing to the photoelectric effect and to incoherent scattering, which must occur. Darwin did this by introducing an absorption factor h, so that the transmitted amplitude at each plane was $(1-h-iq_0)$ of the incident amplitude, instead of $(1-iq_0)$, as we have assumed. Prins has pointed out that the absorption may be allowed for more easily by carrying out the calculation as we have done here, assuming, however, that q, $\bar{q}$, and q_0 may be complex; that is to say, by assuming the scattering of the radiation to be accompanied by an appropriate change of phase, which will just give the absorption. We shall consider Prins's method in detail below, but shall, however, first assume the absorption coefficient to be negligible in comparison with the reflection coefficient. This case was considered in detail by Darwin, and it is in fact approximately realised when crystals of low absorption coefficient reflect radiation of wave-lengths that are not too long.

(*e*) *Crystal with negligible absorption:* It is more convenient in this case to introduce a quantity ϵ, such that $\epsilon=q_0+v$, or, by (*2*.72),

$$\kappa a\sin\theta=m\pi-q_0+\epsilon. \tag{2.76}$$

If θ_1 is the angle of reflection corresponding to the modified Bragg law, by (2.63),

$$\kappa a(\sin\theta - \sin\theta_1) = \epsilon, \text{ or } \kappa a \cos\theta_1 \,\Delta\theta_1 = \epsilon, \tag{2.77}$$

where $\Delta\theta_1 = \theta - \theta_1$. The quantity ϵ is thus proportional to the deviation of the angle of incidence under consideration from that corresponding to the modified reflection law deduced in II, § 3(*c*). In terms of ϵ, equation (2.75) becomes

$$\frac{S_0}{T_0} = \frac{-q}{\epsilon \pm \sqrt{\epsilon^2 - q\bar{q}}}. \tag{2.78}$$

Since there is assumed to be no absorption, q is equal to $\bar{q}$, and both are real; so that if $-q < \epsilon < +q$, (2.78) becomes

$$\frac{S_0}{T_0} = \frac{-q}{\epsilon \pm i\sqrt{q^2 - \epsilon^2}}, \tag{2.79}$$

and $$|S_0/T_0|^2 = 1.$$

This means that, over the narrow angular range defined by (2.79), the reflection from a perfect crystal with no absorption should be total. Outside this range, the reflection diminishes rapidly and symmetrically on either side. For positive ϵ the positive value of the square-root in (2.78) must be taken, and for negative ϵ the negative value, if the condition that the reflection coefficient must be less than unity is to be satisfied.

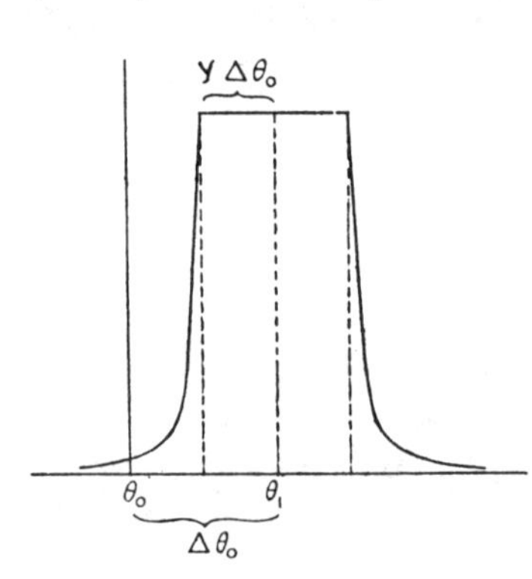

FIG. 25. The reflection curve for a perfect crystal without absorption

The reflection curve has the form shown in fig. 25.

The ordinates are the values of the reflection coefficient $|S_0/T_0|^2$, or $R(\theta)$, as defined by (2.42a), and the abscissae are glancing angles of incidence or reflection. The middle of the range of total reflection lies at $\epsilon = 0$, or $\theta = \theta_1$, the modified Bragg angle. The range of total reflection extends from $\epsilon = +q$ to $\epsilon = -q$, or, by (2.77), from

$$\theta = \theta_1 + q/\kappa a \cos\theta_1 \text{ to } \theta = \theta_1 - q/\kappa a \cos\theta_1, \tag{2.80}$$

and is therefore proportional to the *amplitude* reflected by a single plane.

It is interesting to see how this angular range of total reflection is situated with respect to the angle θ_0 given by the original Bragg reflection law. By (2.60) and (2.65), we have

$$\Delta\theta_0 = \theta_1 - \theta_0 = \delta/\sin\theta_0 \cos\theta_0 = -q_0/\kappa a \cos\theta_0. \tag{2.81}$$

Since the whole angular range in which we are interested is only of the order of a few seconds, we can use θ_0 or θ_1 indifferently in the angular functions. By (2.80) and (2.81), therefore, we see that the semi-angular width of the range of total reflection is $\Delta\theta_0|q|/|q_0|$. The value of the ratio $|q|/|q_0|$, which we shall denote by γ, depends on the state of polarisation of the incident radiation. Let γ_1 be its value when the electric vector is perpendicular to the plane of incidence, γ_2 when it is parallel to it. Then

$$\gamma_1=\frac{|f(2\theta)|}{|f(0)|},\ \gamma_2=\frac{|f(2\theta)|}{|f(0)|}\cos 2\theta. \tag{2.82}$$

If the atom scatters as a point, $|f(2\theta)|=|f(0)|$, and $\gamma_1=1$. One limit of the range of total reflection for the corresponding state of polarisation then lies at the Bragg angle θ_0, and if, as in the usual cases, δ is positive, it will be the lower limit. For all real atoms $|f(2\theta)|<|f(0)|$ for any finite value of the angle θ, and $\gamma_1<1$. The range of total reflection then contracts, and lies wholly above the angle θ_0. The value of γ_2 is always less than unity for any angle greater than zero. If the radiation were unpolarised, we should have two curves of different widths superposed, and this would have the effect of rounding off the corners of the curve in fig. 25.

(*f*) *The integrated reflection from the face of a perfect crystal:* The reflecting power of the crystal is by definition equal to $|S_0/T_0|^2$, so that the integrated reflection from the large perfect crystal with neglible absorption is equal to the area lying beneath the curve in fig. 25. The area beneath the two portions of the curve lying outside the range of total reflection has to be obtained by integrating the expression for $|S_0/T_0|^2$ obtained from (2.78) from $\epsilon=+q$ to $\epsilon=+\infty$ and from $\epsilon=-q$ to $\epsilon=-\infty$. The integrations are of a simple type, q and q_0 being, of course, treated as constants. It is found that one quarter of the whole area is contributed by these parts of the curve. The whole area is thus 4/3 times the area beneath the range of total reflection, that is to say, to 4/3 times its angular breadth, since the ordinate is here unity. Thus the integrated reflection is given by

$$\int R(\theta)\,d\theta=\tfrac{4}{3}(2\gamma\Delta\theta_0)=\tfrac{8}{3}|\gamma||\delta|/\sin\theta_0\cos\theta_0. \tag{2.83}$$

We suppose the radiation to be unpolarised, and so put

$$|\gamma|=\tfrac{1}{2}(|\gamma_1|+|\gamma_2|),$$

and, using the value of δ from (2.60), get

$$\int R(\theta)\,d\theta=\frac{8}{3\pi}\frac{N\lambda^2}{\sin 2\theta_0}|f(2\theta)|\frac{e^2}{mc^2}\frac{1+|\cos 2\theta_0|}{2}, \tag{2.84}$$

a formula first obtained by Darwin.

If we compare this result with (2.53), the corresponding formula for a mosaic crystal, we see that they differ greatly. The integrated reflection from the mosaic is proportional to the square of the number of atoms per unit volume, and to the square of the scattering factor f; that for the perfect crystal is proportional to the first powers of these quantities. The numerical values for the same reflection given by the two formulae differ very greatly, that from the mosaic formula being much the larger. We shall have to discuss in Chapter VI what experiment has to say in helping to decide which formula is applicable, but before doing so we must consider one or two further points of theory.

(*g*) *Primary extinction:* When the angle of incidence θ has a value lying within the range of total reflection the energy of the primary beam is wholly diverted into the secondary, or reflected, beam. Under these conditions, the primary beam diminishes rapidly in intensity as it passes through the crystal, which behaves as if it had a large absorption coefficient, even although the ordinary absorption coefficient is, as we have assumed, negligible. We can see in a general way how such diminution may come about if we remember that there is a phase-lag of $\pi/2$ at each reflection, so that the twice-reflected waves, which again travel in the direction of the incident wave, will now be out of phase with it, and will reduce, and not increase, its amplitude. From the equations of the last section we can easily calculate the magnitude of the effect.

From (2.71), it will be seen that ξ represents absorption, the amplitude of the primary beam T_r being reduced by a factor approximately $e^{-\xi}$ at each plane, and its intensity, assuming ξ to be real, by $e^{-2\xi}$. Thus 2ξ is the absorption coefficient per plane. This effect, which must not be confused with ordinary absorption, we shall refer to as *extinction*, or more accurately, following Darwin, as *primary* extinction, in order to distinguish it from the essentially different process of secondary extinction, which was discussed in II, § 2(*j*) in connection with mosaic crystals.

By (2.73) and (2.76), $\xi = \sqrt{q^2 - \epsilon^2}$, and within the reflecting range $q^2 > \epsilon^2$, so that ξ is real. Within this range ϵ varies from $-q$ to $+q$ and $\bar{\xi}$, the mean value of ξ, is $\pi|q|/4$. We can therefore take for the mean value of the extinction coefficient, $2\bar{\xi}$, or $\pi|q|/2$. Thus

$$2\bar{\xi} = \frac{\pi N \lambda a}{2 \sin \theta} |f| \frac{e^2}{mc^2}, \tag{2.85}$$

or, for an nth order spectrum,

$$2\bar{\xi} = \frac{\pi a^2 N}{n} |f| \frac{e^2}{mc^2}. \tag{2.85a}$$

We have here taken $\lambda/\sin \theta$ as constant over the reflecting range, and equal to $2a/n$, the value given by the Bragg law, which is quite accurate enough, since the range of reflection is very small, and occurs at an angle differing only slightly from the Bragg angle.

If β is the amplitude absorption-factor per plane due to ordinary absorption, and μ_0 is the linear absorption coefficient,

$$2\beta = \mu_0 a \operatorname{cosec} \theta, \tag{2.86}$$

since a distance $a \operatorname{cosec} \theta$ is traversed by the wave in passing between each pair of crystal planes.

For MoKα radiation ($\lambda = 0{\cdot}71$A), μ_0 for rock salt is 16·4. For 200 and 600 we may take F = 21·5 and 7·0 respectively.* We therefore get, using the known values of a and N for rock salt,

$$\text{For } 200,\ 2\bar{\xi} = 3{\cdot}39 \times 10^{-4},\ 2\beta = 3{\cdot}66 \times 10^{-6}.$$

$$\text{For } 600,\ 2\bar{\xi} = 3{\cdot}66 \times 10^{-5},\ 2\beta = 1{\cdot}22 \times 10^{-6}.$$

For diamond, the corresponding figures are †

$$\text{For } 111,\ 2\bar{\xi} = 1{\cdot}22 \times 10^{-4},\ 2\beta = 2{\cdot}91 \times 10^{-7}.$$

From these figures, the decrease in the intensity of the primary beam after passing through any number of planes may be calculated.

Some results are shown in fig. 26. The ordinates show the intensity of the transmitted beam, the incident intensity being taken as unity

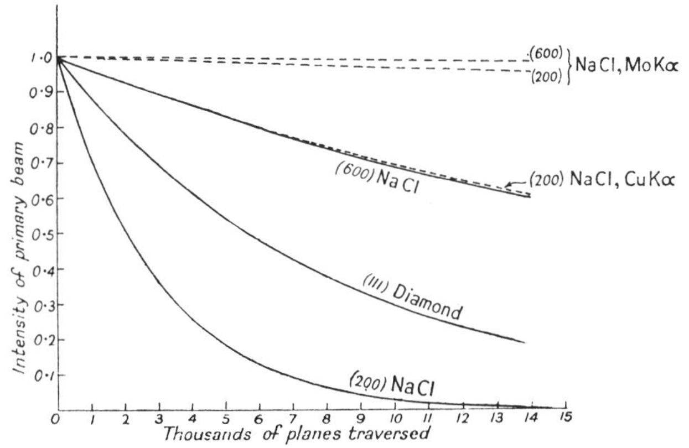

FIG. 26. Comparison of extinction and absorption for rock salt and diamond with MoKα radiation

the abscissae, the number of thousands of planes passed through. The dotted curves show the effect of ordinary absorption which, with these crystals, and with MoKα radiation, is nearly negligible in a thickness of

* The formulae of II, §§ 2(d), 2(e), and 2(f) were worked out assuming a simple lattice. They apply to the compound lattice if the structure factor F is used instead of f, provided that the crystal is not polar (see p. 65). The values of F used for NaCl are based on those of James and Firth,[16] and those for C on the tables of James and Brindley.[17]

† The plane in the case of diamond 111 is the composite pair, a being the identity period.

10,000 planes or so. On the other hand, we see that owing to extinction, shown by the full curves, the primary beam falls off rapidly in intensity with increasing depth. In the case of 200 NaCl, only about 1/100 of the incident intensity penetrates to a depth greater than 12,000 planes, and only one half to a distance of about 2000 planes, a distance about equal to the wave-length of orange light. The whole process of reflection, and of dynamical interchange between the primary and secondary beam, is over in a layer in which the intensity of the incident beam would not be reduced by the ordinary processes of absorption by more than a few per cent. It is evident, too, that the assumption that the intensity reflected from a block of crystal 10,000 planes thick is proportional to the volume irradiated, which we might safely make so far as absorption alone is concerned, becomes quite invalid owing to extinction; for only rather less than one fifth of the incident energy penetrates as much as half way through such a block.

With the spectrum 600, conditions are rather different. The absorption per plane is still smaller than for 200, but extinction is now very much smaller, only diminishing the intensity of the primary beam by about thirty per cent. in a thickness of 10,000 planes. A thickness some ten times as great would be required for complete dynamical interchange between primary and secondary beam to be set up. The extinction coefficient is directly proportional to the amplitude reflected from a single plane, and so falls off rapidly for the higher orders both on account of increase in θ and decrease in f.

With diamond, the curves are very similar. The ordinary absorption is very small. The intensity is reduced by extinction to 1/100 of its incident value in a depth of about 30,000 planes. For crystals such as these, with relatively penetrating radiation, the condition of negligible absorption, assumed by Darwin, upon which the formulae we have so far obtained depend, is evidently justifiable. But with less penetrating radiation, or with crystals containing heavier elements, matters are very different. For CuKα radiation ($\lambda = 1{\cdot}539$A), for example, $\mu_0 = 160$ in rock salt, and the curve for ordinary absorption for 200 falls very close to the curve for extinction for 600. The assumption that absorption is negligible in comparison with extinction is here evidently not justifiable. The next step, therefore, must be to take absorption into account in the equations for reflection from the perfect crystal.

(*h*) *The perfect crystal with absorption—Prins's method:* In applying the correction for absorption in the perfect crystal, we shall follow the method of Prins [15] by supposing the scattering coefficients q_0, q and $\bar{q}$ to be complex. It is easy to see that a complex scattering coefficient is formally equivalent to an absorption; for if q_0 is complex, so is the refractive index, and we can write instead of $n = 1 - \delta$,

$$n = 1 - \delta - i\beta. \tag{2.87}$$

Suppose now that a wave is travelling in a direction x in a medium whose refractive index is n. Then to get the displacement at x, given that at $x=0$, we have to multiply by a phase factor $e^{-i\kappa n x}$ ($\kappa=2\pi/\lambda$), or in the case we are considering, where the refractive index is complex, by $e^{-i\kappa(1-\delta)x}\,e^{-\kappa\beta x}$. The real part of the refractive index, $1-\delta$, is thus concerned with the phase-lag, the imaginary part, β, corresponds to an absorption, that is to say, to a diminution of the amplitude of the wave with increasing x. Since the ordinary linear absorption coefficient, μ_0, refers to intensity, and not to amplitude, we have

$$\mu_0=2\kappa\beta=4\pi\beta/\lambda. \tag{2.88}$$

Prins's method is to substitute $\delta+i\beta$ for δ in Darwin's formula (2.75), and to determine β so as to agree with the observed absorption coefficient of the crystal. Darwin's method of treating absorption by including the factor h is equivalent to making q_0 complex. Prins's treatment differs from Darwin's in that q is also complex, since the absorption is now represented by an appropriate phase-lag in the scattered radiation.

We have now to modify (2.75) to include the complex scattering factors. The relation between δ and q_0 is given by equation (2.60), which we may now write as

$$-(\delta+i\beta)=\frac{\sin\theta_0}{\kappa a}\,q_0. \tag{2.89a}$$

Similarly, we may put for the corresponding equations for q and $\bar{q}$

$$-(\mathrm{A}_1+i\mathrm{B}_1)=\frac{\sin\theta_0}{\kappa a}\,q,$$

$$-(\mathrm{A}_2+i\mathrm{B}_2)=\frac{\sin\theta_0}{\kappa a}\,\bar{q}. \tag{2.89b}$$

If θ_0 is the Bragg angle, we have, by (2.72), $\kappa a(\sin\theta-\sin\theta_0)=v$, which, since $\theta-\theta_0$ in the cases which interest us is very small, say $\Delta\theta$, gives

$$\cos\theta_0\sin\theta_0\,\Delta\theta=(\sin\theta_0/\kappa a)\,v. \tag{2.89c}$$

Substitution of the last three equations in (2.75) gives at once

$$\frac{\mathrm{S}_0}{\mathrm{T}_0}=\frac{\mathrm{A}_1+i\mathrm{B}_1}{\mathrm{C}\pm\sqrt{\mathrm{C}^2-(\mathrm{A}_1+i\mathrm{B}_1)(\mathrm{A}_2+i\mathrm{B}_2)}}, \tag{2.90}$$

$$\text{where } \mathrm{C}=\cos\theta_0\sin\theta_0\Delta\theta-\delta-i\beta,$$

an equation for $\mathrm{S}_0/\mathrm{T}_0$ in terms of $\Delta\theta$, the angle between the direction of incidence θ and that corresponding to the Bragg angle θ_0. If the crystal is not polar, we need not distinguish between q and $\bar{q}$, and can put $\mathrm{A}_1+i\mathrm{B}_1=\mathrm{A}_2+i\mathrm{B}_2=\mathrm{A}+i\mathrm{B}$, and in the following discussion we shall do this.

The equation takes a rather simpler form if we refer the angle not to θ_0 as origin, but to the corrected Bragg angle θ_1. By (2.81),

$$\theta_1 - \theta_0 = \Delta\theta_0 = \delta \sec \theta_0 \operatorname{cosec} \theta_0,$$

and so, if $\Delta\theta_1$ is the small angle $\theta - \theta_1$, $\Delta\theta = \Delta\theta_1 + \delta \sec \theta_0 \operatorname{cosec} \theta_0$. The half-width of the range of perfect reflection corresponding to zero absorption is, by (2.82), $\gamma\Delta\theta_0$, or $A \sec \theta_0 \operatorname{cosec} \theta_0$, since, for no absorption, $B = \beta = 0$, and $\gamma = q/q_0 = A/\delta$. If then we express the angle $\Delta\theta_1$ as a fraction of the half-width of total reflection for no absorption, putting $\Delta\theta_1 = \eta\gamma\Delta\theta_0$, (2.90) becomes

$$\frac{S_0}{T_0} = \frac{1 + iB/A}{\eta - i\beta/A \pm \sqrt{(\eta - i\beta/A)^2 - (1 + iB/A)^2}}, \tag{2.91}$$

and the reflection coefficient $R(\eta)$ is given by the modulus of this expression. This form of the expression is given by Renninger,[18] and an analogous form has been used by Allison.[19] For no absorption it reduces to

$$\frac{S_0}{T_0} = \frac{1}{\eta \pm \sqrt{\eta^2 - 1}} = \frac{1}{\eta \pm i\sqrt{1 - \eta^2}}, \tag{2.92}$$

a modified form of the Darwin equation (2.78). Equation (2.92) shows plainly that the limits of total reflection correspond to $\eta = \pm 1$.

It is not easy to see without detailed discussion how the modulus of (2.91) behaves when B and β are not zero. The Darwin curve for $B = \beta = 0$ is a limiting curve, and all curves for B and $\beta > 0$ lie beneath it. The flat region of total reflection disappears. The discussion of the relationship of B, β, and A, and the possible methods of determining their values, will be postponed until Chapter VI, where the agreement between the various theoretical formulae and the results of observation will be considered.

Examples of reflection curves calculated from (2.91) by Renninger for the spectra 200, 400 and 600 of rock salt are shown in fig. 27. The method of calculation will be found in § 3(*j*) of Chapter VI. Their exact form depends on the assumptions made with regard to the relationship between B and β, but with any assumption that is reasonable the curves are asymmetric, and their centres of gravity do not lie at $\theta = \theta_1$, as for the case of zero absorption, but at rather smaller angles. It will be seen that the effect of absorption increases with increasing order of spectrum, and is greater when the electric vector in the incident radiation is parallel to the plane of incidence than when it is perpendicular to that plane.

The abscissae of the curves in fig. 27 are values of η, that is to say, the breadth of the Darwin curve for total reflection, assuming no absorption, is taken as unity in all cases. The curves for all orders are therefore referred to the one Darwin curve, shown dotted in the figure. In actual angular measure, the breadth of the limiting Darwin

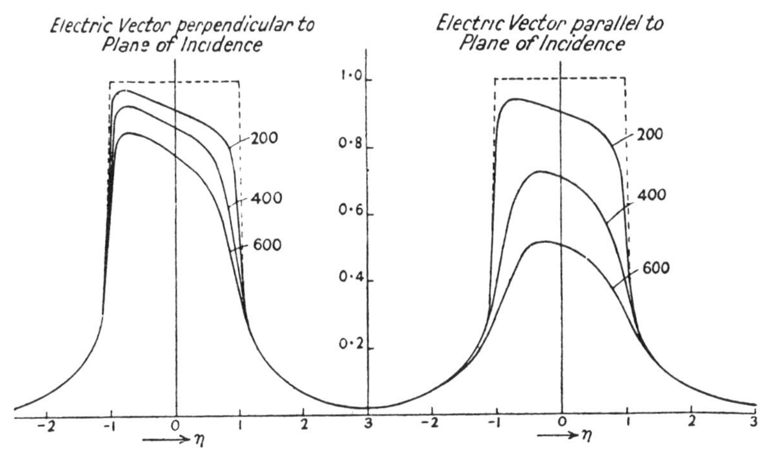

FIG. 27. Reflection curves for a perfect crystal with absorption (Renninger) [18]

curve varies with the order of the spectrum. The breadth of the range of total reflection in radians is D or D cos θ_0, according to whether the electric vector is perpendicular or parallel to the plane of incidence, where

$$D = (2N\lambda^2/\pi \sin 2\theta_0)\,|F|\,e^2/mc^2, \qquad (2.93)$$

F being the structure factor of the spectrum concerned, and N the number of crystal units in unit volume. When the electric vector is perpendicular to the plane of incidence, the angular breadths of the ranges of total reflection for 200, 400, and 600 in the case under consideration are respectively 7·6″, 2·7″, and 1·6″; for the other direction of polarisation the corresponding figures are 6·5″, 1·1″, and 0·5″.

(*i*) *The composite lattice in the perfect crystal:* As a rule, those crystals for which a direct comparison between the theoretical and observed reflecting powers can be made do not consist of atoms of one kind arranged on a simple space-lattice, but have a composite unit cell, so that each atomic plane in our discussions of the preceding sections must be replaced by a group of planes, one for each atom of the unit cell. We cannot without further enquiry allow for this simply by substituting the structure factor F for the atomic scattering factor *f* in the formulae we have derived; for the perfect-crystal formulae are based on the idea of multiple reflections of waves within the crystal, some of which will take place from one side of the groups of planes and some from the other. Now, as we have seen in II, § 1(*d*), if the crystal is polar with respect to the set of planes concerned, the structure factor depends upon the particular side of the group of planes from which the reflection takes place. The difference is only appreciable in those special cases in which the frequency of the incident radiation is in the immediate neighbourhood of an absorption edge for one of the atoms in the crystal unit. The appropriate form of the equation for the reflection coefficient is then (*2*.90).

The effect in any case occurs only when the crystal is polar, even when the condition of nearness of the frequency of the incident radiation to an absorption edge of a crystal atom is fulfilled, and in most actual cases it can be neglected. For the formulae relating to a crystal with zero absorption we have then simply to replace f by F, the structure factor of the crystal unit for the reflection considered.

In the expression for the absorbing crystal, we have to use instead of $\mathrm{A}+i\mathrm{B}$,

$$\sum_k (\mathrm{A}_k + i\mathrm{B}_k) e^{2\pi i a_k / a}, \tag{2.94}$$

where $\mathrm{A}_k + i\mathrm{B}_k$ is the value of $\mathrm{A}+i\mathrm{B}$ for the kth atom of the unit cell, and a_k is the distance of the atom from the reflecting plane chosen to correspond to the zero of phase.

The fraction of the quantity β, as determined from the absorption coefficient of the crystal, which is to be assigned to any particular atom is proportional to the contribution of that atom to the absorption coefficient.

4. Ewald's Dynamical Theory

(*a*) *Introductory:* We shall conclude this chapter on the intensity of reflection of X-rays by crystals with some account of Ewald's theory, which deals essentially with the same aspect of the problem of diffraction as that treated by Darwin—the dynamical interaction of the scattered waves and the crystal lattice. It arose, however, quite independently of the work of Darwin, and its scope is rather more general. Its method of approach, moreover, differs entirely. Space does not allow a full account of the theory to be given here, and an outline of the very elegant methods used must suffice. For mathematical details, the reader must consult Ewald's three papers in the *Annalen der Physik*.[20] The work had its origin in an attempt to express the laws of crystal optics in terms of the discrete lattice, instead of in terms of the continuum. This work was published by Ewald in 1912 as his doctorate thesis under Sommerfeld. Meantime the discovery of the diffraction of X-rays by crystals had been made, and Ewald was able to extend his work to include the case of these much shorter waves.

The underlying idea of the theory may be stated briefly as follows. Each lattice-point of the crystal is supposed to be occupied by a dipole which can be set into oscillation by the radiation field of any electromagnetic wave passing through the crystal. The oscillating dipoles themselves emit radiation, which produces a radiation field.

The lattice is first of all thought of as being of unlimited extent, and the dipoles as being in a state of oscillation expressible in terms of a plane wave advancing through the lattice with a certain speed. It is important to be clear that this 'dipole-wave' refers only to the state of oscillation of the dipoles themselves, and is not an electromagnetic wave passing through the medium in which they are situated. This initially arbitrary state of motion of the dipoles sets up electromagnetic waves, and these waves interact with the dipoles and affect their motion. The first problem to be solved may now be stated as follows. We suppose the oscillation of the dipoles described by the dipole-wave to be due to an electromagnetic field. If we know this field we can calculate the state of motion of the dipoles, and from this, we can again calculate the field to which it gives rise. We wish the whole system to be dynamically self-contained, so that the field calculated from the dipole oscillations is just that which would produce those oscillations. All this refers to an infinite crystal, in which the state of motion is supposed to have been set up in some way. In all practical cases, the crystal is of finite extent and a wave enters from outside. The second, and more difficult, part of the work deals with the problem of such a finite crystal.

(*b*) *The electromagnetic field due to a dipole-wave in a crystal lattice*: The essential problem to be solved at the outset concerns the kind of electromagnetic field that will be set up by a dipole-wave such as we have considered above. In order to understand the principles involved, we shall discuss briefly a simple case given by Ewald himself in a later paper.[21] We consider a set of lattice-planes numbered 0, 1, 2, ... p, fig. 28, equally spaced at a distance a apart. The dipole-wave passes through the lattice in a direction normal to these planes and in the direction of increasing p, or downwards in fig. 28, so that all the dipoles

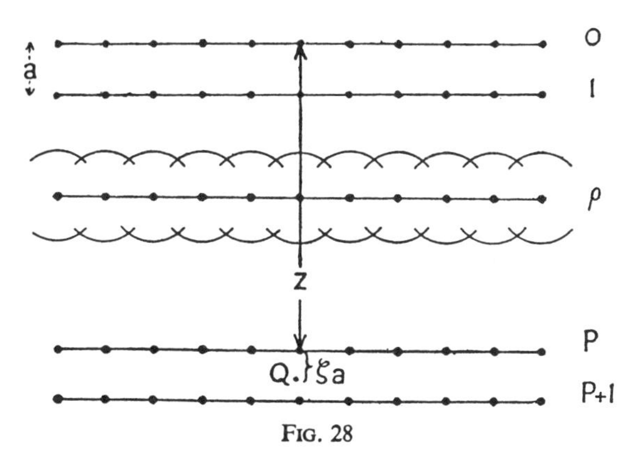

FIG. 28

in any one plane oscillate in phase. The spherical wavelets which each dipole emits run together at a short distance from the planes to form a set of plane waves whose fronts are parallel to those of the dipole-waves. There will be two such sets of waves from each plane, one travelling up, the other down, and the resultant field in the lattice is obtained by summing the waves from all the planes. These scattered waves travel with the velocity of light, c, which is not the velocity of the dipole-waves. We shall consider the waves in terms of their circular frequency ω, the same for both the dipole- and the scattered waves, and of their wave-numbers, or reciprocal wave-lengths, which we shall denote by K for the dipole-wave, and k for the scattered waves. If v is the velocity of the dipole-waves, we have

$$2\pi c = \omega/k, \quad 2\pi v = \omega/\mathrm{K}.$$

If the oscillations of the dipoles on the uppermost plane, $p=0$, are given at a time t by $e^{-i\omega t}$, those of the dipoles on the pth plane are given by $e^{-i\omega t}\, e^{2\pi i \mathrm{K} pa}$. Let us calculate the field at a point Q between the Pth and (P + 1)th planes, say at a distance $z = (\mathrm{P} + \zeta)\, a$ below the surface, where ζ is a positive proper fraction. The resultant at Q of the waves from all the planes above Q is, apart from a time factor,

$$\sum_{p=0}^{p=\mathrm{P}} e^{2\pi i \mathrm{K} pa}\, e^{2\pi i k(\mathrm{P}+\zeta-p)a}, \tag{2.95}$$

the sum being taken over all the lattice-planes above Q; for the wave from the pth plane has to travel a distance $(\mathrm{P} + \zeta - p)a$ to get to Q, and does so with velocity c, and wave number k, but it starts with a phase-lag $2\pi \mathrm{K} pa$. The sum (2.95) is a geometrical progression, and it can be written in the form

$$\frac{e^{2\pi i k z}}{1 - e^{2\pi i(\mathrm{K}-k)a}} - \frac{e^{2\pi i(\mathrm{K}-k)(1-\zeta)a}}{1 - e^{2\pi i(\mathrm{K}-k)a}}\, e^{2\pi i \mathrm{K} z}. \tag{2.96}$$

We notice that the resultant wave consists of two parts, the first, a wave travelling downwards with a velocity c, corresponding to k; the second, a wave travelling, on the average, at the same speed as the dipole-wave, wave-number K. It is, however, a rather unusual kind of wave. Since ζ grows from 0 to 1 as the wave passes between any two planes, we see that immediately before passing through any plane the phase of the wave is that corresponding to transmission with the velocity of the dipole-wave. On passing through the plane there is a jump in phase, so that the wave in between two planes has a phase differing somewhat from that corresponding to the dipole-wave. It has, however, resumed the correct value by the time the next plane is reached, so that, on the average, the phase of the wave is equal to that of the dipole-wave. It is this wave, which accompanies the dipole-wave (Ewald's *gleichlaufende Welle*), which is essential in the discussion of the unbounded crystal.

A fuller treatment shows that the wave travelling with the velocity c depends on the crystal boundary. It is a 'boundary wave' in the sense in which we shall use that term later on, and the resultant field due to it at any point in the crystal depends on the exact form of the boundary of the crystal, but does not necessarily become zero, however large the crystal.* For the time being, however, we shall consider only the wave which accompanies the dipole-wave. We shall be interested only in dipole-waves which travel with velocities very nearly equal to that of light. $K-k$ will thus always be small, and the amplitude of the wave in the second term of (2.96) can be written $i/2\pi a(K-k)$, by expanding the exponential in the denominator.

We have also to take into account the waves coming from the planes below Q, and this is done in the same way. We again get a wave travelling at the same speed as the dipole-wave, as well as a boundary wave; and for the total wave which accompanies the dipole-wave we find the amplitude $2ik/2\pi a(K^2-k^2)$, which, since K and k are nearly equal, may again be written $i/2\pi a(K-k)$. The unit of amplitude is here, of course, the amplitude of the wave emitted by a single plane of dipoles, and this we may easily obtain by the method given in II, § 2(*b*) for calculating q, the amplitude scattered by a single plane of atoms. In the formulae there given, $\sin\theta$ is now unity, and the amplitude becomes $-in\lambda_0$ times the amplitude at unit distance due to a single dipole.

If p is the electric moment of a dipole, it is a well-known result of electromagnetic theory that the amplitude of the wave that it scatters is, at unit distance, $p_\perp\omega^2/c^2$, where $p_\perp$ is the component of the moment at right angles to the wave-vector.

Now $\omega/c=2\pi/\lambda_0=2\pi k$, and n, the number of dipoles per unit area of the planes, is equal to a/V where V is the volume of the lattice-cell. We find, therefore, for the amplitude of the wave scattered by a single plane $-i4\pi^2akp_\perp/V$, and for that which accompanies the dipole-wave

$$\frac{2\pi}{V}p_\perp\frac{k}{K-k}=\frac{2\pi}{V}\frac{1}{\epsilon}p_\perp, \tag{2.97}$$

where ϵ is given by

$$K=k(1+\epsilon). \tag{2.98}$$

For this wave, the quantity ϵ is equivalent to δ, the defect from unity of the 'refractive index' for the wave considered.

We have thus arrived at the important result that a dipole-wave in the lattice is accompanied by a wave travelling with the same speed, whose amplitude is inversely proportional to the difference between K and k. This means that the wave in question has an appreciable amplitude only when K is very nearly equal to k; that is to say, the dipole-wave does not excite an electromagnetic wave of appreciable

* Such 'boundary waves' occur, strictly speaking, in all problems in which the transmission of light in a medium is considered, and are due to the reflection of waves from the boundaries of the medium. They are, however, commonly neglected, and this can be justified by considering the average field in a portion of medium whose boundaries vary slightly and irregularly with the time over a range of the order of the wave-length of the radiation. The average field at any point due to the boundary waves is then zero, but the main transmitted wave is unaffected.

amplitude unless its own velocity is very nearly the velocity of light. We shall see that, dynamically, no dipole-wave can travel with the velocity of light, so that the amplitude of the wave which is set up never becomes infinite, even in an infinite crystal.

The excitation of the wave has strong formal analogies with the phenomenon of resonance. As the dipole-wave travels through the crystal with a given velocity it excites at each plane of atoms a wave which travels with a slightly different velocity. The different waves so excited do not build up to form a wave of appreciable amplitude unless the speed of exciting waves and excited wave are very nearly equal. In virtue of the analogy with resonance, Ewald speaks of the quantity ϵ as the 'resonance error' (*Resonanzfehler*). For a rigid deduction of the properties of the waves, Ewald's papers must be consulted.

(*c*) *The alternative methods of description of a dipole-wave in a lattice:* Let the lattice-points of an infinite crystal, at each of which a dipole is situated, be defined by the vector

$$\mathbf{r} = u\mathbf{a} + v\mathbf{b} + w\mathbf{c},$$

u, v, w, being integers, and **a**, **b**, and **c**, the primitive translations of the lattice. Let the wave-vector of the dipole-wave be **K**, a vector of magnitude $1/\lambda$, whose direction is that of the normal to the wave-fronts of the dipole-waves.

If $e^{-i\omega t}$ defines the state of oscillation of the dipole at the origin O, that of a dipole situated at a point A defined by the vector **r** is

$$e^{-i\omega t}\, e^{2\pi i \mathbf{K}\cdot\mathbf{r}}; \tag{2.99}$$

for the scalar product $2\pi\mathbf{K}\cdot\mathbf{r}$ is equal to $2\pi/\lambda$ multiplied by the length of the normal from the origin onto the wave-front passing through A, and thus gives the phase difference between the dipole oscillations at O and A.

The fact that the lattice-points are described by a set of discrete values of the vector **r** has as a consequence that the description of the dipole oscillations given by the dipole-wave (*2*.99) is not unique. It is easy to show that **K** can be replaced by any one of an infinite set of other wave-vectors each of which describes exactly the same set of vibrations. These wave-vectors are the set $\mathbf{K}+\mathbf{r}^*$, where $\mathbf{r}^*$ is any vector in the reciprocal lattice. This may be seen at once as follows. We have

$$(\mathbf{K}+\mathbf{r}^*)\cdot\mathbf{r} = \mathbf{K}\cdot\mathbf{r} + \mathbf{r}^*\cdot\mathbf{r},$$

and the scalar product of a lattice-vector and a vector of the corresponding reciprocal lattice is always an integer. (See equation (*2*.6a), p. 28). Thus, since the exponential of $2\pi i$ times any integer is always unity, we have

$$e^{2\pi i \mathbf{K}\cdot\mathbf{r}} = e^{2\pi i (\mathbf{K}+\mathbf{r}^*\cdot\mathbf{r})}, \tag{2.100}$$

so that the wave-vectors $\mathbf{K}$ and $\mathbf{K}+\mathbf{r}^*$ define the same set of dipole oscillations.

This result can be expressed geometrically in a very simple way. Let O, fig. 29, be the origin of the reciprocal lattice, and let AO be the vector $\mathbf{K}_0$, the wave-vector of the assumed dipole-wave. Let OB be the reciprocal-lattice vector $\mathbf{r}_m^*$ to the point (hkl) or (m) say; then $\mathbf{K}_m$, the vector $\overrightarrow{AB}$ is equal to $\mathbf{K}_0+\mathbf{r}_m^*$, and is also, as we have just seen, the wave-vector of a dipole-wave that describes exactly the same set of dipole oscillations as the wave-vector $\mathbf{K}_0$. It is clear, moreover, that the vector from the point A to *any* point of the reciprocal lattice will be a possible wave-vector for describing the same set of oscillations. These different dipole-waves have different velocities; for the velocity is proportional to the reciprocal of the wave-vector.

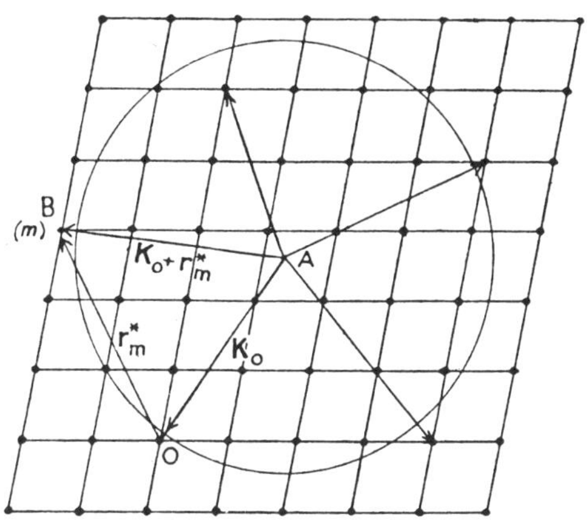

FIG. 29. The wave-vectors of the different equivalent dipole-waves, and the sphere of reflection

The principle underlying this idea of the alternative description of dipole-waves may perhaps be made clearer by a very simple special

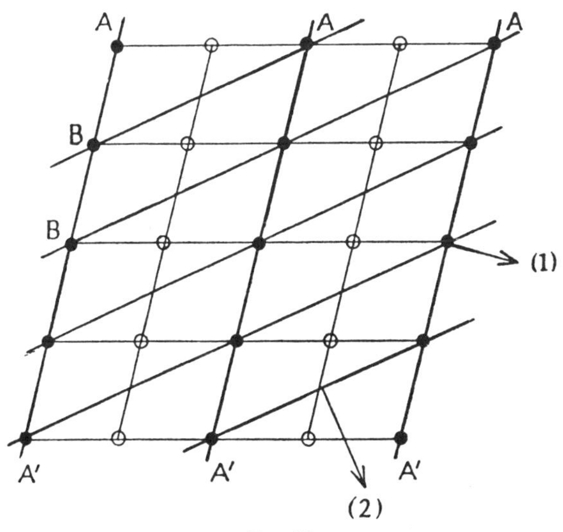

FIG. 30

case. In fig. 30, let the circles and dots represent lattice-points, and let all the points denoted by dots contain dipoles in the same phase, say at crests of the wave, at a given instant; all those denoted by circles will then lie at troughs. The lines marked AA′ represent wave-fronts of a wave travelling in the direction of the arrow (1), say. It is clear, however, that lines such as AB are also the loci of points in the same phase, and, so far as the oscillations of the dipoles are concerned, may be considered as wave-fronts of a wave that travels in the direction of the arrow (2), its velocity and wave-length being different from, but the frequency, of course, the same, as those of the wave considered first.

(*d*) *The relationship of the interference wave-field to the dipole-waves:* Let us suppose now that the wave-vector AO, fig. 29, which we may denote by **K**(000), or simply by $\mathbf{K}_0$, since it is drawn to the origin of the reciprocal lattice, corresponds to a dipole-wave having very nearly the velocity of light. Then, as we have seen in II, § 4(*b*), it will be accompanied by an electromagnetic wave in the medium, which travels with the same speed, and has an appreciable amplitude if this speed is near enough to c, the velocity of light. Now, among the infinite number of other waves which are alternative descriptions of the dipole-wave, there may be some whose velocities are also nearly equal to c. If there are any such dipole-waves, we must suppose that they too will be accompanied by appreciable electromagnetic waves. The existence of one train of waves in the crystal will in fact in general entail the possible and indeed necessary existence of others, and here we begin to see the emergence of the different interference beams.

Suppose now that with centre A (fig. 29) and radius k corresponding to the wave-vector of light of the frequency ω in free space, a sphere is described. This sphere, which is the sphere of reflection of I, § 1(*b*), will pass near, but not through, the point O. We see, in fact, from (2.98) that it will pass a distance ϵk from it. The amplitude of the wave which will accompany the dipole-wave will therefore be inversely proportional to the distance of O from the sphere. Now the sphere may pass near other lattice-points. The line joining A to such a point (m) is the wave-vector $\mathbf{K}_m$ of a possible dipole-wave, and this will be accompanied by an electromagnetic wave of amplitude inversely proportional to the distance of the point from the sphere.

We see at once the relationship of these different waves to the interference beams in the elementary lattice theory discussed in Chapter I. In the corresponding construction for the diffracted waves described in I, § 1(*b*), the point O would lie exactly on the sphere, since all the waves are supposed to travel with the velocity of light. Diffracted beams derived from the primary beam occur only if the corresponding lattice-points lie exactly on the sphere. In the case we are now considering no lattice-point can lie exactly on the sphere, for it would correspond to a dipole-wave travelling with exactly the velocity of light, which we shall see to be impossible when we consider the matter dynamically. But we see that to each point lying near enough to the sphere there corresponds a dipole-wave travelling nearly with the velocity of light, which will be accompanied by a corresponding electromagnetic wave whose amplitude varies inversely as the distance of the point in question from the sphere. All such electromagnetic waves taken together constitute the interference wave-field corresponding to the arbitrarily assumed set of dipole vibrations. We have no right to call one of the beams the primary beam in preference to any other; for to assume the existence of any one of them implies the existence of all the rest.

(*e*) *The dynamical problem:* We have now to pass from the purely kinematical aspect of the problem, which has occupied us so far, to the dynamical aspect. We have to calculate what set of interference waves and dipole-waves can coexist, so that the system is dynamically self-contained. We start by assuming a dipole-wave of velocity very nearly that of light, in a certain definite direction. We do not know the exact velocity until the problem is solved. We know, however, that it must be very nearly that of light, so that we can draw the sphere of reflection, and determine which, if any, of the lattice-points lie near enough to it to give waves of amplitude great enough to contribute appreciably to the wave-field. Let AO, fig. 31, be the vector $\mathbf{K}_0$ representing the wave initially chosen, and suppose that, in all, n points lie near enough to the sphere to be considered, the wave-vectors drawn to them from A being $\mathbf{K}_0$, $\mathbf{K}_1$, ... $\mathbf{K}_m$ and the corresponding values of ϵ, ϵ_0, ϵ_1, ... ϵ_m. We have now to determine the position of the point A, and the corresponding n values of ϵ which satisfy the dynamical conditions. It is convenient to have a name for the point A, and we shall call it the 'wave-point'.* Each of the n waves gives a certain component to the total electric field at any point, and we are now going to suppose that the oscillations of the dipoles are produced by this field.

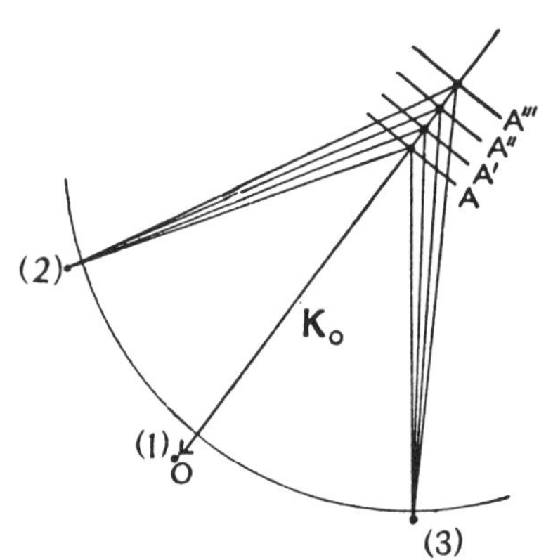

FIG. 31. The wave-points and dispersion surfaces

We first determine the relation between the moment of a dipole and the field in which it is situated. Let the vector $\mathbf{p}$ denote the electric moment at the time t of the dipole at the point defined by the vector $\mathbf{r}$.

* Ewald calls it the *Anregungspunkt*, a term not easy to translate suitably.

The dipole* is supposed to be due to the small mutual displacement of charges $+e$ and $-e$, of mass m. If ω_0 is the circular frequency of the free dipole vibrations, the equation of motion, neglecting damping, is

$$\ddot{\mathbf{p}} + \omega_0{}^2\mathbf{p} = \frac{e^2}{m}\,\mathbf{E}, \tag{2.101}$$

$\mathbf{E}$ being the electric field acting on the dipole. Since $\mathbf{p}$ oscillates with the frequency ω of the dipole-waves, $\ddot{\mathbf{p}} = -\omega^2\mathbf{p}$, and the solution of (2.101) for a steady state is

$$\mathbf{E} = -\frac{m}{e^2}(\omega^2 - \omega_0{}^2)\,\mathbf{p}, \tag{2.102}$$

which gives the dipole moment $\mathbf{p}$ in terms of the field $\mathbf{E}$.

We now have to calculate the field $\mathbf{E}$, which is that due to the n waves, in terms of the dipole moments. This we can write down at once from (2.97), which gives

$$\mathbf{E} = \frac{2\pi}{\mathrm{V}} \sum_m \mathbf{p}_{[m]} \frac{1}{\epsilon_m} e^{2\pi i \mathbf{K}_m \cdot \mathbf{r}}. \tag{2.103}$$

There is one term for each of the n waves that have appreciable amplitude, and $\mathbf{p}_{[m]}$ is the component perpendicular to the wave-vector $\mathbf{K}_m$ of the dipole-moment at the origin. The phase factor is introduced because the field at the lattice-point defined by $\mathbf{r}$ is required. By (2.100), II, § 4(*c*), this phase factor is the same for all the n waves, and $\mathbf{p}$ in (2.102) of course contains the same factor. If then we write $\mathbf{b}$ for the maximum moment of the dipole, we get on substituting (2.103) in (2.102)

$$\mathbf{b} = \frac{1}{\Omega} \sum_m \mathbf{b}_{[m]} \frac{1}{\epsilon_m}, \tag{2.104}$$

where

$$\Omega = -\frac{m}{2\pi e^2}(\omega^2 - \omega_0{}^2)\,\mathrm{V}. \tag{2.105}$$

The quantity Ω of equation (2.105) is related in a simple way to the refractive index of the crystal for radiation of frequency ω. By equation (2.61) of II, § 3(*b*), the refractive index n for wave-length λ of a medium containing N atoms per unit volume, each of scattering power $f(0)$ in the direction in which the wave is travelling, is given by

$$n = 1 - \delta = 1 - \frac{\lambda^2}{2\pi}\frac{e^2}{mc^2}\mathrm{N}f(0).$$

* We have retained the idea that the atoms of the lattice scatter as dipoles, which was used by Ewald, since it is simple, and illustrates the principles involved as well as any more complicated assumption. It must be remembered, however, that a dipole scatters symmetrically about its axis, and effectively, therefore, we are assuming f not to depend on the angle of scattering. A more detailed discussion for this general case will be found in Chapter VIII.

For a dipole of natural frequency ω_0, $f(0) = \omega^2/(\omega^2 - \omega_0^2)$, and the refractive index of the crystal lattice composed of dipoles is therefore, from (2.105), given by

$$n = 1 + \frac{2\pi}{V}\frac{e^2}{m}\frac{1}{\omega_0^2 - \omega^2} \qquad (2.106)$$

$$= 1 + 1/\Omega,$$

since $N = 1/V$, and $\lambda/c = 2\pi/\omega$.

Thus

$$\delta = -\frac{1}{\Omega}. \qquad (2.107)$$

The refractive index refers to a wave of frequency ω travelling through the crystal when no other wave is excited. We cannot consider the crystal as having a single definite refractive index when the composite wave-field we have been discussing in the last few paragraphs is excited, for each wave-train will travel with a slightly different speed.

In most materials the refractive index for X-rays is less than unity, so that δ is positive and Ω negative, which corresponds to the case in which the natural frequency ω_0 of the dipole oscillators is smaller than ω the frequency of the radiation.

(*f*) *The dispersion equation and the dispersion surface:* Equation (2.104) is a vector equation, and the sum on the right-hand side contains n terms. A typical one, the mth for example, contains $\mathbf{b}_{[m]}$ the component of $\mathbf{b}$ perpendicular to the vector $\mathbf{K}_m$ from the wave-point A to the reciprocal-lattice point (m). The direction of $\mathbf{K}_m$ being given, $\mathbf{b}_{[m]}$ is therefore fixed if its components parallel to two directions, say in and perpendicular to some plane containing $\mathbf{K}_m$, are given. For a given position of A, $2n$ components are therefore required to express the sum of the vectors on the right-hand side of (2.104). The equation can thus be written as $2n$ linear equations involving these $2n$ components.

The coefficients in these equations are multiples of $1/\epsilon_0$, $1/\epsilon_1$ ... $1/\epsilon_m$. These quantities are not independent of one another. If, for example, ϵ_0 is known, all the rest are also known. For ϵ_0 fixes the position of the wave-point A for a given direction of the vector $\mathbf{K}_0$, and once A is known the sphere of reflection can be drawn, and the values of the remaining quantities follow at once from the distance of the corresponding reciprocal-lattice points from it (see fig. 29). Thus, in effect, the coefficients of the $2n$ linear equations contain ϵ_0 only. The equations have a solution only if the determinant formed by their coefficients vanishes, and the vanishing of this determinant yields an equation of degree $2n$ for ϵ_0, which has $2n$ solutions. Each solution gives a value of ϵ_0, and thus a position for the wave-point A, and hence all the other wave-vectors $\mathbf{K}_m$, and the whole set of waves constituting the dynamically self-contained wave-field.

For any particular direction of the wave-vector $\mathbf{K}_0$, there are $2n$ possible positions of the wave-point A, and so $2n$ possible wave-fields that satisfy the dynamical conditions. Moreover, for the infinite lattice, the superposition in any proportion of any number of these independent solutions is itself a solution. Ewald calls the equation of degree $2n$ for ϵ_0 the 'dispersion equation'.

Let the points A, A′, ... fig. 31, on the line OA be the $2n$ possible positions of the wave-point, as determined from the $2n$ values of ϵ_0 which satisfy the dispersion equation for the particular direction of $\mathbf{K}_0$. From these points we may draw the $2n$ vectors $\mathbf{K}_m$, $\mathbf{K}'_m$, $\mathbf{K}''_m$ ... to any lattice-point (m) which determines a wave. There will then be $2n$ radiating sets of wave-vectors, each of which defines a dynamically possible wave-field. Each wave-point is the centre of an appropriate sphere of radius k, the sphere of reflection, and the distance of the point (m) from such a sphere gives ϵ_m which determines the amplitude and velocity of the corresponding wave.

Now $\mathbf{K}_0$ might have been chosen over a small range of directions without altering the number of lattice-points which lie near enough to the wave-sphere to give interference waves of an appreciable amplitude. For each of these directions there are $2n$ possible positions of A. If we suppose the direction of $\mathbf{K}_0$ to vary over the permitted range, the $2n$ points A, A′, ... will describe a surface of $2n$ sheets, the dispersion surface, which is the locus of all positions of the wave-point A that give dynamically self-contained sets of interference waves. The dispersion equation is the equation to this surface.

(*g*) *The dispersion surface for the case of two waves:* As an example, we shall consider the practically important case in which there are only two sets of dipole-waves giving rise to appreciable electromagnetic waves, so that only two points of the reciprocal lattice lie near enough to the sphere of reflection to be considered. Let these two points be denoted by (o) and (m) in fig. 32, and let the corresponding wave-vectors be $\mathbf{K}_0$ and $\mathbf{K}_m$. Suppose first that the dipole vibrations are perpendicular to the plane containing these two wave-vectors, so that $\mathbf{b} = \mathbf{b}_{[m]}$. The dispersion equation (2.104) then becomes

$$\frac{1}{\epsilon_0} + \frac{1}{\epsilon_m} = \Omega, \qquad (2.108)$$

and we shall now show that the wave-point must lie on a certain hyperboloid.

With the points (o) and (m), fig. 32, as centres let circles be described having radii $k(1 + 1/\Omega)$, k being the reciprocal wave-length in free space of the radiation having the frequency ω of the dipole-waves. As we have seen in II, § 4(*f*), the quantity $(1 + 1/\Omega)$ is the refractive index of the crystal for a single wave-train of frequency ω. Let these circles intersect at the point Q. Since $|\mathbf{K}_0|$ and $|\mathbf{K}_m|$ will differ only

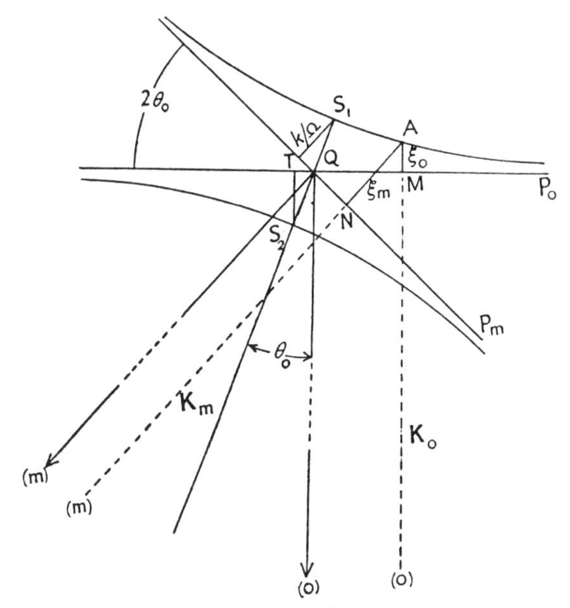

FIG 32. The dispersion surface for two waves

slightly from k, the wave-point A will lie at a distance from Q small in comparison with the distances Q(o) and Q(m); so that the circles in the region with which we are concerned may be taken as coinciding with their tangents at the point Q, the straight lines QP_o, QP_m, perpendicular to Q(o) and Q(m) respectively.

The distances of the wave-point A from (o) and (m) are $|\mathbf{K}_0|$ and $|\mathbf{K}_m|$, which are equal to $k(1+\epsilon_0)$ and $k(1+\epsilon_m)$. Since the distance QA is small in comparison with the distances Q(o) and A(o), the lines Q(o) and A(o) are nearly parallel, and AM, the distance of A from the line QP_o, is quite nearly enough equal to A(o) – Q(o); or, if we write ξ_0 for AM,

$$\xi_0 = k(1+\epsilon_0) - k(1+1/\Omega) = k(\epsilon_0 - 1/\Omega).$$

Similarly, if ξ_m is the distance of A from the line QP_m,

$$\xi_m = k(\epsilon_m - 1/\Omega).$$

From the last two equations

$$\epsilon_0 = \xi_0/k + 1/\Omega, \quad \epsilon_m = \xi_m/k + 1/\Omega. \tag{2.109}$$

The dispersion equation (2.108) can be written in the form

$$\Omega\epsilon_0\,\epsilon_m = \epsilon_0 + \epsilon_m,$$

and on substituting the values of ϵ_0 and ϵ_m from (2.109) we obtain

$$\xi_0\xi_m = k^2/\Omega^2. \tag{2.110}$$

Equation (2.110) is that of a hyperbola whose asymptotes are the lines QP_o, and QP_m, and whose axis is the line bisecting the angle between the lines Q(*o*) and Q(*m*). The apses of the hyperbola, S_1 and S_2, are the points which are equidistant from the two asymptotes; so that $\xi_0 = \xi_m = \pm k/\Omega$, and the diameter of the hyperbola, the distance S_1S_2, is therefore equal to $2k/\Omega \cos\theta$, where 2θ is the angle between the directions of Q(*o*) and Q(*m*).

The distance of the apse S_2 from the point (*o*) is equal to $Q(o) - S_2T$, fig. 32, or to k, since $Q(o) = k(1 + 1/\Omega)$, and $S_2T = k/\Omega$. Thus S_2 lies at the point that would be the centre of the sphere of reflection if $S_2(o)$ and $S_2(m)$ were the primary and diffracted rays on the simple Laue theory. Ewald calls this point the Laue point, and denotes it by L. The single point L of the elementary theory, is, on the dynamical theory, replaced by the two branches of the hyperbola, or more accurately by the two sheets of the hyperboloid formed by supposing the figure to rotate through a small angle about an axis joining the points (*o*) and (*m*). This small rotation allows a small variation in the direction of incidence. Any point on the hyperboloid so produced is a possible wave-point, when the radiation is polarised so that the electric vector is perpendicular to the plane containing Q(*o*) and Q(*m*).

If the electric vector of the radiation lies in the plane containing these two directions the dispersion equation is still that of a hyperbola, which has the same asymptotes as that considered above, but a diameter that is smaller in the ratio $\cos 2\theta_0 : 1$. This is easily proved by a method very similar to that given above, and we shall not deduce it here.

As we saw in II, § 4(*f*), Ω is negative in most cases that are of practical importance, so that Q, the centre of the hyperbola, lies nearer to the

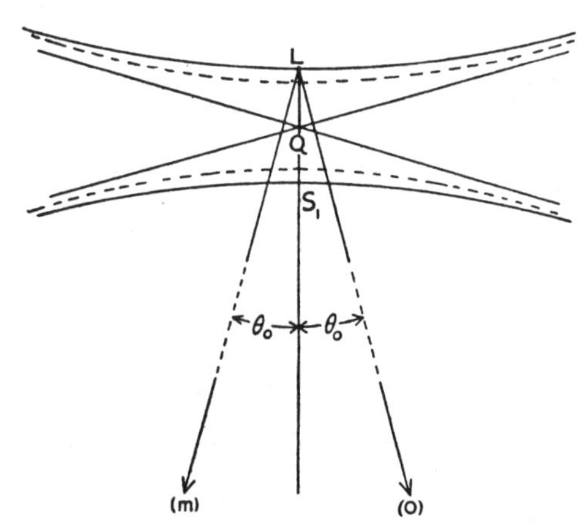

FIG. 33. The dispersion hyperboloids for two waves for the two states of polarisation, drawn for the case Ω negative. L is the Laue point, equidistant $1/\lambda$ from the reciprocal-lattice points (*o*) and (*m*)

points (o) and (m) than the Laue point L. The hyperbolas are shown for this case in fig. 33, in which the dotted curves correspond to the state of polarisation with the electric vector in the plane Q(o)(m), and the full curve to that with the electric vector perpendicular to this plane.

Since $1/\Omega = -\delta$, the diameters of the two hyperbolas can be written $2k\delta \sec\theta_0$ and $2k\delta \sec\theta_0 \cos 2\theta_0$ for the two states of polarisation, $1-\delta$ being the refractive index of the crystal for the radiation concerned. For X-rays, δ for most substances is a number of the order of 10^{-5}, so that the diameter LS_1, fig. 33, is of the order 10^{-5} of the distances L(o), L(m).

(*h*) *The transition from the case of a single wave to that of two waves:* If there is only one wave, the dispersion surface is a sphere of radius $k(1+1/\Omega)$, or $k(1-\delta)$, round the lattice-point (o) as centre, corresponding to a wave travelling with a velocity appropriate to the refractive index $1-\delta$. Suppose now that spheres of this radius are described about both points (o) and (m), and let the dotted circles in fig. 34 be the traces of these spheres in the plane containing the rays Q(o) and Q(m). Their intersection Q is the centre of the two hyperbolas that we have considered in the last paragraph, whose asymptotes coincide with the circles in the immediate neighbourhood of Q. If the wave-point A lies in the neighbourhood of this intersection there will be two waves of appreciable strength corresponding to each position of the wave-point, and for each direction of $\mathbf{K}_0$ two possible wave-points for each state of polarisation, corresponding to the intersections of $\mathbf{K}_0$ with the hyperbolas.

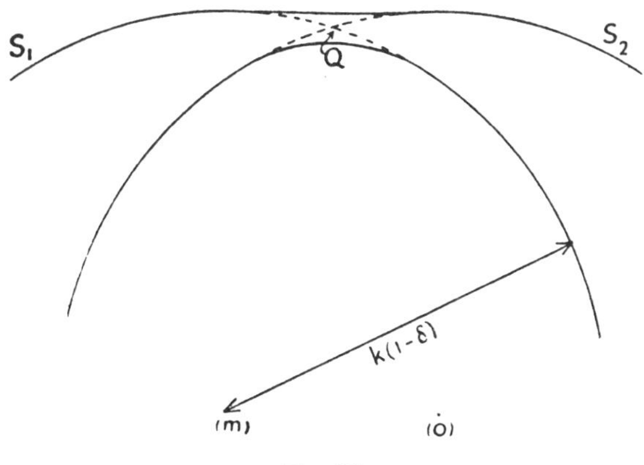

FIG. 34

It is clear that as the wave-point moves away from the region of the intersection we get only one wave; because only one point will lie near enough to the sphere of reflection having the wave-point as centre to give a wave of appreciable amplitude. In such a case the wave-point must lie on one or other of the circles, and we can speak of the crystal as having a definite refractive index. As the wave-point approaches the region of the intersection of the circles, two waves have appreciable strength, interaction occurs between them, and the dispersion surface breaks up into two sheets of which the hyperboloid considered above forms the central portion. If both states of polarisation are considered, there are four branches. When the wave-point is in this region the crystal cannot be said to have a definite refractive index. With unpolarised radiation, for any given direction of $\mathbf{K}_0$, four waves travelling in slightly different directions may co-exist, and their velocities again will depend on the exact directions.

(*i*) *The field due to a dipole-wave in a semi-infinite crystal. The boundary waves:* We have so far considered the processes taking place inside a crystal of infinite extent, and by so doing have been able to neglect the effect of the boundaries. In practice, however, we are always interested in cases in which a wave enters the crystal from outside. The relation between the processes taking place inside and outside the crystal is, in fact, the really important matter, and we must consider all that we have done so far only as a step towards the solution of that problem. We shall therefore now consider a semi-infinite crystal bounded by a plane surface.

We cannot suppose that the conditions for dynamical self-consistency which we have obtained for the infinite crystal will suffice for a crystal with a definite boundary. We saw in the elementary discussion in II, § 4(*b*), p. 67, that, in addition to the wave accompanying the dipole-wave, there was another wave, the boundary wave, travelling with the velocity of light. Discussion of the case in which the dipole-wave travels in the semi-infinite crystal in a direction inclined to the surface shows that to each internal wave, as we may call the wave which accompanies a dipole-wave through the crystal, there correspond two waves travelling with velocity c, one, the inner boundary wave, travelling inside the crystal, the other, the outer boundary wave, travelling in the medium outside. In what follows we shall only quote results, referring the reader to Ewald's papers for proofs.

The boundary waves are given in terms of the reciprocal-lattice construction as follows. Let AO, fig. 35, be a wave-vector corresponding to an internal wave, and let the wave-sphere of radius k be drawn with A as centre. The sphere does not pass through, but only near, the point O. Let a line through O, normal to the crystal surface, cut the sphere in C, near O. Then $\overrightarrow{AC}$ is the wave-vector of the inner boundary wave which corresponds to the internal wave-vector AO. If OC prolonged cuts the sphere again at D, $\overrightarrow{AD}$ is the wave-vector of the outer boundary wave. If $\mathbf{K}_0$ is the wave-vector of the internal wave, $\overrightarrow{AC}$, the inner boundary wave-vector, will be denoted by $\mathbf{k}_0$.

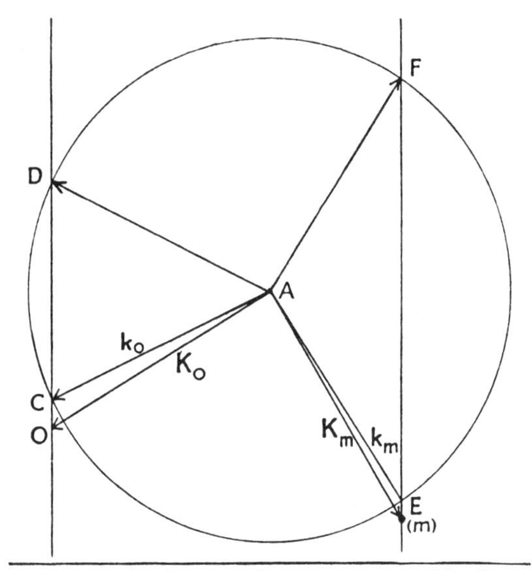

FIG. 35. The relation between the wave-vectors of the internal wave and the inner and outer boundary waves

In exactly the same way we can construct the boundary wave-vectors corresponding to any other internal wave $\mathbf{K}_m$. $\overrightarrow{AE}$ and $\overrightarrow{AF}$ are two such vectors in fig. 35, and are obtained from the intersections E and F with the sphere of a line normal to the crystal surface through the lattice-point (m) that determines the internal wave-vector $\mathbf{K}_m$. It is evident from the construction that these wave-vectors have the right magnitude k, corresponding to the velocity of light. We see also that the direction of the outer boundary wave is the reflection at the surface of the direction of the inner boundary wave. Ewald shows that the amplitude of the outer boundary wave is equal to, and that of the inner boundary wave equal and opposite to, the amplitude of the corresponding internal wave.

Now it is evident that the condition of dynamical self-consistency for the semi-infinite crystal cannot be satisfied by the set of waves derived from a single wave-point A; for, as we have seen, these conditions are satisfied by the internal waves taken alone, and we have now to add to these, which alone would maintain the dipole vibrations, the field due to the inner boundary waves. In the corresponding optical problem there is only one internal wave, and the solution is obtained by superposing on the whole wave-field, inside and outside the crystal, a wave whose direction is that of the inner boundary wave, and whose amplitude is equal and opposite to it. Inside the crystal this wave exactly destroys the inner boundary waves; the field inside is therefore just that internal field which satisfies the dynamical conditions. Outside the crystal the superposed wave remains as the 'incident wave'; the outer boundary wave is the reflected wave, and the internal wave is the refracted wave.

For the X-ray problem we cannot proceed on exactly the same lines; for in general there is more than one internal wave, and so more than one inner boundary wave. The directions of these waves differ, and they cannot, therefore, be neutralised throughout the crystal by a single incident wave. A solution on similar lines is, however, made possible by the fact that there are $2n$ independent solutions of the dynamical problem for the infinite crystal, n being the number of internal waves. It is possible to superpose these $2n$ solutions in such a way that all the inner boundary waves except one, say $\mathbf{k}_0$, have a resultant which is zero. $\mathbf{k}_0$ can then be neutralised by the incident wave, and we have, inside the crystal, a dynamically self-consistent wave-field composed of $2n$ suitably superposed internal wave systems, which together constitute the interference field corresponding to the given incident wave.

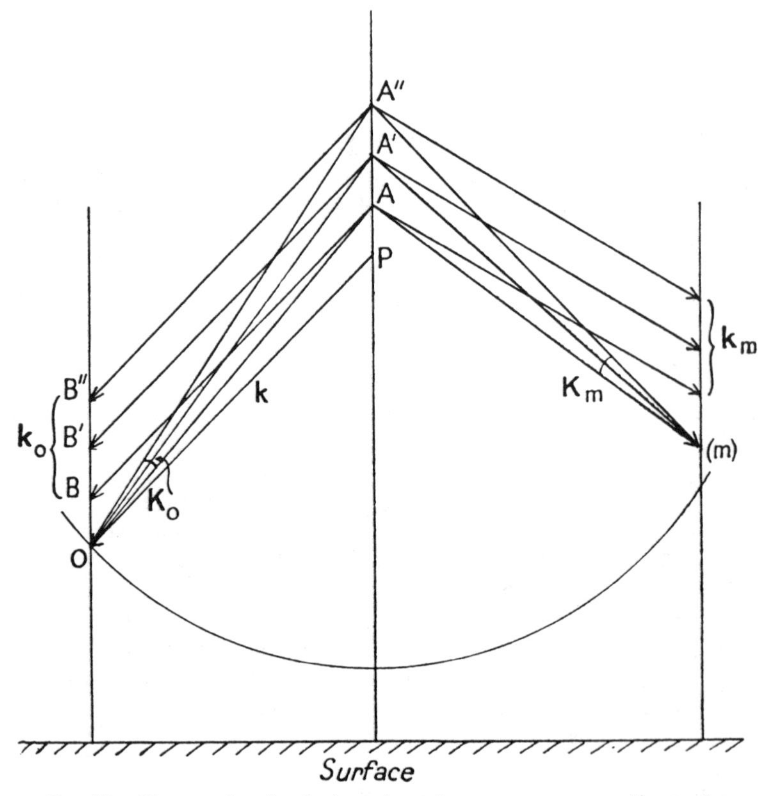

FIG. 36. Construction for the inner boundary waves corresponding to the internal wave-vectors drawn from a series of wave-points

The resultant internal wave corresponding to any lattice-point (m) is due to the superposition of $2n$ waves whose vectors are $\mathbf{K}_m$, $\mathbf{K}'_m$, ... all differing slightly in direction and magnitude. But the directions of the $2n$ corresponding inner boundary waves must all be the same, or else they could not cancel one another everywhere. This gives the key to the solution.

Let PO, fig. 36, be a vector of length k in the direction of the incident wave-normal and therefore fixed. Through P draw a line PA′A″ ... perpendicular to the crystal surface, cutting the $2n$ sheets of the dispersion surface corresponding to the problem in the points A, A′, A″ .... Then AO, A′O, A″O, ... give $\mathbf{K}_1$, $\mathbf{K}'_1$, ... the $2n$ possible internal wave-vectors. With A, A′, A″, ... as centres, describe spheres of radius k, and through O draw OBB′B″, ... parallel to PAA′ ... , cutting these spheres in B, B′, ... Then AB, AB′, ... are, by the construction of fig. 35, the $2n$ inner boundary-wave vectors $\mathbf{k}_0$, $\mathbf{k}'_0$, ... and these are evidently all parallel to each other and to the incident wave-vector. It is evident moreover that an analogous construction gives the inner boundary-wave vectors $\mathbf{k}_m$, $\mathbf{k}'_m$, ... and that these are also all parallel. The necessity of the parallelism of the inner wave-vectors corresponding to the different solutions, which is the first condition to be satisfied if they are to cancel one another, fixes the wave-points A, A′, A″, ... for any angle of incidence. The amplitudes must then be adjusted so that this mutual cancellation takes place for all the waves except the one which is chosen as the incident wave.

(*j*) *The relation between the external and internal waves considered in terms of the boundary conditions:* The method of dealing with the bounded crystal outlined in II, § 4(*i*), is due to Ewald, and was that by which the solution of the problem was first obtained. There is, however, an alternative method of considering the matter, which is perhaps easier to follow because it is more directly physical. It was given by Laue [22] in a very interesting paper in which the dynamical theory is considered in terms of a three-dimensionally periodic distribution of electric charge-density, instead of in terms of a lattice consisting of dipoles. We shall consider this theory in some detail in Chapter VIII, after we have discussed the representation of the crystal by means of Fourier series, but we may anticipate some of the results here.

Let us suppose a plane wave-train, whose wave-vector is $\mathbf{k}_0$, falls on the plane surface of the semi-infinite crystal from outside. The wave-field set up inside the crystal when a steady state is reached will be of the type already discussed, but there will be certain necessary boundary conditions to be fulfilled.

First of all, if $\mathbf{K}_0$ is the wave-vector of that wave inside the crystal which is to be considered as the primary wave, and as derived directly from the incident wave $\mathbf{k}_0$, the components of $\mathbf{k}_0$ and $\mathbf{K}_0$ parallel to the surface must be equal; for only so can the necessary continuity of phase at the surface be attained. This is the condition that gives Snell's law of refraction in optics.

Let SS′, fig. 37, be the trace in the plane of the paper of the surface of the crystal, and let A(o) be the vector $\mathbf{K}_0$ drawn to the reciprocal-lattice point (o). Let P(o) be a vector $\mathbf{k}_0$ drawn to the same point. Then if PP′ is the normal to the surface SS′, A must lie on PP′, for only then will the components of $\mathbf{k}_0$ and $\mathbf{K}_0$ parallel to the surface be equal. Since A must also lie on the dispersion surface, this fixes at once the $2n$ possible positions of the wave-point, A, A′, A″, ...; for these must lie at the points of intersection of PP′ with the $2n$ sheets of this surface.

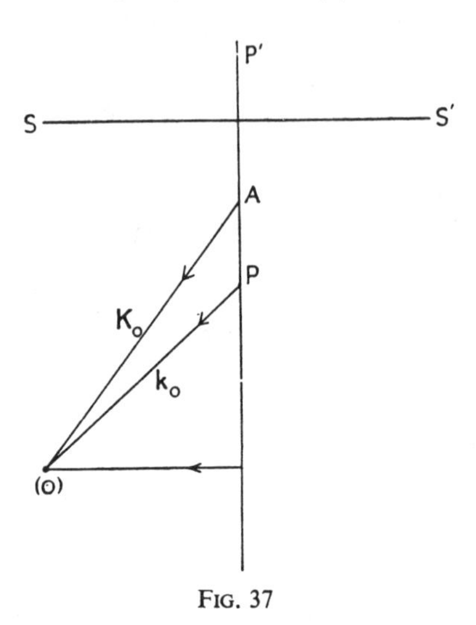

FIG. 37

The $2n$ wave-points determine $2n$ possible wave-fields in the crystal, but these are no longer independent, for certain conditions have to be satisfied in their superposition. These conditions we have discussed in II, § 4(i), in the form in which they were given by Ewald, but they may also be introduced in the form of the usual boundary conditions to be satisfied by the electric fields inside and outside the crystal—the continuity of the tangential component of the electric field, and of the normal component of the electric induction. This method of solving the problem is physically, although not formally, identical with that given by Ewald. The boundary waves are not present in the final wave-field, and make their appearance only as a formal step in the solution of the problem.

(k) *One primary and one secondary wave in the finite crystal:* (i) We shall consider first the case in which both primary and secondary wave-normals are directed into the crystal. Let $\mathbf{K}_0$ and $\mathbf{K}_m$ be the wave-vectors of the primary and secondary waves in the crystal, and let $\mathbf{k}_0$ be that of the incident wave. Both $\mathbf{K}_0$ and $\mathbf{K}_m$ make acute angles with the inward-drawn normal to the crystal surface, which we shall suppose to be the direction of the z axis. In fig. 38, let (o) and (m) be the reciprocal-lattice points concerned, and let L be the Laue point, so that L(o) and L(m) each have magnitude $|\mathbf{k}|$.

Let the radiation be polarised with the electric vector perpendicular to the plane L(o)(m). The two branches of the dispersion hyperbola for this case are shown in the figure. It is to be remembered that in any actual case the distance of L from the lattice-points (o) and (m) is

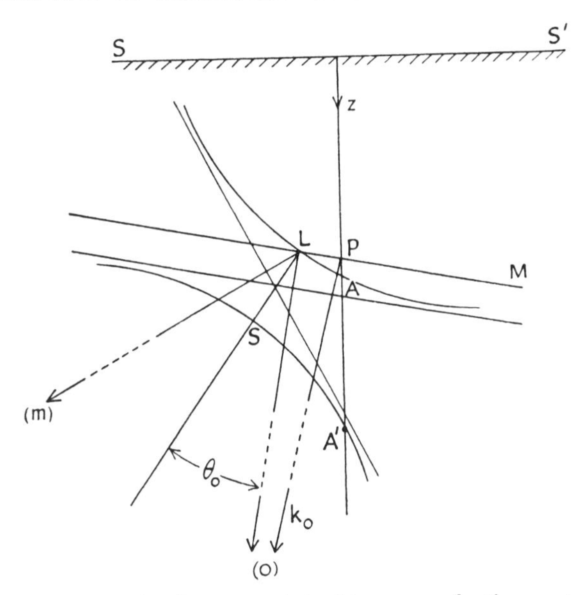

FIG. 38. Construction for wave-points of two waves, for the case in which the secondary wave is directed into the crystal

of the order of 10^5 times the diameter of the hyperbola. With the point (*o*) as centre, let a circle be described of radius k. This circle passes through L, and within the region of the figure can without sensible error be replaced by its tangent at that point, the straight line LM. The line joining any point P on this line to the point (*o*) has length k and can be taken to represent the wave-vector $\mathbf{k}_0$ of the incident radiation. As the angle of incidence varies, P moves along the line LM. When it coincides with L, the glancing angle of incidence on the crystal planes that are normal to the reciprocal vector (*o*)(*m*) is θ_0, that corresponding to the Bragg angle of reflection at those planes.

Let SS′ be the crystal surface, and let the line PAA′ be drawn through P normal to the surface. Then, as we have seen in II, §§ 4(*i*) and 4(*j*), A and A′, the intersections of this line with the hyperbola, give the possible positions of the wave-point for the angle of incidence corresponding to the position of P. Each point gives a possible wave-field. The wave-vectors $\mathbf{K}_0$ and $\mathbf{K}_m$, corresponding to the point A, are the lines A(*o*) and A(*m*), and $\mathbf{K}'_0$ and $\mathbf{K}'_m$, corresponding to the point A′, are the lines A′(*o*) and A′(*m*). To avoid confusion these lines are not actually drawn in fig. 38.

The relative amplitudes of the waves (*o*) and (*m*) for any position of of A are determined by that position; for A determines the magnitudes of the vectors $\mathbf{K}_0$ and $\mathbf{K}_m$, and these determine ϵ_0 and ϵ_m, to which the amplitudes of the waves are inversely proportional. The relative amplitudes of the two fields as a whole determined by A and A′ are, however, given by the boundary conditions, which, in Ewald's form, are here (*a*) that the inner boundary waves corresponding to the point (*m*) must cancel each other, and (*b*) that the incident beam should cancel the sum of the inner boundary waves corresponding to the point (*o*).

In the case we are considering, A and A′ must always lie on different branches of the hyperbola, and the two waves which constitute each wave-train will have slightly different velocities and directions. Interchange of energy occurs between the primary and secondary beams, analogous to that which takes place between a pair of coupled pendulums whose frequencies are nearly equal, and a detailed investigation shows that the primary beam has all the energy at the surface. At a certain depth this state of affairs is reversed, and so the alternation goes on. The depth of the layer in which a complete alternation takes place is the greater the more nearly the velocities of the two waves approach, but this closeness is limited by the diameter of the hyperbola, since A and A′ always lie on different branches; so that interchange must always occur. It is, however, improbable that it would ever be possible to detect it experimentally, for the interchange depth is small, and the effect would be averaged out by any imperfection in the crystal.

(ii) We may now consider the case of symmetrical reflection at a crystal face. The crystal surface then bisects the angle between L(*o*) and L(*m*), and these conditions are represented by fig. 39, in which the lettering corresponds to that in fig. 38.

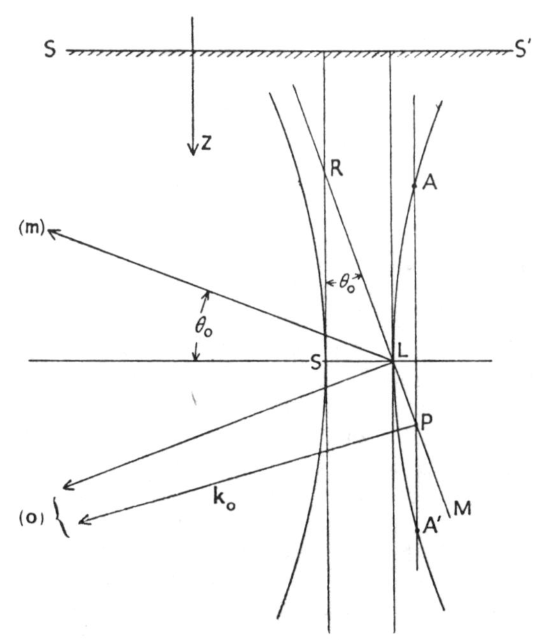

FIG 39. Construction for the wave-points for the case of symmetrical reflection at a crystal face

The line through P normal to the surface now cuts the hyperbola in points A and A′ which lie on the same branch. The solution should still, however, be of the interchange type that we have discussed above, since there are still two wave-fields between which interaction can take place. This would indeed be true for a crystal strictly without absorption, but it has been shown by Kohler[23] that if the crystal has the smallest absorption, so small that the equations need not be modified, but large enough to ensure that no energy gets through a thick slab, only *one* of the points A and A′ gives rise to an appreciable wave-field. There is thus never more than one wave-field operative at a time in the reflection experiment in any real case, and so no interchange. A discussion of this point will be found in Chapter VIII, § 1(*h*).

As the point P moves towards L, that is to say, as the glancing angle of incidence increases, A and A′ move closer together, finally coinciding at L. If the angle of incidence becomes still greater, the point P passes into the region between the tangents to the two branches of the hyperbola which are parallel to z, the normal to the crystal surface, and the line through P parallel to z cannot then cut the hyperbola. As the angle of incidence increases still further, P moves out of this region again, and the solution once more becomes of the type we have already discussed. We have now to enquire more closely into the meaning of the absence of any intersection with the hyperbola of the line through P, and normal to the surface.

In such a case, the z components of the imaginary points of intersection are conjugate complex quantities, the x and y components remaining real. The wave-vectors $\mathbf{K}_0$ and $\mathbf{K}_m$ corresponding to any such imaginary point of intersection are complex.

It is also a necessary consequence of the relation

$$\mathbf{K}_m = \mathbf{K}_0 + \mathbf{r}_m^*,$$

$\mathbf{r}_m^*$ being the vector in the reciprocal lattice from the point (o) to the point (m), that if $\mathbf{K}_0$ and $\mathbf{K}_m$ are complex, the imaginary parts of each must be equal, since their difference is equal to $\mathbf{r}_m^*$, a real quantity. The two wave-vectors must thus have equal imaginary components parallel to z. Now, as we have seen in II, § 3(*h*), a complex wave-vector implies either an increase or a decrease in the amplitude of the wave with distance travelled. The amplitudes of the waves (o) and (m) must therefore either both increase or both decrease with increase of depth in the crystal. It is evident that if energy is to be conserved they can only decrease. Both primary and secondary waves thus die away exponentially to zero as the depth in the crystal increases, and this state of affairs clearly corresponds to a deflection of the whole energy of the primary beam into the secondary, or in other words, to total reflection of the incident energy. Ewald's theory, like Darwin's, thus leads to the result that, over a certain small range of angles of incidence, reflection from the ideally perfect crystal with negligible absorption will be total.

It is easy to show that the two theories agree as to the angular range of total reflection. From fig. 39, this range is the angle turned through by the line P(o) while P moves from L to R, which is equal to LR/k, and LR = LS cosec θ_0, LS being the diameter of the hyperbola, which, by II, § 4(g), is $2k/\Omega \cos\theta_0$.

Thus the angular range of total reflection is 2 sec θ_0 cosec θ_0/Ω, or 2δ sec θ_0 cosec θ_0, in agreement with the value obtained from Darwin's theory in II, § 3(e), if we assume that $f(2\theta) = f(0)$, which is true for the simple classical dipole upon which the theory is based, for the state of polarisation that we have assumed. For the simple dipole oscillators here considered, the lower limit of the range of perfect reflection is the Bragg angle θ_0. If, as in any actual case, $f(2\theta)$ is not equal to $f(0)$, the diameter of the hyperbola, and so the breadth of total reflection, are reduced in the ratio $f(2\theta)/f(0)$. The centre of the hyperbola remains in the same position, and the angle of incidence corresponding to the middle of the range of total reflection is unaltered. This applies to radiation polarised with the electric vector perpendicular to the plane of incidence. In II, § 4(g), we have seen that if the electric vector lies in the plane of incidence, the diameter of the hyperbola, and hence the range of total reflection, are reduced in the ratio cos $2\theta_0$: 1, again in agreement with the results of Darwin's theory. The proofs of these results in terms of the dynamical theory will be found in Chapter VIII, § 1(g).

(l) *Unsymmetrical reflection*: In II, § 4(k) above, we considered the case of symmetrical reflection in which the crystal planes are parallel to the surface. It is easy to modify our construction to suit the case in which the surface is inclined to the crystal planes, as shown in fig. 40b. In fig. 40a, L is again the Laue point, and L(o) and L(m) the directions of incidence and reflection corresponding to the Bragg reflection condition, just as in fig. 39. C is the centre of the hyperbola, which, to avoid confusion, we have not drawn. LQ is parallel to the crystal surface and is inclined at an angle ϕ to the lattice-planes taking part in the reflection, and at an angle ψ to L(o), the direction of incidence corresponding to the Bragg condition; so that $\psi = \theta_0 - \phi$, where θ_0 is the Bragg angle.

Let S be the intersection of CS, the line through the centre of the hyperbola and perpendicular to the crystal surface, and LS the line perpendicular to L(o). Then, by the construction used in fig. 39, the line S(o) joining S to the reciprocal-lattice point (o) is the direction of incidence corresponding to the middle of the range of total reflection. The difference, $\Delta\theta_0$, between the angular settings of the crystal for the middle of the range of total reflection and for reflection at the Bragg angle is thus equal to the angle between S(o) and L(o), or

$$\Delta\theta_0 = \mathrm{LS}/\mathrm{L}(o) = \mathrm{LS}/k. \qquad (2.111)$$

From fig. 40a,

$$LS = LC \sin LCS/\sin CSL$$
$$= LC \cos \phi/\sin \psi = LC \cos(\theta_0 - \psi)/\sin \psi, \quad (2.112)$$

and LC, the distance from the Laue point to the centre of the hyperbola, is by II, § 4(*g*), $-k/\Omega \cos \theta_0$, so that by (2.111) and (2.112),

$$\Delta\theta_0 = -\cos(\theta_0 - \psi)/\Omega \cos \theta_0 \sin \psi$$
$$= -(\cot \psi + \tan \theta_0)/\Omega, \quad (2.113)$$

a formula first obtained by Ewald.[24]

If the crystal is cut so that the incident beam is nearly parallel to the surface when reflection takes place, ψ is small, and, by (2.113), $\Delta\theta_0$

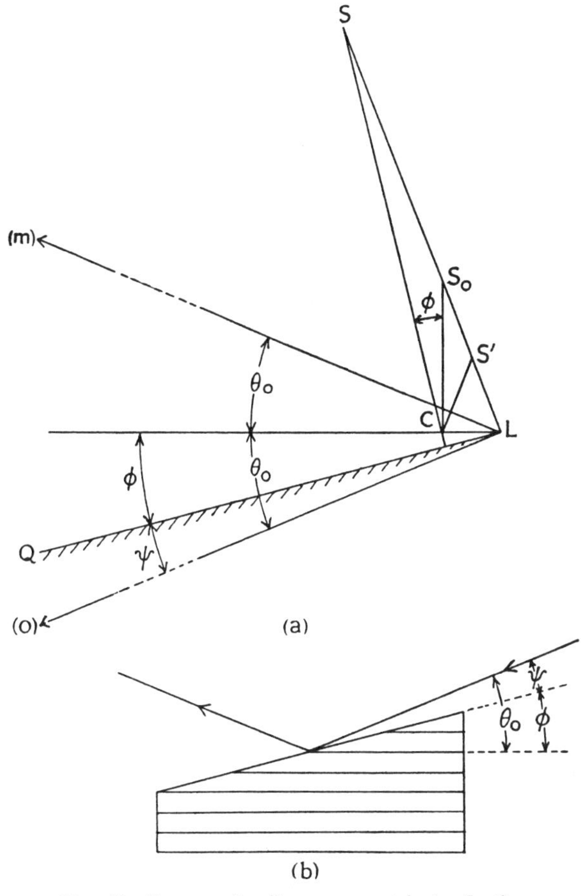

FIG. 40. Construction for unsymmetrical reflection

becomes large. The same result is evident geometrically from fig. 40a, for as ψ diminishes, the distance of S from L rapidly increases. For symmetrical reflection, $\psi = \theta_0$, and S becomes S_0 in fig. 40a; $\Delta\theta_0$ then reduces to the usual value for symmetrical reflection

$$\Delta\theta_0 = -1/\Omega \sin\theta_0 \cos\theta_0.$$

That this normal value of the deviation from the Bragg angle can be greatly increased by cutting the crystal so that its face is nearly parallel to the incident beam in the position of reflection was first pointed out by Bergen Davis,[25] and was used by him and his collaborators to measure the refractive indices of crystals.

Let us now suppose the crystal to be cut so that the *reflected* beam leaves the crystal nearly parallel to the surface; that is to say, that ψ in fig. 40a is increased until it approaches $2\theta_0$. By (*2*.113), $\Delta\theta_0$ then becomes $-1/2\Omega \sin\theta_0 \cos\theta_0$, or half the value for symmetrical reflection, a result that is also obvious from fig. 40a; for CS must now be normal to L(*m*) instead of to L(*o*), and intersects LS at S′, half-way between L and S_0.

We shall return to these formulae in Chapter IV, when discussing the measurement of the refractive index, but it is convenient to derive them here, since they are deduced more readily in terms of the dynamical theory than by treating the problem as a simple case of refraction with a refractive index $1-\delta$, where $\delta = -1/\Omega$, as was done by Stenström, Bergen Davis and others. It is in fact not really correct to do this, since, as we have seen in the discussion of the dynamical theory, we cannot strictly think of the crystal, when it is reflecting, as having a refractive index that is a definite and constant property of its material. Ewald has shown, however, that the errors made in deducing $\Delta\theta_0$ on the assumption that the problem is simply one of refraction are of the order of δ^2, and so are within the limits of experimental error.

(*m*) *Conclusion:* The account of the dynamical theory given in this chapter is based on Ewald's original treatment, and an attempt has been made to keep the main lines of the argument clear, and not to overload it with mathematical detail. The theory, as developed, assumes the lattice to be composed of scattering dipoles, situated at the lattice-points, and in Chapter VIII we shall return to the consideration of the case in which the scattering matter is distributed continuously throughout the volume of the cell, and in which absorption cannot be neglected. We shall find that the results already deduced in this chapter by means of the Darwin-Prins theory also follow from the dynamical theory, which, however, also enables us to calculate the details of the wave-field within the crystal. It will be advantageous to postpone the more detailed discussion of the theory until certain other aspects of the subject have been dealt with; but because of the importance of the dynamical theory in the development of the treatment of diffraction by crystals, and because of the intrinsic beauty of the methods by which Ewald attacked and solved the problem, a preliminary account has been introduced at this stage that will serve to give an idea of the main content of the theory.

The discussion has shown what an essentially complicated matter the diffraction of short waves by a crystal lattice is, if rigidly treated, but it has also shown that in so far as the directions of the spectra are concerned, the results of the dynamical theory differ from those of the simple theory of Laue and Bragg by seconds of arc only. For many purposes, such as the analysis of crystal structure, they are thus insignificant. But with the intensities, matters are different; the divergence between the two theories is very great, and we shall have to discuss at some length in Chapter VI what experiment has to say in deciding when one theory or the other has to be used. It is not, however, possible to do this until we have discussed the atomic scattering factor, f, which occurs in all the intensity formulae, and the effect of the temperature motion of the lattice. The next three chapters will accordingly be devoted to these topics.

References

1. J. J. Thomson, *Conduction of Electricity through Gases*, 2nd Ed. p. 321.
2. G. Friedel, *Comptes Rendus*, **157**, 1533 (1913); W. L. Bragg, *Physikal. Zeit.*, **15**, 77 (1914); M. v. Laue, *Ann. d. Physik*, **50**, 433 (1916); P. P. Ewald, *Physica*, **5**, 363 (1925); Kolkmeyer, Karssen and Bijvoet, *Physica*, **6**, 336 (1925); P. P. Ewald, and C. Hermann, *Zeit. f. Krist.*, **65**, 251 (1927).
3. D. Koster, K. S. Knol and J. A. Prins, *Zeit. f. Physik*, **63**, 345 (1930).
4. C. G. Darwin, *Phil. Mag.*, **27**, 315 (1914) (I); **27**, 675 (1914) (II).
5. A. H. Compton, *Phys. Rev.*, **9**, 29 (1917).
6. W. L. Bragg, R. W. James and C. H. Bosanquet, *Phil. Mag.*, **41**, 309, (1921).
7. W. L. Bragg, R. W. James and C. H. Bosanquet, *Phil. Mag.*, **42**, 1 (1921).
8. R. W. James, G. King and H. Horrocks, *Proc. Roy. Soc.*, **153**, A, 230 (1935).
9. J. A. Bearden, *Phys. Rev.*, **29**, 20 (1927).
10. R. J. Havighurst, *Phys. Rev.*, **28**, 882 (1926).
11. W. H. Bragg, *Phil. Mag.*, **27**, 881 (1914).
12. C. G. Darwin, *Phil. Mag.*, **43**, 800 (1922).
13. W. Stenström, *Lund Dissertation* (1919).
14. P. P. Ewald, *Physikal. Zeit.*, **21**, 617 (1921); *Zeit. f. Physik*, **2**, 332 (1920); **30**, 1 (1924).
15. J. A. Prins, *Zeit. f. Physik*, **63**, 477 (1930).
16. R. W. James and E. M. Firth, *Proc. Roy. Soc.*, **117**, A, 62 (1927).
17. R. W. James and G. W. Brindley, *Phil. Mag.*, **12**, 81 (1931); *Zeit. f. Krist.*, **78**, 470 (1931).
18. M. Renninger, *Zeit. f. Krist.*, **89**, 344 (1934).

19. S. K. Allison, *Phys. Rev.*, **41**, 1 (1932).
20. P. P. Ewald, *Ann. d. Physik*, **49**, 117 (1916); **49**, 1 (1916); **54**, 519 (1917).
21. P. P. Ewald, *Fortschritte d. Chemie*, **18**, 494 (1925).
22. M. v. Laue, *Ergebnisse d. exacten Naturwiss.*, **10**, 133 (1930).
23. M. Kohler, *Ann. d. Physik*, **18**, 265 (1933).
24. P. P. Ewald, *Zeit. f. Physik*, **30**, 1 (1924).
25. Bergen Davis and H. M. Terrill, *Proc. Nat. Acad. Sci.*, **8**, 357 (1922).

CHAPTER III

THE ATOMIC SCATTERING FACTOR

1. Introductory

At the beginning of the last chapter, in II, § 1(*b*), we defined provisionally a quantity *f*, the atomic scattering factor, which is a measure of the amplitude scattered by an atom when radiation of a given amplitude falls upon it. It is expressed in terms of the amplitude scattered by a single classical electron under the same conditions, that is to say, by an electron scattering according to the Thomson formula (*2*.8). We have seen in the last chapter how all expressions for the intensity of X-ray spectra contain this quantity *f*, whose existence we have, however, so far merely assumed in a formal way. At this stage it is necessary to deal with it in much greater detail, and to discuss to what extent it is possible to calculate it theoretically in terms of atomic properties. Such a discussion is of the greatest importance to crystal analysis, for in the detailed determination of structures, comparison between the observed intensities of the spectra and those calculated from some assumed atomic arrangement plays an essential part; and such calculations cannot be made unless the *f*-factors of the atoms concerned in the structure are known.

The importance of the atomic scattering factor has been realised from the beginning. Laue introduced it formally into his equations, and Darwin used it in his early papers, and indicated clearly that it must depend on the spatial arrangement of the electrons in the atoms. In their first papers on structure analysis, W. L. and W. H. Bragg took the amplitude scattered by an atom as proportional to the atomic weight, but Sir William Bragg [1] in his Bakerian Lecture of 1915 expressed clearly the importance of taking into account the effect of the distribution of electrons over a region whose dimensions are comparable with the wave-length of the radiation employed. About the same time, Debye [2] pointed out that atoms irregularly arranged, as in a gas, must still produce diffraction effects with X-rays, because of the grouping of the electrons which are responsible for the scattering about the atomic nuclei; and in a very important paper, which laid the foundations of subsequent work on the scattering of X-rays by gases, he showed how to calculate the scattering factors with certain assumptions as to the electronic arrangements. A. H. Compton [3] in 1917 calculated the scattering factors for certain simple configurations of electrons, and used them to estimate the dimensions of the atoms in calcite and rock salt by comparison of observed and calculated intensities of spectra. Similar estimates were also made by Debye and Scherrer [4] in 1918.

In all these papers the classical theory of the scattering of electrons by X-rays had of necessity to be used as a basis for calculation, for the older quantum theory provided no means of calculating the scattering of coherent radiation by electrons. On the other hand, the existence of the diffraction of X-rays by crystals showed definitely that such coherent radiation forms an important part of the total scattering. The situation was theoretically very unsatisfactory; for while in the time just preceding the development of the newer quantum mechanics, the gradual accumulation of accurate intensity measurements was making it clear that calculations based on the classical theory of scattering gave results of the right general type and order of magnitude if an electron distribution in harmony with the Bohr atomic theory was assumed, yet there was no proper theoretical sanction for using the classical formulae. The difficulties have been largely, although not entirely, removed since the advent of the new quantum mechanics and the development of the theory of scattering in the hands of Wentzel, Waller, and others; for the quantum-mechanical and classical formulae are found to be formally nearly identical, so that the general agreement between experiment and theory based on the mixture of classical and old quantum ideas, which had to be used in calculating the scattering factors, was satisfactorily explained.

2. The Classical Calculation of Scattering Factors

(*a*) *General considerations:* We shall first discuss the scattering of radiation by atoms according to the classical theory, which forms a useful introduction to the treatment in terms of the quantum theory. The assumptions which we shall make in this section may be summed up as follows:

(1) We shall suppose the atom to contain electrons which are distributed throughout a volume comparable with the atomic dimensions, and with the wave-length of the radiation which falls on it.

(2) Each electron is supposed to be so loosely bound in the atom that it scatters according to the Thomson formula as a 'free' electron. If we suppose the electron to be bound in the atom with such a force that it oscillates with a natural frequency ω_0 when disturbed, we are in effect supposing that ω, the frequency of the incident radiation, is very large compared with ω_0, but yet is not so large that relativity corrections become important. The electron will then scatter exactly out of phase with the incident radiation, with an amplitude given by the Thomson formula.

(3) We also assume that any orbital motion of the electrons within the atom takes place so slowly that there is no appreciable alteration of the configuration during a large number of complete alternations of the field-vector of the incident radiation. Using these assumptions, we can deduce in a general way certain properties of the scattering factor.

In the forward direction there will be no path difference between the waves scattered by electrons in different parts of the atom. Since the electrons are all supposed to be loosely bound, any change of phase on scattering is the same for all of them, and so the amplitude scattered in a forward direction will be Z times that due to a single electron, Z being the total number of electrons in the atom. On the other hand, in a direction making a finite angle with the direction of the incident radiation there will be path differences between the waves scattered from electrons in different parts of the atom, which will therefore interfere, and produce a resultant amplitude smaller than Z times that due to a single electron. The phase differences that may be attained depend on the angle of scattering, the wave-length, and the volume throughout which the electrons are distributed. The scattering factor f will thus approach Z for small angles of scattering, and will fall away with increasing angle at a rate that, for a given wave-length, is determined by the distribution of electrons within the atom.

We can see then from these very general considerations that it is not sufficient to take the amplitude scattered by an atom as proportional to the atomic number Z at all angles. The ratio between the amplitudes scattered by two atoms of different kinds will, moreover, not be constant, since f for each atom will, in general, depend in a different way on the angle of scattering;* and this may be of great importance in crystal analysis. We see too that if it is possible by any means to determine f experimentally as a function of the angle of scattering we can get direct information concerning the distribution of the electrons within the atom. Both the above aspects of the subject have been important in its development.

In calculating scattering factors by classical methods, we obtain different results according to whether we consider the atoms as forming part of a crystal lattice, or as irregularly arranged in a gas. In either case, we must in practice deal with an average effect due to a very large number of atoms, and the method of averaging depends on the conditions of the problem. We shall consider the point in a little detail, since the principles involved are important, and are not always clearly understood.

(*b*) *Scattering by atoms arranged in a crystal lattice:* Suppose that every point of a lattice is occupied by an atom of the same kind, the centres of the atoms coinciding with the lattice-points. The atomic electrons then form a group of scattering points associated with each lattice-point, and the problem to be solved is evidently similar to that of calculating the structure factor for a composite crystal unit. The scattering factor that we have to calculate is in fact the structure factor of the atom. There is, however, a difference. We can no longer

* A very direct demonstration of this for the atoms of calcium and fluorine is given by the intensities of the spectra from the crystal fluorite. See, R. W. James and J. T. Randall, *Phil. Mag.*, 1, 1202 (1926).

assume the groups at the lattice-points to be identical. They will be identical on the average, but, since the electrons may be moving in the atoms, at any given instant the electronic configurations about the different lattice-points may differ. We recognise here, in the assumption of an instantaneous configuration of point electrons, just that type of concept which characterises the classical point of view, and which the quantum theory declares to be without physical meaning. For the time being, however, we shall use it.

Suppose to begin with that each atom contains only one electron, the position of which with respect to the centre of the atom, and so to the corresponding lattice-point, is determined for one of the atoms by a vector **r**. The phase difference at a distance large compared with the atomic dimensions between the wave scattered by this electron and by one supposed coincident with the atomic centre is $\kappa\mathbf{S}\cdot\mathbf{r}$, by (*1*.6). The vector **S** is normal to the plane that would reflect the direction of incidence into the direction of scattering, and its magnitude is $2 \sin \theta$, where 2θ is the angle of scattering; κ is equal to $2\pi/\lambda$.

Let us now suppose the direction of scattering to be that corresponding to a spectrum, so that **S** is normal to the corresponding lattice-planes. The radiation scattered by electrons supposed coincident with the lattice-points will now be in phase from all the atoms. In calculating the amplitude scattered by any configuration of electrons we have therefore simply to take into account the phase differences due to the departures of the electrons from the atomic centres. The amplitude due to a given configuration will then be $\Sigma e^{i\kappa\mathbf{S}\cdot\mathbf{r}}$ times that due to a single electron, the sum being taken over all the electrons in all the atoms. Now this amplitude is just the same as would be scattered by a single atom containing all the electrons, fixed by the corresponding vectors radiating out from the common centre. If N, the total number of atoms, is large, the probability that an electron lies within any given element of volume of this composite atom is N times the probability that one lies within the corresponding element of volume of a single atom chosen at random. Let us denote by $\rho(\mathbf{r})\,d\tau$ the probability that an electron in any atom lies within an element of volume $d\tau$, whose mid-point is defined by the vector **r**. Then we can write for the amplitude scattered by the composite atom considered above

$$\mathrm{N}\int \rho(\mathbf{r})\, e^{i\kappa\mathbf{S}\cdot\mathbf{r}}\, d\tau,$$

the integral being taken throughout the whole volume of a single atom. Thus the amplitude scattered by the whole lattice in the direction of the spectrum is the same as it would be if each lattice-point were occupied by an atom scattering f times the amplitude scattered by a single electron, where

$$f = \int \rho(\mathbf{r})\, e^{i\kappa\mathbf{S}\cdot\mathbf{r}}\, d\tau. \tag{3.1}$$

Let **r** make an angle α with the direction of **S**. Then

$$\kappa\mathbf{S}\cdot\mathbf{r} = 2\kappa \sin\theta\, r\cos\alpha = \mu r\cos\alpha,$$

where

$$\mu = 2\kappa\sin\theta = 4\pi(\sin\theta)/\lambda. \tag{3.2}$$

Now suppose that $\rho(\mathbf{r})$ depends only on the magnitude of **r**, and not on its direction, so that the average atom is spherically symmetrical. Then we may take as a convenient element of volume an annulus of a spherical shell of radius r, thickness dr, and width $d\alpha$, having **S** as axis. (See fig. 41.)

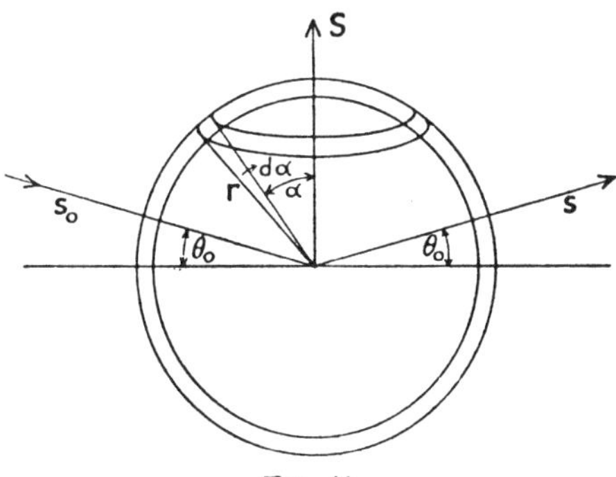

FIG. 41

Thus $d\tau = 2\pi r^2 \sin\alpha\, d\alpha\, dr$, so that if we put $\mu r\cos\alpha = x$, and $-\mu r\sin\alpha\, d\alpha = dx$, (3.1) gives

$$f = \int_0^\infty \frac{2\pi r^2}{\mu r}\rho(r)\,dr \int_{-\mu r}^{+\mu r} e^{ix}\,dx = \int_0^\infty 4\pi r^2\rho(r)\frac{\sin\mu r}{\mu r}\,dr. \tag{3.3}$$

If $U(r)dr$ is the probability that an electron lies between radii r and $r+dr$ in any atom, $U(r) = 4\pi r^2\rho(r)$, so that (3.3) becomes

$$f = \int_0^\infty U(r)\frac{\sin\mu r}{\mu r}\,dr, \tag{3.4}$$

where $\mu = 4\pi(\sin\theta)/\lambda$. The scattering factor is spherically symmetrical and real. The calculation is very similar if, instead of a single electron, each atom contains a number of electrons, say Z. For generality, suppose the motions of all the electrons are of different types, having different probability functions $\rho_1, \rho_2, \dots \rho_z$. Again we shall suppose that all directions of the radius are equally likely for any of the electrons, so that we can define radial probability functions $U_1(r) \dots U_z(r)$, giving the probability that, in any atom, electrons of any given type should lie between the radii r and $r+dr$. Then it is evident that (3.4) becomes

$$f = \int_0^\infty \{U_1(r) + U_2(r) + \dots + U_z(r)\}\frac{\sin\mu r}{\mu r}\,dr = \int_0^\infty U(r)\frac{\sin\mu r}{\mu r}\,dr, \tag{3.5}$$

where now $U(r)dr = \sum_n U_n(r)dr$, and is the total probability that an electron will be found between radii r and $r+dr$ in any atom. The relation $\int_0^\infty U(r)dr = Z$, where Z is the number of electrons in a single atom, must of course hold good.

The average taken here is that over the instantaneous positions of the electrons in all the atoms in the crystal, but it will usually be assumed that this is the same as the time-average of the distribution in a single atom. The orientation of the atom has been taken as a random one, so that the average atom has spherical symmetry. The possibility of taking the average in this way depends on the fact that the centres of the atoms are fixed and that we are considering the reflection of a spectrum, and not of the general background, so that there is a regular phase relationship between the radiation scattered from corresponding points in the different atoms. We can therefore add the *amplitudes* scattered by the electrons in different atoms.

(c) *Scattering by atoms distributed at random:* Let us suppose now that the same atoms are irregularly distributed, as in a gas. There is no longer any regular phase relationship between the radiation scattered from corresponding points of different atoms. We have now, therefore, to average the *intensities* scattered by a large number of independent atoms, which can have any orientation. This is equivalent to taking the intensity scattered by any configuration of electrons in one atom, and averaging over all possible configurations and orientations.* Let there be Z electrons, fixed by vectors $\mathbf{r}_1, \ldots \mathbf{r}_n, \mathbf{r}_m, \ldots$. Then the amplitude scattered by the atom in the configuration given by these vectors is $\sum_n e^{i\kappa \mathbf{S}\cdot\mathbf{r}_n}$, the sum being over all the Z electrons. The intensity for this particular configuration is the square of the modulus of this expression. The measured intensity is the average of this intensity over all possible configurations. Thus, in terms of the intensity scattered by a Thomson electron, we have

$$I = \overline{\left| \sum_n e^{i\kappa \mathbf{S}\cdot\mathbf{r}_n} \right|^2} = \overline{\sum_n \sum_m e^{i\kappa \mathbf{S}\cdot\mathbf{r}_n - \mathbf{r}_m}}. \qquad (3.6)$$

The double summation in (3.6) contains terms involving all possible pairs of electrons in the atom. Those terms for which $n=m$ are all equal to unity on averaging, so that they contribute in all Z to the intensity. To get the average values of the terms for which $n \neq m$ we proceed as follows. Let $\rho_n(\mathbf{r}_n)d\tau_n$, $\rho_m(\mathbf{r}_m)d\tau_m$, be respectively the probabilities that the electrons n and m lie within the small elements of

* This procedure neglects external interference effects, produced by the relationship of electrons in different atoms, and this is justifiable if the average distance between atoms is large in comparison with the wave-length. (See Chapter IX, § 3(a).)

volume $d\tau_n$ and $d\tau_m$ fixed by the vectors $\mathbf{r}_n$ and $\mathbf{r}_m$. The product $\rho_n(\mathbf{r}_n)\rho_m(\mathbf{r}_m)\,d\tau_n\,d\tau_m$ gives the probability that electron n lies in $d\tau_n$ while at the same time electron m lies in $d\tau_m$, and this gives the weight to be assigned to the term in the double sum involving n and m in taking the average. To get the average value of such a term we multiply it by this weight, and integrate over the whole volume, both for $d\tau_n$ and $d\tau_m$. We get in this way for the average intensity

$$\mathrm{I} = \mathrm{Z} + \sum_n^{n \neq m} \sum_m \int \rho_m(\mathbf{r}_m) e^{-i\kappa \mathbf{S}\cdot\mathbf{r}_m} d\tau_m \int \rho_n(\mathbf{r}_n) e^{i\kappa \mathbf{S}\cdot\mathbf{r}_n} d\tau_n. \tag{3.7}$$

We have here supposed the positions of the electrons to be independent. so that the position of m does not depend in any way on that of n, The two integrals can therefore be evaluated independently, and so by (3.1)

$$\mathrm{I} = \mathrm{Z} + \sum_n^{n \neq m} \sum_m f_n f_m, \tag{3.8}$$

where f_n and f_m are the scattering factors that would have been obtained for the individual electrons had the atoms been arranged in a crystal lattice.* In what follows we shall always use f in this sense. Since there are Z electrons altogether, we may write (3.8) in the form

$$\mathrm{I} = \sum_n (1 - f_n^2) + (\sum_n f_n)^2 = \sum_n (1 - f_n^2) + f^2, \tag{3.9}$$

where $f = \Sigma f_n$, and is the scattering factor for the whole atom, as defined by (3.5). The method followed in the calculation just given is based in principle on the work of Debye.[2] Compton [5] and Raman [6] have given essentially similar treatments. Compton, however, assumed f_n to be the same for each atomic electron, so that it could be taken as equal to f/Z, where f is the scattering factor for the whole atom, and Z is the total number of electrons in the atom. Equation (3.9) becomes on this assumption

$$\mathrm{I} = \mathrm{Z}\{1 - (f/\mathrm{Z})^2\} + f^2 = \mathrm{Z} - f^2/\mathrm{Z} + f^2. \tag{3.10}$$

The assumption of equal f's for all the atomic electrons is not justifiable, as we shall see later, and the more exact expression (3.9) is therefore preferable to (3.10), although the numerical difference between them is not in fact great. The expression for the classical scattering in its more exact form was first given by Woo,[7] and, independently, by James.[8]

We may now compare equation (3.5), which refers to atoms arranged in a crystal lattice, and (3.9), which refers to the same atoms arranged as a gas. Suppose that in each case there are N atoms in all. The intensity scattered by the gas will be N times the average intensity scattered by a single atom, or $\mathrm{N}\{\Sigma(1 - f_n^2) + f^2\}$. On the other hand, for the direction of the lattice spectrum the total amplitude is $\mathrm{N}f$ and the

* In the formulae of this paragraph the f's are supposed to be real With complex values of f the correct form of the product in (3.8) would be $f_m^* f_n$.

intensity N^2f^2. The two formulae have, moreover, quite different dependences on angle of scattering. We can see this most easily if we take the case of an atom containing a single electron. Equation (*3*.9) then reduces to unity, which means that each atom scatters as a Thomson electron, as would be expected, since the electrons in the different atoms form a random arrangement. If the radiation were polarised with the electric vector perpendicular to the plane containing the directions of incidence and scattering the scattered intensity in this plane would be the same at all angles. The single-electron atom in the crystal, however, scatters with a factor f which depends on the angle of scattering, decreasing as it increases.

The difference between the two cases can be put briefly in the following way. The first formula assumes that the whole statistical distribution of charge in the average atom scatters at once, so that the *amplitudes* of the waves from each element of the distribution may be added. The regular arrangement of atoms allows this to be done. In the second case all the elements of the statistical distribution are supposed to scatter independently, corresponding to the idea that they represent entirely independent configurations of the same or of different atoms, with no regular phase relationship between them. We have therefore to average the *intensities* scattered by the different configurations. One is essentially a statement about amplitudes, the other about intensities

(*d*) *Coherent and incoherent radiation:* The classical treatment that we have just given assumes the scattered radiation all to have the same wave-length as the incident radiation, while actually a very considerable fraction of it is scattered with an increase in wave-length. This incoherent or Compton scattering can take no part in the formation of the interference maxima, and in these we are concerned only with the coherent radiation. Incoherent radiation occurs only in the background, and forms a very small fraction of the intensity of a strong spectrum. On the other hand, the radiation scattered from a gas contains a very much larger fraction of incoherent radiation, since the coherent radiation is not now reinforced by the agreement in phase of the contributions from the different atoms.

Equation (*3*.9) shows that according to the classical theory the total radiation scattered per atom by a monatomic gas is

$$I = \sum_n (1 - f_n^2) + f^2.$$

Now the *amplitude* scattered per atom when the atoms are arrayed in a crystal lattice is, for the directions of the interference maxima, f, and here we must suppose only the coherent radiation to take part. If the formula could be taken over into quantum theory, the second term f^2 might be interpreted as giving the intensity per atom of the coherent radiation, and the first term $\sum_n (1 - f_n^2)$ as giving that of the incoherent radiation. This was suggested as an interpretation by A. H. Compton.

The quantum theory of scattering, as we shall see, does in fact suggest that formally this is very nearly correct. The scattering from an assemblage of electrons, calculated according to the classical theory, is equal to the total scattering, coherent and incoherent, calculated according to the quantum theory, if interaction between the electrons is neglected, or if it is taken into account in certain approximate manners, and if the exchange terms are neglected. Nevertheless, although it appears to be formally correct to assign the first and second terms of (*3*.9) to incoherent and coherent radiation respectively, it seems at this stage to be rather arbitrary to do so; for the whole calculation is based on interference methods, and assumes no change of wave-length. These difficulties emphasise the need for a treatment of the theory of scattering based on quantum principles, and we shall outline such a treatment in the third part of this chapter.

(*e*) *Numerical calculation of scattering factors on classical basis:* The first serious attempt to evaluate the radial probability function $U(r)$ on which the f-factor depends was made by Hartree[9] in 1925, on the basis of Bohr's orbit theory of the atom. He estimated the dimensions of the orbits from the X-ray and optical term values, and deduced the necessary time-averages, and so obtained $U(r)$. The scattering factor could then be calculated from (*3*.5) by numerical integration. Hartree's f-curves obtained in this way were of great value in crystal analysis, where a general idea of the way in which scattering depends on the angle of scattering may often be enough, particularly if the relative values for different atoms are fairly correct. Absolute comparisons with f-curves obtained experimentally by W. L. Bragg, James and Bosanquet,[10] and others, showed important systematic deviations between theory and experiment. The Hartree curves always showed a hump for moderate angles of scattering, which was not found experimentally. Williams,[11] and Jauncey[12] tried to account for this in terms of incoherent scattering; but since these curves have been entirely superseded by the much more accurate ones subsequently obtained by Hartree himself on a quantum-mechanical basis, using his method of the self-consistent field, we shall not discuss them further, and shall proceed at once to the consideration of the quantum theory of scattering.

3. The Treatment of Scattering by the Methods of Wave-Mechanics

(*a*) *Introductory:* In the classical treatment of the scattering of radiation by electrons it is assumed that we have the right to employ such ideas as the exact position and velocity of an electron at a given instant.

Now from the point of view of the quantum theory it is impossible in principle to determine exactly the simultaneous position and velocity of a particle; for any exact determination of position involves the interaction of the particle with radiation, and this will alter its momentum, and so the whole of its subsequent motion. We cannot consider ourselves to be outside a system, impartially noting the positions and velocities of each of its particles, without in any way altering these, which is in effect what classical physics assumes.

Although we cannot determine exact configurations in this way, we possess in the methods of wave-mechanics a calculus enabling us to determine the *probability* of finding an electron in a stated region, given the field of force acting on it. To each mechanical problem there corresponds a certain equation, the wave-equation, whose solution, known as the wave-function, we shall denote by u. It is a function of the co-ordinates and of the time, which is in general complex. Let $dv = dx\,dy\,dz$ be a small element of volume in the neighbourhood of the point (x, y, z). Then $|u(x, y, z, t)|^2\,dv$ is to be interpreted as the probability that, under the conditions of experiment to which the wave-equation applies, the electron will be found at time t in the element of volume dv. In a certain sense therefore $e|u|^2$ or euu^*, where u^* is the conjugate of u, may be thought of as the average charge-density at the point (x, y, z), an interpretation first given to it by Schrödinger. It is better, however, to use the probability interpretation, since we are led into difficulties in many cases in using the idea of the charge-density. The justification for using the idea of the charge-density is that $e|u|^2$ occurs in some expressions, notably, as we shall see, in that for the scattering factor, exactly in the way in which charge-density occurs in the corresponding classical case. For a number of electrons the probability interpretation has to be extended as follows. Suppose the co-ordinates $x_1, y_1, z_1, \ldots x_n, y_n, z_n$ refer to the electrons (1) ... (n). Then u will be a function of all these co-ordinates and of the time, and the probability that at time t electron (1) lies within the element of volume dv_1 in the neighbourhood of (x_1, y_1, z_1), electron (n) within the element of volume dv_n in the neighbourhood of (x_n, y_n, z_n), and so on is

$$|u(x_1, y_1, z_1, \ldots x_n, y_n, z_n, t)|^2\,dv_1 \ldots dv_n.$$

The problem that interests us in this section is the calculation of the amplitude of the scattered radiation when an electromagnetic wave falls on an atom whose wave-functions when undisturbed by the radiation are known. This problem has not yet been fully and satisfactorily solved. To deal with it, it would appear to be necessary to consider the atom and the total radiation field as forming a single quantised system, a method that has been used by Dirac, but which is entirely beyond the scope of the present treatment.* There exist, however, other methods which, while less satisfactory logically than

* See for example, Fermi, *Reviews of Modern Physics*, Vol. 4, p. 87, 1932.

the fuller treatment, give the results that we require much more simply, and have in fact generally been used in such problems. The calculation of the coherent and incoherent radiation from electrons under the influence of an electromagnetic wave was first carried out by Wentzel,[13] and by Waller.[14] The work is an extension of the dispersion formula of Kramers and Heisenberg[15] to waves of lengths comparable with the atomic dimensions. The dispersion formula had already been obtained by the methods of wave-mechanics for long waves by Schrödinger[16].

In the brief outline of the methods which follows, we shall consider the case of an atom containing a single electron. We have already seen that the quantity $e|u|^2$ can then be considered in a certain sense as the average charge-density in the atom at the point where the wave-function has the value u. In ordinary electrical theory, ρ the charge-density, and $\mathbf{j}$ the current-density at any point are connected by the equation of continuity,

$$\frac{\partial \rho}{\partial t} + \operatorname{div} \mathbf{j} = 0. \qquad (3.11)$$

Schrödinger showed that it is possible to derive from the wave-equation an equation of the same form as (3.11), in which ρ is represented by $e|u|^2$. The quantity in the equation corresponding to $\mathbf{j}$ may therefore be interpreted as a current-density. We shall refer to it in what follows as the Schrödinger current-density.

The simplest method of treating radiation problems is to assume that the electromagnetic field due to an atom whose wave-function is known is to be calculated from the Schrödinger current-density in just the same way that the field due to a distribution of currents is calculated according to the ordinary classical electromagnetic theory. So long as the atom is in a stationary state, the Schrödinger charge and current-density do not depend on the time, and the atom will not radiate, but if an electromagnetic wave falls on the electron, the field acting on it now varies periodically with the time. We may treat the field due to the incident wave as a perturbation of the atomic field, and calculate the resulting perturbation of the wave-function. The corresponding perturbed current-density now varies periodically with the time, and gives rise to radiation, which is to be calculated according to the known laws of classical electrodynamics for a periodically varying current distribution. This method is only approximate, for it neglects altogether the reaction of the atomic field on the radiation field; but so long as we can treat the incident electric field as a small perturbation of the atomic field, and this is justifiable in the case to which we shall apply the method, it yields results that agree with those of the more detailed theory.

(*b*) *The wave-equation:* We have first of all to write down the wave-equation appropriate to the problem, which will be that for an electromagnetic field varying with the time. We shall use the non-relativistic form of the equation, although for the heavier atoms relativity effects will be appreciable for the inner electrons.

The usual method of obtaining the wave-equation is as follows. We write the total energy of the electron in the Hamiltonian form, as a function of its momenta and co-ordinates, and in the Hamiltonian replace the momentum by the operator $(\hbar/i)$ grad.* The Hamiltonian is thus transformed into an operator, which is made to operate on the wave-function u, and the whole is then equated to Wu, W being the total energy. Since we are dealing with a field varying with the time, W is not a constant, and it must be replaced by the operator $i\hbar\,\partial/\partial t$, a transformation that is arrived at as follows. For a stationary state of energy W the wave-function u may be written in the form

$$u = \psi(x, y, z)e^{-\frac{i}{\hbar}\mathrm{W}t}, \tag{3.12}$$

where ψ is a function of the co-ordinates only. From this equation it is clear that $i\hbar\,\partial u/\partial t = \mathrm{W}u$, so that W is equivalent to the operator $i\hbar\,\partial/\partial t$, and this is now assumed to hold generally.

The wave-equation is thus

$$\mathrm{H}u = \mathrm{W}u = i\hbar\frac{\partial u}{\partial t}, \tag{3.13}$$

H being the Hamiltonian operator.

For an electron of mass m in a field of vector-potential **A** and scalar potential ϕ, the appropriate form of the Hamiltonian is†

$$\mathrm{H} = \frac{1}{2m}(\mathbf{p} - e\mathbf{A}/c)^2 + e\phi. \tag{3.14}$$

The vector **p** is related to the ordinary mechanical momentum $m\mathbf{v}$ by the equation

$$\mathbf{p} = m\mathbf{v} + e\mathbf{A}/c, \tag{3.15}$$

and it may be shown that it is **p** that has to be used in place of the ordinary mechanical momentum if the equations of motion of the charge are to take the Lagrangian form. We may therefore infer that in writing down the wave-equation according to the rules given above, it is **p**, as defined by (3.15), that must be replaced by the operator $(\hbar/i)$ grad.

The electric intensity **E** is connected with **A** and ϕ by the relation

$$\mathbf{E} = -\frac{1}{c}\frac{\partial \mathbf{A}}{\partial t} - \operatorname{grad}\phi. \tag{3.16}$$

* It is convenient to use $\hbar = h/2\pi$, where h is the usual Plank's constant, since this materially simplifies the formulae by getting rid of a large number of factors 2π.

† See for example, W. Heitler, *Quantum Theory of Radiation*, 2nd Edn., Chap. I, §§(2) and (6), or C. Schaefer, *Einführung in die theoretische Physik*, Vol. 3, pt. 2, §46, p. 294.

We now suppose the field acting on the electron to be that due to the nucleus of the atom, together with that due to an incident plane electromagnetic wave, which we shall further assume to be plane-polarised. The wave-field is to be treated as a small perturbation, so that terms involving the square of the vector potential **A** may be neglected. The scalar potential of the wave-field is zero. Using these results, we obtain for the wave-equation, according to the rules given above,

$$\left(-\frac{\hbar^2}{2m}\nabla^2+e\phi_a\right)u-\frac{\hbar e}{imc}(\mathbf{A}\cdot\mathrm{grad})\,u=i\hbar\frac{\partial u}{\partial t},\tag{3.17}$$

$$\mathrm{H}_a u-\frac{\hbar e}{imc}(\mathbf{A}\cdot\mathrm{grad})\,u=i\hbar\frac{\partial u}{\partial t},\tag{3.17a}$$

or

$$\mathrm{H}_a u+\mathrm{H}'u=i\hbar\frac{\partial u}{\partial t},\tag{3.17b}$$

where H_a is the Hamiltonian operator for the unperturbed atom. If u_n is a solution of the equation for the unperturbed atom,

$$\mathrm{H}_a u_n=i\hbar\frac{\partial u}{\partial t}=\mathrm{W}_n u_n,\tag{3.18}$$

and the various solutions of (3.18), u_1, u_2, ... u_n, are the characteristic wave-functions corresponding to the different stationary states of energies W_1, W_2, ... W_n. These states may be discrete, corresponding to negative energy values, and to the closed orbits of the older quantum theory; or they may belong to the continuum of positive energy states, the analogues of the hyperbolic orbits of the older theory. The second term of (3.17b), $\mathrm{H}'u$, contains the effect of the perturbing field, and our problem is to solve (3.17), treating it as a perturbation of (3.18).

(*c*) *The calculation of the perturbed wave-function:* Only a bare outline of the solution of (3.17) can be given here, and for details the reader is referred to books on wave-mechanics. Whatever the solution u may be, it can always be expressed as a linear combination of wave-functions satisfying the unperturbed equation (3.18), so that we can write

$$u=\sum_m a_m u_m=\sum_m a_m\psi_m e^{-\frac{i}{\hbar}\mathrm{W}_m t},\tag{3.19}$$

where $|a_m|^2$ is the probability that as a result of the perturbation the atom is in the state m. Our aim is thus to determine the coefficients a_m. To do this we substitute (3.19) in (3.17b), and using (3.18) obtain at once

$$\sum_m a_m\mathrm{H}'u_m=i\hbar\sum_m u_m\frac{\partial a_m}{\partial t}.\tag{3.20}$$

In order to determine how any coefficient, say a_k, varies with the time we first of all multiply both sides of (3.20) by u_k^*, the conjugate of u_k, and then integrate over all space.

The wave-functions are supposed to be orthogonal and normalised, so that

$$\int u_k^* u_m \, dv = \begin{cases} 0 & \text{if} \quad m \neq k \\ 1 & \text{if} \quad m = k. \end{cases} \tag{3.21}$$

The integration thus gives

$$i\hbar \frac{\partial a_k}{\partial t} = \sum_m a_m \int u_k^* \mathrm{H}' u_m \, dv. \tag{3.22}$$

Equation (3.22) is exact, but it cannot be used directly to determine a_k, since the right-hand side contains all the unknown coefficients a, and to proceed further we must approximate. Suppose the atom to be initially in the state n, so that to begin with we know that $a_n = 1$, while all the other coefficients are zero. We may then, as a first approximation, suppose that all the coefficients except a_n remain small, and can neglect them in (3.22), and at the same time may put $a_n = 1$. This gives as a first approximation, which will suffice for our purpose,

$$i\hbar \frac{\partial a_k}{\partial t} = \int u_k^* \mathrm{H}' u_n \, dv = e^{\frac{i}{\hbar}(\mathrm{W}_k - \mathrm{W}_n)t} \, \mathrm{H}'_{nk}, \tag{3.23}$$

where

$$\mathrm{H}'_{nk} = \int \psi_k^* \mathrm{H}' \psi_n \, dv. \tag{3.24}$$

The determination of the perturbed wave-function thus involves the calculation of the quantities H'_{nk}, which are called the *matrix elements* of the perturbing field. These contain the vector potential of the incident wave, and we must write down an expression for this.

Suppose the electric field in the incident wave at a point whose distance from the origin is given by the vector **r** is

$$\mathbf{E} = \mathbf{E}_0 \cos(\omega t - \kappa \mathbf{s}_0 \cdot \mathbf{r}), \tag{3.25}$$

where $\mathbf{s}_0$ is the unit vector in the direction of the wave-normal. Then by (3.16), since $\phi = 0$,

$$\mathbf{A} = -\frac{c\mathbf{E}_0}{\omega} \sin(\omega t - \kappa \mathbf{s}_0 \cdot \mathbf{r}),$$

or in the more convenient complex form,

$$\mathbf{A} = -\frac{c\mathbf{E}_0}{2i\omega} e^{i\omega t - i\kappa \mathbf{s}_0 \cdot \mathbf{r}} + \text{conjugate term}. \tag{3.26}$$

Now H′, by (3.17), contains the scalar product of **A** and a gradient, which is equal to the magnitude of **A** multiplied by the component of the gradient in the direction of the electric field-vector in the incident wave. Let us denote this direction by x_0. Then we can write for H'_{nk}

$$H'_{nk} = -\frac{\hbar e}{2m\omega}|\mathbf{E}_0|\left\{e^{i\omega t}\int \psi_k^* e^{-i\kappa \mathbf{s}_0\cdot\mathbf{r}} \frac{\partial \psi_n}{\partial x_0}\,dv - e^{-i\omega t}\int \psi_k^* e^{i\kappa \mathbf{s}_0\cdot\mathbf{r}} \frac{\partial \psi_n}{\partial x_0}\,dv\right\}. \quad (3.27)$$

If we denote the first integral in (3.27) by B^0_{nk}, and if B^{0*}_{kn} is its conjugate with n and k interchanged, then it is easy to show by partial integration, remembering that $\mathbf{s}_0$ is perpendicular to x_0, that the second integral in (3.27) is equal to $-B^{0*}_{kn}$. Thus (3.27) becomes

$$H'_{nk} = -\frac{\hbar e|\mathbf{E}_0|}{2m\omega}\left\{B^0_{nk}e^{i\omega t} + B^{0*}_{kn}e^{-i\omega t}\right\}. \quad (3.28)$$

This value of H'_{nk} must now be substituted in (3.23), and the integration for a_k carried out, which by (3.19), and on the assumption that $a_n = 1$, gives for the perturbed wave-function u

$$u = \psi_n e^{-\frac{i}{\hbar}W_n t} + \frac{e|\mathbf{E}_0|}{2m\omega}\sum_k{}' \left[\frac{B^0_{nk}e^{\frac{i}{\hbar}(\hbar\omega - W_n)t}}{\omega_{kn}+\omega} + \frac{B^{0*}_{kn}e^{-\frac{i}{\hbar}(\hbar\omega + W_n)t}}{\omega_{kn}-\omega}\right]\psi_k, \quad (3.29)$$

where the summation is to be taken over all values of k except n, and is to include integration over the continuum of positive energy states. The quantity ω_{kn} in (3.29) is defined by

$$\hbar\omega_{kn} = W_k - W_n,$$

so that it is the circular frequency corresponding to the transition of the atom from the state k to the state n. Owing to the action of the perturbing field, there is a certain probability, given by the square of the modulus of the expression in the square brackets in (3.29), that the atom, initially in the state n, will be found in another state k, and this probability can evidently be calculated if the wave-functions are known. For X-rays, for which ω is large, the probability that the atom is in a state other than n will be small, unless the frequency of the incident radiation is nearly equal to one of the possible transition frequencies of the atom, but in this case the corresponding coefficient may be considerable.

(d) The calculation of the current-density and of the corresponding scattered radiation : If we multiply equation (3.17) for u by u^*, and the conjugate equation for u^* by u, and subtract one from the other, we obtain

$$\frac{\hbar^2}{2m}(u\nabla^2 u^* - u^*\nabla^2 u) + \frac{i\hbar e}{mc}\mathbf{A}\cdot(u^*\,\mathrm{grad}\,u + u\,\mathrm{grad}\,u^*) = i\hbar\frac{\partial}{\partial t}(uu^*).$$

Since div $\mathbf{A} = 0$ for a plane electromagnetic wave, this may be written in the form

$$\mathrm{div}\left\{\frac{i\hbar e}{2m}(u\,\mathrm{grad}\,u^* - u^*\,\mathrm{grad}\,u) - \frac{e^2}{mc}\mathbf{A}\,uu^*\right\} + \frac{\partial}{\partial t}(euu^*) = 0.$$

If now we compare this equation with the equation of continuity, (*3*.11), remembering that euu^* is a charge-density, we see that $\mathbf{j}$, the Schrödinger current-density, is given by

$$\mathbf{j}=\frac{i\hbar e}{2m}(u \operatorname{grad} u^* - u^* \operatorname{grad} u) - \frac{e^2}{mc}\mathbf{A}\, uu^*. \tag{3.30}$$

The coherent radiation scattered by the atom, having the same wavelength as the incident radiation, is now to be calculated by treating this current-density as if it were an ordinary classical current-density obeying the laws of electrodynamics, and employing for u in (*3*.30) the perturbed wave-function (*3*.29).

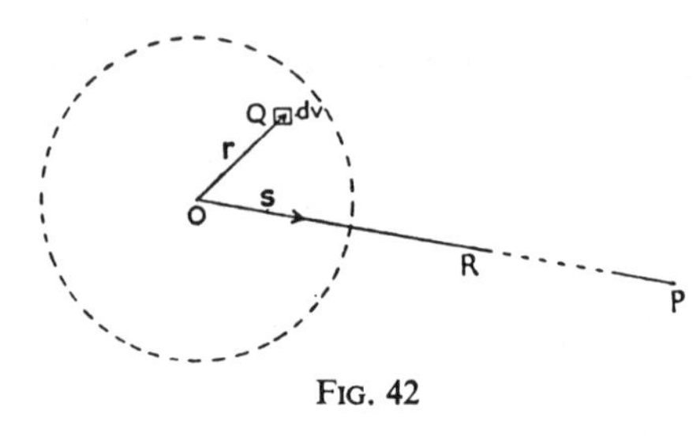

FIG. 42

Suppose now that the current-density at a point Q, fig. 42, whose position relative to the origin O is fixed by the vector $\mathbf{r}$, is $\mathbf{j}$, and let $\mathbf{j}$ oscillate with circular frequency ω. Let the field at a point P distant R from O be required, R being large in comparison with the dimensions of the region about O within which the current-density is appreciable, and let $\mathbf{s}$ be the unit vector in the direction of OP. Then the result, from classical electrodynamics, that we shall use is that the amplitude of the scattered radiation at P may be derived from an electric moment M at O, given by

$$\mathrm{M} = \pm\frac{1}{i\omega}\int j_p\, e^{\pm i\kappa \mathbf{s}\cdot\mathbf{r}}\, dv. \tag{3.31}$$

In this expression, j_p is the component of $\mathbf{j}$ perpendicular to the unit vector $\mathbf{s}$ in the direction of scattering, and is supposed to contain the time through a factor $e^{\pm i\omega t}$. The integration is to be taken over the whole of the volume of the current distribution. If $\mathbf{j}$, the current-density, is real, the real part of (*3*.31) is of course to be understood as the moment. The electric *field* at P is found in the usual way by multiplying the moment M by ω^2/c^2, and by the appropriate phase and amplitude factors corresponding to the distance of P from O. The exponential term in the integrand of (*3*.31) allows for the effect on the phase of the scattered wave of the varying distance of the element of volume dv from O.

(*e*) *The scattering of radiation of frequency large compared with the atomic absorption frequencies:* (i) Coherent scattering: If ω the frequency of the incident radiation is large in comparison with any of the atomic transition frequencies ω_{kn}, we can neglect the terms under the summation sign in (*3*.29), since they are then small and, moreover, only appear in the expression for the current-density in terms of order E_0^2 which is itself a small quantity, since the field is to be considered as a small perturbation. The first term in (*3*.30) does not then contain the time, and contributes nothing to the scattering, and we are left with the time-dependent current-density $-(e^2/mc)\mathbf{A}|u|^2$, or $-(e^2/mc)\mathbf{A}|\psi_n|^2$ if the atom is in the state n.

By (*3*.26), **A** is equal to the real part of

$$-\frac{c\mathbf{E}_0}{i\omega}e^{i(\omega t-\kappa \mathbf{s}_0\cdot\mathbf{r})},$$

so that the current-density vector is proportional to and in the direction of the incident electric field vector **E**. If we substitute this value of **A** in (*3*.31) we obtain for the scattering moment M the real part of

$$-\frac{e^2}{m\omega^2}E_{0p}e^{i\omega t}\int|\psi_n|^2e^{i\kappa(\mathbf{s}-\mathbf{s}_0\cdot\mathbf{r})}\,dv, \qquad (3.32)$$

and for the scattered coherent intensity at unit distance in the direction **s**

$$I_{\text{coh}}=\left(\frac{e^2}{mc^2}\right)^2|E_{0p}|^2\left|\int|\psi_n|^2e^{i\kappa\mathbf{S}\cdot\mathbf{r}}\,dv\right|^2. \qquad (3.33)$$

Here E_{0p} is the component of $\mathbf{E}_0$ perpendicular to **s**. If **E** is perpendicular to the plane containing both $\mathbf{s}_0$ and **s**, $E_{0p}=|\mathbf{E}_0|$; if it lies in this plane $E_{0p}=|\mathbf{E}_0|\cos 2\theta$, where 2θ is the angle between the directions of incidence and scattering. In (*3*.33) we have again written S for $\mathbf{s}-\mathbf{s}_0$.

The intensity of the coherent scattering of unpolarised radiation by an atom in the state n at unit distance is therefore

$$I_{n,n}=|\mathbf{E}_0|^2\left(\frac{e^2}{mc^2}\right)^2\frac{1+\cos^2 2\theta}{2}|f_{n,n}|^2, \qquad (3.34)$$

where

$$f_{n,n}=\int|\psi_n|^2e^{i\kappa\mathbf{S}\cdot\mathbf{r}}\,dv. \qquad (3.35)$$

Equation (*3*.34) shows that the intensity of coherent scattering by the atom is $|f_{n,n}|^2$ times the intensity scattered under the same conditions by a Thomson electron (cf. equation (*2*.11)): so that $f_{n,n}$ is the atomic scattering factor for coherent radiation when the atom is in the state n.

If we compare (*3*.35) with (*3*.1) of III, § 2(*b*), which gives the classical scattering factor for atoms arranged in a crystal lattice, we see that the two expressions become identical if we put $\rho(\mathbf{r})$ equal to $|\psi_n|^2$, that is to say, if we employ for the probability that an electron is in any given region of the atom the probability $|\psi_n|^2$ calculated according to the methods of wave-mechanics. In other words, to obtain the coherent scattering we may treat $e|\psi_n|^2$, the Schrödinger charge-density, as a classically scattering charge-density, ψ_n being so normalised that $\int|\psi_n|^2\,dv=1$, so that the probability of finding the electron somewhere is unity. The radial charge-density $U(r)$ of (*3*.4) is equal to $4\pi r^2|\psi_n|^2$.

There is now, of course, no difference between atoms in a crystal and in a gas. The difference between the two cases depended on ideas about electron configurations that are not valid in quantum mechanics. Formula (3.34) gives the *coherent* scattering in either case. This important result was obtained independently by Wentzel,[13] and by Waller [14] in 1927. It is to be remembered that it applies only when the incident frequency is considerably greater than the natural frequencies of the atom, and yet is not so great that relativity effects become important.

(ii) Incoherent scattering: The coherent scattering is associated with processes in which the electron remains in the same energy state. We can bring the incoherent scattering into the same scheme by assuming it to be associated with transitions from an initial state n to other final states m. The natural extension of the method that we have already used is to take the current-density associated with such a transition under the influence of the perturbing field to be

$$\mathbf{j}_{n,m}=\frac{i\hbar e}{2m}(u_n \operatorname{grad} u_m^* - u_m^* \operatorname{grad} u_n) - \frac{e^2}{mc}\mathbf{A}\, u_n u_m^*. \tag{3.36}$$

We assume the incident frequency to be subject to the same restrictions as before. The first term in (3.36) now depends on the time, but is associated with the spontaneous emission from the atom; the effect of the perturbation is contained in the second term. Using methods exactly analogous to those discussed above, we find for the intensity associated with the transition from n to m

$$\mathrm{I}_{n,m}=\left(\frac{e^2}{mc^2}\right)^2\left(\frac{\omega'}{\omega}\right)^2\left|\int \psi_n\psi_m^* e^{i\kappa\mathbf{S}\cdot\mathbf{r}}\,dv\right|^2|\mathrm{E}_{0p}|^2, \tag{3.37}$$

where $\omega'=\omega-\omega_{mn}$, and is the frequency of the scattered radiation, which is now no longer equal to that of the incident radiation. The total incoherent scattering is to be obtained by summing (3.37) over all values of m except n; and the total scattered intensity, including that of the coherent radiation, by extending the summation over all values of m, including n. Now we are assuming ω to be much greater than ω_{mn} for all possible values of m, and thus, to a close approximation, we can put $(\omega'/\omega)^2=1$. This gives for I, the total scattered intensity, expressed now in terms of $\mathrm{I_T}$, the scattering by a Thomson electron,

$$\mathrm{I}=\sum_m\left|\int \psi_m^*\psi_n e^{i\kappa\mathbf{S}\cdot\mathbf{r}}\,dv\right|^2\mathrm{I_T}. \tag{3.38}$$

We may now show that, with the assumptions as to frequency we have made, $\mathrm{I}=\mathrm{I_T}$ for an atom containing a single electron. The proof is as follows. Any function may be expanded as a linear combination of a set of characteristic functions ψ_k. Let us then put

$$\psi_n e^{i\kappa\mathbf{S}\cdot\mathbf{r}}=\sum_k \mathrm{C}_{nk}\psi_k. \tag{3.39}$$

Any coefficient such as C_{nm} is determined by multiplying each side of (3.39) by $\psi_m^* dv$, and integrating over all space. By the orthogonality and normalisation conditions obeyed by the wave-functions, $\int \psi_m^* \psi_k \, dv$ vanishes for $k \neq m$, and is unity for $k = m$, and so we get

$$\int \psi_m^* \psi_n e^{i\kappa \mathbf{S}\cdot\mathbf{r}} \, dv = C_{nm}.$$

By (3.38), therefore, the total scattering is $\sum_m |C_{nm}|^2 I_T$, and we have to show that $\sum_m |C_{nm}|^2$ is unity.

Now by (3.39)

$$|\psi_n|^2 = |\sum_m C_{nm} \psi_m|^2. \tag{3.40}$$

If we multiply each side of (3.40) by dv, and integrate over all space, the orthogonality conditions give

$$\int |\psi_n|^2 \, dv = \sum_m |C_{nm}|^2 \int |\psi_m|^2 \, dv,$$

which proves the proposition, since each integral is equal to unity by the normalisation conditions.

For the single electron, within the limits of frequency imposed, we have therefore

$$I_{coh} = |f|^2 I_T; \quad I_{inc} = (1 - |f|^2) I_T. \tag{3.41}$$

We have here written f for the scattering factor $f_{n,n}$ of equation (3.35), and shall in general do this when, as will usually be the case, it is assumed that the atom is in the ground state initially.

It is interesting to compare the results obtained in this section with those derived from the purely classical theory of scattering. According to the latter, an atom containing a single loosely bound electron should scatter as a single Thomson electron. Apart from a polarisation factor, the scattered intensity should be the same in all directions, and all the scattered radiation should have the wave-length of the incident radiation. According to the quantum theory, however, the intensity of the coherent radiation, having the same wave-length as the incident radiation, should have the classical value only for very small angles of scattering, and should fall off rapidly with increasing angle. This falling off is to be calculated by supposing the point charge to be replaced by a continuous charge-distribution which has everywhere a density proportional to $|\psi|^2$, and determining the scattering from this distribution by the ordinary methods of interference, making due allowance for the phase difference between the contributions from different parts of it. As the intensity of the coherent scattering falls off, that of the incoherent scattering, which has wave-lengths slightly greater than that of the incident radiation and extending over a small range, rises, but the sum of the coherent and incoherent intensities in any direction is equal to the classical value. This result is due to Wentzel.[13]

The whole of the discussion in this section refers to an atom containing a single electron. We are naturally more interested in atoms containing a number of electrons, and in a later section we shall discuss how the results that we have obtained can be extended to such cases. Before going on to this, however, we shall complete the discussion of the hydrogen-like atom by considering briefly the case in which the frequency of the incident radiation is nearly equal to one of the transition frequencies ω_{kn} of the atom.

(*f*) *The dispersion terms :* As ω approaches ω_{kn}, the corresponding term in the summation of (3.29) becomes large, and we can no longer neglect it in calculating the scattering. We shall expect the value of f for radiation of frequency not far from that of an atomic absorption edge to differ appreciably from the value calculated from (3.35). We should, of course, expect the same thing on purely classical grounds; for when the incident frequency approaches that of one of the free vibrations of the system we can no longer use the formula for scattering by a free electron, and resonance effects occur. The effect is closely related to anomalous dispersion, for we have seen that the scattering factor determines the refractive index, so that anomalous dispersion means anomalous scattering. We shall discuss this in detail in Chapter IV, but it will be convenient at this stage to calculate M, the scattering moment for coherent radiation, when the dispersion terms are taken into account.

For brevity, we shall denote the summation in (3.29) by Σ, and its conjugate by Σ^*. In calculating the current-density from (3.30) we may neglect those terms in $-(e^2/mc)\mathrm{A}|u|^2$ which contain Σ or Σ^*; for they will be at least of the second order in E_0, which is a small perturbing field. We have to consider the products of u_n and u_n^* with the gradients of Σ^* and Σ respectively; and of Σ and Σ^* with the gradients of u_n^* and u_n. This, after a certain amount of algebra, leads to the following expression for the scattering moment for coherent radiation for an atom containing a single electron in a state n, in the direction of the unit vector s

$$\mathrm{M} = -\frac{e^2}{2m\omega^2}\left[\mathrm{E}_{0p} f_{n,n} - \frac{2\hbar}{m}|\mathbf{E}_0| \sum_k{}' \frac{\omega_{kn}}{\omega_{kn}^2 - \omega^2} \mathrm{B}_{nk}^0 (\mathrm{B}_{nk}^*)_p\right]. \quad (3.42)$$

Here $f_{n,n}$ is the scattering factor as defined in (3.35), and E_{0p} is the component of the electric vector in the incident radiation perpendicular to s, the direction of scattering. The sum is to be taken over all values of k except n, and includes integration over the continuum of positive energy states.

The factor B_{nk}^* is given by

$$\mathrm{B}_{nk}^* = \int \psi_k e^{i\kappa \mathbf{s}\cdot\mathbf{r}} \operatorname{grad} \psi_n^* \, dv, \quad (3.43)$$

and the suffix p in equation (3.42) denotes that it is only the component of the gradient in a direction perpendicular to $\mathbf{s}$, the direction of scattering, that is effective. This fact has been used in reducing (3.43) to the form given above; and it has also been assumed that the wave-functions ψ_n and ψ_k are real. This will be the case in such numerical applications as we shall make of the formula, and it has the effect of materially simplifying the expression for M, the moment.

(*g*) *The general form of the scattering factor for the many-electron atom:* We shall now discuss briefly how the ideas developed in the earlier part of this section may be extended to the case of an atom containing a number of electrons. The appropriate form of the wave-equation is now

$$\sum_k \left\{ \left(-\frac{\hbar^2}{2m} \nabla_k{}^2 + e\phi_k \right) u + \frac{i\hbar e}{mc} (\mathbf{A}_k \cdot \text{grad}_k\, u) \right\} = i\hbar \frac{\partial u}{\partial t}. \qquad (3.44)$$

The suffix k refers to the co-ordinates of the electron k, and the sum is to be taken over all the electrons in the atom. The wave-function u is now a function of the time, and of all the electron co-ordinates, 3Z in number if the atom contains Z electrons.

The wave-functions corresponding to the stationary states of the unperturbed atom are of the type

$$u_n = \Psi_n(\mathbf{r}_1, \mathbf{r}_2, \ldots \mathbf{r}_z) e^{-\frac{i}{\hbar} W_n t}, \qquad (3.45)$$

where Ψ_n is a function of the co-ordinates of all the electrons, which are supposed to be given by the vectors $\mathbf{r}_1, \ldots \mathbf{r}_k, \ldots \mathbf{r}_z$, and W_n is the energy of the state n. These wave-functions are solutions of (3.44) when the perturbing field is zero.

We have first to find a quantity to represent the average charge-density for the different electrons. The quantity $|u|^2 dv$, where $dv = dv_1\, dv_2 \ldots dv_z$, gives the probability that electron (1) lies within the element of volume dv_1, at the distance $\mathbf{r}_1$, electron (2) within dv_2 at $\mathbf{r}_2$, and so on. Suppose $|u|^2 dv$ to be integrated over all the electron co-ordinates *except* those of electron (k). This integral gives the probability that the electron (k) lies within the element of volume dv_k at $\mathbf{r}_k$ while the remaining electrons are anywhere at all, and so is in a sense a measure of the average charge-density at $\mathbf{r}_k$ due to the electron (k). We can thus write for ρ_k, the charge-density associated with the electron (k),

$$\rho_k = e \int |u|^2 dv'_k, \qquad (3.46)$$

where dv'_k denotes that the integration is over all electron co-ordinates *except* those of (k). The charge-density ρ_k, integrated over the co-ordinates of electron (k), is, as it should be, equal to the electronic charge, since $\int |u|^2 dv = 1$, the wave-function u being normalised to unity.

By a method analogous to that given in § 3(*d*) of this chapter, it can be shown that ρ_k, as defined in (*3*.46), obeys the equation

$$\operatorname{div} \mathbf{j}_k + \frac{\partial \rho_k}{\partial t} = 0,$$

where $\mathbf{j}_k$, the current-density associated with the electron (k), is the expression (*3*.30) integrated over all the electron co-ordinates, except those of (k); and this confirms the identification of ρ_k with the charge-density associated with electron (k). Accepting this identification, we may infer from § 3(*e*) of this chapter that the appropriate scattering factor for coherent radiation from an atom in the nth state is, by analogy with (*3*.35),

$$f_{n,n} = \sum_k \int \left\{ \int |u_n|^2 \, dv'_k \right\} e^{i\kappa \mathbf{S}\cdot\mathbf{r}_k} dv_k = \int |\Psi_n|^2 \sum_k e^{i\kappa \mathbf{S}\cdot\mathbf{r}_k} dv, \qquad (3.47)$$

where $\mathbf{r}_k$ is the vector to the element of volume dv_k, and the integration is over all the 3Z electron co-ordinates.

The corresponding expression for the total scattered intensity, coherent and incoherent, when the frequency of the radiation is much greater than ω_{mn} is by analogy with (*3*.38)

$$\mathrm{I} = \sum_m \left| \int \Psi_m^* \Psi_n \sum_k e^{i\kappa \mathbf{S}\cdot\mathbf{r}_k} dv \right|^2 \mathrm{I_T}. \qquad (3.48)$$

Here again, the atom is supposed to be initially in the state n, and the sum is to be taken over all the possible wave-functions Ψ_m of the atom. The term $m = n$ in the sum gives, of course, the coherent scattering.

By a process very similar to that given in § 3(*e*) ii, using the orthogonality and normalisation conditions of Ψ_n and Ψ_m, we can transform (*3*.48) into the simpler expression

$$\mathrm{I} = \int |\Psi_n|^2 \left| \sum_k e^{i\kappa \mathbf{S}\cdot\mathbf{r}_k} \right|^2 dv \; \mathrm{I_T}, \qquad (3.49)$$

which reduces again to $\mathrm{I_T}$ if there is only one electron.

(*h*) *Approximate forms of the expressions for coherent and total scattering for a many-electron atom. The exclusion principle:* The exact solution of the wave-equation for an atom containing a number of electrons is in general impossible, and we have to be content with approximations of varying degrees of closeness. We shall consider the form taken by the expressions for the scattering for some of the more important of these.

For certain approximations to the total wave-function, Ψ_n may be expressed as a product of Z individual electronic wave-functions. This is, for example, permissible if we can neglect the interaction energy of the electrons, or if the interaction is taken into account by certain approximate methods to be considered below. We can then write

$$\Psi_n = \psi_1(\mathbf{r}_1)\, \psi_2(\mathbf{r}_2) \ldots \psi_k(\mathbf{r}_k) \ldots \psi_z(\mathbf{r}_z). \qquad (3.50)$$

With this approximation, (3.47) reduces at once to

$$f_{n,n}=\sum_k\int|\psi_k|^2e^{i\kappa\mathbf{S}\cdot\mathbf{r}_k}dv_k=\sum f_k; \tag{3.51}$$

or the f-factor for the whole atom is the sum of the f-factors for the individual electronic wave-functions, each calculated according to equation (3.35), a result that is very important in the numerical calculation of f-factors.

With the same approximation, equation (3.49) for the total intensity expressed in terms of the Thomson scattering I_T as unity reduces at once to

$$I=Z+\sum_j{}'\sum_k{}' f_j f_k^*, \tag{3.52}$$

where f_j and f_k are the f-factors for the electronic wave-functions ψ_j and ψ_k, and each sum is taken over all the values of j and k except those for which $j=k$. It is assumed that the individual electronic wave-functions are normalised. This expression, it will be seen, is formally identical with (3.8) of III, § 2(*c*), which we obtained by a purely classical argument. We have thus the interesting result that the total scattering from an atom, calculated classically, is formally identical with that calculated according to wave-mechanics if we may express the total wave-function of the atom as the product of the wave-functions of its individual electrons. The true wave-function can never be written in this simple form, and we shall discuss later what sort of approximation is made by doing so, but we must first consider another point.

Even if we retain the products of the individual electronic wave-functions as a basis for the total wave-function, the form (3.50) for Ψ_n is too simple, for it associates a specified electron with each wave-function. For example, electron (k) is supposed to occupy the wave-function ψ_k, and so on. Now this goes beyond any knowledge that we can possibly obtain about the electrons. We have no means of identifying particular electrons, and every possible permutation of the electrons amongst the Z wave-functions of (3.50) will correspond to the same energy. The correct wave-function to use, if we are to keep to the product form, is a linear function of Z! products such as (3.50), representing all the possible permutations of the Z electrons amongst the Z wave-functions. Such a total wave-function does not assume any particular distribution of the individual electrons, but assigns an equal probability to all such distributions.

The total wave-function must, however, satisfy a further condition. By Pauli's exclusion principle, no two electrons can occupy identical wave-functions, and this also must be shown by the form of the total wave-function. This is attained by writing Ψ_n as the determinant of Z rows and Z columns whose successive rows are

$$\psi_1(1),\ \psi_1(2),\ \ldots\ \psi_1(z);\quad \psi_2(1),\ \psi_2(2),\ \ldots\ \psi_2(z);$$

and so on, the suffixes denoting the different electronic wave-functions, and the numbers in the brackets the electrons. This determinant is a linear function of products of Z wave-functions, involving all possible permutations of the electrons amongst the wave-functions. It also vanishes if any two of the wave-functions become identical; for then two rows of the determinant will be identical. The vanishing of the total wave-function symbolises the fact that the probability of two electrons occupying identical wave-functions is zero. The total wave-function in this form thus expresses (*a*) that we cannot identify particular electrons, and (*b*) that no two electrons can occupy identical wave-functions.

If we use a wave-function of this type in (*3*.47) and (*3*.49), we find that the scattering factor for coherent radiation $f_{n,n}$ takes the same form as for the simple expression (*3*.50). It is still the sum of the scattering factors corresponding to the individual wave-functions which the electrons can occupy. The expression for the total scattering I is, however, slightly modified. We find

$$\mathrm{I}=\mathrm{Z}+\sum_{j}\sum_{k}^{j\neq k} f_j f_k^* - \sum_{j}\sum_{k}^{j\neq k} |f_{jk}|^2. \qquad (3.53)$$

The new quantity f_{jk} that appears in the expression is defined by

$$f_{jk}=\int \psi_j^*(p)\,\psi_k(p)\,e^{i\kappa \mathbf{S}\cdot\mathbf{r}_p}\,dv_p, \qquad (3.54)$$

the integration being over all the co-ordinates of any one electron (p). The quantity f_{jk} will usually be small, since ψ_j is in general small in those regions where ψ_k is large, and *vice versa*, but the correction may be appreciable in some cases. Equation (*3*.53) for the total scattering was first given by Waller and Hartree.[17]

Since $|\sum_k f_k|^2=\sum_k |f_k|^2+\sum_j\sum_k^{j\neq k} f_j f_k^*$, equation (*3*.53) gives for the coherent and incoherent parts of the scattered intensity

$$\mathrm{I}_{\mathrm{coh}}=|\sum_k f_k|^2, \qquad (3.55)$$

$$\mathrm{I}_{\mathrm{inc}}=\sum_k (1-|f_k|^2)-\sum_j\sum_k^{j\neq k}|f_{jk}|^2. \qquad (3.56)$$

Physically, the last term owes its existence to the fact that, in the many-electron atom, a number of electron levels are occupied. Transitions which involve the passage of electrons to such occupied levels are forbidden, and this is taken into account by the particular form of wave-function used.

In using (*3*.56) or (*3*.53), it is important to notice that the extra term $-\sum_{jk}|f_{jk}|^2$, due to exchange, appears only when the two electrons associated with the wave-functions have the same spin. The integrations considered above refer only to the space wave-functions. If the spins of the two electrons are the same, they cannot occupy the same space wave-function, but if their spins are opposite, there is nothing to prevent their doing so, since Pauli's principle does not apply to the space wave-functions considered alone, but to the whole wave-function, including the spin. For electrons with the same spin, the space wave-function must be written in the determinant, or *antisymmetric*, form considered above, which leads to the extra term in the scattering formula; but if the electrons have opposite spins, the wave-function is symmetrical in the electron co-ordinates, and the extra term does not then appear. Thus, the summation in the last term in (*3*.53) or (*3*.56) is to be taken over all pairs of different electrons with the *same* spin. Equations (*3*.55) and (*3*.56) are the best approximations to the scattered radiation to be obtained within the limits of frequency that have been assumed, using a non-relativistic wave-equation, and a total wave-function based on the products of the individual electronic wave-functions.

4. Numerical Methods

(*a*) *The method of the self-consistent field:* The wave-function Ψ for the whole atom could be expressed as a linear function of the products of the individual electronic wave-functions if there were no interaction between the electrons. The problem of an atom with Z electrons then becomes virtually Z independent three-dimensional problems, instead of a single problem in 3Z-dimensional space, as it is if interaction is taken into account. In practice we cannot, of course, neglect the interaction of the electrons. Nevertheless, it is possible to treat the problem as a set of Z 3-dimensional problems, and at the same time to allow for the interaction of the electrons to a fair degree of approximation. The method that we shall describe was introduced by Hartree,[18] and it has the very great merit that it leads to numerical estimates of the atomic wave-functions.

An electron in an atom is acted on by a central field due to the nucleus, together with the field due to all the other electrons, which will not, in general, be a central field. Let us suppose, however, that we may represent its average effect as a central field. To calculate the wave-function of this electron we have then the problem of a single electron in a central field, a 3-dimensional problem. If this process is repeated for each electron, we have finally Z wave-functions, all of which may be treated as independent, since they already contain an allowance for the interaction of the electrons. Ψ, the total wave-function for the atom, can then be built up from their product, and the scattering from the atom can be calculated from them using the equations developed in the last section. Moreover, it can be shown that the approximation so obtained is the best possible if Ψ is to keep the simple product form.

Before discussing the method in greater detail, it will be necessary to quote a few results from the wave-mechanical theory of an electron in a central field. Each negative energy state of such an electron is characterised by four quantum numbers, n, l, m, and s. Of these, n is the total quantum number, which mainly determines the energy of the state; l is a measure of the total angular momentum of the electron in the state concerned, and m of the possible components of that angular momentum parallel to a specified direction. For any value of l there are $2l+1$ possible values of m, ranging from $-l$ to $+l$. The fourth quantum number, s, represents the electron spin, and can have two opposite values. The group of wave-functions with given values of n and l, which we shall in future call an (n, l) group, thus consists in all of $2(2l+1)$ different states. States for which $l=0, 1, 2, 3, \ldots$ are known respectively as $s, p, d, f, \ldots$ states. A complete s-group contains two states, a p-group 6, a d-group 10, and so on. A complete group is denoted by putting the symbol for the type of electron in brackets, and writing as an index the number of electrons in the group. For example, $(2s)^2$, $(2p)^6$, $(3d)^{10}$. Now $l \lesseqgtr n-1$, so that the possible groups are:— $1s$, containing in all two states and constituting the K shell when occupied by electrons in an atom; $2s$ and $2p$, containing in all 8 states, constituting the L shell; $3s$, $3p$, and $3d$, containing 18 states, constituting the M shell; and so on.

Each wave-function of an (n, l) group has $n-l-1$ radial nodes, which occur at the same radii for all the functions of the group, but, except for the s-states, the individual wave-functions have not spherical symmetry. The total probability function, $\Sigma|\psi|^2$ for all the wave-functions of a completed (n, l) group has, however, spherical symmetry. All the atomic electrons, with the exception of the outermost ones, will usually be members of such closed groups. If the outer electrons are in s-states, the total charge distribution of the atom will have spherical symmetry, and it will be necessary to consider only the radial variation of the wave-functions. We shall assume this to be so in what follows, although the outermost electrons will often be in states that are not spherically symmetrical. For these we must be content with a radial average.

For such a spherically symmetrical atom the radial probability function, $U(r)$, which occurs in equation (3.4) of III, § 2(b), for f, has according to the wave-mechanical theory of scattering to be replaced by $4\pi r^2\Sigma|\psi|^2$. We have therefore to calculate $r\psi$ which, following Hartree, we shall denote by P. In terms of P the radial wave-equation for an electron in a central field becomes

$$\frac{d^2\mathrm{P}}{dr^2}+\left[\frac{2m}{\hbar^2}\{\mathrm{E}-\mathrm{V}(r)\}-\frac{l(l+1)}{r^2}\right]\mathrm{P}=0. \tag{3.57}$$

$V(r)$ is here the potential of the central field at a point distant r from the centre of the atom, and E is the total energy of the electron.

If the electron is a member of a closed (n, l) group, $V(r)$ will be made up of the potential due to the nuclear charge, together with that due to the electrons of the other closed groups and to the other electrons of its own group. Let us assume that we know the charge distribution in the whole atom. We can then calculate the field at any distance r due to it. This, however, will not be the field acting on the electron; for it includes the field due to the density distribution of that electron itself, which must therefore be subtracted from the total field to give the field acting on the electron. The density distribution due to the group (n, l) will no longer be spherically symmetrical when the contribution to it from a single electron has been subtracted. Hartree, however, as an approximation, takes this residual field to be spherically symmetrical, and obtains it simply by subtracting from the total field due to the group $1/2(2l+1)$ of its value. In this way, starting with an assumed charge distribution in the atom based on our general knowledge of atomic structure, we make as good an estimate as possible of the field $V(r)$ acting on the electron. We shall call this the *initial field.* It will not, of course, be known as an analytical function, but as a series of numerical values, tabulated for different values of r.

With this assumed initial field, equation (*3*.57) must be integrated numerically, subject to the necessary conditions that P must be everywhere finite, and must be zero for $r=0$ and $r=\infty$. Solutions satisfying these conditions can be found only for certain discrete values of the energy parameter E. The energy of the state for which we are carrying out the integration will be known approximately from the X-ray or optical term values, and this approximate value of E is used for the first integration. For the actual method of carrying out the numerical integration the reader must consult the papers of Hartree. The integration is carried out with an assumed value of the energy parameter E. If the right value of E has been chosen, the solution dies away exponentially to zero for large r, and the value of E must be adjusted by trial until this condition is satisfied. The integration carried out with this correct value of E gives the radial wave-function P corresponding to the initial $V(r)$. Since (*3*.57) is a homogeneous equation, the absolute value of P is not determined, and this must be fixed for the problem under consideration by appropriate normalisation. P is to represent the radial charge-density for the closed (n, l) group, and we must therefore have

$$\int_0^\infty P^2\,dr = 2(2l+1), \qquad (3.58)$$

which gives the normalisation factor.

From the radial charge-distribution so determined we can calculate the contribution of the (n, l) group concerned to the total potential

V(r). If we have started initially with the correct atomic field, this calculated contribution must be equal to the assumed contribution of the group to the intial field. If this is so, then the field is said to be *self-consistent*, and we may assume that the correct wave-function for the group (n, l) has been found. If the initial and final fields differ, as they usually will, the initial field must be modified, and the calculation must be repeated until self-consistency is attained. Unless the error is large, it is not necessary to rework the whole integration, for appropriate variation methods may be used. This process must be carried out for

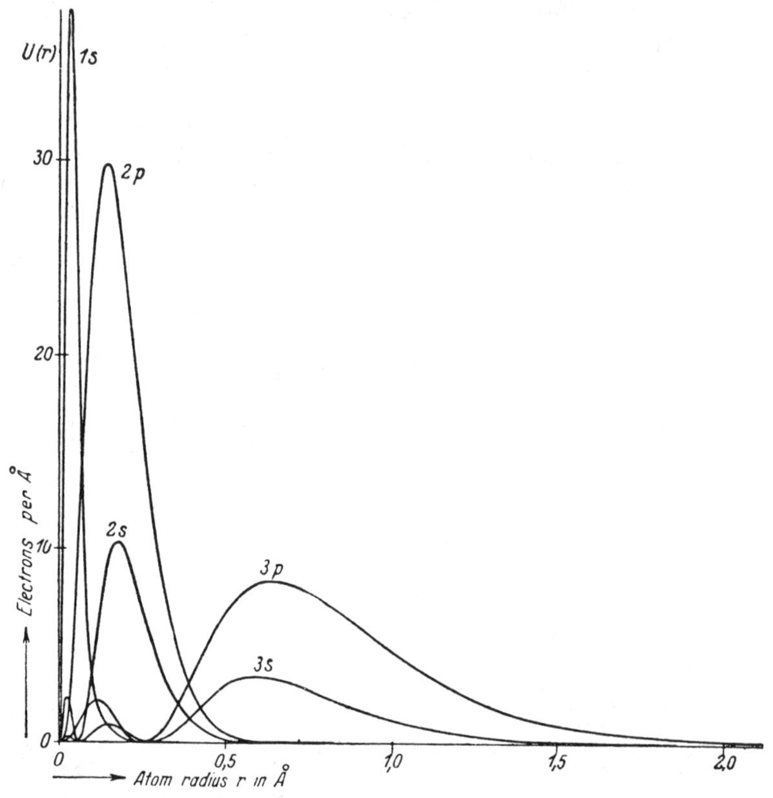

FIG. 43. Radial charge distribution for the different electron groups of K^+
(James, *Ergebnisse der technischen Röntgenkunde*, vol. III, 1933)

each electron group, until complete self-consistency for the whole atomic field is reached.

In applying the method, one is greatly helped by the fact that there is not much overlapping of the density distributions of the different electron shells. This will be apparent from fig. 43 in which the normalised values of $|P|^2$ for the different groups of electrons in the potassium ion, K^+, are shown. Because most of the charge due to the L and M shells lies outside the region of the K shell, it gives only a small contribution to the field acting on the K electrons. The field of the latter can therefore be adjusted to self-consistency first of all, and will need only slight changes, even if the subsequent work shows the initial fields assumed for the L and M electrons to need considerable modification. After the K electrons, the L are dealt with, and then the M, and so on. The method involves a considerable amount of labour, although it is not, in practice, so great as might be supposed, and self-consistent fields have been calculated for a number of the lighter atoms, and for a few of the heavier ones. The labour is much greater for the heavier atoms, owing to the greater number of electron groups involved.

(*b*) *The self-consistent field including exchange—Fock's equation:* It can be shown that so long as the total atomic wave-function is expressed, as in (*3*.50), as a product of a set of wave-functions for single electrons in central fields, the method of the self-consistent field, as outlined in III, § 4(*a*), leads to the best solution. A rather closer approximation to the wave-function takes into account the fact that individual electrons cannot be identified, and, as we saw in §3(*h*), this is allowed for by writing the wave-function in the determinant form there discussed. Slater [19] and Fock [20] independently suggested the modification of the method of the self-consistent field to include this exchange principle, and Fock and Petrashen [21] have given a numerical solution for the case of sodium. D. R. Hartree and W. Hartree [22,39,42] have obtained solutions for beryllium and for certain other atoms.

It is not possible to give details of the solutions here. The method makes use of the variation principle. It can be shown that if the wave-equation is written in the form

$$\mathrm{H}\Psi = \mathrm{W}\Psi \tag{3.59}$$

for a state of energy W, then

$$\mathrm{W} = \int \Psi^* \mathrm{H} \Psi \, dv \Big/ \int \Psi \Psi^* \, dv, \tag{3.60}$$

and that if Ψ is not exactly, but approximately, a solution of (*3*.59), then that approximation is the closest that makes W, as defined by (*3*.60), a minimum. If Ψ is written as a simple product of wave-functions, application of this principle leads to the equation of the self-consistent field, in which the field acting on any one electron is that due to nucleus and to the statistical charge-density of the remaining electrons. The method of Slater and Fock consists in putting Ψ in the determinant form that takes account of exchange, and determining the best value by the variation method. This leads to a set of simultaneous differential equations for the electronic wave-functions, whose numerical solution by a method analogous to that of the self-consistent field is a matter of considerable complexity, for the details of which we must refer those interested to the original papers.

Fig. 44, plotted from data taken from the paper of Hartree and Hartree, shows the radial electron-density for the $3p$ wave-function of Cl^-, as calculated by the method of the self-consistent field, and from

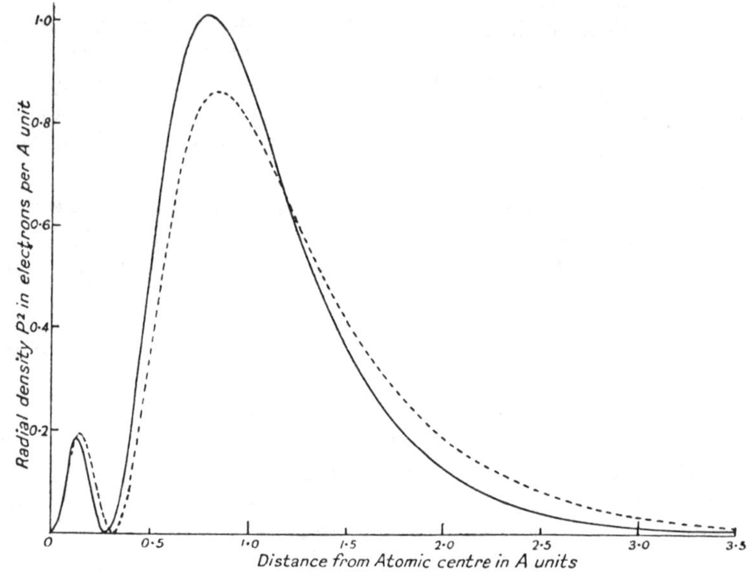

FIG. 44. Curves showing P^2 for the $3p$ electron group of Cl^-, calculated by the method of the self-consistent field. The full line is calculated with allowance for exchange, the dotted line without (Hartree & Hartree)

Fock's equation. It will be seen that for this electron group the difference is considerable, and that there is a tendency for the Fock charge-distribution to be drawn in towards the centre of the atom, as compared with that calculated without exchange. The effect of exchange is much smaller for the inner electrons, and the change in f is therefore not very large, although appreciable. In fig. 45, the full and dotted lines show respectively the f-curves for Cu^+ calculated with and without allowance for exchange. For certain other atomic properties, however, such, for example, as the diamagnetic susceptibility, which depends on $\overline{r^2}$ and so is sensitive to the radial charge-distribution, the effect is greater, and agreement between theory and experiment is materially improved by taking exchange into account.

Both methods that have been discussed neglect spin and relativity effects, which could probably be taken into account with a considerable amount of labour. A more important approximation is that inherent in the use of central-field wave-functions—the neglect of the interaction between individual pairs of electrons, and its replacement by an averaged

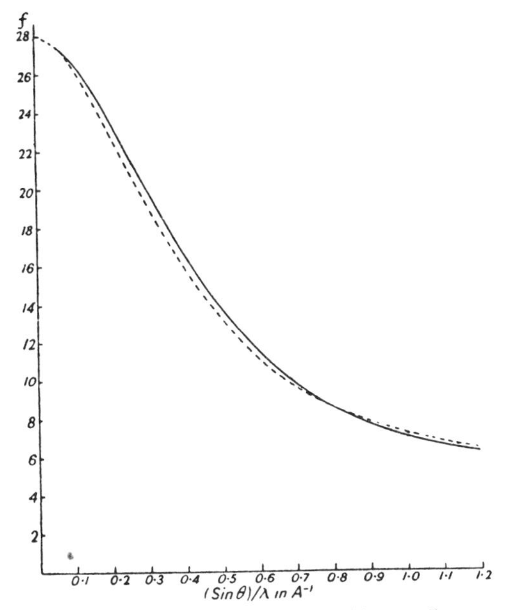

FIG. 45. f-curves for Cu^+, with and without allowance for exchange. Full line with exchange, dotted line without.

effect. It is not easy to see how this modifies the wave-functions, and little quantitative work has yet been done on these lines. For the purpose of calculating f-curves for use in the analysis of crystals the approximations so far made are, however, probably in nearly every case quite good enough.

(*c*) *The Pauling-Sherman method:* Pauling [23] has calculated approximate atomic wave-functions by assuming each electron of the atom to move in a hydrogen-like field, that of the nucleus reduced by an appropriate screening constant, different for each electron group. Extending this work, Pauling and Sherman [24] have determined a complete set of f-curves for all atomic numbers. Pauling's method is simpler than that of the self-consistent field, but its results are almost certainly less accurate, since the electronic fields must depart rather widely from the simple Coulomb type. The method tends to accentuate the maxima and minima in the radial wave-function. We shall discuss the effect of this on the scattering factor in § 4(*e*) of this chapter.

(*d*) *The Thomas-Fermi method:* An approximate method of calculating atomic fields and charge distributions that does not depend on the direct application of wave-mechanics was developed independently by Thomas [25] and Fermi. [26] The atomic electrons are treated as a degenerate gas obeying Fermi-Dirac statistics and the Pauli principle, the energy of the atom in its ground state being in fact simply the zero-point energy of this gas. Electrons are supposed to be distributed in the 6-dimensional phase-space corresponding to the 3 position co-ordinates x, y, z, and the 3 momentum co-ordinates, p_x, p_y, p_z, of an electron, at the rate of two in each h^3 of volume. Thus, if there are N electrons in all, we may write

$$\mathrm{N} = \int \frac{2}{h^3} dx\, dy\, dz\, dp_x\, dp_y\, dp_z, \tag{3.61}$$

where the integration is over the whole of the phase-space occupied by electrons. Now the range of the momentum co-ordinates is limited by the assumption that the electron cannot escape from the atom, so that if V is the potential at any point, we must always have $|p| < \sqrt{2me\mathrm{V}}$. The integration over the momentum co-ordinates can therefore be replaced by the volume of a sphere having a radius $\sqrt{2me\mathrm{V}}$. Equation (3.61) then becomes

$$\mathrm{N} = \int \frac{8\pi}{3h^3} (2me\mathrm{V})^{\frac{3}{2}}\, dx\, dy\, dz, \tag{3.62}$$

and, if ρ is the average charge-density at the point x, y, z,

$$\rho = -\frac{8\pi e}{3h^3} (2me\mathrm{V})^{\frac{3}{2}}. \tag{3.63}$$

Now, by Poisson's equation, $\nabla^2 \mathrm{V} = -4\pi\rho$, which, since V is a function of r only, can, by (3.63), be written in the form

$$\frac{1}{r^2}\frac{d}{dr}\left(r^2 \frac{d\mathrm{V}}{dr}\right) = \frac{32\pi^2 e}{3h^3} (2me\mathrm{V})^{\frac{3}{2}}. \tag{3.64}$$

This equation has now to be solved with the conditions that $\mathrm{V} \to 0$ as $r \to \infty$, and $\mathrm{V}r \to \mathrm{N}e$ as $r \to 0$, N being the total number of electrons in the atom. Thomas solved this equation numerically for the case of caesium,* $\mathrm{N} = 55$, obtaining V, and hence, by (3.63), the charge-density ρ, as a function of the radius r; and once the charge-density is known the f-curve can be calculated, since $\mathrm{U}(r)$ of (3.5) is equal to $4\pi r^2 \rho$.

The Thomas-Fermi charge distributions for different atoms bear the following relation to one another. If a total charge Z_0, including the nuclear charge, lies within a radius r_0 for an atom of atomic number N_0, then a charge $\mathrm{Z} = \mathrm{Z}_0(\mathrm{N}/\mathrm{N}_0)$ lies within a radius $r = r_0(\mathrm{N}_0/\mathrm{N})^{\frac{1}{3}}$ for an atom of atomic number N. If we notice that the radial charge-density $\mathrm{U}(r)$ is equal to $-d\mathrm{Z}/dr$, we may easily deduce the following rule for determining the f-curve for an atom of atomic number N, given that for one of atomic number N_0. To obtain the value of f for an atom of atomic number N at $(\sin\theta)/\lambda = x$, take the value of f from the curve

* An extended numerical solution of the Thomas-Fermi equation for caesium has been given by Bush and Caldwell, who used the Bush differential analyser. *Phys. Rev.*, **38**, 1898 (1931).

for the atom of atomic number N_0 for $(\sin\theta)/\lambda = x(N_0/N)^{\frac{1}{3}}$, and multiply it by N/N_0. In this way, given the f-curve for any standard atom, it is easy to calculate the f-curves for all other atoms.

Heisenberg [27] has shown how to calculate the incoherent scattering from an atom, using the Thomas field, and Bewilogua [28] has deduced the necessary numerical tables for applying Heisenberg's method to the rapid calculation of the incoherent scattering from any atom. (Cf. Chapter IX, § 1(*b*).)

(*e*) *The calculation of f-curves from the charge distributions:* We shall consider in this section how to calculate the scattering factor f for coherent radiation when the radial charge-density distribution $U(r)$ is known, and also how the character of this distribution determines the way in which f depends on the angle of scattering, and on the wave-length of the scattered radiation. The determination of f involves the evaluation, graphical or numerical, of the integral

$$f = \int_0^\infty U(r) \frac{\sin \mu r}{\mu r} dr, \tag{3.65}$$

where $\mu = 4\pi(\sin\theta)/\lambda$. The wave-length λ and the angle θ affect f only in the combination $(\sin\theta)/\lambda$, so that f has the same value for a given order of spectrum from a given crystal whatever the wave-length of the radiation, provided of course that this lies within the range over which the theory so far developed is valid.

In discussing the evaluation of (*3*.65), it is most convenient to take a definite example, and we shall consider the case of the potassium ion, K^+, for which $U(r)$ has been determined by Hartree [29] by the method of the self-consistent field. This method allows us to determine f separately for the different electron groups of the atom, and a knowledge of these separate f-curves is often useful. They are required, for example, in the calculation of the incoherent scattering, according to equations (*3*.9) or (*3*.56), and it is important, too, from many points of view, to know which parts of the atom contribute most strongly to f.

The normalised values of P^2, which is the same as $U(r)$, are shown in fig. 43. Curve A of fig. 46 shows again the value of P^2 for the $3p$ group of ionised potassium, plotted as a function of r, and the other curves in the same figure show $(\sin \mu r)/\mu r$, also plotted as a function of r, for a series of values of $(\sin\theta)/\lambda$. To find the value of f for the $3p$ group for any particular value of $(\sin\theta)/\lambda$, we draw another curve, whose ordinates are the products of corresponding ordinates of the P^2 curve, A, and of the $(\sin \mu r)/\mu r$ curve for the value of $(\sin\theta)/\lambda$ concerned. The area of this curve is equal to the integral (*3*.65), and so to f*. A series of such curves is shown in fig. 47, and it is instructive to study

* The integration can, of course, be carried out numerically instead of graphically, and this is often more convenient in practice.

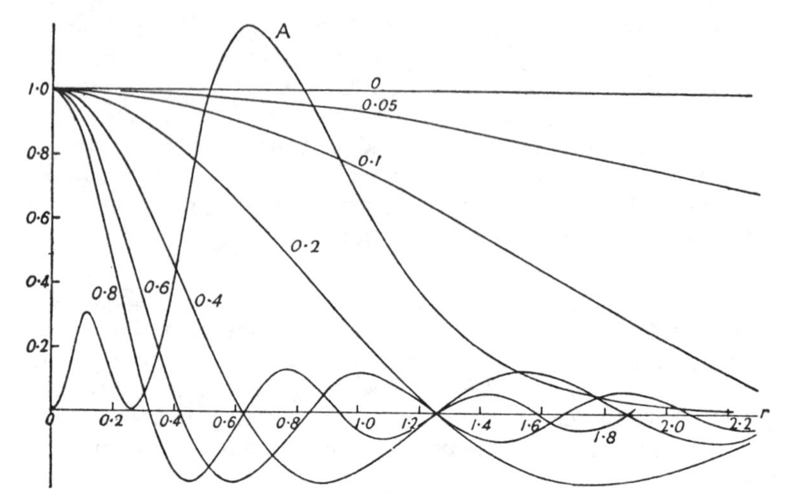

FIG. 46. The radial charge-density P^2 for the $3p$ group of K^+, curve A, and curves of $(\sin \mu r)/\mu r$ for different values of $(\sin \theta)/\lambda$, plotted as functions of r

the way in which they vary with increasing $(\sin \theta)/\lambda$. The curve for $(\sin \mu r)/\mu r$ has a main maximum, of value unity, for $\mu r = 0$, and falls to zero at $r = \pi/\mu$, or $\lambda/4 \sin \theta$. For larger values of r, the curve is alternately positive and negative, but the subsidiary maxima and minima are all considerably smaller than the main maximum, and fall off in amplitude rather slowly. If $(\sin \theta)/\lambda$ is small, the whole effective part

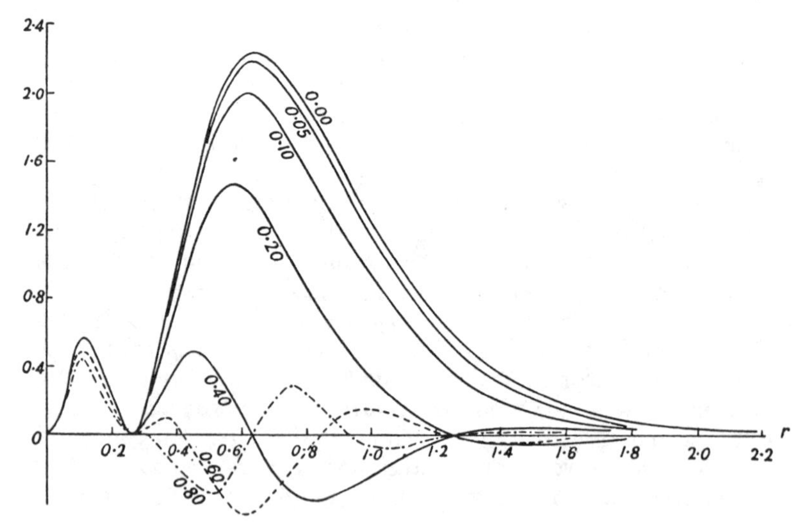

FIG. 47. Products of P^2 for $3p$ group of K^+ and $(\sin \mu r)/\mu r$ for different values of $(\sin \theta)/\lambda$ plotted as functions of r

of the curve for P^2 lies within the main maximum of the curve $(\sin \mu r)/\mu r$; the curve of products then differs only slightly from the P^2 curve itself, and f is nearly equal to the total number of electrons in the group. As $(\sin \theta)/\lambda$ increases, the main maximum of $(\sin \mu r)/\mu r$ becomes narrower, and the area beneath the curve of products becomes smaller. The curves in fig. 47 show clearly how the outer parts of the charge-density rapidly become ineffective in producing scattering as $(\sin \theta)/\lambda$ increases. For $(\sin \theta)/\lambda = 0{\cdot}4$, for example, there is a considerable negative area beneath the curve of products, due to the first minimum of $(\sin \mu r)/\mu r$, which now lies in regions of r occupied by the main bulk of the charge distribution P^2. This negative area nearly neutralises the positive area due to the main maximum, and the total area is very small. At a rather larger value of $(\sin \theta)/\lambda$ the total area, and so f, actually becomes negative, although the negative value never becomes greater than about 0·03 per electron. At a still larger value of $(\sin \theta)/\lambda$, f becomes positive once more, but never exceeds about 0·03 per electron. and it thereafter dies away slowly without again becoming negative, The f-curve for the $3p$ group is shown in fig. 48, together with the curves for the other groups of electrons in the potassium ion.

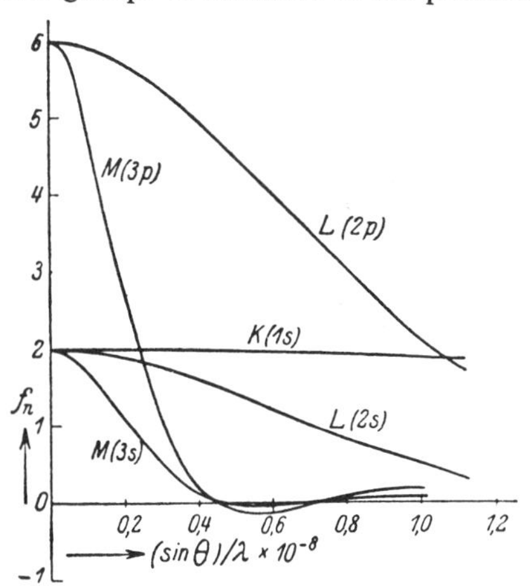

FIG. 48. f-curves for the individual electron groups of K^+
(James, *Ergebnisse der technischen Röntgenkunde*, vol. **III**, 1933)

The behaviour of f that has been described is typical of a charge distribution of this kind. Once f has fallen to zero it never again reaches an appreciable value, either positive or negative, and this is primarily due to the broad main maxima of the charge distribution P^2. When $(\sin \theta)/\lambda$ has become so large that the first zero value of $(\sin \mu r)/\mu r$ lies at a smaller radius than the main bulk of the charge distribution, the breadth of the latter includes not one but several of the alternately positive and negative subsidiary maxima and minima of $(\sin \mu r)/\mu r$. The curve of products thus oscillates, and its area is always small.

It is evident that the narrower the main maxima of P^2, the more pronounced will be the oscillatory character of f after it has once fallen to zero. An infinitely narrow maximum, that is to say, a spherical shell, would give for f simply the curve $(\sin \mu r_0)/\mu r_0$, where r_0 is the radius of the shell. The definite humps on the f-curves obtained from the orbit model of the atom had their origin in the circular orbits, which, on the average, are equivalent to such spherical shells of charge. With the much more diffuse distributions characteristic of the wave-mechanical treatment of the problem, these humps are smoothed out, and this, as we shall see, is entirely in accordance with experiment.

It will be clear from fig. 46 that the smaller the radius at which the main charge distribution of the group lies, the more slowly will f for that group decrease as $(\sin \theta)/\lambda$ increases. The contribution to f from the K electrons, for example, falls off very slowly for all but the lightest atoms. There will in fact be little falling off in the f-curve with increasing angle of scattering if the main bulk of the charge distribution lies within a sphere whose radius is small compared with the wave-length of the radiation. It is interesting to compare the f-curves shown in fig. 48 for the $1s$, $2s$, and $3s$ groups in potassium, each of which contains the same number of electrons, and to correlate them with the corresponding charge distributions, which are shown in fig. 43.

In carrying out the calculation of f by the method described above, the necessity for plotting a series of $(\sin \mu r)/\mu r$ curves, one for each value of μ, may be avoided by the use of a device due to Mr. A. Baxter. Let the curve of $(\sin r)/r$ be plotted against $\log r$ as abscissae. Then the curve of $(\sin \mu r)/\mu r$ may be derived from this simply by moving the origin of $\log r$ through a distance $\log \mu$. The curve of $U(r)$ is plotted also as a function of $\log r$, on a piece of tracing paper, and by superimposing this curve on that for $(\sin r)/r$ the relative positions of the two curves can be obtained for any value of μ by a simple displacement parallel to the axis of $\log r$. To obtain the product of $U(r)(\sin \mu r)/\mu r$ for different values of r for any given value of μ, we make the abscissae of $\log r$ for the $U(r)$ curve coincide with those of $\log r + \log \mu$ for the $(\sin r)/r$ curve. The products required are then those of corresponding ordinates of the two curves.

In fig. 49 the $(\sin r)/r$ curve and the $U(r)$ curve for the $3p$ group of Cl^- are shown plotted as functions of $\log r$. The relative positions of the curves correspond to $(\sin \theta)/\lambda = 0{\cdot}6$. The dotted lines show the positions of the main maximum of the $U(r)$ curve for a series of values of $(\sin \theta)/\lambda$. This method of plotting shows very clearly how rapidly the value of f falls away with increasing $(\sin \theta)/\lambda$ for this electron group, for its value is small unless a considerable part of the main maximum of $U(r)$ lies within the main positive region of $(\sin r)/r$.

In general, the uncertainties in the calculation of the wave-functions by the method of the self-consistent field are greatest for the outer electrons of the atoms. On the other hand, as we have just seen, these

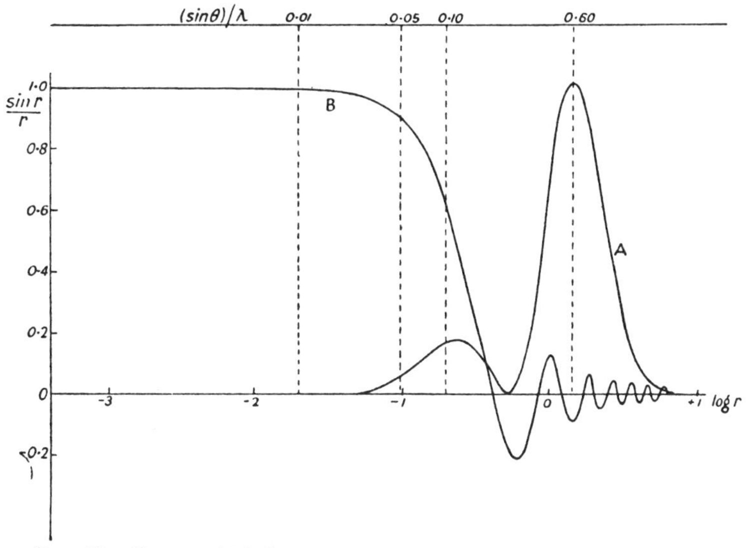

FIG. 49. Curves of P^2 for $3p$ group of Cl^-, and of $(\sin r)/r$, plotted as functions of $\log r$, for $(\sin \theta)/\lambda = 0{\cdot}6$

electrons make a comparatively small contribution to f, except for quite small values of $(\sin \theta)/\lambda$. The Hartree method will thus give f-curves which are likely to be fairly accurate for large values of $(\sin \theta)/\lambda$, but less trustworthy for the smaller values, a limitation, however, that it shares with all other methods yet devised.

(f) Comparison of f-curves based on different atomic models. A discussion of the available tables of f-values : The methods available for the calculation of f may be summarised as follows: (1) The method of the self-consistent field, with or without exchange. (2) Approximate methods, such as those of Pauling and Sherman, in which the atomic wave-functions are treated as hydrogen-like, with appropriate effective nuclear charges. (3) The method of Thomas and Fermi, in which the atomic electron-cloud is treated as a degenerate electron-gas.

Of these, the Hartree method (1) is almost certainly the best, especially if the effects of exchange are included. At present, however, Fock's equation has been solved numerically only for a few atoms, and even the simpler calculation of the self-consistent field has been made for relatively few, and these mostly among the lighter ones. The labour involved in the calculation, especially for the heavier atoms, is such that it will probably be some time before anything like a complete table of f-values calculated by this method will be available. Moreover, for the heavier atoms, the dispersion and relativity effects are in any case considerable, and the f-values calculated from them not very reliable. It is possible to do something towards filling up the gaps by interpolation. James and Brindley have given tables of f calculated by a method of interpolation based on the self-consistent fields of ten different atoms, and these should be fairly accurate for atoms with atomic numbers less than about 25. The values so calculated are given in Table XIV, Appendix IV, Vol. **1**, p. 328. At the time the tables were constructed data were lacking on which to base reliable interpolations for the heavier atoms. The interpolations were based, not on the values of f for the atoms as a whole, but on those for the individual electron groups, and James and Brindley [30] give in their paper tables from which the f-curves of such groups can readily be calculated. Self-consistent fields have now been calculated for a number of atoms not included in the tables, notably for caesium, tungsten, gold, and mercury, among the heavier atoms.

Pauling and Sherman,[24] using method (2), have calculated complete tables of f-curves for all atomic numbers. While the curves they obtain are in fair general agreement with those obtained by the Hartree method, there are certain systematic deviations. The Pauling method tends to accentuate the maxima and minima of the radial charge distributions. Now, as we have seen on p. 128, a narrowing of a maximum in the radial distribution corresponding to any group causes a more marked oscillation of the f-curve for that group with varying $(\sin \theta)/\lambda$, and this oscillation will leave its impress on the f-curve for the atom as a whole. The f-curves calculated from the old orbit model of the atom showed marked humps, corresponding to the sharp maxima in the radial charge-distribution due to the circular orbits. The smearing out of the charge-distribution which is characteristic of the wave-mechanical models of the atom tends to remove these humps from the curves. Vestiges of them remain, however, both on the Hartree and on the Pauling-Sherman curves, but they are much more marked on the latter.

As we have seen in III, § 4(d), the Thomas-Fermi method (3) allows us to calculate very easily the f-curves for all atoms if the curve for any atom is known, simply by an appropriate change of scale of ordinates and abscissae. Curves obtained in this way were first given by W. L. Bragg [31] in his report to the Solvay Conference in 1927, and in a paper with J. West [32] he soon afterwards gave a short table of Thomas f-values. More complete tables, taken from a paper by James and Brindley,[33] will be found in Appendix IV, Vol. **1**, p. 332, Table XV; and also in Tables I & II, § XI(b), p. 571 of the International Crystal Tables,* from calculations by Pauling.

The Thomas field, apart from a change of scale, has the same general form for all atoms, and so it cannot reproduce any of the individual peculiarities of the fields of particular atoms. It is essentially an average field, and the maxima and minima in the charge-density due to

* *Internationale Tabellen zur Bestimmung von Kristallstrukturen*, Borntraeger, Berlin (1935).

the different electron groups are entirely smoothed out. The f-curves calculated from the Thomas distribution show no trace of the slight oscillations that appear on the Pauling-Sherman curves, and to a lesser extent on the Hartree curves. We should, on the whole, expect the Thomas curves to be a better approximation for the heavier atoms than for the lighter ones, but not to be so good an approximation as the Hartree curves. In figs. 50 and 51 the Hartree, Pauling-Sherman, and Thomas curves are plotted for a number of atoms for comparison. It will be seen that although all three methods agree as to the general magnitude of f at all angles, yet there are very considerable differences in detail, amounting for some of the heavier atoms to several units. Moreover, it does not seem possible to give any simple general rules for expressing the nature of the differences between the curves of various types. For the heavier atoms it would apppear to be best to use the

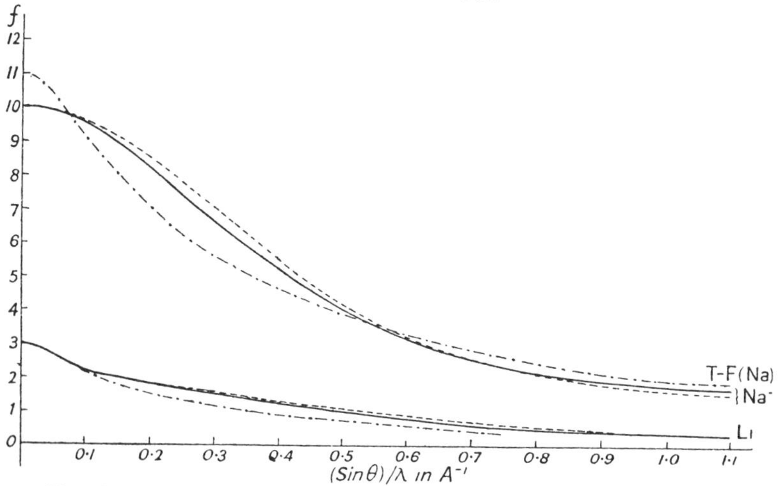

FIG. 50. Comparison of f-curves for Na, Na^+ and Li calculated from Hartree, Pauling-Sherman and Thomas-Fermi charge distributions.
——— Hartree, – – – – Pauling-Sherman, –·–·–· Thomas-Fermi

Hartree curves if they are available, and, in their absence, the Thomas curves. It is possible that a more general survey of the effects of exchange on the Hartree curves for the heavier atoms might invalidate this conclusion; but it is not very likely since the effect of exchange on the inner electron groups of the heavier atoms, which give the main contribution to f, is not large.

For neutral atoms, the Thomas-Fermi charge-distribution does not fall away exponentially with increasing radius, as a wave-mechanical distribution would, but more slowly; for a positive ion, all its charge lies within a finite radius, and for a negative ion, no distribution can be obtained. Lenz [34] and Jensen [35] have sought to avoid these disadvantages, while retaining the advantage possessed by the Thomas-Fermi

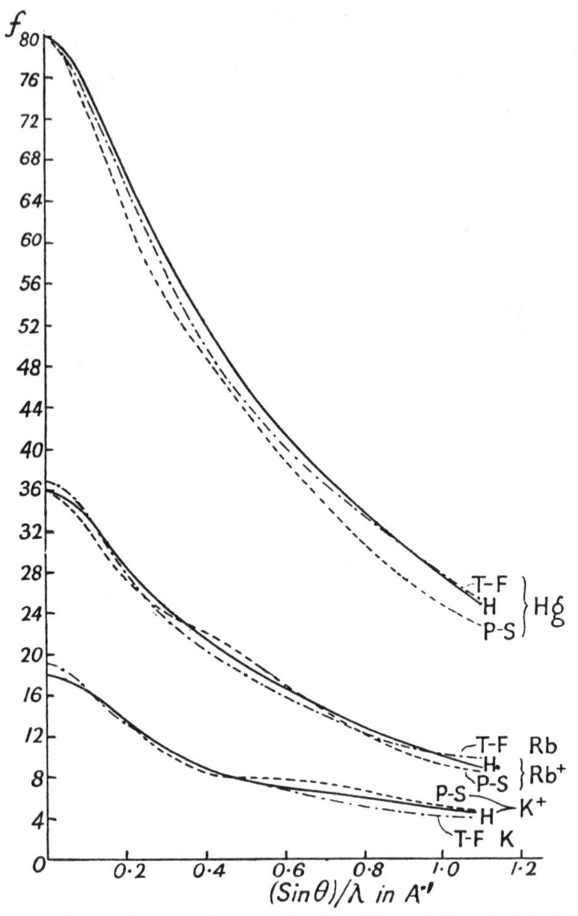

FIG. 51. Comparison of f-curves for Hg, Rb and K calculated from Hartree ———, Pauling-Sherman - - - - and Thomas-Fermi -·-·-· charge distributions.

field that it can be applied to any atom, and have applied a wave-mechanical variation principle, starting with the Thomas-Fermi equation. Tables of f-values based on the Lenz-Jensen model have been published by v. Sz. Nagy,[36] which do in fact agree rather more closely with the Hartree curves, in those examples where comparisons can be made, than do the Thomas curves. The method does not, however, appear to be very sound in principle, although it may nevertheless be a fairly good empirical approximation.

(g) *Limitations of applicability of the tabulated f-curves:* In this chapter, we have dealt with the theoretical calculation of the scattering factors of atoms for X-rays on the basis of the quantum theory. We shall see later that the agreement between the values so calculated and the observed values is fairly satisfactory, but this comparison must be left until certain other matters have been discussed. It is, however, advisable at this stage to emphasize once more the limiting assumptions that have been made throughout the discussion in this chapter. First of all, it has been assumed that the incident radiation has a frequency considerably higher than any absorption frequency of the atom. It will be evident that for the X-rays generally employed in work on crystal structures this condition does not hold for the heavier atoms, although it is fairly well satisfied for the lighter ones. The f-values for the heavier atoms given in the tables that we have discussed must therefore be regarded as limiting values for high frequencies, and care must be exercised in using them. We shall discuss in the next chapter the effect of using radiation of frequency comparable with the atomic absorption frequencies, and shall see there that very considerable corrections have to be made to the tabulated f-curves on this account.

Secondly, the non-relativistic form of the wave-equation has been used throughout. We must not therefore expect the results that we have obtained to apply when the frequency of the incident radiation is too large; nor can we expect relativity effects to be negligible for the inner electrons of the heavier atoms. At present no numerical data are available to make an estimate of the magnitude of this effect at all easy, but it is probably by no means negligible and, because it affects the inner electrons, it will influence the whole range of the f-curves. We shall deal briefly with the scattering of high-frequency radiation at a later stage, in considering the scattering of X-rays by gases, but it has little application to the problems of crystal analysis.

References

1. W. H. Bragg, *Phil. Trans. Roy. Soc.*, **A**, **215**, 153 (1915).
2. P. Debye, *Ann. der Physik*, **46**, 809 (1915).
3. A. H. Compton, *Phys. Rev.*, **9**, 29 (1917).
4. P. Debye and P. Scherrer, *Physikal Zeit.*, **19**, 474 (1918).
5. A. H. Compton, *Phys. Rev.*, **35**, 925 (1930).
6. C. V. Raman, *Indian Journal of Physics*, **3**, 357 (1928).
7. Y. H. Woo, *Phys. Rev.*, **41**, 21 (1932).
8. R. W. James, *Ergebnisse d. tech. Röntgenkunde*, **3**, 32 (1933).
9. D. R. Hartree, *Phil. Mag.*, **50**, 289 (1925).
10. W. L. Bragg, R. W. James and C. H. Bosanquet, *Phil. Mag.*, **44**, 433 (1922); *Zeit. f. Physik*, **8**, 77 (1921).
11. E. J. Williams, *Phil. Mag.*, **2**, 657 (1926).
12. G. E. M. Jauncey, *Phys. Rev.*, **29**, 605 (1927).
13. G. Wentzel, *Zeit. f. Physik*, **43**, 1 (1927).

14. I. Waller, *Phil. Mag.*, **4**, 1228 (1927); *Nature*, **120**, 155 (1927); *Nature*, **15**, 969 (1927); *Zeit. f. Physik*, **51**, 213 (1928).
15. H. Kramers and W. Heisenberg, *Zeit. f. Physik*, **31**, 681 (1925).
16. E. Schrödinger, *Ann. d. Physik*, **81**, 109 (1926).
17. I. Waller and D. R. Hartree, *Proc. Roy. Soc.*, **A**, **124**, 119 (1929).
18. D. R. Hartree, *Proc. Camb. Phil. Soc.*, **24**, 89 and 111 (1928).
19. J. C. Slater, *Phys. Rev.*, **35**, 210 (1929); **34**, 1293 (1929).
20. V. Fock, *Zeit. f. Physik*, **61**, 126 (1930).
21. V. Fock and M. Petrashen, *Physikal. Zeit. Sowjet.*, **6**, 368 (1934).
22. D. R. Hartree and W. Hartree, Be, *Proc. Roy. Soc.*, **A**, **150**, 9 (1935); Cl^-, **A**, **156**, 45 (1936); Cu^+, **A**, **157**, 490 (1936)
23. L. Pauling, *Proc. Roy. Soc.*, **A**, **114**, 181 (1927).
24. L. Pauling and J. Sherman, *Zeit. f. Krist.*, **81**, 1 (1932).
25. L. H. Thomas, *Proc. Camb. Phil. Soc.*, **23**, 542 (1926).
26. E. Fermi, *Zeit. f. Physik*, **48**, 73 (1928).
27. W. Heisenberg, *Physikal. Zeit.*, **32**, 737 (1932).
28. L. Bewilogua, *Physikal. Zeit.*, **32**, 740 (1932).
29. D. R. Hartree, *Proc. Roy. Soc.*, **A**, **143**, 506 (1934).
30. R. W. James and G. W. Brindley, *Phil. Mag.*, **12**, 81 (1931).
31. W. L. Bragg, 'L'intensité de reflexion des rayons-X'. *Solvay Conference Reports*, 1927.
32. W. L. Bragg and J. West, *Zeit. f. Krist.*, **69**, 118 (1928).
33. R. W. James and G. W. Brindley, *Zeit. f. Krist.*, **78**, 470 (1931).
34. W. Lenz, *Zeit. f. Physik*, **77**, 713 (1932).
35. H. Jensen, *Zeit. f. Physik*, **77**, 722 (1932).
36. B. v. Sz. Nagy, *Zeit. f. Physik*, **91**, 105 (1934).

The following papers give numerical values of the self-consistent fields of atoms in addition to those already quoted:

37. O, O^+, O^{++}, O^{+++}, D. R. Hartree and M. M. Black, *Proc. Roy. Soc.*, **A**, **139**, 311 (1933).
38. Cl^- and Cu^+; K^+ and Cs^+; F^-, Al^{+3}, Rb^+, D. R. Hartree, *Proc. Roy. Soc.*, **A**, **141**, 282 (1933); **A**, **143**, 506 (1934); **A**, **151**, 96 (1935).
39. Be, Be^+; Ca, Ca^{++}; Hg, Hg^{++}, D. R. Hartree and W. Hartree, *Proc. Roy. Soc.*, **A**, **149**, 210 (1935).
40. Ag^+, M. M. Black, *Memoirs Manchester Lit. and Phil. Soc.*, **79**, 29 (1935).
41. W. M. G. Manning and J. Millman, *Phys. Rev.*, **49**, 848 (1936).
42. D. R. Hartree and W. Hartree, Ca^{++}, Ca^+, Ca, *Proc. Roy. Soc.*, **A**, **164**, 167 (1938); K^+ and Ar, *Proc. Roy. Soc.*, **A**, **166**, 450 (1938); Na^- and K^-, *Proc. Camb. Phil. Soc.*, **34**, 550 (1938).

CHAPTER IV

THE ANOMALOUS SCATTERING AND DISPERSION OF X-RAYS

1. Theoretical

(*a*) *Introductory:* In calculating the atomic scattering factor, we have hitherto assumed the frequency ω of the incident radiation to be large in comparison with any natural absorption frequency ω_{kn} of the scattering atom. This assumption has enabled us to neglect the terms involving the frequency in formula (*3*.42), III, § 3(*f*), for the scattering moment, and to use only the first term in calculating the *f*-curves for the atoms. Such procedure gives a good approximation to the scattering of radiations such as those usually employed in crystal analysis by the lighter atoms; but it is nevertheless unjustifiable in many cases of practical importance. To take only one example: in the investigation of the structure of alloys containing metals such as copper, zinc, iron, or manganese, Cu Kα radiation is often employed, the frequency of which is of the same order of magnitude as the K absorption frequencies of these elements, being higher for some and lower for others. The assumption that $\omega \gg \omega_{kn}$ cannot therefore be justified in such cases, and we must make some estimate of the importance of those terms in the formula for the scattering moment that have so far been neglected. Reference to equation (*3*.42) will show that the additional terms are not very important unless ω is nearly equal to ω_{kn}. Their values may then, however, be considerable; so that for radiation of frequencies differing only slightly from that corresponding to the absorption edge of the scattering atom *f* may depend markedly on the frequency. There will in fact be an anomalous scattering which is very closely connected with the anomalous dispersion of the radiation in the material of the scatterer. Even for frequencies differing considerably from that of an absorption edge, the effect of dispersion is not entirely negligible.

The most direct method of dealing with the problem, and formally the most correct one, is to discuss the coefficients of the second term in formula (*3*.42), which involve the atomic wave-functions. This has been done by Hönl, but we shall postpone the consideration of his work. We prefer here to begin by making use of the close formal similarity that is found to exist between the classical and quantum-mechanical formulae for scattering, and to approach the subject from the classical point of view, showing afterwards the correspondence between the formulae so obtained and those of quantum mechanics.

(*b*) *Scattering by a classical dipole-oscillator:* The classical theory of dispersion supposes the atoms to scatter as if they contained electric dipole-oscillators having certain definite natural frequencies, which are identified with the absorption frequencies of the atoms. Such oscillators may be considered as produced by the simple harmonic vibration of charges $\pm e$ with a relative displacement x at time t. For definiteness, suppose the oscillator to consist of an electron of mass m vibrating about, a massive positive charge, which we may consider to be at rest. On it falls an electromagnetic wave of circular frequency ω, whose electric vector at the position of the dipole is given at time t by $\mathbf{E} = \mathbf{E}_0 e^{i\omega}$. The equation of motion of the electron under the action of the wave is

$$\ddot{\mathbf{x}} + k\dot{\mathbf{x}} + \omega_s^2\mathbf{x} = \frac{e\mathbf{E}_0}{m} e^{i\omega t}. \qquad (4.1)$$

In this equation, a damping factor k, proportional to the velocity of the displaced charge, has been introduced, and ω_s is the natural circular frequency of the dipole if the charge is displaced, and allowed to oscillate without either applied field or damping.* The action of the magnetic field of the incident wave on the motion of the dipole, which is small, has been neglected.

When a steady state has been reached, the dipole executes forced oscillations of frequency ω under the action of the incident wave, its moment at time t being given by

$$\mathbf{M} = e\mathbf{x} = \frac{e^2}{m} \frac{\mathbf{E}_0 e^{i\omega t}}{\omega_s^2 - \omega^2 + ik\omega}, \qquad (4.2)$$

a result easily verified by direct substitution in (4.1). The axis of the resultant dipole oscillator is parallel to the direction of the incident electric vector $\mathbf{E}$.

The oscillating dipole is the source of an electromagnetic wave of the same frequency. At distances great in comparison with the amplitude of oscillation this wave is spherical, and its electric vector at a distance $\mathbf{r}$ at a time t at a point in the equatorial plane of the dipole is $\omega^2/c^2|\mathbf{r}|$ times the moment of the dipole at the time $t - |\mathbf{r}|/c$. The amplitude A of the scattered wave at unit distance in the equatorial plane is therefore

$$\mathrm{A} = \frac{e^2}{mc^2} \frac{\omega^2 \mathrm{E}_0}{\omega_s^2 - \omega^2 + ik\omega}. \qquad (4.3)$$

Now f, the scattering factor of the dipole, is defined as the ratio of the amplitude scattered by the oscillator to that scattered by a free classical electron under the same conditions. The scattering by a free electron is obtained by putting $\omega_s = 0$ and $k = 0$ in (4.3); for in the limit, when the restoring force on the electron is zero, the natural frequency is also zero. For the free electron, the scattered amplitude at unit distance in the equatorial plane is therefore $\mathrm{A}' = -(e^2/mc^2)\mathrm{E}_0$.

* If the damping is due purely to the classical radiation damping, $k = 2e^2\omega^2/3mc^3$, where ω is the frequency of the oscillation. See, for example, Lorentz, *Theory of Electrons*, p. 48; Compton and Allison, *X-Rays in Theory and Experiment*, p. 266.

The negative sign means that the wave scattered by the free electron in the forward direction has a phase opposite to that of the primary wave. Using this result, we get for f, the scattering factor of a dipole oscillator of natural frequency ω_s, for waves of frequency ω,

$$f = \frac{\omega^2}{\omega^2 - \omega_s{}^2 - ik\omega}. \tag{4.4}$$

It is to be emphasised that, with the definition of f used above, the scattered wave is *opposed* in phase to the primary wave when the scattering factor f is *positive.*

If $\omega \gg \omega_s$, f for the oscillator is unity. If $\omega \ll \omega_s$, f is negative, and the dipole then scatters a wave in phase with the primary wave, the amplitude of which is proportional to the square of the incident frequency. The difference between the values of f for very high and very low frequencies is thus unity. In the two cases we have discussed, the term $ik\omega$ in the denominator of (4.4) could be neglected. This is no longer so when ω is nearly equal to ω_s, and the scattering factor then becomes complex, which means that both the amplitude and the phase of the scattered radiation depend on the incident frequency.

(*c*) *The relation between the scattering and the absorption and refractive index in a medium containing dipoles:* Suppose now that a plane incident wave passes through a medium containing a large number N of similar dipoles per unit volume, each of which scatters a wave that we shall assume to be in phase with the primary wave in the direction of incidence. That is to say, we shall assume f for the dipoles to be negative. Then, by the method of Fresnel zones, the waves scattered by the dipoles lying in any thin sheet of the medium parallel to the primary wave-front may be shown to combine to form a resultant wave whose phase lags $\pi/2$ behind that of the primary wave. The successive additions of these small components $\pi/2$ out of phase with the primary wave cause the resultant transmitted wave to travel through the medium with a phase-velocity which is less than c, the velocity of light in free space. In other words, the medium has a refractive index for the waves that differs from unity. We have seen in II, § 3(*b*), p. 54, that for X-rays the refractive index n is given in terms of the scattering factor f by

$$n = 1 - \frac{\mathrm{N}\lambda^2}{2\pi}\frac{e^2}{mc^2}f = 1 - \frac{2\pi \mathrm{N}e^2}{m\omega^2}f. \tag{4.5}$$

In the case we have considered, in which f is negative, corresponding to $\omega \ll \omega_s$, the refractive index of the medium will be greater than unity, and the phase-velocity of the wave less than c. If, however, $\omega \gg \omega_s$, so that f is positive, the refractive index is less than unity, and the phase velocity of the transmitted wave is greater than c. In neither case is there any appreciable absorption of the radiation by the medium; the effect of the scattering is merely to change the phase-velocity of the transmitted wave. If, however, ω is comparable with ω_s, f will be complex. By (4.5), the medium will then also have a complex refractive index, and this, as we saw in II, § 3(*h*), p. 62, means that it absorbs the radiation.

Let us put

$$f = f' + if'' \tag{4.6}$$

$$n = 1 - \alpha - i\beta, \tag{4.7}$$

so that $$\alpha = (2\pi Ne^2/m\omega^2)f', \quad \beta = (2\pi Ne^2/m\omega^2)f''. \tag{4.8}$$

Then by (2.88), p. 63, the medium will have a linear absorption coefficient μ_0 for waves of frequency ω, given by

$$\mu_0 = 2\omega\beta/c = (4\pi Ne^2/m\omega c)f'', \tag{4.9}$$

where f'' is the imaginary part of the expression (4.4) for f.

From (4.4), we find for f' and f''

$$f' = \frac{\omega^2(\omega^2 - \omega_s^2)}{(\omega^2 - \omega_s^2)^2 + k^2\omega^2}, \quad f'' = \frac{k\omega^3}{(\omega^2 - \omega_s^2)^2 + k^2\omega^2}, \tag{4.10}$$

giving for $\mu_a(\omega)$, the absorption per dipole of the medium for a wave whose frequency is exactly ω,

$$\mu_a(\omega) = \frac{4\pi e^2 k}{mc} \frac{\omega^2}{(\omega^2 - \omega_s^2)^2 + k^2\omega^2}. \tag{4.11}$$

It is instructive to consider the question of absorption in a slightly different way. By (4.10), f'' is always positive. The imaginary part of f, if'', thus denotes a component of the scattering by the dipole the phase of which lags $\pi/2$ *behind* that of the primary wave. We have seen that the dipoles lying in any thin sheet parallel to the primary wave-front produce a resultant wave whose phase is retarded $\pi/2$ behind that of the waves scattered by the individual dipoles in the sheet. If this phase already lags $\pi/2$ behind that of the primary wave, the retardation of the resultant scattered wave will be π, so that it opposes the primary wave. The effect of the imaginary part of f will be to add to the resultant scattered wave a component that at each point of the medium is out of phase with the primary wave. There is thus a progressive diminution in the amplitude of the transmitted wave, or, in other words, an absorption.

(*d*) *The breadth of the absorption lines:* The expression (4.11) for the absorption per dipole has a maximum value for a frequency very nearly equal to ω_s, and this maximum is sharper the smaller the damping coefficient k. If a train of waves consisting of a continuous range of frequencies, which includes ω_s, passes through the medium, only those frequencies very near to ω_s will be appreciably absorbed. The medium thus exhibits line absorption, and the breadth of the absorption lines is measured by k, which is easily shown to be the breadth at half-value of the maximum in $\mu_a(\omega)$ expressed in terms of ω.

Let $E(\omega)d\omega$ be the energy passing per sq. cm. per second in the incident wave-train associated with frequencies between ω and $\omega+d\omega$. Each dipole in the medium absorbs a fraction of this energy equal to $\mu_a(\omega)$, and if the train of waves contains frequencies extending over a continuous range from ω_1 to ω_2, which includes ω_s, the total energy absorbed per dipole per second from the wave-train will be

$$E(\text{abs}) = \int_{\omega_1}^{\omega_2} \mu_a(\omega)\, E(\omega)\, d\omega. \tag{4.12}$$

If k is small, the breadth of the maximum in $\mu_a(\omega)$ is also small. For the purpose of the integration, therefore, $E(\omega)$ can be taken as constant, and equal to $E(\omega_s)$, its value in the neighbourhood of ω_s. For the same reason, the range of integration is not important so long as it includes ω_s, and may therefore be taken from $\omega=0$ to $\omega=\infty$. Using (*4*.11), we may therefore write (*4*.12) in the form

$$E(\text{abs}) = E(\omega_s)\frac{4\pi e^2 k}{mc}\int_0^{\infty}\frac{\omega^2 d\omega}{(\omega^2-\omega_s{}^2)^2+k^2\omega^2}. \tag{4.13}$$

Because of the small breadth in ω of the integrand, we can replace ω^2 by $\omega_s{}^2$ in the small term $k^2\omega^2$, and can also put $\omega^2=\omega\omega_s$ in the numerator of the integrand. This gives

$$E(\text{abs}) = E(\omega_s)\frac{2\pi e^2 k\omega_s}{mc}\int_0^{\infty}\frac{2\omega\, d\omega}{(\omega^2-\omega_s{}^2)^2+k^2\omega_s{}^2}. \tag{4.14}$$

The integral is now a standard form, and its value is $\pi/k\omega_s$, so that

$$E(\text{abs}) = \frac{2\pi^2 e^2}{mc}E(\omega_s), \tag{4.15}$$

that is to say, each dipole absorbs from the continuous radiation a fraction $2\pi^2e^2/mc$ of the energy density at a frequency ω_s equal to the natural frequency of the dipole. If instead of the circular frequency ω we use the true frequency $\nu(=\omega/2\pi)$, we have

$$E(\text{abs}) = \frac{\pi e^2}{mc}E(\nu_s), \tag{4.16}$$

where $E(\nu_s)$ is now the intensity per unit frequency range in the incident radiation in the neighbourhood of $\nu=\nu_s$. This formula was first obtained by Ladenburg.[1]

Equation (*4*.15), giving the energy absorbed by the oscillator, does not contain the damping coefficient k. Integral (*4*.14), from which (*4*.15) is obtained, assumes k to be small, and the final result of the integration does not then depend on its actual value. Equation (*4*.15) gives in fact the absorption by an oscillator with negligibly small damping. Houston [2] has shown that the same expression may be obtained by considering the work done by the incident electric field on the oscillating charge. In the limit, when the damping becomes negligibly small, and the frequency of the incident radiation coincides with the natural frequency of the oscillator, the amplitude of the oscillations increases indefinitely with the time. This must be considered as the classical analogue of the photoelectric effect, and (4.15) may be used, as we shall see below, to calculate the photoelectric absorption coefficient. The more exact expression (4.13), which approximates to (4.14) only in the limiting case of very small damping, includes loss of energy by scattering as well. This may be seen as follows. Let us suppose that ω differs greatly from ω_s. Then, by (4.11) and (4.12), $\delta\mathrm{E}$, the energy absorbed from the radiation between frequencies ω and $\omega+d\omega$, is

$$\delta\mathrm{E}=\frac{4\pi e^2 k}{mc}\,\mathrm{E}(\omega)\,\frac{d\omega}{\omega^2}\,.$$

Now if k is due only to the electromagnetic damping of the oscillations its value (see footnote, p. 136) is $2e^2\omega^2/3mc^3$, which gives

$$\delta\mathrm{E}=\frac{8\pi}{3}\left(\frac{e^2}{mc^2}\right)^2\mathrm{E}(\omega)\,d\omega,$$

and this is the Thomson expression for the scattering by a free electron.

(e) *Comparison between the classical and quantum-theory expressions for the scattering factor. The oscillator strength:* Equation (4.10) shows that unless ω is very nearly equal to ω_s, and so long as the damping is small, we can write for f', the real part of the scattering factor for a single dipole for radiation of frequency ω,

$$f'=\frac{\omega^2}{\omega^2-\omega_s{}^2}\,. \tag{4.17}$$

From (4.9), f'', the imaginary part of the scattering factor, is given by

$$f''=\frac{mc}{4\pi e^2}\,\omega\mu_a(\omega), \tag{4.18}$$

where $\mu_a(\omega)$ is the absorption coefficient per dipole for radiation of frequency ω. This absorption coefficient is extremely small for an oscillator with small damping unless ω is very nearly equal to ω_s, and for the time being we shall neglect it and consider only the real part, f'.

Let the atom contain a number of oscillators, say $g(1)$ of frequency ω_1, ... $g(s)$ of frequency ω_s. Then the total real part of the atomic scattering factor is

$$f'=\sum_s\frac{g(s)\,\omega^2}{\omega^2-\omega_s{}^2}, \tag{4.19}$$

which may be written in the form

$$f'=\sum_s g(s)-\sum_s\frac{g(s)\,\omega_s{}^2}{\omega_s{}^2-\omega^2}\,. \tag{4.20}$$

We shall call $g(s)$ the oscillator strength corresponding to the natural frequency ω_s. Expression (*4*.20) for the scattering power of an atom must now be compared with that deduced in the last chapter by the methods of wave-mechanics. We note first of all that with the assumption made throughout this chapter, that the polarisation factor is unity, which means that we are considering only the scattering in the plane perpendicular to the electric vector of the incident radiation, the scattering from an atom containing dipole oscillators is independent of the angle of scattering. As we saw in the last chapter, this is not true of the scattering from a real atom unless the region of the atom within which the charge-density has appreciable magnitude has dimensions small in comparison with the wave-length of the radiation; and it is only in such a case that any real comparison can be made between (*4*.20) and the expression for f obtained from equation (*3*.42), on p. 112. Let us for the time being, however, assume that we are dealing with radiation the wave-length of which is large in comparison with the atomic dimensions. For our present purpose, the investigation of the dispersion terms, this assumption is less of a limitation than might at first sight be supposed. Those dispersion terms corresponding for example to the K absorption edge are only important for frequencies of the incident radiation near the K absorption frequency, and the corresponding wave-lengths are considerably greater than the linear dimensions of the K charge-distribution. The corresponding statements for the other absorption edges are also true; so that for the dispersion terms the assumption that we have made gives a very good approximation.

With this assumption in mind, let us compare equation (*4*.20) with the value of f derived from equation (*3*.42) of III, § 3 (f), p. 112. It is to be remembered that (*3*.42) applies to coherent scattering from an atom containing a single electron in the state n. Moreover, in deducing it, no allowance was made for any form of damping or loss of energy from the atomic system by radiation. The value of f deduced from (*3*.42) therefore corresponds to the *real* part, f', of the scattering factor. To obtain it we put E_0 equal to unity in the square bracket in (*3*.42), getting

$$f' = f_{n,n} - \sum_k{}' \left\{ \frac{2\hbar}{m\omega_{kn}} B^0_{nk} (B^*_{nk})_p \frac{\omega^2_{kn}}{\omega^2_{kn} - \omega^2} \right\}. \qquad (4.21)$$

It will be seen that the two expressions (*4*.20) and (*4*.21) have the same form. To each natural dipole frequency ω_s in the classical expression there corresponds in the quantum expression a frequency ω_{kn}, which is the Bohr frequency associated with the transition of the atom from the energy state k to the state n in which it is supposed to remain during the scattering. To each oscillator strength $g(s)$ there corresponds a coefficient $g(k,n)$, where

$$g(k,n) = \frac{2\hbar}{m} B^0_{nk} (B^*_{nk})_p / \omega_{kn}. \qquad (4.22)$$

The term $f_{n,n}$ in (4.21) is the scattering factor for the atom in the state n for coherent radiation of frequency large in comparison with any transition frequency ω_{kn}; and this has to be identified with the sum of the oscillator strengths for the whole atom. If the incident wave-length is large in comparison with the linear dimensions of the atom, $f_{n,n}$ is equal to the number of electrons in the atom—in the case of the one-electron atom that we are considering, unity. Thus, if the two expressions (4.20) and (4.21) are to be identified, we must have for the one-electron atom

$$\sum_s g(s) = 1. \tag{4.23}$$

Let us now return to the consideration of the quantities $g(k, n)$, the analogues of the oscillator strengths of the classical expression. With the assumption that we are making as to wave-length, (4.22) reduces to a simpler form. Suppose the electric vector $\mathbf{E}_0$ to be parallel to the x direction. Since $\lambda \gg |\mathbf{r}|$, we can put $e^{-i\kappa \mathbf{s}_0 \cdot \mathbf{r}}$ equal to unity in (3.27)* and so get

$$\mathrm{B}^0_{nk} = \int \psi_k^* \frac{\partial \psi_n}{\partial x} \, d\tau, \quad (\mathrm{B}^*_{nk})_p = \int \psi_k \frac{\partial \psi_n^*}{\partial x} \, d\tau.$$

From the properties of the wave-equation it is then not difficult to show that

$$\mathrm{B}^0_{nk} = \frac{m}{\hbar} \omega_{kn} x_{nk}, \quad \mathrm{B}^*_{nk} = \frac{m}{\hbar} \omega_{kn} x_{kn}, \tag{4.24}$$

where
$$x_{nk} = \int \psi_k^* x \psi_n \, d\tau.$$

Equation (4.22) then becomes

$$g(k, n) = \frac{2m}{\hbar} \omega_{kn} |x_{nk}|^2. \tag{4.25}$$

The oscillator strength $g(k, n)$ is thus proportional to the square of the co-ordinate matrix-element, x_{nk}, and so to the probability of the spontaneous transition of the atom from the state k to the state n with the emission of radiation of frequency ω_{kn} polarised parallel to x, the direction of the incident electric vector.

Since x represents a direction fixed relative to the atom, and since, on the average, the atom may present any aspect to the incident radiation, equation (4.25) is more correctly written

$$g(k, n) = \frac{2m\omega_{kn}}{3\hbar} \{|x_{nk}|^2 + |y_{nk}|^2 + |z_{nk}|^2\}. \tag{4.26}$$

Now it can be shown that with $g(k, n)$ defined as in (4.25) or (4.26)

$$\sum_k g(k, n) = 1 \tag{4.27}$$

* In other words, we are neglecting the phase differences between the contributions to the scattering from the different parts of the atom.

for a one-electron atom.* This result is known as the Thomas-Reiche-Kuhn [3] summation rule, and we see that it leads to the result that we had already inferred from the comparison of (*4*.20) and (*4*.21).

With the assumption we have made as to wave-length, the formal relationship between the classical and quantum formulae for the scattering of coherent radiation is now complete for the one-electron atom. The atom scatters as if it contained a number of dipole oscillators, one for each of the possible transition frequencies from states k to the state n in which the atom is supposed to be, usually of course the ground state. The strength of each such virtual dipole oscillator is to be taken as proportional to the transition probability between the two states concerned, and its frequency as equal to the corresponding Bohr frequency. The total oscillator strength is equal to unity, the number of electrons the atom contains. The states k include, of course, not only the discrete states of negative energy, but also the continuum of positive energy states, and, for these, the summations in (*4*.20) and (*4*.21) will become integrations.

The quantum dispersion formula was first given by Kramers and Heisenberg [4] on the basis of the correspondence principle. It was first obtained in terms of the concepts of wave-mechanics by Schrödinger [5] and was extended to include X-ray frequencies by Waller.[6]

(*f*) *Extension to the many-electron atom :* The dispersion formula as we have deduced it in (*3*.42) applies only to an atom containing a single electron, and we are as a rule more interested in atoms containing many electrons. To the extent to which we can regard the atomic electrons as being independent, or their interactions as being taken into account by some method such as that of the self-consistent field, we can think of the dispersion formula for the atom as a sum of a set of formulae such as (*4*.21), one for each electron, or group of electrons, such as the K, L, or M groups: but here certain difficulties arise which are absent when a single electron is considered. They are most easily discussed by taking a special case, say the electrons of the K shell. The K state is then the state n of (*3*.42) or (*4*.21), and ω_{kn} is a frequency corresponding to the transition from the ground state K to another state k. Now we have seen that the strength of the virtual oscillator associated with this frequency is proportional to the probability of a transition between these two states. If this probability is zero the corresponding oscillator strength is also zero. In a one-electron atom, the electron may pass from its lowest state to one of a number of discrete states, each transition corresponding to a definite frequency in the line absorption spectrum

* The law is true for a many-electron atom if we define x_{nk} as $\int \psi_k^* \Sigma x_i \psi_n \, d\tau$, where x_i is the co-ordinate of the ith electron, the integration being then over the configuration space of all the electrons. The sum of the oscillator strengths is then equal to Z, the total number of electrons in the atom. See Bethe; Geiger and Scheel's *Handbuch der Physik*, Vol. **24,** Part I, 2nd Edn.

of the atom; but many such transitions to discrete energy states possible for the single electron cannot take place if the electron is a member of the K shell of a many-electron atom. A number of these states will already be filled, and it would be necessary to remove an outer electron in order to create a vacancy to which the K electron could pass; for by Pauli's principle an electron cannot pass to a state already occupied by another electron. The probability of transition to these states is therefore zero; and so are the corresponding oscillator strengths in the dispersion formula. The electron can pass only to the outermost unoccupied discrete states, or to one of the states of positive energy, corresponding to the removal of the electron from the atom—a hyperbolic orbit on the older quantum theory.

The matrix elements corresponding to the transitions of the K electrons to the outer discrete states are small, for they depend on the product of the wave-functions of the two states concerned, and the wave-function of an inner state is small in those regions where the wave-function of an outer state is appreciable, and *vice versa*. In effect, therefore, only the outer states of positive energy are important. These form a continuous set of energy levels, and not a discrete set, and correspond to a region of continuous absorption, and not to line absorption. The virtual oscillators by which the scattering from the K electrons can be represented thus also form a continuous set, having every possible frequency, from the K absorption frequency, corresponding to the energy $\hbar\omega_K$ just sufficient to remove the electron from the atom, upwards. We have now to discuss the way in which the distribution of the oscillators in this continuum depends on the frequency.

(*g*) *The oscillator-density for the continuum of positive energy states:* Let $(dg/d\omega)d\omega$ be the number of virtual oscillators having frequencies lying between the limits ω and $\omega+d\omega$. We shall refer to $dg/d\omega$ as the oscillator-density at frequency ω. For the K electrons, it has a lower limit at $\omega=\omega_K$, the K absorption frequency. The total oscillator strength associated with the K electrons, in so far as the continuum of energy states is concerned, is therefore

$$g_K=\int_{\omega_K}^{\infty}\left(\frac{dg}{d\omega}\right)_K d\omega. \tag{4.28}$$

The value of this integral will in general be considerably less than 2, the total number of K electrons. Sugiura [7] has shown that for a hydrogen-like electron in the $1s$ state the integral over the positive energy states has the value 0·437, instead of unity as it would have if the sum over the discrete states were also included. The K electrons in actual atoms are not hydrogen-like, and we cannot expect the oscillator strength to be 0·437 per electron; but we are at all events prepared to find it considerably less than unity.

Similar considerations apply to the L electrons. Again, the

only possible transitions are to the positive energy states. The absorption edge is more complicated, because of the different types of L level, but the principles are the same. We must note, however, that amongst the forbidden transitions are not only those to the outer electron groups, but also those to the K levels having bigger negative energies. Now the oscillator strength corresponding to a transition from L to K must be considered as having the opposite sign to that for the corresponding transition from K to L. That is to say, we assume generally $g(k, n) = -g(n, k)$. The justification for this may be seen if the matter is considered statistically. The one transition corresponds to an absorption of radiation, the other to an emission, and the net result of equal numbers of the two types of transition will be zero.

Algebraically, therefore, the L oscillators, because of the forbidden transitions L to K, are in excess of their normal strength by just as much as the K oscillators are deficient because of the forbidden transitions K to L. We can extend this argument to the whole atom, and arrive at the result that the total oscillator strength for it is still equal to the total number of electrons it contains. Formula (*4*.20) therefore becomes

$$f' = Z + \sum_{K} \int_{\omega_K}^{\infty} \frac{\omega^2 (dg/d\omega)_K}{\omega_i^2 - \omega^2} d\omega, \tag{4.29}$$

where ω_i is now the frequency of the incident radiation, and $(dg/d\omega)_K$ is the oscillator-density at frequency ω corresponding to the K continuum. The sum is to be taken over all the electron groups K, L, M, … . The integral $\int_{\omega_K}^{\infty} (dg/d\omega)_K d\omega$ will not in general be equal to the number of electrons in the corresponding group, but to a smaller number.

Formula (*4*.29), like all the formulae of the last few paragraphs, applies only if the wave-length of the incident radiation is large in comparison with the atomic dimensions. This, as we have seen, is always true for those wave-lengths for which any given dispersion term is appreciable. We can thus as a good approximation to the total real part of f for an actual atom use (*4*.29) for the dispersion terms, and replace Z by f_0, the scattering factor for frequencies high in comparison with any natural frequency of the atom. As a very useful approximation for the real part of the scattering factor for incident radiation of frequency ω_i we may therefore write

$$f' = f_0 + \Delta f', \tag{4.30}$$

where

$$\Delta f' = \sum_{K} \int_{\omega_K}^{\infty} \frac{\omega^2 (dg/d\omega)_K}{\omega_i^2 - \omega^2} d\omega, \tag{4.30a}$$

without restriction as to wave-length, except that it must not be so short that relativity corrections become appreciable. We have discussed the numerical calculation of the scattering factor f_0 in the concluding sections of Chapter III, and its values are to be found tabulated. It becomes equal to Z for the limiting case of long wave-lengths or small angles of scattering.

(*h*) *The determination of the oscillator-density from the photoelectric absorption :* In principle, the oscillator-density can be determined if the atomic wave-functions are known, so that the necessary matrix elements can be calculated, and we shall discuss this method in due course. We propose, however, first of all to consider a simpler method, which uses the connection between the oscillator-density and the atomic absorption coefficient.

We have seen by (*4*.15) that each oscillator of frequency ω absorbs in unit time a quantity of energy $(2\pi^2 e^2/mc) . \mathrm{E}(\omega)$, where $\mathrm{E}(\omega)$ is the energy-density in the incident radiation at frequency ω. Suppose now that radiation containing frequencies extending over a narrow range $d\omega$, which is yet large in comparison with the natural absorption-width of an oscillator, falls on the atom.

The total energy incident per sq. cm. per sec. is $\mathrm{E}(\omega)d\omega$; and if $(dg/d\omega)$ is the oscillator-density in the neighbourhood of ω, there will be in each atom $(dg/d\omega)d\omega$ oscillators that absorb the radiation. The atomic absorption coefficient $\mu_a(\omega)$ for frequency ω is therefore given by

$$\mu_a(\omega) = \text{Energy absorbed per atom/Energy incident per sq. cm.}$$
$$= (2\pi^2 e^2/mc)\,\mathrm{E}(\omega)(dg/d\omega)\,d\omega/\mathrm{E}(\omega)\,d\omega,$$

or $$(dg/d\omega) = (mc/2\pi^2 e^2)\mu_a(\omega). \tag{4.31}$$

The distribution of oscillators and the variation of the photoelectric absorption with frequency thus follow the same law, apart from a numerical factor, and the oscillator-density can therefore be determined empirically from the experimental values of the photoelectric absorption coefficient.

Dispersion formulae based on this connection have been given by Kronig,[8] by Kallmann and Mark,[9] and by Bothe.[10] Kallman and Mark assumed the total number of oscillators associated with the K absorption edge to be two, the number of K electrons, and taking $\mu_a(\omega)$ as varying inversely as ω^3, proceeded to calculate the absorption coefficients, and the dispersion formula. The values they obtained in this way were of the right order of magnitude. We have seen, however, that the total K oscillator strength is less than two, and we shall therefore reverse the argument, and use the measured values of the absorption coefficient to calculate the oscillator strength, a method that has been used by Williams [11] and others.

Prins,[12] from measurements of the refractive index of iron for X-rays of frequency near that of the K absorption edge for iron, concluded that g_K for iron should be about 1·3 instead of 2, a result that is supported by the calculations of Williams from the photoelectric absorption, and by those of Hönl from the wave-functions. An explanation on the lines that we have discussed above of this apparent deficiency in oscillator strength was shortly afterwards given by Kronig and Kramers.[13]

The variation of the atomic absorption coefficient with frequency for the K edge is fairly well represented by the empirical formulae

$$\mu_a(\omega)=\frac{\mathrm{A}}{\omega^n}=\left(\frac{\omega_{\mathrm{K}}}{\omega}\right)^n\mu_a(\omega_{\mathrm{K}}) \quad \text{for } \omega>\omega_{\mathrm{K}},$$

$$\mu_a(\omega)=0 \quad \text{for } \omega<\omega_{\mathrm{K}}, \tag{4.32}$$

where n has a value not very different from 3, ω_{K} is the frequency of the K absorption edge, and $\mu_a(\omega_{\mathrm{K}})$ the absorption coefficient due to the K electrons for that frequency. Assuming a formula of this type, we get, by (*4*.28) and (*4*.31), for the oscillator strength associated with the K edge

$$g_{\mathrm{K}}=\frac{mc\mathrm{A}}{2\pi^2e^2}\int_{\omega_{\mathrm{K}}}^{\infty}\frac{d\omega}{\omega^n}=\frac{mc}{2\pi^2e^2}\frac{\omega_{\mathrm{K}}}{n-1}\mu_a(\omega_{\mathrm{K}}). \tag{4.33}$$

From this formula the oscillator strength g_{K} can be calculated if n and $\mu_a(\omega_{\mathrm{K}})$ are determined from experiments on absorption.

From (*4*.30a) we get for $\Delta f'_{\mathrm{K}}$, the contribution to the dispersion terms in f' from the K electrons, for radiation of frequency ω_i,

$$\Delta f'_{\mathrm{K}}=\frac{mc}{2\pi^2e^2}\mu_a(\omega_{\mathrm{K}})\,\omega_{\mathrm{K}}^n\int_{\omega_{\mathrm{K}}}^{\infty}\frac{\omega^2d\omega}{(\omega_i^{\,2}-\omega^2)\,\omega^n}, \tag{4.34}$$

and there will be similar expressions for each of the other absorption edges.

We give below the values of (*4*.34) for $n=2$, 2·5, and 3.

$$n=2, \qquad \Delta f'_{\mathrm{K}}=g_{\mathrm{K}}\frac{1}{2x}\log_e\frac{|x-1|}{x+1}, \tag{4.35}$$

$$n=2{\cdot}5, \qquad \Delta f'_{\mathrm{K}}=g_{\mathrm{K}}\frac{3}{2x^{\frac{3}{2}}}\left\{\frac{\pi}{2}-\cot^{-1}x^{\frac{1}{2}}-\tfrac{1}{2}\log_e\frac{x^{\frac{1}{2}}+1}{|x^{\frac{1}{2}}-1|}\right\}, \tag{4.36}$$

$$n=3, \qquad \Delta f'_{\mathrm{K}}=g_{\mathrm{K}}\frac{1}{x^2}\log_e|x^2-1|, \tag{4.37}$$

where $x=\omega_i/\omega_{\mathrm{K}}$, and g_{K} is the total oscillator strength for the K edge, which can be determined from (*4*.33).

For a single oscillator of strength g_{K} and frequency ω_{K}, the corresponding value of $\Delta f'_{\mathrm{K}}$ would be

$$\Delta f'_{\mathrm{K}}=g_{\mathrm{K}}/(x^2-1). \tag{4.38}$$

All these expressions become infinite for $x=1$, that is to say, when $\omega_i=\omega_{\mathrm{K}}$; but it must be remembered that they all cease to be valid over a short range on either side of the absorption edge,* and the proper expression could be calculated only if the damping could be allowed for accurately.

(*i*) *The imaginary part of the scattering factor:* We have not yet considered f'', the imaginary part of the scattering factor, corresponding to a scattered wave whose phase in the forward direction lags $\pi/2$ behind that of the primary wave. So long as we are dealing with scattering by a system containing only a discrete set of oscillators with definite frequencies, f'' is important only when the incident frequency lies very close to one of the natural oscillator frequencies. Provided that such frequencies are avoided, the total scattering from such a system may be taken as f'. Matters are, however, different when, as in the case of an actual atom, the virtual oscillators have frequencies extending over a continuous range. Suppose, for example, we are considering the electrons of the K shell. Then if ω_i, the frequency of the incident radiation, is greater than ω_K, the K absorption frequency, ω_i must always coincide with the frequency of some oscillator of the K continuum, and f'' will have a value not in general negligible. If we denote by $i\Delta f''_K$ the contribution of the K electrons to the imaginary part of f, we see from equation (4.9) of IV, § 1 (*c*) that

$$\Delta f''_K = \frac{mc}{4\pi e^2}\omega_i \mu_K(\omega_i), \tag{4.39}$$

where $\mu_K(\omega_i)$ is the contribution of the K electrons to the atomic absorption coefficient for frequency ω_i. Since $\mu_K(\omega_i)=0$ if $\omega_i<\omega_K$, $\Delta f''_K$ will differ from zero only if $\omega_i>\omega_K$.

Using the relation (4.31) between the absorption coefficient and the oscillator-density, we obtain from (4.39)

$$\Delta f''_K = \frac{\pi}{2}\omega_i (dg/d\omega)_{K_i}. \tag{4.40}$$

If we assume an inverse power law of the type (4.32) for the variation of the absorption coefficient with frequency, we can express $\Delta f''_K$ in terms of g_K, the total oscillator strength of the K continuum. Thus, from (4.39) and (4.32),

$$\begin{aligned} \Delta f''_K &= \frac{mc}{4\pi e^2}\omega_i \left(\frac{\omega_K}{\omega_i}\right)^n \mu_a(\omega_K) \\ &= \frac{\pi}{2}\frac{n-1}{x^{n-1}} g_K, \quad \text{if } x>1, \end{aligned} \tag{4.41}$$

by (4.33), where $x=\omega_i/\omega_K$, as before.

* It has been assumed in the calculation that led to these formulae that μ changes inappreciably with a change of frequency of the order of k, the natural width of an absorption line, *i.e.* that $\frac{k}{\mu}\frac{d\mu}{d\omega} \ll 1$, and this condition is not fulfilled at the absorption edge where μ changes suddenly.

(*j*) *The total scattering factor f*: To sum up the last few paragraphs, let us write for f, the total atomic scattering factor for frequency ω_i,

$$f = f_0 + \Delta f' + i\Delta f'', \tag{4.42}$$

where f_0 is the atomic scattering factor for frequencies high in comparison with any atomic absorption frequency, and is independent of the incident frequency, while $\Delta f'$ and $\Delta f''$ are the real and imaginary parts of f that depend on the frequency. Then

$$\Delta f' = \sum_{\mathrm{K}} \Delta f'_{\mathrm{K}} = \sum_{\mathrm{K}} \int_{\omega_{\mathrm{K}}}^{\infty} \frac{(dg/d\omega)_{\mathrm{K}}\,\omega^2\,d\omega}{\omega_i^2 - \omega^2}, \tag{4.43}$$

$$\Delta f'' = \sum_{\mathrm{K}} \Delta f''_{\mathrm{K}} = \sum_{\mathrm{K}} \frac{\pi}{2}\omega_i (dg/d\omega)_{\mathrm{K}_i} = \sum_{\mathrm{K}} \frac{mc}{4\pi e^2}\omega_i \mu_{\mathrm{K}}(\omega_i), \tag{4.43a}$$

$\Delta f''_{\mathrm{K}}$ being zero for $\omega_i < \omega_{\mathrm{K}}$.

The atomic scattering factor that occurs in the formulae for the intensity of reflection of X-rays by crystals developed in Chapter II is the modulus of f as given by (*4*.42), or $|f|$, and is given by

$$|f|^2 = (f_0 + \Delta f')^2 + (\Delta f'')^2. \tag{4.44}$$

The difference between $|f|$ and f_0, the value of f calculated from the atomic charge-distribution by the methods outlined in Chapter III on the assumption that the frequency of the radiation is large in comparison with any atomic absorption frequencies, is denoted by δf, and is a convenient measure of the part of f that depends on the frequency.

(*k*) *The variation of the scattering factor with frequency*: In order to get a definite idea of the way in which the scattering factor varies with frequency according to the theory we have outlined, we may consider the case of an atom containing a K electron group whose contribution to the atomic absorption coefficient varies inversely as the cube of the frequency of the incident radiation. In this case, by (*4*.37) and (*4*.41),

$$\Delta f'_{\mathrm{K}} = g_{\mathrm{K}} \frac{1}{x^2} \log_e |x^2 - 1|,$$

$$\Delta f''_{\mathrm{K}} = g_{\mathrm{K}} \frac{\pi}{x^2}, \; x > 1; \quad = 0, \; x < 1. \tag{4.45}$$

The curve A of fig. 52 shows $\Delta f'_{\mathrm{K}}/g_{\mathrm{K}}$, the curve B, $\Delta f''_{\mathrm{K}}/g_{\mathrm{K}}$, plotted against x, the ratio of the incident frequency to the K absorption frequency. For very large values of x, that is to say, for frequencies large in comparison with the K absorption frequency, both $\Delta f'_{\mathrm{K}}$ and $\Delta f''_{\mathrm{K}}$ tend to zero, and this means that, so far as any effect of the K electrons is concerned, the scattering factor is then f_0, the calculation of which was considered in Chapter III.

For very low frequencies, on the other hand, $\Delta f'_{\mathrm{K}}$, the real contribution to Δf_{K}, tends to $-g_{\mathrm{K}}$, while the imaginary contribution $\Delta f''_{\mathrm{K}}$ is zero. For frequencies much lower than that of the K absorption

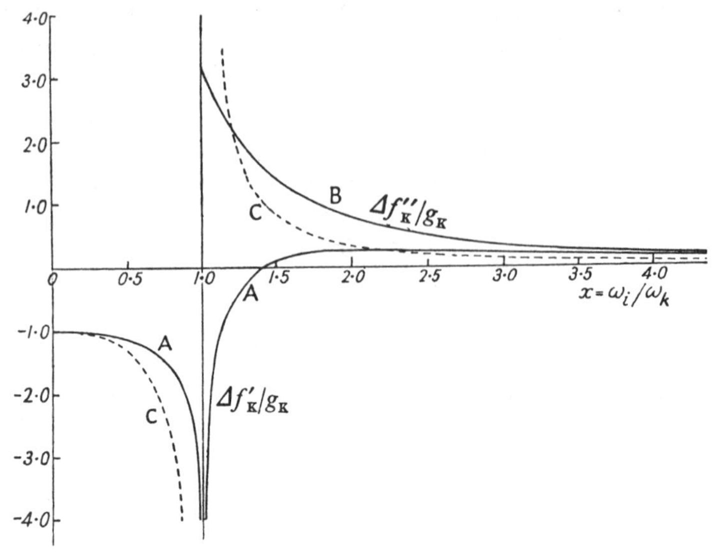

FIG. 52. Corrections to the scattering factor on account of dispersion for a K electron whose contribution to the atomic absorption coefficient varies as the inverse cube of the frequency of the radiation. Curve A shows $\Delta f'_K/g_K$ and curve B $\Delta f''_K/g_K$. Curve C is the corresponding curve for a simple harmonic oscillator

edge, therefore, the standard scattering factor f_0 has to be reduced by g_K, the oscillator strength of the K continuum.*

In terms of the classical picture, this corresponds physically to the fact that an oscillator forced into vibration by a wave of frequency much smaller than its own natural frequency acquires a negligibly small amplitude. The K electrons for these low frequencies contribute nothing to the scattering factor. If we confine our attention for the moment to radiation scattered in the direction of the incident wave-normal, so that f_0 is equal to Z, the total number of electrons in the atom, we might expect that the elimination of the scattering from the two K electrons would diminish the value of f by 2. Actually the diminution is only g_K, which is always less than 2, and may be as low as 1·2 or 1·3. This apparent contradiction may be explained as follows. For scattering in the forward direction, the total oscillator strength is Z. Because transitions of the K electrons to the occupied L, M, ... states are forbidden, g_K is less than 2, as we have already seen. But g_L, the oscillator strength of the L continuum, is in excess because of the forbidden L to K transitions by exactly the amount by which g_K is too

* We are considering here only the variation of the contribution of the K shell with frequency, and are neglecting the variation of the contributions from the other groups, which must also occur. We are considering a state of affairs corresponding fairly well to a frequency small in comparison with that of the K edge, but large in comparison with that of the L edge.

small because of the forbidden K to L transitions. The total oscillator strength of the atom thus remains Z; but when the K electrons cease to contribute it is diminished by g_K, and not by 2.

As x increases from very small values, $\Delta f'_K$ decreases, slowly at first, but rapidly as the value $x=1$ is approached, tending to an infinitely great negative value at $x=1$ itself, although, as we have seen, the approximation to the dispersion terms that we are using is not valid in the immediate neighbourhood of $x=1$, owing to the fact that damping has been neglected.

For $x<1$ the scattering factor is less than f_0 by at least g_K, and for radiation just on the long wave-length side of the absorption edge by considerably more.

When x is just greater than unity, or the incident frequency just greater than the K absorption frequency, $\Delta f'_K$ is again a large negative quantity. It rises rapidly as x increases, becomes positive, and reaches a maximum of about $0{\cdot}28 g_K$ for x about 2·2, thereafter slowly decreasing towards zero for large values of x. The height and position of the maximum depends on the value of n, the inverse power according to which the absorption varies with the frequency. The maximum is higher, and occurs at a smaller value of x, the greater n. Williams has shown that the locus of the maxima is the curve for $\Delta f'_K$ corresponding to a single dipole oscillator of frequency ω_K, and strength g_K. Such a curve represents the limiting case for which $n=\infty$. The curve for such an oscillator is shown by the dotted line C in fig. 52. It will be seen that the curve for the single oscillator tends to infinite positive values as x tends to unity from larger values. Curves such as that for $n=3$ may be considered as produced by the superposition of an infinite set of such dipole curves, corresponding to the continuum of oscillators, each successive curve being displaced relative to its predecessor in the direction of increasing x, and correspondingly diminished in ordinates. This superposition of curves produces large negative values on both sides of $x=1$.

(*l*) *The effect of the imaginary component of f on the amplitude and phase of the scattered wave:* The imaginary part of Δf_K has values differing from zero only when $x>1$. For the case considered in IV, § 1(*k*), its value is πg_K for $x=1$, and it decreases slowly to zero as x becomes large. It represents a component of the scattered radiation from the atom having a phase that lags $\pi/2$ behind that of the primary wave. The phase-lag of the resultant wave, which is π for very high frequencies (large x), thus diminishes as the frequency of the incident radiation approaches that of the absorption edge from the short wave-length side, and an increasingly important component with a phase-lag of $\pi/2$ is added; and it may differ very appreciably from π if the frequency of the incident radiation is only just greater than that of the absorption edge.

On the long wave-length side of the absorption edge on the other hand, apart from any effects due to damping, or to the other absorption edges, which will always be small, the scattered wave will have a phase-lag π if the resultant f for the whole atom is positive.

In order to discuss the effect of the imaginary component, it is best to take a numerical example. We shall consider an idealised iron atom, containing K electrons for which $\Delta f'_K$ and $\Delta f''_K$ are given by (4.45). The total number of electrons is 26, and only the K electrons are supposed to contribute anything to the dispersion and absorption. The oscillator strength of the K continuum, g_K, we shall take as 1·3, in agreement with the observations of Prins [12] for iron. The curve giving f_0 for the atom as a function of $(\sin\theta)/\lambda$ we shall take to be that calculated from the Thomas-Fermi charge distribution, as tabulated by James and Brindley. These assumptions will serve to give a general idea of the numerical magnitude of the effect we are considering.

We shall take two cases, (1) when $(\sin\theta)/\lambda = 0$, and $f_0 = 26$, and (2) when $(\sin\theta)/\lambda = 1{\cdot}0\text{A}^{-1}$, $f_0 = 6{\cdot}3$. Using these values we calculate the value of $|f|$ from the equation

$$|f|^2 = (f_0 + \Delta f'_K)^2 + (\Delta f''_K)^2,$$

with the values of $\Delta f'_K$ and $\Delta f''_K$ given by equation (4.45). This assumes that the scattering by the K electrons is independent of $(\sin\theta)/\lambda$ which is nearly enough true for an atom such as iron over the range in $(\sin\theta)/\lambda$ under discussion.

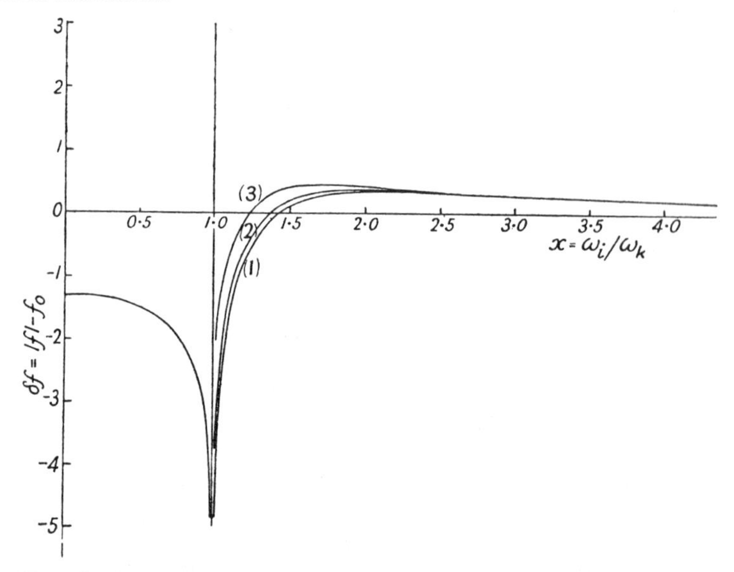

FIG. 53. Illustrating the effect of an imaginary component $\Delta f''$ of the scattering factor. δf is plotted as a function of ω_i/ω_K for the K electron of the iron atom. In curve (1), $\Delta f''$ is neglected, in curves (2) and (3) it is included for $(\sin\theta)/\lambda$ equal to 0 and 1A^{-1} respectively

With the values of $|f|$ so determined we calculate $\delta f = |f| - f_0$, which is the correction to be added to f_0 to give $|f|$, the total scattering factor.

Curve (1) of fig. 53 shows again $\Delta f'_K$, plotted as a function of x. This is the value of δf neglecting the imaginary component $\Delta f''_K$. Curves (2) and (3) are those for δf including the effect of the imaginary term, for $(\sin\theta)/\lambda = 0$, and $(\sin\theta)/\lambda = 1$ respectively. The effect of the imaginary component is much greater for the larger value of $(\sin\theta)/\lambda$, because the scattering factor for the electrons other than the K electrons has there fallen off very greatly, and the imaginary component makes relatively a far greater contribution to the total scattering factor than it does for $(\sin\theta)/\lambda = 0$. If the scattering from the rest of the atom is large, $(\Delta f''_K)^2$ is nearly negligible in comparison with $(f_0 + \Delta f'_K)^2$.

The phase of the scattered wave lags behind that of the primary wave by an amount $\pi - \phi$, where

$$\tan\phi = \frac{\Delta f''_K}{f_0 + \Delta f'_K}.$$

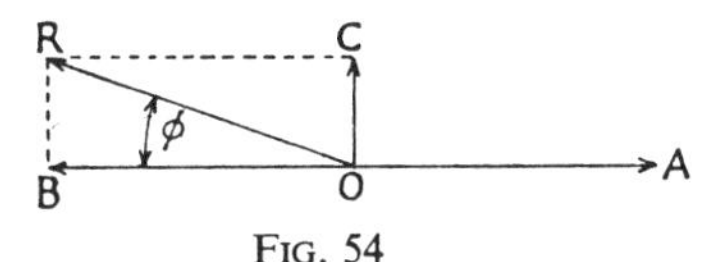

Fig. 54

This may be understood from fig. 54, in which the direction OA represents the phase of the primary wave, and OB and OC the phases of the real and imaginary parts of the scattered wave, lagging respectively π and $\pi/2$ behind the phase of the primary wave. If OB and OC also represent the magnitudes of the real and imaginary components, OR gives the resultant magnitude, and the angle AOR the phase-lag of the scattered wave.

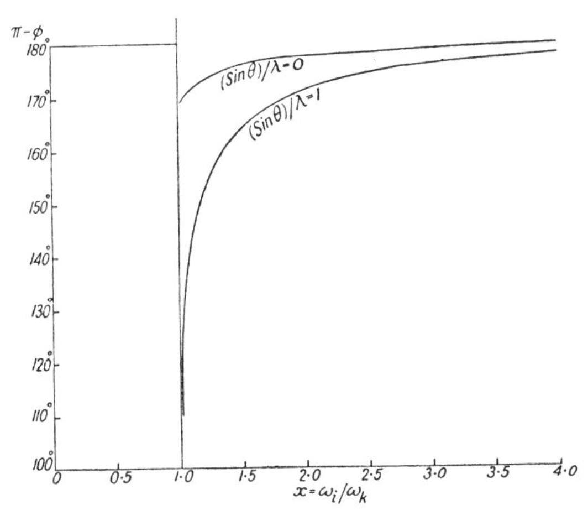

Fig. 55. The phase-lag of the scattered radiation from iron as a function of $x = \omega_i/\omega_K$ for different values of $(\sin\theta)/\lambda$

In fig. 55 this phase-lag $\pi - \phi$ is plotted as a function of x for $(\sin\theta)/\lambda = 0$ and 1. It will be seen once more that the phase variation with x is much more important for the large values of $(\sin\theta)/\lambda$ than for the small, and the reason for this will be obvious from what has been said above. The importance of the variation of $|f|$ and the phase-lag with $(\sin\theta)/\lambda$ was first clearly pointed out by Coster, Knol and Prins.[14]

It is clear from the discussion of the last two sections that there is in general a sudden change both in the value of f and in the phase of the scattered wave when the frequency of the incident radiation passes through a value equal to that of the absorption edge. The usual formulae for the structure factors of crystal units containing atoms of different kinds are deduced on the assumption that all the atoms scatter in the same phase, and are therefore valid only when the frequency of the radiation differs considerably from any atomic absorption frequency.

It will also be evident that the numerical values of f, as usually tabulated, which are always values of f_0, cannot be applied indiscriminately to the scattering of X-rays of any wave-length. If the wave-length λ is greater than that corresponding to an absorption edge of the scattering atom, the tabulated value of f_0 must be reduced by at least the oscillator strength of the corresponding electron group. With the heavier atoms, and with the wave-lengths usually employed in the analysis of crystal structures, it may easily happen that λ is greater than both λ_K and λ_L, so that the proper value of f to use may be perhaps seven or eight units below the tabulated value. With the data at present available as to oscillator strengths, it is not easy to apply accurate corrections; but it is not difficult, and it is certainly worth while, to make rough estimates of the kind we have given, based on such oscillator strengths as are known.* The corrections will be practically independent of the angle of scattering, for the reasons we have considered above.

(*m*) *The influence of damping of the oscillators on f:* We have so far entirely neglected the damping coefficient k in the calculation of f. In this section we shall consider its influence, and it will be found that this is very small, except in the immediate neighbourhood of the absorption edge, so that no serious error is made in neglecting it. From equation (*4*.4) of IV, § 1 (*b*), we see that the complex value of f for a dipole oscillator of natural frequency ω_s for incident radiation of frequency ω_i is

$$f=\frac{\omega_i^2}{\omega_i^2-\omega_s^2-ik\omega_i}.$$

If the oscillator-density of the K continuum in the region of frequency ω is $(dg/d\omega)$, the total complex value of f for the K continuum is

$$f=\omega_i^2\int_{\omega_K}^{\infty}\frac{(dg/d\omega)\,d\omega}{\omega_i^2-\omega^2-ik\omega_i}. \tag{4.46}$$

Suppose that, as in the last two sections, $(dg/d\omega)=A/\omega^3$. Then,

* See Appendix III for corrections in the case of the K electrons.

since the total oscillator-density of the K continuum is g_K, $A=2g_K\omega_K{}^2$, and (4.46) may be written

$$f=2g_K\omega_i{}^2\omega_K{}^2\int_{\omega_K}^{\infty}\frac{d\omega}{\omega^3(\omega_i{}^2-\omega^2-ik\omega_i)}. \tag{4.47}$$

We now make the substitutions $x=\omega_i/\omega_K$, $\xi=\omega/\omega_K$, $\kappa=k/\omega_K$, and (4.47) becomes

$$f=2x^2g_K\int_1^{\infty}\frac{d\xi}{\xi^3(x^2-\xi^2-i\kappa x)}. \tag{4.48}$$

Prins [12] has given the value of the integral in (4.48). His results may be written in the form

$$f=\frac{x^2\log_e(1-x^2+i\kappa x)+x^2-i\kappa x}{(x^2-i\kappa x)^2}. \tag{4.49}$$

If κ is so small that its square and higher powers may be neglected, Prins shows that the separation of the real and imaginary parts of (4.49) yields approximate formulae, which in our notation may be written as follows:

(*a*) For $x<1$ $(\omega_i<\omega_K)$,

$$\Delta f'_K=g_K\frac{1}{x^2}\log_e(1-x^2),$$

$$\Delta f''_K=g_K\frac{\kappa}{x}\left[\frac{2-x^2}{1-x^2}+\frac{2}{x^2}\log_e(1-x^2)\right]. \tag{4.50a}$$

(*b*) For $x>1$ $(\omega_i>\omega_K)$,

$$\Delta f'_K=g_K\left\{\frac{1}{x^2}\log_e(x^2-1)-\frac{2\pi\kappa}{x^3}\right\},$$

$$\Delta f''_K=g_K\left[\frac{\kappa}{x}\left\{\frac{x^2-2}{x^2-1}+\frac{2}{x^2}\log_e(x^2-1)\right\}+\frac{\pi}{x^2}\right]. \tag{4.50b}$$

These formulae are not valid in the immediate neighbourhood of $x=1$, although for the values of κ that occur in practice they cover most of the necessary range. Glocker and Schäfer [15] have given the closed forms of these expressions, but we shall only quote their results for the special case $x=1$, when their formulae take the form

$$\Delta f'_K/g_K=\frac{(1-\kappa^2)\log_e\kappa-\pi\kappa+1+\kappa^2}{(1+\kappa^2)^2},$$

$$\Delta f''_K/g_K=\frac{2\kappa\log_e\kappa+\frac{\pi}{2}(1-\kappa^2)+\kappa(1+\kappa^2)}{(1+\kappa^2)^2}. \tag{4.51}$$

For very small damping, these formulae reduce to the approximate forms

$$\Delta f'_K/g_K=1+\log_e\kappa,$$
$$\Delta f''_K/g_K=\pi/2, \tag{4.52}$$

which are useful for estimating the lowest value that f can take in the neighbourhood of an absorption edge.

From experiments on the natural breadth of spectral lines, it may be inferred that κ is of order of magnitude 10^{-3}. This is greater than the value calculated from the classical damping factor, $k = 2e^2\omega^2/3mc^3$, which gives for the wave-length of the absorption edge, λ_K, $\kappa = k/\omega_K = (4\pi/3)(e^2/mc^2)/\lambda_K$. For the K absorption edge of iron, $\lambda_K = 1{\cdot}74$A, and $\kappa = 7 \times 10^{-5}$.

On evaluating the formulae (*4*.50) and (*4*.51) with κ equal to 10^{-3}, we find that the curve for $\Delta f'_K$, plotted as a function of x, instead of approaching minus infinity as x approaches unity, has the finite value $-6{\cdot}91g_K$ for $x = 1$. Except, however, for values of x within about 0·005 of unity the curve follows so closely that of fig. 52, which was plotted neglecting damping, that the two curves could not be shown separately on the scale of the diagram.

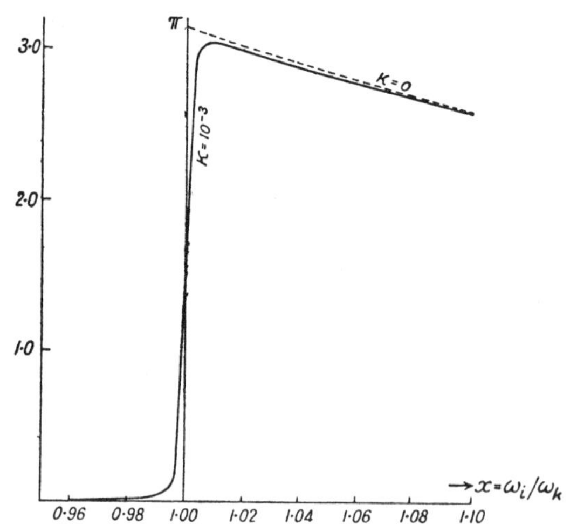

FIG. 56. The effect of damping on the imaginary part of the scattering factor in the neighbourhood of $\omega_i/\omega_K = 1$

In fig. 56, the curve for $\Delta f''_K/g_K$ as a function of x is shown for values of x in the neighbourhood of unity. The full curve was drawn assuming $\kappa = 10^{-3}$, and the dotted curve assuming $\kappa = 0$. The curve for zero damping meets the axis of ordinates at $\Delta f''_K/g_K = \pi$, but the curve for finite damping reaches a maximum for a value of x slightly greater than unity, and then falls very rapidly, crossing the axis at

$$\Delta f''_K/g_K = 1{\cdot}56.$$

It does not become zero immediately for $x < 1$, but approaches the x-axis gradually.

For very small damping, the intersection of the curve with the axis of ordinates occurs at $\Delta f''_K/g_K = \pi/2$, and the fall of the curve becomes sharper as κ becomes smaller, until in the limit, for $\kappa = 0$, it follows the axis.

The results of this section show that it is in general unnecessary to consider the effects of damping in comparing theoretical and experimental dispersion curves. The errors introduced in neglecting it are certainly smaller than the experimental errors, or than our uncertainty concerning other quantities, such as the oscillator strengths, that occur in the formulae.

(*n*) *The calculation of oscillator strengths from the atomic wave-functions :* Another method of approach to the problem of determining the oscillator strengths, in principle less empirical than that discussed above, consists in using such information as we have concerning the atomic wave-functions to calculate them directly from formulae such as (*4*.22) or (*4*.26). This method has been used by Sugiura[7] and, more recently, by Hönl.[16]

Only for the hydrogen-like atom are the wave-functions known accurately enough for a solution of the problem along these lines to be more than an approximation. For the K electrons, the approximation is probably as good as that obtained from the measured absorption coefficients; for the L electrons it is probably worse.

We shall first suppose the wave-length of the incident radiation to be long in comparison with the dimensions of the K shell, so that equations (*4*.25) or (*4*.26) may be used for the oscillator strengths. Sugiura showed that, for a single electron bound in a Coulomb field in its ground state 1*s*, the oscillator-density $dg/d\omega$, for frequency ω in the K oscillator-continuum, is given by

$$\frac{dg}{dz}=\frac{2^7}{3}\frac{g(z)}{z^4}, \tag{4.53}$$

where $z=\omega/\omega_0$, ω_0 being the K absorption frequency; and

$$g(z)=e^{-\frac{4}{\sqrt{z-1}}\tan^{-1}\sqrt{z-1}}\Big/\left(1-e^{-2\pi/\sqrt{z-1}}\right). \tag{4.54}$$

By graphical integration over the K continuum, from $z=1$ to $z=\infty$, he finds that the total oscillator strength corresponding to the continuum of positive energy states for a single hydrogen-like electron is 0·437.

Hönl has shown that, with the limits $1\leqslant z\leqslant 4$, $g(z)$ may be replaced to a close approximation by the much simpler expression

$$\bar{g}(z)=\frac{e^{-4}}{3}(4z-1). \tag{4.55}$$

Using this, instead of $g(z)$, in (*4*.53), we get for the total strength of the continuum of oscillators for a single electron

$$\begin{aligned} g &= \frac{2^7}{9}e^{-4}\int_1^\infty\left(\frac{4}{z^3}-\frac{1}{z^4}\right)dz \\ &= \frac{2^7}{27}\cdot 5e^{-4}=0{\cdot}434. \end{aligned} \tag{4.56}$$

According to this result, therefore, the total oscillator strength for the

K electrons of the atom, assuming them to be hydrogen-like, and that only transitions to positive energy states are possible, ought to be about 0·87, a conclusion that had already been reached by Kronig and Kramers.[13] Now calculations based on absorption measurements, and on experiments such as those of Prins on total reflection, suggest that a value of the order of 1·3 is more nearly correct for most atoms. We have now to consider the reasons for this large discrepancy, and possible means of correcting it.

The two main corrections that have to be made to the calculations of Sugiura to make them applicable to electrons bound in an atom are discussed by Hönl. They depend on the fact that the K electrons in real atoms form part of a complex electronic structure. The influence of the other electrons has therefore to be considered. To a fair degree of approximation the K electrons can, as a matter of fact, be treated as hydrogen-like; but, owing to the screening effect of one K electron on the other, the effective nuclear charge must not be taken as Z, but as $(Z-s)$, where s may be taken as approximately constant and equal to about 0·3. The value obtained by Hylleraas [17] for helium is 5/16. In addition to this, however, the outer electrons, although they produce only a very small field in the region of the K shell, give a nearly constant potential V_0. The effective potential $V(r)$ at distance r in the region of the K shell may therefore to this degree of approximation be written

$$V(r) = -(Z-s)e/r + V_0.$$

The negative energy of the ground state will thus be decreased by an amount eV_0, and we shall have for the ionisation energy I_K,

$$I_K = (Z-s)^2 Rhc - eV_0, \qquad (4.57)$$

where R is Rydberg's constant; or if ω_0 is the frequency of the absorption edge for a hydrogen-like atom of nuclear charge $Z-s$, and ω_K that for the actual atom, then

$$\omega_K = \omega_0(1-\delta_K), \quad \text{where} \quad \delta_K = \frac{(Z-s)^2 - I_K/Rhc}{(Z-s)^2}. \qquad (4.58)$$

For the heavier elements, terms in

$$(Z-s)^4\,\alpha^2,\ (Z-s)^6\,\alpha^4, \text{ and } (Z-s)^8\,\alpha^6,$$

where α is the fine-structure constant, the value of which is $7{\cdot}3\times10^{-3}$, must be added to $(Z-s)^2$ in (4.58). I_K, the ionisation energy, in terms of the wavelength of the K absorption edge, is equal to hc/λ_K, and since $10^8/R = 911$, the expression for δ_K can be written in the form

$$\delta_K = (A - 911/\lambda_K)/A, \qquad (4.58a)$$

where λ_K is expressed in Ångström units, and

$$A = Z_1^2 + 1{\cdot}33\times10^{-5}Z_1^4 + 3{\cdot}55\times10^{-10}Z_1^6 + 11{\cdot}7\times10^{-15}Z_1^8, \qquad (4.58b)$$

where $Z_1 = Z - 0{\cdot}3$.

The values of the matrix elements that determine the oscillator-density depend on the values of the wave-functions of the ground and excited states of the K electron in the region where both have appreciable magnitude; that is to say, in the region of the K shell. Now in this region, as we have just seen, the wave-functions may still be considered to be hydrogen-like. The oscillator-density for any frequency ω of the K continuum may thus still be calculated from equation (4.53). If we express the oscillator-density in terms of ω_K, the actual frequency of the absorption edge, we obtain for the two K electrons, since $\omega_K = \omega_0(1 - \delta_K)$,

$$\left(\frac{dg}{d\omega}\right)_K = \frac{2^8 e^{-4}}{9\omega_K}\left\{\frac{4}{(1-\delta_K)^2}\left(\frac{\omega_K}{\omega}\right)^3 - \frac{1}{(1-\delta_K)^3}\left(\frac{\omega_K}{\omega}\right)^4\right\}. \tag{4.59}$$

The oscillator-density so derived is the sum of two terms, each of which is of the form A/ω^n, and over the relevant frequency range it approximates fairly closely to a single term varying inversely as $\omega^{2\cdot 7}$.

Although the oscillator-density remains to a good approximation the same as that for a hydrogen-like atom, the total strength of the continuum will be altered. For, as we have seen, the ionisation energy is reduced by the presence of the outer electrons, or, in other words, the K continuum of positive energy states extends to a lower energy value. To get the total oscillator strength, we have to take as a lower limit of integration for (4.56) a value of z corresponding to $\omega = \omega_K$, instead of $z = 1$, which corresponds to $\omega = \omega_0$. The integration must thus extend from $z = 1 - \delta_K$ to $z = \infty$, and this gives at once for the two K electrons,

$$g_K = \frac{2^8 e^{-4}}{9}\left\{\frac{2}{(1-\delta_K)^2} - \frac{1}{3(1-\delta_K)^3}\right\}. \tag{4.60}$$

A virtually identical explanation of the difference between the values of g_K for a hydrogen-like electron and an electron forming part of an atom was given independently by E. J. Williams.[11]

The figures in Table IV. 1 are taken from Hönl's paper, and give the values of δ_K and g_K for a number of atoms, as calculated from (4.58) and (4.60) using the known values of the K absorption edges. Some

TABLE IV. 1

K Oscillator strengths (Hönl)

Element	δ_K	g_K
14 Si	0·276	1·53
20 Ca	0·240	1·41
24 Cr	0·218	1·34
26 Fe	0·215	1·33
30 Zn	0·205	1·31
42 Mo	0·182	1·24
60 Nd	0·154	1·17
74 W	0·143	1·15
92 U	0·131	1·12

values obtained by Williams from atomic absorption coefficients, as tabulated by Jönsson,* are shown in Table IV. 2.

TABLE IV. 2

K and L Oscillator strengths (Williams)

Element	g_K	g_L
13 Al	1·40	—
26 Fe	1·45	—
29 Cu	1·35	5·2
47 Ag	1·29	5·6
82 Pb	—	5·0

$\Delta f'_K$ and $\Delta f''_K$ can now be calculated at once from (*4*.43) and (*4*.43a) of IV, § 1(*j*), if the value of the oscillator-density given by (*4*.59) is used. The result is

$$\Delta f'_K = \frac{2^7 e^{-4}}{9}\left\{\frac{4}{(1-\delta_K)^2}\frac{1}{x^2}\log_e|x^2-1| - \frac{1}{(1-\delta_K)^3}\left(\frac{2}{x^2}+\frac{1}{x^3}\log_e\left|\frac{x-1}{x+1}\right|\right)\right\},$$

$$\Delta f''_K = \frac{2^7 e^{-4}}{9}\pi\left\{\frac{4}{x^2(1-\delta_K)^2} - \frac{1}{x^3(1-\delta_K)^3}\right\} \text{ if } x>1,$$

$$= 0 \text{ if } x<1, \qquad (4\cdot61)$$

where $x = \omega_i/\omega_K$, as before.

Hönl makes similar calculations for the L electrons, which are not, however, on so sound a theoretical basis as those for the K electrons, since the underlying assumption of a hydrogen-like L electron is not a good approximation. Some values of the L oscillator strengths so calculated are given in Table IV. 3. It will be seen that they differ appreciably from those calculated by Williams, and show a greater variation with the atomic number of the element.

TABLE IV. 3

L Oscillator strengths (Hönl)

Element	40 Zr	46 Pd	60 Nd	74 W	92U
$2g_{2s}$	1·53	1·50	1·42	1·35	1·28
$6g_{2p}$	5·14	4·70	3·93	3·55	3·22
g_L	6·67	6·20	5·35	4·90	4·50

Wheeler and Bearden [18] have also calculated the oscillator strengths from the wave-functions for a number of atoms by a method that has the advantage of relative simplicity, and that does not make the assumption of hydrogen-like electrons. It would appear, moreover, to be applicable to the L as well as to the K electrons.

* *Uppsala Dissertation*, 1928

TABLE IV. 4

K Oscillator strengths (Wheeler and Bearden)

Element	g(K, L)	g(K, M)	g(K, N)	g(K, O)	g_K
2 He					2·0
8 O *	0·162				1·78
17 Cl †	0·247	0·013			1·48
29 Cu †	0·320	0·030			1·30
55 Cs †	0·370	0·054	0·011	0·003	1·12

Let m be an occupied wave-function of an atom, other than a K wave-function. Then, by the summation rule, g_K, the oscillator strength for the K electrons, is given by

$$g_K = 2\{1 - \sum_m g(K, m)\}, \tag{4.62}$$

where $g(K, m)$ is the oscillator strength of the virtual oscillator associated with the transition from the state K to the state m. The sum has now to be taken over the relatively few occupied discrete states, instead of over the continuum, and avoids the extrapolation of the continuum of the hydrogen-like atom that is used in the Hönl method. The oscillator strength $g(K, m)$ is given by (*4*.26), and is determined numerically from the Hartree wave-functions calculated by the method of the self-consistent field. As Wheeler and Bearden themselves point out, it is not really correct to use these functions in (*4*.62), since the summation rule does not properly apply to them. The self-consistent wave-functions, although of course normalised, are not strictly orthogonal. They consider, however, that the errors introduced by this are probably smaller than those due to uncertainty in the values of the wave-functions. The general agreement between the values of g_K as calculated in this way and as calculated by Hönl's method is in fact remarkable.

(*o*) *The quadrupole terms in the scattering factor:* So far, our calculations of the dispersion terms have been based on the assumption that λ, the wave-length of the incident radiation, may be considered as very large in comparison with the dimensions of the electron shell concerned. To conclude the theoretical discussion of dispersion, we must see what errors are introduced by this assumption. This problem also has been dealt with by Hönl.[19] It involves detailed discussion of the matrix-elements of equation (*4*.22), which determine the oscillator strengths, and we must be content here with a very brief outline of the results. For particulars, reference must be made to the original paper.

The effects due to the K shell are considered, and if a is the mean

* D. R. Hartree and M. Black, *Proc. Roy. Soc.*, **A, 139**, 311 (1932).

† D. R. Hartree, *Proc. Roy. Soc.*, **A, 141**, 282 (1933); **A, 143**, 506 (1934).

radius of the shell,* a/λ is still assumed to be small, although not now negligible, so that its increasing powers decrease rapidly in magnitude. This assumption can always safely be made in practice.

The quantities to be evaluated are

$$\mathbf{B}^0_{nk} = \int \psi_k^* \, e^{i\kappa \mathbf{s}_0 \cdot \mathbf{r}} \, \text{grad} \, \psi_n \, dv,$$

and $\mathbf{B}^*_{nk}$, which is the same integral, with $\mathbf{s}$, the unit vector in the direction of scattering replacing $\mathbf{s}_0$, that in the direction of incidence. The gradients in $\mathbf{B}^0_{nk}$ and $\mathbf{B}^*_{nk}$ are to be taken parallel to the direction of the incident electric vector and perpendicular to the direction of scattering respectively.

The wave-function ψ_n in the case we are discussing is that corresponding to the ground state, 1*s*, of a hydrogen-like electron; while ψ_k corresponds to one of the states of positive energy of the same atom. Such a wave-function may be written as the product of two functions; one, which is not quantised, and determines the energy of the state, depending only on *r*, the radius from the centre of the atom; the other, a function only of the angular co-ordinates, a spherical harmonic of the type $P_l^m(\cos\theta)\begin{matrix}\cos m\phi\\ \sin m\phi\end{matrix}$, determined by the two quantum numbers *l* and *m*. For every energy-value of the continuum of energy states there are still a number of different wave-functions, corresponding to the different values of *l* and *m*.

It can be shown that $\mathbf{B}^0_{nk}$ differs from zero only if $m=1$; but *l* may still take any value except zero, which is excluded, since $l \nless m$. For each value of the energy of the excited state there is thus a series of values of the matrix-elements, one for each value of *l*. For any given value of *l*, we may write

$$(\mathbf{E}_0 \cdot \mathbf{B}^0_{nk})\mathbf{B}^*_{nk}/|\mathbf{E}_0| = |A_l|^2 \Phi_l, \tag{4.63}$$

where A_l depends on the energy of the excited state, and Φ_l is an angular function, depending on the direction of polarisation of the incident radiation, and the angle of scattering 2θ.

By (*4*.22), therefore, for any given frequency ω of the K continuum, the oscillator-density is proportional to $\sum_l |A_l|^2 \Phi_l/\omega$, and the oscillator strength will be the sum of a set of integrals, one for each value of *l*.

Hönl shows that the angular factor Φ_l differs in form for different values of *l*. If $l=1$, Φ_l is unity when the incident electric vector is perpendicular to the plane containing the directions of incidence and scattering, and $\cos 2\theta$ if it lies in this plane. This is the angular dependence on the direction of polarisation characteristic of a dipole oscillator, and we may call the terms for which $l=1$ the *dipole terms*.

* More precisely, a is the parameter in the hydrogen-like wave-function for the K electron, $\psi_0 \sim e^{-r/a}$, and has the value $a_0/(Z-s)$, a_0 being the radius of the first Bohr hydrogen orbit, 0·532A.

When $l=2$, the values of Φ_l for the perpendicular and parallel directions of the electric vector are $\cos 2\theta$ and $\cos 4\theta$ respectively, and these values are characteristic of a quadrupole oscillator. The terms for which $l=2$ may therefore be called the *quadrupole terms*; and similarly, $l=3$ gives *octopole terms*. In practice, as we shall see, even the quadrupole terms are very small, and those for higher values of l are quite negligible.

We have now to consider the quantities $|A_l|^2$. Hönl shows that they may be expanded in ascending powers of κ^2, where $\kappa=2\pi a/\lambda$, provided that κ^2 is small. In his notation

$$\begin{aligned} |A_1|^2 &= f_1^{(0)} + \kappa^2 f_1^{(2)} + \kappa^4 f_1^{(4)} + \ldots && \text{Dipole terms} \\ |A_2|^2 &= \kappa^2 f_2^{(2)} + \kappa^4 f_2^{(4)} + \ldots && \text{Quadrupole terms} \\ |A_3|^2 &= \kappa^4 f_3^{(4)} + \ldots && \text{Octopole terms} \end{aligned} \tag{4.64}$$

The largest quadrupole term is thus proportional to κ^2, and the largest octopole term to κ^4, so that in comparison with the largest dipole term they are of small importance.

For the coefficients in (*4*.64) Hönl gives the values

$$\begin{aligned} f_1^{(0)} &= \frac{m}{\hbar^2}\frac{2^6}{3}\frac{1}{z^3}g(z) \\ f_1^{(2)} &= \frac{m}{\hbar^2}\frac{2^8}{15}\frac{z-2}{z^5}g(z) \\ f_2^{(2)} &= \frac{m}{\hbar^2}\frac{2^8}{15}\frac{4z-3}{z^5}g(z), \end{aligned} \tag{4.65}$$

where z and $g(z)$ have the same significance as in (*4*.53). For the limiting case of very long waves, all the terms in (*4*.64) vanish except $f_1^{(0)}$, the first of the dipole terms in $|A_1|^2$, and the oscillator-density must in this case reduce to that given by equation (*4*.53). By comparing (*4*.65) and (*4*.53), we see that $z f_1^{(0)}$ is, apart from the polarisation factor Φ_1, proportional to dg/dz, the oscillator-density for the case of long waves. If we include in the expression for dg/dz the terms involving κ^2 as well, we obtain from (*4*.65) for the two K electrons:

(1) *Dipole terms*

$$\frac{dg}{dz}=\frac{2^8}{3}\frac{g(z)}{z^4}\Phi_1+\kappa^2\frac{2^{10}}{15}\frac{z-2}{z^6}g(z)\Phi_1; \tag{4.66a}$$

(2) *Quadrupole terms*

$$\frac{dg}{dz}=\kappa^2\frac{2^{10}}{15}\frac{4z-3}{z^6}g(z)\,\Phi_2. \tag{4.66b}$$

If the approximate value for $g(z)$ from (*4*.55) is used, the terms involving κ^2, which are the correction to be added to the oscillator-density applicable to very long waves, are:

Dipole terms

$$\left(\frac{dg}{dz}\right)_1 = \frac{2^{10}e^{-4}}{45}\left(\frac{4}{z^4} - \frac{9}{z^5} + \frac{2}{z^6}\right)\Phi_1\kappa^2;$$

Quadrupole terms

$$\left(\frac{dg}{dz}\right)_2 = \frac{2^{10}e^{-4}}{45}\left(\frac{16}{z^4} - \frac{16}{z^5} + \frac{3}{z^6}\right)\Phi_2\kappa^2. \qquad (4.67)$$

These expressions apply to the hydrogen-like electron. To allow for the influence of the other electrons, we proceed exactly as in IV, § 1(*n*), p. 159, expressing the oscillator-density at the frequency ω in the K continuum in terms of ω and ω_K, the actual K absorption frequency, by writing $z=(1-\delta_K)\omega/\omega_K$ in (4.67). Similarly, to obtain the additions to the total oscillator strength, $(\delta g_K)_1$ and $(\delta g_K)_2$, on account of the new terms, we integrate (4.67) with respect to z between the limits $1-\delta_K$ and infinity. It is not necessary to give the resulting expressions explicitly; they are numerically very small in all practical cases. Suppose, for example, the scattering atom to be iron. From Table IV. 1, p. 159, $\delta_K=0{\cdot}215$, and on performing the necessary integration, and substituting this value, we find

$$(\delta g_K)_1 = -0{\cdot}762\Phi_1\kappa^2 \quad ; \quad (\delta g_K)_2 = +1{\cdot}044\Phi_2\kappa^2.$$

For Fe, Z=26, and $a=0{\cdot}0207$A. For Mo Kα($\lambda=0{\cdot}71$A) we then have $\kappa^2=0{\cdot}033$; while for wave-lengths in the region of the K absorption edge for iron ($\lambda=1{\cdot}74$A), where the other factors in the correcting terms are by far the largest, we find $\kappa^2=0{\cdot}0056$; so that the error made by neglecting these terms in the oscillator strength, and consequently in f, will be much smaller than the present limits of uncertainty in the determination, either theoretical or experimental, of these quantities.

The correction to $\Delta f'_K$, the real part of the dispersion terms, is obtained by performing the integration (4.43) using the additional terms in the oscillator-density. The integrals are all of a simple type, but the resulting expressions are rather cumbersome, and we shall refer the reader to Hönl's paper for particulars. The general order of magnitude of the change in $\Delta f'_K$ is roughly $\Delta f'_K \delta g_K/g_K$, which again is of the order of $\kappa^2\Delta f'_K$, a quantity of little importance in the present state of our knowledge.

In addition to the correcting terms in the real part of Δf_K, there will be corresponding terms in the imaginary part $\Delta f''_K$. These may be calculated at once from (4.67), by means of (4.43a), which gives

$$\Delta f''_K = \frac{\pi}{2} z_i\, dg/dz_i,\ (z_i=\omega_i/\omega_K) \text{ if } z_i>1;\ \text{ and } \Delta f''_K=0, \text{ if } z_i<1.$$

(*p*) *The total scattering factor, including quadrupole terms:* We may now calculate the total scattering factor, including both the real and imaginary parts of the dispersion terms, and the polarisation factors. Although the terms involving κ^2 are very small, as we have seen, we shall retain them in order to show the effect of such terms on the form of the total scattering factor. For the whole scattering factor, assuming any direction of polarisation, let us write

$$f=(f_0+\xi_1^{(0)}+\kappa^2\xi_1^{(2)})\Phi_1+\kappa^2\xi_2^{(2)}\Phi_2+i\{(\eta_1^{(0)}+\kappa^2\eta_1^{(2)})\Phi_1+\kappa^2\eta_2^{(2)}\Phi_2\}. \tag{4.68}$$

We have here followed in part Hönl's notation for the dispersion terms. $\xi_1^{(0)}$ and $\eta_1^{(0)}$ are the quantities we have previously denoted by $\Delta f'$ and $\Delta f''$. They are the sums for all the absorption edges of the real and imaginary parts of the dispersion terms respectively, assuming the incident wave-length to be very large in comparison with the dimensions of the K shell. The terms $\kappa^2\xi_1^{(2)}$ and $\kappa^2\xi_2^{(2)}$ are the corresponding additions to the real part of the scattering factor due respectively to the dipole and quadrupole terms of (4.67); $\kappa^2\eta_1^{(2)}$ and $\kappa^2\eta_2^{(2)}$ are the corresponding imaginary terms.

Let f_1 and f_2 be the scattering factors when the incident electric vector is respectively perpendicular and parallel to the plane of scattering. The values of Φ_1 and Φ_2 are, for f_1, unity and $\cos 2\theta$; and for f_2, $\cos 2\theta$ and $\cos 4\theta$. Then, from (4.68),

$$\begin{aligned} f_1&=f_0+\xi_1^{(0)}+\kappa^2\xi_1^{(2)}+\kappa^2\xi_2^{(2)}+i(\eta_1^{(0)}+\kappa^2\eta_1^{(2)}+\kappa^2\eta_2^{(2)}\cos 2\theta)\\ f_2&=(f_0+\xi_1^{(0)}+\kappa^2\xi_1^{(2)})\cos 2\theta+\kappa^2\xi_2^{(2)}\cos 4\theta\\ &\qquad+i\{(\eta_1^{(0)}+\kappa^2\eta_1^{(2)})\cos 2\theta+\kappa^2\eta_2^{(2)}\cos 4\theta\}. \end{aligned} \tag{4.69}$$

For unpolarised radiation, the total *intensity* factor for scattering is $\frac{1}{2}\{|f_1|^2+|f_2|^2\}$, and this is equal to $\frac{1}{2}|f|^2(1+\cos^2 2\theta)$, if f is the scattering factor defined in the usual way, in terms of the scattering by a classical electron.

From (4.69), retaining the terms in κ^2, we obtain

$$|f|^2\frac{1+\cos^2 2\theta}{2}=(\mathrm{A}+\kappa^2\mathrm{B})\frac{1+\cos^2 2\theta}{2}+2\kappa^2\mathrm{C}\cos^3 2\theta, \tag{4.70}$$

where

$$\begin{aligned} \mathrm{A}&=(f_0+\xi_1^{(0)})^2+(\eta_1^{(0)})^2.\\ \mathrm{B}&=2(f_0+\xi_1^{(0)})\xi_1^{(2)}+2\eta_1^{(0)}\eta_1^{(2)}.\\ \mathrm{C}&=(f_0+\xi_1^{(0)})\xi_2^{(2)}+\eta_1^{(0)}\eta_2^{(2)}. \end{aligned} \tag{4.71}$$

From (4.69), (4.70) and (4.71), after a certain amount of reduction, treating products of the quantities ξ and η as small, and retaining terms up to κ^2, we get for δf, the difference between the total scattering factor $|f|$ and f_0, the limiting value of f for very short waves,

$$\delta f=|f|-f_0=\Delta f'+\frac{1}{2}\frac{(\Delta f'')^2}{f_0+\Delta f'}+\kappa^2\left(\xi_1^{(2)}+\xi_2^{(2)}\frac{2\cos^3 2\theta}{1+\cos^2 2\theta}\right). \tag{4.72}$$

In equation (4.72), we have again written for $\xi_1^{(0)}$ and $\eta_1^{(0)}$ the values

$\Delta f'$ and $\Delta f''$, the real and imaginary parts of the dispersion terms for very long waves. The first two terms give the value of δf, assuming long waves. It was this that was plotted in fig. 53. The correction that has to be applied if the dimensions of the electronic system concerned are not negligible in comparison with λ is contained in the term depending on κ^2, which represents the whole difference between the result of the simpler considerations of IV, § 1(n), and the detailed discussion of the present section.

(q) *The dependence of* δf *on the angle of scattering :* The quantity δf represents that part of the total scattering factor $|f|$ which depends on the frequency of the incident radiation. From (*4*.72), it will be seen that this correction does not depend very greatly on the angle of scattering 2θ.

The first term $\Delta f'$, which is by far the most important, does not depend on 2θ. The second term, which contains $\Delta f''$, that part of the scattering factor due to radiation scattered $\pi/2$ out of phase with the incident wave, will depend on the angle of scattering; for it contains in the denominator f_0, which diminishes as 2θ increases. This term will therefore increase with increasing angle of scattering. We have already discussed this from a rather different point of view in IV, § 1(l), and the difference between the values of δf for two angles of scattering for a typical case is shown in fig. 53. The effect is appreciable but not very large, amounting perhaps in some cases to half a unit in f, between small and large angles of scattering.

The quadrupole term involving $\kappa^2 \xi_2^{(2)}$ also depends on 2θ, but is in general very small indeed in comparison with $\Delta f'$. It will have its largest values for wave-lengths in the neighbourhood of the critical absorption wave-lengths of the atom. The best chance of detecting its existence experimentally would be given by examining the scattering factor for λ just longer than λ_K say; for then the absorption term, which also depends on the angle of scattering, will be zero for the K absorption, and will have only very small values corresponding to the distant L and M absorption edges, while the terms involving κ^2 will have the relatively large values due to the anomalous dispersion effect. Hönl gives the values of the terms in (*4*.72) for Zn Kα radiation scattered by copper ($\lambda_K = 1{\cdot}378$, $\lambda/\lambda_K = 1{\cdot}038$), a case examined by Rusterholz.[20] He finds

$$\delta f = -3{\cdot}50 + \frac{0{\cdot}026}{f_0 - 3{\cdot}50} - 0{\cdot}016 \frac{2\cos^3 2\theta}{1 + \cos^2 2\theta}.$$

Even in this favourable case, the effect of the quadrupole terms is far smaller than the possible experimental error. Rusterholz finds $\delta f = -3{\cdot}68$, which is in excellent general agreement with the theory, but the uncertainties in the measurements probably amount to several tenths of a unit, and evidently there is no possibility of detecting the quadrupole term merely from its variation with the angle of scattering, especially as the variation of f_0 itself with the angle of scattering is not known with the necessary accuracy. At present the quadrupole term may safely be neglected in quantitative work on scattering.

2. Experimental Verification of the Dispersion Formulae

(*a*) *Introductory:* The experiments which provide the material for testing the dispersion formulae that have been developed in the first part of this chapter are of two main types, (1) the measurement of refractive indices of materials for X-rays, and (2) the measurement of the atomic scattering factor for wave-lengths in the neighbourhood of an absorption edge. We shall consider in turn examples of each type of experiment, beginning with the measurement of refractive indices.

We have seen in Chapter II, § 3(*b*), that n, the refractive index of a substance for X-rays of wave-length λ, may be written in the form

$$n = 1 - \delta = 1 - \frac{\lambda^2 e^2}{2\pi mc^2} \sum_a \mathrm{N}_a f_a(0), \tag{4.73}$$

where $f_a(0)$ is the scattering factor for the direction of the incident radiation of an atom of type (a), of which there are N_a in unit volume of the material. The sum is to be taken over all the different types of atom present in the substance. If λ is much shorter than the wave-length of any absorption edge of the atom (a), $f_a(0)$ may be taken as equal to Z_a, the number of electrons in the atom; and if this is true for all the atoms in the substance, δ/λ^2 should have a constant value, independent of the wave-length, which can be calculated from the density and chemical composition of the substance. But as λ approaches an absorption edge, $f(0)$ for the atom in question begins to depend on the frequency, and we have to write for $f(0)$ the more accurate form $\mathrm{Z} + \Delta f$, where Δf is the dispersion correction given by (*4*.43) and (*4*.43a). Equation (*4*.73) then becomes

$$n = 1 - \delta = 1 - \frac{\lambda^2 e^2}{2\pi mc^2} \sum_a (\mathrm{Z}_a + \Delta f_a)\mathrm{N}_a; \tag{4.74}$$

δ/λ^2 will no longer be constant, and from its variation Δf can be determined. Measurements of the refractive index could thus provide a check on the dispersion formulae.

Such measurements require very accurate work, for the refractive indices of most substances for X-rays differ from unity by only a few parts in a million. In all practical cases δ is positive, so that the refractive index is less than unity. It is true that the contribution to Δf from a single electron group may become negative for wave-lengths near an absorption edge, but the negative contribution will not outweigh the positive contributions from Z and from the other groups.

Another point to notice is that δ is in general complex. If we put $\delta = \alpha + i\beta$, then, by (*4*.9) of IV, § 1(*c*),

$$\beta = \frac{\lambda}{4\pi}\mu = \frac{\lambda}{4\pi}\sum_a N_a \mu_a, \qquad (4.75)$$

where μ is the linear absorption coefficient of the material, and μ_a is the atomic absorption coefficient for the atom of type (*a*). Since there is some absorption for all wave-lengths, β will never be zero, although it will not be large unless λ is just less than one of the absorption wave-lengths of an atom of the material. We are thus in general concerned with refraction by an absorbing medium, and it is necessary to be clear as to what is measured in an experiment of any given type.

In strictness, the refractive index of an absorbing medium, as measured by the ratio $\sin i/\sin r$, depends on the angle of incidence, and Snell's law is not exactly obeyed.* With α and β as small as they are for X-rays, however, the variations of the effective refractive index from a constant value are of order α^2 and β^2, and may be neglected. Experiments that deal with angular deviations give effectively the real part of the refractive index, $1 - \alpha$, and determine $\Delta f'$, the real part of Δf. The absorption is, however, more important when the refractive index is determined by measuring the critical angle; for it influences considerably the sharpness of the limit of total reflection, and reduces the possible accuracy obtainable by such methods.

(*b*) *Measurement of the refractive index from the deviations from Bragg's law:* That the phase velocity of an X-ray wave in a crystal should differ from *c*, its velocity in free space, was shown by Darwin as early as 1913, as we have already seen in Chapter II. Ewald's dynamical theory, published independently of, and soon after, Darwin's work, gives the same result. Both these writers showed that the Bragg reflection law, $2d \sin\theta = m\lambda$, is only approximately true, and obtained a corrected reflection condition, which may be written in the form

$$m\lambda = 2d(1 - \delta/\sin^2\theta_m)\sin\theta_m. \qquad (4.76)$$

These theoretical predictions seem, however, to have been lost sight of, and the deviation from the Bragg law was rediscovered experimentally in Lund in 1919 by Stenström,[21] who, in accurate measurements of X-ray wave-lengths of the order of several Ångström units by reflection from crystals, found that the wave-length determined was smaller the higher the order of the spectrum used in the determination. This variation he correctly ascribed to the refraction of the X-rays by the crystal, but the differences depended on angular measurements of the order of a few seconds of arc, and no really reliable value of δ could be deduced from them.

Stenström's results were soon confirmed by Duane and Patterson [22]

* See for example, C. Schaefer, *Einführung in die theoretische Physik*, Vol. 3, Part I, p. 431.

and by Hjalmar,[23] with different wave-lengths and crystals. Ewald [24] then pointed out that such results were to be expected theoretically, and showed that the deviations observed by Hjalmar were of the right order of magnitude.

Qualitatively, the results may easily be understood from equation (*4*.76). According to this, the value of θ is greater for all orders than that given by Bragg's law, but the deviation decreases with increasing order; so that the observed wave-length also decreases. A detailed account of this early work will be found in Siegbahn's book, *Spektroskopie der Röntgenstrahlen*, *2nd* Edn., 1930.

In all these experiments, the crystal surface was parallel to the reflecting planes. The incident beam, when the crystal was set to reflect, was therefore inclined at an angle of several degrees to the surface, and the bending due to refraction as it entered the crystal was very small. Davis and Terrill,[25] in 1922, pointed out that the deviation could be greatly increased by grinding the surface at such an angle to the reflecting planes that the incident beam was inclined at only a very small angle to it when the crystal was in the reflecting position.

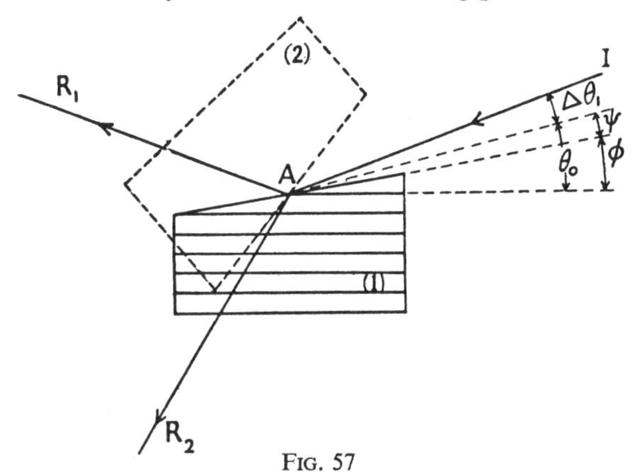

FIG. 57

The arrangement may be understood by referring to fig. 57. The surface of the crystal is inclined at an angle ϕ to the reflecting planes, and in position (1) of the crystal, the incident beam IA makes the smaller angle, and the reflected beam AR_1 the larger angle with it. Let ψ be the angle between the surface and the direction of the incident beam predicted by the Bragg law, so that $\psi+\phi=\theta_0$, the Bragg angle; and let $\Delta\theta_1$ be the angle between this direction and the actual direction of incidence when reflection occurs. We have already dealt with the theory of such unsymmetrical reflections in Chapter II, § 4(*l*), when discussing the dynamical theory of reflection. By equation (*2*.113),

$$\Delta\theta_1=\delta(\cot\psi+\tan\theta_0)=\delta\{\cot(\theta_0-\phi)+\tan\theta_0\}, \qquad (4.77)$$

and it is plain that $\Delta\theta_1$ increases rapidly with decreasing ψ.

In order to measure the deviation, Davis and von Nardroff [26] used the following device. Suppose the crystal to be rotated about an axis through A, and normal to the plane of incidence, into position (2), in which it again reflects, the reflected beam being now AR_2. In this position, the incident beam makes the larger angle with the surface, and the reflected beam the smaller one. $\Delta\theta_2$, the deviation of the setting of the crystal from that appropriate to the Bragg angle, is now given by

$$\Delta\theta_2 = \delta\{\cot(\theta_0 + \phi) + \tan\theta_0\}. \tag{4.78}$$

The angle between the two settings (1) and (2) of the crystal is

$$180° - (2\theta_0 + \Delta\theta_1 + \Delta\theta_2),$$

and this angle was measured to within about half a second of arc by Davis and von Nardroff, who used a specially accurate X-ray spectrometer. To deduce the values of θ_0 and δ from the measurements, we may use the following method, due to Ewald,[28] which is substantially equivalent to that used by Davis and von Nardroff.

Let us put $\alpha_\phi = \frac{1}{2}(\Delta\theta_1 + \Delta\theta_2) + \theta_0$ and let α_0 be the value of α for symmetrical reflection, so that

$$\alpha_0 - \theta_0 = 2\delta/\sin 2\theta_0. \tag{4.79}$$

From equations (*4*.77), (*4*.78), and (*4*.79) it is then easy to show that

$$(\alpha_\phi - \alpha_0)\tan^2\theta_0 = (\alpha_\phi - \theta_0)\tan^2\phi, \tag{4.80}$$

which can be solved for θ_0 by successive approximations, since the value of θ_0 is already known with considerable accuracy. Once θ_0 has been found, the value of δ may be calculated from (*4*.77) and (*4*.78).

Some values obtained by von Nardroff [27] for the reflection of Mo $K\alpha_1$ ($\lambda = 0{\cdot}7078$A) from the cube planes of a crystal of iron pyrites, FeS_2, are given in Table IV. 5. The increase of the deviation as ψ approaches θ_0 is plain. Similar experiments with copper radiation reflected by iron pyrites gave $\delta = (17{\cdot}6 \pm 0{\cdot}5).10^{-6}$.

TABLE IV. 5

Mo $K\alpha_1$ ($\lambda = 0{\cdot}7078$A) reflected from FeS_2

ϕ	α_ϕ	$\frac{1}{2}(\Delta\theta_1 + \Delta\theta_2)$	$\Delta\theta_1$	$\delta \times 10^6$
0° (Natural face)	7° 31′ 28″	5″		
6° 31′ 57·5″	7° 31′ 43·5″	20·6″	39″	3·3
7° 18′ 39″	7° 32′ 48·5″	85·6″	169″	3·03

$\theta_0 = 7° \; 31' \; 23{\cdot}0''$

A variation of the method, in which the crystal is rotated about an axis perpendicular to the reflecting planes, has been used by Hatley [29] with a calcite crystal reflecting Mo $K\alpha_1$. He obtained

$$\delta = (2{\cdot}03 \pm 0{\cdot}09).10^{-6}.$$

Field and Lindsay [42] have recently used the method of unsymmetrical reflection in measuring the refractive index of cerussite ($PbCO_3$) for X-rays over the ranges of wave-length 2·5 to 3·6A and 4·8 to 5·4A. Within these limits of wave-length lie the five M absorption edges of lead. Instead of grinding a surface on the crystal they used a natural face that made a suitable angle with the atomic planes at which reflection was observed. The results they obtained, while in general agreement with theory, do not give any very definite information about the anomalous dispersion associated with the M absorption of lead.

(*c*) *Comparison with theory :* The K absorption edges of Fe and S are at 1·739A and 5·009A respectively. With Mo Kα radiation ($\lambda = 0{\cdot}708$A) the dispersion corrections will be very small. From Hönl's values, given in Table IV. 1, we estimate the value of $\Delta f'$ to be 0·2 for Fe and 0·1 for S, giving for $Z + \Delta f'$ for FeS_2 the value 58·4. The lattice spacing of the cube planes is 2·703A, and the cube having this edge contains half a molecule of FeS_2; so that $N = 2{\cdot}520 \times 10^{23}$. Substitution of these values in (*4*.74) gives $\delta = 3{\cdot}31 \times 10^{-6}$, in good agreement with the values obtained experimentally.

The value calculated for Cu Kα_1 radiation in the same way is $15{\cdot}2 \times 10^{-6}$. Here there is a rather bigger dispersion correction, since λ/λ_K for Fe with Cu Kα radiation is 0·88, which gives $\Delta f' = -1{\cdot}4$. In the case of Hatley's measurements with calcite, the dispersion corrections are negligible, and the calculated value of δ is $1{\cdot}93 \times 10^{-6}$. The general agreement between theory and experiment is on the whole as good as can be expected, considering the difficulty of the measurements. Such experiments are, however, not very well suited to the investigation of the dispersion formulae, for which it is necessary to measure the refractive indices for a variety of wave-lengths, including those near the absorption edge.

The methods that we have discussed in this section depend essentially on the crystalline nature of the refracting medium, and require also that the interference conditions should be fulfilled during the measurements. Now matter refracts X-rays whether it is crystalline or not. In fact, as we saw when dealing with the dynamical theory of interference, it is strictly only when the interference conditions are *not* fulfilled that we may consider the crystal as having a definite refractive index. The two remaining methods that we shall discuss apply to non-crystalline materials, such as glass, or to micro-crystalline materials, as well as to single crystals.

(*d*) *Refractive indices by total reflection :* In 1922, A. H. Compton [30] pointed out that if the refractive index of a substance for X-rays is less than unity it ought to be possible to obtain total *external* reflection from a smooth surface of it, since the X-rays, on entering the substance from the air, are going into a medium of smaller refractive index. If we assume the substance not to absorb the rays, so that δ is real, the laws of refraction lead at once to the result that X-rays should be totally reflected from a plane surface at all glancing angles smaller than a certain critical angle θ_c, given by

$$\cos \theta_c = 1 - \delta,$$

which, since δ, and hence θ_c, are small, gives, on expanding the cosine, quite nearly enough

$$\theta_c = \sqrt{2\delta}. \tag{4.81}$$

For example, if δ for the material is 3×10^{-6}, $\theta_c = 2{\cdot}4 \times 10^{-3}$ radians, or about 8 minutes of arc. The glancing angles of total reflection therefore, although small, are easily observable, and Compton at once verified his prediction experimentally, showing that the tungsten L line, $\lambda = 1{\cdot}279$A, was totally reflected from a sheet of plate glass at an angle of $10'$, and from a sheet of silver at $22{\cdot}5'$, corresponding closely to the angles to be expected from the theoretical refractive indices. Some account of Compton's experiments has already been given in Appendix I of the first volume of *The Crystalline State*, and we need not consider them in greater detail here. We are more concerned with the applicability of the method to the testing of the dispersion formula, for which it seems at first sight to be well suited, since a wide choice of substances is available for reflectors, and the angular deviations to be expected are greater than those to be obtained by pure refraction at a surface. There are, however, difficulties which to a large extent discount these advantages.

As was pointed out by Prins,[12] the sharpness of the limit of total reflection is much affected by absorption in the material; and since we are mainly interested in the refractive index for wave-lengths in the neighbourhood of the absorption edge, this is rather serious. The effect is well known from ordinary optical theory. Let us assume the Fresnel laws of reflection to apply also to the case of the reflection of X-rays. If the electric vector in the incident radiation is perpendicular to the plane of incidence, and A_0 and A_r are the amplitudes of the incident and reflected beams respectively, Fresnel's formula gives

$$A_r = \frac{\sin \theta_1 - n \sin \theta_2}{\sin \theta_1 + n \sin \theta_2} A_0, \tag{4.82}$$

where n is the refractive index, and θ_1 and θ_2 are respectively the glancing angles of incidence onto, and refraction into, the medium.

Now $\cos \theta_1 = n \cos \theta_2$, so that if $n = 1 - \delta$, (*4*.82) can be written

$$\frac{A_r}{A_0} = \frac{\sin \theta_1 - \sqrt{n^2 - \cos^2 \theta_1}}{\sin \theta_1 + \sqrt{n^2 - \cos^2 \theta_1}} = \frac{\theta_1 - \sqrt{\theta_1{}^2 - 2\delta}}{\theta_1 + \sqrt{\theta_1{}^2 - 2\delta}}, \tag{4.83}$$

since δ and θ_1 are small. The reflection coefficient for intensity is the square of the modulus of this expression, which becomes unity if

$2\delta > \theta_1^2$, provided that δ is real. This corresponds to total reflection. We have seen, however, that δ is in general complex, and equal to $\alpha + i\beta$. Equation (4.83) then gives for the ratio of the incident and reflected intensities

$$\frac{I_r}{I_0} = \left| \frac{\theta_1 - \sqrt{\theta_1^2 - 2\alpha - 2i\beta}}{\theta_1 + \sqrt{\theta_1^2 - 2\alpha - 2i\beta}} \right|^2, \tag{4.84}$$

which can be evaluated by putting

$$\sqrt{\theta_1^2 - 2\alpha - 2i\beta} = p + iq,$$

from which p and q are found by squaring and equating the real and imaginary parts. This gives

$$\frac{I_r}{I_0} = \frac{(\theta_1 - p)^2 + q^2}{(\theta_1 + p)^2 + q^2}, \tag{4.85}$$

where

$$p^2 = \tfrac{1}{2}\{\sqrt{(\theta_1^2 - 2\alpha)^2 + 4\beta^2} + \theta_1^2 - 2\alpha\},$$

$$q^2 = \tfrac{1}{2}\{\sqrt{(\theta_1^2 - 2\alpha)^2 + 4\beta^2} - \theta_1^2 + 2\alpha\}. \tag{4.86}$$

If the electric vector of the incident radiation lies in the plane of reflection, an expression slightly different in form from (4.85), but giving the same numerical values within the limits of experimental error, is obtained for the reflection coefficient. Similar expressions have been given by Forster,[31] Jentzsch,[32] and others.

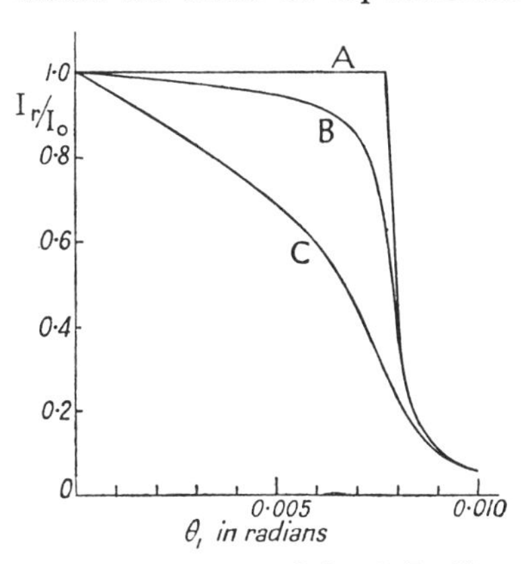

FIG. 58. Reflection coefficient I_r/I_0 of iron for X-rays as a function of the glancing angle. A, negligible absorption. B, weak absorption (long-wave side of Fe absorption edge). C, stronger absorption (short-wave side of Fe absorption edge). (After Prins)

The types of reflection curve given by (4.85) for zero, small, and large absorption are given in fig. 58, which is taken from the paper by Prins. For all three curves α is taken as 30×10^{-6}. For curve A, $\beta = 0$, and the absorption is zero. This corresponds to reflection from a completely transparent medium, and in this case there is a sharp limit of total reflection. For curve B, $\beta = 10^{-6}$, and for curve C, $\beta = 7 \times 10^{-6}$. Curve B thus applies approximately to the case of the reflection from iron of a wave-length rather longer than that of the K absorption edge of this element, and curve C to a wave-length rather shorter than that of the K edge. It is plain that when the absorption is large there is no definite critical angle.

These conclusions were verified by Prins [12] in a very interesting series of experiments in which X-rays of various wave-lengths were reflected from an iron mirror. In some of the experiments, continuous radiation was reflected from the mirror, which was caused to rotate so as to vary the angle of incidence. So long as the absorption is small, for each wave-length there is a certain critical angle of glancing incidence below which no reflection occurs. The reflected beam falls on a crystal, the plane of incidence for the latter being normal to that for the mirror, and thence on to a photographic plate. The method is essentially one of 'crossed spectra', analogous to Newton's well-known method of crossed prisms. The first spectrum is produced by the mirror, which gives a different limit of total reflection for each wave-length. The second spectrum is produced by the crystal, which spreads the reflected beam into a spectrum. The limit of total reflection appears on the photographic plate as a dispersion curve.

The difference in the sharpness of total reflection for wave-lengths on the two sides of the K absorption edge of the iron is also well shown in some of the photographs reproduced by Prins in his paper. The experiments gave definite evidence of the existence of anomalous dispersion, and Prins concluded from his results that the total oscillator strength for the K continuum of iron was 1·3, and not 2. That this number should be nearer one than two had already been suggested by Houston,[2] Kronig,[8] and others; the work of Prins brought new evidence of this from experiments of a different kind. We have already discussed the reason for the lower values of the oscillator strength; the lines along which the explanation had to be sought appear to have been pointed out first by Kronig and Kramers.[13]

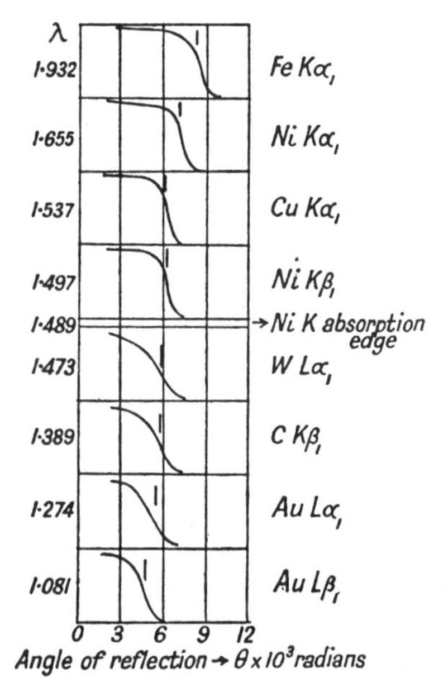

FIG. 59. Reflection curves from a nickel mirror for various wave-lengths (H. Kiessig). The short vertical lines denote $\theta_c = \sqrt{2\delta}$, the critical angle of reflection without absorption

Perhaps the most complete set of observations on total reflection in the neighbourhood of an absorption edge are those of Kiessig,[33] who used nickel mirrors deposited by evaporation on glass. The reflection curves were measured by means of an ionisation spectrometer, and a number of them are reproduced in fig. 59. The abscissae give the angles of reflection in radians (1 radian = $3{\cdot}44 \times 10^3$ minutes of arc), and the ordinates the reflection coefficients. The difference in the general shape of the curves for wave-lengths on the two sides of the K absorption edge of nickel is very well shown; but it is evident that even on the long wave-length side, where the absorption due to the K electrons is zero, the absorption due to the remaining electrons causes a very noticeable rounding off of the reflection curves.

In order to compare the results with theory, Kiessig assumed a value for β equal to $\mu\lambda/4\pi$, μ being the linear absorption coefficient of nickel for wave-length λ, and then chose the value of α so as to give the best fit between the experimental curve and that calculated from* (*4*.85). The values of α he obtained are given in Table IV. 6. The third column shows α/λ^2, which, in the absence of a dispersion effect, should be constant. The 'anomalous dispersion', indicated by the fall in the value of α/λ^2 on either side of the absorption edge, is very plainly shown.

In comparing the experimental values with theory, one difficulty is to know the value of the density of the nickel film. Kiessig found the best agreement if he assumed the density of the film to be 7·5, instead of 8·8, the usual value for nickel. He found the low value of the density to be confirmed by direct measurements of the thickness and weight of nickel films on glass.

TABLE IV. 6

Values of α from total reflection by nickel (Kiessig)

λ	$\alpha \times 10^6$	$(\alpha/\lambda^2) \times 10^{10}$(obs.)	$(\alpha/\lambda^2) \times 10^{10}$(Calc.)
1·922 A	34·5	9·25	8·75
1·655	25·0	9·13	8·56
1·537	18·7 †	7·92 †	8·22
1·497	18·7	8·35	7·64
1·489	absorption edge		
1·473	17·5	8·07	7·92
1·389	17·0	8·80	8·72
1·274	14·8	9·11	9·12
1·081	11·05	9·40	9·40

If thin enough films are used, the reflection curves show maxima and minima, which are due to interference between the rays reflected from the outer surface, and those reflected from the nickel-glass surface inside the film. Since δ is greater for nickel than for glass, the refractive index of the nickel for the rays is less than that of glass. At the nickel-glass surface, we therefore have the normal case of reflection from an optically denser medium. Because of the very oblique angle which the rays make with the film, the path difference between the beams reflected from the upper and lower surfaces is small enough to give interference,

* The formula used by Kiessig, although slightly different in form, is equivalent to (4.85).

† Made with a different nickel mirror.

even with the relatively short X-rays used in these experiments. Maxima and minima in the reflection curve were observed as the angle of incidence was altered, and from the spacing of these it was possible to determine the thickness of the film, and hence, by weighing the total amount of nickel, the average density. Values of the order of 7·9 were found. In the actual experiments to determine the refractive index, disturbances due to interference were avoided by using films so thick that no appreciable amount of radiation was reflected from the lower surface.

In the last column of Table IV. 6, the values of α/λ^2 for nickel, calculated from Hönl's dispersion formula, are given. It is not possible to give absolute values, owing to the uncertainty as to the density of the film, and so the observed and calculated values for the shortest wave-length have been made equal. This corresponds to a density of about 7·3. Actually, the longer wave-lengths should give more reliable values, because of the sharper limit to the total reflection, but the experimental figures seem to be more variable for this side of the absorption edge. The variations can hardly be real, and it is evident that the errors of experiment are still considerable. The difference between the values

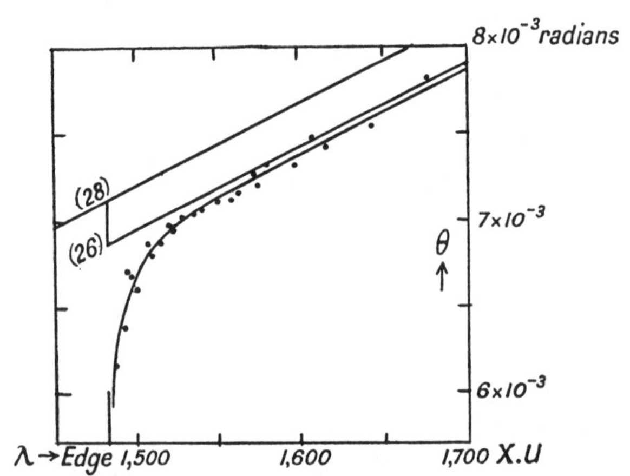

FIG. 60. Critical angle of total reflection as a function of the wave-length. Reflector, nickel ($\lambda_K = 1{\cdot}438$A). The straight lines are the theoretical curves for 26 and 28 free electrons (Lameris and J. Prins)
(*Physica*, **I**, 881 (1934))

of α/λ^2 for long and short-waves is evidently rather smaller according to these experiments than the theoretical work predicts. All that can be said is that the effect is of the right general order, and that anomalous dispersion is very clearly shown. Considering the difficulty of the work, the agreement between theory and experiment is probably as close as can be expected.

Lameris and Prins [34] have also made observations with nickel mirrors, using the method of crossed spectra described on p. 174. The curve which they obtain, giving the relation between the critical angle of total reflection and the wave-length, is reproduced in fig. 60. The variations that they obtain are smaller than those observed by Kiessig, and they attribute some of the discrepancies in his results to the fact that he used more than one mirror, and that the densities of the different mirrors may not have been the same. The values of α/λ^2 calculated by Lameris and Prins from their curve are, however, rather greater than those deduced from Hönl's theory.

The experiments described in this section show in a very interesting way the essential similarity between the properties of X-rays and those of light. They show examples of total reflection from an optically rarer medium, ordinary reflection from an optically denser medium, and interference in thin films. They show too that the Fresnel laws for the intensity of reflection at a surface are with a fair degree of exactness quantitatively true. This similarity is still further exemplified by the next method to be described.

(*e*) *Refractive indices from the deviations produced by prisms:* The first to determine the refractive indices of X-rays by directly measuring the deviation of a beam of rays by a prism were Larsson, Siegbahn, and Waller,[35] in 1924, and the method has since been used by Larsson,[36] Stauss,[37] Bearden,[38] and others. In principle it is very simple. A beam of X-rays falls on a right-angled prism, A, fig. 61, of the material to be investigated, at a small glancing angle of incidence, θ, which is

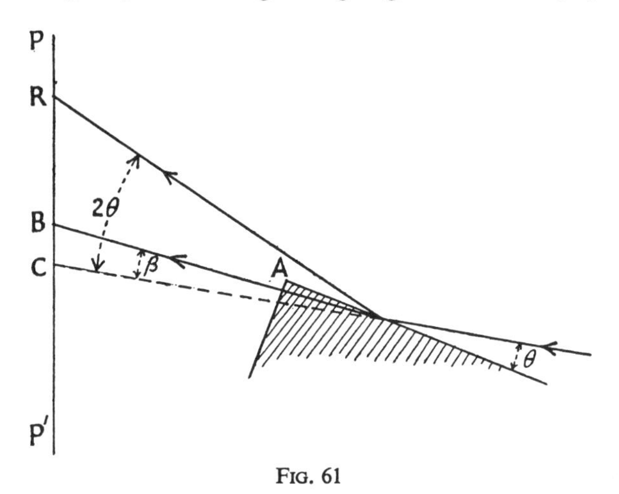

FIG. 61

just greater than the critical angle for the wave-length for which the refractive index is to be measured. Suppose this wave-length to be present in the incident radiation as a characteristic line spectrum, together with a certain amount of continuous background. The angle

θ will be less than the critical angle for some of the longer waves of this continuous spectrum, and so there will be a totally reflected beam, which falls on the photographic plate PP′ at R, fig. 61. The characteristic radiation is refracted on entering the prism, through a relatively large angle, because of the very small glancing angle of incidence. It meets the opposite face of the prism nearly normally, so that the deviation due to refraction is here negligible, and falls on the plate at B. Let C be the point at which the continuation of the incident beam meets the plate. Then RC gives 2θ, and BC gives β, the deviation of the incident beam on entering the prism. The real part of the refractive index, $1-\alpha$, is therefore given by

$$1-\alpha=\cos\theta/\cos(\theta-\beta),$$

or, since θ and β are small,

$$\alpha=\beta(\theta-\beta/2).$$

In practice, owing to the absorption of the X-rays, only the extreme edge of the prism can be used, and great care has to be taken to ensure that this edge is really sharp. Because of the strong absorption of the beam, however, that part of the radiation that just passes over the edge of the prism makes a very convenient record on the plate of the direct beam C.

Fig. 62 (Pl. II) is an example of a spectrogram obtained in this way by Bearden. The line-spectrum of small dispersion due to several components in the incident beam is clearly shown. In Table IV. 7, the values of α for calcite, obtained for a series of wave-lengths by Larsson, are given. The range of wave-lengths includes the K absorption edge for calcium, 3·064A, and the series of measurements is very suitable for testing the dispersion formula.

Column (4), Table IV. 7, shows the values of $Z+\Delta f'$, calculated from the experimental values by means of equation (*4*.74), which can be put in the more convenient form

$$Z+\Delta f'=2\pi\frac{W}{\rho F}\frac{m}{e}\frac{\alpha}{\lambda^2},$$

where F is Faraday's constant, 9648·9 abs. e.m.u., ρ is the density of the material, and W its molecular weight. The specific charge of the electron, e/m, is then to be expressed in e.m.u. We have used the value $1{\cdot}761\times10^7$ for this constant. The differences between the value of Z, which is 50 for $CaCO_3$, and the numbers in column (4) give the values of the total dispersion correction, $\Delta f'$, for the whole molecule. Most of this is due to the calcium atom, but the effect of oxygen is by no means negligible, and some allowance has been made for it by assuming Hönl's formula to hold for oxygen, in order to get the value of the correction. For the three oxygen atoms this correction is sometimes as high as half a unit in $\Delta f'$, even although the absorption edge of oxygen is at 23·5A. This is due to the rise in the dispersion curve to positive

PLATE II

FIG. 62. Spectrum of CuKα radiation produced by a prism of quartz. D, direct beam, R, reflected beam (Bearden)
(*Phys. Rev.*, **39,** 1 (1932))

Plate II — See other side

TABLE IV. 7

Refractive indices and anomalous dispersion of calcite for X-rays. Values of $\alpha = 1 - n$ (Larsson)

λ	$1/x = \lambda/\lambda_K$	$\alpha \times 10^6$	$Z + \Delta f'$	$\Delta f'$ (Ca)
1·539	0·502	8·80	50·43	0·26
1·934	0·631	13·89	50·39	0·16
2·498	0·815	22·37	48·66	− 1·68
2·509	0·819	23·26	50·12	− 0·23
2·774	0·905	27·05	47·71	− 2·70
2·931	0·956	28·54	45·09	− 5·35
3·025	0·989	29·34	43·52	− 6·94
3·040	0·992	28·67	42·11	− 8·35
3·070	1·002	30·18	43·46	− 7·00
3·083	1·006	32·02	45·65	− 4·82
3·218	1·050	35·98	47·16	− 3·34
3·378	1·102	39·57	47·07	− 3·46
3·447	1·125	41·87	47·83	− 2·72
3·734	1·218	49·19	47·88	− 2·74

values of $\Delta f'$ on the short-wave side of the absorption edge. Column (5) in the table gives the value of $\Delta f'$ for calcium, and the values so obtained are plotted in fig. 63 for comparison with the theoretical curve derived from Hönl's formula. It is evident that the general agree-

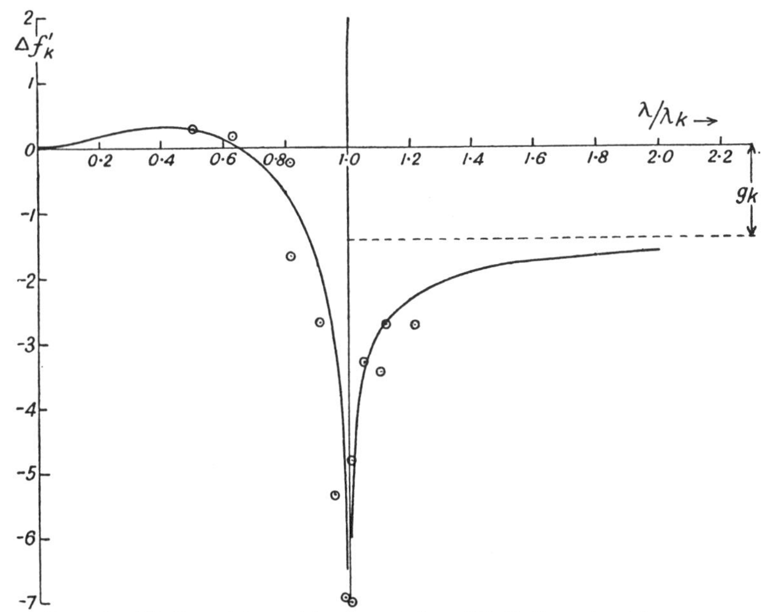

FIG. 63. $\Delta f'_K$ for calcium. The curve is from Hönl's formula. The points are from Larsson's experiments, corrected for the dispersion of oxygen and carbon

ment between theory and experiment is as close as can be expected. It is to be remembered that what is derived from the formula is not $\Delta f'$, but $50 + \Delta f'$; so that no great significance can be given to the absolute values within perhaps a tenth of a unit. Nevertheless, the points as plotted have not been brought into agreement with the curve at any place, and the fit, such as it is, is absolute. Since no allowance has been made for damping, the curve is not valid in the immediate neighbourhood of the absorption edge. The experiments show in a very striking way the essential correctness of the ideas underlying the theory of dispersion.

Bearden [39] has recently made very accurate determinations of the refractive index of quartz for Cu Kα and Kβ radiations, using this method. He obtains for Cu Kα, $\alpha = (8{\cdot}553 \pm 0{\cdot}006).10^{-6}$, which leads to 30·28 for $Z + \Delta f'$, Z for SiO_2 being 30. Bearden indeed uses these values to determine the wave-length of the radiation in order to decide between the value obtained from grating measurements, and that obtained from crystal measurements; but this would appear to be pushing belief in the dispersion formula rather far.

Another method of using direct refraction by a prism is due to Davis and Slack,[40] who interposed a prism in the path of the X-ray beam between the two crystals in a double-crystal spectrometer, and measured the resulting displacement of the maximum of reflection, which was of the order of 5 or 6 seconds of arc. Good values of the refractive index were obtained in this way, and the method has also been used by Bearden and Shaw,[41] with quartz. The values they obtained for the refractive index were in very close agreement with the values obtained earlier by Bearden by the direct refraction method described above.

It has not been possible to include in this account of the experiments on refractive indices of X-rays more than a fraction of the work that has been done in this field. The examples chosen are such as serve to illustrate the theory of anomalous dispersion, which is the main topic of this chapter. The reader is referred to the excellent account of the work on refractive indices given in Siegbahn's *Spektroskopie der Röntgenstrahlen.* Reference to a number of papers not explicitly quoted in the text will be found in the bibliography at the end of the chapter.[43]

(*f*) *Tests of the dispersion formulae from the measurement of atomic scattering factors:* In this section, we shall examine the evidence that has been obtained for the existence of anomalous dispersion from direct measurements of the atomic scattering factors of elements for wave-lengths in the neighbourhood of their critical absorption wave-lengths. Atomic scattering factors may be determined by measuring the intensities of X-ray spectra from crystals, and making use of one or other of the formulae deduced in Chapter II; and this would appear at first sight to be a simple method of testing the accuracy of the dispersion formulae considered in the first part of this chapter. In practice, however, the method has its difficulties, for it is not easy to make absolute measurements of the intensities of X-ray spectra, and the interpretation of such measurements is not always simple. We shall postpone to Chapter VI the consideration of the methods of measuring atomic scattering factors, but, for the sake of completeness, we must anticipate some of the results of such measurements here.

We have seen that it is the real part of the scattering factor that is determined from measurements of the refractive index. In measurements of the scattering factor based on determinations of the intensities of X-ray spectra, on the other hand, it is $|f|$, the modulus of the total scattering factor f, as defined in § 1(j) of this chapter. If f_0 is the scattering factor for waves so short that the dispersion terms are negligible, we can take

$$\delta f = |f| - f_0$$

as a measure of the dispersion correction as determined from intensity measurements. By equations (4.44) and (4.74), if $\Delta f'_K$ and $\Delta f''_K$ are the real and imaginary parts respectively of the dispersion terms associated with the K absorption edge, we can write as a good enough approximation for most purposes

$$\delta f_K = \Delta f'_K + \tfrac{1}{2}\frac{(\Delta f''_K)^2}{f_0 + \Delta f'_K}. \qquad (4.87)$$

$\Delta f''_K$ is zero if λ, the wave-length of the scattered radiation, is greater than λ_K, the K absorption wave-length of the scatterer. We shall use equation (4.87) in comparing the results of theory and experiment.

The first to obtain direct qualitative evidence of anomalous scattering in the neighbourhood of an absorption edge from the intensities of X-ray spectra were Mark and Szilard,[44] who in 1925 reflected the Kα radiation of strontium from the (111) face of a crystal of rubidium bromide (RbBr). The structure of the crystal is of the rock-salt type, and the structure factors of the spectra of odd indices are equal to the difference of the scattering factors of Br and Rb. Now the atoms may be assumed to be ionised in the crystal, and Rb^+ and Br^- contain the same number of electrons. The difference between the scattering factors for wave-lengths far from the absorption edges of either element will therefore be so small that spectra such as 111 or 333 should not appear. Mark and Szilard found them to be absent for the Kα radiations of Cu, Fe, Co, and Zn; with Sr Kα radiation, however, both 111 and 333 were quite measurable.

The K absorption edge of Rb lies at 0·814A, and that of Br at 0·918A, while the wave-length of Sr Kα radiation is 0·877A, and so lies between the two absorption edges. The scattering factor of each element will thus be considerably affected by anomalous dispersion. From a rough calculation based on Hönl's formula (4.61), we find the value of δf to be about $-2{\cdot}7$ for Br, and $-4{\cdot}0$ for Rb, so that there will be a great enough difference between the two scattering factors for Sr Kα radiation to give difference spectra for low orders.

The experiments of Coster, Knol, and Prins [45] on the difference between the reflecting powers of opposite octahedral faces of the polar zincblende crystal, which have already been described in Chapter II, § 1(*d*), were also evidence of the existence of anomalous scattering.

In 1929, Miss Armstrong [46] published a series of measurements of the f-factors of copper and iron for the Kα radiations of copper and molybdenum. The metals were investigated in the form of powder by Havighurst's method (Chapter II, § 2(*i*), equation (*2*.49)), and the values of the scattering factors were expressed in absolute measure by comparison with a rock-salt standard. The values of $|f|$ for both copper and iron were lower for Cu Kα radiation than for Mo Kα; the difference was greatest for copper, and amounted to about three units in $|f|$, being roughly independent of the angle of scattering.

The value of λ/λ_K, the ratio of the wave-length of the scattered radiation to that of the K absorption edge of the scatterer, is 0·515 for Mo Kα radiation and 1·12 for Cu Kα radiation scattered from copper. According to Hönl's formula, the scattering factor of copper for the Cu Kα radiation should be lowered by about 2·7 units, while that for Mo Kα should be raised by about 0·25 units with respect to the scattering factor for short-wave radiation. The author simply records her results, and does not ascribe the difference between the f-factors for the two radiations to the effects of dispersion.

Measurements such as those just considered are subject to some uncertainty, owing to the difficulty of allowing for absorption in the powder. Wyckoff,[47,48] shortly afterwards extended the work of Miss Armstrong by measuring the scattering factors of nickel and oxygen for the Kα radiations of Mo, Cu, and Ni, using powdered nickel oxide (NiO) as the scatterer. Nickel oxide is a crystal of the rock-salt type; the structure factors of its spectra are therefore of two kinds, $F_S = f_{Ni} + f_O$, and $F_D = f_{Ni} - f_O$, and if each is determined as a function of the angle of scattering it is possible, by taking half the sum and half the difference of the ordinates of the curves for F_S and F_D, to calculate the f-curves for nickel and oxygen separately. The curves for oxygen with the three radiations were very much the same, but those for nickel differed considerably, the curve given by molybdenum radiation being the highest, and that given by copper radiation the lowest. The curve given by nickel radiation occupied an intermediate position, and the differences between the three values of f were very nearly equal at all angles. The f-curves obtained by Wyckoff in these experiments are shown in fig. 64.

For oxygen as scatterer, λ/λ_K is 0·030 for Mo Kα, 0·065 for Cu Kα, and 0·070 for Ni Kα. For all these radiations therefore the value of δf is inappreciable. For nickel, the corresponding values of λ/λ_K for the three radiations are 0·478, 1·037, and 1·116. On reference to the curve for $\Delta f'$ for nickel in fig. 66,* it will be seen that δf for Mo Kα is about $+0{\cdot}25$, while for Cu Kα it is about $-3{\cdot}5$, and for Ni Kα about $-2{\cdot}7$. The results obtained by Wyckoff are thus in general accord

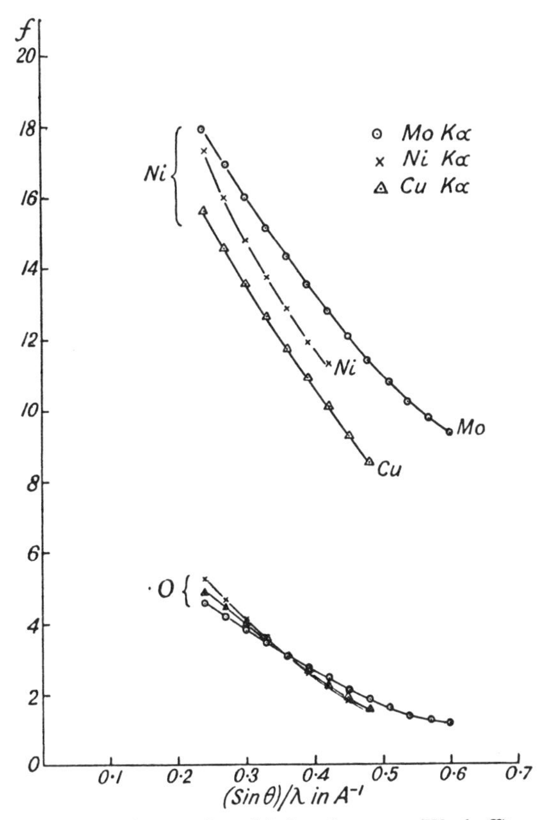

FIG. 64. f-curves for nickel and oxygen (Wyckoff).

with the theoretical expectations, although the observed values of δf are rather smaller than the calculated ones.

Wyckoff, in these experiments, used as a control the fact that $|f|$ for oxygen is found to be about the same for three radiations, all of which are far from its absorption frequency. The comparison of the results for oxygen with those for nickel is free from uncertainties due to absorption, because the nickel and oxygen are constituents of the same crystal, so that the absorption correction must be the same for both. On the other hand, because the two atoms are in the same crystal, the correction for the effect of temperature is difficult, as we shall see in the next chapter. This correction is, however, unlikely to be very

* The curve for Ni is not actually plotted. It lies so close to that for Fe that it could hardly be plotted separately on the scale of the figure.

large within the range of $(\sin\theta)/\lambda$ used in these experiments. The comparison with rock salt, on which the absolute value of $|f|$ depends, is of course still subject to the uncertainties due to absorption.

Further experiments by the same general methods, and with similar results, were made by Wyckoff,[49] using powders of metallic nickel, copper, and iron, and the K radiations from Mo, Cu, Ni, and Fe.

Bradley and Hope [50] used as scatterers powdered crystals of the alloy of iron and aluminium FeAl, the structure of which is of the caesium-chloride type. The intensities of the spectra were determined using the Kα radiations from Mo, Cu, Co, Fe, and Cr. The separate scattering factors for iron and aluminium were determined from the structure factors in a manner analogous to that outlined above for the case of nickel oxide. Absolute values were obtained by assuming the scattering factor for aluminium to be known from the work of James, Brindley, and Wood (cf. Chapter VI), in which the measured value of f for this element was shown to agree closely with the theoretical value, calculated by Hartree's method of the self-consistent field. In this way, the scattering factor for iron was determined as a function of the wave-length. The results for the 110 spectrum of iron are given in column (3) of Table IV. 8. The figures originally given by Bradley and Hope

TABLE IV. 8

The atomic scattering factor for iron. $|f_{110}|$

Radiation	λ/λ_K	Bradley and Hope	Wyckoff *	Glocker and Schäfer
Mo Kα	0·41	17·2	16·3	16·8
Cu Kα	0·89	16·1	11·8	11·2
Ni Kα	0·95	—	10·1	—
Co Kα	1·01	14·1	—	—
Fe Kα	1·10	15·7	13·8	13·6
Cr Kα	1·31	17·2	—	—

include no allowance for the dispersion terms in the scattering factor of aluminium, which are not entirely negligible for the wave-lengths used, and the figures in the table are taken from a later paper by Bradley and Rodgers,[51] and have been corrected for this effect. The fall in $|f|$ for wave-lengths in the neighbourhood of the absorption wave-length for iron is well shown, but the absolute agreement with theory is not very good. The high value for Cr Kα ($\lambda/\lambda_K = 1·31$) is almost certainly due to experimental error. It is not to be expected on any likely theory, and is not confirmed by the results obtained by other workers.

The work of Bradley and Hope is of interest both on account of the ingenuity of the methods used, and because it was undertaken with a

* Corrected for temperature.

view to applying the results to a problem of practical importance. Many alloys contain metals such as copper, zinc, and manganese, whose scattering powers are so similar for most wave-lengths that it is extremely difficult to distinguish between them in determining the crystal structures of the alloys. The magnetic Heusler alloys, for example, contain copper and manganese, and since the magnetic properties appear to be connected with the formation of a super-lattice, it is important in investigating these alloys to be able to distinguish between the positions of the copper and manganese atoms. Bradley and Rodgers [51] were able to do this by comparing the X-ray powder photographs obtained from the alloys with iron, copper, and zinc radiations. They found that the relative intensities of the weak spectra, in which the scattering factor of manganese nearly balanced that of copper, varied with the wave-length of the radiation employed, owing to the way in which the scattering powers of the two elements varied on account of anomalous dispersion; and from these variations in intensity the structure was determined.

Measurements of the scattering power of iron for different wave-lengths in the neighbourhood of the absorption edge were also made by Glocker and Schäfer.[52] They used powdered iron mixed with powdered aluminium, which served as a standard. The values they obtained were, however, very much lower than was to be expected theoretically, expecially on the short-wave side of the absorption edge. There is little doubt that their results were vitiated by an incorrect allowance for absorption in the mixed powder. Very great care is necessary in using this method, in which one powder is taken as the standard substance; and unless the powder has at least a certain degree of fineness, wholly misleading results may be obtained by assuming the effective absorption coefficient of the mixed powder to be that calculated from the relative proportions of the two constituents. Glocker and Schäfer themselves realised the cause of the discrepancies in their first results, and a discussion of it is given by Schäfer [53] in a later paper. Their amended values are given in column (5) of Table V. 8. The technique of the method of mixed powders has been developed by Brentano,[54] by Rusterholz,[55] and by Brindley and Spiers,[56] and if proper precautions are taken it can give very good results. The question is discussed in rather greater detail in Chapter VI, § 4(*d*).

Rusterholz [55] has measured the scattering factor of copper for a series of wave-lengths on the long-wave side of the absorption edge. He also used the method of mixed powders, with aluminium as a standard. His values, which are corrected for temperature, are given in Tables IV. 9, and IV. 10. Table IV. 9 gives the values of $|f|$ for Zn Kα radiation, $\lambda/\lambda_K = 1{\cdot}038$, for a series of values of $(\sin\theta)/\lambda$. Column (2) of the same table gives the values of f_0 for copper calculated from the values of the wave-functions for Cu^+ given by D. R. and W. Hartree, who used the method of the self-consistent field including exchange.

TABLE IV. 9

Scattering factors for copper

$(\sin\theta)/\lambda$	f_0	Zn Kα radiation Rusterholz. $\lvert f\rvert$	δf	Cu Kα radiation Brindley and Spiers $\lvert f\rvert$	δf
0·240	21·9	18·3	−3·6	18·9	−3·0
0·277	20·6	15·6	−5·0	17·2	−3·4
0·393	16·6	12·6	−4·0	13·5	−3·1
0·459	14·6	10·6	−4·0	11·5	−3·1
0·604	11·3	7·55	−3·8	8·4	−2·9
0·620	11·0	7·15	−3·8	8·0	−3·0
			$\delta f_{(calc.)}$ −3·4		$\delta f_{(calc.)}$ −2·6

These values differ somewhat from those given in Table 2 of Rusterholz's paper, which include no allowance for electron exchange. Column (4) shows the values of δf.

Table IV. 10 shows the values of $|f|$ and δf for the 220 reflection from copper ($(\sin\theta)/\lambda = 0{\cdot}393$) for a series of wave-lengths. These

TABLE IV. 10

$|f_{(220)}|$ for copper for different wave-lengths

Radiation	λ	λ/λ_K	$\lvert f\rvert$	f_0	δf
Os Lα	1·389A	1·008	11·35	16·6	−5·25
Zn Kα	1·432	1·038	12·5		−4·1
Cu Kα	1·537	1·11	13·2		−3·4
Ni Kα	1·655	1·20	13·6		−3·0
Co Kα	1·785	1·29	13·7		−2·9
Cr Kα	2·285	1·66	13·9		−2·7

values are shown on the curve in fig. 66 for comparison with the values of δf for copper calculated from Hönl's theory. The experimental values of δf are rather larger than the theoretical ones. It must be remembered, however, that in the absence of a value for a wave-length well on the short-wave side of the absorption edge, the absolute value of δf in the column for the experimental figures includes any error in the calculation of f_0, and is not entirely an experimental value. The agreement with theory has not been improved by using the revised calculated values for the scattering factor of copper.

More recent determinations of this kind are those of Brindley and Spiers,[57] who measured the scattering factors of copper and nickel for copper radiation, using the method of mixed powders. They obtained the values of δf by subtracting the theoretical value of the scattering factor for short waves, f_0, from the measured values $|f|$, corrected for temperature by the methods described in Chapter V. Their values agreed closely with those to be expected from Hönl's theory. The agreement is, however, not quite so good when the revised figures for f_0, including the effects of electron exchange, are used. The revised values are given by Brindley and Ridley,[58] and are included in Table IV. 9, for comparison with the results obtained by Rusterholz, although it should be noted that the results are obtained for different wave-lengths. The theoretical values for the two wave-lengths, from Hönl's equation, are also shown in the table.

A very careful determination of the scattering factor of tungsten for wave-lengths in the neighbourhood of its L absorption edges has been made by Brentano and Baxter.[59] They used the method of mixed powders, with aluminium as the standard powder, and took great precautions to avoid the errors to which the use of the method is liable. They made comparative measurements for wave-lengths on both sides of the absorption edge, using the same powder specimens. The value of δf they expressed in approximately absolute measure by assuming the value of $|f|$ obtained for the 200 spectrum from tungsten with the $K\alpha$ radiation from tin to be unaffected by dispersion, and to have the value calculated by the Thomas-Fermi method. The difference between the values of $|f|$ obtained using two wave-lengths, one much longer and the other much shorter than the wave-length of the L absorption edge, should be considerably greater than the corresponding difference for the K absorption edge; for there are eight L electrons, and only two K electrons. Brentano and Baxter found this larger difference, and estimated from their measurements a value of 4·5 for the total oscillator strength associated with the eight L electrons. Hönl's theoretical estimate gives the values 1·35 for L_I, and 3·55 for $L_{II}+L_{III}$; a total of 4·9.

The values of $|f|$ and δf obtained by Brentano and Baxter are given in Table IV. 11, and are shown graphically by the circles in fig. 65. No reliable data exist for the calculation of a theoretical curve for the

TABLE IV. 11

Values of δf for L absorption edge of tungsten (Brentano and Baxter)

Radiation	λ	$\lambda/\lambda_{L_{III}}$	δf
Sn $K\alpha$	0·492A	0·405	0·0
Ag $K\alpha$	0·560	0·461	0·3
Rh $K\alpha$	0·614	0·506	−0·4
Mo $K\alpha$	0·710	0·585	−1·8
Zr $K\alpha$	0·786	0·648	−3·1
Zn $K\beta$	1·292	1·065	−10·1
Zn $K\alpha$	1·434	1·180	−7·5
Cu $K\alpha$	1·539	1·267	−6·3
Fe $K\alpha$	1·934	1·593	−5·8
Cr $K\alpha$	2·229	1·836	−5·9

Absorption edge L_{III} for tungsten, $\lambda_{L_{III}}$, is 1·213A.

L edge. The curve in the figure is that given by Brentano and Baxter, and was calculated by them from such approximate data as were available. There is good general agreement between this curve and the experimental points on the long-wave side of the absorption edge. On the short-wave side the agreement is less good, and the theoretical curve rises to high values again for wave-lengths much nearer to the absorption edge than is indicated by the experimental points. It must be remembered, however, that considerable uncertainty attaches to the theoretical curve in this case, and also that the absorption edge is not

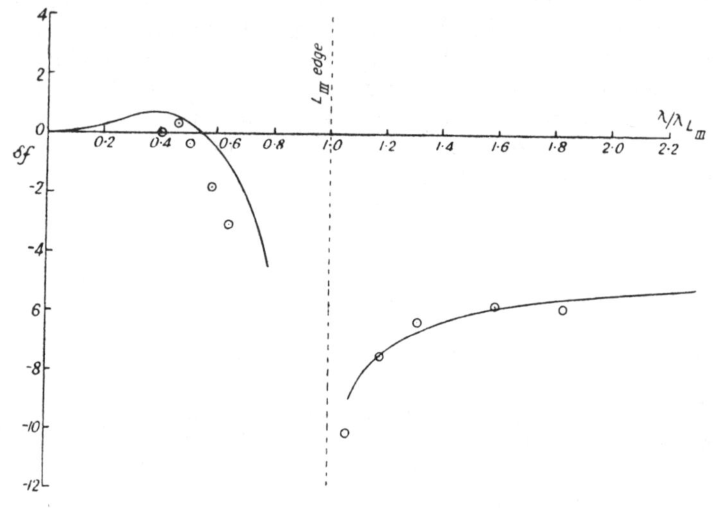

FIG. 65. Anomalous scattering for L absorption of tungsten (Brentano and Baxter)

simple. The values of λ/λ_L are expressed in terms of L_{III} in drawing the curve.

Brentano and Baxter also made measurements with thallium chloride, in which the chlorine in the same crystal was taken as a standard scatterer in deducing the f-curve for thallium. The f-curves obtained with Mo Kα and Cu Kα radiations, whose wave-lengths lie on opposite sides of the L absorption edge for thallium, differed by a constant amount, independent of the angle of scattering, which is in accordance with theory, since the quadrupole terms are very small, and the dipole terms in the dispersion formula should vary only very slightly with the angle of scattering.

(g) *General conclusions:* It is not very easy to draw definite conclusions from the existing experimental work on dispersion that is based on the measurement of scattering factors. As we have seen, all workers agree in finding a marked decrease in the value of $|f|$ for wave-lengths in the immediate neighbourhood of an absorption edge of the scatterer. The fall of $|f|$ on both sides of the absorption edge, as predicted by theory, is also found experimentally; and most observers agree in finding the absolute decrease in $|f|$ for a given scatterer and a given radiation to be nearly independent of the angle of scattering. Measurements of the same quantity made by different observers differ so much among themselves, however, that it is clear that difficulties of experimental technique are still so great as to prevent any very exact test of the dispersion formulae by this method.

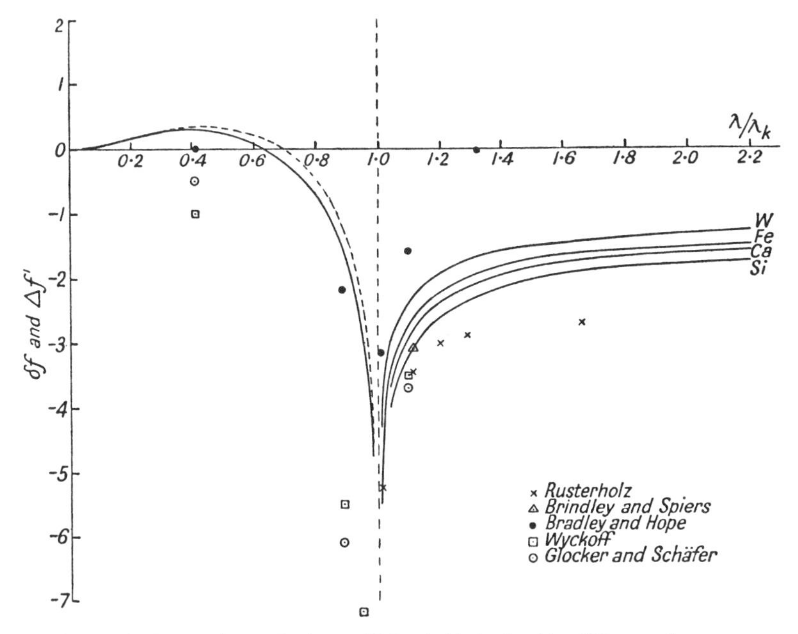

FIG. 66. Comparison of values of δf and Δf obtained by different observers with Hönl's theory

We have tried to summarise the results of the observations for the K edge in fig. 66. In this figure, for λ/λ_K greater than unity, the curves giving $\Delta f'_K$ according to Hönl's theory are drawn for silicon, calcium, iron, and tungsten. For large values of λ/λ_K, these approach asymptotically to the values $-1{\cdot}53$, $-1{\cdot}41$, $-1{\cdot}33$, and $-1{\cdot}15$, the K oscillator strengths for the four elements. For λ/λ_K less than unity, the full curve in the figure shows Hönl's curve for $\Delta f'_K$ for iron only. The dotted curve shows δf_K, as distinct from $\Delta f'_K$ for the 110 spectrum of iron. This serves to give an idea of the effect of the imaginary part of the scattering factor on the total scattering factor. Measurements of δf for iron have been made, as we have seen, by Wyckoff, by Bradley and Hope, and by Glocker and Schäfer. The points obtained in these three sets of experiments are shown in the figure by squares, by black dots, and by circles, respectively. The observed values of $|f|$ are those for the 110 spectrum of iron (Table IV. 8), and to get δf we have subtracted the tabulated value of f_0, calculated by the Thomas-Fermi method, 17·3 for Fe(110), from the observed value of $|f|$. This procedure is not very satisfactory, since there may well be an error of as much as half a unit in the calculated value of f_0. Another method of making the comparison would have been to put the points for Mo Kα onto the curve, assuming them to be correct. This would have improved the agreement in some cases, and made it worse in others. All the points for $\lambda < \lambda_K$ lie too low, but whether this indicates a systematic error in the theory, or whether it is due to the difficulty of making accurate measurements of intensities for wave-lengths just shorter than the critical absorption wave-length, which excite fluorescent radiation in the scatterer, it is not easy to say. It may perhaps be significant that Brentano and Baxter find a similar deviation for the L absorption edge.

The remaining observations to be considered are those of Rusterholz, and of Brindley and Spiers, who used copper as a scatterer. The dispersion curves for iron and copper are so similar that they could hardly be shown as separate on the scale of fig. 66, and the curve drawn there for iron will serve for the comparison of these results with theory. In neither set of experiments was a value obtained for a wave-length shorter than λ_K, and to get δf all that can be done is to subtract the theoretical value of f_0 from the measured values of $|f|$. The theoretical value is here that calculated from the Hartree atomic field, with exchange. The points obtained from Rusterholz's numbers are shown by crosses; Brindley and Spiers obtain only one value, and it is shown by a triangle. The points would have been raised by about half a unit had the value of f_0 without exchange been used, but this value should be less accurate than that including exchange.

The 'scatter' of the points is fairly wide, although the scale on which they are plotted is such as to show it up rather badly. It cannot be said either that the results contradict the theory, or that they support it, except in a general way. Taking into account the evidence obtained from the measurement of refractive indices, we may, however, probably regard the theory as supported by experiment to an extent sufficient to allow us to use it with some confidence in calculating the dispersion corrections to be applied to the scattering factors for use in the analysis of crystals. In Appendix III we give tables, based on Hönl's theory, to aid in the rapid calculation of $\Delta f'$ and $\Delta f''$ for the K absorption edge. The correction for the K edge may be quite important for wave-lengths and scatterers that are commonly used in such work. For the heavier elements, the L dispersion terms will be of importance also, but it is not very easy to obtain numerical values for the corrections, although from Hönl's values for the oscillator strengths, tabulated in Table IV. 3, some estimate may still be made. With molybdenum radiation, such corrections will scarcely ever be necessary; with copper radiation, they could be neglected for all elements lighter than the rare earths.

References

1. R. Ladenburg, *Zeit. f. Physik*, **4**, 451 (1921).
2. R. A. Houston, *Phil. Mag.*, **2**, 512 (1926).
3. W. Thomas, *Naturwiss.*, **13**, 627 (1925); W. Kuhn, *Zeit. f. Physik*, **33**, 408 (1925); F. Reiche and W. Thomas, *Zeit. f. Physik*, **34**, 510 (1925).
4. H. A. Kramers and W. Heisenberg, *Zeit. f. Physik*, **31**, 681 (1925).
5. E. Schrödinger, *Ann. d. Physik*, **81**, 109 (1926).
6. I. Waller, *Phil. Mag.*, **4**, 1228 (1927).
7. Y. Sugiura, *Jour. de Physique et le Radium*, **8**, 113 (1927).
8. R. de L. Kronig, *Jour. Opt. Soc. Amer.*, **12**, 547 (1926).
9. H. Kallman and H. Mark, *Naturwiss.*, **14**, 648 (1926); *Ann. d. Physik*, **82**, 585 (1927).
10. W. Bothe, *Zeit, f. Physik*, **40**, 653 (1927).
11. E. J. Williams, *Proc. Roy. Soc.*, **A**, **143**, 358 (1934).
12. J. A. Prins, *Zeit. f. Physik*, **47**, 479 (1928).
13. R. de L. Kronig and H. A. Kramers, *Zeit. f. Physik*, **48**, 174 (1928).
14. D. Coster, K. S. Knol and J. A. Prins, *Zeit. F. Physik*, **63**, 345 (1930); D. Coster and K. S. Knol, *Zeit. f. Physik*, **75**, 340 (1932); D. Coster and K. S. Knol, *Proc. Roy. Soc.*, **A**, **139**, 459 (1933).
15. R. Glocker and K. Schäfer, *Zeit. f. Physik*, **73**, 289 (1931).
16. H. Hönl, *Zeit. f. Physik*, **84**, 1 (1933).
17. E. A. Hylleraas, *Zeit. f. Physik*, **48**, 469 (1928).
18. J. A. Wheeler and J. A. Bearden, *Phys. Rev.*, **46**, 755 (1934).
19. H. Hönl, *Ann. d. Physik*, **18**, 625 (1933).
20. A Rusterholz, *Zeit. f. Physik*, **82**, 538 (1933).
21. W. Stenström, *Lund Dissertation*, 1919.
22. W. Duane and R. A. Patterson, *Phys. Rev.*, **16**, 532 (1920).
23. E. Hjalmar, *Zeit. f. Physik*, **1** 439 (1920); **3**, 262 (1920).
24. P. P. Ewald, *Zeit. f. Physik*, **2**, 332 (1920).
25. Bergen Davis and H. M. Terrill, *Proc. Nat. Acad. Sci.*, **8**, 357 (1922).
26. Bergen Davis and R. von Nardroff, *Proc. Nat. Acad. Sci.*, **10**, 60, 384 (1924).
27. R. von Nardroff, *Phys. Rev.*, **24**, 143 (1924).
28. P. P. Ewald, *Zeit. f. Physik*, **30**, 1 (1924).
29. C. C. Hatley, *Phys. Rev.*, **24**, 486 (1924).
30. A. H. Compton, *Phil. Mag.*, **45**, 1121 (1923).
31. R. Forster, *Helv. Phys. Acta.*, **1**, 18 (1927); *Naturwiss.*, **15**, 969 (1927).
32. F. Jentzsch, *Physikal. Zeit.*, **30**, 268 (1929); F. Jentzsch and E. Nähring, *Naturwiss.*, **17**, 980 (1929); M. Schön, *Zeit. f. Physik*, **58**, 165 (1929); J. Thibaud, *Phys. Rev.*, **35**, 1452 (1930); E. Dershem and M. Schein, *Zeit. f. Physik*, **75**, 395 (1932); F. Jentzsch and H. Steps, *Zeit. f. Physik*, **91**, 151 (1934).
33. H. Kiessig, *Ann. der Physik*, **10**, 715, 769 (1931).

43. A. J. Lameris and J. A. Prins, *Physica*, **1**, 881 (1934).

35. A. Larsson, M. Siegbahn and I. Waller, *Naturwiss.*, **12**, 1212 (1924).
36. A. Larsson, *Uppsala Dissertation*, (1929).
37. H. E. Stauss, *Phys. Rev.*, **36**, 1101 (1930); *Jour. Opt. Soc. Amer.*, **20**, 616 (1930).
38. J. A. Bearden, *Phys. Rev.*, **38**, 835 (1931).
39. J. A. Bearden, *Phys. Rev.*, **39**, 1 (1932).
40. Bergen Davis and C. M. Slack, *Phys. Rev.*, **25**, 881 (1925); C. M. Slack, *Phys. Rev.*, **27**, 691 (1926).
41. J. A. Bearden and C. H. Shaw, *Phys. Rev.*, **46**, 759 (1934).
42. J. E. Field and G. A. Lindsay, *Phys. Rev.*, **51**, 165 (1937).
43. The following papers, not explicitly quoted in the text, deal with the total reflection of X-rays (the list is not exhaustive):

    F. Wolfers, *Comptes Rendus*, **176**, 1385 (1923); **177**, 32 (1923).
    P. Kirkpatrick, *Nature*, **113**, 98 (1924).
    H. Stauss, *Nature*, **114**, 88 (1924).
    M. de Broglie and J. Thibaud, *Comptes Rendus*, **181**, 1034 (1925).
    W. Linnik and W. Laschkarew, *Zeit. f. Physik*, **38**, 659 (1926).
    R. L. Doan, *Phil. Mag.*, **4**, 100 (1927).
    H. W. Edwards, *Phys. Rev.*, **30**, 91 (1927).
    W. E. Laschkarew and S. D. Herzrücken, *Zeit. f. Physik*, **52**, 739 (1928).
    H. E. Stauss, *Phys. Rev.*, **31**, 491 (1928).
    E Dershem, *Phys. Rev.*, **33**, 659 (1929).
    K. Kellerman, *Ann. d. Physik*, **4**, 185 (1930).
    C. B. O. Mohr, *Proc. Roy. Soc.*, **A**, **133**, 292 (1931).
    S. W. Smith, *Phys. Rev.*, **40**, 156 (1932).
44. H. Mark and L. Szilard, *Zeit. f. Physik*, **33**, 688 (1925).
45. D. Coster, K. S. Knol and J. A. Prins, *Zeit. f. Physik*, **63**, 345 (1930).
46. Alice H. Armstrong, *Phys. Rev.*, **34**, 931 (1929).
47. R. W. G. Wyckoff and Alice H. Armstrong, *Zeit. f. Krist.*, **72**, 319 (1929).
48. R. W. G. Wyckoff, *Phys. Rev.*, **35**, 215, 583 (1930);
49. R. W. G. Wyckoff, *Phys. Rev.*, **36**, 1116 (1930).
50. A. J. Bradley and R. A. H. Hope, *Proc. Roy. Soc.*, **A**, **136**, 272 (1932).
51. A. J. Bradley and J. W. Rodgers, *Proc. Roy. Soc.*, A, **144**, 340 (1934).
52. R. Glocker and K. Schäfer, *Zeit f. Physik*, **73**, 289 (1931).
53. K. Schäfer, *Zeit. f. Physik*, **86**, 738 (1933)
54. J. C. M. Brentano, *Phil. Mag.*, **4**, 620 (1927); **6**, 178 (1928); *Zeit. f. Physik*, **70**, 74 (1931).
55. A. A. Rusterholz, *Zeit. f. Physik*, **82**, 538 (1933).
56. G. W. Brindley and F. W. Spiers, *Proc. Phys. Soc. Lond.*, **46**, 841 (1934).
57. G. W. Brindley and F. W. Spiers, *Phil. Mag.*, **20**, 865 (1935); **21**, 778 (1936).
58. G. W. Brindley and P. Ridley, *Proc. Phys. Soc.*, **50**, 96 (1938).
59. J. C. M. Brentano and A. Baxter, *Zeit. f. Physik*, **89**, 720 (1934).

CHAPTER V

THE INFLUENCE OF TEMPERATURE ON THE DIFFRACTION OF X-RAYS BY CRYSTALS

1. Theoretical

(*a*) *Introductory :* In Chapter I we considered, simply as a geometrical problem in diffraction, the influence on the intensities of the diffraction maxima of small and random displacements of the lattice-points from their mean positions. We found that such displacements diminished the intensities of the spectra, leaving their sharpness unaltered, while at the same time the general scattering in directions other than those of the spectra was increased. The intensities of the diffraction spectra were reduced by a factor e^{-2M}, where

$$M = 8\pi^2 \overline{u_s^2}(\sin^2 \theta)/\lambda^2, \tag{5.1}$$

$\overline{u_s^2}$ being the mean-square displacement of a lattice-point in a direction perpendicular to the reflecting planes, that is to say, in the direction of the vector S, as defined by equation (*1.5*), p. 3, of Chapter I. The atoms in actual crystals are subject continually to small displacements on account of their thermal motions, and in this chapter we shall consider in some detail the influence of the temperature of the crystal and of its elastic properties, both on the intensities of the diffraction maxima and on the diffuse scattering.

Very soon after the discovery of the diffraction of X-rays by crystals this problem was attacked, and in so far as the effect on the diffraction maxima was concerned, successfully solved, by Debye. In his earlier papers, Debye [1] supposed the atoms to execute independent simple-harmonic oscillations about positions of equilibrium, but in an important paper published in 1914 [2] he applied the methods he had recently developed in his theory of specific heats,[3] for approximating to the modes of vibration of discontinuous solids, to obtain numerical estimates of the mean-square displacements of the atoms in simple cubic crystals. Since the publication of Debye's work on specific heats, the problem of the modes of vibration of lattices consisting of interconnected particles had been dealt with by Born and v. Kármán,[4] and the influence of this work is to be seen in Debye's 1914 paper.

The complete expression for the temperature factor in terms of the normal modes of the lattice was first given by Waller in 1923,[5] and was published by him in greater detail in his Uppsala dissertation in 1925.[6] Meanwhile, Born's well-known work [7] on the dynamics of crystal lattices had appeared. Important contributions to the subject were also made by Faxén,[8] who in 1918 pointed out that the method of averaging used by Debye was not strictly correct, and gave the necessary modifications. These, however, affect the general scattering, and not the intensities of the interference maxima. Ideas of a similar nature were put forward by Brillouin [9] in his work on the scattering of light. The earlier work was based on classical dynamics, and classical scattering theory, supplemented by the theory of the Planck oscillator. More recent treatments, in which the principles of the quantum theory are used, are those of Waller,[10] Ott,[11] Born,[12] and Born and Sarginson.[13,14] The account of the subject to be given in this chapter is based on the methods of the earlier papers, although it does not follow any of them in detail, but the results obtained are in agreement with those given by the later, and more rigid, treatments.

We shall start from equation (*1*.29) of § 3 (*b*), Chapter I, which, for a lattice composed of identical atoms of scattering factor f_0, we may now write in the form

$$\bar{I} = |f_0|^2 \sum_n \sum_{n'} e^{i\kappa \mathbf{S}\cdot\mathbf{r}_n - \mathbf{r}_{n'}} e^{-\frac{1}{2}\overline{p^2_{n,n'}}}, \tag{5.2}$$

where

$$\overline{p^2_{n,n'}} = \overline{\{\kappa \mathbf{S}\cdot\mathbf{u}_n - \mathbf{u}_{n'}\}^2}. \tag{5.3}$$

Here, $\mathbf{u}_n$ and $\mathbf{u}_{n'}$ are the vector displacements of the atoms n and n' from their position of equilibrium in the undisturbed lattice, and the sums are to be taken over all the N lattice-points. The average $\overline{p^2_{n,n'}}$ is to be taken over a time very long in comparison with the period of oscillation of the lattice-points.

In equation (*5*.2), and in all subsequent equations unless the contrary is stated, the intensity is expressed as a multiple of that scattered by a single classical electron under the same conditions. If the incident radiation is unpolarised, we must therefore multiply (*5*.2) by a factor C, given by

$$C = (e^2/mc^2)^2(1 + \cos^2 2\theta)/2R^2, \tag{5.4}$$

to obtain the actual intensity, 2θ being the angle of scattering, and R the distance of the point of observation from the crystal.

In Chapter I, we assumed the vibrations of the atoms situated at the different lattice-points to be entirely independent of one another, so that the mean value of $p^2_{n,n'}$ was the same for any pair of points n and n'. It was therefore possible to take a common factor $e^{-\frac{1}{2}\overline{p^2}}$ outside the summation signs, and this led directly to equation (*1*.35) for the total scattered intensity of unmodified * wave-length.

In a real crystal, the molecules or atoms associated with the lattice-points are linked to one another by the interatomic forces. The displacements of one atom must, in principle, affect those of its neighbours, and on the whole, adjacent atoms are more likely to be displaced in the same direction than in opposite directions. The average value

* It should be noted that because the radiation is scattered from moving atoms there must be a very slight modification of the scattered wave-length, which, however, lies below the limits of observation in the case of X-rays.

of $\kappa \mathbf{S} \cdot \mathbf{u}_n - \mathbf{u}_{n'}$ is therefore not the same for all pairs of atoms, and the assumptions made in Chapter I are thus far too simple. We must now consider what effect the interdependence of the lattice-points will have on the methods to be used in averaging, and on the scattered intensity. We shall find that the results of Chapter I remain true in so far as the influence of lattice vibrations on the main diffraction maxima is concerned, but that the expression for the general scattering is greatly modified, and now indicates the existence of broad and diffuse maxima in the directions of the lattice spectra, a result that was first predicted by Faxén, and considered later in more detail by Waller.

These diffuse maxima were first definitely observed by Laval [15] in 1938. They appear as a matter of fact on many oscillation photographs and Laue photographs taken long before this, and had been ascribed to various causes, or neglected altogether. Since the publication of Laval's results, a large amount of work has been done on the subject, some of which we shall deal with in a later section. It is becoming plain that much interesting information as to the elastic properties of crystals, and also as to their actual structure, is to be obtained by studying the diffuse maxima. It is indeed hardly too much to say that an effect which for many years had been regarded as of secondary importance in comparison with the reduction of the intensity of the main maxima by thermal movement has now become that most likely to repay detailed study.

(*b*) *The normal co-ordinates of the lattice vibrations:* The problem of the vibrations of a lattice of interconnected particles was dealt with by Born and v. Kármán,[4] and has been discussed in great detail by Born [7] in his book *Dynamik der Kristallgitter*, and we shall first outline the theory briefly for the case of a simple lattice, consisting of particles all of one kind. The particles are coupled, and the displacement of any one of them, in principle, affects all the others, so that the equations of motion of any one point involve the displacements of all the other points.

Let us assume for a moment that the lattice is of infinite extent. Such a lattice can be traversed by plane waves, and it can be shown that any oscillation of the lattice so small that the potential energy can be expressed as a quadratic function of the displacements and velocities of the lattice-points can be expressed as the sum of the displacements due to a number of such plane waves, having various amplitudes and wave-lengths. That is to say that the actual displacement $\mathbf{u}_n$ of any point n can be expressed as the sum of its displacements due to the individual waves.

If the displacements are small, so that the principle of superposition applies, each wave travels through the lattice independently of any of the others. The whole configuration of the lattice may be expressed in terms of the displacements due to this set of independent waves, just as well as in terms of the co-ordinates of the individual lattice-points, and the waves can therefore be considered as themselves constituting a set of co-ordinates, which are independent, and to be preferred to the ordinary geometrical co-ordinates of the lattice-points for the treatment of vibration problems. Such co-ordinates are known as *normal co-ordinates.*

(*c*) *Plane waves in a lattice :* Let $\mathbf{r}_n$ be the vector from the origin to the lattice-point n of a rectangular lattice whose translations parallel to the axes of x, y, z, are $\mathbf{a}$, $\mathbf{b}$, $\mathbf{c}$, and let $\mathbf{u}_n(k)$ be the displacement of the point n due to a plane wave whose wave-vector is $\mathbf{k}$. The magnitude of $\mathbf{k}$ is equal to the number of wave-lengths in a distance 2π, and its direction is that of the wave-normal. If ω_k is the circular frequency, we can write

$$\mathbf{u}_n(k) = \mathbf{a}_k \cos\{\omega_k t - \mathbf{k}\cdot\mathbf{r}_n - \delta_k\}, \tag{5.5}$$

where $\mathbf{a}_k$ is a vector giving the amplitude and direction of the displacement. Waves travelling in a discrete lattice have the peculiarity that there is a definite lower limit to their possible wave-length in any direction. This may be seen as follows. The lattice-vector $\mathbf{r}_n$ must have one of a discrete set of values given by

$$\mathbf{r}_n = n_1\mathbf{a} + n_2\mathbf{b} + n_3\mathbf{c},$$

with n_1, n_2, n_3 integers. The scalar product $\mathbf{k}\cdot\mathbf{r}_n$ can therefore be written

$$\mathbf{k}\cdot\mathbf{r}_n = n_1 k_x a + n_2 k_y b + n_3 k_z c, \tag{5.6}$$

where k_x, k_y, k_z are the components of $\mathbf{k}$, and so the number of wave-lengths in length 2π parallel to x, y, z respectively. Thus $k_x a$, $k_y b$, $k_z c$ or ϕ_x, ϕ_y, ϕ_z, are the phase differences due to the waves between the displacements of neighbouring atoms in rows parallel to the three axes.

It is convenient to consider ϕ_x, ϕ_y, ϕ_z as the components of a vector $\boldsymbol{\varphi}$, which defines the wave, and which for a cubic lattice is in the direction of the wave-vector $\mathbf{k}$; and n_1, n_2, n_3 as the components of a vector $\mathbf{n}$, which defines a lattice-point. We can then write

$$\mathbf{k}\cdot\mathbf{r}_n = \boldsymbol{\varphi}\cdot\mathbf{n}. \tag{5.7}$$

All possible values of ϕ_x, ϕ_y, ϕ_z for a lattice of discrete points lie between the limits $\pm\pi$. To see what this means physically, we may consider a plane longitudinal wave propagated in the direction of the a axis. A phase difference $\phi_x = +\pi$ or $-\pi$ means that, in the direction of a, successive points are moving either towards or away from one another, the distinction between the positive and negative sign of ϕ_x being simply one of the direction of propagation of the wave. The wave-length is $2a$, and no shorter wave is possible ; for an increase in ϕ_x beyond π simply repeats the lattice displacements produced by longer waves already included in the range $\phi_x < \pi$. We have therefore the conditions

$$-\pi < \phi_x,\ \phi_y,\ \phi_z < \pi. \tag{5.8}$$

The vibrations of the adjacent points in a lattice for one particular vibration direction when the phase differences between successive points are a maximum are shown in fig. 67.

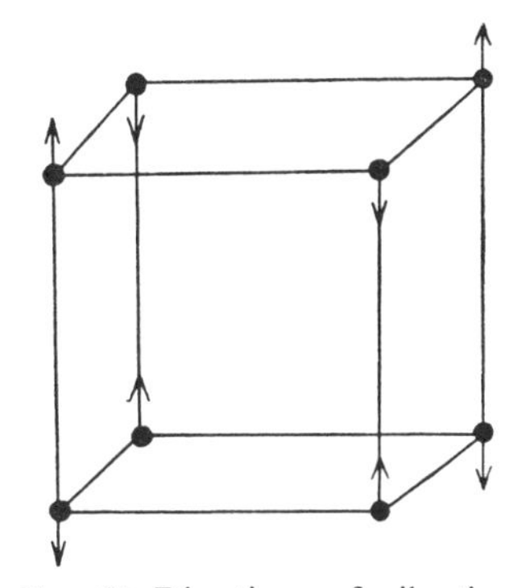

FIG. 67. Directions of vibration of neighbouring lattice-points for maximum phase difference

We see then that there is a definite upper limit to the frequency, and so a lower limit to the wave-length, of the waves that can be propagated in any given direction through a lattice consisting of discrete points.

(*d*) *Waves in a finite lattice:* In a lattice of finite dimensions there is not only a lower limit but also an upper limit to the possible wave-lengths of the lattice waves. If the finite block of lattice contains N points, its configuration can be uniquely defined by 3N co-ordinates, and since, as we have seen above, we may consider each lattice wave as an independent co-ordinate, we are led to infer that only 3N waves will be possible. The lattice waves in a finite block of crystal may be expected to form a discrete set.

The existence of this finite number of independent waves is connected with the boundary conditions that must be fulfilled at the surfaces of the block, and it can be shown that the total number of normal modes is independent of the exact boundary conditions. Let the crystal be the rectangular parallelepiped discussed in Chapter I, having N_1, N_2, N_3 points parallel to the edges, a, b, c, and so $N = N_1N_2N_3$ points in all. One possible set of boundary conditions would be that all the surface points should remain undisplaced. The vibrations of the lattice could then be considered as represented by a set of independent stationary modes of vibration, with nodal planes parallel to the surfaces of the block; and, as is well known, such stationary modes can be built up from suitable progressive waves.

Another method, and one that we shall use here, is to suppose the block $N_1N_2N_3$ to form part of an infinite lattice, in which the motion repeats itself exactly in blocks of this size, so that all motions are periodic in distances N_1a, N_2b, N_3c parallel to a, b, and c respectively. We can consider any one such block as a finite lattice subject to certain definite boundary conditions, and the periodicity of the motion gives for the possible phases of the waves the conditions

$$N_1\phi_x = 2\pi p, \quad N_2\phi_y = 2\pi q, \quad N_3\phi_z = 2\pi r, \tag{5.9}$$

p, q, and r being integers. Thus the possible values of ϕ_x, ϕ_y, ϕ_z are

$$\phi_x = \frac{2\pi}{N_1}p, \quad \phi_y = \frac{2\pi}{N_2}q, \quad \phi_z = \frac{2\pi}{N_3}r. \tag{5.10}$$

We may very conveniently represent these possible values by points in a rectangular lattice whose translations parallel to the three axes are $2\pi/N_1$, $2\pi/N_2$, $2\pi/N_3$. The vector to any lattice-point (p, q, r) in this lattice gives a possible wave-vector $\boldsymbol{\varphi}$ for a wave in the crystal lattice. The limiting values of p, q, r are $p = \pm N_1/2$, $q = \pm N_2/2$, $r = \pm N_3/2$, and the phase-lattice therefore contains $N_1N_2N_3(=N)$ points. There are thus N possible wave-vectors, counting the positive and negative directions of propagation separately, and to each wave-vector there correspond three independent directions of vibration; so that there are, as we should expect, 3N independent lattice waves.

The total volume of the phase-lattice is $(2\pi)^3$, and it contains N points, uniformly distributed. The density of points in the lattice is therefore $N/(2\pi)^3$, and is very large for crystals of more than microscopic dimensions.

(e) The calculation of $\overline{\{\kappa\mathbf{S}\cdot(\mathbf{u}_n - \mathbf{u}_{n'})\}^2}$ *in terms of the normal coordinates :* Let us now consider the case of thermal vibrations in a lattice composed of atoms all of one kind. We may represent the displacements of the lattice-points as the sum of their displacements due to the 3N elastic waves discussed in the last paragraph, all of which are entirely independent. In the actual crystal lattice, the phases of these waves vary rapidly and arbitrarily with the time, owing to the damping that must occur. Each wave is in effect continually being stopped and started afresh with an arbitrary phase, after a time absolutely very short, but long enough to contain many periods of vibration of the wave. The average energy to be associated with any wave has to be determined from the principles of thermal equilibrium between the different modes of vibration, and will be that appropriate to the frequency of the wave.

It must be observed that the picture here given of the lattice vibrations is still an approximation. Each normal mode consists of a harmonic wave, and under such conditions the lattice will not expand with increasing amplitude of the waves. The existence of a coefficient of thermal expansion in real crystals shows that terms other than linear combinations of the displacements come into the expressions for the restoring forces, and that the true representation of the motion in terms of a set of harmonic waves is impossible. It is, however, a justifiable approximation for many purposes, if the temperature is not too high.

Assuming its validity, we may write for the vector displacement $\mathbf{u}_n$ of the lattice-point n

$$\mathbf{u}_n = \sum_{\phi j} \mathbf{e}_{\phi j} a_{\phi j} \cos\{\omega_{\phi j} t - \mathrm{n}\cdot\boldsymbol{\varphi} - \delta_{\phi j}\}. \tag{5.11}$$

In (5.11), $\mathbf{e}_{\phi j}$, is a unit vector in one of the three independent directions of vibration of the lattice $(j = 1, 2, 3)$ associated with the wave $\boldsymbol{\varphi}$. These three directions are always at right angles; $\omega_{\phi j}$ is the circular frequency, and $\delta_{\phi j}$ a phase, which may change many times in an arbitrary way in any time long enough for an observation to be made; $a_{\phi j}$ is the amplitude of the wave; the sum, which has 3N terms, is to be taken over the N values of $\boldsymbol{\varphi}$ and the three values of j.

The quantity $p^2_{n,n'}$ of (5.3), whose time-average is required, may therefore be written

$$p^2_{n,n'} = \{\kappa\mathbf{S}\cdot\mathbf{u}_n - \mathbf{u}_{n'}\}^2$$

$$= \left[\sum_{\phi j} (\kappa\mathbf{S}\cdot\mathbf{e}_{\phi j}) a_{\phi j}\{\cos(\omega_{\phi j} t - \mathbf{n}\cdot\boldsymbol{\varphi} - \delta_{\phi j}) - \cos(\omega_{\phi j} t - \mathbf{n}'\cdot\boldsymbol{\varphi} - \delta_{\phi j})\}\right]^2. \tag{5.12}$$

The square of the series on the right-hand side of (5.12) contains squares of the terms corresponding to each pair of values $\boldsymbol{\varphi}$, j, and products of terms having different values. These products vanish on averaging over the time, since their phases are entirely independent, and we are left with the squares of terms such as that in the curly brackets. Of these, each cosine-squared term gives 1/2 on averaging, while twice the product of the two cosines gives the cosine of the sum, which vanishes on averaging, plus the cosine of the difference, $\cos\{\boldsymbol{\varphi}\cdot(\mathbf{n}-\mathbf{n}')\}$. Thus, the mean value of $(p^2_{n,n'})/2$ can be written

$$\tfrac{1}{2}\overline{p^2_{n,n'}} = \tfrac{1}{2}\sum_{\phi j} (\kappa\mathbf{S}\cdot\mathbf{e}_{\phi j})^2 \overline{a^2_{\phi j}} \left[1 - \cos\{\boldsymbol{\varphi}\cdot(\mathbf{n}-\mathbf{n}')\}\right]. \tag{5.13}$$

The first term of (5.13) is equal to $8\pi^2(\sin^2\theta/\lambda^2)\sum_{\phi j}\overline{(a^2_{\phi j})_S}$, where the suffix S denotes that component of $a_{\varphi j}$ which is parallel to the vector $\mathbf{S}$, and 2θ is the angle of scattering. The mean-square displacement due to a wave is half the mean-square amplitude; and since the waves are entirely independent, the total mean-square displacement $\overline{u_s{}^2}$ is equal to the sum of the mean-square displacements due to the individual waves. Thus, the first term of (5.13) is equal to $16\pi^2\overline{u_s{}^2}(\sin^2\theta)/\lambda^2$, or to 2M, the value obtained for the exponent of the temperature factor by the simple method of averaging used in Chapter I, and is the same for any pair of atoms. The effect of the coupling of the atoms is contained in the second term of (5.13), which represents the modification introduced by the more accurate method of averaging used in this chapter. The existence of this term was first pointed out by Faxén.[8]

(*f*) *The relation between the mean-square amplitude and the temperature :* The total kinetic energy of the lattice, E_{kin}, is given by

$$E_{kin} = \tfrac{1}{2}\sum_n m\dot{\mathbf{u}}_n^2 \tag{5.14}$$

m being the mass of a lattice-point. By (5.11),

$$\dot{\mathbf{u}}_n = -\sum_{\phi j} \mathbf{e}_{\phi j} a_{\phi j} \omega_{\phi j} \sin(\omega_{\phi j} t - \mathbf{n}\cdot\boldsymbol{\varphi} - \delta_{\phi j}),$$

and the mean-square value of this velocity is

$$\overline{\dot{\mathbf{u}}_n^2} = \tfrac{1}{2}\sum_{\phi j} \overline{a_{\phi j}^2}\,\omega_{\phi j}^2, \tag{5.15}$$

since the products of the sines corresponding to different values of $\omega_{\phi j}$ vanish on averaging, because of the independence of the phases of the waves, and the average of each square term is 1/2, while $\mathbf{e}_\varphi$, is a unit vector.

Now $\overline{\mathrm{E}}$, the mean total energy, is twice the mean kinetic energy, and so, by (5.14) and (5.15),

$$\overline{\mathrm{E}} = \tfrac{1}{2}\mathrm{N}m\sum_{\phi j} \omega_{\phi j}^2 \overline{a_{\phi j}^2}, \tag{5.16}$$

N being the total number of atoms.

If $\overline{\mathrm{E}}_{\phi j}$ is the average energy associated with the wave ϕ, j,

$$\overline{\mathrm{E}} = \sum_{\phi j} \overline{\mathrm{E}}_{\phi j}, \tag{5.17}$$

and by comparing (5.16) and (5.17) we obtain for the mean-square amplitude due to the mode ϕ, j

$$\overline{a_{\phi j}^2} = \frac{2}{m\mathrm{N}}\frac{\overline{\mathrm{E}}_{\phi j}}{\omega_{\phi j}^2}. \tag{5.18}$$

From the quantum theory of the harmonic oscillator,

$$\overline{\mathrm{E}}_{\phi j} = (\bar{n}_{\phi j} + \tfrac{1}{2})\hbar\omega_{\phi j}, \tag{5.19}$$

where $\bar{n}_{\phi j}$ is the mean value of the quantum number of the harmonic oscillator in thermal equilibrium, and this, by Planck's theorem, is given by

$$\bar{n}_{\phi j} = 1\Big/\left(e^{\frac{\hbar\omega_{\phi j}}{k\mathrm{T}}} - 1\right), \tag{5.20}$$

k being Boltzmann's constant, and T the absolute temperature.

The last three equations express $\overline{a_{\phi j}^2}$ as a function of the temperature, and of the frequency of the corresponding wave.

Equation (5.13) may now be written

$$\tfrac{1}{2}\overline{p_{n,n'}^2} = \sum_{\phi j} \mathrm{G}_{\phi j}[1 - \cos(\boldsymbol{\varphi}\cdot\mathbf{n} - \mathbf{n}')], \tag{5.21}$$

where

$$\mathrm{G}_{\phi j} = \frac{\hbar}{m\mathrm{N}}\frac{(\kappa\mathbf{S}\cdot\mathbf{e}_{\phi j})^2(\bar{n}_{\phi j} + \tfrac{1}{2})}{\omega_{\phi j}}$$

$$= \frac{\hbar}{2m\mathrm{N}}\frac{(\kappa\mathbf{S}\cdot\mathbf{e}_{\phi j})^2}{\omega_{\phi j}}\coth\left(\frac{1}{2}\frac{\hbar\omega_{\phi j}}{k\mathrm{T}}\right). \tag{5.22}$$

In terms of our previous notation therefore,

$$\sum_{\phi j} \mathrm{G}_{\phi j} = 2\mathrm{M}. \tag{5.23}$$

For high enough temperatures ($kT \gg \hbar\omega_{\phi j}$),

$$(\bar{n}_{\phi j} + \tfrac{1}{2})\hbar\omega_{\phi j} \to kT, \tag{5.24}$$

and in this case,

$$G_{\phi j} = \frac{kT}{mN}\frac{1}{\omega^2_{\phi j}}(\kappa \mathbf{S}\cdot\mathbf{e}_{\phi j})^2, \tag{5.25}$$

a form that is often accurate enough.

It will be seen that the accurate calculation of M involves a knowledge of the lattice frequencies $\omega_{\phi j}$, and this involves a knowledge of the interatomic forces that we do not possess, even in the simplest case. In the absence of such knowledge, we shall have to be content with approximate estimates of M, such for example as can be made for simple cubic crystals by Debye's method.

(*g*) *The calculation of the average scattering from a lattice in thermal movement :* We now introduce the value of $\frac{1}{2}\overline{p^2_{n,n'}}$, as given by (5.21), nto (5.2), the expression for the average intensity $\bar{I}$, which is a function of S/λ, and obtain

$$\bar{I}(S/\lambda) = |f_0|^2 \sum_n \sum_{n'} e^{i\frac{2\pi}{\lambda}\mathbf{S}\cdot\mathbf{r}_n - \mathbf{r}_{n'}} \exp.\left[-\sum_{\phi j} G_{\phi j}\{1 - \cos(\boldsymbol{\varphi}\cdot\mathbf{n} - \mathbf{n}')\}\right], \tag{5.26}$$

or using (5.23),

$$\bar{I}(S/\lambda) = |f_0|^2 e^{-2M} \sum_n \sum_{n'} e^{i\frac{2\pi}{\lambda}\mathbf{S}\cdot\mathbf{r}_n - \mathbf{r}_{n'}} \exp.\left\{\sum_{\phi j} G_{\phi j}\cos(\boldsymbol{\varphi}\cdot\mathbf{n} - \mathbf{n}')\right\}.$$

Since the exponent in the last factor is small, this may be written to a degree of approximation good enough for many purposes in the form

$$\begin{aligned}\bar{I}(S/\lambda) &= |f_0|^2 e^{-2M} \sum_n \sum_{n'} e^{i\frac{2\pi}{\lambda}\mathbf{S}\cdot\mathbf{r}_n - \mathbf{r}_{n'}} \\ &\quad + |f_0|^2 e^{-2M} \sum_n \sum_{n'} \sum_{\phi j} G_{\phi j} e^{i\frac{2\pi}{\lambda}\mathbf{S}\cdot\mathbf{r}_n - \mathbf{r}_{n'}} \cos(\boldsymbol{\varphi}\cdot\mathbf{n} - \mathbf{n}') \\ &= |f_0|^2 e^{-2M} I_0(S/\lambda) + I_2(S/\lambda).\end{aligned} \tag{5.27}$$

The first term gives the ordinary interference maxima. The function $I_0(S/\lambda)$ is the intensity scattered by the undisturbed lattice when the atoms are supposed to scatter as points, and the effects of the polarisation of the incident radiation are left out of account. It is the interference function for the undisturbed lattice, as defined in I, § 1(*c*). The atomic scattering factor for the atom at rest is $|f_0|$. On account of the thermal motions, the intensities of the spectra given by the undisturbed lattice are reduced by the factor e^{-2M}, but their sharpness is not thereby altered. The second term $I_2(S/\lambda)$ gives the general scattering. We shall see that this too has maxima in directions corresponding to the spectra, but that these are much less sharp than the diffraction maxima, and remain diffuse however large the crystal.

The summation over n and n' in I_2 may be carried out as follows.

Put
$$\frac{2\pi}{\lambda}\mathbf{S}\cdot\mathbf{r}_n - \mathbf{r}_{n'} = x, \quad \boldsymbol{\varphi}\cdot\mathbf{n} - \mathbf{n}' = y.$$

Then
$$I_2(S/\lambda) = \tfrac{1}{4}|f_0|^2 e^{-2M} \sum_{\phi j} G_{\phi j} \sum_{n} \sum_{n'} \{e^{i(x+y)} + e^{i(x-y)}\}. \qquad (5.28)$$

By (5.7),
$$\boldsymbol{\varphi}\cdot\mathbf{n} = \mathbf{k}\cdot\mathbf{r}_n = 2\pi\mathbf{g}\cdot\mathbf{r}_n,$$

where
$$\mathbf{k} = 2\pi\mathbf{g}, \qquad (5.29)$$

so that $\mathbf{g}$ is a vector in the direction of the wave-normal of the lattice wave $\boldsymbol{\varphi}$, whose magnitude is equal to the reciprocal wave-length. We may therefore put

$$x + y = 2\pi(\mathbf{S}/\lambda + \mathbf{g})\cdot(\mathbf{r}_n - \mathbf{r}_{n'}), \quad x - y = 2\pi(\mathbf{S}/\lambda - \mathbf{g})\cdot(\mathbf{r}_n - \mathbf{r}_{n'}),$$

and substituting these values in (5.28), we obtain at once

$$I_2(S/\lambda) = \tfrac{1}{4}|f_0|^2 e^{-2M} \sum_{\phi j} G_{\phi j}\{I_0(\mathbf{S}/\lambda - \mathbf{g}) + I_0(\mathbf{S}/\lambda + \mathbf{g})\}, \qquad (5.30)$$

where $I_0(\mathbf{S}/\lambda \pm \mathbf{g})$ is the same function of $\mathbf{S}/\lambda \pm \mathbf{g}$ that $I_0(\mathbf{S}/\lambda)$ is of $\mathbf{S}/\lambda$, and therefore, by § 1 (*c*) of Chapter I, so long as the crystal contains a large number of units, has appreciable values only when $\mathbf{S}/\lambda \pm \mathbf{g}$ is a vector in the reciprocal lattice.

(*h*) *The scattering function considered with the aid of the reciprocal lattice :* We can understand the significance of equations (5.27) and (5.30) most easily by returning to the methods outlined in Chapter I, §§ 1 (*b*) and 1 (*c*). With each point of the reciprocal-lattice space is associated a certain scattering function $I(\mathbf{S}/\lambda)$, the value of which corresponding to the point lying at the extremity of the vector $\mathbf{S}/\lambda$ drawn from the lattice origin gives the intensity scattered by the crystal when the directions of incidence and scattering correspond to the vector $\mathbf{S}$. The intensity is expressed in terms of that scattered by a single classical electron under the same conditions, that is to say, that in representing the intensity in the reciprocal-lattice space we put $|f_0|^2 = 1$ in (5.27) and (5.30), which is equivalent to supposing the atoms to scatter as points. The distribution of the scattering function that was specially considered in Chapter I was that of the interference function $I_0(\mathbf{S}/\lambda)$, which is given by (1.21a) when the crystal has the form of a parallelepiped. The details of the distribution will vary with the external form of the crystal in a manner that will be discussed at greater length in Chapter X; but in all cases the interference function has maxima at the reciprocal-lattice points themselves, that is to say, when the vector $\mathbf{S}/\lambda$ is a vector in the reciprocal lattice, and so long as the crystal contains a large number of unit cells, the intensity of the distribution is negligible except in the immediate neighbourhood of these points. We have now to consider how the distribution is modified when the lattice is subject to thermal vibrations.

To the first term in (5.27) there corresponds a distribution which is that of the interference function I_0 multiplied by a factor e^{-2M}, which is itself a function of S/λ, by (5.25). The more distant the reciprocal-lattice point from the origin, that is to say the higher the order of the spectrum to which it corresponds, the more the intensity associated with it is diminished; but there is no change in the sharpness of the maxima, in so far as the first term of (5.27) is concerned. This is the ordinary Debye effect.

To the second term in (5.27), $I_2(S/\lambda)$, there also corresponds a distribution in the reciprocal-lattice space. This gives the diffuse scattering, and it is everywhere zero when the lattice is at rest. According to (5.30), $I_2(S/\lambda)$ is made up of independent contributions from each state of polarisation j of each lattice wave φj. Let us assume for the moment that there is one wave only, φ, with a wave-vector $\mathbf{g}$, and reciprocal wave-length $|\mathbf{g}|$, in a state of polarisation j. Keeping in mind the form of the interference function I_0, we see from (5.30) that to this wave there will correspond two maxima in the scattering distribution, of peak value $\frac{1}{2}e^{-2M}\, G_{\varphi j}$, each of which will have the same form and distribution as the main interference maxima. They will not, however, be situated at the reciprocal-lattice points, but will be symmetrically disposed on either side of them at vector distances $\pm\mathbf{g}$. This corresponds to the condition that $I_0(S/\lambda \pm \mathbf{g})$ has a maximum only when the argument is a vector from the origin to a reciprocal-lattice point. The intensities of these maxima are governed by the value of $G_{\phi j}$, and are much less than those corresponding to the main maxima. Each of the three independent states of polarisation, j, of the wave gives an independent pair of maxima situated in the same positions, and the total value of the scattering function corresponding to them is the sum of the values corresponding to the individual distributions.

The actual motion of the lattice we must think of as made up of a huge number, 3N, of lattice waves, and to each of these there corresponds a pair of maxima in the scattering function. Taken all together, these independent maxima will produce in effect a continuous distribution about each reciprocal-lattice point. The shorter the lattice wave, the greater will be the distance of the corresponding maxima from any given reciprocal-lattice point; and since the shortest acoustic waves in a crystal lattice have lengths of the order of twice the lattice spacing, it will be seen that this intensity distribution will be much more diffuse than those corresponding to the ordinary lattice spectra.

The picture we now have of the distribution of the interference function corresponding to a lattice in thermal motion is somewhat as follows. Around each reciprocal-lattice point is a compact and exceedingly sharp maximum, corresponding to a possible lattice spectrum. These maxima are reduced in intensity by the thermal movement, the reduction being greater the greater the distance of the reciprocal-lattice point from the origin, but they are not made more diffuse. The sharp maxima form the nuclei of much weaker and much more diffuse distributions, which correspond to the diffuse scattering. These again have their maxima at the reciprocal-lattice points, but the distribution is not the same about each, even for a point lattice, and depends both on the order of the spectrum and on the elastic behaviour of the crystal.

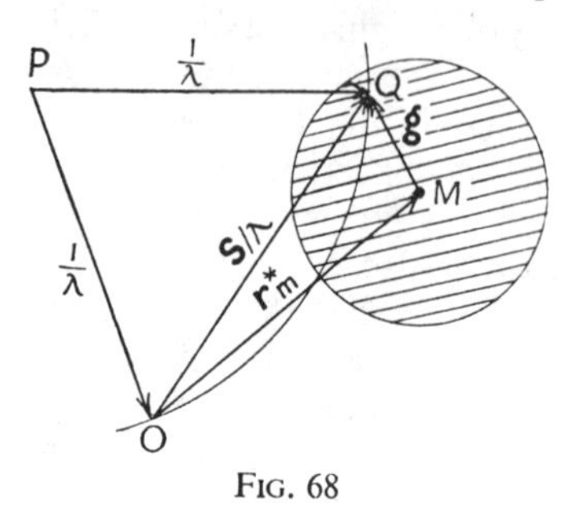

FIG. 68

It will perhaps be helpful to consider the matter from a slightly different point of view with the aid of a diagram. In fig. 68, let $\overrightarrow{PO}$ be a vector of length $1/\lambda$ in the direction of incidence, and let the circle be the trace of the corresponding sphere of reflection. Let M be the reciprocal-lattice point $m(hkl)$, so that $\overrightarrow{OM}$ is the reciprocal-lattice vector $\mathbf{r}^*_m$. Suppose we require the intensity scattered in the direction PQ, Q being also on the sphere of reflection, so that $\overrightarrow{OQ} = \mathbf{S}/\lambda$. According to (*5.30*), the lattice waves travelling in the directions MQ or QM, whose reciprocal wave-lengths are equal to MQ($=|\mathbf{g}|$), are those that give the main contribution to the scattered intensity in the direction PQ. For $\overrightarrow{OQ} - \overrightarrow{MQ} = \overrightarrow{OM}$, and is a vector in the reciprocal lattice. In principle, of course, every lattice wave makes some contribution at Q, but because of the form of the function I_0, it is only those whose wave-vectors lie in the immediate neighbourhood of $\overrightarrow{MQ}$ that make any appreciable contribution.

The fact that that part of the scattering function corresponding to I_2 has diffuse maxima around the reciprocal-lattice points leads at once to an important conclusion. As we have seen in Chapter I, § 1(*b*), it is only when the sphere of reflection passes directly through M that we shall expect the lattice reflection (*m*) to appear in any appreciable intensity; but from fig. 68, it is plain that we may expect weaker and much more diffuse reflections in positions not very different from those appropriate to those of the lattice reflections to make their appearance under conditions of incidence that should definitely exclude these. This of course assumes that the intensity of the diffuse maxima is great enough for the effects to be observed, and as we shall see, this is often true. We shall discuss the methods by which these maxima may be observed in Section 3 of this chapter, but must first complete the theoretical discussion.

(*i*) *The diffuse maxima considered as optical 'ghosts':* We are indebted to Professor R. Peierls for the following simple and instructive qualitative method of explaining the diffuse scattering. Each elastic wave in the lattice may be considered as imposing on it a periodic error in the spacings, which is the analogue in three dimensions of the one-dimensional periodic error in ruling often found in optical diffraction gratings. The effect of such a periodic error in an optical grating is to cause it to give faint subsidiary spectra on each side of the main maxima, which are termed 'ghosts'. If there are G rulings in unit length of the grating, while the periodic error repeats itself g times in unit length, the ghosts that accompany the mth order reflection occur in positions corresponding to first order reflections from gratings having $mG \pm g$ rulings in unit length. In the crystal, each set of elastic waves produces its own ghosts, and the assemblage of them round each lattice spectrum constitutes the diffuse scattering.

The following treatment gives the positions of the ghosts corresponding to any elastic wave. Let $\rho_0(x, y, z)$ be the density of scattering matter at the point (x, y, z) in the undisturbed crystal. As we shall see in Chapter VII, this density can be expressed as a triple Fourier series, each component of which is a plane wave-form, the planes of constant phase being parallel to one of the sets of lattice-planes of the crystal, and the wave-lengths being the spacings of these planes. To each pair of spectra $h, k, l, \bar{h}, \bar{k}, \bar{l}$ there corresponds one such Fourier wave, in the sense that a scattering distribution represented by the wave-form could produce these spectra, and these only. In terms of the reciprocal lattice, we may write (cf. VII, § 1(*g*), p. 348),

$$\rho_0(x, y, z) = \sum_m A(m) \cos\{2\pi(\mathbf{r}\cdot\mathbf{r}_m^*) + \alpha_m\}, \qquad (5.31)$$

where $\mathbf{r}_m^*$ is the reciprocal-lattice vector to the point m, and the sum is to be taken over all the reciprocal-lattice points. The amplitude factor $A(m)$ is proportional to the amplitude of the corresponding spectrum, and α_m is a phase term, the value of which does not affect the present argument. Through the crystal we shall suppose an elastic wave to run whose vector displacement at the point (x, y, z) is given by

$$\mathbf{u} = \mathbf{a}\cos(\omega t - 2\pi\mathbf{r}\cdot\mathbf{g}). \qquad (5.32)$$

If u_x, u_y, u_z are the components of $\mathbf{u}$ parallel to the axes of x, y, z, and if they are small, the modified density $\rho(x, y, z)$ may be written

$$\rho(x, y, z) = \left(\rho_0(x, y, z) - u_x\frac{\partial\rho_0}{\partial x} - u_y\frac{\partial\rho_0}{\partial y} - u_z\frac{\partial\rho_0}{\partial z}\right)\left(1 - \frac{\partial u_x}{\partial x} - \frac{\partial u_y}{\partial y} - \frac{\partial u_z}{\partial z}\right). \qquad (5.33)$$

The first bracket represents the result of a uniform displacement, the second that of the compressions and rarefactions due to the variation of the displacement with the co-ordinates. Since the displacements are small, their products with one another and with their derivatives may be neglected, and (5.33) then becomes

$$\rho(x, y, z) = \rho_0(x, y, z) - \frac{\partial}{\partial x}(\rho_0 u_x) - \frac{\partial}{\partial y}(\rho_0 u_y) - \frac{\partial}{\partial z}(\rho_0 u_z). \qquad (5.34)$$

By (5.31) and (5.32),

$$\rho_0 u_x = \tfrac{1}{2}\sum_m A(m) a_x [\cos\{\omega t - 2\pi(\mathbf{r}_m^* + \mathbf{g}\cdot\mathbf{r}) + \alpha_m\} + \cos\{\omega t - 2\pi(\mathbf{r}_m^* - \mathbf{g}\cdot\mathbf{r}) + \alpha_m\}],$$

with similar expressions for $u_y\rho_0$ and $u_z\rho_0$. Differentiating $\rho_0 u_x$, $\rho_0 u_y$, and $\rho_0 u_z$ with respect to x, y, and z respectively, and adding, we obtain, using (5.34),

$$\rho(x, y, z) = \rho_0(x, y, z) + \pi\sum_m A(m)[(\mathbf{a}\cdot\mathbf{r}_m^* + \mathbf{g})\sin\{\omega t - 2\pi(\mathbf{r}_m^* + \mathbf{g}\cdot\mathbf{r}) + \alpha_m\} + (\mathbf{a}\cdot\mathbf{r}_m^* - \mathbf{g})\sin\{\omega t - 2\pi(\mathbf{r}_m^* - \mathbf{g}\cdot\mathbf{r}) + \alpha_m\}]. \qquad (5.35)$$

We see therefore that, associated with each Fourier wave m, of wave-vector $\mathbf{r}_m^*$, of the original distribution, there are in the modulated distribution waves of vectors $\mathbf{r}_m^* + \mathbf{g}$ and $\mathbf{r}_m^* - \mathbf{g}$. These new periodicities, like the Fourier components of the original distribution, may be supposed to reflect X-rays according to Bragg's law. The fact that the modulations in (5.35) are represented by progressive waves does not affect the present argument. The velocity of any elastic wave is so small in comparison with that of a light wave that, in a diffraction problem, the elastic wave may be considered as being at rest in calculating the directions of the diffracted beams.

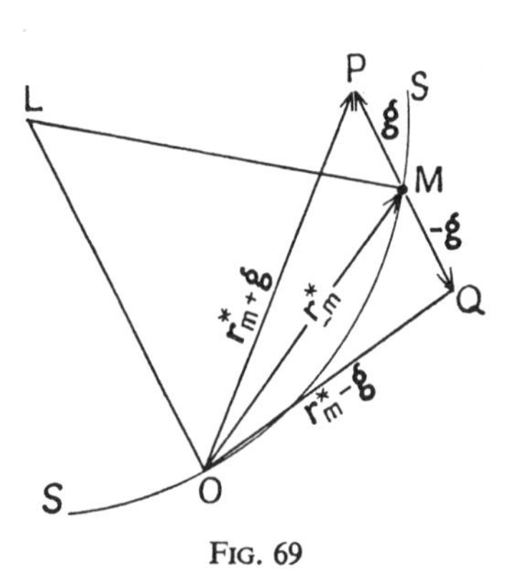

FIG. 69

Consider the Fourier wave corresponding to the vector $\mathbf{r}_m^*$ or $\overrightarrow{OM}$ in fig. 69. This set of waves, whose planes of constant phase are perpendicular to $\overrightarrow{OM}$, may be supposed to produce the mth order spectrum by reflecting LO into LM. Now let the lattice be distorted by the elastic wave whose wave-vector is $\mathbf{g}$. Let MP be the vector $\mathbf{g}$ and MQ the vector $-\mathbf{g}$. Then OP and OQ are respectively the vectors $\mathbf{r}_m^* + \mathbf{g}$ and $\mathbf{r}_m^* - \mathbf{g}$, and are the wave-vectors of the extra periodicities produced in the lattice by the elastic wave $\mathbf{g}$. To the points P and Q correspond the optical ghosts produced by this wave, in the sense that the corresponding extra spectra will appear when the sphere of reflection SS is displaced so as to pass through either P or Q. To every elastic wave propagated in the crystal there corresponds a pair of points such as P and Q, and the whole assemblage forms a diffuse cloud round each reciprocal-lattice point.

Equation (5.35) shows that the amplitudes of the modulations, and so of the corresponding spectra, are proportional to $\mathbf{a}\cdot(\mathbf{r}_m^* \pm \mathbf{g})$, that is to say, to the components of the displacement due to the waves normal to their wave-fronts. This result is to be expected physically, and we shall see that it is also a consequence of the Faxén-Waller theory.

(*j*) *The distribution of intensity in the diffuse maxima :* We must now return to the discussion of equation (5.30), which for this purpose may conveniently be written in the form

$$I_2(S/\lambda) = |f_0|^2 e^{-2M}\sigma, \tag{5.36}$$

where

$$\sigma = \tfrac{1}{2}\sum_{\phi j} G_{\phi j}\{I_0(S/\lambda + \mathbf{g}) + I_0(S/\lambda - \mathbf{g})\}. \tag{5.37}$$

If N, the total number of unit cells in the crystal, is large, the sum over ϕ in (5.37) may be replaced by an integration over the volume of the phase-cell considered in § 1(*c*) of this chapter. There are $N\,d\phi/(2\pi)^3$ waves represented by points lying within an element of volume $d\phi = d\phi_x\,d\phi_y\,d\phi_z$ of the phase-cell, having phases lying between ϕ_x and $\phi_x + d\phi_x$, ϕ_y and $\phi_y + d\phi_y$, ϕ_z and $\phi_z + d\phi_z$. We therefore write (5.37) in the form

$$\sigma = \frac{1}{2}\frac{N}{(2\pi)^3}\sum_j \int G_{\phi j}\{I_0(S/\lambda + g) + I_0(S/\lambda - g)\}\,d\phi. \tag{5.38}$$

For a cubic lattice, of spacing a, $\mathbf{r}_n = a\mathbf{n}$, and so, by (5.29), $\boldsymbol{\varphi} = 2\pi a\mathbf{g}$. The integration over the phase-cell can therefore be replaced by an integration in the reciprocal-lattice space, $\mathbf{g}$ being a vector drawn from a reciprocal-lattice point as origin. Let dv be an element of volume in the reciprocal-lattice space at the extremity of the vector $\mathbf{g}$. Then $d\phi = (2\pi a)^3\,dv$, and

$$\sigma = \tfrac{1}{2}Na^3 \sum_j \int G_{\phi j}\{I_0(S/\lambda + \mathbf{g}) + I_0(S/\lambda - \mathbf{g})\}\,dv. \tag{5.39}$$

During the integration, S/λ remains constant. The limits of integration for $|\phi|$ are $\pm\pi$, and so those for each component of $\mathbf{g}$ are $\pm 1/2a$, or $\pm a^*/2$; that is to say, the integration is to extend through the complete reciprocal-lattice cell.

Suppose for example we are calculating the scattering in the direction PQ of fig. 68. Then $S/\lambda - \mathbf{g}$ is a vector in the reciprocal lattice if $\mathbf{g}$ is the vector $\overrightarrow{MQ}$, and for this value of $\mathbf{g}$, $I_0(S/\lambda - \mathbf{g})$ has a sharp maximum. If the vector $\overrightarrow{MQ}$ lies within the limits of integration, the vector from any other reciprocal-lattice point to Q, which would also correspond to a maximum of the function, must lie outside these limits. The function I_0 has very sharp maxima, and negligible values in regions away from these maxima, while the function $G_{\phi j}$ varies relatively slowly with $\mathbf{g}$, so that in performing the integration we may take it as constant, with the value corresponding to **g**, the vector $\overrightarrow{\mathrm{MQ}}$, and we have then merely to evaluate the integral $\int \mathrm{I}_0(\mathbf{S}/\lambda - \mathbf{g})\, dv$ with $\mathbf{S}/\lambda$ constant. If the crystal is a rectangular block, and if N is large, the value of this integral over a single maximum for a cubic crystal is N/a^3 (cf. Chapter II, § 2(*f*), equation (2.39)). The first term in (5.39) gives the same contribution as the second, for $\mathrm{I}_0(\mathbf{S}/\lambda + \mathbf{g})$ has a maximum within the limits of integration when $\mathbf{g} = -\overrightarrow{\mathrm{MQ}} = \overrightarrow{\mathrm{QM}}$. Equation (5.39) therefore becomes

$$\sigma = \mathrm{N}^2 \sum_j (\mathrm{G}_{\phi j})_g, \tag{5.40}$$

where $(\mathrm{G}_{\phi j})_g$ is the value of $\mathrm{G}_{\phi j}$ corresponding to the wave-vector **g**.

The value of $\mathrm{G}_{\phi j}$ is given by equation (5.22), but for our purpose it will usually be enough to use the approximate form (5.25), corresponding to temperatures that are not too low. If $v_{\phi j}$ is the velocity of a lattice wave ϕj, and **g** its wave-vector, then $v_{\phi j} = \omega_{\phi j}/2\pi|\mathbf{g}|$, so that (5.25) may be written

$$(\mathrm{G}_{\phi j})_g = \frac{k\mathrm{T}}{4\pi^2 m\mathrm{N}|\mathbf{g}|^2} \frac{1}{v_{\phi j}^2} (\kappa\mathbf{S}\cdot\mathbf{e}_{\phi j})^2, \tag{5.41}$$

and equation (5.40) becomes

$$\sigma = \frac{k\mathrm{TN}}{4\pi^2 m|\mathbf{g}|^2} \sum_j \frac{1}{v_{\phi j}^2} (\kappa\mathbf{S}\cdot\mathbf{e}_{\phi j})^2, \tag{5.42}$$

which gives the value of σ corresponding to a point in the scattering distribution at a vector distance **g** from a reciprocal-lattice point. The value of the scattering function corresponding to this point is determined by the nature of the elastic waves of wave-vector **g**.

The vibrations in the three states of polarisation j are independent, and

$$\begin{aligned} \kappa\mathbf{S}\cdot\mathbf{e}_{\phi j} &= (4\pi/\lambda) \sin\theta \cos\alpha_{sj} \\ &= 2\pi(|\mathbf{S}|/\lambda) \cos\alpha_{sj}, \end{aligned} \tag{5.43}$$

where α_{sj} is the angle between the vector **S** and the direction of vibration of the component j. Then, by (5.42) and (5.43),

$$\sigma = \frac{k\mathrm{TN}}{m|g|^2} \left(\frac{|\mathrm{S}|}{\lambda}\right)^2 \sum_j \frac{\cos^2\alpha_{sj}}{v_{\phi j}^2} \tag{5.44}$$

$$= \frac{k\mathrm{TN}}{m} \frac{\mathrm{R}^2}{r^2} \sum_j \frac{\cos^2\alpha_{sj}}{v_{\phi j}^2}, \tag{5.45}$$

where R is the distance of the point considered from the lattice origin, and r its distance from the relevant reciprocal-lattice point. The summation is to be taken over the three independent directions of polarisation of the lattice wave ϕ. Equation (5.45) gives the value of σ for a point at a distance r from a reciprocal-lattice point in the direction of the wave-vector **g**. The magnitude of the relevant value of $\boldsymbol{\varphi}$ is given by $|\boldsymbol{\varphi}| = 2\pi a|\mathbf{g}| = 2\pi ar$.

The last factor in (5.45) is strongly directional. The velocities of the differently polarised components corresponding to a given direction of propagation in general differ considerably, and even in a cubic crystal the velocity of a given type of wave depends on the direction of propagation relative to the crystal axes, while pure longitudinal and pure transverse waves can travel only in certain special directions.

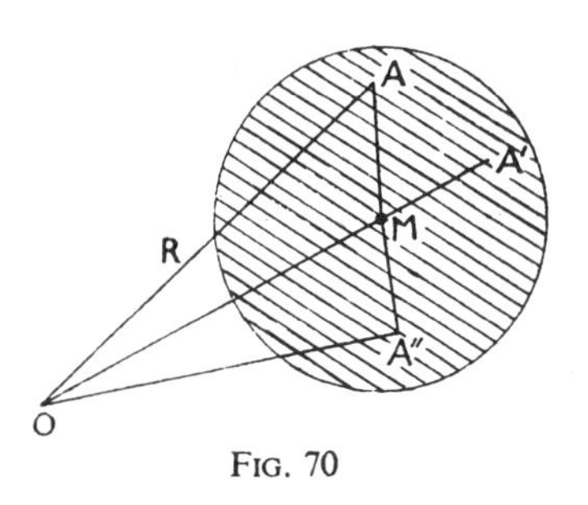

FIG. 70

The factor $\cos^2 \alpha_{sj}$ also exerts a strongly directional effect, the nature of which may be seen in the following way. Let A, fig. 70, be a point in the neighbourhood of the reciprocal-lattice point M, at which the value of the scattering function is required, so that the lengths MA and OA correspond respectively to r and R in (5.45), and the vector $\overrightarrow{OA} = \mathbf{S}/\lambda$. Then the vector $\overrightarrow{MA} = \mathbf{g}$, and is the wave-vector of the elastic wave contributing to the scattering function at A. The longitudinal component adds a contribution proportional to $\cos^2$(OAM) and a transverse wave lying in the plane of the paper a contribution proportional to $\sin^2$(OAM), although the velocity factor in the two cases will in general be different. A transverse component perpendicular to the plane of the paper will give no contribution, since it is perpendicular to $\overrightarrow{OA}$, or to $\mathbf{S}/\lambda$. At A′, the longitudinal component would give the maximum contribution, and the transverse components none at all, while at A″, the longitudinal component of the appropriate wave, which is now travelling in the direction MA″, would give no contribution. We shall deal later in more detail with this question, but must first investigate in a general way the diffuseness of the maxima.

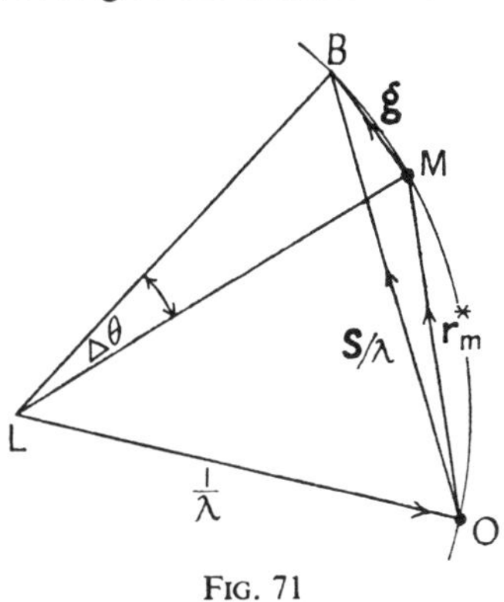

FIG. 71

Suppose the crystal to be set so as to reflect the mth order spectrum. The sphere of reflection in fig. 71 then passes through M, the reciprocal-lattice point. The intensity in a direction LB, making an angle $\Delta\theta$ with LM, the direction of the spectrum, is determined by the wave-vector $\overrightarrow{MB} (= \mathbf{g})$. The angle $\Delta\theta = |\mathbf{g}|/\text{LB} = |\mathbf{g}|\lambda$, and so, from (5.44), we see that if the directional factors are left out of account, the diffusely scattered radiation falls off inversely as the square of the angular distance $\Delta\theta$ from the direction of the spectrum. It must be remembered that we can never use exactly defined directions of incidence, such as have been assumed in the discussion, and that the slight range of angles of incidence inevitable in practice involves an integration over the corresponding range of $\mathbf{S}/\lambda$; and also that $|\mathbf{g}|$, being the reciprocal wave-length in a finite lattice, can never be zero. Equation (5.44) will not therefore apply for the exact direction of the lattice spectra, and σ will not become infinite. The sharp principal maximum and the diffuse maximum are illustrated diagrammatically in fig. 72.

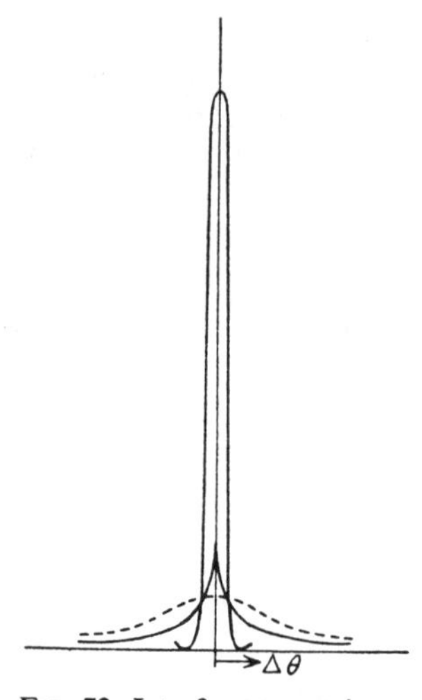

FIG. 72. Interference maximum and diffuse scattering

So far as equation (5.45) is concerned, we see that the greater the value of R, that is to say the higher the order of the spectrum, the greater the intensity of the diffuse cloud. Before comparing the formula with the results of observation, we must, however, as (5.36) shows, multiply the value of σ given by (5.45) by the polarisation factor, the square of the structure factor, and the Debye factor e^{-2M}, all of which will in general cause a falling off of the intensity with increasing order. It will, however, remain true that strong diffuse maxima corresponding to very low orders of spectra are not to be expected. Throughout this discussion the effects of extinction have been neglected, and it has been assumed that the crystal under consideration reflects as a mosaic. Most of the experimental results refer to organic crystals, which are soft and easily distorted, and for these the assumption is probably justifiable. A treatment of the temperature scattering in terms of the dynamical theory would be a matter of considerable complexity, but it is unlikely that the results would be qualitatively very different from those obtained in this discussion.

(*k*) *The surfaces of equal diffusion for a cubic crystal:* Jahn [16] has shown how to calculate the approximate form of the surfaces surrounding each reciprocal-lattice point over which σ has a constant value. We may call these surfaces *surfaces of equal diffusion.** His treatment, adapted to the notation used in this chapter, is as follows.

Let L, M, N be the direction cosines of the vector $\mathbf{r}^*_{lll}$ from the origin

* The term *iso-diffusion surface* appears to be getting into the literature, but with the good Latin word *equal* available, this seems to be an unnecessarily unpleasant hybrid, and we shall not use it here.

to the reciprocal-lattice point $m(h, k, l)$, l, m, n * those of the wave-vector $\mathbf{g}$, and A_j, B_j, C_j those of the direction of vibration j of the lattice wave, all referred to the axes of the reciprocal lattice.

Since $$\mathbf{S}/\lambda = \mathbf{r}_m^* + \mathbf{g},$$

$$\kappa\mathbf{S}\cdot\mathbf{e}_{\phi j} = 2\pi|\mathbf{r}_m^*|(\mathrm{L}A_j + \mathrm{M}B_j + \mathrm{N}C_j) + 2\pi|\mathbf{g}|(lA_j + mB_j + nC_j). \quad (5.46)$$

As an approximation, we may assume $|\mathbf{r}_m^*| \gg |\mathbf{g}|$, which is justifiable in many cases, and we can then neglect the terms involving $|\mathbf{g}|$ in taking the square of the right-hand side of (5.46). In a cubic crystal, for the reciprocal-lattice point $m(hkl)$,

$$\mathrm{L}|\mathbf{r}_m^*| = ha^*, \quad \mathrm{M}|\mathbf{r}_m^*| = ka^*, \quad \mathrm{N}|\mathbf{r}_m^*| = la^*,$$

and so, by (5.42), to the degree of approximation assumed,

$$\sigma = \frac{k\mathrm{TN}}{m} \frac{(a^*)^2}{|\mathbf{g}|^2} \sum_j \frac{(hA_j + kB_j + lC_j)^2}{v_{\phi j}^2}. \quad (5.47)$$

If r is the distance from the reciprocal-lattice point $m(hkl)$ of a point on the surface of equal diffusion, we may write, since $r = |\mathbf{g}|$,

$$r^2 \sim \sum_j \frac{(hA_j + kB_j + lC_j)^2}{v_{\phi j}^2}. \quad (5.48)$$

To the degree of approximation used, the surface of equal diffusion is centro-symmetrical about the recriprocal-lattice point. The more accurate expression, (5.46), would not give this result.

The three velocities $v_{\phi j}$ corresponding to the three independent components of the elastic wave $g(lmn)$ in a cubic crystal of density ρ are given in terms of the three elastic constants c_{11}, c_{12}, c_{44} by the three solutions of the simultaneous equations

$$\left.\begin{aligned} \{c_{11}l^2 + c_{44}(m^2+n^2) - \rho v_j{}^2\}\,A_j + (c_{12}+c_{44})(lmB_j + lnC_j) &= 0,\\ \{c_{11}m^2 + c_{44}(n^2+l^2) - \rho v_j{}^2\}\,B_j + (c_{12}+c_{44})(mnC_j + mlA_j) &= 0,\\ \{c_{11}n^2 + c_{44}(l^2+m^2) - \rho v_j{}^2\}\,C_j + (c_{12}+c_{44})(nlA_j + nmB_j) &= 0. \end{aligned}\right\} \quad (5.49)$$

In using these equations, we are assuming the diffuse maxima to depend only on the acoustic waves, and not on the vibrations of atom against atom in a complex unit; and we are also neglecting the effects of dispersion, or the dependence of the velocity of the waves on the frequency. For obtaining a general picture of the effect, and of its dependence on the elastic properties of the crystal, this procedure is justifiable, but the limitations of the results so derived must be borne in mind.

It is not difficult to determine the radii of the surfaces of equal diffusion along a number of specified directions, for which the equations take simple forms, and so to map out their shape. For example, if the waves travel parallel to the cubic axis [100], $l = 1$, $m = 0$, $n = 0$, and equations (5.49) reduce to

$$(c_{11} - \rho v_j{}^2)A_j = (c_{44} - \rho v_j{}^2)B_j = (c_{44} - \rho v_j{}^2)C_j = 0,$$

* Confusion can hardly be caused by the occurence of l with different meanings in the direction cosines l, m, n, and the indices of the spectra.

which have the solutions

$$\left.\begin{array}{llll} v_1^2 = c_{11}/\rho; & A_1 = 1, & B_1 = 0, & C_1 = 0, \\ v_2^2 = c_{44}/\rho; & A_2 = 0, & B_2 = 1, & C_2 = 0, \\ v_3^2 = c_{44}/\rho; & A_3 = 0, & B_3 = 0, & C_3 = 1, \end{array}\right\} \tag{5.50}$$

corresponding to a longitudinal wave of velocity $v_1 = \surd(c_{11}/\rho)$, and two independent transverse waves with equal velocities $\surd(c_{44}/\rho)$ vibrating in the directions [010] and [001]. In general, c_{44} is considerably greater than c_{11}.

Substituting the results (5.50) in (5.48), we obtain for the radius of the surface of equal diffusion in the direction [100] for the spectrum hkl

$$r^2[100] = h^2/c_{11} + (k^2 + l^2)/c_{44}. \tag{5.51}$$

The corresponding radii for the directions [010] and [001] are given by

$$r^2[010] = k^2/c_{11} + (l^2 + h^2)/c_{44}; \; r^2[001] = l^2/c_{11} + (h^2 + k^2)/c_{44}. \tag{5.51a}$$

For waves propagated along the two-fold axis [110], and for the corresponding radii of the surfaces of equal diffusion parallel to this direction, the following equations apply. For the velocities and directions of vibration,

$$\begin{array}{llll} 2\rho v_1^2 = c_{11} + c_{12} + 2c_{44}; & A_1 = B_1 = 1/\surd 2, & C_1 = 0, \\ 2\rho v_2^2 = c_{11} - c_{12} \quad ; & A_2 = -B_2 = 1/\surd 2, & C_2 = 0, \\ \rho v_3^2 = c_{44} \quad ; & A_3 = B_3 = 0 \quad , & C_3 = 1, \end{array}$$

the first being a pure longitudinal wave, and the other two pure transverse waves with different velocities and frequencies; and for the radii of the surfaces of equal diffusion,

$$r^2[110] = \frac{(h+k)^2}{c_{11} + c_{12} + 2c_{44}} + \frac{(h-k)^2}{c_{11} - c_{12}} + \frac{l^2}{c_{44}}. \tag{5.52}$$

The results for the directions of the five other two-fold axes can be written down at once by appropriate changes of indices and signs in (5.52).

For the trigonal axis [111],

$$r^2[111] = \frac{(h+k+l)^2}{c_{11} + 2c_{12} + 4c_{44}} + \frac{2\{(h^2+k^2+l^2) - (hk+kl+lh)\}}{c_{11} - c_{12} + c_{44}}, \tag{5.53}$$

from which again the values for the other trigonal axes can be obtained by the appropriate interchanges. Reference may be made to Table 1 in Jahn's paper for the intercepts, in the special directions considered, for a number of other types of reciprocal-lattice points.

The distribution about the origin point (000), that is to say for directions of scattering making very small angles with the direction of the primary beam, must be considered separately. In this case, the first term in (5.46) vanishes, since $|\mathbf{r}_m^*| = 0$, and the term hitherto neglected becomes the only one. Jahn shows that when, as before, dispersion of the elastic waves is neglected, and only acoustic frequencies are considered, the intensity distribution is a function of (l, m, n) only, which means that the contours of the surfaces of equal diffusion surrounding the point (000) are radial straight lines.

Even in a cubic crystal, the waves propagated in a general direction are not pure transverse and longitudinal, and the expression for the surfaces of equal diffusion becomes somewhat complicated. For a treatment of this case, reference may be made to the work of Jahn (*loc. cit.*).

(*l*) *The shapes of the surfaces of equal diffusion in some particular cases:* The form of the surfaces of equal diffusion about any given reciprocal-lattice point may vary considerably for different types of cubic crystal. For crystals such as sodium and lead, for example, $c_{11}-c_{12}<2c_{44}$, while for others, such as rock salt and sylvine, $c_{11}-c_{12}>2c_{44}$, and this affects the form of the surfaces materially. For example, Jahn has shown that the radius of the section of a surface of equal diffusion about the reciprocal-lattice point (100) by a plane passing through the point, and perpendicular to [001], is given by

$$r^2(\phi)=\frac{c_{44}\cos^2\phi+c_{11}\sin^2\phi}{c_{11}c_{44}+\frac{1}{4}(c_{11}+c_{12})(c_{11}-c_{12}-2c_{44})\sin^2 2\phi}, \tag{5.54}$$

ϕ being the angle between r and the direction [100]. For crystals of the sodium type, the second term in the denominator of (5.54) is negative, and this leads to relatively large values of r when $\sin^2 2\phi=1$, that is to say, when $\phi=45°$, 135°, 225°, and 315°; while for crystals of the sylvine type both terms in the denominator are always positive, and no prominent maxima in r appear.

Fig. 73 has been drawn for the case of sodium, from the values of the elastic constants quoted by Jahn, $c_{12}=0{\cdot}760\,c_{11}$, $c_{44}=0{\cdot}914\,c_{11}$; for fig. 74, the results of Voigt for rock salt were used, $c_{12}=0{\cdot}284\,c_{11}$, $c_{44}=0{\cdot}279\,c_{11}$. Curve (*a*) in each figure shows a section through the surface surrounding the reciprocal-lattice point (200) by a plane passing through the point and perpendicular to [001]. The sections by a plane perpendicular to [010] are the same as those shown in the figure, while the section by a plane perpendicular to [100] is in each case a circle. The surface for sodium has well-marked projections in the directions $\pm[110]$, $\pm[1\bar{1}0]$, $\pm[101]$, $\pm[\bar{1}01]$; the surface for rock salt shows no such projections.

In both figures, curves (*b*), (*c*), and (*d*) are sections through the reciprocal-lattice point (220). In each case, (*b*) is a section by a plane perpendicular to [001], and (*c*) the section by a plane perpendicular either to [100] or [010]; while the left-hand side of (*d*) shows one half of the section by a plane perpendicular to [110], and the right-hand side one half of the section by a plane perpendicular to $[1\bar{1}0]$. Corresponding points of the surface are lettered in the same way in the sections (*b*), (*c*), and (*d*).

The sensitiveness of the form of the surface to the elastic properties of the crystal suggests that a study of the diffuse scattering may become an important method of investigating them, particularly for crystals whose elastic properties depend in a more marked way on direction than do those of cubic crystals, and we shall return to this aspect of

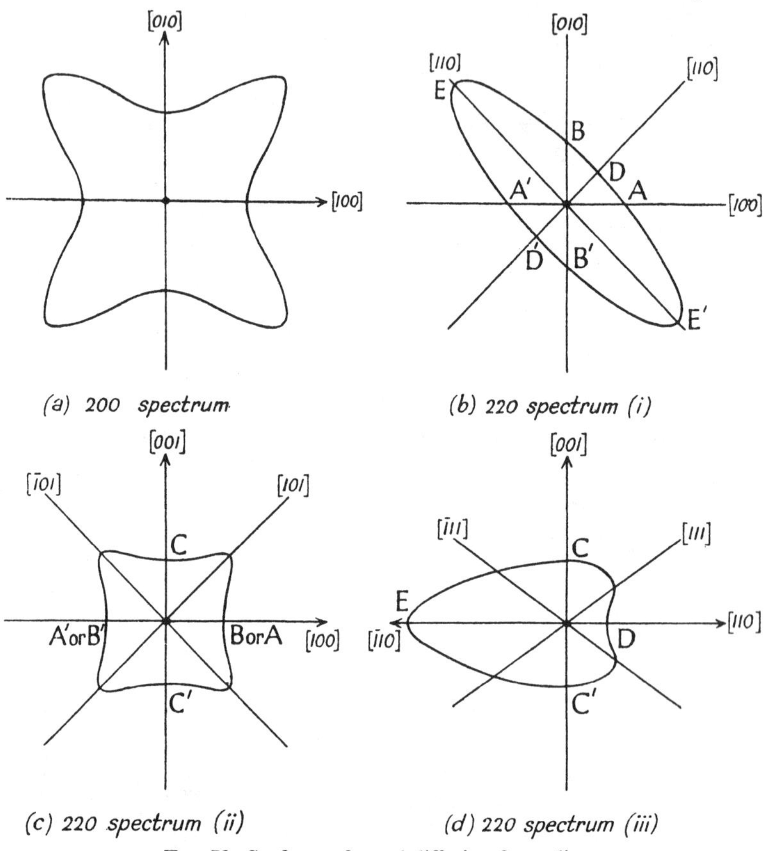

(*a*) 200 *spectrum*

(*b*) 220 *spectrum* (*i*)

(*c*) 220 *spectrum* (*ii*)

(*d*) 220 *spectrum* (*iii*)

FIG. 73. Surfaces of equal diffusion for sodium

the matter in discussing the experimental work on diffuse scattering in Section 3 of this chapter.

Figs. 73 and 74 are drawn to illustrate the form of the surfaces, rather than their relative magnitudes, but it is plain from the form of the equations for $r^2[lmn]$ that the radius of the surface corresponding to a given intensity of scattering is, *ceteris paribus*, greater the smaller the elastic constants of the crystal, and that so far as order of magnitude is concerned the radii are inversely as the square-roots of the elastic constants. We may therefore expect to get well-marked effects due to diffuse scattering from crystals that are not too hard or too rigid; and as a matter of fact they are particularly well shown by crystals of organic compounds, which usually fulfil this condition, although they are by no means confined to such crystals, and are exhibited even by diamond.

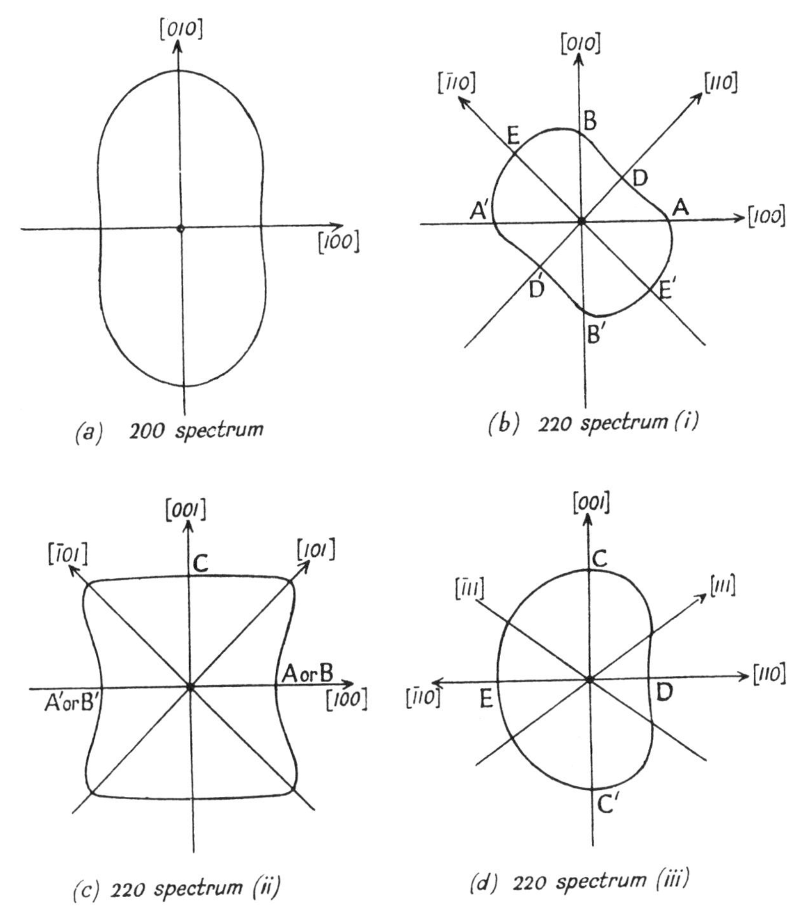

FIG. 74. Surfaces of equal diffusion for rock salt

For a cubic crystal consisting of scattering points arranged on a primitive lattice, the radii of corresponding surfaces are proportional to the order of the spectrum $\surd(h^2+k^2+l^2)$, but in a real crystal the atomic scattering factor, the Debye factor, and the polarisation factor will tend to reduce the intensities for the higher orders. We shall discuss this point in considering the experimental work.

(*m*) *The numerical evaluation of the temperature factor* e^{-2M} *for a cubic crystal by Debye's method:* In V, § 1(*f*), we obtained for 2M, the exponent of the temperature factor for the lattice spectra, the expression

$$2M = \sum_{\phi j} G_{\phi j} = \frac{\hbar}{mN} \sum_{\phi j} \frac{(\kappa \mathbf{S} \cdot \mathbf{e}_{\phi j})^2 (\bar{n}_{\phi j} + \frac{1}{2})}{\omega_{\phi j}}. \tag{5.55}$$

In the absence of detailed knowledge of the lattice frequencies, $\omega_{\phi j}$, it is not possible to evaluate this expression exactly in any actual case. It will be seen, however, that the determination of M involves a summation over all the lattice frequencies, and does not require an explicit knowledge of the individual frequencies, such as was required for calculating the diffuse scattering in a given direction. For this reason, it is possible in certain cases to use approximate methods that lead to numerical values of M probably not very far from the truth. We shall consider in detail the method introduced by Debye, which is applicable to a cubic crystal composed of atoms all of one kind.

Most of the lattice waves are long in comparison with the dimensions of the unit cell, and for such waves the crystal may to a good approximation be considered as a continuous solid. Debye assumed a cubic crystal to behave to a first approximation as an *isotropic* continuous solid, in which all waves of a given type would be propagated with the same speed, whatever their lengths and directions, although the speeds of longitudinal and transverse waves would of course differ.

In a discontinuous lattice, it is not true that waves of different lengths travel with the same speed, even in the same direction; and we have already seen that it is by no means true that all waves of the same length travel with the same speed in different directions. Indeed, pure transverse and pure longitudinal waves are possible only in certain directions, even in a cubic crystal. Nevertheless, by assuming the crystal to behave with regard to the transmission of elastic waves as an isotropic solid, without structure, and suitably choosing the average values of the velocities of the longitudinal and transverse waves, it is possible to obtain a fair approximation to M.

As we have already seen, a solid block of the lattice containing N atoms can transmit only 3N independent waves. We shall suppose these to be the N longitudinal, and 2N transverse waves represented by the N points of the phase-lattice of V, § 1(*d*); but we shall suppose all the waves of any one type to be transmitted with the same speed as they would be in an isotropic solid whose elastic properties approximate to those of the crystal block for long waves. We shall take the three values of the index j as referring to the one longitudinal and the two transverse components of the waves.

If v, ω, $\boldsymbol{\varphi}$, and $\mathbf{k}$ are corresponding values of velocity, frequency, phase-vector, and wave-vector, we have for a cubic lattice

$$\boldsymbol{\varphi} = \mathbf{k}a, \quad \omega = v/|\mathbf{k}|, \quad |\boldsymbol{\varphi}| = a\omega/v. \tag{5.56}$$

The assumption that we have made above is that v is constant for a given type, j, of wave.

There are $N/(2\pi)^3$ points in each unit volume of the phase-lattice. For the isotropic solid, the number of independent waves of type j having a value of $|\boldsymbol{\varphi}|$ less than a given maximum value ϕ_m is equal to the number of phase-lattice points inside a sphere of radius ϕ_m; and if ϕ_m is the maximum value of ϕ permitted by the lattice, this number must be equal to N. Thus

$$\frac{4\pi}{3}\phi_m^3\frac{N}{(2\pi)^3}=N, \tag{5.57}$$

or, if ω_{mj} is the maximum frequency corresponding to waves of type j permitted by the lattice, our assumption as to the constancy of velocity allows us to write

$$\omega_{mj}^3=\frac{3}{4\pi}(2\pi)^3\left(\frac{v_j}{a}\right)^3, \tag{5.58}$$

where v_j is the velocity of elastic waves of type j. We see from this that the value of ω_{mj} depends upon whether the waves are longitudinal or transverse.

Debye, in his theory of specific heats, assumes a mean maximum frequency ω_m defined in terms of the velocities v_t and v_l of the transverse and longitudinal waves by the equation

$$\omega_m^3\left(\frac{1}{v_l^3}+\frac{2}{v_t^3}\right)=\frac{9}{4\pi}\frac{(2\pi)^3}{a^3}, \tag{5.59}$$

but for the moment we shall consider the longitudinal and transverse waves separately.

Since the density of points in the phase-lattice is large if N is large, the summation over ϕ in (5.55) may be replaced by an integration over the volume of the phase-lattice. The number of phase-lattice points corresponding to values of $|\boldsymbol{\varphi}|$ between $|\boldsymbol{\varphi}|$ and $|\boldsymbol{\varphi}|+d|\boldsymbol{\varphi}|$ is $4\pi|\boldsymbol{\varphi}|^2d|\boldsymbol{\varphi}|\,N/(2\pi)^3$, which may be used as the element of volume in the integration. It is, however, more convenient to change the integration variable to ω, putting $4\pi|\boldsymbol{\varphi}|^2d|\boldsymbol{\varphi}|=4\pi(a/v)^3\omega^2d\omega$, by (5.56), and taking the limits of integration from 0 to ω_{mj}.

Now $(\kappa\mathbf{S}\cdot\mathbf{e}_{\phi j})^2$ in (5.55) is κ^2S^2 times the square of the projection of the unit vector $\mathbf{e}_{\phi j}$ on the direction of S, and for a wave of a given type j, $\mathbf{e}_{\phi j}$ takes all directions with equal probability during the integration over the sphere for a given value of ω. The mean square of its projection in any one direction such as S is therefore 1/3; and since the mean-square amplitude of the wave is the same whatever the direction of j, for the isotropic medium, we can replace $(\kappa\mathbf{S}\cdot\mathbf{e}_{\phi j})^2$ by its mean value $\kappa^2S^2/3$ in the integration with respect to ω. Equation (5.55) may therefore be written

$$2M=\frac{4\pi}{3}\kappa^2S^2\frac{\hbar}{m}\left(\frac{a}{2\pi}\right)^3\sum_j\frac{1}{v_j^3}\int_0^{\omega_{mj}}(\bar{n}_\omega+\tfrac{1}{2})\,\omega\,d\omega, \tag{5.60}$$

where ω_{mj} is the maximum frequency for waves of type j.

Again changing the integration variable, putting

$$\xi = \hbar\omega/kT, \quad x = \hbar\omega_m/kT, \tag{5.61}$$

and using for $\bar{n}_\omega$ the value given in (5.20), we obtain for the integral in (5.60)

$$\int_0^{\omega_m} (\bar{n}_\omega + \tfrac{1}{2})\,\omega\, d\omega = \left(\frac{kT}{\hbar}\right)^2 \int_0^x \left\{\frac{\xi}{e^\xi - 1} + \tfrac{1}{2}\xi\right\} d\xi$$

$$= \omega_m^2 \{\Phi(x)/x + 1/4\}, \tag{5.62}$$

where

$$\Phi(x) = \frac{1}{x}\int_0^x \frac{\xi\, d\xi}{e^\xi - 1}. \tag{5.63}$$

On substituting for $(a/v_j)^3$ in (5.60) from (5.56), we get finally

$$2M = \frac{\kappa^2 S^2 \hbar}{m} \sum_j \frac{1}{\omega_{mj}} \{\Phi(x_j)/x_j + 1/4\}. \tag{5.64}$$

The function $\Phi(x)$ was evaluated numerically by Debye, who gives a table of its values in his paper. A table will also be found in the *International Crystal Tables*,* Table IV, p. 574, Vol. 2.

In equation (5.64), the temperature is contained in the parameter x_j. According to Debye's theory of specific heats, the value of C_v, the specific heat at constant volume, is given as a function of the temperature for all monatomic cubic elements by the same expression, provided that the temperature is expressed as a multiple of a certain temperature Θ characteristic of the substance. This 'characteristic temperature', as Debye calls it, is defined by the equation

$$\hbar\omega_m = h\nu_m = k\Theta, \tag{5.65}$$

where ν_m is the maximum frequency of the elastic vibrations of the solid. Thus, by (5.61),

$$x = \Theta/T. \tag{5.66}$$

The characteristic temperature for a given crystal determines, roughly speaking, what temperatures may be considered as 'high' in dealing with the crystal. If $T > \Theta$, or $x < 1$, quantum phenomena become relatively unimportant, and we can assume as a good approximation equipartition of energy among the different modes of vibration of the lattice.

Strictly speaking, since Θ depends on ω_{mj}, which in its turn depends on the velocity of the waves, different values of Θ should be used for longitudinal and transverse vibrations. We cannot of course separate them, and a mean value must be used. Debye's Θ is defined in terms of the mean ω_m of equation (5.59); we take here a rather different mean value, Θ_M, assuming a mean of the expression under the sum-

* *Internationale Tabellen zur Bestimmung von Kristallstrukturen*, Bornträger, Berlin, 1935.

mation sign in (5.60) for different values of j, and obtain, since $\kappa^2 S^2 = 16\pi^2(\sin^2\theta)/\lambda^2$,

$$
\begin{aligned}
2M &= \frac{3\kappa^2 S^2 \hbar^2}{mk\,\Theta_M}\left\{\frac{\Phi(x)}{x}+\frac{1}{4}\right\} \\
&= \frac{12h^2}{mk\,\Theta_M}\left\{\frac{\Phi(x)}{x}+\frac{1}{4}\right\}\frac{\sin^2\theta}{\lambda^2},
\end{aligned} \tag{5.67}
$$

Θ_M being a characteristic temperature defined in terms of a rather different maximum frequency from that of Debye, but not differing greatly from it numerically. We shall return to this point in the next section. In (5.67), we have put $\hbar = h/2\pi$, h being the usual form of Planck's constant.

The value of M obtained by Debye in 1914 was of the same form as (5.67), but of only half the magnitude. The necessity for the extra factor of 2 was pointed out by Waller, and is due to the method used in counting the normal modes of the vibrating lattice.

The term 1/4 in the brackets comes from the so-called zero-point energy of vibration of the lattice, which is a necessary consequence of the quantum theory of the harmonic oscillator. The term also appears in Debye's expression, although this dates from a time before the development of wave-mechanics. Its necessity had been inferred by Planck, from the form taken by the quantum expression for the mean energy of an oscillator at high temperatures.

Equation (5.67) may also be written in the form

$$M = \frac{6h^2 T}{mk\,\Theta_M^2}\left\{\Phi(x)+\frac{x}{4}\right\}\frac{\sin^2\theta}{\lambda^2}, \tag{5.68}$$

which is very convenient, since for $x<1$, or T higher than the characteristic temperature, $\Phi(x)+x/4$ differs only slightly from unity. Table V. 1 gives values of this quantity as a function of x, taken from a paper by Zener.[17]

TABLE V. 1

x	$\Phi(x)+x/4$	x	$\Phi(x)+x/4$
0·0	1·000	1·2	1·040
0·2	1·001	1·4	1·054
0·4	1·004	1·6	1·069
0·6	1·010	1·8	1·087
0·8	1·018	2·0	1·107
1·0	1·028	2·5	1·164

Since M is equal to $8\pi^2\overline{u_s^2}(\sin^2\theta)/\lambda^2$, where $\overline{u_s^2}$ is the mean-square displacement of a lattice-point parallel to **S**, we find by comparing this expression with (5.68) the following value for the mean-square displacement in terms of the temperature and physical constants for the cubic crystal

$$\overline{u_s^2} = \frac{3h^2\mathrm{T}}{4\pi^2 mk\,\Theta_{\mathrm{M}}^2}\left\{\Phi(x) + \frac{x}{4}\right\}. \tag{5.69}$$

(*n*) *Determination of the characteristic temperature :* The temperature factor can be determined numerically for the simple cubic crystals to which the above treatment applies if the characteristic temperature is known. There are three principal methods by which it can be estimated, (*a*) from measurements of the variation of the specific heat with temperature, (*b*) from measurements of the elastic constants of the crystal, and (*c*) from optical measurements of the selective frequencies of absorption and reflection of infra-red radiation by the crystal, the last method applying to ionic crystals, such as rock salt.

The most reliable method is probably the first. According to Debye's theory, the specific heats of all simple cubic monatomic elements at constant volume are given by the same function* of $x(=\Theta/\mathrm{T})$. If Θ is properly chosen for each substance, the curve giving the specific heat as a function of T/Θ should be the same for all of them. At very low temperatures, C_v should vary as $(\mathrm{T}/\Theta)^3$, and this is closely confirmed by experiment. The values of Θ giving the closest agreement can be found by comparing the variations of the specific heats of a number of substances at very low temperatures. Values of Θ obtained in this way for a number of elements are given in Table V. 2. These values were very kindly prepared for us by Dr. F. Simon, F.R.S., to whom we are greatly indebted. The values for certain simple diatomic compounds whose crystals have cubic symmetry, and which obey Debye's law fairly closely, have also been included in the table.

We have seen that Θ will differ according to whether it is based on the maximum frequency for longitudinal or for transverse waves. A mean, Θ_{D}, is used in the theory of specific heats, which is related to Θ_l and Θ_t, the values for longitudinal and transverse waves by the equation

$$3\,\Theta_{\mathrm{D}}^{-3} = \Theta_l^{-3} + 2\,\Theta_t^{-3}. \tag{5.70}$$

Now, by (5.68), we see that the appropriate mean value of Θ for the temperature factor, Θ_{M}, will be given by

$$3\,\Theta_{\mathrm{M}}^{-2} = \Theta_l^{-2} + 2\,\Theta_t^{-2}, \tag{5.71}$$

so long as the temperature is not too low. This difference has been pointed out by Zener and Bilinsky,[18] who show that Θ_{M} is always larger than Θ_{D} by a few per cent. The way in which the difference may be estimated will be best understood by considering the relation of Θ to the elastic constants.

The elastic constants determine the velocities of the longitudinal and

* Often known as Debye's function.

TABLE V. 2

Values of the characteristic temperature, Θ, based on specific heats

Substance	Θ in degree abs.	Substance	Θ in degree abs.
H_2	(105)	Ta	245
D_2	(97)	Bi	(100)
		Cr	485
He	25–35	Mo	380
Ne	64	W	310
A	80		
Kr	63	Fe	430
Xe	55	Co	410
		Ni	400
Cu	320		
Ag	210	Pd	275
Au	175	Ir	285
		Pt	230
Be	(900)		
Mg	(320)		
Ca	230		
Zn	(220)		
Cd	(155)	NaCl	281
Hg	(95)	KCl	230
		KBr	177
Al	390	CaF_2	474
Tl	(93)	FeS_2	645
		Diamond	1860
Sn(wh)	(130)		
Pb	88		

The values in brackets refer to elements not crystallising in the cubic system, but give nevertheless a fair representation of the average energy of the lattice vibrations.

transverse waves. Let K be the compressional elasticity, n the rigidity, and σ Poisson's ratio, the ratio of the lateral contraction to the longitudinal extension when a rod of the substance under consideration is stretched. The rigidity modulus, n, determines the velocity of the transverse waves; for longitudinal waves in which no lateral motion occurs, the appropriate elastic modulus is $K+4n/3$. If v_t and v_l are the velocities of transverse and longitudinal waves, and ρ is the density of the substance,

$$v_t^2 = n/\rho, \quad v_l^2 = (K+4n/3)/\rho, \tag{5.72}$$

and n, K, and σ are connected by the relations

$$n = \frac{3K(1-2\sigma)}{2(1+\sigma)}, \quad K+\frac{4n}{3} = \frac{3K(1-\sigma)}{1+\sigma}. \tag{5.73}$$

If $\nu_{mj}(=\omega_{mj}/2\pi)$ is the maximum frequency, and v_j is the corresponding velocity,

$$\nu_{mi}=\left(\frac{3N}{4\pi}\right)^{\frac{1}{3}}\frac{v_j}{\rho^{\frac{1}{3}}M^{\frac{1}{3}}}, \tag{5.74}$$

by (5.58), where N is Avogadro's number, $6{\cdot}03\times10^{23}$, and M is the atomic weight of the material, since a^3, the volume of unit cell of the lattice, is equal to $M/\rho N$. Using (5.72), (5.73), and (5.74), we find, since $h\nu_m=k\Theta$,

$$\Theta_l=\frac{h}{k}\left(\frac{3N}{4\pi}\right)^{\frac{1}{3}}\frac{K^{\frac{1}{2}}}{M^{\frac{1}{3}}\rho^{\frac{1}{6}}}\left\{\frac{3(1-\sigma)}{1+\sigma}\right\}^{\frac{1}{2}},$$

$$\Theta_t=\frac{h}{k}\left(\frac{3N}{4\pi}\right)^{\frac{1}{3}}\frac{K^{\frac{1}{2}}}{M^{\frac{1}{3}}\rho^{\frac{1}{6}}}\left\{\frac{3(1-2\sigma)}{2(1+\sigma)}\right\}^{\frac{1}{2}}. \tag{5.75}$$

The average value of Θ appropriate to the specific heat theory, Θ_D, given by (5.70), is therefore

$$\Theta_D=\frac{h}{k}\left(\frac{9N}{4\pi}\right)^{\frac{1}{3}}\frac{K^{\frac{1}{2}}}{M^{\frac{1}{3}}\rho^{\frac{1}{6}}f^{\frac{1}{3}}(\sigma)}, \tag{5.76}$$

where

$$f(\sigma)=\left\{\frac{1+\sigma}{3(1-\sigma)}\right\}^{\frac{3}{2}}+2\left\{\frac{2(1+\sigma)}{3(1-2\sigma)}\right\}^{\frac{3}{2}},$$

while Θ_M, the value most appropriate to the temperature factor, is given by

$$\Theta_M=\frac{h\sqrt{3}}{k}\left(\frac{3N}{4\pi}\right)^{\frac{1}{3}}\frac{K^{\frac{1}{2}}}{M^{\frac{1}{3}}\rho^{\frac{1}{6}}f'^{\frac{1}{2}}(\sigma)},$$

where

$$f'(\sigma)=\frac{1+\sigma}{3(1-\sigma)}+\frac{4(1+\sigma)}{3(1-2\sigma)}. \tag{5.77}$$

Table V. 3, taken from the paper of Zener and Bilinsky,[18] gives the ratio Θ_M/Θ_D for a series of values of Poisson's ratio.

TABLE V. 3

σ	0·25	0·3	0·35	0·4	0·45	0·5
$\dfrac{\Theta_M}{\Theta_D}$	1·025	1·027	1·032	1·040	1·051	1·070

The values of Θ calculated from the elastic constants are in good general agreement with those obtained from the theory of specific heats, an indication of the general applicability of the theory. On account of the difficulties and uncertainties inherent in the measurement of the elastic constants of crystals, it is probable, however, that the thermal values are more reliable. Equations (5.76) and (5.77) show that the characteristic temperature of a crystal is to a large extent determined by its compressibility. Crystals with a large compressibility (i.e. small K) have a low characteristic temperature and a large temperature factor of X-ray reflection; those with a low compressibility have a high characteristic temperature and a small temperature factor. In general, we may say that the harder the crystal the smaller will be its temperature factor. Other things being equal, the lower the density and atomic weight the larger Θ, and the smaller the temperature factor. Diamond, which combines great hardness with low atomic weight and density, has the highest characteristic temperature of any known substance.

(*o*) *The contribution of the different frequency ranges of the elastic spectrum to M:* According to equations (5.67) and (5.68), M consists of two parts, one, M_0 independent of the temperature, having its origin in the zero-point energy of the lattice, the other, M_T, depending on the temperature. M_0 and M_T are given by

$$M_0 = \frac{3}{2}\frac{h^2}{mk\Theta}\frac{\sin^2\theta}{\lambda^2} = \frac{3}{2}\frac{h}{m\nu_m}\frac{\sin^2\theta}{\lambda^2}, \tag{5.78}$$

$$M_T = \frac{6h^2T}{mk\Theta^2}\Phi(x)\frac{\sin^2\theta}{\lambda^2} = \frac{6kT}{m\nu_m^2}\Phi(x)\frac{\sin^2\theta}{\lambda^2}, \tag{5.79}$$

where ν_m is the maximum frequency of the elastic spectrum, and $x = h\nu_m/kT$. We shall consider first the temperature-dependent part, M_T, the only part of M that can be determined by direct experiment. M_T is not simply proportional to T, for $\Phi(x)$ is also a function of T.

The contributions of the different portions of the elastic spectrum to M_T are determined by $\Phi(x)$. Fig. 75 shows $\xi/(e^\xi - 1)$, the integrand of the integral in $\Phi(x)$, as defined by (5.63), plotted as a function of the frequency, ν, A separate curve must be drawn for each value of T, but all such curves may be derived from the curve for one temperature by the proper change of scale of the abscissae. The value of $\Phi(x)$ for a crystal for which the maximum frequency is ν_m is the area between the curve and the axis of ν up to the limit ν_m, divided by ν_m. In fig. 75, curves are drawn for a series of temperatures ranging from 25° abs. up to 900° abs. The limiting frequencies, ν_m, are also shown for NaCl($\Theta = 281°$, $\nu_m = 59 \times 10^{11}$ sec^{-1}), and for aluminium ($\Theta = 398°$, $\nu_m = 83 \times 10^{11}$ sec^{-1}).

It will be seen that, for all values of ν_m, $\Phi(x)$ approaches unity for very large values of T. For a given value of ν_m, the lower the temperature the smaller $\Phi(x)$, and the smaller the contribution of the higher frequencies to it, and so to M_T. It will be seen for example, that at 900° abs. all frequency ranges possible in the crystal make approxi-

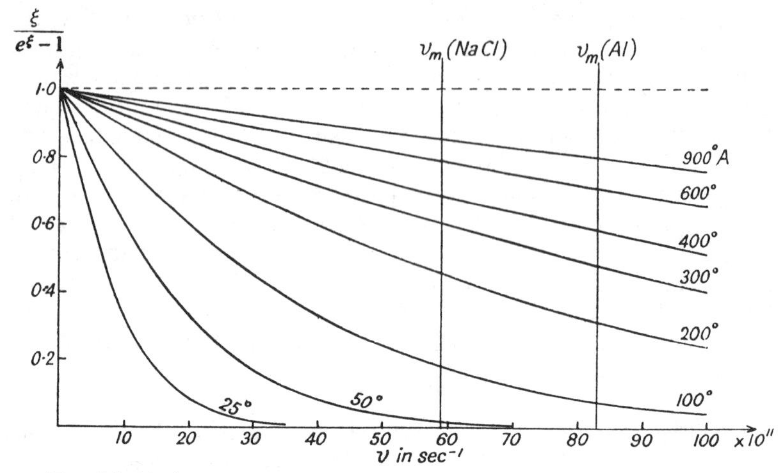

FIG. 75. $\xi/(e^\xi - 1)$ as a function of the frequency ν, for different values of the absolute temperature

mately equal contributions to $\Phi(x)$ for NaCl, so that the conditions are nearly those of equipartition of energy between the different normal modes of the lattice; whereas at 25° abs. the higher frequencies make a negligible contribution. It is for this reason that the Debye formula for the specific heat agrees best with experiment at low temperatures; for then only the long waves make any appreciable contribution, and, for these, the assumption that the solid behaves as a continuum is justified.

The value of $\Phi(x)$ at any temperature is the ratio of the area beneath the corresponding curve up to the limit ν_m to the area of the rectangle of height unity and base ν_m. At any given temperature therefore $\Phi(x)$ is smaller the larger ν_m, or the higher Θ. In the case of diamond for example, for which $\Theta = 1860°$ abs., and $\nu_m = 390 \times 10^{11}$ sec^{-1}, $\Phi(x)$ is considerably less than unity for any reasonable temperature. Moreover, we see from (5.79) that M_T is inversely proportional to ν_m^2, even apart from the effect of ν_m on $\Phi(x)$; so that for substances such as diamond M_T is not very appreciable at any temperature at which measurements of the intensities of X-ray spectra can be made. This is in agreement with the observed result that temperature has no measurable effect on the intensity of reflection of X-rays by diamond.

M_0, the part of M depending on the zero-point vibrations, is inversely proportional to ν_m, while M_T, as we have just seen, is inversely proportional to ν_m^2 at high temperatures, and to a rather higher power of ν_m at lower temperatures. For crystals with small ν_m, such as rock salt, M_T is the more important part of M, but as ν_m becomes larger, the relative importance of M_0 increases, and for very large ν_m it may be much the larger part of M. The formulae (5.78) and (5.79) give for rock salt at room temperature, 290° abs., if λ is expressed in Ångström units,

$$M_T = 1{\cdot}206\ (\sin^2\theta)/\lambda^2, \quad M_0 = 0{\cdot}347\ (\sin^2\theta)/\lambda^2.$$

For diamond, on the other hand,

$$M_T = 0{\cdot}020\ (\sin^2\theta)/\lambda^2, \quad M_0 = 0{\cdot}127\ (\sin^2\theta)/\lambda^2.$$

The part of M depending on the temperature is inappreciable, but the effect of the zero-point vibration on the intensity of reflection of X-rays will be by no means negligible for large values of $(\sin\theta)/\lambda$.

(*p*) *The limitations of the Debye-Waller formula for M:* In deducing the numerical value of M by Debye's method, a number of approximations have been made to which attention has already been drawn; we shall here summarise them briefly, and attempt to estimate their importance.

(i) Debye's theory assumes the volume of the crystal to remain constant; it refers throughout to the specific heat at constant volume, whereas all measurements that we can make on the reflection of X-rays by crystals are carried out under conditions of constant pressure, the volume of the crystal changing with changing temperature.

The existence of this volume change with temperature shows that it is not sufficient to express the energy of the crystal simply as a quadratic function of the displacements of its atoms, which is the assumption underlying the expression of the lattice vibrations in terms of normal co-ordinates, but that higher powers of the displacements must be involved. Waller [19] has made calculations to investigate the effect of such terms on the temperature factor. The results are qualitatively of the right kind, but it is not possible to estimate their magnitude numerically with any certainty.

Zener and Bilinsky [18] have attempted to allow numerically for the expansion of the lattice by treating the characteristic temperature, Θ, as a function of the temperature of the lattice, instead of as a constant, and we shall return to their work when discussing the agreement between theory and experiment.

(ii) Another fundamentalassumption in Debye's theory is that waves of all lengths and frequencies and directions are propagated with the same speed, so that the actual crystal is replaced by a model isotropic continuum, the discrete nature of the crystal being allowed for by terminating the series of waves propagated by the continuum at a definite upper limit. The work of Born and v. Kármán shows that this assumption of a constant velocity cannot be justified for the discrete lattice, and that waves of the higher frequencies travel with speeds depending on their length. The law that follows from the assumption of the continuum, that $\rho(\nu)d\nu$ the number of waves having frequencies between ν and $\nu + d\nu$ is proportional to $\nu^2 d\nu$, cannot be correct. Recent work by Blackman [20] shows that even in a lattice composed of atoms all of one kind there will not be a single maximum of $\rho(\nu)$ at the highest frequency as there should be according to the continuum theory, but several maxima, some of which are at much lower frequencies.

(iii) The Debye theory has been applied not only to cubic crystals composed of atoms of one kind, but also to diatomic crystals, such as KCl and NaCl, whose specific heats can be fairly well represented by the Debye formula with an appropriate value of the characteristic temperature. A discussion of the normal modes of vibration of crystals of this type, such as was given by Born and v. Kármán, shows that the frequencies of the elastic spectrum do not form a continuous series, but are separated into two sets with a finite interval between them. One group, comprising the lower frequencies, corresponds to the ordinary elastic vibrations of the lattice as a whole, in which neighbouring atoms of different kinds oscillate together; while the other group, comprising the higher frequencies, corresponds on the whole to oscillations of neighbours against one another, or, if we consider the two kinds of atom lying on similar lattices, to vibrations of one lattice against the other. The higher frequencies are sometimes known as the "optical frequencies", and the lower ones as the "acoustic frequencies".

If the two kinds of atoms are oppositely charged ions, the optical frequencies are associated with periodic changes in the electric polarisation in the lattice, and so with emission and absorption of radiation. They are responsible for the selective absorption and reflection of the infra-red radiation shown by so many crystals, the so-called *Reststrahlen.*

For crystals such as KCl and NaCl, in which the two kinds of atom do not differ greatly in mass, there is no large separation between the acoustic and the optical frequencies, and the latter may, to a good approximation, be regarded as forming the limit of the elastic spectrum, and as defining the characteristic temperature. If, however, the crystal is more complicated, and contains groups such as molecules or complex ions, the vibrations of the atoms within these groups will have frequencies much higher than those of the elastic vibrations of the lattice as a whole. The acoustic and optical frequencies will then differ greatly, and the latter will contribute little to the temperature factor, since the mean-square amplitude of the interatomic vibrations will be small in comparison with those of the elastic waves.

(*q*) *Formal representation of the temperature factor for a complex crystal:* The effect of the temperature on the intensities of the lattice spectra from a complex crystal cannot be taken into account simply by multiplying the structure factor F, as defined in Chapter II, equation (2.14), by a temperature factor of the type e^{-M}. Formally, we must consider the mean-square vibration of each kind of atom in the unit cell separately, and if f_j is the scattering factor of an atom of type j

when at rest, we must use instead of f_j in the expression for the structure factor F, $f_j e^{-M_j}$, where $M_j = 8\pi^2 \overline{u_{js}^2} (\sin^2\theta)/\lambda^2$, $\overline{u_{js}^2}$ being the mean-square displacement of atoms of type j in the direction of S. Then

$$F(hkl) = \sum_j f_j e^{-M_j} e^{2\pi i \left(\frac{hx_j}{a} + \frac{ky_j}{b} + \frac{lz_j}{c}\right)}, \qquad (5.80)$$

x_j, y_j, z_j being the mean co-ordinates of the atom j in the unit cell.

The formal proof of this expression is easily obtained by an extension of that given for the simple lattice in Chapter I, § 3(*b*). We need not carry out the more elaborate process of averaging given in this chapter, since, as we have seen, it affects only the general scattering. In the expression (*2*.1), Chapter II, for the amplitude, we replace $\boldsymbol{\rho}_1$, $\boldsymbol{\rho}_2$, ... the vectors fixing the atoms in the unit cell, by $\boldsymbol{\rho}_1 + \mathbf{u}_{n1}$, $\boldsymbol{\rho}_2 + \mathbf{u}_{n2}$, ... , $\mathbf{u}_{n1}$, $\mathbf{u}_{n2}$, ... being the displacements of the atoms 1, 2, ... from their mean positions in the unit cell n. The structure-amplitude must then be included under the summation over n. If we form the square of the modulus of this sum, and take its mean value for all possible displacements, assuming that the mean-square displacement of an atom of the same type is the same in all the unit cells, and that products of the displacements of different atoms vanish on averaging, we obtain without difficulty equation (*5*.80) for the structure factor. It is not necessary to go through the proof, as no new principles are involved.

It will be evident from (*5*.80) that if M differs considerably for the different atoms of the unit we cannot use a single temperature factor of the type e^{-M} multiplying the whole structure factor. There exists no method of calculating $\overline{u_j^2}$ directly from theoretical considerations, although it has been determined experimentally in some simple cases, as we shall see later.

It will be seen that the effect of thermal movement cannot be allowed for with any certainty for any but the simplest types of crystal. It is only because most of the lattice waves that have an important influence on the temperature factor lie in those regions of frequency where the unit cell may be considered as vibrating as a whole that formulae such as we have developed do apparently apply fairly well in many cases.

(*r*) *The temperature factor for crystals not having cubic symmetry:* With the exception of the last paragraph, the whole discussion has so far dealt with crystals having cubic symmetry, for which it has been possible to make the assumption that $\overline{u_s^2}$ is the same at a given temperature, whatever the direction of S relative to the crystal, so that M is a function of $(\sin\theta)/\lambda$ alone, and not of the indices of the spectrum, except in so far as these determine $\sin\theta$.

For crystals of lower symmetry, this will not in general be true. The forces acting on an atom when it is displaced from its mean position may well differ greatly in different directions, and the amplitude of the atomic vibrations for a given average energy cannot be independent of the direction of displacement. To take a rather extreme case, it is not to be expected that in a crystal such as mica the mean amplitude of the atomic vibrations will be the same for directions lying in the cleavage planes as for directions perpendicular to them.

Very little progress has been made towards the theoretical calculation of M for anisotropic crystals, but it is possible to state certain very general rules that must be obeyed. Let us consider only the elastic vibrations, leaving the optical frequencies out of account for the moment. Then, as a result of the general theory of elasticity, it must be possible to express the mean-square amplitude of vibration $\overline{u_s^2}$ in any direction in terms of the mean-square displacements $\overline{u_x^2}$, $\overline{u_y^2}$, $\overline{u_z^2}$ in three mutually perpendicular directions, provided that these three directions, x, y, z, coincide with the principal axes of elasticity of the crystal. We can express $\overline{u_s^2}$ in the form

$$\overline{u_s^2} = l^2\overline{u_x^2} + m^2\overline{u_y^2} + n^2\overline{u_z^2}, \tag{5.81}$$

where l, m, n are the direction cosines of the direction of S with respect to the principal axes.

If the crystal has no symmetry, or only a centre of symmetry, there is nothing in its external form to determine the directions of the principal axes; but, if it has a two-fold axis, one principal axis must lie along this; and if it has an axis of symmetry higher than two-fold, one principal axis must lie along this while the other two may be chosen in any two mutually perpendicular directions at right angles to it. The mean-square displacement in all directions at right angles to the symmetry axis will then be the same.

In a trigonal, tetragonal, or hexagonal crystal, let the z axis be the principal symmetry axis. Then $\overline{u_x^2} = \overline{u_y^2} = \overline{u_p^2}$, say, and if ϕ is the angle that the direction of S makes with the symmetry axis,

$$\overline{u_s^2} = \overline{u_p^2} \sin^2 \phi + \overline{u_z^2} \cos^2 \phi. \tag{5.82}$$

The temperature factor for the reflection at a set of planes in any direction can, if (5.82) is true, be expressed in terms of those for a set perpendicular to, and a set parallel to the axis of symmetry. Suppose that $M_z = B_z(\sin^2 \theta)/\lambda^2$, and $M_p = B_p(\sin^2 \theta)/\lambda^2$ are the exponents of the temperature factor determined from planes perpendicular to and parallel to the axis respectively. Then M_s, for planes perpendicular to S, is given by

$$M_s = (B_p \sin^2 \phi + B_z \cos^2 \phi)(\sin^2 \theta)/\lambda^2, \tag{5.83}$$

a formula that should apply to trigonal, tetragonal, or hexagonal crystals containing one kind of atom. Zener [17] has attempted to calculate M_s numerically for zinc and cadmium, both hexagonal crystals, in terms of the elastic constants, but the values he obtained are not at all in good agreement with the measurements of Brindley [21] for zinc, although Brindley finds that he can express his results in the form (5.83). As we shall see in Section 3 of this chapter, it is probably easier to obtain information as to the vibrations in non-cubic crystals by studying the diffuse scattering than by studying the reduction of the intensities of the spectra.

(*s*) *Rotations of molecules and groups of atoms in crystals :* Vibrational motions, such as we have discussed in this chapter, are not the only kind of thermal movement that may occur in a crystal structure. The possibility that molecules, or groups of atoms, held in a crystal structure may rotate has been discussed by Pauling.[22] According to the ideas of quantum mechanics, there is always a small probability at any temperature that a group of atoms held in a position of equilibrium in a crystal structure may turn end for end, even although classically it may possess insufficient energy to do so. This probability increases with increasing temperature, and the oscillations of the group about a stable position of equilibrium may pass over into more or less free rotation.

Pauling investigates the criterion for vibration or rotation in certain simple cases, and finds that rotation may occur at quite moderate temperatures. He concludes that some of the transitions of crystals to allotropic forms are connected with the setting in of rotations of some of their constituent groups.

In this chapter, we are concerned not with the thermodynamics and mechanics of such rotations, but with their effect on the intensities of the diffraction spectra given by the crystal with X-rays. No general rules for the calculation of the effect can be given; it must depend on the type of group, and on the nature of the rotation. It is perhaps worth while, however, to illustrate the principles involved by considering a simple case.

Let a group consist of a central atom, whose scattering factor is f_0, surrounded by a number of atoms, whose scattering factors are f_1, f_2, ... f_s, at vector distances $\boldsymbol{\rho}_1$, $\boldsymbol{\rho}_2$, ... $\boldsymbol{\rho}_s$, from it, and let the magnitudes of these vectors be constant. Let the central atoms of a number N of such groups occupy the points of a simple space-lattice, and remain fixed while the groups rotate about them. If the rotation is a thermal effect, its rate will be slow in comparison with the frequency of the incident radiation, and we can consider any instantaneous configuration of the groups of the lattice as persisting unchanged during a very large number of oscillations of the radiation field. Let $\boldsymbol{\rho}_{n1}$, $\boldsymbol{\rho}_{n2}$, ... $\boldsymbol{\rho}_{ns}$, be the vectors defining the instantaneous configuration of the group occupying the lattice-point n, which is distant $\mathbf{r}_n$ from the origin. Then the amplitude scattered by the crystal when the directions of incidence and scattering are defined by the vector $\mathbf{S}$ is proportional to

$$\mathrm{A} = \sum_n f_0 e^{i\kappa \mathbf{S}\cdot\mathbf{r}_n} + \sum_n \Big(\sum_s f_s e^{i\kappa \mathbf{S}\cdot(\mathbf{r}_n + \boldsymbol{\rho}_{ns})}\Big).$$

For the direction of a spectrum, $e^{i\kappa \mathbf{S}\cdot\mathbf{r}_n}=1$, and so

$$A=\sum_n f_0+\sum_n\left(\sum_s f_s e^{i\kappa \mathbf{S}\cdot\boldsymbol{\rho}_{ns}}\right). \qquad (5.84)$$

If all directions of the group are equally probable during the rotation, and N, the number of lattice-points, is large, each sum over n for a given value of s in (5.84) corresponds to a uniform distribution of scattering matter over the surface of a sphere; for all directions of $\boldsymbol{\rho}_{ns}$ are equally likely. We have already dealt with the diffraction due to a spherical shell in Chapter III, § 2(*b*), and can at once write A in the form

$$A=N\left\{f_0+\sum_s f_s\frac{\sin\phi_s}{\phi_s}\right\}, \qquad (5.85)$$

where $\phi_s=4\pi|\boldsymbol{\rho}_s|(\sin\theta)/\lambda$. The rotation of the groups can thus be taken into account by ascribing to each an effective scattering factor, equal to the expression in the brackets in (5.85), and using this in the structure factor, the corresponding co-ordinates being those of the centre of the group. If, for example, the group in question were the NO_3^- group of the nitrates, the effective scattering power would be $f_N+3f_0(\sin\phi_0)/\phi_0$. The radius of the spherical shell would be about 1·25A, and the effect of the oxygen atoms would thus fall off very rapidly, becoming zero for $(\sin\theta)/\lambda=0{\cdot}2$, and then becoming negative up to $(\sin\theta)/\lambda=0{\cdot}4$. The effect of the rotation is that the oxygen contributes little to any spectra except those of quite low order. We have assumed all orientations of the group to be equally probable during the rotation, so that the effective group for scattering has spherical symmetry. This is evidently the simplest assumption that can be made, but it is not likely to be true in many actual cases. It would be possible to solve the problem when certain orientations were preferred, but the calculation would as a rule be one of some complexity, and no such case will be considered here.

Evidence, based on the diffraction of X-rays, that rotation actually occurs, has been given by a number of workers. Kracek, Hendricks, and Posnjak [23] have investigated the crystal forms of NH_4NO_3 at different temperatures, and find that there are five modifications. The one that is stable above 125° C. is cubic, and has one molecule of the salt in the unit cell, a structure that is not possible unless each group has spherical symmetry, at all events statistically. They suggest that both NH_4^+ and NO_3^- rotate freely about their central atoms. The NO_3^- group will then behave, in so far as its scattering power for X-rays is concerned, like a spherical atom with an f-factor equal to that deduced above. The authors also suggest that similar rotations may occur in the crystalline states of the nitrates of calcium, barium, strontium, and lead, for which the agreement between the intensities of the observed and calculated X-ray spectra is unsatisfactory.

Certain cases are known of crystals having a higher apparent space-group symmetry than can be accounted for from their constitution and the number of molecules in the unit cell, and this may perhaps be due to rotations of certain of their constituent groups to give an average group with a higher effective symmetry than that possessed by the group at rest. There can be little doubt that rotations such as we have discussed do occur, but it would be very desirable to test some such case by making proper absolute measurements of the intensities of the X-ray spectra, and comparing them with those to be expected on the hypothesis of rotation. Caution must in any case be used in explaining away awkward inconsistencies between observed and calculated intensities in this manner.

Another very interesting case is that of the paraffins. Müller [24] has shown that when crystals of the paraffins are heated to within a few degrees of their melting-points certain of their X-ray spectra disappear in a manner that could be explained by assuming the long chains to rotate about their length. Hendricks,[25] in an investigation of the long-chain amine halides at ordinary temperatures, obtained results indicating that the aliphatic chains were straight; but, on repeating the experiments at the temperature of liquid air, he found evidence of the usual zig-zag chain. These results could be explained by assuming rotation of the chain to set in about the direction of its length at the higher temperature. It is not very easy to calculate quantitatively the effect of such a rotation on the intensities of the spectra, but it is plain that the spectra due to the spacing along the length of the chain should persist, while all evidence of the zig-zag nature of the chain must disappear. Bernal [26] has also obtained evidence of rotation in the crystals of the long-chain alcohol dodecylol, based on optical observations of the change from a higher to a lower symmetry on cooling. The higher symmetry appears to be consistent only with rotation of the molecules about their long axes.

Such rotations represent the first stages of the breakdown of the crystalline structure, and of the approach to the amorphous state, but the fact that the mean position of the centre or axis of the group or molecule retains its regular position on the lattice involves certain fixed phase-relationships between the waves scattered from the different groups and molecules, and makes the treatment of the diffraction problem differ from that for the case of a gas consisting of molecules of the same kind. The square of the scattering factor given by (5.85) for the group is not proportional to the intensity per molecule scattered by the same group as a gas, as we shall see in a later chapter.

2. Experimental Tests of the Formulae for the Debye Temperature Factor

(*a*) *Introductory:* Debye's prediction that the intensities of the diffraction spectra obtained from a crystal by means of X-rays should decrease as the temperature of the crystal increases was soon verified by W. H. Bragg,[27] who, in 1914, carried out experiments with rock salt and sylvine (KCl). With rock salt he made measurements at 15° C. and 370° C., and demonstrated both the diminution of the intensity of the interference maxima at the higher temperature, and the change in their position due to the expansion of the lattice. He concluded that the effect was of the order of magnitude to be expected from Debye's theory, but the methods of measuring the intensities of the spectra were at that time not well enough developed to make a really accurate test possible. Backhurst,[28] in 1922, made measurements of the temperature factor for aluminium, corundum, carborundum, diamond, and graphite. He found only very general agreement with the theory of Debye. Of the crystals he used, however, only diamond and aluminium are cubic. For diamond, Backhurst found a negligible temperature factor, as was to be expected from its hardness and high characteristic temperature.* For aluminium, the results were not in agreement with the theory, but were probably much affected by recrystallisation and by annealing due to the heating, which would influence the primary and secondary extinction, an effect not at the time well understood. The same may probably be said of the results of Collins,[29] who also worked with aluminium.

In 1925, James [30] made measurements with rock salt, using both molybdenum and rhodium radiation. About thirty reflections were investigated, at high temperatures ranging from 290° abs. to 900° abs. These experiments showed that the exponent M of the temperature factor varied as $(\sin^2 \theta)/\lambda^2$, as the theory demands, but that the magnitude of the effect was very considerably greater than that to be expected from Debye's formula, as given in his paper of 1914. James concluded that his results were best represented by a formula of the type $M = BT^2(\sin^2 \theta)/\lambda^2$, B being a constant. According to the theory, however, M should vary very approximately as T over the range of temperature used in the experiments, and as T^2 only at temperatures below about 50° abs.

In all the experiments so far considered, the crystal had been heated above the temperature of the room. Now, as we have already seen in § 1(*p*) of this chapter, it is not to be expected that the theoretical formula will be valid if the temperature is so high that the energy of the crystal cannot be expressed as a quadratic function of the displacements and velocities of its atoms. It appeared therefore to be desirable to make measurements at low temperatures, where the amplitudes of the vibrations remain small. There is also much less chance of recrystallisation, and consequent change of extinction, if the temperature of the crystal is varied, say, between that of the room and that of liquid air, than there is if the temperature is raised. Extinction effects of this kind

* This result was confirmed for low temperatures by Ehrenberg, Ewald and Mark (*Zeit. f. Krist.*, **66**, 547 (1928)).

may be very marked, and can easily vitiate the results completely unless care is used (see, for example, James, *loc. cit.*, p. 592).

In 1927, James and Miss Firth [31,32] extended the measurements with rock salt to the temperature of liquid air. In the meantime Waller had published his modified form of the Debye theory, which doubled the value of the exponent in the temperature factor. From the temperature of liquid air up to about 500° abs. the results of James and Miss Firth agreed well with the amended theory, but, at higher temperatures, the decrease in intensity of the spectra with increasing temperature was still far too rapid. Entirely similar results were obtained by James and Brindley [33] with sylvine, which, because of the similarity in mass of the atoms of potassium and chlorine, approximates very closely to the ideal cubic lattice of the Debye theory; while James, Brindley, and Wood [34] found close agreement with theory in the case of aluminium, when it was cooled to the temperature of liquid air.

It may be concluded that the Debye-Waller formula represents the temperature factor for cubic crystals of simple type with fair accuracy, so long as the temperature is not too high. At high temperatures, however, the increasing expansion of the crystal causes changes in its elastic properties, and the assumption of a constant characteristic temperature in the formula for M is no longer justifiable; nor can the amplitudes of vibration of the atom any longer be regarded as small. At high temperatures, therefore, the simple theory breaks down.

(*b*) *The temperature factor for sylvine* (*KCl*): We shall take as an example for more detailed discussion the case of sylvine, which, as we have seen, approximates closely to the type of simple cubic lattice for which the theory was developed.

If ρ_0 and ρ_T are the intensities* of a given spectrum at a standard temperature T_0, and at T respectively,

$$\rho_T/\rho_0 = e^{-2M_T}/e^{-2M_0}, \tag{5.86}$$

where M_0 and M_T are the values of M of (5.67) for the temperatures T_0 and T. If we neglect the small variation in the glancing angle θ due to the change in the lattice spacing with temperature, and assume the characteristic temperature to be independent of T, we obtain

$$\left(\frac{\lambda}{\sin\theta}\right)^2 \log_e\left(\frac{\rho_T}{\rho_0}\right) = \frac{12h^2}{mk\Theta}\left\{\frac{\Phi(x_0)}{x_0} - \frac{\Phi(x)}{x}\right\}, \tag{5.87}$$

where $x = \Theta/T$, and $x_0 = \Theta/T_0$.

In the experiments of James and Brindley on sylvine, T_0 was the

* For the intensity of a spectrum, the integrated reflection, as defined in §2(*e*), Chapter II, is to be understood. The assumption is made here, and throughout this chapter, that the appropriate formulae are those for the mosaic crystal, and not those of the dynamical theory. The evidence supporting this will be found in the papers quoted, and will be discussed in the next chapter.

laboratory temperature, 290° abs., and the measurements were made for values of T ranging from the temperature of liquid air, 86° abs., to about 900° abs. The curve A in fig. 76 shows $(\lambda/\sin\theta)^2 \log_{10}(\rho_T/\rho_0)$ plotted as a function of T, with $\Theta = 230°$, the value deduced from the

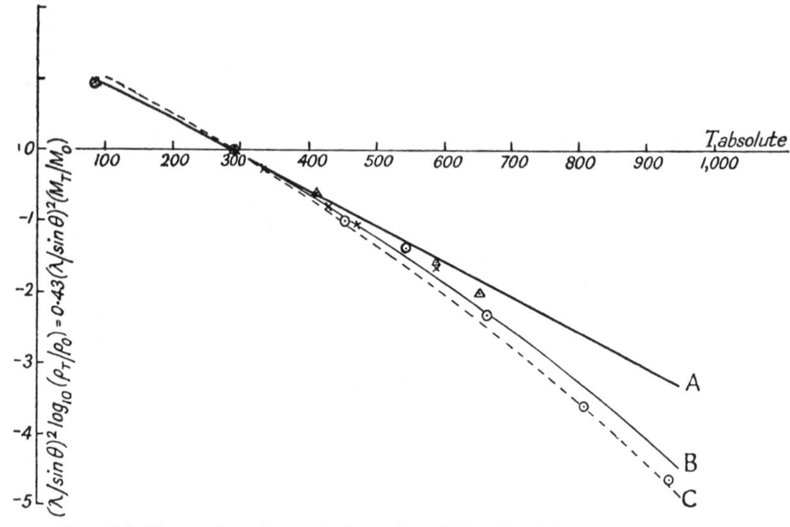

FIG. 76. Illustrating the variation of $M(\lambda/\sin\theta)^2$ with temperature. Experimental points from 400, ⊙, from 600, △, and from 800, ×. Theoretical curves: A, from the Debye-Waller theory, B and C from the same theory as modified by Zener and Bilinsky

variation of the specific heat at low temperatures. For m in (5.87), the mean mass of an atom of potassium and an atom of chlorine is used. The circles, triangles, and crosses represent the experimental values obtained respectively from the 400, the 600, and the 800 spectra. It will be seen that there is very fair agreement between theory and experiment up to about 400° abs., but that above this temperature the observed decrease in intensity of the spectra becomes very much greater than that given by the Debye-Waller theory. We have already seen that agreement between experiment and the simple theory is hardly to be expected at high temperatures, since the conditions assumed in the theory are manifestly not fulfilled.

The Debye theory implies that the volume of the crystal remains constant as the temperature changes, and Θ is deduced from the specific heat at constant volume, C_v. Now in all experiments on the temperature factor it is the pressure to which the crystal is subjected that remains constant, while its volume changes, owing to the thermal expansion. The elastic properties must also vary with the temperature, and in so far as the properties of the crystal may be expressed in terms of a characteristic temperature, we must suppose this to be not a constant, but a function of the temperature of the crystal. Zener and Bilinsky [18] have given a method for estimating the change of Θ with temperature, and have applied it to correct the Debye-Waller formula for this effect.

$\Theta(T)$, the characteristic temperature at temperature T, may be defined by the equation

$$C_v(T) = C_D\left\{\frac{\Theta(T)}{T}\right\}, \tag{5.88}$$

where C_D is the Debye function of the theory of specific heats, and $C_v(T)$ is the specific heat at constant volume at temperature T. In principle, it would be possible to determine $\Theta(T)$ from the variation of C_v with temperature, but the method is not practicable, because of the insensitiveness of C_D to the value of Θ when T is greater than Θ. Zener and Bilinsky therefore employ another method, which consists in comparing the observed entropy of the crystal at any temperature T_1, which is given by

$$S_{obs}(T_1) = k\int_0^{T_1}\frac{C_p\,dT}{T}, \tag{5.89}$$

with the entropy calculated from a Debye model, in which the crystal is supposed to be first expanded at absolute zero to a volume equal to its volume at T_1, and then heated from zero to T_1 at constant volume. During this process, the characteristic temperature may be treated as constant, and equal to its value at T_1. Assuming this model crystal to obey the Debye law, we have for its entropy at T_1

$$S_D(T_1) = k\int_0^{T_1}\frac{C_D\{\Theta(T_1)/T\}}{T}\,dT, \tag{5.90}$$

since the change in volume at absolute zero involves no entropy change. Now the two values of the entropy given by (5.89) and (5.90) must be the same at all temperatures. Using this fact, Zener and Bilinsky obtain the equation

$$\frac{d}{dT}(\log_e\Theta) = KC_p,$$

where

$$K = (C_p - C_v)/TC_pC_v, \tag{5.91}$$

a quantity that is very nearly constant for simple solids over the whole range of temperatures concerned.

Equation (5.91) leads on integration to

$$\Theta_T/\Theta_0 = e^{-KH(T)},$$

where

$$H(T) = \int_0^T C_p\,dT. \tag{5.92}$$

Θ_0 is the value of Θ as deduced from the behaviour of the specific heat at very low temperatures.

Zener and Bilinsky give in their paper tables of the numerical values of $\log_e(\Theta_T/\Theta_0)$ for rock salt and sylvine, and curves B and C of fig. 76 are drawn from these data, using equation (5.74) in the form

$$\left(\frac{\lambda}{\sin\theta}\right)^2 \log_e \frac{\rho_T}{\rho_0} = \frac{12h^2}{mk}\left\{\frac{\Phi(x_0)}{x_0\Theta_{290}} - \frac{\Phi(x)}{x\Theta_T}\right\}, \tag{5.93}$$

with $x_0 = \Theta_{290}/T$, $x = \Theta_T/T$. For curve B, a value of Θ_0 equal to 240° has been used, and for curve C a value equal to 230°. It will be seen that these curves reproduce in a general way the observed decrease in intensity with rising temperature. Except for those corresponding to the two highest temperatures, the experimental points are fitted better by the higher value of the characteristic temperature, 240°, than by 230°, the value deduced from the specific heats; but, as we have already seen in § 1(*n*) of this chapter, we should expect Θ_M, the appropriate value of Θ for the temperature factor, to be greater than Θ_D, the value deduced from the theory of specific heats.

(*c*) *The mean-square amplitude of the atomic vibrations:* We have seen that M, the exponent of the temperature factor, can be written in the form

$$M = 8\pi^2 \overline{u_s^2}(\sin^2\theta)/\lambda^2, \tag{5.94}$$

where $\overline{u_s^2}$ is the mean-square displacement of the atoms of the lattice in the direction S, and perpendicular to the planes at which the spectrum may be considered as formed by reflection at a glancing angle θ. If therefore M can be determined experimentally, we can at once calculate the mean-square displacements of the atoms from equation (5.94). As a matter of fact, we cannot determine M by experiment, but only the difference between its values at two different temperatures. According to equation (5.67), M contains a term not depending on the temperature, but due to the zero-point vibrations of the atoms, which is cancelled in taking the difference of M at two temperatures. This term can, of course, be included, and the mean-square displacement calculated accordingly; but it is to be remembered that this goes beyond direct experiment. All that can be determined directly by comparing the intensities of a spectrum at two temperatures is the difference between the mean-square displacements at those temperatures.

Estimates of the mean-square displacements of the atoms in a crystal were first made by James and Miss Firth[32] for rock salt. More accurate calculations from the same experimental data were made shortly afterwards by Waller and James,[35] who were able to determine separately the values of $\overline{u^2}$ for the atoms of sodium and chlorine.

For the calculation of $\overline{u^2}$ it is convenient to express M in a rather different form. From (5.63),

$$\Phi(x) = \frac{1}{x}\int_0^x \frac{\xi\, d\xi}{e^\xi - 1}.$$

The integrand may be expanded in terms of Bernoulli's numbers if $|\xi|<2\pi$, giving

$$\frac{\xi}{e^{\xi}-1}=1-\tfrac{1}{2}\xi+\tfrac{1}{12}\xi^2-\tfrac{1}{720}\xi^4+\dots .$$

On integration, and division by x^2, we obtain

$$\frac{\Phi(x)}{x}=\frac{1}{x}-\tfrac{1}{4}+\tfrac{1}{36}x-\tfrac{1}{3600}x^3+\dots, \tag{5.95}$$

an expansion that is valid for $T>\Theta/2\pi$, and so over most of the range of experiment for crystals such as rock salt and sylvine.

We now substitute this expression in equation (5.67) for M. The term 1/4 in M due to the zero-point energy is cancelled by the term $-1/4$ in (5.95), and since $x=h\nu_m/kT$, we obtain, using (5.69),

$$\overline{u_s^2}=\beta T+\gamma/T+\delta/T^3+\dots, \tag{5.96}$$

where

$$\beta=\frac{3k}{4\pi^2 m\nu_m^2}=\frac{3h^2}{4\pi^2 mk\Theta_m^2},\quad \gamma=\frac{1}{12}\left(\frac{h}{2\pi}\right)^2\frac{1}{mk},\quad \delta=-\frac{1}{1200}\left(\frac{h}{2\pi}\right)^2\frac{\Theta_m^2}{mk}. \tag{5.97}$$

The coefficient γ contains only known constants, and δ is small, and can be estimated nearly enough from approximate values of Θ. The values of γ and δ being known, β can be determined from the difference of M at two temperatures. If M_1 and M_2 refer to the temperatures T_1 and T_2,

$$M_1-M_2=\frac{8\pi^2\sin^2\theta}{\lambda^2}\left\{\beta(T_1-T_2)+\gamma\left(\frac{1}{T_1}-\frac{1}{T_2}\right)+\delta\left(\frac{1}{T_1^3}-\frac{1}{T_2^3}\right)+\dots\right\}$$
$$=\tfrac{1}{2}\log_e(\rho_1/\rho_2), \tag{5.98}$$

where ρ_1 and ρ_2 are the integrated reflections of the same spectrum at T_1 and T_2 respectively. Using the calculated values of γ and δ, and the observed value of ρ_1/ρ_2, we can at once calculate β from (5.98), and so $\overline{u_s^2}$ from (5.96). If the crystal is cubic, the mean-square displacement parallel to all directions is the same, and the mean-square of the total displacement, $\overline{u^2}$, is equal to $3\overline{u_s^2}$.

These equations apply to a crystal containing one kind of atom, and matters become a little more complicated for a crystal such as rock salt, for which the mean-square displacements of the two kinds of atom will not in general be the same. The spectra given by such a crystal are of two types, having their indices either all even or all odd. For any spectrum with even indices, the structure factor is the sum of the scattering factors of the two atoms, for one with odd indices, their difference. If the suffixes 1 and 2 refer respectively to Na and Cl, we may write for the sum and difference spectra

$$F_s=f_1e^{-M_1}+f_2e^{-M_2},\quad F_D=f_2e^{-M_2}-f_1e^{-M_1}.$$

By plotting the measured values of F_s and F_D against $(\sin\theta)/\lambda$, and taking half the sum and half the difference of corresponding ordinates of the two curves so obtained, we can find $f_1 e^{-M_1}$ and $f_2 e^{-M_2}$ separately for any given value of $(\sin\theta)/\lambda$. This was done both for the temperature of the laboratory, and for that of liquid air, and this gave $M_{290} - M_{86}$ for each kind of atom. Equation (5.98) then gave β_1 and β_2, and hence, from (5.96), the mean-square displacements of the two atoms at any temperature. By this means Waller and James [35] found for rock salt $\beta_{Na} = 6{\cdot}56 \times 10^{-21}$, $\beta_{Cl} = 5{\cdot}31 \times 10^{-21}$, and for the root-mean-square displacements at 290° abs.,

$$\sqrt{\overline{u_{Cl}^2}} = 0{\cdot}22\text{A}, \qquad \sqrt{\overline{u_{Na}^2}} = 0{\cdot}24\text{A}.$$

By the same method, James and Brindley [33] found the root-mean-square displacements to be 0·25A and 0·15A for KCl, at 290° abs. and 86° abs. respectively; while for metallic aluminium, James, Brindley and Wood [34] found for the corresponding values 0·17A and 0·11A. Shonka,[36] from an extensive series of measurements on sodium fluoride, found a mean value for the two atoms of 0·18A at laboratory temperatures; and Brindley,[37] by comparing the f-curves calculated from Hartree's method of the self-consistent field with the measured f-curves obtained from powdered crystals by Havighurst, found values for the root-mean-square displacements for NaCl and NaF in fair general agreement with those given above.

These numbers give some idea of the displacements due to thermal vibrations in such relatively soft crystals as those considered above. The distances of closest approach of the atoms in NaCl, KCl, and Al are 2·814A, 3·138A and 2·786A respectively, and the root-mean-square displacements at 290° abs. are 0·082, 0·081 and 0·061 of these distances.

(*d*) *The relation between β and the elastic constants:* It is possible to obtain a check between the theoretical and experimental values of β. Although for a crystal such as rock salt the values of β for the two atoms separately cannot be calculated, Waller has shown that if m_1 and m_2 are the masses of the two atoms, and ρ and a the density and lattice spacing of the crystal, then

$$\bar{\beta} = \frac{m_1\beta_1 + m_2\beta_2}{m_1 + m_2}$$
$$= \frac{k}{12(m_1+m_2)}\left\{\frac{\sqrt[3]{3\pi^2}}{\pi^2}\rho a^2 \frac{c_{44}(2c_{11}+c_{44}) + \frac{1}{5}b_1(c_{11}+c_{12})}{c_{11}c_{44}^2 + \frac{1}{5}b_1(c_{11}+c_{12})c_{44} + \frac{1}{105}b_1{}^2 b_2} + \frac{3}{\pi v_0{}^2}\right\}, \tag{5.99}$$

where $b_1 = c_{11} - c_{12} - 2c_{44}$, $b_2 = c_{11} + 2c_{12} + c_{44}$; c_{11}, c_{12} and c_{44} being the elastic constants of the crystal in the notation of Voigt. Voigt gives for rock salt the values

$$c_{11} = 4{\cdot}560 \times 10^{11}, \quad c_{12} = 1{\cdot}294 \times 10^{11}, \quad c_{44} = 1{\cdot}270 \times 10^{11} \text{ dynes/cm}^2.$$

For ν_0, the proper frequency corresponding to the *Reststrahlen* from rock salt, the value corresponding to a wave-length 61μ may be used. These values give $\bar{\beta}=5{\cdot}7\times10^{-21}$, which may be compared with the value $5{\cdot}81\times10^{-21}$, calculated from the experimental results, the agreement being satisfactorily close.

3. The Experimental Study of Diffuse Scattering

(*a*) *Introductory:* The experiments so far discussed have dealt only with the effect of temperature on the interference maxima, that is to say, with the first term of equation (*5.27*). The second term, $I_2(S/\lambda)$, which gives the diffuse scattering, has been discussed theoretically at considerable length in Section 1 of this chapter, and we have now to see how the conclusions reached agree with observation. The important work on diffuse scattering done by the American school under the leadership of Jauncey was started before diffuse maxima had been observed experimentally, and before the implications of the Faxén-Waller theory were generally understood, and it is based mainly on a scattering formula given by Jauncey,[38] and amplified by Jauncey and Harvey,[39] and by Woo.[40] We shall discuss this formula, and some of the work that has been based on it, in a later paragraph. It gives the average diffuse scattering, and predicts no maxima and minima, and the methods used are in fact such as to exclude their observation.

Diffuse scattering maxima must have been observed many times without their nature having been understood. Indeed radial streaks were observed by Friedrich [41] as early as 1913, and it was suggested by Faxén in 1923 that they were due to thermal movement of the atoms of the lattice. It is possible also that some part of the anomalous reflections observed by Clark and Duane [42] in 1922 and 1923 and ascribed by them to the emission of characteristic X-rays by the atoms of the crystal itself may have been due to extra reflections given by the white radiation in the primary beam. The first to realise fully the experimental implications of the Faxén-Waller theory appears to have been Laval,[15] who in 1939 published accounts of a remarkable series of observations made with the ionisation spectrometer, which confirmed fully the main predictions of the theory. Laval's work was published in a periodical not usually easily accessible to physicists, and so did not at once become generally known, and independent observations of the phenomenon were made in the United States of America, in India, and in England. In 1938, Wadlund [43] found unexplained radial streaks and spots on Laue photographs from rock salt and sylvine, taken with radiation from a tungsten target, and these were explained by Zachariasen [44] in 1940 as being due to thermal agitation. Zachariasen put forward a simplified form of the Faxén-Waller theory, which was tested by Siegel and Zachariasen,[45] Siegel,[46]

Jauncey and Baltzer,[47] Gregg and Gingrich,[48] and Kirkpatrick.[49] Zachariasen [50] in his first paper assumed all elastic waves to travel with the same speed in the crystal, an assumption which is, as we have already seen, far too simple, but in a later paper he gives a more rigorous calculation for the case of the cubic lattice. Jauncey and Baltzer,[51] found their results to be in general agreement with the more rigorous formula. A later careful study of the positions and intensities of the maxima for KCl was made by J. H. Hall.[52]

In India, Raman and Nilakantan [53] in 1940 found extra reflections on Laue photographs from diamond. These they supposed to be due to the excitation of lattice vibrations by the primary X-rays, which were then assumed to be reflected from the planes of varying density set up by the vibrations. They did not consider the thermal vibrations of the lattice to be an adequate explanation of the effects they observed and, in support of this, pointed out that the extra reflections were not diffuse, but relatively sharp,[54] and that they did not depend markedly on the temperature. In considering their work, it is important to realise that extra reflections may sometimes be produced by imperfections in the crystal due to surface disorder, or to strains. This has been pointed out by Kirkpatrick,[49] and Teague,[55] among others. Such extra reflections will be sensitive to changes in crystal structure or texture, but not to changes in temperature, unless these also produce changes in structure. Lonsdale and Smith [56,57] have shown that the extra reflections observed by Raman and Nilakantan are structure sensitive, and they also show that diamonds of the rare type, classed as Type II by Robertson, Fox, and Martin,[58] do not show these reflections at all, but that *all* diamonds show weak and more diffuse reflections which are sensitive to changes of temperature. Raman [59,60] and his colleagues have obtained extra reflections from a number of other substances, and maintain that they cannot be explained in terms of the Faxén-Waller theory. Raman's theory has been criticised by Born,[61] and by Lonsdale and Smith.[56,57] Most of the observed phenomena are explicable in detail, and in quite an unforced manner, in terms of the theory of lattice vibrations, and it would not yet appear to be necessary to intro duce an explanation of an entirely new type.

In England, Preston [62] in 1938 and 1939 observed streaks on Laue photographs from an alloy of 4 per cent. of copper in aluminium, which changed in intensity as the alloy hardened with age at room temperature. When radiation from a copper or silver target was used, spots appeared on some of the streaks, which were due to the characteristic radiation. Preston found that the streaks could be explained as being due to segregation of the copper atoms into small groups, but he found that some traces of them were visible on photographs taken within one hour of quenching the alloy. To see if these were due to segregation taking place very rapidly, he took Laue photographs of the crystal at 500° C., at which temperature there should be no segregation.

He found that the effects ascribed to ageing did disappear, but that new diffuse spots appeared, which were also found on photographs from pure aluminium, where no question of segregation could arise. The diffuse pattern was in this case more intense at high temperatures than at low, and the effects were completely reversible with temperature. Preston at first attempted to explain the effects by the existence in the crystal of small groups of atoms, a point of view that was later developed by Sir William Bragg.[63] Preston also obtained diffuse-spot patterns from rock salt and magnesium oxide.

As a result of the observation that Laue photographs from benzil showed an intricate pattern of diffuse spots and streaks, even with quite moderate exposures, Mrs. Lonsdale and H. Smith undertook an extensive investigation into the conditions under which such diffuse patterns are produced. A preliminary account of the work was given in 1940 by Lonsdale, Knaggs and Smith.[64] A full account by Lonsdale and Smith [65] contains a remarkable series of very beautiful photographs, mostly from organic crystals, some of which will be considered in more detail below. Mrs. Lonsdale has taken a prominent part in the work on diffuse reflection, and in its exposition, and reference should be made to two summarising reports by her,[66, 67] in which the whole subject is discussed very fully, and in which a detailed bibliography will be found. It is not our aim here to attempt a complete review of the experimental work, but to deal with a few of the investigations that seem to illustrate particularly clearly the main principles discussed in the theoretical section of this chapter. The treatment will not be historical. We shall consider first some of the more recent work, because much of the earlier work is more easily understood when seen in true perspective in the light of later discoveries.

(*b*) *The study of diffuse patterns on Laue photographs:* The method that has mainly been used in the investigation of diffuse maxima is to take a Laue photograph, either with the direct radiation from a tube giving characteristic X-rays, or with characteristic radiation that has first been made monochromatic by a preliminary reflection from a crystal. The diffuse maxima are very much weaker than the corresponding lattice spectra, and accordingly fairly large exposures, measured by the product of intensity and time, are necessary, especially when monochromatic radiation is used. The best detailed studies by this method are those of Mrs. Lonsdale [64] and her collaborators, who have mainly employed the 5 kW tube at the Davy-Faraday Laboratory, with anticathodes of copper, silver, or molybdenum. For the copper target, the tube was operated at about 34 kV with a current of 140 mA. With such a tube, the exposure time for an ordinary Laue photograph is a few minutes, and is still quite moderate even with monochromatised radiation. The crystals used by Lonsdale and Smith were mostly those of organic substances, which, because of their softness, show a relatively large temperature effect, and correspondingly intense diffuse patterns. Diffuse spots occur indeed on many Laue and oscillation photographs from organic crystals taken with quite moderate exposures, and had been repeatedly observed before their nature was recognised. The photographs are usually taken on flat films, placed normal to the direction of the incident beam at a distance of 3 to 4 cm. from the crystal. Quite small crystals, such as are suitable for ordinary oscillation photographs, are used. For organic crystals, Cu $K\alpha$ radiation is very suitable, but for simple inorganic crystals, which have as a rule smaller spacings than organic crystals, Mo $K\alpha$ is more suitable.

Suppose then a photograph is taken with direct radiation from a copper anticathode, the crystal being at rest. A continuous range of wave-lengths, the 'white radiation', accompanies the characteristic radiation from the target, and most of the spots on the photograph are produced by this. They form the Laue pattern, and are situated on a series of ellipses, all of which pass through the point where the direct beam intersects the plate. It may happen that one or more of these spots is very intense, because by chance the crystal is so set that it is formed by the characteristic radiation, but the main pattern is due to the white radiation. If the exposure is considerable, the Laue pattern may be accompanied by a pattern of diffuse spots and radial streaks. The shape and appearance of the diffuse spots is quite different from those of the Laue spots. If a number of them occur, it will be found that they lie either at or very near the positions at which the spots on an oscillation photograph would appear if it were taken with the characteristic radiation from the target, and with an oscillation of the crystal over a range of a few degrees on either side of the setting used for the Laue photograph. This is well illustrated by fig. 77 (Pl. III), a Laue photograph from sylvine (KCl) figured by Lonsdale and Smith.[65] Unfiltered radiation from a silver target was used. The direction of incidence was normal to [100], and inclined at an angle of 20° to [010]. The typical oscillation pattern of diffuse spots is clearly seen, superimposed on the Laue pattern.

The origin of the diffuse pattern is easily understood if we consider the distribution of the intensity function in the reciprocal-lattice space. As we have seen in V, § 1 (*h*), there will be a diffuse region of intensity about each reciprocal-lattice point. These regions are indicated by shaded circles around the points of fig. 78 (Pl. III). The actual diffuse domains are not spherical, and the figure in this respect is purely diagrammatic. It represents any plane of the reciprocal lattice perpendicular to the [100] axis. The lattice is assumed to be simple cubic, which, for the purpose of representing intensities, is a very close approximation, because of the close equality in scattering power of K and Cl. The line PO, lying in the plane through the origin, makes an angle of 20° with [010], and the circles (0), (1), and (2) are respectively the traces of the sphere of reflection in the plane through the origin and in two

PLATE III

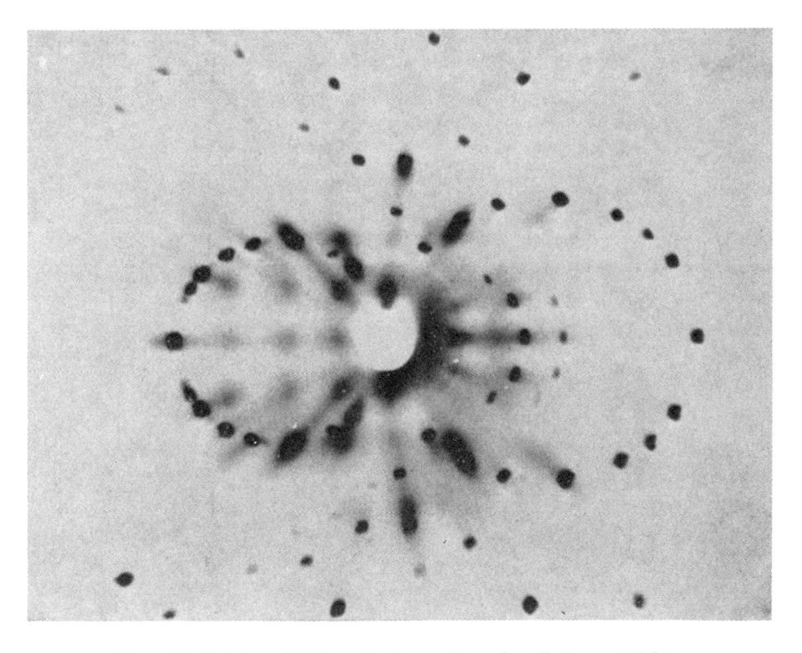

FIG. 77. Sylvine, [100] vertical: unfiltered radiation at 20° to the [010] axis (Lonsdale and Smith)
(*Proc. Roy. Soc.*, **A, 179,** (1942))

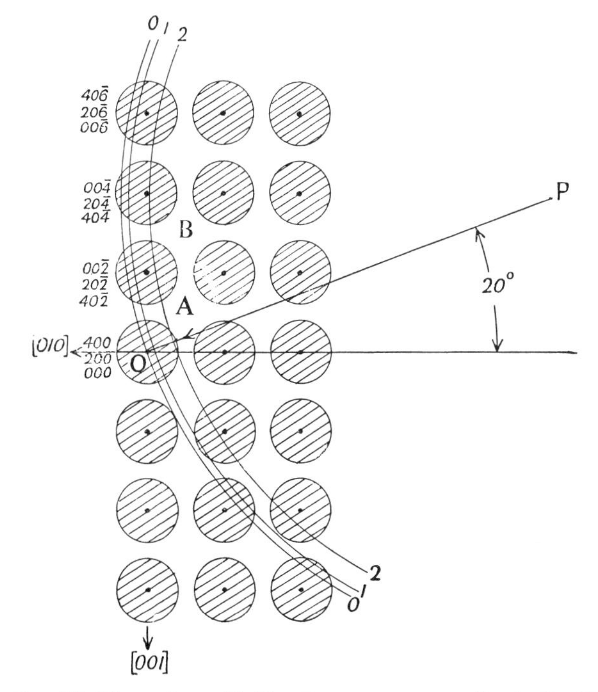

FIG. 78. The reciprocal-lattice diagram corresponding to fig. 77

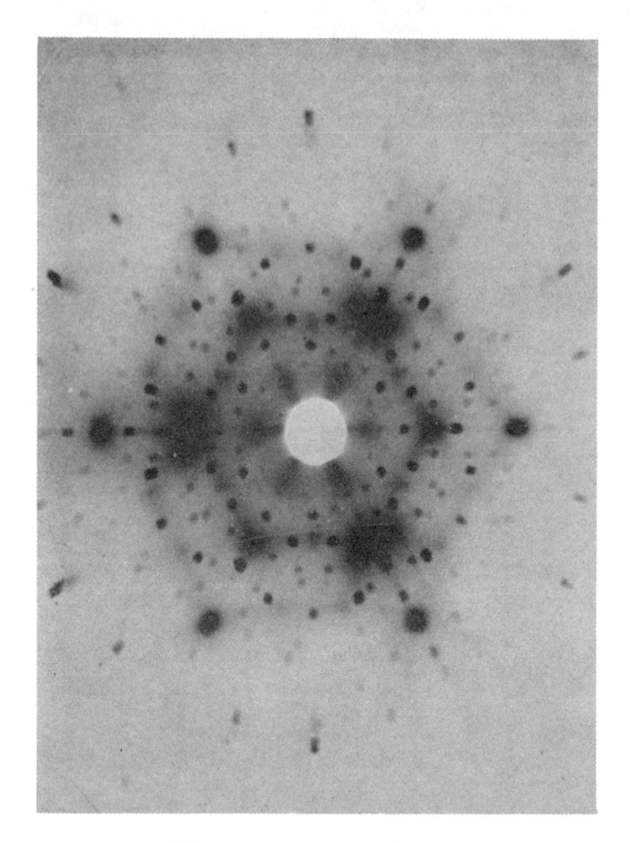

FIG. 79. Benzil, [11$\bar{2}$0] vertical: unfiltered Cu radiation along [0001] (Lonsdale and Smith, *ibid.*)

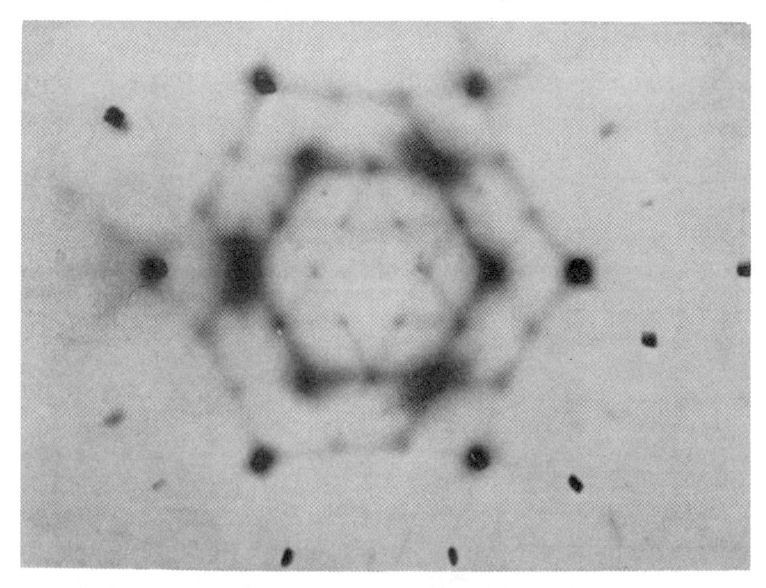

FIG. 80. Benzil, [11$\bar{2}$0] vertical: in monochromatic radiation, Cu Kα, very nearly along [0001] (Lonsdale and Smith, *ibid.*)

succeeding planes, drawn for the case of Ag Kα radiation. The diagram thus corresponds to the conditions under which fig. 77 was obtained.

If the lattice is allowed to perform a small oscillation about an axis through O perpendicular to the plane of the paper, the passage of the lattice-points through circle (0) gives the spectra on the equatorial layer line, and their passage through circles (1) and (2), the spectra on the first and second layer lines respectively. With the fixed direction of incidence, no spots lie actually on the sphere of reflection, but the diffuse domains surrounding a number of the points are intersected by the sphere, and the corresponding diffuse spots occur in the photograph, nearly in the positions in which the spectra produced by a small oscillation of the crystal would occur. The upper portion of fig. 78 corresponds to the left-hand side of the photograph in fig. 77. Point A, for example, corresponds to the first row of diffuse spots to the left of the centre of fig. 77, $00\bar{2}$, $20\bar{2}$ and $40\bar{2}$. The spot $00\bar{2}$ on the equatorial layer line is the weakest of the three and, as we see from fig. 78, the circle (0) cuts the corresponding diffuse domain near its edge. Circles (1) and (2) cut it nearer the centre, and at roughly equal distances from it, and the spots on the first and second layer lines are correspondingly much stronger, and approximately equally intense. Again, in the second row to the left, corresponding to the point B in the diagram, the spot $00\bar{4}$ on the equatorial layer line, and the spot $20\bar{4}$ on the first layer line are weak, while the spot $40\bar{4}$ on the second layer line is very strong, and nearly coincides with a very intense Laue spot. Reference to fig. 78 shows that circle (2) passes very near the reciprocal-lattice point ($40\bar{4}$), so that the conditions for selective reflection are very nearly fulfilled. Even this rough diagram, in which no attempt is made to represent the actual forms of the different domains, enables us to predict with some accuracy the general distribution of intensity in the diffuse spots.

That the diffuse reflections are formed by monochromatic radiation was shown by the work of Laval [15] and Preston,[62] and photographs obtained by Lonsdale and Smith [65] illustrate this point clearly. When radiation that has been made monochromatic by a preliminary reflection from a crystal is substituted for the direct radiation from the target, the Laue pattern, which is due to the continuous radiation, disappears, but the diffuse pattern remains almost unchanged. An excellent example of this is given by photographs obtained from the hexagonal crystal benzil, with the radiation passing along [0001]. The first, fig. 79 (Pl. IV), was obtained with unfiltered copper radiation; the second, fig. 80 (Pl. IV), with monochromatic Cu Kα radiation. Except for a few spots, which happen to be formed by Cu radiation, the Laue pattern completely disappears, and the details of the diffuse pattern are beautifully shown. In this photograph, the diffuse spots are often connected by streaks, which are non-radial, but the radial streaks that are often seen on photographs taken with direct radiation, such for example as fig. 77, are absent.

The non-radial streaks, which are formed by monochromatic radiation, may be explained by supposing the diffuse domains around neighbouring points in the reciprocal lattice to overlap to a certain extent. Suppose the interference function to have a distribution similar to that shown, purely diagrammatically, in fig. 81. If SS′ is the trace of the sphere of reflection, it is clear that the diffuse spots corresponding to the points A and B and C will be joined by a connecting streak. That such overlapping of the domains of neighbouring spots does sometimes occur was shown very directly by Laval. Photographs such as fig. 80 show in a rather striking way a section of the reciprocal lattice of the crystal.

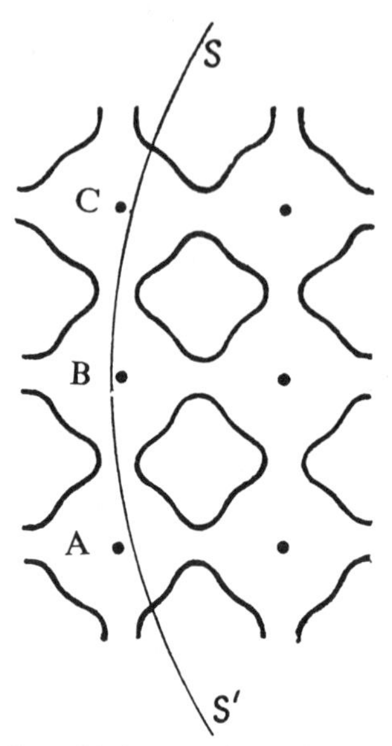

FIG. 81. Explanation of non-radial streaks with monochromatic radiation

(*c*) *The radial streaks:* The radial diffuse streaks, like the Laue spots, are formed by the continuous radiation. We may also study their formation by means of the reciprocal lattice, but for this purpose it is convenient to draw the diagram in such a way that for radiation of wave-length λ the translations of the reciprocal lattice are given by

$$\mathbf{a}\cdot\mathbf{a}^* = \mathbf{b}\cdot\mathbf{b}^* = \mathbf{c}\cdot\mathbf{c}^* = \lambda$$

instead of by

$$\mathbf{a}\cdot\mathbf{a}^* = \mathbf{b}\cdot\mathbf{b}^* = \mathbf{c}\cdot\mathbf{c}^* = 1,$$

the relation used elsewhere in this book. The radius of the sphere of reflection will then be unity, independent of the wave-length, but the linear dimensions of the reciprocal lattice are proportional to the wave-length.

When radiation from an anticathode such as copper is used, a continuous band of radiation is present in addition to the strong characteristic copper radiation. Let the dots in fig. 82 represent the points in one plane of the reciprocal lattice for the characteristic wave-length. We suppose for the sake of argument that wave-lengths down to a lower limit of one half of the characteristic wave-length are present in the radiation, and the crosses represent the reciprocal-lattice points for this wave-length. In the diagram, for example, A and A′ are the points corresponding to (110) for the two wave-lengths, and the points for intermediate wave-lengths lie between A and A′ on the line AA′, which is radial with respect to the origin. Points corresponding to longer wave-lengths lie on the same line, beyond A, which will coalesce with the line for shorter wave-lengths through (220). The appropriate reciprocal-lattice diagram thus consists of a series of radiating lines, one in the

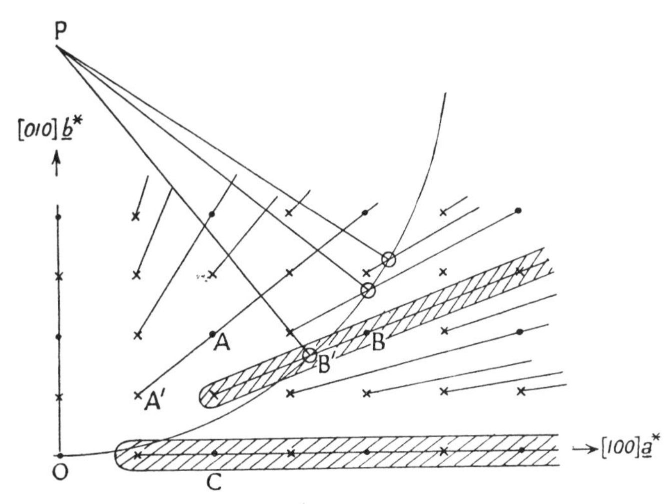

FIG. 82. Explanation of Laue spots and radial streaks

direction of each reciprocal-lattice vector from the origin. Each such line terminates at a distance from the origin corresponding to the shortest wave-length present. If the crystal contains a large number of units, and if there are no lattice vibrations, we must suppose these lines to be very narrow. They will be weighted as regards intensity in the neighbourhood of the points corresponding to the characteristic wave-length, and generally, in accordance with the structure factors of the lattice spectra, with the values of which we may suppose the reciprocal-lattice points themselves to be weighted.

If the sphere of reflection, now of radius unity, is drawn for any given direction of incidence, the intersections of this sphere with the radial lines determine the directions of the Laue reflections. All that is necessary is to join P, the centre of the sphere, to the point of intersection. The complete diagram must, of course, be thought of as in three dimensions. The radial lines corresponding to all the planes belonging to the same zone lie in the same plane, which passes through the origin. The intersection of this plane with the sphere of reflection defines a circular cone, having P as vertex, on which the Laue reflections belonging to the same zone must lie. The corresponding spectra on the photographic plate will lie on the ellipse formed by the intersection of this cone with the plane of the plate, which passes through the centre of the photograph.

If the thermal motions of the lattice are such that each characteristic spot is accompanied by a diffuse domain, it is clear that each radial line will also be accompanied by a diffuse domain, more or less cylindrical in form, with a more intense axial core, which is the original radial line. Examples of two such diffuse regions are shown diagrammatically in fig. 82, corresponding to the points B (210), and C (100). From the way in which the sphere of reflection cuts the domain B, it is plain that the Laue spot corresponding to B′ will be accompanied by a radial diffuse streak. We see from the diagram also that the longer radial streaks will tend to lie near the centre of the pattern, since the elongated domains are more nearly parallel to the surface of the sphere of reflection when the angle of scattering is small.

If monochromatic radiation is used, the Laue spot B′, and the radial streak disappear, and only the diffuse spot due to the intersection of the sphere with the monochromatic diffuse cloud around the characteristic point B remains.

(*d*) *The relation between the characteristic, Laue, and diffuse reflections:* The diffuse reflection is strongest when the setting of the crystal is such as to give the characteristic reflection. Laue spot and characteristic reflection then coincide, and the diffuse spot appears as a background to them. As the orientation of the crystal is changed from this setting, the Laue spot is rapidly displaced. The diffuse reflection is displaced very much less than the Laue spot, but usually in the same direction, and it also becomes weaker, although in certain cases it may persist over a range of a number of degrees from the setting for maximum intensity.

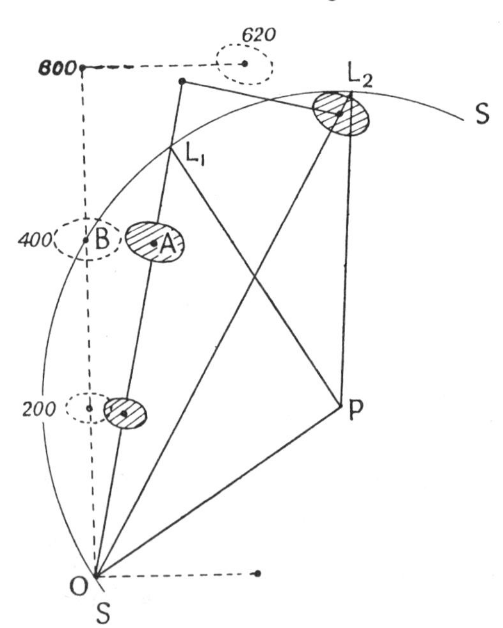

FIG. 84. Reciprocal-lattice diagram corresponding to fig. 83 (Pl. V)

These displacements are illustrated by some photographs obtained by Jauncey and Baltzer[51] from rock salt, with the direct radiation from a copper target, reproduced in fig. 83 (Pl. V). They show the equatorial reflections from the cleavage planes for a number of settings. The departure from the correct setting for the Bragg reflection 400 is 12°

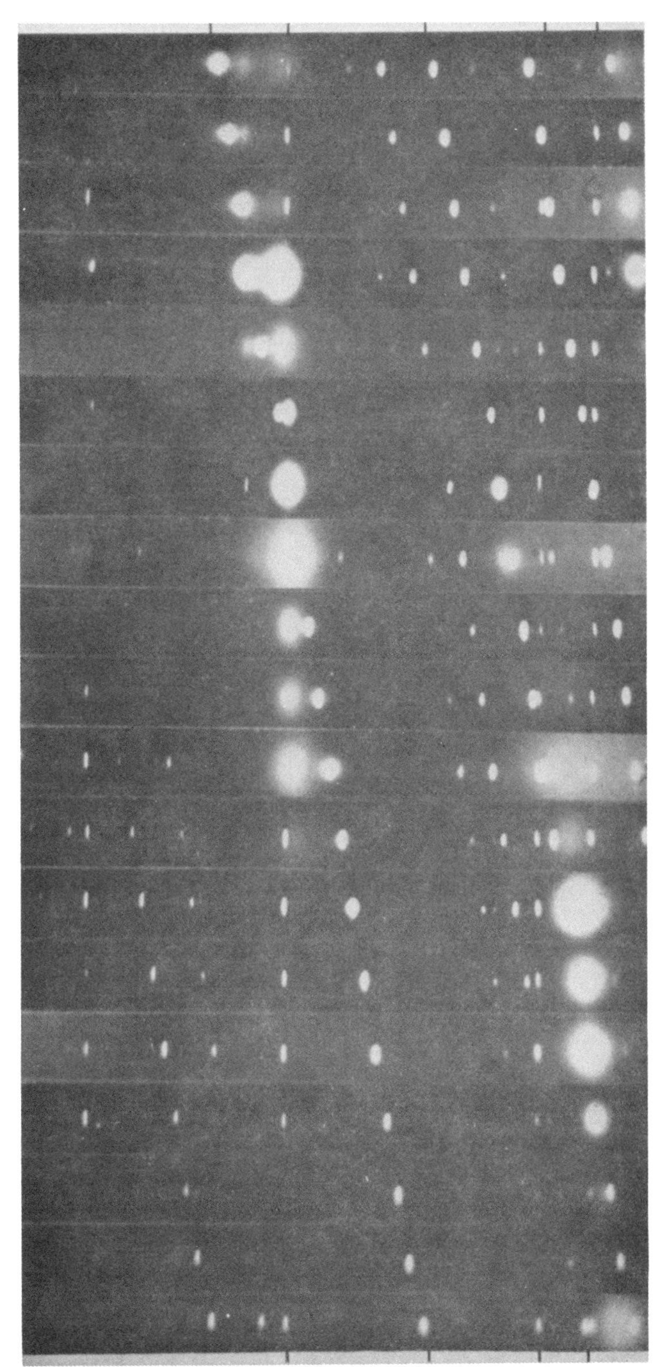

FIG. 83. Photograph illustrating the displacement of the Laue spots and the diffuse spots with varying angle of incidence. Equatorial reflections 400 and 620 from rock-salt cleavage planes. The departure from the correct setting for 400 is 12° for the bottom strip, and decreases by 1° for each successive strip (Jauncey)

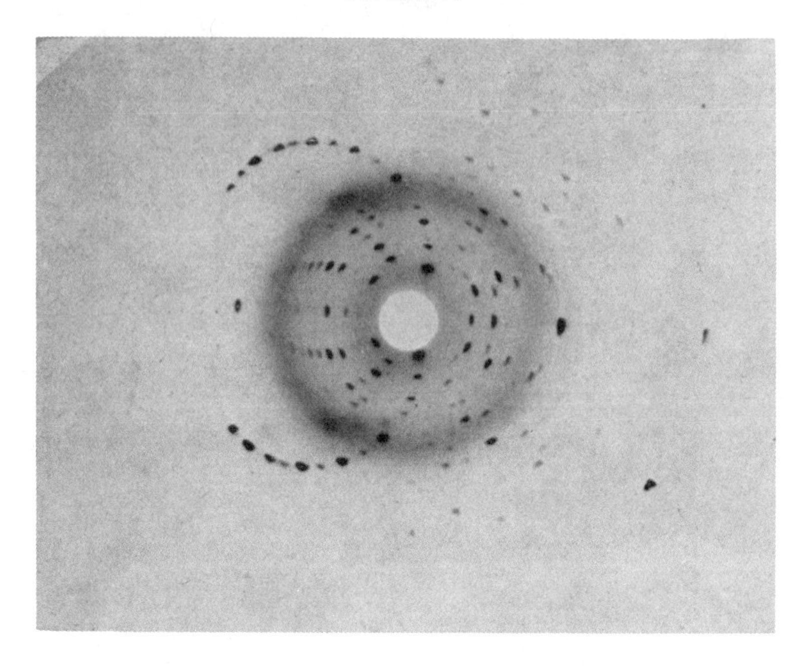

FIG. 85 (*a*). Sorbic acid: filtered Cu radiation, $17\frac{1}{2}$ min. exposure at the temperature of liquid air (Lonsdale and Smith, *ibid.*)

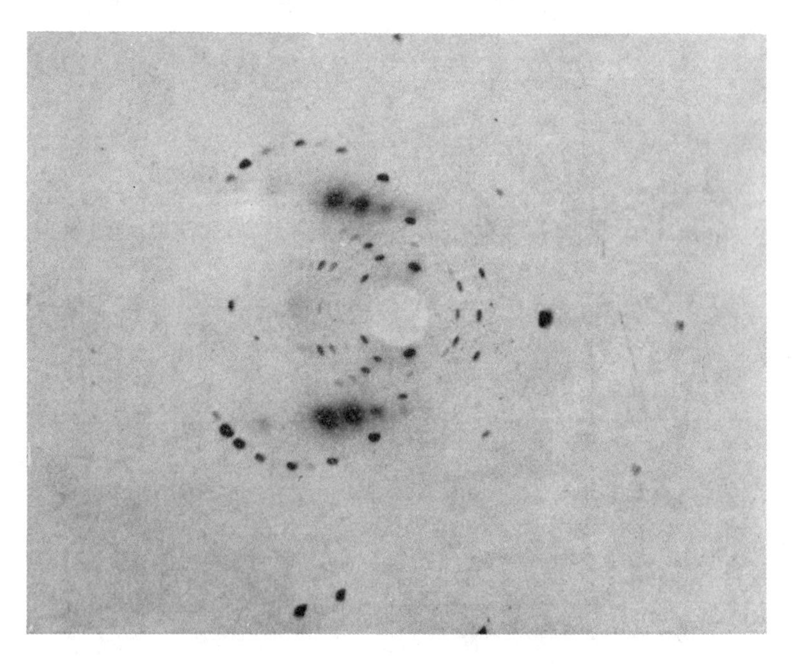

FIG. 85 (*b*). Sorbic acid: setting as in (*a*); 10 min. exposure at room temperature (Lonsdale and Smith, *ibid.*)

(P. 2

for the bottom strip, and decreases by 1° for each successive strip. The positions of the 400 and 620 reflections for Cu Kα are marked on each strip to serve as reference points. The relevant reciprocal-lattice diagram is shown, true to scale, in fig. 84. The initial position of the reciprocal-lattice point (400) corresponding to the wave-length Cu Kα is shown by the point A. P is the centre of the sphere of reflection, and L_1, the intersection of the [100] axis, OA, with the sphere of reflection, defines the direction PL_1 of the Laue reflection $h00$. Similarly, PL_2 is the direction of the Laue reflection 620. Suppose now that the lattice rotates in a counter-clockwise direction about an axis through O perpendicular to the plane of the paper. The point L_1 moves along the circle SS, and the Laue spot approaches the position of the characteristic spot, coinciding with it when A reaches the position B on the sphere of reflection. Before this setting is reached, however, the diffuse domain about A intersects the sphere of reflection in a region lying between B and L_1. A faint diffuse reflection then appears, slightly displaced from the position of the characteristic 400 reflection towards that of the Laue spot. As A moves towards B, the diffuse reflection grows stronger, and its displacement from the characteristic position decreases, and finally, Laue spot, characteristic spot, and diffuse maximum all coincide when A reaches B. If A rotates beyond B, the Laue spot is displaced on the other side of the characteristic spot. The various stages can readily be followed on fig. 83, for both the 400 and 620 reflections. A similar series of photographs is given by Lonsdale and Smith for urea nitrate. Fig. 84 explains why it is that the positions of the diffuse maxima are much less sensitive to changes of crystal setting than those of the Laue spots.

(*e*) *The qualitative effect of temperature on the diffuse maxima:* That the diffuse scattering does depend on the temperature of the crystal was shown by Laval [15] in the series of experiments to be discussed later, and also by Preston.[62] The intensity of the diffuse maxima was found to increase with the temperature. Lonsdale and Smith [65] have compared photographs taken at room temperature with others from the same crystal in the same orientation, taken at the temperature of liquid air. At the lower temperature, the diffuse pattern disappears almost completely, except for a few very strong spots. The Laue pattern, on the other hand, becomes more intense, and many more spots are usually to be seen. A pair of photographs from sorbic acid, fig. 85 (*a*) & (*b*) (Pl. VI), show this well. The halo in the photograph at the lower temperature is due to scattering from the liquid air used to cool the crystal. The temperature effect is completely reversible, the maxima reappearing in their original strength when the temperature of the crystal rises to that of the room. In the case of diamond, the effect of temperature on the true diffuse maxima is less marked, but still quite definite. There can be no doubt that the diffuse maxima are a thermal effect.

(*f*) *The shape of the diffuse maxima and its dependence on crystal structure:* According to the theory developed in the first section of this chapter, the form of the diffuse maxima is governed mainly by equation (5.45),

$$\sigma = \frac{kTN}{m} \frac{R^2}{r^2} \sum_j \frac{\cos^2 \alpha_{sj}}{v_{\phi j}^2} .$$

If we leave out of account the effects of the polarisation factor and the structure factor, this equation gives the intensity of the diffuse domain at a vector distance **r** from a reciprocal-lattice point, **R** being the vector distance from the origin of the point at which the intensity is considered. For many purposes, it is a good enough approximation to take **R** as the distance of the reciprocal-lattice point itself from the origin. In fig. 86, let P be a point at a distance r from a reciprocal-lattice point M. The diffuse intensity at P is due to the lattice waves

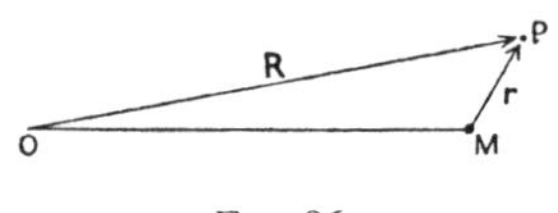

FIG. 86

travelling in the direction MP whose reciprocal wave-lengths are equal to the distance MP. The intensity contributed by the wave is inversely proportional to the square of its velocity, so that a large intensity will in general be associated with a small elasticity. It is, however, only the component parallel to OP of the elastic vibration associated with the wave that makes any contribution to the diffuse scattering at P. Since P is usually much nearer to M than it is to O, it is often accurate enough when estimating the general magnitude of the effect, to say that it is only the components of the elastic vibrations parallel to the reciprocal-lattice vector $\overrightarrow{OM}$ that make any appreciable contribution to the diffuse scattering associated with the reciprocal-lattice point M. Although it is possible to calculate the detailed distribution of the intensity function about the lattice-point only for very simple crystals, the considerations outlined above enable us to draw several important conclusions that can be tested experimentally.

Consider for example a crystal in which the atoms are arranged in a series of layers, within each of which they are relatively firmly linked. Structures of this kind are found in mica, graphite, and many other crystals. Such a layer lattice will yield most readily to displacements perpendicular to the layer planes. Vibrations in directions parallel to the planes will be of relatively small amplitude, whatever the direction of propagation of the wave with which they are associated, but, equally, the component vibration perpendicular to the planes will be relatively large for all waves. Suppose for example the layer planes are the planes (001). The vector from the origin to the reciprocal-lattice point (001) is then always parallel to the direction of maximum vibration, whatever the direction of the wave, and the point (001) will be surrounded by a large and approximately spherical domain of diffuse intensity. The portion A of the distribution illustrated in fig. 87 is due to the longi-

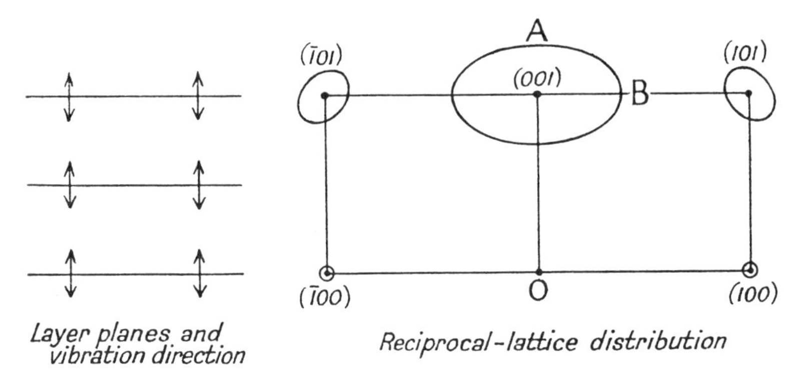

FIG. 87. Reciprocal-lattice diagram illustrating diffuse distribution for a layer lattice

tudinal component of waves travelling perpendicular to the layer planes, while the portion B is due to the transverse components, perpendicular to the layer planes, of waves travelling parallel to the planes in the direction [100]. The diffuse domains will not be truly spherical, because the speed of the waves in different directions will not be the same.

Around the points (100), on the other hand, the diffuse domain will be small and weak; for the lattice vector to the point (100) is perpendicular to the only direction in the crystal in which there is an appreciable amplitude of vibration. Points such as (101) show a diffuse distribution smaller than that about (001) because only the component of the vibration parallel to [101] is effective. An example of a very large diffuse maximum due to a layer structure is shown in fig. 88 (Pl. VII), obtained from urea nitrate by Lonsdale and Smith. The maximum in this case is plainly visible when the setting of the crystal is more than 12° away from that appropriate to the corresponding spectrum.

Another interesting case is that of the chain structure. Suppose a crystal to consist of chain-like molecules, arranged with their lengths parallel to a single direction. According to Müller, the compressibility of such substances is several times smaller along the chain direction than at right angles to it, and they will not therefore transmit *longitudinal* vibrations of appreciable amplitude along the length of the chain axis. Transverse waves travelling in directions at right angles to the chains can be transmitted, since these will be concerned with displacements of chain relative to chain along the direction of their own length, and transverse vibrations travelling along the directions of the chains can also be transmitted. Suppose the chains to be parallel to [001]. Then the diffuse domains surrounding the reciprocal-lattice point (001) will have no extension parallel to [001], since such extensions would be associated with waves both travelling and vibrating parallel to the chain direction; but they will have an extension at right-angles to [001], corresponding to transverse waves travelling in directions perpendicular to the chains. The (001) diffuse domains will therefore be disc-shaped. The domains for points such as (100), which have the chain direction as zone axis, show a more or less uniform extension in all directions. The distributions are represented diagrammatically in fig. 89.

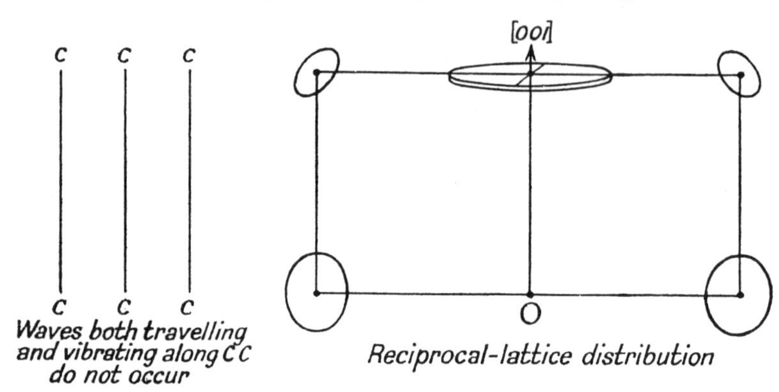

FIG. 89. Recipocal-lattice diagram illustrating diffuse distributions due to a chain structure

The sphere of reflection will intersect such a disc-shaped domain along a narrow strip, and the corresponding diffuse reflections on actual photographs will appear as narrow streaks. An interesting example of this is given by photographs from sorbic acid, obtained by Lonsdale, Robertson and Woodward,[68] one of which is seen in fig. 90 (Pl. VII), which show both the broad diffuse spots characteristic of the layer lattice, and the narrow streaks characteristic of the chain structure. The sorbic acid molecules are chain-like, and are arranged in parallel rows, but their arrangement is also such as to produce a series of layers. By studying the relative positions of the two types of spot, it is possible to determine with considerable accuracy the orientation of the molecules in the crystal structure. It is clear from this example that the diffuse reflections may sometimes be of great help in guessing a preliminary structure in crystal analysis.

(*g*) *The shapes of the diffuse maxima for cubic crystals :* In §§ 1 (*k*) and 1 (*l*) of this chapter, we discussed the approximate form of the surfaces of equal diffusion for cubic crystals. The details of the surfaces depend on the elastic nature of the crystal, and figs. 73 and 74 show sections of some of the surfaces calculated from Jahn's formula for sodium and rock salt, two crystals that belong to elastically dissimilar classes.

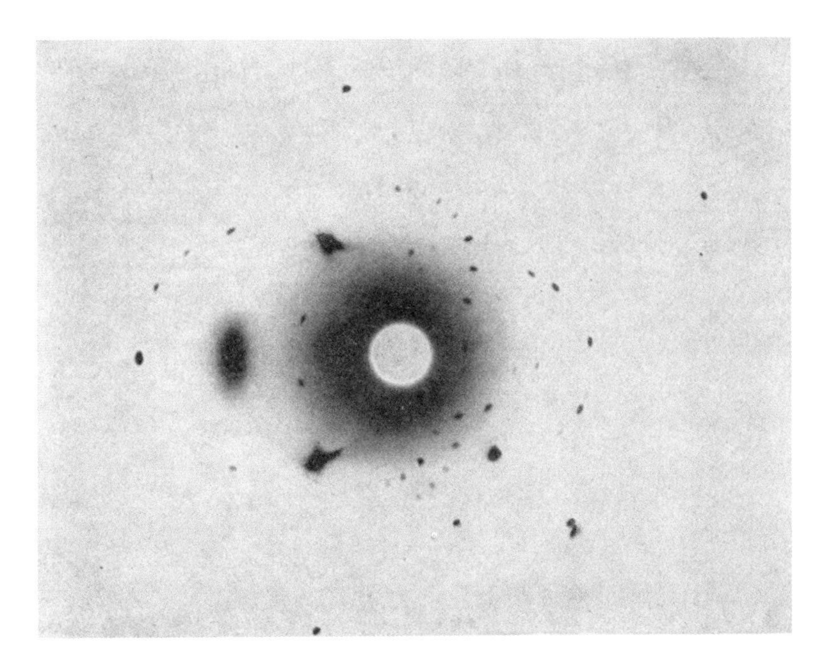

FIG. 88. A large diffuse maximum due to a layer structure, urea nitrate (Lonsdale and Smith, *ibid.*)

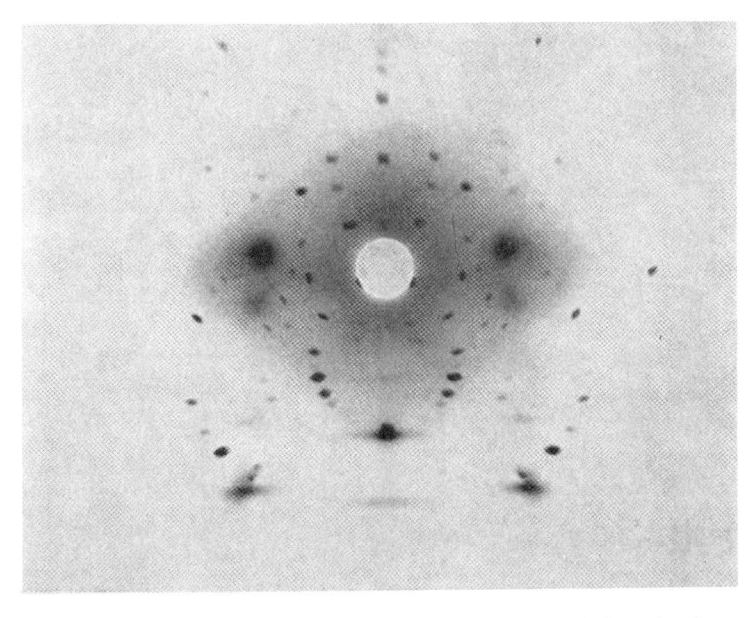

FIG. 90. Photograph from sorbic acid with filtered Cu radiation, showing broad and narrow diffuse spots (Lonsdale, Robertson and Woodward) (*Proc. Roy. Soc.*, **A, 178,** 43 (1941))

PLATE VIII

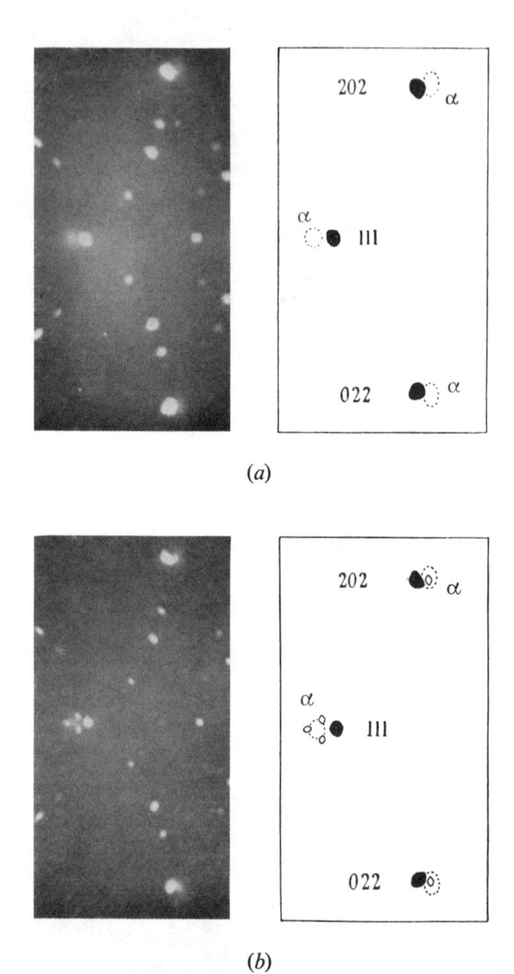

FIG. 92. Unfiltered Cu radiation, $\theta_{111} = -23{\cdot}8$. Film normal to undeviated incident beam. ● Laue spot, ◌ primary, ◠ secondary spot. (*a*) Type II diamond, showing no secondary reflections ; (*b*) Type I diamond, showing strong secondary reflections (Lonsdale)

(*Proc. Roy. Soc.*, **A, 179,** 315 (1942))

Each is elastically anisotropic, but for sodium $c_{11}-c_{12}<2c_{44}$, while for rock salt, $c_{11}-c_{12}>2c_{44}$. Lonsdale and Smith,[56] and Jahn and Lonsdale,[69] have made a photographic study of the forms of the surfaces for some cubic crystals. Their method is to study the shapes and intensities of the diffuse spots, as recorded on Laue photographs for different settings of the crystal near that giving the characteristic reflection. The spot as recorded gives a section of the diffuse distribution by the sphere of reflection, and this can be compared with theory.

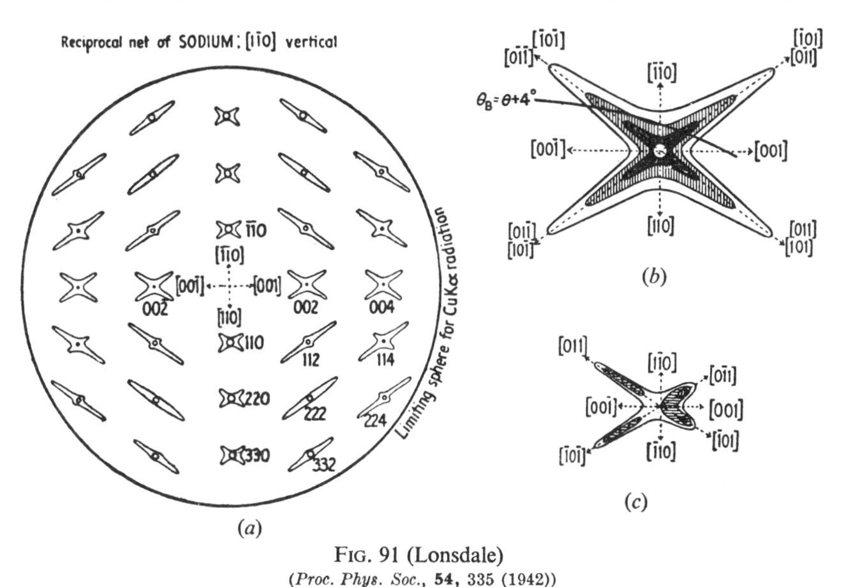

FIG. 91 (Lonsdale)
(*Proc. Phys. Soc.*, **54**, 335 (1942))

Fig. 91 (*a*) shows the intensity distribution in the reciprocal-lattice space for sodium; fig. 91 (*b*) is an enlargement of the neighbourhood of the point (002), showing the trace of the sphere of reflection for a setting 4° from the reflecting position, and fig. 91 (*c*) shows the section of the domain by the sphere of reflection. This section has a scissor-shaped form, and this is shown very clearly in photographs obtained by Lonsdale and Smith, for example, in that reproduced in Plate 4 (*a*) in Mrs. Lonsdale's report to the Physical Society.[66]

Lonsdale and Smith find that sodium and lead give similar distributions, although sodium has a body-centred structure and lead a face-centred one. This similarity is, however, entirely in accord with theory, for it is the elastic behaviour that should principally determine the diffuse scattering, and lead and sodium are elastically similar. The diffuse reflections from tungsten, on the other hand, are weak, small, and more or less elliptical in form, and disappear quickly as the setting of the crystal departs from that appropriate to the Bragg reflection. The elasticity of tungsten is large and isotropic, with $c_{11}-c_{12}=2c_{44}$, and we should therefore expect a distribution of this kind, quite different from the large distributions given by the soft anisotropic metals, with their prominent extensions in certain directions. It will be seen from equation (5.54), p. 213, that these extensions depend largely on the fact that for elastically anisotropic crystals $c_{11} - c_{12} - 2c_{44}$ differs from zero. The experimental results strongly support the view that it is the elastic nature of the crystal, and therefore the acoustical rather than the optical frequencies, that govern the diffuse scattering.

(*h*) *Primary and secondary extra reflections:* We have referred already to the existence of extra reflections that are structure-sensitive, but not directly temperature-sensitive, which are due to strain or imperfection in the crystal, and which were first observed by Raman. Such reflections have been called by Mrs. Lonsdale *secondary* extra reflections, to distinguish them from the *primary* extra reflections, which are the diffuse reflections due to thermal motion. An interesting study of the two types of reflection given by diamond has been made by Lonsdale and Smith,[56,57] and some mention has already been made of this work, but we shall here consider it in a little more detail.

All diamonds show the primary diffuse reflections. They are single and nearly circular, and are observed when the setting of the crystal is not more than from 3° to 5° from the setting that gives the corresponding Bragg reflection. With the $[1\bar{1}0]$ axis of the crystal vertical, and perpendicular to the incident beam, primary diffuse spots were observed associated with $\{111\}$, $\{220\}$, $\{113\}$, $\{004\}$ and $\{331\}$. All these reflections are markedly temperature-sensitive.

Secondary reflections are given only by certain diamonds. The rare diamonds classed by Robertson, Fox, and Martin as being of Type II do not show them. These diamonds are unusually transparent to certain infra-red and ultra-violet radiations. The secondary reflections are structure-sensitive, and vary in intensity from diamond to diamond, whereas the primary diffuse reflections do not vary appreciably.

It was observed that the diamonds giving intense secondary reflections were those that showed *large* extinction for the ordinary X-ray reflections, whereas Type II diamonds, and those Type I diamonds giving only faint secondary reflections, showed little extinction, and gave extremely strong ordinary spectra.

The primary spots are explicable in terms of the Faxén-Waller theory if the regions of diffuse intensity distribution around the reciprocal-lattice points are assumed to be roughly spherical, and of very limited extent, in accordance with the high elasticity of the crystal. The positions of the secondary spots on the other hand, correspond geometrically to extensions of the intensity domains along the cube axes, as Raman was the first to point out.[59] In a photograph taken with $[1\bar{1}0]$ vertical, the secondary reflections associated with 111 form a triangle with the primary diffuse reflection as a background. Fig. 92(*a*) (Pl. VIII) shows a Laue photograph from a Type II diamond, taken with unfiltered copper radiation. The primary diffuse spot 111 is plainly shown, but there are no secondary spots. Fig. 92(*b*) was taken under the same conditions, but with a Type I diamond, and it shows strong secondary reflections as well as primary diffuse spots. The triangular arrangement of the secondary spots in the case of 111 is well shown, but the diffuse background to the triangle, which is the primary diffuse spot, might easily be overlooked.

These secondary extra reflections are not properly speaking diffuse reflections, and have a quite different appearance. They are almost certainly due to strains, and analogous rather to the spots sometimes observed on photographs from cleaved or cold-worked metals. No fully satisfactory explanation of them has yet been given. On the other hand, they would seem to be no more easily explained in terms of Raman's theory than on the thermal theory. Pisharoty [70] has suggested that the 'forbidden' 222 reflection from diamond is an extra reflection, but there is no evidence in support of this, and Lonsdale and Smith have found that the 222 reflection from both Type I and Type II diamonds has all the characteristics of a true Bragg reflection, and none of those of an extra reflection.

(*i*) *Laval's investigation of diffuse scattering :* We have left until last the discussion of the pioneer work of Laval [15] in this field, because it is probably easier to follow the results of the photographic investigations than those obtained with the ionisation spectrometer. Laval was, however, the first to see clearly how observations of diffuse scattering could be made, and his work is still quantitatively as complete as any. In most of his work, Laval used a crystal of sylvine (KCl), mounted on the table of an ionisation spectrometer in such a way that it could be rotated into any orientation. Monochromatic radiation, either Mo Kα or Cu Kα, was obtained by reflecting the primary beam from the surface of a curved quartz crystal. The ionisation chamber was filled with argon, and was connected to a very sensitive quadrant electrometer. The quantity measured by Laval, and called by him the scattering power, P, is defined in the following way. Let I_0 be the intensity of the primary beam, C the scattering power of a single Thomson electron placed in the same beam, *dm* the mass of the crystal scattering radiation, and *di* the intensity scattered by it into the solid angle $d\omega$. Then

$$di = PI_0C\,dm\,d\omega. \qquad (5.100)$$

The crystal was first set with its reflecting face normal to the line bisecting the directions of incidence and scattering. If it was then rotated about this bisector as axis, while the angle between the directions of incidence and scattering, 2θ, remained constant, the variation of the scattering with azimuth could be studied; and by changing the angle of scattering, and rocking the crystal, it was possible to study the distribution of diffuse scattering. It is easiest to consider the matter, as before, in terms of the reciprocal lattice.

Suppose, for example, that the reflection 002 is to be investigated. Let OX, OY, OZ, fig. 93, be the three reciprocal axes, and M the point

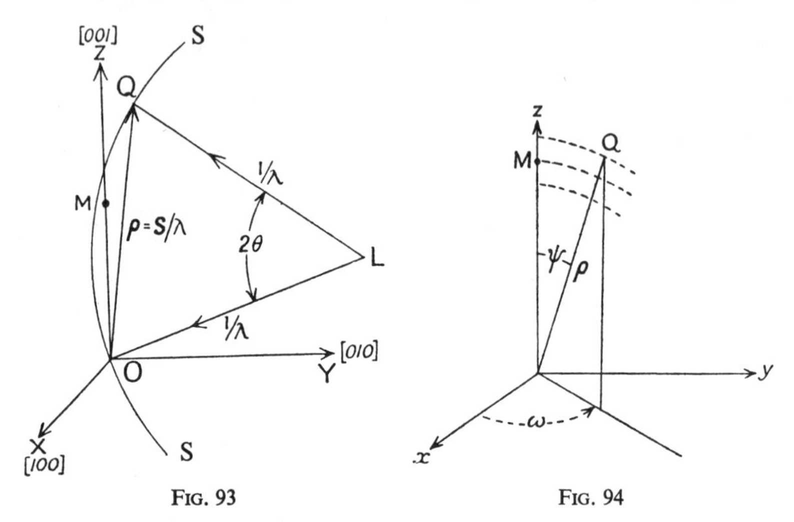

FIG. 93 FIG. 94

(002). SS represents the sphere of reflection for a certain direction of incidence LO. Let LQ be the direction of scattering, Q being on the sphere. Then OQ is the vector S/λ, which we shall here denote by $\boldsymbol{\rho}$. Its magnitude is $2(\sin\theta)/\lambda$. By measuring the scattering power P for the particular direction of incidence corresponding to LO, with the angle of scattering 2θ, we obtain the value corresponding to the point Q at the extremity of the vector $\boldsymbol{\rho}$ in the reciprocal-lattice space. By varying the length and direction of $\boldsymbol{\rho}$ it is possible to map out the scattering domain about the lattice-point. The point Q may be fixed, as in fig. 94, by the radial distance $|\boldsymbol{\rho}|$, the azimuth ω, which is the angle between the xz plane and the plane containing $\boldsymbol{\rho}$ and OZ, and the angle ψ between OZ and $\boldsymbol{\rho}$. The azimuth ω is altered by rotating the crystal about the normal to the (001) face. The length of $|\boldsymbol{\rho}|$ is changed by varying the angle of scattering, and the angle ψ by rotating the crystal about an axis lying in the reflecting face, and perpendicular to the meridian plane containing $\boldsymbol{\rho}$, which will in practice be the plane of incidence and scattering. When $\psi = 0$, the normal to the reflecting plane bisects the angle between the directions of incidence and scattering.

In this way, Laval was able to map out the domains of diffuse reflection for a number of reciprocal-lattice points. Some of his results for 004, with $\omega = 45°$, are shown in fig. 95. The ordinates give the values of P, and the abscissae values of ψ. The curves are drawn for a series of values of ρ/ρ_{004}. Each curve is drawn for a constant value of 2θ, and the corresponding values of $\theta - \theta_B$, where θ_B is the Bragg angle, are shown in the figure. Any one curve thus gives a section of the diffuse domain for a constant angle of mis-setting of the crystal. When

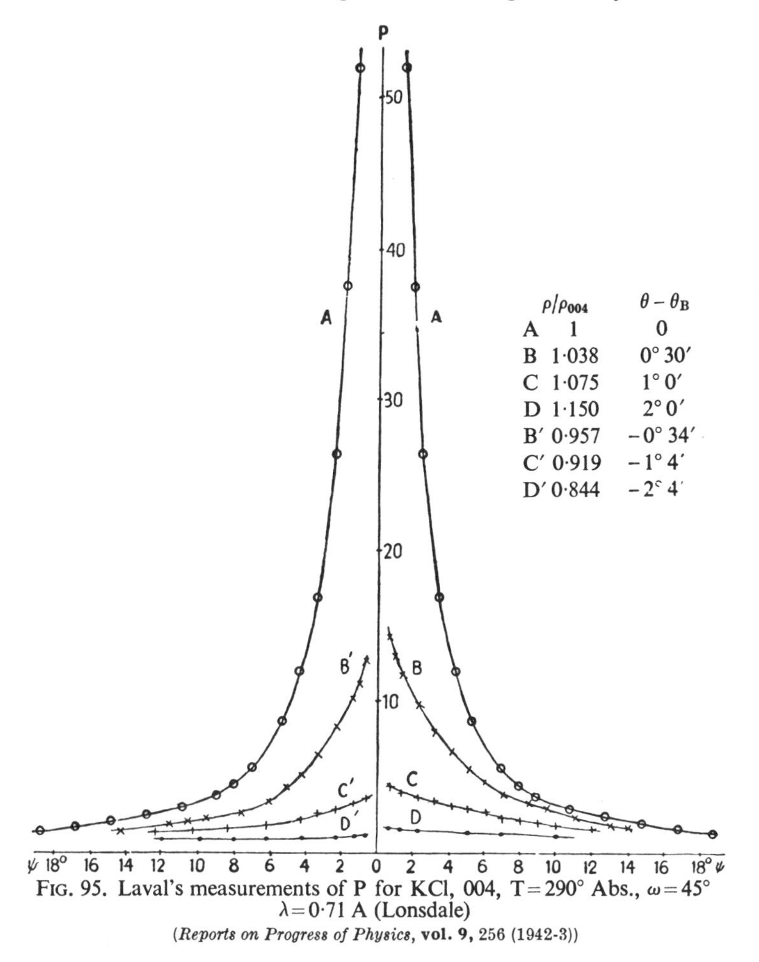

FIG. 95. Laval's measurements of P for KCl, 004, T=290° Abs., $\omega=45°$ $\lambda=0{\cdot}71$ A (Lonsdale)

(*Reports on Progress of Physics*, **vol. 9**, 256 (1942-3))

$\rho/\rho_{004}=1$, the value $\psi=0$ would correspond to the ordinary 004 spectrum, and the curve would rise to very high values of P for small values of ψ. It is not therefore possible to investigate the diffuse reflection in regions too near the reciprocal-lattice point.

The curves show very clearly the existence of a very broad maximum of scattering about the reciprocal-lattice point, and it will be seen that for a large enough value of ψ, all the curves tend to a steady value, a background scattering γ appropriate to the region between the reciprocal-lattice points. The value of this minimum scattering for KCl is about half the average unmodified background scattering given by the

Jauncey formula. It will be seen in fig. 95 that the scattering is not symmetrical about M, but is larger for the larger values of ρ, which is in accord with theory.

For the spectrum 002, Laval found that P was independent of ω for given ρ and ψ, indicating that the diffuse domains are figures of revolution about [001]. This domain is also entirely surrounded by a region of weak scattering; but for the 004 and 006 domains Laval found that P exhibited maxima and minima as ω changed, having its lowest values for $\omega = 45°$, $135°$, $225°$, and $315°$, and maximum values for $\omega = 0°$, $90°$, $180°$, and $270°$. This result may be explained by supposing that along the directions [100], [010], [$\bar{1}$00] and [0$\bar{1}$0] the 004 and 204, and the 006 and 206 domains overlap, so that along these directions the weak background scattering is never reached. Laval's work thus indicates clearly the existence in some cases of connecting bridges between neighbouring domains. As we have already seen, these bridges correspond to the non-radial diffuse connecting streaks often found on Laue photographs taken with monochromatic radiation, examples of which are shown in fig. 80. One of Laval's diagrams for the contours of equal diffuse scattering for the domains 006 and 206 is reproduced in fig. 96. The values of ρ_{006} and ρ_{206} are respectively 0·957A and

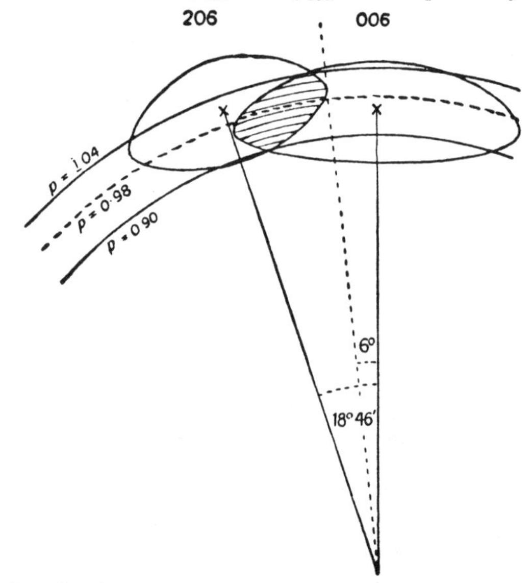

FIG. 96. Laval's diagram of 006 and 206 contours of equal diffusion for KCl, $\lambda = 0{\cdot}71$ $\omega = 0°$, showing overlap between $\rho = 0{\cdot}90$ and $\rho = 1{\cdot}04$ for values of $\psi > 6°$ (Lonsdale)

(*Reports on Progress of Physics*, **vol. 9**, 271 (1942-3))

1·016A. Laval considered the boundary of a domain to be reached when the extrapolated maximum of a curve such as D in fig. 95 differed from its steady background value by less than 1/10 of that value.

Working in this way, Laval mapped the positions and the extent of a number of domains, and obtained a series of contours of equal diffusion showing a general similarity to those given by Jahn's calculations, although his readings are not taken at close enough intervals to reproduce the finer details.

He also measured the peak intensities of the scattering curves for a constant angle of mis-setting ($\theta - \theta_B = 2° \ 30'$). He found, using Mo K$\alpha$ radiation at 290° abs., that the intensities fell off for a series of orders from the (001) face in the following way:

$$002 : 004 : 006 : 008 : 0010 = 100 : 50 : 33 : 21 : 4.$$

This falling off is less rapid than that of the integrated reflections for the same spectra, for which James and Brindley [33] found the sequence 100 : 22 : 5·3 : 2·3 : 0·40. It is, however, not at all easy to interpret the intensity measurements without a more exact knowledge of the experimental conditions. Laval's results have been discussed from this point of view by Mrs. Lonsdale,[67] who concludes that the actual domains are nearly identical, both in absolute intensity and distribution of intensity, and that although this equality is to some extent fortuitous, it is certain that the decrease of integrated intensity for the diffuse maxima is very much less than for the Bragg reflections. Indeed, if there is any falling off at all, it is not very marked. This result is on the whole in accord with theory, for equations (5.30) and (5.44) show that while both the diffuse and the Bragg spectra are affected in the same way by the structure factor and the Debye factor, the expression for the diffuse intensity contains in addition the factor S^2, or $4 \sin^2 \theta$, which increases with increasing angle of scattering. Further quantitative work is, however, needed before any very definite statement can be made.

(*j*) *The quantitative dependence of the intensity of the diffuse scattering on temperature :* Laval also investigated in some detail the dependence of the intensity of the diffuse scattering on the temperature of the crystal. His results have been compared with the predictions of the lattice-vibration theory by Born and Lonsdale.[71]

Equations (5.30) and (5.42) of the theoretical discussion in Section 1 of this chapter indicate that if the temperature is not too low ($T > \Theta$), the diffuse scattering power P at any given point in the reciprocal-lattice space is proportional to Te^{-2M}. With the same limitation as to temperature, and for a cubic crystal, we may write, using (5.68),

$$2M = \frac{12h^2T}{mk\Theta^2}\frac{\sin^2\theta}{\lambda^2} = \frac{3h^2T}{mka^2\Theta^2}(h_1{}^2 + h_2{}^2 + h_3{}^2), \tag{5.101}$$

where a is the lattice spacing, Θ Debye's characteristic temperature, and h_1, h_2, h_3 are the indices of the spectrum. This result is expressed by Born [72] in the form

$$P \propto Te^{-T/T_h}, \tag{5.102}$$

where
$$T_h = \frac{mka^2\Theta^2}{3h^2} \frac{1}{h_1^2+h_2^2+h_3^2} = \frac{T_0}{h_1^2+h_2^2+h_3^2}. \tag{5.102a}$$

Equation (5.102), as it stands, indicates that T_h is an inversion temperature, above which the intensity of the diffuse scattering will decrease with increasing temperature. Too much stress must not, however, be placed on this, for in the deduction of (5.30) it is assumed that 2M is small enough for the approximation $e^{-2M} = 1 - 2M$ to be valid, and this is certainly not true when $T = T_h$, so that $2M = 1$. Table V. 4 is due to Born and Lonsdale, and gives the values of Θ and T_0 for a number of crystals. In the case of KCl, studied by Laval, T_h will be a high temperature, above the melting point of the crystal, for the low-order spectra, but for the higher orders it may be in the neighbourhood of room temperature. The formula thus indicates a different type of temperature dependence for high and low-order spectra. Laval measured the scattering power at 290°, 550° and 665° abs., for the 002 and 004 domains of KCl, for a number of different values of ψ, that is to say at a variety of distances from the reciprocal-lattice points for $\theta = \theta_B$. His values of ψ range from 2° to 11° in the case of 002, and from 1° 45′ to 16° 45′ in the case of 004. His results show that P_T/P_{290} is nearly constant for different values of ψ, which means that the distribution of intensity in the domains does not depend appreciably on the temperature. Born and Lonsdale have compared his figures with the formula $P \propto Te^{-T/T_h}$, and find agreement within the limits of experimental error, which is about 10 per cent.

Experiments were made by Baltzer [73] with rock salt at a temperature of 100° abs., where the relationship (5.102) cannot be expected to hold. He finds that reasonable agreement is to be obtained by using the Debye-Jauncey formula for the diffuse scattering, $P \propto 1 - e^{-2M}$.

TABLE V. 4

°K	Face-centred Cu	Pb	Al	Body-centred Li	Na	Pseudo-simple NaCl	KCl	KBr	Diamond C
Θ	315	88	398	510	202	281	230	177	2340
T_0	14450	6880	12260	3800	2970	12850	13670	14230	146400

(*k*) *Jauncey's scattering formula and its investigation:* We have so far discussed only the more recent work on diffuse scattering, which has been done since the existence of the diffuse maxima was experimentally demonstrated by Laval, and by Preston in 1939, but, before this, a large amount of careful work on diffuse scattering from crystals had been done in America, under the leadership of Jauncey. The experimental arrangements in all this work were such as expressly to exclude scattering in the regions of the lattice spectra, interest being directed mainly to the background scattering.

The theoretical formula which is used in the discussion of this work, derived by Jauncey [38,39] and Harvey, rests on a purely classical basis. The scattering of a number of electrons is treated in a manner analogous to that used by Compton,[74] and by Raman,[75] in their derivation of the total scattering from a monatomic gas. This has been discussed in detail in Chapter III, § 2(*c*). Jauncey extends Compton's calculation by assuming the electrons to be collected together in groups of Z each about the points of a simple cubic lattice, and he then supposes these groups to have a temperature motion about the mean position of the lattice-points. The simplifying assumptions made by Jauncey and Harvey in deducing their formula correspond exactly to the assumption made by Debye in his original derivation of the temperature factor, that the vibrations of any atom are independent of those of its neighbours. As we saw in Chapter I, § 3(*b*), Debye's assumption gives for I_d, the diffuse scattering in any direction, the expression

$$I_d = N f_0^2 (1 - e^{-2M}). \tag{5.103}$$

The scattering crystal contains N atoms, each with a scattering factor f_0 when at rest. M is the usual exponent of the temperature factor for the crystal. Both f_0 and M are functions of $(\sin\theta)/\lambda$. I_d is as usual expressed in terms of the scattering by a single Thomson electron. This term in the scattering formula is part of the interference pattern produced by the crystal lattice, and represents the diffusely scattered *unmodified* radiation, but the atoms will also scatter incoherently, with change of wave-length, and the total diffuse scattering in any direction will be the sum of the modified and unmodified contributions.

In discussing Compton's formula for the total scattering, from a gas, we saw that his classical derivation leads to an expression that is formally identical with the *total* scattering, modified and unmodified, by a gas consisting of single atoms, as given by the methods of quantum mechanics, assuming the absence of interaction between the electrons of the atoms,* and neglecting exchange and relativity effects. The total scattering per atom from a monatomic gas is then given by equation (3.9) of Chapter III.

Since Jauncey's method of calculating the total scattering from a crystal is an application of Compton's method to that case, we should expect it to give a formula identical with that obtained by adding the unmodified diffuse scattering, as obtained by Debye, to the incoherent scattering of § 2(*d*), Chapter III. This addition gives

$$I_d/N = f^2(1 - e^{-2M}) + \sum_n (1 - f_n^2), \tag{5.104}$$

where $f = \sum_n f_n$, f_n being the scattering factor for a single electron of the atom, and the sum being taken over all the electrons in the atom.

* Or, more strictly, that the wave-functions of the atoms may be expressed as products of the wave-functions of their individual electrons.

Equation (*5*.104) is in fact equivalent to the formula of Jauncey and Harvey, which is given in terms of the scattering per electron, whereas (*5*.104) gives the scattering per atom. If we denote by S the scattering per electron, (*5*.104) gives

$$S = I_d/NZ = f^2(1 - e^{-2M})/Z + 1 - \sum_n f_n^2/Z. \qquad (5.105)$$

The formula of Jauncey and Harvey is

$$S = 1 + (Z - 1)f'^2/Z^2 - F^2/Z. \qquad (5.106)$$

Here, F is the atomic scattering factor as affected by the heat motion, and so is equivalent to fe^{-M} in our notation, while the quantity f' is defined by Jauncey as follows:

$$f'^2 = f^2 + (f^2 - Z\sum_n f_n^2)/(Z - 1), \qquad (5.107)$$

so that f' reduces to f if $f_n = f/Z$, that is to say, if the average value of f_n is the same for all the electrons of the atom. By substituting the values of F and f' in (*5*.106), we obtain at once (*5*.105). In a later paper, Jauncey [76] introduces another quantity f'', which in our notation is defined by

$$f''^2 = Z\sum_n f_n^2, \qquad (5.108)$$

so that f''/Z is the root-mean-square of the f's for the individual atomic electrons. By means of these equations, the formulae of Jauncey may be expressed in the notation employed in this book.

In Woo's modification of the Jauncey formula, the Breit-Dirac [77,78] correcting factor is introduced into the expression for the modified radiation, the assumption being made that this correction, deduced originally for the case of scattering by free electrons, also applies to electrons bound in atoms. The formula for the diffuse scattering then becomes

$$S = f^2(1 - e^{-2M})/Z + (1 - \sum_n f_n^2/Z)/R^3, \qquad (5.109)$$

where $\qquad R = 1 + h(1 - \cos 2\theta)/mc\lambda.$

Equation (*5*.109) represents the diffuse scattering from a simple crystal lattice on the assumption that the atoms are all independent, so that there is no coupling between neighbouring atoms. If such coupling exists, as it of course must in an actual crystal, the Debye formula for the unmodified part of the general scattering is, as we have already seen, far too simple. The simple addition to the unmodified scattering of the modified radiation from the atoms of the crystal considered as a gas in which the atoms are all separate assumes, too, the independence of the atoms when in the crystal state, an assumption that cannot be justified for the outer electrons, and the similar assumption that the atoms oscillate as a whole, without distortion of their electron shells, is involved in the use of fe^{-M} for the scattering factor as modified by temperature. The agreement between theory and experiment in the case of the temperature factor for the interference maxima shows, however, that it is unlikely that much error will be introduced by this assumption.

The first reliable measurements of diffuse scattering were made by Jauncey and May,[79] who used rock salt, and their method has been used, with slight modifications, by Harvey and others in later investigations. The crystal is in the form of a thin slice, about 1 mm. in thickness, whose surface is parallel to the (100) face, and is mounted on the table of an X-ray spectrometer. Radiation falls on this slice in a direction making an angle θ with the normal to its face, and the ionisation chamber is set so as to receive the radiation scattered through an angle ϕ. If $\phi = 2\theta$, the directions of incidence and scattering make equal angles with the normal to the crystal slice, and all the rays that enter the chamber have travelled the same distance, $t \sec \theta$, in the crystal. A simple formula, due to Crowther,[80] then gives the relation between the scattering coefficient, the thickness of the slice, and the intensity of the incident and transmitted radiation. When $\phi = 2\theta$, however, the crystal will in general reflect X-rays, since not monochromatic radiation, but a broad band of wave-lengths is used, obtained by filtering the incident radiation. Some wave-lengths of this band will fulfil the reflection condition, and if the scattering is determined for a series of values of ϕ that includes 2θ, there will be a peak at this angle corresponding to an interference maximum. Jauncey and May draw a curve through the points on the graph giving the scattering as a function of ϕ on either side of the maximum, and so obtain by interpolation the value of the scattering for $\phi = 2\theta$ in the absence of the interference maximum. It will be seen that this method tends to eliminate from the measurement any hump on the curve of general scattering in the region of the interference maximum. The early work of Jauncey and May showed general qualitative agreement with theory, but did not give absolute values.

Harvey,[81] in a series of experiments on sylvine, has tested the scattering formula at temperatures ranging from that of liquid air, 86° abs., to 1020° abs. The method he used was in principle that of Jauncey and May, but the scattering was expressed in absolute measure by means of the following device. Jauncey and Harvey have shown that Dirac's formula gives accurately the shape of the scattering curve for paraffin wax. Harvey assumes that it also gives the correct absolute value, and compares the scattering by sylvine with that given by a slab of paraffin wax at an angle of 90°. Making due allowances for the relative densities of the two substances, he is able to express the scattering by sylvine in absolute measure. In applying Crowther's formula, a number of corrections have to be made, including that for the different absorption of the modified and unmodified radiation in the ionisation chamber, but reference must be made to the original paper for details.

In fig. 97, which is taken from Harvey's paper, the curves for S, the scattering per electron, at a series of temperatures are shown plotted against $(\sin\theta)/\lambda$. The full curves show the calculated values obtained from equation (5.105). In making the calculations, the values of fe^{-M}

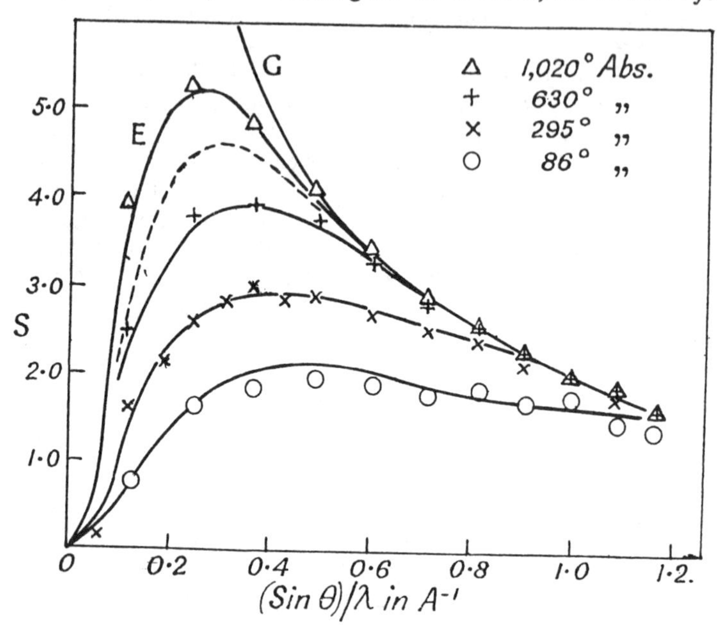

FIG. 97. Diffuse scattering of X-rays from sylvine at different temperatures (Harvey)
(*Phys. Rev.*, **44**, 133 (1933))

for sylvine, as observed by James and Brindley,[33] were used, and not the values calculated from the Debye-Waller formula. The dotted curve is that calculated for 1020° abs. from the Debye-Waller formula, while curve E is that calculated for the same temperature using an extrapolated value of fe^{-M} based on the results of James and Brindley. The diffuse scattering thus confirms the high values of M observed at high temperatures by these workers.

Harvey uses f^2/Z for $\sum_n f_n^2$, in the absence of information concerning the values of f for the different electron groups. This information is now available, and the recalculation of the formulae using it would affect the results slightly, although probably not very materially.

The curve G gives the scattering from a monatomic gas, supposed to consist of equal numbers of atoms of K and Cl, as calculated from the Compton formula, that is to say, by putting $e^{-2M}=0$ in (5.109). It will be seen that the scattering curve for the crystal approaches more and more closely to that for the gas as the temperature increases, a result that is of course to be expected. The approximation at any given temperature is closest at large angles of scattering, since the value of M for a given mean-square displacement of the atoms is proportional to $(\sin^2\theta)/\lambda^2$.

Harvey, Williams, and Jauncey [82] have compared the measurements of Williams [83] on the diffuse scattering from sodium fluoride with theory, taking into account the exchange terms in the formula for the modified scattering, as given in equation (3.56), Chapter III. The modified scattering, apart from the Breit-Dirac factor, is, according to Waller and Hartree [84],

$$I_{inc} = \sum_n (1 - f_n^2) - \sum_j \sum_k^{j \neq k} |f_{jk}|^2. \tag{5.110}$$

Harvey, Williams, and Jauncey use the tables of James and Brindley [85] in order to calculate f_n and f_{jk} for the different electrons. For the unmodified part of the scattering, fe^{-M} is taken from the experimental results of Shonka [36] on sodium fluoride. Fig. 98 shows their results graphically. The circles give the experimental points, taken from the work of Williams. Curve III shows S as a function of $(\sin\theta)/\lambda$, cal-

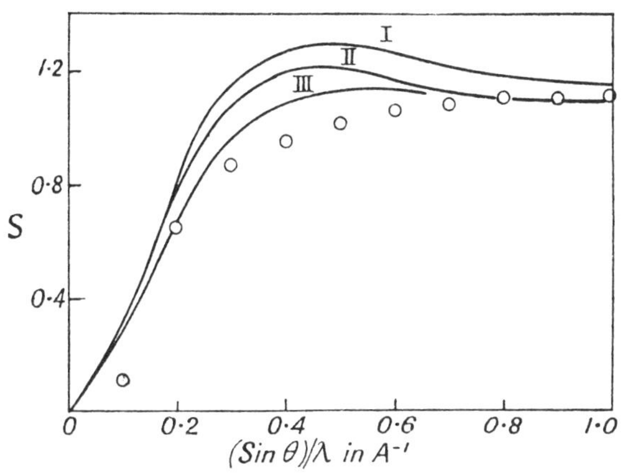

FIG. 98. Diffuse scattering from sodium fluoride. Curves after Harvey, Williams and Jauncey [82]; experimental points from Williams [83]

culated using (5.110) for the modified scattering. In Curve II, the exchange terms were neglected, and in Curve I, the approximation f^2/Z was used for $\sum_n f_n^2$, which is the same as assuming the contribution to the total f to be the same for all the electrons of the atom. The degree of approximation involved in this may be seen from the tables given by Jauncey and Pennell,[86] which were based on the tables of James and Brindley.

The experimental results are certainly best represented by Curve III, as we should expect, but it must be admitted that the possibility of cumulative error in the theoretical calculation of S is considerable, and too much must not be based on the details of the agreement.

In later experiments, a considerable amount of work on diffuse scattering from zinc crystals was done by Jauncey,[87] Claus, Bruce,[88] McNatt,[89,90,91] and others. The anisotropy of the scattering distribution was checked, and the two principal characteristic temperatures, corresponding to vibrations parallel to and perpendicular to the hexagonal axis, were measured by Jauncey and Bruce. McNatt [92] has also tested Hönl's dispersion formula by measuring the diffuse scattering from zinc and copper for radiation in the neighbourhood of the absorption frequencies.

In all these experiments, the regions of the lattice spectra were avoided, and moreover, fairly wide incident beams, and a continuous range of wave-lengths from the white radiation from a tungsten target, were usually employed. The conditions were such as to smooth out discontinuities in the background, and to avoid any diffuse maxima associated with the lattice spectra. The discussion in the earlier part of this chapter, in which the distribution of the intensity function in the reciprocal-lattice space was considered, shows that in the case of KCl continuous regions of weak scattering surround the diffuse maxima completely only near the origin, corresponding to small angles of scattering. In the regions of the reciprocal-lattice space more remote from the origin there is considerable overlapping of the domains of diffuse scattering surrounding the lattice-points, and the regions of true weak scattering become smaller. Laval's work shows that the true weak unmodified scattering for KCl is no more than half that indicated by the Jauncey formula, but the experiments made to test that formula were never such as to measure the weak scattering, and always give an average effect. One case indeed of an abnormally low scattering was observed by McNatt,[93] and was ascribed by him to a dip in the f-curve for zinc, due to some internal distortion of the atom. Such an explanation is in itself improbable, and Mrs. Lonsdale [67] has pointed out that it is not necessary. Conditions in this particular experiment were such that what was measured approximated to the true weak scattering half-way between the 0002 and 0004 reciprocal-lattice points.

Jauncey's formula, because of its method of derivation, which neglects interaction between atoms, gives only an average effect, and predicts no maxima in the diffuse scattering. The experimental methods used in testing it are, however, such as to give an average effect, and so fair agreement between theory and experiment has been found as regards the absolute magnitude of the general scattering. The experiments have indeed tested this in a very thorough way, and have had an important place in the development of the subject; but future work is likely to be directed to the details of the background pattern of the diffuse scattering, which are not given by formulae such as (5.109).

References

1. P. Debye, *Verh. d. Deutsch. Phys. Ges.*, **15**, 678, 738, 857 (1913).
2. P. Debye, *Ann. d. Physik*, **43**, 49 (1914).
3. P. Debye, *Ann. d. Physik*, **39**, 789 (1912).
4. M. Born and T. v. Kármán, *Physikal. Zeit.*, **13**, 297 (1912); **14**, 15 (1913).
5. I. Waller, *Zeit. f. Physik*, **17**, 398 (1923).
6. I. Waller, ' Theoretische Studien zur Interferenz-u. Dispersionstheorie der Röntgenstrahlen.' *Uppsala Dissertation*, 1925.
7. M. Born, *Dynamik der Kristallgitter*, Leipzig and Berlin, 1915; *Enzykl. der math. Wissenschaft.*, Bd. V. 3, 527.
8. H. Faxén, *Ann. d. Physik*, **54**, 615 (1918); *Zeit. f. Physik*, **17**, 266 (1923).
9. L. Brillouin, *Ann. d. Physique*, **17**, 88 (1922).
10. I. Waller, *Zeit. f. Physik*, **51**, 213 (1928).
11. H. Ott, *Ann. d. Physik*, **23**, 169 (1935).
12. M. Born, *Proc. Roy. Soc.*, **A**, **180**, 397 (1942); *Proc. Phys. Soc.*, **54**, 362 (1942); *Rep. Prog. Phys.* (*Physical Society*), **9**, 294 (1942-3).
13. M. Born and K. Sarginson, *Proc. Roy. Soc.*, **A**, **179**, 69 (1941).
14. K. Sarginson, *Proc. Roy. Soc.*, **A**, **180**, 305 (1942).
15. J. Laval, *C. R. Acad. Sci. Paris*, **207**, 169 (1938); **208**, 1512 (1939); *Bull. Soc. franc. Miner.*, **62**, 137 (1939).
16. H. Jahn, *Proc. Roy. Soc.*, **A**, **179**, 320 (1942); **A**, **180**, 476 (1942).
17. C. Zener, *Phys. Rev.*, **49**, 122 (1936).
18. C. Zener and S. Bilinsky, *Phys. Rev.*, **50**, 101 (1936).
19. I. Waller, *Ann. d. Physik*, **83**, 153 (1927).
20. M. Blackman, *Proc. Roy. Soc.*, **A**, **148**, 365, 384 (1935); A, **149**, 117, 126 (1935); *Rep. Prog. Phys.*, **8**, 11 (1941).
21. G. W. Brindley, *Phil. Mag.*, **21**, 790 (1936); *Proc. Leeds Phil. Soc.*, **3**, 200 (1936).
22. L. Pauling, *Phys. Rev.*, **36**, 430 (1930).
23. F. C. Kracek, S. B. Hendricks and E. Posnjak, *Nature*, **128**, 410 (1931).
24. A. Müller, *Proc. Roy. Soc.*, A, **127**, 417 (1930).
25. S. B. Hendricks, *Zeit. f. Krist.*, **67**, 106, 119, 465 (1928); **68**, 189 (1928); *Nature*, **126**, 167 (1930).
26. J. D. Bernal, *Zeit. f. Krist.*, **83**, 153 (1932).
27. W. H. Bragg, *Phil. Mag.*, **27**, 881 (1914).
28. I. Backhurst, *Proc. Roy. Soc.*, **A**, **102**, 340 (1922).
29. E. H. Collins, *Phys. Rev.*, **24**, 152 (1924).
30. R. W. James, *Phil. Mag.*, **49**, 585 (1925).
31. R. W. James, *Manchester Memoirs*, **71**, 9 (1926-7).
32. R. W. James and E. M. Firth, *Proc. Roy. Soc.*, **A**, **117**, 62 (1927).
33. R. W. James and G. W. Brindley, *Proc. Roy. Soc.*, **A**, **121**, 155 (1928).

34. R. W. James, G. W. Brindley and R. G. Wood, *Proc. Roy. Soc.*, **A, 125,** 401 (1929).
35. I. Waller and R. W. James, *Proc. Roy. Soc.*, **A, 117,** 214 (1927).
36. J. J. Shonka, *Phys. Rev.*, **43**, 947 (1933).
37. G. W. Brindley, *Phil. Mag.*, **9**, 193 (1930).
38. G. E. M. Jauncey, *Phys. Rev.*, **37**, 1193 (1931).
39. G. E. M. Jauncey and G. G. Harvey, *Phys. Rev.*, **37**, 1203 (1931).
40. Y. H. Woo, *Proc. Nat. Acad. Sci.*, **17**, 467 (1931); *Phys. Rev.*, **38**, 6 (1931); **39**, 555 (1932); **41**, 21 (1932).
41. W. Friedrich, *Physikal. Zeit.*, **14**, 1079 (1913).
42. G. L. Clark and W. Duane, *Phys. Rev.*, **21**, 379 (1923).
43. A. P. R. Wadlund, *Phys. Rev.*, **53**, 843 (1938).
44. W. H. Zachariasen, *Phys. Rev.*, **57**, 597 (1940).
45. S. Siegel and W. H. Zachariasen, *Phys. Rev.*, **57**, 795 (1940).
46. S. Siegel, *Phys. Rev.*, **59**, 371 (1941).
47. G. E. M. Jauncey and O. J. Baltzer, *Phys. Rev.*, **58**, 1116 (1940).
48. R. Q. Gregg and N. S. Gingrich, *Phys. Rev.*, **59**, 619 (1941).
49. P. Kirkpatrick, *Phys. Rev.*, **59**, 452 (1941).
50. W. H. Zachariasen, *Phys. Rev.*, **59**, 207, 766, 860, 909 (1941).
51. G. E. M. Jauncey and O. J. Baltzer, *Phys. Rev.*, **59**, 699 (1941).
52. J. H. Hall, *Phys. Rev.*, **61**, 158 (1942).
53. C. V. Raman and P. Nilakantan, *Nature*, **145**, 667, 860; **146**, 523, 686 (1940).
54. C. V. Raman and P. Nilakantan, *Phys. Rev.*, **60**, 63 (1941).
55. D. S. Teague, *Phys. Rev.*, **62**, 179 (1942).
56. K. Lonsdale and H. Smith, *Nature*, **148**, 112, 257 (1941); **149**, 402 (1942); *Phys. Rev.*, **60**, 617 (1941).
57. K. Lonsdale, *Proc. Roy. Soc.*, **A, 179**, 315 (1942).
58. R. Robertson, J. J. Fox and A. E. Martin, *Phil. Trans.*, **A, 232**, 463 (1934); *Proc. Roy. Soc.*, **A, 157**, 579 (1936).
59. C. V. Raman, *Proc. Roy. Soc.*, **A, 179**, 289, 302 (1941-2).
60. C. V. Raman and P. Nilakantan, *Proc. Indian Acad. Sci.*, Bangalore, **11, A**, 379, 389; **12, A**, 141 (1941); *Nature*, **146**, 523 (1940).
61. M. Born, *Nature*, **149**, 402 (1942); *Proc. Phys. Soc.*, **54**, 362 (1942).
62. G. D. Preston, *Proc. Roy. Soc.*, **A, 167**, 526 (1938); **A, 172,** 116 (1939).
63. W. H. Bragg, *Proc. Roy. Soc.*, **A, 179**, 51, 94 (1941-2); *Proc. Phys. Soc.*, **54**, 354 (1942).
64. K. Lonsdale, I. Knaggs and H. Smith, *Nature*, **146**, 332 (1940).
65. K. Lonsdale and H. Smith, *Proc. Roy. Soc.*, **A, 179**, 8 (1941-2).
66. K. Lonsdale, *Proc. Phys. Soc.*, **54**, 314 (1942).
67. K. Lonsdale, *Rep. Prog. Phys.* (*Phys. Soc.*), **9**, 256 (1942-3).
68. K. Lonsdale, J. M. Robertson and I. Woodward, *Proc. Roy. Soc.*, **A, 178**, 43 (1941).

69. H. Jahn and K. Lonsdale, *Phys. Rev.*, **61**, 375 (1942).
70. P. R. Pisharoty, *Proc. Indian Acad. Sci.*, Bangalore, **14**, **A**, 56, 377, 434 (1942).
71. M. Born and K. Lonsdale, *Nature*, **150**, 490 (1942).
72. M. Born, *Rep. Prog. Phys.*, **9**, 294 (1942-3).
73. O. J. Baltzer, *Phys. Rev.*, **60**, 460 (1941).
74. A. H. Compton, *Phys. Rev.*, **35**, 925 (1930).
75. C. V. Raman, *Indian Jour. Phys.*, **3**, 357 (1928).
76. G. E. M. Jauncey, *Phys. Rev.*, **42**, 453 (1932).
77. G. Breit, *Phys. Rev.*, **27**, 362 (1926).
78. P. A. M. Dirac, *Proc. Roy. Soc.*, **A**, **111**, 405 (1926).
79. G. E. M. Jauncey and H. L. May, *Phys. Rev.*, **23**, 128 (1924).
80. J. A. Crowther, *Proc. Roy. Soc.*, **A**, **86**, 478 (1912).
81. G. G. Harvey, *Phys. Rev.*, **38**, 593 (1931); **43**, 591, 707 (1933); **44**, 133 (1933).
82. G. G. Harvey, P. S. Williams and G. E. M. Jauncey, *Phys. Rev.*, **46**, 365 (1934).
83. P. S. Williams, *Phys. Rev.*, **46**, 83 (1934).
84. I. Waller and D. R. Hartree, *Proc. Roy. Soc.*, **A**, **124**, 119 (1929).
85. R. W. James and G. W. Brindley, *Phil. Mag.*, **12**, 81 (1931).
86. G. E. M. Jauncey and Ford Pennell, *Phys. Rev.*, **43**, 505 (1933).
87. G. E. M. Jauncey and W. D. Claus, *Phys. Rev.*, **46**, 941 (1934).
88. G. E. M. Jauncey and W. A. Bruce, *Phys. Rev.*, **50**, 408, 413 (1936); **51**, 1062, 1067 (1937).
89. G. E. M. Jauncey and E. M. McNatt, *Phys. Rev.*, **55**, 498 (1939).
90. R. D. Miller and E. S. Forster, *Phys. Rev.*, **50**, 417 (1936).
91. O. J. Baltzer and E. M. McNatt, *Phys. Rev.*, **55**, 237 (1939).
92. E. M. McNatt, *Phys. Rev.*, **56**, 406 (1939).
93. E. M. McNatt, *Phys. Rev.*, **57**, 621 (1940).

CHAPTER VI

EXPERIMENTAL TESTS OF THE INTENSITY FORMULAE

1. Primary and Secondary Extinction

(*a*) *Introductory*: The first five chapters of this volume have been devoted to a discussion of the quantitative theory of the diffraction of X-rays by crystals. At this stage it is necessary to consider how closely the formulae that have been deduced represent the intensities of the reflections from actual crystals, and the present chapter will be devoted to a discussion of this question.

In principle, the most direct method of testing this point would appear to be to measure the absolute values of the integrated reflection from the face of a large crystal of simple structure, preferably a crystal of an element. Using the theoretical values of the scattering factors, calculated by one or other of the methods described in Chapter III, and making due allowance for the temperature factor, one could then compare the observed value with that calculated from the formulae of Chapter II. The first quantitative tests of the reflection formulae, some of which will be described in greater detail a little later, were made in this way. Rock salt was generally used in these experiments, since it can be obtained in large pieces, and has a very simple structure, which is known with certainty.

The principle of such a determination is as follows. The crystal is mounted on the table of an X-ray spectrometer, in such a way that it can be rotated about an axis lying in the reflecting face. The slit of the spectrometer is illuminated by a beam of X-rays made monochromatic by an initial reflection from a fixed crystal. It is not enough to use radiation direct from the anticathode, for this will contain a continuous range of wave-lengths as well as the monochromatic characteristic radiation. The crystal to be experimented on has a face large enough to receive the whole of the incident beam from the spectrometer slit, and the slit of the ionisation chamber is wide enough to receive the whole of the beam reflected from the crystal. The crystal is then rotated about the axis of the instrument at a known uniform angular rate through the whole range of angles over which it reflects the radiation with appreciable intensity. The total ionisation produced during rotation is measured with a suitable electrometer, connected to the ionisation chamber. Let this total ionisation be denoted by E, and let I be the ionisation produced in the chamber when the crystal is removed, and the chamber is set so that the incident beam enters it directly for one second. Then, if ω is the angular velocity with which the crystal rotates, the integrated reflection, ρ, is given, as we saw in Chapter II, §§ 2(*c*), 2(*d*) and 2(*e*), by

$$\rho = \frac{E\omega}{I}. \tag{6.1}$$

In Chapter II, we derived formulae for the integrated reflection from the face of a large crystal, and the observed reflection has now to be compared with the value to be expected from these formulae; but here a difficulty arises, the nature of which has already been indicated in Chapter II. We found there that the formula for the integrated reflection took quite different forms according to whether we assumed the crystal to be perfect, and to consist of an entirely regular array of exactly parallel and equally spaced planes extending through its whole volume, or to be a conglomerate, made up of small independent fragments, nearly but not quite parallel in orientation, each individual fragment being so small that the effects of absorption, and of the multiple reflection of the X-rays within it, could be neglected. We shall call crystals of these two limiting types *perfect* and *mosaic* crystals respectively.

As we showed in Chapter II, § 2(*h*), the integrated reflection from the mosaic crystal for unpolarised radiation is given by

$$\rho = \frac{N^2\lambda^3}{2\mu}|F|^2\left(\frac{e^2}{mc^2}\right)^2\frac{1+\cos^2 2\theta}{2\sin 2\theta} = \frac{Q}{2\mu}, \tag{6.2}$$

μ being the linear absorption coefficient of the crystal for rays of wave-length λ.

If, on the other hand, the crystal is perfect, and if the absorption is not too large, we saw in Chapter II, § 3(*f*), that the integrated reflection is given by

$$\rho = \frac{8}{3\pi}N\lambda^2|F|\frac{e^2}{mc^2}\frac{1+|\cos 2\theta|}{2\sin 2\theta}. \tag{6.3}$$

In both these formulae, θ is the Bragg angle of reflection, and N is the number of scattering units, each of scattering factor F, in unit volume. F is supposed to be corrected for the thermal movement of the lattice, and to have the form of (*5*.80), Chapter V, § 1(*q*).

Now (*6*.2) and (*6*.3) differ very greatly. In the case of a strong spectrum, the integrated reflection calculated on the assumption that the crystal is a mosaic may perhaps be as much as twenty or thirty times as great as that calculated on the assumption that it is perfect. The state of a real crystal will as a rule lie somewhere between these two extremes, and neither formula will apply to it exactly. This is by far the greatest difficulty that is encountered in interpreting quantitative measurements of the intensities of X-ray spectra, and if we are to understand the discussions to be given later of the experimental tests of the formulae, it will be necessary to consider in greater detail the effect of the state of perfection of a crystal on the intensity with which it reflects X-rays. A preliminary discussion has already been given in Chapter II, §§ 2 (*j*) and 3(*g*), which will serve as an introduction to the more detailed account to be given here.

The discrepancies between formulae (*6*.2) and (*6*.3) were quite clearly realised by Darwin as early as 1914; and from the measurements of Moseley and Darwin [1] on the intensity of reflection of white radiation from rock salt, which was fairly well represented by (*6*.2), he concluded that this crystal could not be perfect, but must have a structure approximating to that to which the name mosaic has since been given. Sir William Bragg,[2] in 1914, had found that the integrated reflections for the strong spectra from small pieces of diamond were not proportional to the volume of the reflecting crystal, as they should have been according to equation (*2*.30) of Chapter II, on which equation (*6*.2) is based. He had concluded that diamond must exhibit abnormal absorption for X-rays passing through it at the reflecting angle, and had shown the existence of this enhanced absorption experimentally. W. L. Bragg, James and Bosanquet,[3] in 1921 and 1922, had made absolute measurements of the integrated reflections from rock salt, and had found good quantitative agreement between their results and formula (*6*.2). From their measurements for the stronger spectra they had, however, concluded that, even with rock salt, abnormal absorption occurred at the reflecting angle, and had shown that the effective absorption coefficient was greater than the ordinary absorption coefficient by an amount proportional to the intensity of the spectrum. In a paper written in 1922, Darwin [4] analysed and criticised the method of measuring the effective absorption used by Bragg, James and Bosanquet, and it is to this paper that we owe a proper understanding of the principles involved.

(*b*) *The correction for primary extinction:* Darwin showed clearly that there are two distinct effects to be considered. The first of these, which he calls *primary extinction*, we have already dealt with at some length in Chapter II, Section 3. We saw there that when X-rays fall on the face of a large perfect crystal at the proper angle of incidence they are totally reflected.* Total reflection persists over a small range of angles proportional to the amplitude reflected from a single atomic plane, and for most crystals this range is of the order of seconds of arc. The depth of the crystal in which total reflection is set up is, as a rule, only a few thousand atomic planes, and while total reflection is going on the radiation penetrates to a depth very much smaller than that to which it would penetrate if it were stopped only by the ordinary absorption in the crystal. The lower layers of the crystal thus contribute virtually nothing to the reflection. In primary extinction, the lower layers of a crystal are screened from the radiation at angles at

* We neglect for the moment the correction for absorption, which modifies these statements. See Chapter II, § 3(*h*).

which they would otherwise have reflected it, because of the reflection from the upper layers of the *same crystal.*

Most real crystals are not perfect, and we shall now follow the effect on the intensity of the integrated reflection of introducing imperfection. We shall suppose the perfection of the crystal to be reduced in a way that will allow us to follow out the processes, and to study the principles involved, although, it is true, it corresponds exactly to nothing that occurs in nature.

Let us consider a crystal built up of a number of layers, all of equal thickness, and each in effect an entirely independent perfect crystal. The atomic planes in the different layers we shall assume to be not exactly parallel, but inclined to one another at angles which, although small, are yet larger than the angular range over which a single layer will reflect radiation. The number of layers in the depth to which the radiation can penetrate when only ordinary absorption is operative is assumed to be large. Suppose now that each layer is so thick that an appreciable amount of primary extinction takes place in it when it reflects. When radiation falls on the top layer at an angle within the appropriate range, it is mostly reflected by that layer, and very little of it penetrates to the lower layers; and if these belonged to the same perfect crystal they would never have an opportunity to reflect X-rays, because they would always be screened by the top layer. In the case we are considering, however, the lower layers are not exactly parallel to the top layers, and when the angle of incidence is altered so that the top layer no longer reflects, and the incident radiation again penetrates into the crystal, it can be reflected by a lower layer when the planes in this layer make a proper angle with it. In this way, as the crystal turns, each layer has its chance to reflect, and the integrated reflection, which is the total energy reflected as the crystal turns through the whole range over which it can reflect, is much greater than it would have been had the crystal been entirely perfect. It must be borne in mind throughout this discussion that we are considering the integrated reflection, and not the reflecting power when the crystal is set so as to reflect most strongly. This will always be greatest for the most perfect crystal, whereas, as we have just seen, the integrated reflection is increased by the introduction of imperfections.

If the crystal layers are so thin that the effects of multiple reflections within a single one of them can be neglected, the integrated reflection from one layer is just proportional to the volume of that layer intercepted by the incident beam, being equal to that volume multiplied by the quantity Q, as defined in equation (*2*.31) of Chapter II, § 2(*d*). The total integrated reflection from the crystal is then that given by equation (*6*.2). If, on the other hand, the layer is thick enough to show appreciable primary extinction, the integrated reflection from it cannot be taken as proportional to the volume of the layer irradiated, because the reflection is due almost entirely to the first few thousand planes of the layer, the lower ones contributing very little. Darwin, in the paper under discussion, investigates the integrated reflection from a layer of perfect crystal having p planes, and finds the contribution to it from a volume V of the layer to be Q′V, where

$$Q' = Q(\tanh pq)/pq, \tag{6.4}$$

and q is the amplitude reflected from a single plane of atoms when a plane wave of unit amplitude falls on it. As we have already seen in Chapter II, § 2(*b*),

$$|q| = (Na\lambda/\sin\theta)|F|(e^2/mc^2), \tag{6.5}$$

N being the number of diffracting units of scattering power F in unit volume of the crystal, and a the spacing of the planes.

From a simple application of the method of § 2(*h*), Chapter II, it then follows immediately that the integrated reflection from a large crystal composed of independent layers such as we have been considering is given by

$$\begin{aligned} \rho &= Q'/2\mu \\ &= (Q/2\mu)(\tanh pq)/pq, \end{aligned} \tag{6.6}$$

p being the number of reflecting planes in each layer, supposed the same for all. The absorption coefficient μ in this formula is assumed to be the ordinary linear absorption coefficient of the crystal when it is not reflecting. This is the correct coefficient to use only if no two layers are reflecting simultaneously, and we shall have to consider shortly the modification to be introduced when this is not true.

The correction for primary extinction has here been introduced for a crystal consisting of a number of independent strata, and crystals of this particular structure do not in fact exist. A closer approximation to an actual crystal mosaic, although one that is still rather idealised, is obtained by supposing the layers to be broken up into small blocks with varying orientation, and random distribution in depth in the crystal. We may infer that, so long as the individual blocks are not so small that the widths of the diffraction maxima produced by them become appreciable in comparison with the angular deviations of the atomic planes in the blocks themselves from a mean direction, formula (6.6) will still apply to a close approximation.

The correction for primary extinction may be expressed in a rather different form, which involves the actual thickness of the layers instead of the number of planes they contain. Let d be the thickness of a homogeneous layer, so that $d = pa$. Then, since

$$q = NF(e^2/mc^2)\lambda a \operatorname{cosec}\theta, \quad Q = \lambda^3 N^2 F^2 (e^2/mc^2)^2 \operatorname{cosec} 2\theta,$$

it follows that

$$p^2 q^2 = (2Qd^2 \cot\theta)/\lambda, \tag{6.7}$$

which allows us to express (6.6) in terms of the actual thickness of the layers, or of the blocks, in the mosaic. The correction cannot, of course, be applied numerically unless we know the actual thickness of the blocks in the crystal, and then not unless they are all of the same size; so that it has not been found possible to use it effectively for applying a correction for primary extinction in actual experimental work. We may, however, at this stage use it with advantage to calculate how large the homogeneous fragments in a crystal mosaic such as we are considering may be without the correction for primary extinction becoming appreciable.

Let us take the case of rock salt. At the temperature of the laboratory the values of q for the 200, 400 and 600 reflections from this crystal for Mo Kα radiation are respectively $2{\cdot}16 \times 10^{-4}$, $5{\cdot}8 \times 10^{-5}$ and $2{\cdot}3 \times 10^{-5}$. In Table VI. 1, we give the values of $(\tanh pq)/pq$, the correction factor for primary extinction, for a series of values of p, the number of planes in a homogeneous layer, expressed in thousands. The same results are shown graphically in fig. 99, in which a scale of

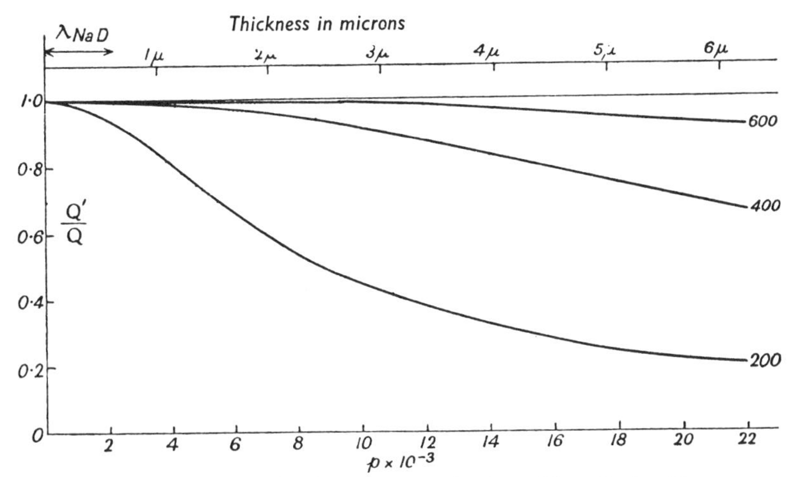

FIG. 99. The correction for primary extinction, $Q'/Q = (\tanh pq)/pq$, for rock salt 200, 400 and 600

thickness expressed in microns ($1\mu = 10^{-4}$ cm.) is also given. It will be clear that, for a given thickness of block, the correction for primary extinction is much more important for the strong spectra than for the weak ones. For rock salt 200, a very strong spectrum, primary extinction cannot be neglected unless the individual blocks of the conglomerate contain fewer than about 1000 planes, whereas blocks of ten times that thickness show no appreciable primary extinction for the 600 spectrum. The actual thickness of the blocks when primary extinction ceases to be appreciable for 200 is about $0{\cdot}3\mu$, or of the order of the wave-length of the longer ultra-violet light waves; whereas, for 600, the correction is still less than 10 per cent. for blocks 6μ in thickness.

TABLE VI. 1

Values of the correction for primary extinction, (tanh pq)/pq, for rock salt at 290° abs. (values of q from experiments of James and Firth [5]) for Mo Kα radiation ($\lambda = 0{\cdot}710$A)

p (in thousands of planes)	(tanh pq)/pq		
	(200)	(400)	(600)
0·5	1·00	1·00	1·00
1	0·99	1·00	1·00
2	0·94	1·00	1·00
4	0·81	0·98	1·00
6	0·66	0·96	0·99
8	0·54	0·93	0·99
10	0·45	0·90	0·98
12	0·38	0·88	0·97
14	0·33	0·83	0·97
16	0·28	0·78	0·96
18	0·25	0·75	0·95
20	0·23	0·71	0·91

For strong spectra, the primary extinction may be very important. For 200, for example, a block 6μ in thickness behaves as if only about 0·21 of its volume were reflecting X-rays. The ordinary linear absorption in this thickness of rock salt when Mo Kα radiation is used is nearly negligible, rather more than 99 per cent. of the radiation being transmitted.

Even for blocks as minute as those considered here, the angular width of the diffraction spectra is still very small; for it is always of the order of the wave-length of the radiation divided by the linear dimensions of the block producing the diffraction, and this, with Mo Kα radiation, $\lambda = 0{\cdot}71$A, is of the order of 20 seconds of arc for a block one micron in thickness.

(*c*) *Secondary extinction :* Let us return to the consideration of the crystal composed of a number of layers, but suppose now that each layer is so thin that primary extinction may be entirely neglected. If the individual layers, although independent, have their atomic planes exactly parallel, the integrated reflection will not be given correctly by equation (*6.2*), even in the absence of primary extinction. Consider a layer at a depth z in the crystal. If I_0 is the intensity of the incident beam, the intensity arriving at a depth z would be reduced to $I_0 e^{-\mu z \operatorname{cosec} \theta}$ by the ordinary absorption; but when the crystal is reflecting X-rays, the reduction is greater than this, since a certain fraction of the radiation is reflected away by each layer through which it passes.

The radiation reflected from any layer is deflected into the reflected beam, and a certain fraction of this reflected radiation will also be deflected back by a second reflection into the direction of the incident beam. There will thus be an interchange of radiation between the incident and reflected beams similar at first sight to that considered in Darwin's theory of the perfect crystal, which led to the idea of primary extinction. There is, however, an essential difference between the two processes. Since the different layers are optically independent, there is no regular phase relationship between the rays reflected from them, and to obtain the resultant effect of a large number of layers we have to add *intensities*, and not amplitudes, as we did in Darwin's theory. This weakening of the incident radiation within the crystal by reflection from optically independent members of the same conglomerate is called by Darwin *secondary extinction.* It is equivalent to an increase in the effective absorption coefficient at the reflecting angle, and can be allowed for to a good approximation simply by an appropriate increase in μ, the absorption coefficient in equation (6.2).

It is easy to get an idea of the correction for a crystal composed of independent strata, in all of which the reflecting planes are parallel. As a first approximation, suppose we consider only the first reflection in both incident and reflected beams, neglecting multiple reflections. A beam of unit cross-section and intensity I, falling on the crystal face at a glancing angle θ_0, intercepts a volume cosec $\theta_0\, dz$ of a layer of thickness dz, if θ_0 is the Bragg angle, and reflects away a fraction proportional to Q cosec $\theta_0\, dz$ into the reflected beam. The total diminution in intensity of the beam of strength I in the layer, due to absorption and reflection together, is therefore $-\mathrm{I}(\mu + a\mathrm{Q})$ cosec $\theta_0\, dz$. The crystal behaves as if it had an absorption coefficient $\mu + a\mathrm{Q}$, instead of μ, and the integrated reflection will be $\mathrm{Q}/2(\mu + a\mathrm{Q})$, instead of $\mathrm{Q}/2\mu$, as it would have been in the absence of secondary extinction. The effect of secondary extinction, as well as of primary extinction, is therefore greatest for the strongest spectra.

We have here considered an extreme case, in which the atomic planes of the independent layers are parallel, and so the effect of the secondary extinction very large. The same effect must, however, occur, although in a lesser degree, in a crystal in which the atomic planes of the different layers are not exactly parallel, but are distributed with a certain angular variation about some mean direction. As before, let us consider reflection from a layer at a depth z. If the planes in any layers above it are so nearly parallel to those of the layer we are considering as to reflect X-rays for the same angular setting of the crystal, that is to say, if the angle between the planes is smaller than the angular width of a diffraction spectrum from a single layer, the lower layers must be screened by the upper ones, and secondary extinction must occur. The more nearly parallel the planes in the different layers, the greater the secondary extinction.

Primary and secondary extinction are thus two essentially distinct effects, and both will in general be operative in any real crystal. The line of argument we have used in this paragraph could still be applied if the layers were thick enough for primary extinction to be appreciable, the only necessary modification being the use of Q′, as defined in (6.4), instead of Q in the statement given above.

(*d*) *Correction for secondary extinction in a crystal composed of small independent blocks :* An argument very similar to that used in the last paragraph can be used if the crystal, instead of being made up of independent parallel layers, is made up of independent blocks with slightly varying orientations, and a random distribution in depth. We shall begin by supposing the blocks to be so small that primary extinction may be neglected, and the range of angles over which their orientation varies to be considerably greater than the angular width of the diffraction spectrum reflected from a single block.

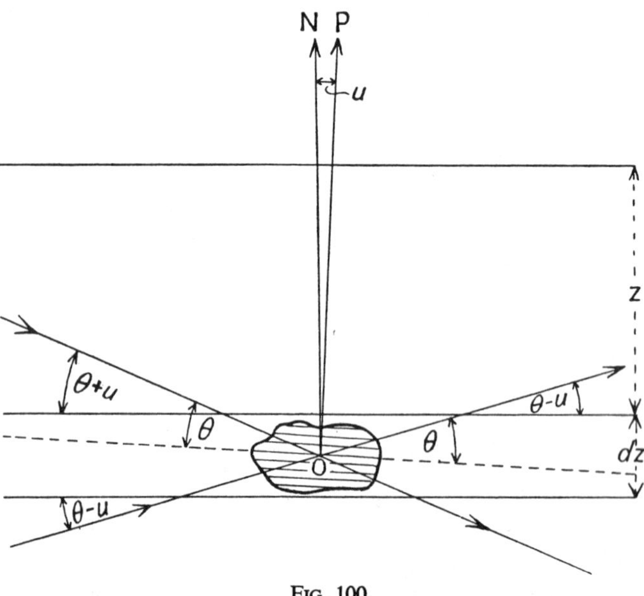

FIG. 100

Consider the blocks in a layer of thickness dz, at a depth z in the crystal. The boundaries of the layer are parallel to the face of the piece of crystal from which the radiation is reflected, but the reflecting planes of the blocks within the layer are not necessarily parallel to the boundaries of the layer. Let OP, fig. 100, the normal to the reflecting planes in a block of volume ΔV, make an angle u with ON, the normal to the crystal face, which we shall take as the direction of reference. For simplicity, we suppose the normals to all the blocks to lie in the plane of incidence. If the slits of the ionisation chamber, or other measuring apparatus, are high enough to receive the whole reflected radiation, small variations of the normals from the plane of incidence make no difference to the integrated reflection.

Let us consider the integrated reflection from the single layer, neglecting for the time being all absorption. The block of volume ΔV contributes an amount $Q\,\Delta V$ to this integrated reflection, and its contribution is made when the angle of incidence of the radiation on the layer is in the neighbourhood of θ_0+u, and the angle of emergence in the neighbourhood of θ_0-u (see fig. 100). Let I be the intensity incident on the layer. When the angle of incidence is θ_0+u, only a fraction of the blocks in any volume V of the layer are set so as to reflect radiation. Let the energy reflected per second from volume V at this angle of incidence be $IG(u)V$. $G(u)$ is thus a measure of the number of blocks in unit volume that are set so as to reflect when the glancing angle of incidence on the crystal face is θ_0+u, and its variation with u is an index of the state of perfection of the crystal. The integrated reflection from the volume V of the layer is obtained by integrating the expression for the reflected energy over all angles of incidence, and dividing by the incident intensity, and this is equal to QV, by equation (2.30), Chapter II, § 2(d). Thus

$$V\int_{-\infty}^{+\infty} G(u)du = QV$$

or

$$\int_{-\infty}^{+\infty} G(u)\,du = Q. \tag{6.8}$$

If the individual blocks are big enough to exhibit primary extinction, and if each contains the same number, p, of reflecting planes, equation (6.8) becomes

$$\int_{-\infty}^{+\infty} G(u)\,du = Q', \tag{6.9}$$

where Q' has the value given by equation (6.4).

We shall now apply these ideas to the calculation of the secondary extinction in the crystal, following the treatment given by Darwin (*loc. cit.*). We suppose the function $G(u)$ to be constant throughout the crystal, and the volume irradiated by the incident beam to be large enough to contain a very large number of independent blocks, so that intensities may be averaged.

Consider a beam falling on an area A of the crystal face at a glancing angle θ_0+u. Let $I_u(z)$ be the energy passing per second, or the *power*, in the incident beam at depth z, when the glancing angle is θ_0+u, so that the intensity is $I_u(z)\operatorname{cosec}(\theta_0+u)/A$. The beam intercepts a volume $A\,dz$ of a layer of thickness dz, and this volume reflects in unit time energy equal to $G(u)A\,dz$ times the incident intensity. Reflection thus reduces the power of the beam by $G(u)I_u(z)\operatorname{cosec}(\theta_0+u)\,dz$ in its passage through the layer; but it will also be reduced by an amount

$\mu I_u(z) \operatorname{cosec}(\theta_0+u)\,dz$ by ordinary absorption, so that the total reduction in power of the primary beam in the layer, owing to both absorption and reflection, is $I_u(z)\{\mu+G(u)\} \operatorname{cosec}(\theta_0+u)\,dz$.

Now those blocks in the layer responsible for the reflection of the radiation incident on the face of the crystal at θ_0+u are also set so as to reflect back into the primary beam radiation in the reflected beam coming from the lower layers. This radiation falls on the lower surface of the layer dz at a glancing angle θ_0-u (see fig. 100), and the power reflected by the volume $A\,dz$ back into the primary beam will be $G(u)E_u(z) \operatorname{cosec}(\theta_0-u)\,dz$, $E_u(z)$ being the power in the reflected beam at depth z.

If $dI_u(z)$ is the total increase in the power of the primary beam in depth dz, at a depth z below the surface of the crystal,

$$\frac{dI_u(z)}{dz} = -\frac{\mu+G(u)}{\sin(\theta_0+u)} I_u(z) + \frac{G(u)}{\sin(\theta_0-u)} E_u(z), \tag{6.10}$$

while the corresponding equation for the variation of the power of the reflected beam with depth will evidently be

$$-\frac{dE_u(z)}{dz} = -\frac{\mu+G(u)}{\sin(\theta_0-u)} E_u(z) + \frac{G(u)}{\sin(\theta_0+u)} I_u(z). \tag{6.10a}$$

Equations (6.10) and (6.10a) must now be solved with the conditions that both $I_u(z)$ and $E_u(z)$ tend to zero for very large z. The solution is quite straightforward if it is assumed that $G(u)$ does not depend on z, so that the equations become linear. It may be obtained from the substitutions $I_u(z)=Ie^{-\alpha z}$, $E_u(z)=E_u e^{-\alpha z}$, I being the power of the incident beam, and E_u the power of the reflected beam corresponding to an angle of incidence θ_0+u, as it leaves the crystal. The solution is

$$\frac{E_u}{I} = \frac{\sin(\theta_0-u)}{\sin\theta_0\cos u}\,\frac{G(u)}{\mu+G(u)+\sqrt{\{\mu+G(u)\}^2-\{G(u)\}^2(1-\cot^2\theta_0\tan^2 u)}}. \tag{6.11}$$

From it we have to find the integrated reflection by integrating E_u/I over all values of u. Before doing so, however, it will be necessary to consider the form of (6.11) rather more closely. The first factor, $\sin(\theta_0-u)/\sin\theta_0\cos u$, is due to the inequality of the angles of incidence and emergence, θ_0+u, and θ_0-u, when reflection is taking place from the blocks we are considering. The question of unsymmetrical reflection is of some importance practically, and a separate paragraph will therefore be devoted to it.

(e) *Unsymmetrical reflection from a crystal mosaic:* Let us first consider a case in which the reflecting planes in the crystal mosaic deviate only slightly from a certain standard direction, but suppose that the face of the piece of crystal from which reflection is taking place has been cut so as to make an angle with this standard direction considerably greater than the angular deviation of the mosaic blocks. This is a state of affairs that is often met with in practice. It frequently happens that no natural face is developed on the crystal parallel to the planes from which it is desired to measure the integrated reflection, so that a face has to be ground on the crystal in the proper direction; and even if a natural face does occur, it is often desirable, for reasons connected with the extinction phenomena we have been considering, to grind it before measuring the reflection. Unless considerable care is taken, it is difficult to avoid errors of the order of a degree in grinding such faces, but fortunately it is easy to eliminate the effects of such errors completely.

Let α be the angle between the average direction of the reflecting planes and the crystal face, measured in the plane of incidence. Reflection then takes place either when the angle of incidence is $\theta_0+\alpha$, and the angle of emergence $\theta_0-\alpha$, or when the angle of incidence is $\theta_0-\alpha$, and the angle of emergence $\theta_0+\alpha$. We consider the effect of ordinary absorption and neglect extinction altogether. If the angle of incidence is $\theta_0+\alpha$, an incident beam of cross-section S_0 cuts off a volume $S_0 \operatorname{cosec}(\theta_0+\alpha)\,dz$ from a layer of thickness dz, parallel to the surface of the crystal. If the layer is at a depth z below the surface of the crystal, every ray reflected from it has traversed a distance

$$z\{\operatorname{cosec}(\theta_0+\alpha)+\operatorname{cosec}(\theta_0-\alpha)\}$$

in the crystal, instead of a distance $2z \operatorname{cosec}\theta_0$, as in the case of symmetrical reflection. Equation (2.41a) of § 2(*h*), Chapter II, thus becomes

$$\left(\frac{E\omega}{I_0}\right)_\alpha = QS_0\int_0^\infty dz \operatorname{cosec}(\theta_0+\alpha)\, e^{-\mu\{\operatorname{cosec}(\theta_0+\alpha)+\operatorname{cosec}(\theta_0-\alpha)\}z},$$

which leads at once to

$$\left(\frac{E\omega}{I}\right)_\alpha = \frac{Q}{2\mu}\,\frac{\sin(\theta_0-\alpha)}{\sin\theta_0\cos\alpha}$$
$$=(Q/2\mu)(1-\cot\theta_0\tan\alpha). \qquad (6.12)$$

The integrated reflection is thus less than the true value, and the difference may be very considerable. For $\theta_0=10°$, and $\alpha=1°$, the error is about 10 per cent., and for $\alpha=5°$, 50 per cent. It can be eliminated by interchanging the directions of incidence and reflection, so that the angle of incidence becomes $\theta_0-\alpha$, and the angle of emergence $\theta_0+\alpha$. The expression for the integrated reflection is then

$$\left(\frac{E\omega}{I}\right)_{-\alpha}=(Q/2\mu)(1+\cot\theta_0\tan\alpha), \qquad (6.13)$$

so that the mean of the two values gives the true value for symmetrical reflection.

The physical significance of the difference between the two expressions may easily be understood from fig. 101.

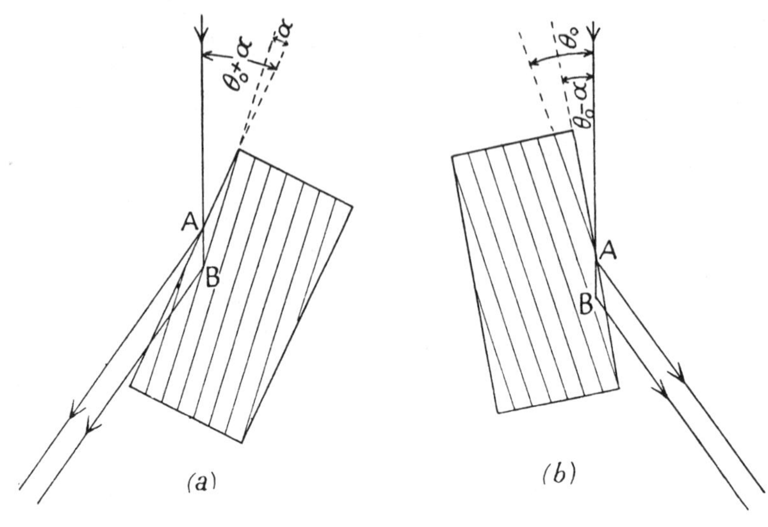

FIG. 101

Consider a narrow pencil of X-rays which penetrates the crystal and irradiates an effective length AB. This length is the same whether the angle of incidence is $\theta_0+\alpha$, as in fig. 101 (*a*), or $\theta_0-\alpha$, as in fig. 101 (*b*). In case (*a*), however, the emergent rays reflected from the irradiated volume have all travelled a greater distance in the crystal than they would have done had the reflection been symmetrical with respect to the surface, while in case (*b*) the distances are correspondingly reduced. The influence of absorption is therefore greater in case (*a*), and less in case (*b*), than it would be for symmetrical reflection.

The importance of this effect, and the method of correcting for it, were first pointed out by Sir William Bragg [2] in 1914. In the determination of the integrated reflection by the method of reflection at a face, measurements should always be taken with the ionisation chamber both right and left, with the crystal settings differing by about $180°-2\theta_0$. If there is any error in the direction of the crystal face, whether owing to inaccurate grinding, or to the presence of a vicinal face, the integrated reflections taken right and left will differ; for in one case the angle of incidence will be larger, and in the other smaller by the same amount, than that for symmetrical reflection. The mean of the two integrated reflections gives the true value for symmetrical reflection. This precaution is particularly necessary for spectra of low order, for which θ_0 is small, since the correcting factor contains the cotangent of θ_0.

Some examples of the effect are given in Table VI. 2, which gives integrated reflections from a single crystal of aluminium on which an octahedral face had been ground. The letters L and R indicate observations made right and left on the spectrometer. The error in the grinding of the face, as estimated from the difference between the left and right readings, must have been rather more than 1½ degrees.

TABLE VI. 2

Integrated reflections from aluminium at 290° Abs., in terms of rock salt 400 as one hundred (James, Brindley and Wood [6])

Spectrum	θ_0	Integrated reflection		
111	8° 43′	(L) 465	(R) 694	Mean 580
222	17° 40′	(L) 133	(R) 155	Mean 144

(*f*) *The integrated reflection from a mosaic of small blocks :* We may now return to the calculation of the integrated reflection from equation (*6*.11). The first factor, as we have just seen, is due to the unsymmetrical orientation with respect to the crystal face of the blocks reflecting when the angle of incidence is $\theta_0 + u$. If the integrated reflections are measured both right and left, as described in the last paragraph, for every value of E_u on one side there corresponds a value E_{-u} at the angle of incidence $\theta_0 - u$, measured on the other side, which differs from the value of E_u, as given by (*6*.11), only in having $\sin(\theta_0 + u)$, instead of $\sin(\theta_0 - u)$, in the first factor. In what follows we shall suppose that means of integrated reflections taken right and left are always considered, and we can therefore put the first factor in (*6*.11) equal to unity in calculating the integrated reflection. We shall also assume that u is always small enough for $\tan^2 u$ to be neglected, and that $G(u)/\mu$ is small enough for us to expand (*6*.11) in powers of it, and to neglect powers higher than the square in the expansion of the square- root.

The integrated reflection from the crystal face can then be written

$$\rho = \int_{-\infty}^{+\infty} \frac{E_u\,du}{I} = \int_{-\infty}^{+\infty} \left\{ \frac{G(u)}{2\mu} - \frac{G^2(u)}{2\mu^2} + \frac{5}{8}\frac{G^3(u)}{\mu^3} \right\} du$$

$$= \frac{Q'}{2\mu} - g_2 \frac{Q'^2}{2\mu^2} + \frac{5}{8} g_3 \frac{Q'^3}{\mu^3}, \qquad (6.14)$$

if we put

$$\int_{-\infty}^{+\infty} G^2(u)\,du = g_2 Q'^2,$$

$$\int_{-\infty}^{+\infty} G^3(u)\,du = g_3 Q'^3, \qquad (6.15)$$

where g_2 and g_3 are constants of the crystal, and Q' has the significance of equation (*6*.4), and is related to $G(u)$ by equation (*6*.9).

Equation (6.14) can now be written in the form

$$\rho = \frac{Q'}{2\left\{\mu + g_2 Q' + (g_2{}^2 - \tfrac{5}{4}g_3)\dfrac{Q'^2}{\mu}\right\}}. \tag{6.16}$$

The correction for primary extinction is here made by using Q' instead of Q, but this involves the assumption that the blocks of the mosaic are all of the same size. The correction for secondary extinction is made by increasing the ordinary linear absorption coefficient of the crystal by an amount that is greater the greater the intensity of the spectrum concerned. If the term in (6.16) involving Q'^2/μ can be neglected, as it probably can for many real mosaic crystals, (6.16) takes the simpler form

$$\rho = \frac{Q'}{2(\mu + g_2 Q')}. \tag{6.17}$$

Equation (6.17) has the form of (6.6) with the linear absorption coefficient increased by a quantity which, for a given crystal, is proportional to Q'.* As we shall see, there are methods of obtaining an estimate of this increase for actual crystals. From the discussion we have just given, however, it is clear, as was pointed out by Darwin, that all such methods correct only for the secondary extinction and leave the primary extinction entirely untouched.

(g) *Reflection curves from actual crystals :* If a crystal mounted on the table of an X-ray spectrometer with the ionisation chamber set at the correct angle to receive a certain spectrum is slowly rotated through the position at which it should reflect according to the Bragg law, it will generally be found that reflection occurs over a range of crystal settings varying from several minutes up to perhaps several degrees. The range of reflection for a perfect crystal, with a parallel incident beam, should be of the order of seconds of arc only, and the observed angular range of reflection, obtained with a slightly divergent beam, such as must be used in practice unless certain special methods to be discussed later are employed, should be almost entirely determined by that divergence.

At this point it is perhaps well to draw attention to the geometry of the process of reflection we are considering here. The angular range of reflection would of course be increased if the actual effective number of co-operating planes were small, so that the reflected beam was widened by diffraction; but, in the experiments we are considering, the width of the diffracted beam itself at any given setting of the crystal is not as a rule greater than can be accounted for by the divergence of the incident beam and the width of the source. As the crystal rotates, radiation

* In some crystals, for example diamond, the parallelism of the optically independent blocks may be such that g_3 is more important than $g_2{}^2$ and is not negligible in comparison with g_1. In (6.17), g_2 is then replaced by $g_2(1 - cQ')$, so that the extinction is relatively more important for the weaker spectra than (6.17) implies.

is always reflected into the same small angular range, but reflection occurs over a range of settings of the crystal. This can be explained only by assuming the crystal to contain regions differing in orientation, but each containing a large number of regularly arranged planes. As the crystal rotates, these are brought in turn into the proper position for reflection. In an extreme case, we may indeed find an apparently single crystal to consist of several separate crystals, differing in orientation, but each in itself fairly perfect. A remarkable example of this is shown in fig. 102(*a*), which was obtained from a crystal of

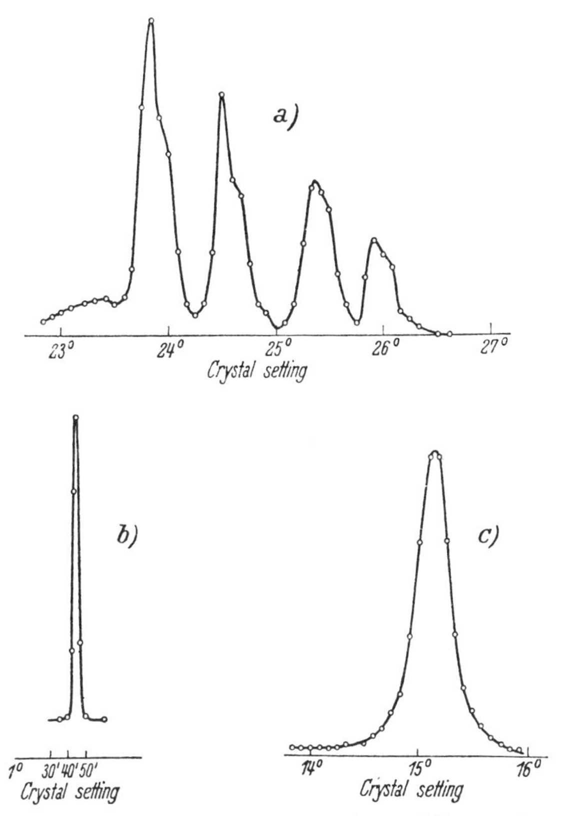

FIG. 102. Examples of reflection curves for specimens of (*a*) corundum, (*b*) β-corundum, (*c*) rock salt; (*a*) and (*b*) with R*h* Kα, (*c*) with Mo Kα (James) (*Zeit. f. Kristal.*, **89**, 295 (1934))

corundum (Al_2O_3). The ordinates give the intensity reflected in the direction of the $3\bar{3}0$ spectrum at the crystal settings given by the abscissae. The region of the crystal irradiated by the incident beam is here seen to consist of at least four distinct parts, each of which gives a reflection sharp enough to show some resolution of the Kα doublet of rhodium, the radiation used in the experiment. Figs. 102(*b*) and

102 (c) show similar curves for crystals of β-corundum, and of rock salt respectively. The area beneath a curve of this kind, which we shall call a *reflection curve*,* is proportional to the integrated reflection from the crystal, and is found to be very nearly constant, for a given spectrum, from different specimens of the same crystal, and to a large extent independent of the shape and width of the reflection curve; whereas the strength of the maximum reflection depends entirely on the particular specimen of the crystal used. It is for this reason that the integrated reflection is taken as a measure of the intensity of a spectrum.

A broad reflection curve is very definite evidence that the crystal giving it is not ideally perfect, but it is by no means easy to get from such curves any accurate idea of the amount of extinction that a particular crystal is likely to show. It is not safe to assume that the form of the reflection curve gives $G(u)$, as defined in § 1 (d) of this chapter, which, as we have seen, defines the secondary extinction. Both primary and secondary extinction depend very much on the fine-scale structure of the crystal, while the reflection curve may be determined by the large-scale structure. The crystal may consist of a number of relatively large fragments with a distribution in orientation that widens the whole reflection curve. The extinction in the crystal will, however, be determined by the texture of the individual large fragments. An extreme case is that shown in fig. 102 (a). Although the reflection curve extends over an angular range of nearly four degrees, it is probable that the individual regions of the crystal are fairly perfect, and would show a large extinction, both primary and secondary—certainly more extinction than would be shown by the rock-salt crystal whose reflection curve is given in fig. 102 (c), and extends over a smaller range of angles.

It is clear in this particular case that the crystal does actually consist of several independent parts, and that the extinction will be determined by the properties of these; but there must be every gradation from this state to one in which there are so many independent fragments that all trace of their individuality disappears in the reflection curve. Caution is therefore needed in drawing any very definite conclusions about extinction from reflection curves. The irregularities that diminish primary extinction, and the lack of perfect orientation of optically independent fragments that diminishes secondary extinction, are most likely to be present in a crystal that has a broad reflection curve, but there is no strict correlation between the breadth of the reflection curve and the amount of the small scale irregularities. Some crystals with comparatively narrow reflection curves show quite small extinction.

A word must be said here as to the correctness of representing a crystal by a mosaic of small perfect blocks. In theoretical work, the assumption of a block structure of this type is practically forced upon us by the difficulties of dealing mathematically with any other type of

* They have been called ' rocking curves ', usually by American authors, and also ' sweep curves ', and neither name is particularly satisfactory.

irregularity, but it is improbable that such a structure is often realised in actual crystals. Imperfection may be due to warping, as for example in the case of rock salt, which often exhibits a large-scale warping clearly visible to the unaided eye; or it may be due to growth in different directions at different parts of the crystal causing misfits and dislocations where the different regions of growth impinge upon one another. It may be due to mechanical impurities causing irregularities of growth, to minute cracks produced by mechanical strains, or to combinations of all these, and probably of many other, causes. We shall not discuss the texture of actual crystals in detail here, but reference may be made to a symposium in the *Zeitschrift für Kristallographie*, vol. **89**, pp. 193–416 (1934), where various authors discuss different aspects of this very complex subject. Here, it is only necessary to point out that anything that prevents exact regularity of arrangement from persisting over more than a few thousand atomic planes will cause the crystal to behave as a mosaic in reflecting X-rays. For a crystal to behave as if it were perfect, exact coherence in phase of the reflected beams from a good many thousand atomic planes is necessary; and when the conditions under which most crystals grow are taken into account, it is not remarkable that perfection of this kind is rarely attained, but rather that it is in fact sometimes attained, as we shall see in our discussion of the reflection of X-rays from actual crystals.

In practice, every gradation, from the perfect crystal to the mosaic occurs, but most crystals approximate more closely to the mosaic than to the perfect type in so far as their properties as reflectors of X-rays are concerned.

It is particularly necessary to guard against the assumption that has sometimes been made that because a crystal appears to be optically perfect, and gives sharp optical reflections from its faces, and good interference fringes from thin slices, it must therefore also be a perfect reflector of X-rays. It must be remembered that X-rays demand a standard of perfection some ten thousand times more exacting than do light waves, because of their much shorter wave-length; and so small dislocations that could have no effect on any optical property are large enough to cause the crystal to be classed as a mosaic when its behaviour as a reflector of X-rays is being considered.

(*h*) *Evidence of crystal imperfection from the absolute intensities of X-ray reflections:* The most reliable evidence as to the state of perfection of a crystal is probably to be obtained by comparing the measured and calculated values of integrated reflections. The difference between the values of the integrated reflection for a strong spectrum as predicted by formulae (*6*.2) and (*6*.3) is so large that if the state of the crystal departs greatly from perfection we should expect the integrated reflections from it to be much greater than those which it could give if it were perfect, on any assumption as to its structure.

As an example, we may here quote some results obtained by James, Brindley and Wood [6] from single crystals of aluminium. These experiments have already been referred to in Chapter V, § 2(*c*), in connection with the measurement of the temperature factor, and the mean-square amplitude of the atomic vibrations. A set of absolute measurements of the integrated reflections from aluminium was made, and at the same time the temperature factor was directly measured. This allows a direct comparison to be made between the observed and calculated intensities. In the calculation of the intensities, Hartree's values for the atomic scattering factor, f, have been used, and these have been corrected for temperature by means of the measured temperature factor, with allowance for zero-point energy. The measurements are absolute, and the observed and calculated figures are independent of one another. The results are shown in Table VI. 3.

TABLE VI. 3

Integrated reflections from aluminium single crystals at 17° C., compared with those calculated from the mosaic and perfect crystal formulae (Mo Kα radiation.)

		Absolute integrated reflections × 10^6		
		Calculated		Observed
Spectrum	θ	Perfect Crystal	Mosaic	
111	8° 43′	19·6	818·3	580
200	10° 6′	16·2	618·6	436
222	17° 40′	6·30	157·9	144
400	20° 32′	4·47	91·3	86·4
333	27° 6′	2·19	28·3	26·2
600	31° 45′	1·31	12·0	12·2
444	37° 23′	0·76	5·14	4·9
800	44° 33′	0·40	2·09	2·1
555	49° 23′	0·37	1·39	1·43

The figures show immediately that the observed intensities are all far greater than those calculated on the assumption that the crystal is perfect. In the case of the 111 reflection, the ratio is thirty to one, and in that of the 800 reflection, five to one. The values for the perfect crystal were calculated without any allowance for ordinary absorption. This is small, but not entirely negligible, and had it been taken into account the calculated values in column (3) would have been still smaller. The observed results for the spectra of high order agree remarkably well with the values calculated on the assumption that the crystal is a mosaic; they are, however, rather low for the strong spectra, and this is evidently due to the existence of a certain amount of secondary extinction in the crystal used in the experiments, which has an appreciable effect only on the strong spectra.

(*i*) *Experimental determination of secondary extinction:* The results quoted in the last paragraph were chosen because they illustrate in the most direct manner possible the point to be made, and give a striking example of an actual crystal that reflects in a manner in no way to be represented by the perfect crystal formula. The first accurate intensity measurements were, however, made with rock salt, which usually approaches very closely to the ideal mosaic type, and shows very little primary extinction. In 1917, A. H. Compton [7] measured the absolute integrated reflection 200 from rock salt, using rhodium $K\alpha$ radiation. His results were probably affected by extinction, as the spectrum is a very strong one, but the integrated reflection was of the general order of magnitude to be expected from equation (*6*.2). In 1921 and 1922, W. L. Bragg, James and Bosanquet [3(I)] made a series of absolute determinations of the integrated reflection from rock salt, with the primary object of determining experimentally the atomic scattering factors of sodium and chlorine. The integrated reflections were measured for 18 different orders, and from these the absolute values of the structure factors $|F|$ for these spectra were determined, assuming formula (*6*.2) to be valid. When these structure factors were plotted against $\sin\theta$, the points were found to lie on two smooth curves, one for structure factors of the type $f_{Cl}+f_{Na}$, the other for those of type $f_{Cl}-f_{Na}$. It was therefore possible to plot f_{Cl} and f_{Na} as functions of $\sin\theta$ by taking half the sum and half the difference of corresponding ordinates of these two curves. For small values of $\sin\theta$, the curves for f_{Cl} and f_{Na} should have approached the limits 18 and 10, the number of electrons in the ions of chlorine and sodium respectively, if we suppose the atoms to be ionised in the crystal. Bragg, James and Bosanquet found this to be true only in so far as the general order of magnitude was concerned. For small values of $\sin\theta$ the curves for both chlorine and sodium were considerably lower than would have been expected, and tended to values more like 14 and 8. This discrepancy they ascribed to extinction, and in their second paper [3(II)] they attempted to correct for this by measuring directly the effective absorption coefficient of the crystal at the reflecting angle.

With this object in view, they measured the absolute value of the integrated reflection not from a crystal face, but from a slice of crystal cut so that the reflecting planes were normal to its faces. The reflected beam in this case passes through the crystal slice, and in Chapter II, § 2(*h*), we saw that if the crystal is a mosaic the integrated reflection is given by

$$\rho = t'Qe^{-\mu t'}, \tag{6.18}$$

where $t'=t\sec\theta$, t being the thickness of the slice, and θ the glancing angle of reflection.

The integrated reflection was measured from a number of slices, cut from the same crystal and differing in thickness. For very thin slices, ρ should be proportional to t', for very thick slices it should die away exponentially with increasing thickness, and it should have a maximum value for $t' = 1/\mu$. It is possible therefore to measure μ by determining the value of t' corresponding to the maximum integrated reflection. Now any radiation contributing to the reflection has passed through the crystal slice at the reflecting angle, and if there is enhanced absorption at that angle it should be possible to detect and to measure it by this method. It is no use trying to determine the effective absorption from the reduction in the intensity of the transmitted beam when the crystal is set so as to reflect most strongly; for the incident beam is always slightly divergent, and not all of it passes through at the reflecting angle when the crystal is in the position for maximum reflection.

Figure 103 shows curves obtained in this way from rock salt for the 200, 220, 400 and 600 spectra. The slices were prepared from the same large crystal, and were ground on a sheet of glass with water, in order to avoid mechanical strain; for it was found that rough grinding made the crystal more imperfect, as indicated by the reflection curves, and so

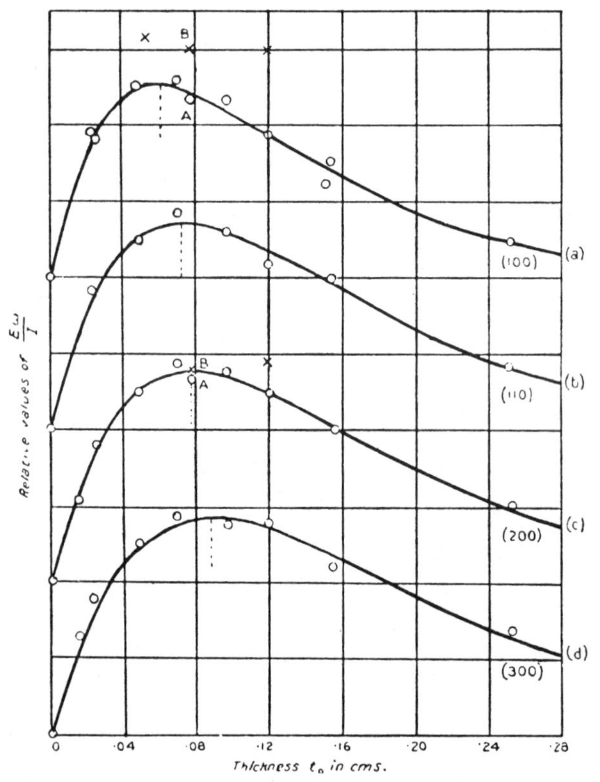

FIG. 103. Integrated reflections measured through a slice of rock salt as a function of the thickness of the slice for different spectra (Bragg, James and Bosanquet)
(*Phil. Mag.*, **42,** 1 (1921))

increased the integrated reflection. A weakness of the method is that different slices have to be used, the mosaic structure of which is not necessarily the same, even although all come from the same crystal. An alternative method is to grind down the same crystal slice, taking the reflection at different thicknesses; but even here the structure may differ at different depths, and may also be altered by the grinding. The secondary extinction is, however, in any case an average property of the crystal, and the points obtained from the different slices lie fairly well on a smooth curve. A few points obtained from slices that had been roughly ground are indicated by crosses in fig. 103, and are evidently abnormal.

It will be seen from fig. 103 that the maximum of the curve falls at a smaller value of t the greater the intensity of the spectrum. The more intense the spectrum the greater therefore is the effective absorption coefficient at the reflecting angle. The crosses in fig. 103 denoting the values obtained from the roughly treated slices, which have a small extinction, all lie well above the corresponding curves, but they lie relatively much higher for 200 than for 400, which again illustrates the fact that extinction is much more important for the stronger spectra.

That the curves of fig. 103 are very closely of the type indicated by (*6*.18) is clearly shown when the results are plotted as in fig. 104,

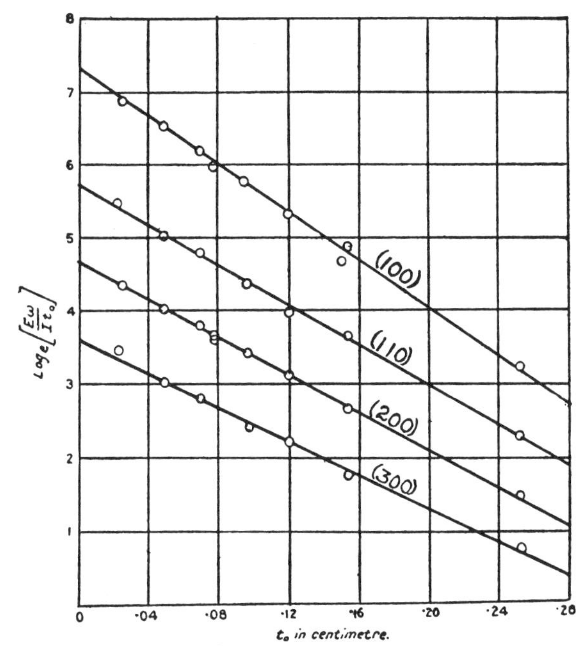

FIG. 104. The logarithm of the integrated reflection measured through a slice of rock salt as a function of the thickness (Bragg, James and Bosanquet) (*Phil. Mag.*, **42**, 1 (1921))

in which the ordinates are $\log_e(\rho/t')$, and the abscissae the actual thickness t. Equation (6.18) gives

$$\log_e(\rho/t') = -\mu t \sec\theta + \text{const.} \tag{6.19}$$

The straightness of the lines in fig. 104 shows therefore that the variation of ρ with t' is of the assumed type, and the slopes of the lines give the values of μ, the effective absorption coefficients for the different reflections, if allowance is made for the factor $\sec\theta$. The difference between the effective absorption coefficient μ, and the ordinary linear absorption coefficient μ_0, we shall denote by ϵ, and call the *secondary extinction.*

In Table VI. 4, the values of μ and ϵ for the particular specimen of rock salt used in the experiments under discussion are shown. It may be remarked that the extinction in this crystal is quite small, and that the value of ϵ relative to μ_0 is often considerably larger. Even so, the absorption coefficient for 200 is increased by 30 per cent. by extinction.

TABLE VI. 4

Secondary extinction, ϵ, for a specimen of rock salt. Rh Kα radiation (W. L. Bragg, James and Bosanquet [3(II)])

Reflection*	Effective absorption coefficient μ	Extinction $\epsilon=\mu-\mu_0$	Integrated reflection (Arbitrary units)
200	16·30	5·60	100
220	13·60	2·90	50·5
400	12·66	1·96	19·9
600	10·72	0·02	4·9

Normal linear absorption coefficient $\mu_0 = 10{\cdot}70$.

The results given in Table VI. 4 indicate that, within the errors of experiment, the extinction is proportional to the integrated reflection, so that, empirically, the absorption coefficient μ is given by

$$\mu = \mu_0 + \epsilon = \mu_0 + k\rho, \tag{6.20}$$

k being constant for the crystal. It is evident, however, that this empirical result may be taken as supporting the expression deduced theoretically by Darwin,

$$\mu = \mu_0 + g_2 Q, \tag{6.21}$$

where g_2 is a constant for the crystal.

It will be clear from the discussion given in §§ 1(*d*) and 1(*f*) of this chapter that the method of Bragg, James and Bosanquet corrects only for secondary extinction, and their results show that in the rock-salt

* The indices used here relate to the true face-centred unit cell of rock salt, containing four molecules of NaCl. Bragg, James and Bosanquet refer their indices to a s naller cell containing half a molecule of NaCl, and this accounts for the apparent discrepancies between the indices used here and those given in the original paper.

crystals used by them there can have been little primary extinction. There are probably few crystals for which this is true, and it must be counted as a fortunate chance that rock salt should have been used in these early experiments, since it allowed the one effect to be isolated.

When the measured values of the scattering factors were corrected for extinction by using the results of these experiments, it was found that they tended much more towards the expected values of 18 and 10 at small angles of scattering, although the actual limiting values could not be determined, since the smallest value of sin θ at which the crystal reflected the radiation used was about 0·1. In this early work, too, no proper correction could be applied for temperature, but this has little influence at the smaller angles of scattering. The curves obtained with and without this correction for extinction are shown in fig. 105.

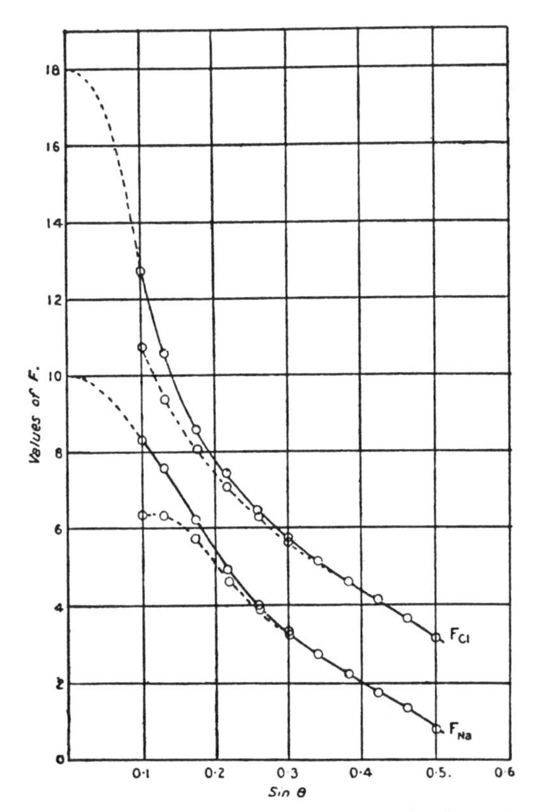

FIG. 105. Measured *f*-curves for sodium and chlorine in rock salt, with and without corrections for secondary extinction. Rh Kα radiation (λ=0·616A) was used in the observations (Bragg, James and Bosanquet, *ibid.*)

The final result of these experiments was to show that the integrated reflection from rock salt was well represented by a formula of the mosaic type, modified as in equation (*6*.17) to allow for secondary extinction. We shall write the formula for the mosaic crystal

$$\rho = \frac{Q}{2(\mu_0 + gQ)}. \tag{6.22}$$

We have here dropped the suffix from g_2 of equation (*6*.17), and shall call g the *secondary extinction coefficient.*

Formula (*6*.22) gave physically likely values for the scattering factors for chlorine and sodium, and led to an approximate count of the number of electrons in these atoms. The observed integrated reflections were in any case far too strong to be represented by a formula of the perfect crystal type (*6*.3). The two crystals discussed here, rock salt and aluminium, both give results in very good agreement with the mosaic formula; but they must be taken as representing an extreme type, and not as exemplifying a general rule. Most crystals show some primary extinction, and some approach closely to the perfect type. So long as a crystal shows only secondary extinction, it is possible to arrive at a fairly reliable correction, but the existence of primary extinction as well makes a reliable correction difficult, if not impossible.

(*j*) *Other methods of estimating secondary extinction:* It must be remembered that the intensity formulae we are discussing were developed primarily to be used as tools in the analysis of crystal structures. What we wish to know are the absolute values of the structure factors, F, for a number of spectra from the crystal, in order that these may be compared with the values calculated from some assumed structure, or used as coefficients in a Fourier series representing the periodic structure of the crystal. We may determine these structure factors from the observed integrated reflections, if we know which formula to use, or what corrections to apply for primary and secondary extinction.

If we can be sure that the crystal shows only secondary extinction it may be possible to make some estimate of its amount, even if the structure of the crystal is not exactly known. If, for example, symmetry considerations tell us that for certain spectra all the atoms must be scattering in phase, we can at once calculate the intensities to be expected, using theoretical values of the scattering factors, and compare these with the measured values. If ρ_0 and ρ are the calculated and measured values respectively of the integrated reflection for the same spectrum, and if g is the secondary extinction coefficient, and if there is no primary extinction, we may write

$$\rho_0 = \frac{Q}{2\mu_0}, \quad \rho = \frac{Q}{2(\mu_0 + gQ)},$$

which leads at once to the equation

$$g = (\rho_0 - \rho)/2\rho\rho_0. \tag{6.23}$$

We may then assume that g is the same for all spectra, and write the corrected integrated reflection ρ_{corr} in terms of the observed reflection ρ in the form

$$\rho_{corr} = \frac{\rho}{1 - 2g\rho}. \tag{6.24}$$

Even if there are no planes for which the atoms scatter all in phase, it is probable that some rather general considerations will give a clue to the approximate structure factor for at least one set of planes. The above calculation can then be carried out using this approximate structure factor, and, if necessary, the extinction may be readjusted when the structure is more accurately known. Devices of this kind are very useful in practice, and examples of their application will be given in the volume of this work devoted to methods.

There is another method which, in principle, will give the secondary extinction coefficient by direct measurement, without there being any need to destroy the crystal by grinding it; and although it is not in general very convenient in application, it is perhaps worth while to describe it briefly, since it might be possible to develop it. Suppose the *absolute* values of the integrated reflection for the same spectrum to have been determined with two different wave-lengths, λ_1 and λ_2. Let ρ_1 and ρ_2, Q_1 and Q_2, μ_1 and μ_2, be the values of ρ, Q, and μ_0 for these two wave-lengths. The extinction coefficient g we assume to be constant for the crystal, and so, if the extinction is entirely secondary,

$$\rho_1 = \frac{Q_1}{2(\mu_1 + gQ_1)}, \quad \rho_2 = \frac{Q_2}{2(\mu_2 + gQ_2)}. \tag{6.25}$$

If we can neglect the dispersion terms in the scattering factors of the atoms of which the crystal is composed for both λ_1 and λ_2, the ratio Q_1/Q_2 is known, since F will then be the same for both wave-lengths. The value of Q_1/Q_2 can therefore be calculated from known quantities. This ratio and the equations (6.25) give in effect three equations to determine Q_1, Q_2, and g. Their solution for g is

$$g = \frac{a - 1}{2(a\rho_2 - \rho_1)}, \tag{6.26}$$

where

$$a = \frac{\rho_1}{\rho_2} \cdot \frac{\mu_1}{\mu_2} \cdot \frac{Q_2}{Q_1}.$$

Unfortunately equation (6.26) is in general rather ill-conditioned since a never differs very greatly from unity, and the method is probably mainly of theoretical interest, particularly as it also depends on the assumption that there is no primary extinction. It was suggested by James and Miss Firth,[5] who employed it to determine the secondary extinction in rock salt, using rhodium and molybdenum radiations. The values they obtained were actually in good agreement with those determined by measuring the reflections through slices of different thickness.

(*k*) *Effects of simultaneous existence of primary and secondary extinction:* The method of estimating secondary extinction described in VI, § 1 (*i*), in which the variation of the integrated reflection taken through a crystal slice is determined for different thicknesses of the slice, is found to work well for some crystals but to break down completely for others. As we have seen, the method corrects only for secondary extinction. If primary extinction is present also, as it usually is in actual crystals,

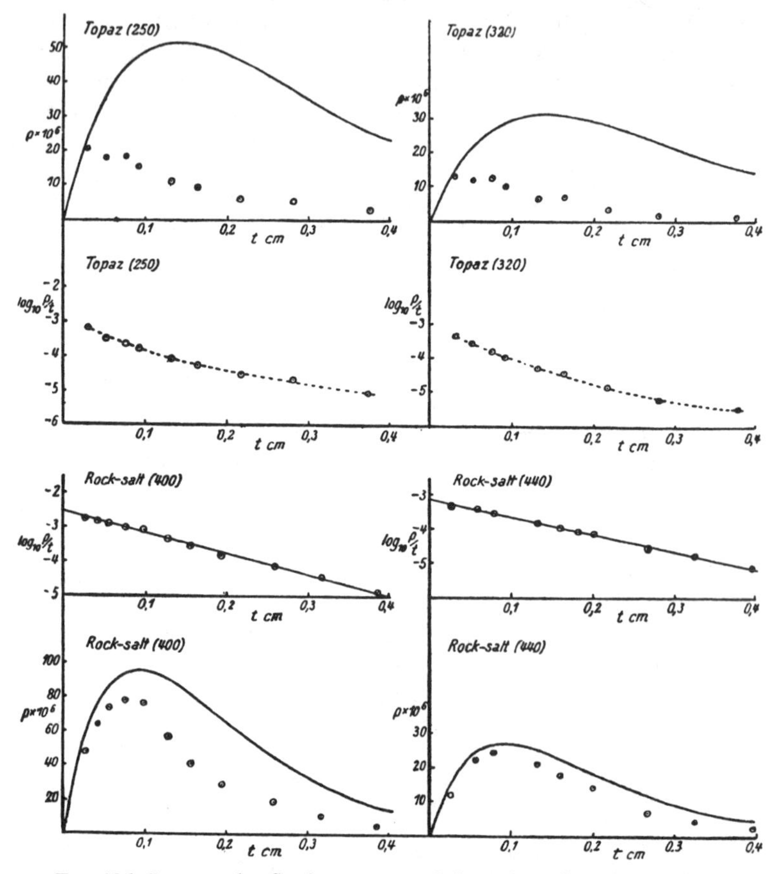

FIG. 106. Integrated reflections measured through a slice of crystal as a function of the thickness for rock salt and topaz (Bragg and West) (*Zeit. f. Krist.*, **69**, 118 (1928))

the arguments that lead to equation (6.19) do not apply, and we shall not expect to get a straight line when $\log(\rho/t')$ is plotted against t. This point is well illustrated by some curves obtained from topaz by W. L. Bragg and J. West,[8] which are shown in fig. 106, together with similar curves obtained by them from rock salt. It will be seen that the curves of $\log(\rho/t')$ against t are straight lines for rock salt, as the earlier work indicated; but the curves for topaz are concave upwards, and no reliable estimate of extinction can be obtained from their slope. The curves of ρ against t are also included in the figure. The full curves indicate the values that should have been obtained in the absence of any extinction, the circles are the experimental points. With topaz, the maxima had not been reached with the thinnest slices that could be cut. Topaz is a relatively perfect crystal, and shows considerable primary extinction. When the surface of a thin slice is ground, we may suppose the surface layers to be converted into a mosaic, the primary extinction in the same regions being largely destroyed. At the same time, the inner layers of the crystal slice remain undisturbed, and retain their primary extinction. Thus, the thinner the slice, the greater the fraction of its volume that shows only secondary extinction. No constant, or even approximately constant, property of the crystal slice is measured by experiments of this kind. The same subject was also investigated independently in some detail by Y. Sakisaka,[9] who used quartz.

(*l*) *Artificial alteration of the state of perfection of crystals:* It was noticed by Bragg, James and Bosanquet [3(I)] that the integrated reflection from a freshly cleaved face of a rock-salt crystal could be greatly increased by grinding the face on fairly coarse emery paper. In one example, the intensity 200 was measured, directly after cleavage, for two different orientations of the crystal face: (*a*) with the edge of the block of crystal upon which the knife had been pressed in cleaving the crystal parallel to the axis of the spectrometer, and (*b*) with the face rotated in its own plane through a right angle from position (*a*). The integrated reflections, in arbitrary units, were (*a*) 12·9 and (*b*) 25·4. The cleavage face was then ground with coarse emery paper until a layer about 1 mm. thick had been removed. After this treatment of the face, the integrated reflections had risen to 100 for both (*a*) and (*b*), and were not appreciably altered by further grinding of the face.

It was also found that the width of the reflection curve could sometimes be increased by grinding a crystal face roughly,* and that faces so prepared always gave relatively high values for the integrated reflection. It seems probable that freshly cleaved rock-salt faces exhibit considerable extinction, probably mainly secondary, which is reduced as the degree of parallelism of the optically independent domains in the neighbourhood of the surface is reduced; but there is, no doubt, a certain amount of primary extinction as well.

In a very interesting paper, Y. Sakisaka [10] examined in some detail the effect of grinding the crystal face. He determined the reflection curves from a number of crystals, using a double-crystal spectrometer, in which the radiation incident on the crystal to be examined had been made nearly parallel by a preliminary reflection from a very good crystal of calcite. As a second reflector, he used in turn calcite, quartz, sodium

* Cf. also Wagner and Kulenkampff, *Ann. d. Physik.*, **68**, 369 (1922).

chlorate, sulphur, iron pyrites, and zincblende. In each case, the reflection from the untouched face was first examined, and the crystal was then ground with carborundum powder of increasing degrees of coarseness, the reflection curve being taken at each stage. The area beneath the reflection curve gave a measure of the integrated reflection. In addition, by examining the distribution of intensity in the cross-sections of the incident and reflected beams, he made an estimate of the depth of the surface layer of the crystal contributing to the reflection. If the intensity is uniform over the cross-section of the incident beam, it will be uniform over that of the reflected beam only if the reflection takes place from an indefinitely thin layer of the crystal. The greater the depth of the crystal contributing to the reflection, the smaller will be the region of the cross-section of the reflected beam over which the intensity is uniform, as very elementary considerations will show.

Sakisaka found that the crystals he examined could be put roughly into the following three classes: (1) The calcite class, for which the width of the reflection curve was small, of the order of 20″ at half-height for good specimens, but increased when the crystal was ground, with a simultaneous decrease in height, until the area under the curve was perhaps doubled. The reflecting layer appeared to be very shallow, and was increased, although not greatly, by the grinding. (2) The quartz class, which included topaz and sodium chlorate, for which the reflecting layer for the untouched face was shallow, and increased only slightly when the crystal was ground. The reflection curve was still fairly narrow, although wider than for good specimens of calcite. Grinding the crystal at first *increased* the height of the reflection curve. Further grinding with coarser abrasives then reduced the height, but increased the width in still greater proportion, and so greatly increased the area. (3) The sulphur class, including also pyrites and zincblende, in which the width of the reflection curve varied greatly from crystal to crystal, but was much wider for the untouched face than for the crystals of classes (1) and (2). The effective layer was thick, the height of the reflection curve was diminished by grinding, and the corresponding increase in area was slight.

In discussing his results, Sakisaka classifies the state of perfection of the crystal in the following way, which, although only approximate, appears to be quite useful. There are two factors to be considered, the first being the size of the perfect blocks or domains in the crystal, corresponding to p in Darwin's formula (*6*.4), the second, the distribution in orientation of the blocks, corresponding to $G(u)$ in Darwin's formula (*6*.8). Corresponding to the first of these factors, we may divide crystals into two classes (*a*) and (*b*). In crystals of class (*a*) the individual domains are large enough to exhibit a considerable amount of primary extinction, in crystals of class (*b*) they are so small as to show little primary extinction. Corresponding to the second factor, we may again divide crystals into two classes, (α), in which the blocks are approximately parallel, so that the reflection will be much influenced by secondary extinction, and (β), in which the blocks are rather irregularly distributed, so that secondary extinction will not be very important. By combining these factors, a rough classification of crystals into four types according to their degree of perfection can be made. These types are denoted by the symbols ($a\alpha$), ($a\beta$), ($b\alpha$) and ($b\beta$). Class ($a\alpha$) would have as its limit the ideally perfect crystal, class ($b\beta$), the ideal mosaic.

In applying a classification of this kind it is necessary to remember that the factors (a) and (b), which are a measure of the degree of primary extinction, might be differently assigned to the same crystal specimen according to the wave-length used in examining it, and to the spectrum examined; for what really determines the primary extinction is not the actual block size, which is proportional to p, but pq, where q is the fraction of the incident radiation reflected by a single plane.

Both the average size of the individual blocks and their relative orientation will be altered by grinding the crystal face. Reduction in the size of blocks that are large enough to show primary extinction will increase the intensity reflected by the crystal at any given angle, and so the height of the reflection curve. If the blocks are already so small that they show little primary extinction, reduction in their size alone will not increase the intensity reflected by the crystal. Grinding, however, may be expected also to decrease the parallelism of the blocks, and so to reduce secondary extinction. The peaks will be broadened, and at the same time reduced in height, but the resultant effect will be to increase the area under the curve if the bilocks were originally fairly parallel, because of the reduction in secondary extinction. An initial *increase* in the height of the reflection curves, such as Sakisaka observed with quartz, must therefore be taken as indicating that the crystal shows much primary extinction, and places it in class (a). The subsequent broadening of the curve, and the increase in its area when the crystal was ground, shows that there was considerable secondary extinction as well.

This type of examination of a crystal appears to have considerable possibilities, but the examples quoted show what an essentially complicated matter the question of crystal perfection is; for, throughout, simplifying assumptions have been introduced. In practice it is not possible to separate primary and secondary extinction as we have done here, and there must be every gradation between the crystal types discussed above.

In the experiments on the temperature factor of rock salt referred to in Chapter V, § 2(a), James [11] found considerable permanent decrease in the intensities of spectra of low orders from rock-salt crystals that had been heated for some time at about 600° C. The intensity of the spectrum 200, for example, in one case fell from 0·71 of the standard to 0·50, after the crystal had been heated to 590° C., and rose only to 0·53 when the crystal had once more cooled to room temperature

The greater part of the decrease was therefore permanent, and so could not be ascribed to the thermal motion of the atoms in the lattice, which would in any case give only a small effect for this particular spectrum. It was almost certainly due to an increase in the primary extinction caused by a recrystallisation of the mosaic at the high temperature. That this explanation was the true one was confirmed by the observation that the permanent decrease in intensity was much less for the spectrum 400, and nearly negligible for 600, although the decrease at high temperatures due to thermal vibrations was quite large for these spectra. In connection with the remarks made in § 1 (*g*) of this chapter concerning the difficulty of drawing any definite conclusions from the reflection curve as to the state of perfection of the crystal, it is interesting to note that the reflection curve of the crystal used in these experiments was not perceptibly altered by the heating.

Some interesting effects due to thermal strain have been described by Sakisaka and Sumoto,[12] who observed that the spots in Laue photographs obtained by transmission of X-rays through a plate of quartz about 5·3 mm. thick, which had been polished on both faces, were all double, the doubling being just such as would be expected if reflection of the X-ray beam took place only in the two surface layers, the contribution of the interior of the crystal plate being in comparison negligible. Such an effect could be explained by assuming that the surface layers had been distorted by the grinding and polishing of the slice of crystal, and had become mosaic in structure, while the still perfect interior of the plate showed large primary extinction. One of the photographs obtained by Sakisaka and Sumoto is shown in Fig. 107 (*a*) (Pl. IX). The corresponding photograph in fig. 107 (*b*) was obtained from the same quartz plate when it was subjected to a thermal strain by setting up a temperature gradient, at right angles to the direction of the incident X-ray beam, of about 167° C. per cm. It is evident that the whole thickness of the plate is now contributing to the reflection. Unequal expansion in different parts of the plate, produced by the temperature gradient, has been enough to destroy the exact regularity in phase of the radiation reflected from different parts of the crystal which is necessary for high primary extinction. The spots became double once more when the crystal was allowed to cool. With *uniform* heating of the plate, the spots remained double, although they became more diffuse.

Nishikawa, Sakisaka and Sumoto [13] have carried out similar experiments with quartz plates set into piezo-electric vibration. The plates were perfect enough to show the double Laue spots described above when not in vibration. The appearance of the spots was modified by the vibration, and in some cases a number of intensity maxima and minima developed between the original doublet, which could be ascribed to the nodes and loops of stationary vibrations set up in the crystal.

The intensity of reflection from crystals may also be increased by a purely mechanical strain. By applying intense pressure locally on one

PLATE IX

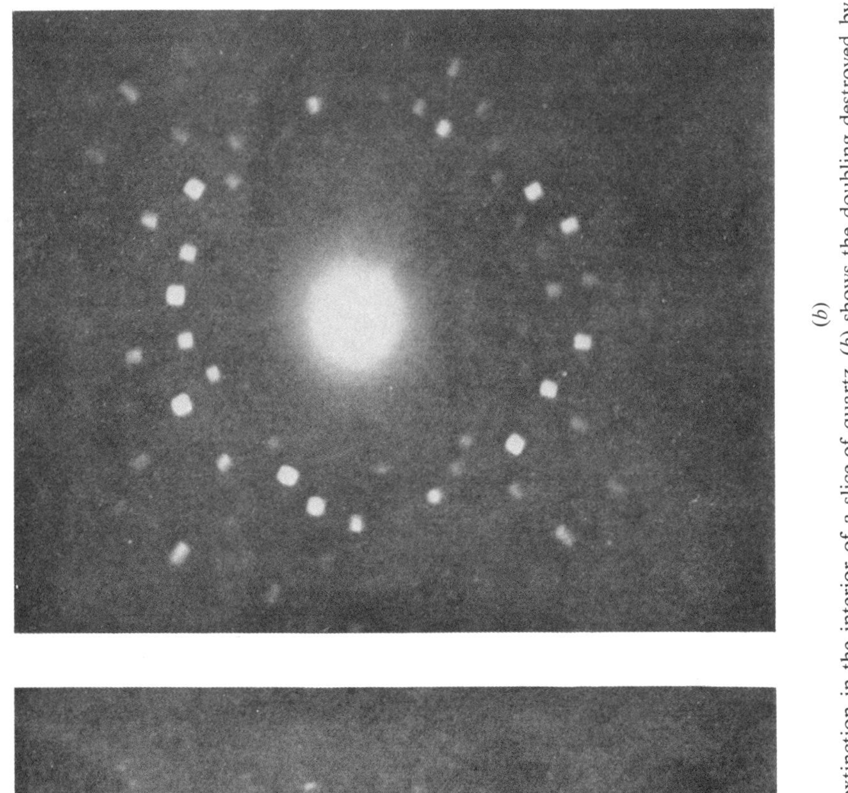

(b)

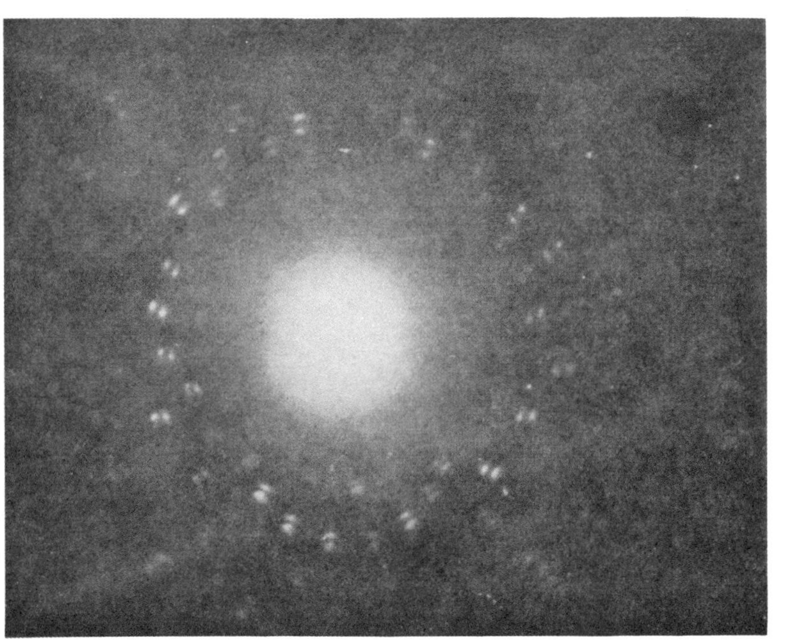

(a)

FIG. 107. (a) shows doubling of Laue spots due to extinction in the interior of a slice of quartz, (b) shows the doubling destroyed by thermal strain due to a temperature gradient set up in the crystal (Sakisaka and Sumoto)

(*Proc. Physico-Mathematical Soc. of Japan*, Ser. 3, Vol. **13**, 211 (1931))

Plate IX — See other side

edge of a plate of quartz E. Fukushima [14] showed that the intensity of reflection of an X-ray beam transmitted through the plate in the region of maximum strain increased to 5 or 6 times its value for the unstrained plate. Experiments of this type, and those employing strain due to unequal heating, or to piezo-electric oscillations, give very direct proof of the interdependence of the reflecting power of a crystal and the state of perfection and regularity of its lattice, the more so in that the effects are reversible, the reflecting power resuming its original value when the strain is removed from the crystal.

2. Quantitative Tests of the Formula for the Mosaic Crystal

(*a*) *Tests from comparison of observed and calculated structure factors :* We have already seen in § 1(*i*) of this chapter that the early work of Bragg, James and Bosanquet showed the reflecting power of a rock-salt crystal to be represented fairly well, so far as general order of magnitude was concerned, by the mosaic formula (*6*.2). When this work was done, it was not possible to allow for the temperature factor with any accuracy, although a provisional allowance was in fact made, based on some measurements by Sir W. H. Bragg.[2] Neither did any reliable means then exist of estimating the atomic scattering factors of the atoms in the crystal, so that no satisfactory quantitative test of the formula was possible. In 1927, James and Miss Firth [5] redetermined the absolute values of the integrated reflections for a number of spectra from the same rock-salt crystal that had been used in the earlier work, using essentially the same experimental method, with improved conditions of steadiness of the X-ray beam. The intensities were measured both at the temperature of the room, and at 86° abs., the temperature of liquid air, so that experimental values of the temperature factor could be obtained, which were found to agree fairly closely with those given by the theoretical formula obtained in the way explained in § 1(*m*) of Chapter V. From these measurements Waller and James [15] deduced the separate temperature factors for the atoms of sodium and chlorine in the rock-salt lattice by the method described in § 2(*c*) of Chapter V.

In formula (*6*.2), the factor $|F|$ is the structure factor of the crystal unit, as defined in Chapter II, §§ 1(*c*) and 1(*e*). If f_1 and f_2 are the atomic scattering factors of chlorine and sodium respectively, when the atoms are at rest, and M_1 and M_2 are the values of the exponents in the temperature factor for the two atoms at the temperature at which the measurements are made, then

$$|F| = |f_1 e^{-M_1} \pm f_2 e^{-M_2}|. \qquad (6.27)$$

The positive sign applies to spectra whose indices, referred to the face-centred cubic lattice, are all even, and the negative sign to spectra whose indices are all odd.

By the time the measurements were completed, theoretical values for

f_1 and f_2, calculated by Hartree by his method of the self-consistent field (Chapter III, § 4(*a*)), were available, and a comparison between the observed integrated reflections and those calculated from formula (*6.*2) was possible. The curves in fig. 108, taken from a paper by James, Waller and Hartree,[16] show the values of the structure factor (*6.*27) plotted as a function of $(\sin\theta)/\lambda$. The actual numbers given as abscissae are values of $\sqrt{h^2+k^2+l^2}$, which is proportional to $(\sin\theta)/\lambda$. Hartree's values for f_1 and f_2, and the values of M_1 and M_2 deduced by Waller and James, based on the measured values of the integrated reflections, were used in calculating F. As pointed out in Chapter V, § 1(*o*), the observed temperature factor gives only that part of M dependent on the temperature. That part of M dependent on any zero-point energy the lattice may possess cannot be observed directly, although it can easily be allowed for theoretically, as we saw in Chapter V, § 2(*c*). Actually, the method used by Waller and James to calculate the values of M_1 and M_2 assumes the existence of zero-point energy, which is therefore included in the comparison between theory and experiment.

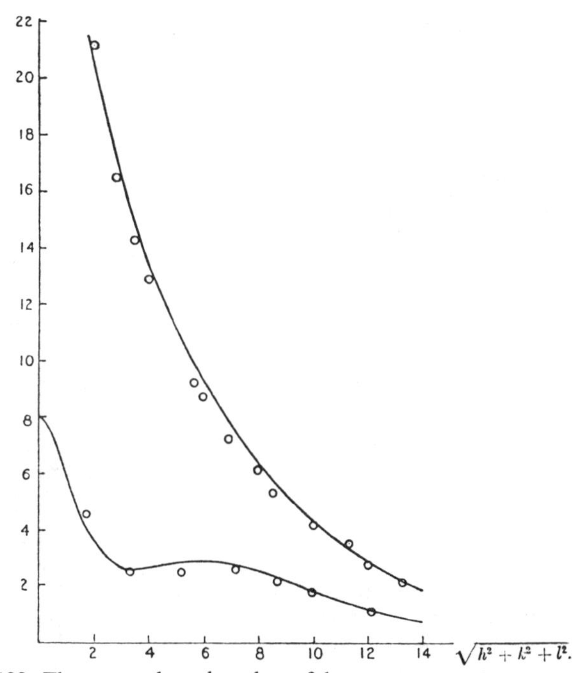

FIG. 108. The curves show the values of the structure amplitude for the two types of spectra from rock salt, calculated from the Hartree model, and corrected to 86° abs. by using the observed temperature factor. The abscissae are values of $\sqrt{h^2+k^2+l^2}$. The existence of zero-point energy is assumed in the calculation. The circles show the observed values at the temperature of liquid air (86° abs.) (James, Waller and Hartree)

(*Proc. Roy. Soc.*, **A**, **118**, 334 (1928))

The curves in fig. 108 were calculated for the temperature of liquid air, 86° abs., and the experimental structure factors at the same temperature were calculated from the observed values of the integrated reflection ρ. On the assumption that the crystal reflects as a mosaic, we have, from (6.2),

$$|F|^2 = \frac{2\mu}{N\lambda^3}\left(\frac{mc^2}{e^2}\right)^2 \frac{2\sin\theta}{1+\cos^2 2\theta}\cdot\rho. \tag{6.28}$$

The value of μ was corrected for secondary extinction by putting $\mu = \mu_0 + gQ$, μ_0 being the ordinary absorption coefficient of the crystal for molybdenum $K\alpha$ radiation, which was used in these experiments. Q has the value given in Chapter II, § 2 (*c*), and the value of g was taken from the measurements of Bragg, James, and Bosanquet, described in § 1 (*i*) of this chapter. The structure factors obtained in this way from the observed values of ρ are shown by the circles in fig. 108. It will be seen that they lie closely on the two curves. The points near the upper curve refer to spectra with indices all even, for which the contributions

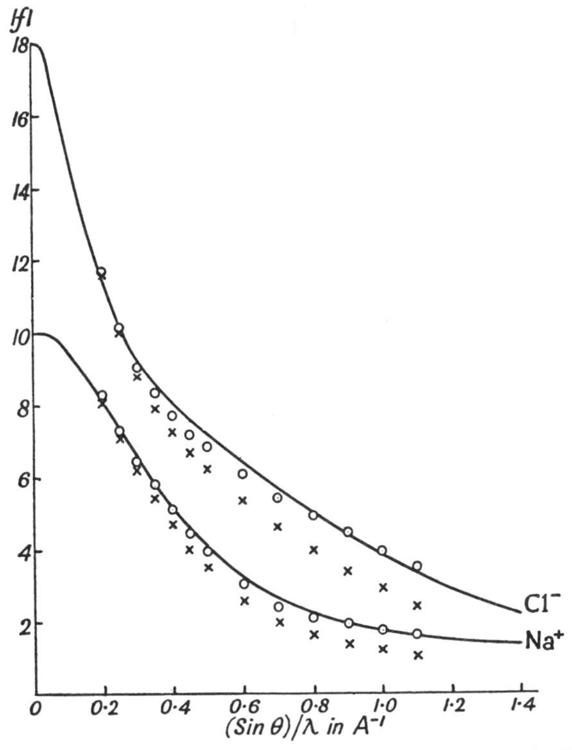

FIG. 109. Theoretical f-curves for Cl^- and Na^+ calculated from the Hartree model for the atom at rest. Circles represent values deduced from observation with allowance for zero-point energy; crosses, values deduced without this allowance

scattered by the two types of atom agree in phase, those near the lower curve refer to spectra with indices all odd, for which the two contributions are always opposite in phase. It will be seen that there is on the whole a very satisfactory agreement between the observed and calculated values of $|F|$, and it should be emphasised that the agreement is absolute, and that no point has been made to fit the theoretical curve.

To get agreement, it was necessary to assume the existence of zero-point energy in the lattice. Had it been neglected, the calculated curves would have fallen considerably above the observed points, and it is interesting to bring this out in another way. In the original work, emphasis was placed on using the observed values of ρ to calculate the atomic scattering factors of sodium and chlorine, so as to test the values of these calculated by the method of the self-consistent field. It is clear that by taking half the sum and half the difference of the ordinates of the two curves drawn through the observed points in fig. 108, experimental values of $f_1e^{-M_1}$, and $f_2e^{-M_2}$ can be determined as functions of $(\sin\theta)/\lambda$. Using the measured values of M_1 and M_2 we may then calculate the values of $|f_1|$ and $|f_2|$ and compare them with Hartree's values. From the observed curves, two values of the scattering factor for the atom at rest may be calculated, according to whether the existence of zero-point energy is assumed or not. In fig. 109, Hartree's theoretical curves for Na^+ and Cl^- are given as functions of $(\sin\theta)/\lambda$. The circles indicate the values of the atomic scattering factors deduced from the observations with allowance for zero-point energy, the crosses the values deduced neglecting it. It is clear that the allowance for zero-point energy is necessary if there is to be agreement between theory and experiment, for the curves deduced when it is neglected differ from the experimental curves by much more than the limits of experimental error. The existence of zero-point energy is, of course, fundamental to modern quantum mechanics, and this result is therefore very satisfactory. Quite recently Brill, Grimm, Hermann, and Peters [17] have repeated the measurements for rock salt, using more sensitive methods of measuring the ionisation, and have confirmed this agreement for spectra up to (10, 10, 0). In correcting for extinction, and temperature, they use the same data as were employed by James, Waller, and Hartree.

These results show clearly that the crystal of rock salt employed in the experiments under discussion reflected very closely according to the mosaic formula (*6.2*). Similar tests were made for crystals of sylvine [18] (KCl), which have already been referred to in Chapter V, § 2(*b*). Again a good agreement between the observed and calculated values of $|F|$ was obtained when allowance was made for zero-point energy. The values obtained in these experiments have recently been somewhat modified by Brindley and Ridley [19] without, however, essentially altering the conclusion to be drawn from them, that sylvine reflects as a mosaic.

The absolute integrated reflections and the temperature factors were measured for single crystals of aluminium by James, Brindley and Wood.[6] The values of the integrated reflection obtained from these crystals at room temperature have already been quoted in Table VI. 3, § 1(h) of this chapter, and the agreement between them and the values indicated by the mosaic formula has been pointed out. In fig. 110, the group of curves A shows the f-curves for Al^{+1}, Al^{+2}, and Al^{+3}, calculated by Hartree's method. For all except very small values of $(\sin\theta)/\lambda$, these curves are so nearly alike that it is not possible to reproduce them separately on the scale of the figure; at small values of $(\sin\theta)/\lambda$, however, they differ appreciably, and approach the limits 12, 11 and 10, respectively, for $(\sin\theta)/\lambda=0$. The curve for Al^{+1},

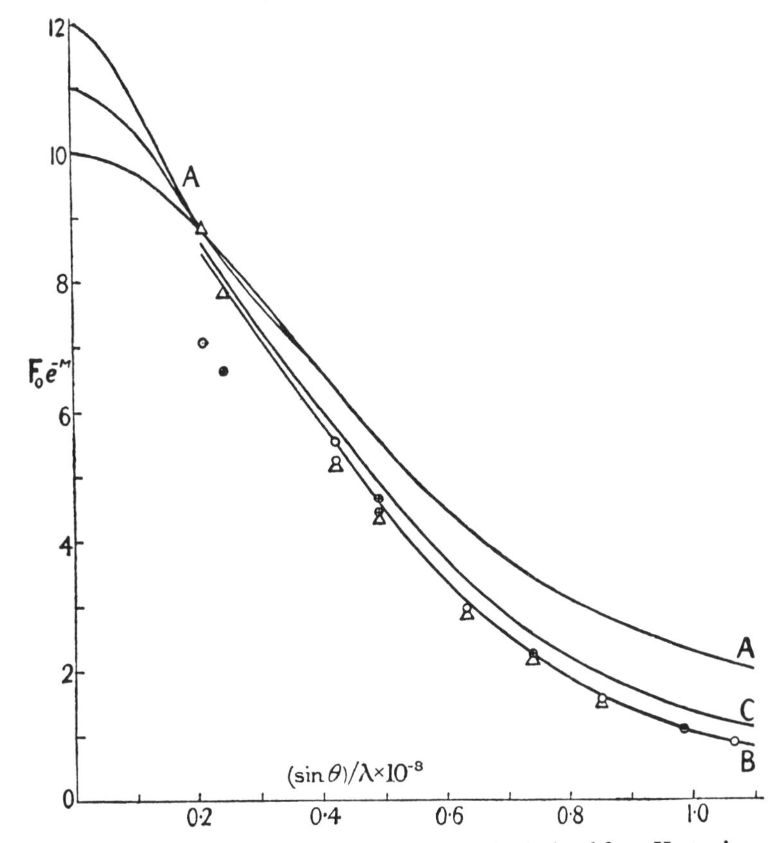

FIG. 110. Curves A, f-curves for Al^{+1}, Al^{+2}, Al^{+3} calculated from Hartree's model for the atom at rest. Curves B and C are the f-curves appropriate to the aluminium crystal at room temperature, calculated from the observed temperature factor; B, assuming zero-point energy, C, assuming no zero-point energy. The circles show observed points obtained from a single crystal of aluminium, and the triangles observations by Bearden from powdered aluminium (James, Brindley and Wood)

(*Proc. Roy. Soc.* **A, 125** 401 (1929))

corrected to a temperature of 290° abs. by means of the observed temperature factor, with allowance for zero-point energy, is shown in the same figure, curve B, curve C being the corresponding curve without allowance for zero-point energy. The experimental values of the f-factor, deduced from the observed values of the integrated reflections by means of (6.28) are shown by the circles. For the higher values of $(\sin\theta)/\lambda$, the agreement, which is again absolute, is remarkably good, although for the smaller values the observed points are low, owing to the existence of secondary extinction, for which no allowance was made in the absorption coefficient. The triangles in the same figure are from an entirely different set of observations made by Bearden [20] at Chicago, who used powdered crystals. These will be discussed later. The curves of fig. 110 show that any attempt to determine the state of ionisation of the atoms in a crystal by means of measurements of the atomic scattering factor is likely to fail, since, for reasons given in Chapter III, § 4(*e*), the curves will differ appreciably only at angles for which no spectra exist.

3. Experimental Tests of the Reflection Formulae for Perfect Crystals

(*a*) *Introductory:* The experimental work considered in the last section showed that the observed integrated reflections from a number of crystals agreed very closely with those to be expected if the crystal had a mosaic structure. We must now consider other experiments, which show that crystals also exist with reflecting properties very nearly those which theory indicates should be characteristic of perfect crystals. As in the theoretical treatment of the subject given in Chapter II, we have here considered imperfect crystals before perfect ones, partly because they were the first to be studied quantitatively with any degree of thoroughness, but mainly because there are certain experimental difficulties inherent in the investigation of perfect crystals which had not to be considered in the case of mosaic crystals.

In discussing the earlier work it is necessary to bear in mind that it was done before the distinction between the mosaic and the perfect crystal had been clearly realised. Quite general considerations, based on elementary optical laws, indicated that the diffraction maxima from crystals should be exceedingly narrow, simply because of the very large number of planes co-operating in the formation of a spectrum; and an elementary treatment, neglecting all multiple scattering, led to the result that the reflected intensity should be proportional to the square of the structure factor. The full theory had indeed already been worked out by Darwin, and by Ewald, but their work was not generally appreciated until more than ten years after its publication, partly because of its very considerable difficulty, but mainly because it appeared during the war of 1914-1918.

In his Bakerian Lecture of 1915, W. H. Bragg [21] noted that the intensity of reflection of X-rays from calcite appeared to be proportional to the first power of the structure factor rather than to its square, and some earlier measurements by him on diamond [2] also pointed in the same direction. In 1917, A. H. Compton [7] made the first absolute measurements of the integrated reflections from rock salt and calcite, and found the value for calcite to be much smaller than was to be expected from formula (*6*.2), which neglects all re-scattering—the mosaic formula as it is now called. Further measurements on diamond by W. H. Bragg in 1921 [22] again indicated clearly that the first power of the structure factor determined the strength of the reflection. These measurements were discussed by Ewald [23] in a paper that brought out clearly the discrepancy between the rival formulae, and introduced the term 'mosaic' for the ideally imperfect crystal. Largely as a result of Ewald's paper, and at his suggestion, an informal conference took place between a number of those interested in the subject, which had the effect of removing many of the difficulties, and of showing clearly for the first time that the two formulae were not in fact rival, but complementary, each applying under appropriate conditions. The results of the discussion were summarised by W. L. Bragg, Darwin and James, in a paper published in the *Philosophical Magazine* in 1926.[24]

An essential point in the theory of reflection by a perfect crystal is the very narrow range of angles of incidence over which it should reflect a parallel beam of X-rays of homogeneous wave-length. Now it is impossible in practice to produce either a strictly parallel or a strictly homogeneous beam of X-rays. The divergence of the beam alone would introduce no particular difficulties, for the crystal would select from it rays incident at those angles at which its reflection coefficient was appreciable, and the angular width of the reflected beam would give a measure of the angular width of the reflection curve of the crystal. A correction for the width of the slit from which the incident beam diverged would of course be necessary, but this would become less important as the distance from the crystal at which observations were made increased. Observations of this kind were made by Ehrenberg and Mark,[25] who allowed radiation from a narrow slit to fall on a diamond, and measured the width of the reflected beam at a distance of 10 metres from the crystal. They found the angular range of reflection to be 5·5″, 13·2″ and 39″ for the spectra 111, 333 and 555 respectively, ranges certainly much smaller than any obtained with mosaic crystals, but nevertheless greater than those predicted by Darwin's formula and, moreover, increasing with increasing order of spectrum, that is to say, with decreasing structure factor, instead of decreasing, as the theory requires.

As Ehrenberg and Mark themselves pointed out, this apparent discrepancy may be ascribed mainly to the lack of homogeneity of the incident radiation. A so-called monochromatic X-ray line has a spectral width of the order of 0·2 to 0·3 X-units, corresponding to a range of angles of reflection considerably greater than the theoretical angular width of the reflection curve of a perfect crystal for monochromatic radiation. The dispersion due to lack of homogeneity of the radiation therefore completely masks the divergence due to diffraction in experiments using direct reflection from a single crystal.

This difficulty may be largely overcome by the use of the double-crystal spectrometer, in which the radiation is reflected in succession from two crystal surfaces. Accurate study of reflection by perfect crystals has been made possible by the development of this instrument, and it will therefore be necessary in the succeeding paragraphs to deal in some detail with its geometrical and optical properties.

(*b*) *The principle of the double-crystal spectrometer:* The double-crystal spectrometer is an ionisation spectrometer in which the incident beam falling on the crystal under examination has first been reflected from another crystal. The principle of employing a preliminary reflection at a crystal surface to produce a monochromatic beam, and to get rid of the continuous radiation always present in the direct rays from the target of an X-ray tube, was employed independently by A. H. Compton,[7] by Bragg, James and Bosanquet,[3(I)] and by Wagner and Kulenkampff,[26] in their measurements of absolute integrated reflections. In 1921, Davis and Stempel [27] used it to study the widths of the reflection curves from freshly cleaved calcite crystals, and examined some whose angular range of reflection was as small as 18″. These experiments were the first in which the value of the double-crystal spectrometer as an instrument of high precision and high resolving power became apparent; but it was not until some time later that the true meaning of the reflection curves obtained with it, and their relationship to the true reflection curve for monochromatic radiation from a single crystal face, was properly understood. The different properties of the two types of setting of the crystal for double reflection were realised by Wagner and Kulenkampff,[26] who noted the absence of dispersion in the parallel setting, which is one of the most important properties of the instrument, and who also pointed out that the polarisation factors for double and single reflection differ. A fuller account of the theory of double-reflection curves was given by Ehrenberg and Mark,[25] Ehrenberg and v. Süsich,[28] and by Davis and Purks.[29] The optical and geometrical principles of the instrument have been dealt with in detail by Schwarzschild,[30] Spencer,[31] Laue,[32] and L. P. Smith.[33] Reference may be made to the detailed account of the theory given in *X-Rays in Theory and Experiment* by A. H. Compton and S. K. Allison. The shorter treatment given here, which, as will be evident, owes much to the account given by these writers, deals only with such points as have a more direct bearing on the topics discussed in this chapter.

Let us assume each crystal to reflect at a glancing angle θ in the same order of spectrum, the incident ray to lie in a horizontal plane, and the reflecting surfaces of the two crystals to lie in vertical planes. Then there are two possible settings for double reflection. In one, illustrated by fig. 111 (*a*), the outward-drawn normals to the two crystal surfaces are inclined to one another at an angle 2θ; in the other, illustrated by fig. 111 (*b*), they are inclined to one another at an angle

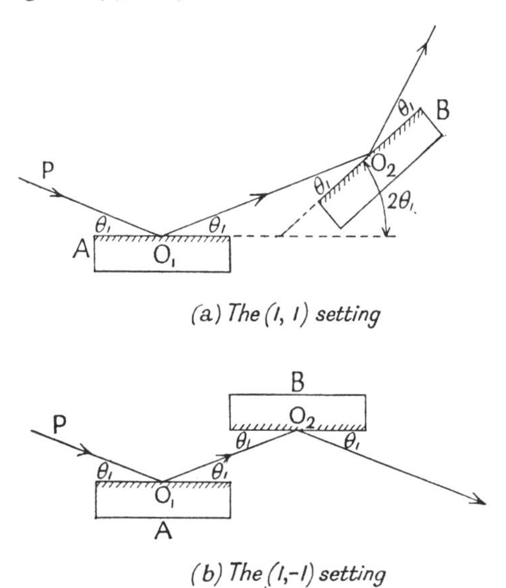

FIG. 111. Illustrating the two types of setting for the double-crystal spectrometer

of 180°. These two types of setting are often referred to as the (1,1), and (1, −1) arrangements respectively, a convenient notation introduced by Allison and Williams,[34] which may be extended to include cases in which reflection does not take place in the same order of spectrum from each crystal. If reflection takes place in the *m*th order from the first crystal, A, and in the *n*th order from the second crystal, B, the two settings are denoted by (m, n), and $(m, -n)$ respectively. If m and n are not equal, the crystal surfaces are no longer parallel for the $(m, -n)$ setting, which cannot therefore be called the parallel setting, as is customary when the order of reflection is the same for both crystals. The essential difference between the two arrangements is that in the (m, n) setting the incident ray PO_1 and the reflected ray from the second crystal lie on the same side of the ray O_1O_2 passing between the two crystals, while in the $(m, -n)$ setting they lie on opposite sides of O_1O_2. In the cases that we shall consider, the two surfaces will usually be parallel, and we shall frequently refer to the (1 −1) setting as the parallel arrangement. There are interesting differences between the optical properties of the two arrangements. From the point of view of the present discussion, the important arrangement is the (1, – 1), or parallel setting, and we shall consider it first.

(*c*) *The parallel arrangement for the double-crystal spectrometer:* Suppose crystal A to remain fixed, with radiation incident on it at such an angle as to give reflection, while crystal B can be rotated about a vertical axis. We wish to know how the radiation reflected by B depends on its angular setting. We take as a standard direction that of a ray exactly horizontal, and incident on A at a glancing angle θ_1. The value chosen for θ_1 is in principle immaterial, but it is better to take it in the neighbourhood of the angle of maximum reflection from the crystal A. To give precision to the discussion, it is advisable to choose a definite angle, and we shall here take θ_1 as being the modified Bragg glancing angle of II, § 3(*c*), corresponding to the middle of the range of total reflection given by the Darwin theory, a convenient direction of reference in discussions involving reflection by perfect crystals.

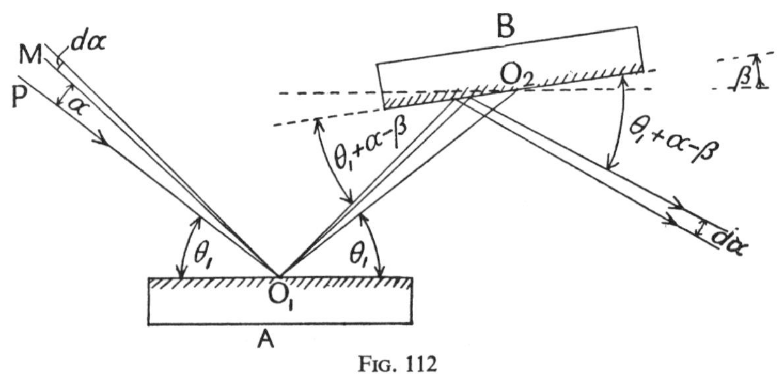

FIG. 112

The standard ray, PO_1 in fig. 112, is reflected along O_1O_2 by crystal A, and falls on the crystal B. When B is in its zero position, its surface is parallel to that of A, and the angle of incidence on it of the ray O_1O_2 is also θ_1, so that if the crystal were ideally perfect, it too would reflect at the middle of its diffraction pattern. The angle of displacement of the crystal B from the zero, or parallel, setting, we denote by β, which is taken as positive when the direction of displacement is anti-clockwise in fig. 112. The angle of incidence of the central ray O_1O_2 on B is then $\theta_1 - \beta$, and unless β is very small there is no appreciable reflection from B. In calculating the way in which the reflection varies with β there are two points to consider, the first, that the incident beam can never be confined to a single direction, but must spread over a finite range of angles; the second, that the incident radiation cannot be strictly homogeneous. For simplicity, we shall consider the two points separately.

We first assume the radiation to be strictly homogeneous, of wavelength λ, but suppose the rays of the incident beam coming from the slit of the instrument to extend over a range of angles, including θ_1, and large in comparison with the angular width of the range of reflection of A, which will be only a few seconds of arc if A is perfect. In any practical case, the range of angles of incidence is likely to be at least one hundred times the range of reflection. Let MO_1, fig. 112, be an incident ray making an angle α with PO_1 in the horizontal plane. This angle we call the horizontal divergence of the ray. In a real incident beam, coming from a cathode spot of finite size, and passing through a slit, there will also be rays diverging from the horizontal plane. If ϕ is the angle that such a ray makes with the horizontal plane, it is easy to show that, so long as ϕ is small, the glancing angle of incidence is diminished by $\frac{1}{2}\phi^2 \tan\theta$. In practice, ϕ can be kept so small that this change may be neglected; but it can in any case be shown that in the parallel arrangement the reflection curve from the second crystal, expressed as a function of β, is independent of the vertical divergence. Accordingly we shall neglect it in what follows, and consider only the horizontal divergence.

Let the power in that part of the incident radiation whose horizontal divergence lies between α and $\alpha + d\alpha$ be $I_0(\alpha)\,d\alpha$. The nature of the function I_0 will depend partly on the form of the slit, and partly on the source of the radiation. Let the crystal A reflect a fraction $R_A(\alpha)$ of this radiation. The power reflected by the crystal A at angles lying between $\theta_1 + \alpha$ and $\theta_1 + \alpha + d\alpha$ is therefore $I_0(\alpha)\,R_A(\alpha)\,d\alpha$. When the setting of the crystal B is β, this power is incident on it at angles lying between $\theta_1 + \alpha - \beta$ and $\theta_1 + \alpha - \beta + d\alpha$, and B therefore reflects a fraction of it equal to $R_B(\alpha - \beta)$, R_B being the reflection coefficient of the crystal B. Both reflection coefficients are expressed as functions of the difference between the actual angle of incidence and the standard angle of incidence θ_1.

The total power, $P(\beta)$, reflected by the crystal B at the setting β, when the horizontal divergence of the incident beam lies between the limits α_1 and α_2, is therefore

$$P(\beta) = \int_{\alpha_1}^{\alpha_2} I_0(\alpha)\,R_A(\alpha)\,R_B(\alpha - \beta)\,d\alpha. \qquad (6.29)$$

In what follows, we shall assume the reflection coefficient R to be the same function of the angle of reflection for each crystal. This condition can be approached in practice by using corresponding portions of two faces produced by one cleavage of a single crystal. The assumption that $R_A = R_B$ in this particular discussion involves, however, the further assumption that the incident radiation is polarised so that its electric vector lies either perpendicular to or parallel to the plane containing the incident and reflected rays. In either case the electric vector makes the same angle with the crystal surface at each reflection.

The coefficient of reflection is, however, different for the two directions of polarisation, and a beam that contains both directions of polarisation will therefore have a different coefficient of reflection from the two crystals, since the weakening of the one component by the successive reflections will be greater than that of the other. The discussion in this paragraph therefore assumes the incident radiation to be polarised in one or other of these two directions, but we shall extend it to include the case of unpolarised radiation a little later.

The effective value of the function R is negligible unless the value of its argument is nearly zero; for the diffraction maximum is very narrow, and lies near θ_1, which has been taken as the zero of reference for the reflection coefficient. The function $I_0(\alpha)$ varies relatively slowly with α, and can be taken as constant over the range of α for which $R(\alpha)$ is appreciable. It may therefore be taken outside the integral, and the limits of integration may at the same time be extended to infinity in either direction; for since the integrand has appreciable values only over a very narrow range, the precise limits are unimportant, so long as they include this range. With these simplifications, we may therefore write

$$P(\beta) = I_0(\alpha) \int_{-\infty}^{+\infty} R(\alpha) R(\alpha - \beta)\, d\alpha. \tag{6.30}$$

By the same argument, the power incident on the crystal B may be written $I_0(\alpha) \int_{-\infty}^{+\infty} R(\alpha)\, d\alpha$. We use this as a normalisation factor, and write for the reflection coefficient at the angle β of the second crystal in the $(1, -1)$ arrangement

$$R_2(\beta) = \int_{-\infty}^{+\infty} R(\alpha) R(\alpha - \beta)\, d\alpha \bigg/ \int_{-\infty}^{+\infty} R(\alpha)\, d\alpha. \tag{6.31}$$

The denominator in (6.31) is the integrated reflection from crystal A for the direction of polarisation considered, and, with the assumption we have made as to the equality of reflection coefficients of the two crystals, from B also. It is therefore constant for the given pair of crystals, and we may write

$$R_2(\beta) = k \int_{-\infty}^{+\infty} R(\alpha) R(\alpha - \beta)\, d\alpha, \tag{6.32}$$

where k is constant. The reflection coefficient at any setting β of the crystal B is therefore determined by the integral of (6.32) with respect to α, β remaining constant.

The properties of this integral may conveniently be studied graphically. In fig. 113(*a*) the two curves $R(\alpha)$, and $R(\alpha - \beta)$ are plotted as functions of α for constant β. The two curves are identical in shape, but the curve $R(\alpha - \beta)$ is displaced relative to the curve $R(\alpha)$ through a distance β in the positive direction of α. The integrand of (6.32) for any given value of α is equal to the product of the ordinates of the two curves, for example, EG and FG, for that value of α. If a series of these products is plotted against the corresponding values of α, the area under the curve is the value of the integral in (6.32). It is at once

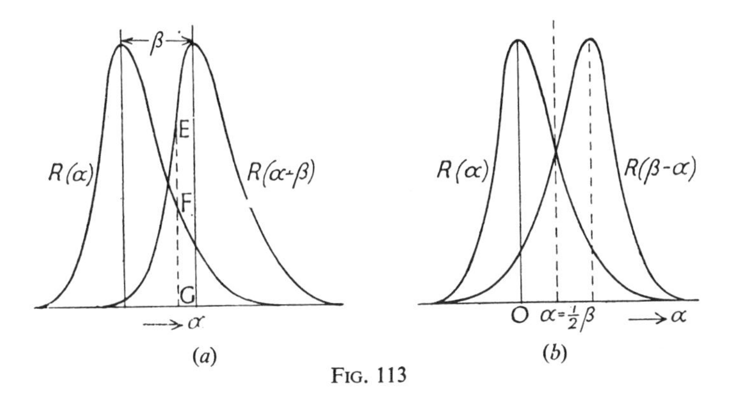

FIG. 113

plain that the value of the integral becomes inappreciable when β is so large that the peaks of the two curves do not overlap, and is a maximum when $\beta=0$, and the crystal faces are parallel, for the two curves then coincide. The maximum reflection coefficient, $R_2(0)$, is then given by

$$R_2(0)=\int_{-\infty}^{+\infty}\{R(\alpha)\}^2 d\alpha \Big/ \int_{-\infty}^{+\infty} R(\alpha)\, d\alpha. \tag{6.33}$$

Suppose the reflected radiation from crystal B at a given setting β is allowed to fall into an ionisation chamber with slits wide enough to receive the whole reflected beam. The ionisation will then be proportional to the reflection coefficient $R_2(\beta)$, and by measuring it for successive settings of the crystal we can map out the reflection curve. It is necessary to be able both to move the crystal B by fractions of a second of arc at a time, and to measure the movement with accuracy; for it is plain that the width of the reflection curve $R_2(\beta)$ must be of the same general magnitude as that of the diffraction curves $R(\alpha)$ from either crystal. The exact relationship between the two will depend on the form of the function $R(\alpha)$, and no general rule can be given. We shall consider a few special cases later.

According to Darwin's theory for a perfect crystal without absorption, reflection should be total over a range of angles of the order of some seconds, and the curve of $R(\alpha)$ should have a flat top. No trace of this flat top will appear in the reflection curve $R_2(\beta)$. This curve has a maximum when $\beta=0$, and the curves of $R(\alpha)$ and $R(\alpha-\beta)$ in fig. 113(*a*) coincide. For any other value of β the curves do not coincide, and the value of the integral must be reduced. Moreover, the maximum reflection coefficient, (6.33), determined by direct integration from

Darwin's formula, is 4/5, and not unity. Direct evidence of total reflection cannot therefore be obtained with the double-crystal spectrometer.

A consideration of fig. 113(*a*) shows that the reflection curve $R_2(\beta)$ is symmetrical about its maximum ordinate at $\beta=0$, so that $R_2(\beta)=R_2(-\beta)$. The overlap of the two curves is evidently exactly the same whether we displace one curve relative to the other by $+\beta$ or $-\beta$ from the position in which they coincide; and this is true however unsymmetrical the reflection curve for the individual crystals. According to Prins's theory, the reflection curve for an absorbing perfect crystal is not symmetrical (see fig. 27, p. 65), but it is not possible to test this asymmetry directly by means of the double-crystal spectrometer. Laue [32] has shown that the shape of the reflection curve $R(\alpha)$ can be determined from that of the double-reflection curve $R_2(\beta)$ only if it is known that $R(\alpha)$ is itself symmetrical. As theory indicates that it probably is not, the best that can be done in the way of checking theory by experiment from the shape of the experimental curves is to assume some definite form for $R(\alpha)$, calculate the corresponding curve for $R_2(\beta)$, and compare it with observation.

(*d*) *Absence of dispersion in the parallel arrangement :* It will now be shown that the $(1, -1)$ arrangement has the very important property of giving the same reflection curve $R_2(\beta)$ for all wave-lengths, provided that the reflection curve $R(\alpha)$ can be taken as independent of the wave-length, which, although not true in general, is very nearly indeed true for such closely neighbouring wave-lengths as those constituting a single characteristic line. Suppose the incident beam to consist of radiation of two separate wave-lengths, λ_1 and λ_2, and let θ_1 and $\theta_1+\Delta\theta$ be the glancing angles of incidence corresponding to the middles of the diffraction patterns for these two wave-lengths. The angle $\Delta\theta$ is supposed to be much smaller than the divergence of the incident beam, so that both wave-lengths will be reflected by the crystal A; and, if B is parallel to A, any rays reflected by A will fall at the same angles of incidence on B, and will be reflected by it also. As the crystal B is rotated it reflects simultaneously, at corresponding angles, for the two wave-lengths, and if the slit of the ionisation chamber is wide enough to take in both reflected beams, the curve giving $R_2(\beta)$ as a function of β, the setting of crystal B, will have the same shape as that for monochromatic radiation; and this will evidently also be true if the incident radiation consists of a continuous band of wave-lengths.

Formally, this result may be expressed as follows. Let λ_0 be the wave-length corresponding to the maximum intensity of the line, and let $J(\lambda-\lambda_0)\,d\lambda$ be proportional to the intensity included in wave-lengths lying between λ and $\lambda+d\lambda$. J_0 is thus the distribution function of intensity in the line, and has appreciable values only over a very small range of wave-lengths in the neighbourhood of λ_0.

Let the ray P_0O_1, fig. 114, making a glancing angle θ_1 with the face of crystal A, be the direction of incidence corresponding to the maximum of the reflection curve, or, for precision, to the modified Bragg angle, for wave-length λ_0. P_1O_1, the corresponding direction for wave-length λ, makes an angle $\theta_1 + \Delta\theta_1$ with the crystal surface, where $\Delta\theta_1 = (\lambda - \lambda_0)\, d\theta/d\lambda$, $d\theta/d\lambda$ being the variation of angle of reflection with wave-length calculated according to Bragg's law. Consider now radiation of wave-length λ, incident along PO_1, fig. 114, which makes an angle $\theta_1 + \alpha$ with the face of crystal A. The argument of the function R for reflection from crystal A for this ray will be $\alpha - \Delta\theta_1$, instead of α, as it would be for wave-length λ_0, and the power reflected by the second crystal in wave-lengths lying between λ and $\lambda + d\lambda$ will be

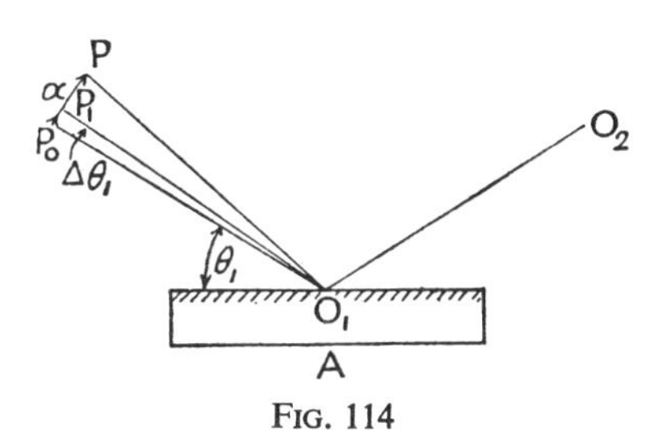

FIG. 114

$$G(\alpha)\, J(\lambda - \lambda_0)\, R(\alpha - \Delta\theta_1)\, R(\alpha - \Delta\theta_1 - \beta)\, d\alpha\, d\lambda. \qquad (6.34)$$

$G(\alpha)$ is here a geometrical factor depending on the distribution of intensity in the incident beam, which may be taken as independent of the wave-length, and $\iint G(\alpha)\, J(\lambda - \lambda_0)\, d\alpha\, d\lambda$ gives the total power in the incident beam. The power reflected by the crystal B is obtained by integrating the expression (6.34) over the whole range of λ and α. In the integration over α, $\Delta\theta_1$ is constant, and, so long as we may take the limits of integration as plus and minus infinity, the reduction of α by a constant value in the integrand makes no difference to the value of the integral. We may therefore write for $P(\beta)$, the power reflected by the crystal B,

$$P(\beta) = G(\alpha) \int_{-\infty}^{+\infty} J(\lambda - \lambda_0)\, d\lambda \int_{-\infty}^{+\infty} R(\alpha)\, R(\alpha - \beta)\, d\alpha \qquad (6.35)$$

$$= k' \int_{-\infty}^{+\infty} R(\alpha)\, R(\alpha - \beta)\, d\alpha, \qquad (6.35a)$$

where k' is a constant for given conditions of working. As before, we have assumed $G(\alpha)$ to vary so slowly with α that it may be taken as constant over the range of α for which $R(\alpha)$ is appreciable.

This expression, considered as a function of β, has exactly the same form as (6.30), which was deduced for a single incident wave-length. In the parallel arrangement, the dispersion is therefore zero, and the finite spectral range of the wave-length employed does not in any way affect the reflection curve. This makes the arrangement of very great value for studying the widths of the reflection curves from crystals, but useless for studying the structure of spectral lines. For this purpose, however, the (1,1) arrangement may be used.

(*e*) *The reflection curve and the dispersion in the* (1,1) *arrangement of the double-crystal spectrometer:* Although the experimental results that we shall discuss later were obtained with the crystals in the (1, −1) arrangement, it is convenient at this point to give briefly, by way of contrast, the properties of the (1, 1) arrangement. Let the crystals be set so that the angle between them is $2\theta_1$. A ray reflected at θ_1 by crystal A will then also be reflected at θ_1 by crystal B. Consider, however, a ray incident on A at $\theta_1+\alpha$. Its angle of incidence on B will now be $\theta_1-\alpha$, instead of also being $\theta_1+\alpha$, as in the parallel position, and the ray will not now necessarily be reflected from the second crystal. This is really the essential difference between the two arrangements. If the second crystal is turned through an angle β from its zero setting, in which it makes an angle $2\theta_1$ with the first crystal, the angle of incidence becomes $\theta_1-\alpha+\beta$. For monochromatic radiation, the expression for $P'(\beta)$, the power reflected by crystal B at the setting β, is therefore

$$P'(\beta)=I_0(\alpha)\int_{-\infty}^{+\infty}R(\alpha)R(\beta-\alpha)\,d\alpha, \qquad (6.36)$$

an equation whose derivation follows exactly the lines of that of equation (6.30) for the parallel arrangement.

The argument of the function R in the second factor of the integrand of (6.36) is $\beta-\alpha$, instead of $\alpha-\beta$, as in (6.30), and this introduces an essential difference. As before, suppose the curves $R(\alpha)$ and $R(\beta-\alpha)$ to be plotted as functions of α (fig. 113(*b*)). The curve for $R(\beta-\alpha)$ cannot be obtained by simple displacement of that for $R(\alpha)$, but is its mirror image about the line $\alpha=\frac{1}{2}\beta$. As may be seen from fig. 113(*b*), if the curve $R(\alpha)$ is symmetrical, so is the curve $P'(\beta)$, which does not then differ from the curve obtained in the parallel arrangement; but if the curve $R(\alpha)$ is unsymmetrical, its overlap with its mirror image $R(\beta-\alpha)$ is different when β has equal positive and negative values, and so the reflection curve from the second crystal, expressed as a function of β, must also be unsymmetrical. Unfortunately, however, the asymmetry of the curve $R(\alpha)$ predicted by Prins cannot be tested in this way, because of the dispersion in the (1, 1) setting, which we shall now consider briefly.

Suppose radiation of two wave-lengths λ_1 and λ_2 to fall on crystal A, and let θ_1 be the angle of incidence corresponding to a reflection maximum for λ_1, and $\theta_1+\Delta\theta$ the analogous angle for λ_2. If the angle between the two crystals is $2\theta_1$ the second crystal will also reflect λ_1 at the maximum; but the radiation of wave-length λ_2 reflected as a maximum from crystal A at an angle $\theta_1+\Delta\theta$ falls at an angle $\theta_1-\Delta\theta$ on crystal B, and will not be reflected by it until it has been turned through an angle $2\Delta\theta$. If the wave-lengths λ_1 and λ_2 are present as two separate lines, such for example as the $K\alpha_1$ and $K\alpha_2$ lines, a peak on the reflection curve will be obtained for each, and the apparatus acts as a spectroscope with a dispersion double that given by reflection from a single crystal. The reflection curves $R_2(\beta)$ given by the individual lines will be broadened, owing to the finite spectral width of each, and the range of reflection due to this cause is in general considerably greater than the width of the monochromatic reflection curve from a good crystal. This method has been much used in studying the widths and fine structure of X-ray lines. Reference may be made to the book of Compton and Allison for an account of work of this kind, which lies outside the scope of the present discussion.

(*f*) *The effect of polarisation of the incident radiation on the double-reflection curve:* We return now to the parallel arrangement. The expressions (*6*.31) and (*6*.32) for the reflection coefficient $R_2(\beta)$ apply when the radiation incident on crystal A is polarised with its electric vector either perpendicular to or parallel to the plane containing the incident and reflected rays. We shall denote these two types of polarisation by the symbols σ and π respectively. The reflection coefficient R has different forms for the two cases, which will be denoted by R_σ and R_π.

The intensity of a beam of unpolarised radiation may be considered as the sum of two equal parts due to independent components in the two directions of polarisation. For unpolarised incident radiation, therefore, (*6*.30) and (*6*.31) become

$$P(\beta) = \tfrac{1}{2} I_0(\alpha) \int_{-\infty}^{+\infty} \{R_\sigma(\alpha) R_\sigma(\alpha - \beta) + R_\pi(\alpha) R_\pi(\alpha - \beta)\}\, d\alpha, \qquad (6.37)$$

and
$$R_2(\beta) = \frac{\int_{-\infty}^{+\infty} \{R_\sigma(\alpha) R_\sigma(\alpha - \beta) + R_\pi(\alpha) R_\pi(\alpha - \beta)\}\, d\alpha}{\int_{-\infty}^{+\infty} \{R_\sigma(\alpha) + R_\pi(\alpha)\}\, d\alpha}. \qquad (6.38)$$

For a perfect crystal with appreciable absorption, such as is considered in the Prins theory, the relation between R_σ and R_π depends in a rather complicated way on the glancing angle θ, and no simple evaluation of the integrals in the above expression is possible, although they can of course be evaluated numerically for any special case.

For a perfect crystal with negligible absorption, Darwin's theory, considered in II, §§ 3(*e*) and 3(*f*), shows that the curve of $R_\sigma(\alpha)$, plotted as a function of α, has the same shape as the curve of $R_\pi(\alpha)$, but that its angular breadth is everywhere greater in the ratio $1 : |\cos 2\theta_0|$, where θ_0 is the Bragg reflection angle. The areas under the curves are therefore in the same ratio, and the denominator in (*6*.38) may be written in the form $(1 + |\cos 2\theta_0|) \int_{-\infty}^{+\infty} R_\sigma(\alpha)\, d\alpha$. When $\beta = 0$, the numerator takes the form $(1+|\cos 2\theta_0|)\int_{-\infty}^{+\infty}\{R_\sigma(\alpha)\}^2\,d\alpha$, so that, in this special case, the reflection coefficient is independent of the state of polarisation, and has the value given in (6.33) which, as we have already stated, is 4/5 for the Darwin curve.

When R_σ and R_π have the Darwin form, the reflection curves from the second crystal for the two standard states of polarisation have the same shape, but $R_2^\pi(\beta)$ is narrower than $R_2^\sigma(\beta)$ in the ratio $|\cos 2\theta_0|:1$. This may be seen from a consideration of fig. 113(*a*); for if the two R-curves are narrower for the state of polarisation π than for the state of polarisation σ in the ratio $|\cos 2\theta_0|:1$, they will overlap for the same fraction of their breadth in the two cases for values of β that are in the same ratio.

(*g*) *Integrated reflections from the double-crystal spectrometer:* We shall now consider how the area of the double-reflection curve, that is to say the integral of $R_2(\beta)$ with respect to β, is related to the integrated reflection from the single crystal.

Suppose equation (6.38) to be integrated with respect to β between the limits plus and minus infinity. Then it is easy to see that for either integral in the numerator

$$\int_{-\infty}^{+\infty} R(\alpha)\,R(\alpha-\beta)\,d\alpha\,d\beta=\left\{\int_{-\infty}^{+\infty} R(\alpha)\,d\alpha\right\}^2; \tag{6.39}$$

for we may put $\alpha-\beta=\phi$ in the integration with respect to β, and $d\beta=-d\phi$, since α is constant. The integral with respect to ϕ is thus identical with that with respect to α, so that (6.39) follows at once. We may therefore write

$$\int R_2(\beta)\,d\beta=\frac{\rho_\sigma{}^2+\rho_\pi{}^2}{\rho_\sigma+\rho_\pi}, \tag{6.40}$$

where $$\rho_\sigma=\int R_\sigma(\alpha)\,d\alpha,\quad\text{and}\quad \rho_\pi=\int R_\pi(\alpha)\,d\alpha, \tag{6.41}$$

the integrated reflections from the single crystal face for radiation having the two standard directions of polarisation. The integrated reflection for unpolarised radiation is $\frac{1}{2}(\rho_\sigma+\rho_\pi)$. For Darwin's formula (6.40) may be simplified. In this case, $\rho_\pi=|\cos 2\theta_0|\,\rho_\sigma$, so that

$$\int R_2(\beta)\,d\beta=\frac{1+\cos^2 2\theta_0}{1+|\cos 2\theta_0|}\rho_\sigma=\frac{2(1+\cos^2 2\theta_0)}{(1+|\cos 2\theta_0|)^2}\rho_D, \tag{6.42}$$

where ρ_D is the integrated reflection for unpolarised radiation as given by equation (2.84) of II, § 3(*f*). From equation (6.42), ρ_D could be calculated from the area of the reflection curve. If Prins's form of the reflection coefficient is used it is not possible to express the integrated reflection so simply in terms of the area of the reflection curve. There is, however, no difference in principle. The integrals (6.41) must be evaluated numerically from the expression given in equation (2.90) of II, § 3(*h*), to obtain the ratio between R_π and R_σ for any desired glancing angle.

The ordinates of the $R_2(\beta)$ curve are equal to the ionisation produced in the chamber for each setting of B, divided by the ionisation produced when the whole beam reflected from A is allowed to enter the ionisation chamber directly, crystal B being removed. The abscissae of the curve are to be measured in radians, and the area of the curve then gives the integrated reflection, uncorrected for polarisation, in absolute measure. In comparing theory with experiment, it will usually be more convenient to take the normalised area of the reflection curve as the measured integrated reflection, and to calculate the value of this from the assumed forms of the diffraction curves than to try to deduce the integrated reflection for unpolarised radiation from the observed curve.

(*h*) *The relation between the widths of the double-reflection and single-reflection curves:* To conclude the discussion of the theory of the double-crystal spectrometer, we must consider how the width at half-maximum, of the observed reflection curve $R_2(\beta)$, which we shall

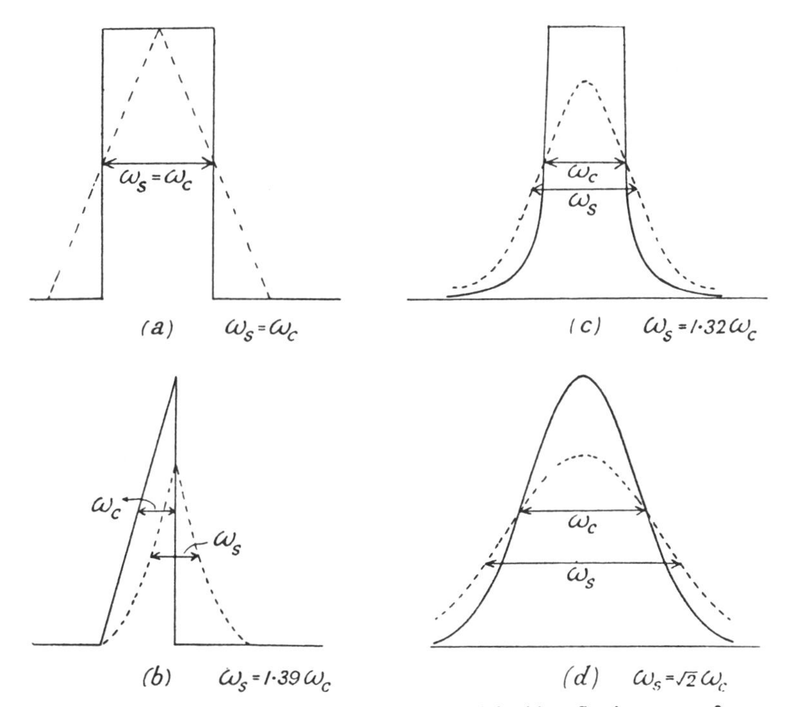

FIG. 115. The relation between the single- and double-reflection curves for different assumed forms of the single-reflection curve

denote by w_s, is related to the corresponding width, w_c, of the reflection curve $R(\alpha)$ for the single crystal. No general rule can be given, since the relationship between the two widths depends on the particular form of the curve $R(\alpha)$. It can, however, be stated that w_s is never less than w_c. It is always of the same order of magnitude, but usually 30 or 40 per cent. greater. We may illustrate this by taking a few special cases. Consider first of all two idealised reflection curves, corresponding to nothing that actually occurs in nature, but nevertheless instructive. In fig. 115(*a*), the crystal is assumed to reflect perfectly over a certain range, and not at all outside it. The reflection curve $R_2(\beta)$ is then an isosceles triangle, its width at half-maximum is equal to the range of perfect reflection, and the maximum reflection coefficient $R_2(0)$ is in this one case unity. In fig. 115(*b*) the reflection curve is assumed to be a triangle with one side vertical, in order to illustrate in a simple way how a very unsymmetrical curve for $R(\alpha)$ gives rise to a symmetrical curve for $R_2(\beta)$. In this case, $w_s = 1{\cdot}39 w_c$, and $R_2(0) = 2/3$. In fig. 115(*c*), the curve $R(\alpha)$ is assumed to be a Darwin curve for radiation polarised in one of the standard directions. Here, $w_s = 1{\cdot}40$ times the range of total reflection, or $1{\cdot}32 w_c$, w_c being the width at half-maximum of the Darwin curve, which is 1·06 times the range of perfect reflection. For unpolarised radiation, these results would, of course, be somewhat modified, since the widths of the curves for the two standard components differ. Finally, in fig. 115(*d*), the curve $R(\alpha)$ is a Gaussian error-curve, and $w_s = \sqrt{2} w_c = 1{\cdot}41 w_c$, and $R_2(0) = 1/\sqrt{2}$, or 0·707.

(*i*) *The diamond as a perfect crystal:* Of all known crystals, the diamond, because of its hardness, and its flawlessness in good specimens, seems to be the most likely to approach perfection of structure; and it was one of the first to be investigated. Experiment shows that good diamonds do in fact reflect X-rays very nearly as perfect crystals should, although no thorough investigation of absolute intensities has so far been made, mainly because the material has usually been available only in the form of small crystals of irregular shape. Reference has already been made in § 3(*a*) of this chapter to the experiments of W. H. Bragg,[2, 22] who found the intensity of reflection from diamond to be proportional to the structure factor, and not to its square; and also to those of Ehrenberg and Mark,[25] who measured directly the widths of the reflection curves from diamond, and found them to be of the order of seconds of arc.

In 1928, Ehrenberg, Ewald, and Mark [35] published the results of measurements of the widths of the reflection curves and of the integrated reflections from diamond, made with the double-crystal spectrometer in the parallel position. Using molybdenum Kα radiation, they found the width at half-maximum of the double-reflection curve $R_2(\beta)$ to be about 4″ for the spectrum 111, and deduced the value 2·8″ for the width of the single-reflection curve $R(\alpha)$, on the assumption that both $R_2(\beta)$ and $R(\alpha)$ are error functions. This value they compare with a theoretical value 4·1″, in the calculation of which, however, they make no allowance for the decrease in scattering power of the carbon atom with increasing angle of scattering. Allowance for this decreases the range of total reflection given by the Darwin theory, which should apply fairly closely to this case, to about 1·8″. It is plain that the crystal used in these experiments was fairly perfect, so far as can be judged from the widths of the reflection curves.

The same workers also made measurements of the relative values of the integrated reflections for a number of spectra from diamond, and compared them with the values to be expected from Darwin's formula (*6*.3), which may conveniently be written in the form

$$\rho = k|F| \cot \theta, \text{ for } \theta < 45°; \quad \rho = k|F| \tan \theta, \text{ for } \theta > 45°,$$

k being a constant for a given crystal and wave-length, and F the structure factor for the spectrum considered. This may be represented graphically in the way shown in fig. 116. The ordinates are values of the integrated reflection. The origin lies at $\cot \theta = \tan \theta = 1$ ($\theta = 45°$); to the right of the origin, the abscissae represent increasing values of $\cot \theta$, to the left, increasing values of $\tan \theta$. The angle of reflection therefore increases from right to left. If the structure factor were the same for all spectra, the graph would consist of two straight lines of equal and opposite inclinations, intersecting at $\cot \theta = \tan \theta = 1$. The cotangent line prolonged would intersect the tangent axis at $\tan \theta = 2$. and *vice versa*.

Two types of spectra occur for the diamond lattice. If the indices hkl of the spectrum are all even, $F = 8f$, unless $h/2$, $k/2$, $l/2$ are all odd, in which case F vanishes;* if all the indices are odd, $F = 4\sqrt{2}f$. The unit cell contains 8 carbon atoms, each of scattering factor f. If f were constant, the integrated reflections from diamond, plotted in the way just described, should therefore lie on two pairs of straight lines, and the slopes of the lines for the spectra of even indices should be $\sqrt{2}$ times those for the spectra of odd indices.

Ehrenberg, Ewald, and Mark found that the observed intensities could be arranged fairly well on two lines with the right relative slopes, but that the absolute slopes were quite different for $\theta < 45°$ and $\theta > 45°$, the slopes for $\theta < 45°$ being very much the greater. They were thus unable to decide definitely whether the intensities were proportional to $|F|$ or to $|F|^2$, although an estimate of the absolute intensity for 111 clearly indicated proportionality to $|F|$. They themselves suggested that the discrepancy of slope was due to neglect of the variation with angle of scattering of the atomic scattering factor of carbon.

* This assumes the atoms to be spherically symmetrical. Actually, a weak but quite definite 222 spectrum occurs, which should be absent according to the rule given.

This point was later tested by Brindley,[36] who used some theoretical values of the f-factor for carbon given by James and Brindley, which, although probably not of great accuracy, give well enough the order of

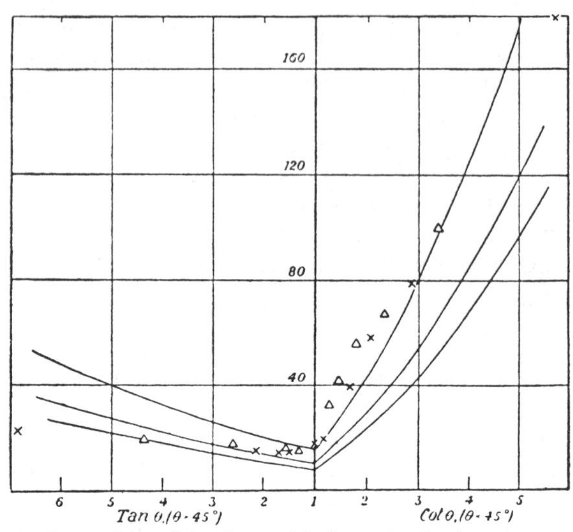

FIG. 116. Comparison of the observed integrated reflections from diamond with the perfect crystal formula (Brindley)
(*Proc. Roy. Soc.*, **A**, **140**, 301 (1933))

the effect, and should be fairly reliable for large values of $(\sin\theta)/\lambda$. The results of Brindley's calculations are given in fig. 116, which is taken from his paper. The curves show the theoretical variation of the intensity with allowance for f, the different curves being drawn for different values of the proportionality factor k. The triangles correspond to intensities of spectra with even indices, the crosses to those of odd indices multiplied by $\sqrt{2}$. It will be seen that this multiplication brings the points of both types more or less on to a single line, indicating at least approximate proportionality of the intensity to the first power of $|F|$. There is a rough general agreement between theory and experiment, although there appears to be quite a systematic divergence, which may be partly due to incorrect allowance for the shape of the crystal, and partly due to the fact that unpolarised radiation seems to have been assumed in the calculations, although, as we saw in VI, § 3(f), the radiation incident on the second crystal is partly polarised by reflection at the first. A rough allowance for this effect does not, however, improve the agreement between theory and experiment.

The effect of the state of perfection of the crystal on the intensity of reflection is well illustrated by some interesting experiments made by Renninger,[37] with a view to measuring the structure factor of the 'forbidden' reflection 222 from diamond, which should be absent if the carbon atoms in the diamond lattice are spherically symmetrical. The existence of this spectrum was first shown by W. H. Bragg in 1921,[22] since when it has been confirmed by a number of observers. The spectrum must have its origin in the binding electrons which will presumably give rise to concentrations of electron density on the lines joining neighbouring carbon atoms. A calculation on the principles of wave-mechanics of the intensity of reflection to be expected has been made by Ewald and Hönl,[38] who give a value for F(222) of about 0·27.

Renninger measured the reflection 222 from octahedral faces of a number of diamonds of varying degrees of perfection, finding a very

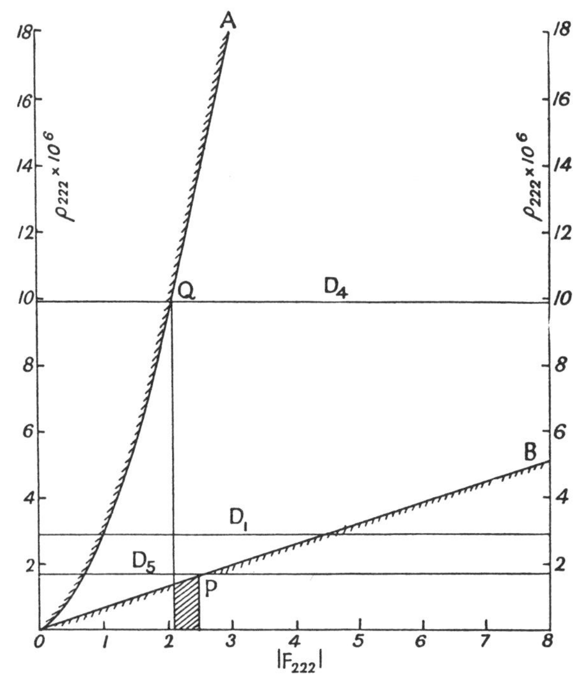

FIG. 117. Illustrating Renninger's method of determining the value of |F| for the 222 spectrum from diamond

considerable variation in intensity. The crystal that appeared most imperfect externally gave a 222 reflection about six times as strong as the one that appeared to be most perfect. He was, however, able to find the structure factor within fairly close limits by an ingenious device. In fig. 117, the abscissae represent assumed values of F(222), the ordinates, integrated reflections. Curve A gives the integrated reflection as a function of the structure factor on the assumption that the crystal has a mosaic structure; for curve B, a perfect crystal, reflecting according to Darwin's formula, is assumed. The points representing the observed integrated reflections must lie somewhere between these two curves. The horizontal lines D_1, D_4, D_5, are drawn at heights corresponding to the observed reflections from three specimens of diamond. Of these, D_5 was very perfect, and it could be reasonably assumed that it reflected nearly as an ideal crystal. The point P where the horizontal line D_5 cuts the lower curve could thus be taken as corresponding very closely to this particular reflection, and gives an upper limit to the structure factor F(222). Similarly, it could be assumed that the crystal D_4, which was outwardly imperfect, and gave a high reflection, reflected nearly as a mosaic, and this gives a lower limit to the structure factor, corresponding to the point Q where the curve A is cut by the line D_4. The two extreme values of F are 2·6 and 2·1, giving a probable value some ten times as great as that calculated theoretically by Ewald and Hönl.*

This method is very interesting, and appears to be capable of extension, although a condition for its applicability is that material of varying degrees of perfection should be available. The limits between which the structure factor may lie will also be much widened if there is very much secondary extinction.

The most recent measurements of the intensity of reflection from diamonds are those of Brill, Grimm, Hermann and Peters,[17] but these were directed mainly to determining the experimental values of the atomic scattering factor for carbon in the diamond crystal, and not to the investigation of the state of perfection of the crystal.

(*j*) *Tests of the Darwin-Prins formula:* Many investigators, the earliest of whom were Davis and Stempel,[27] have shown by means of the double-crystal spectrometer that certain crystals, among them some specimens of calcite, reflect over a range of angles so small as to be of the order predicted by Darwin's theory of reflection by perfect crystals. In this section we shall consider some quantitative tests of the reflection formula for large perfect crystals made with material of this type. Careful measurements have been made by several workers, notably by Allison,[40] and by Parratt,[41] in the United States, and by Renninger [39] in Germany. The observations were as a rule made using as the two reflectors the corresponding surfaces of a freshly cleaved crystal, so that the reflection coefficient $R(\alpha)$ could be considered as the same for each. The experimental part of the work consisted in measuring the double-reflection curve $R_2(\beta)$ (equation (*6*.31) of VI, § 3(*c*)), with the crystals in the parallel setting. The normalising factor in the denominator of equation (*6*.31) is the integrated reflection from the first crystal, which is measured by removing the second crystal and allowing the whole beam reflected from the first crystal to enter the ionisation chamber for a measured time. As we have already

* See, however, note on p. 339, at end of chapter.

seen in VI, § 3(*c*), it is not possible to deduce the form and width of the reflection curve $R(\alpha)$ from the double-reflection curve. The best that can be done is to assume a form for the reflection curve $R(\alpha)$, to calculate from it the corresponding double reflection curve $R_2(\beta)$, and then to compare this with the observed-curve. Again, the observed integrated reflection from the second crystal may be obtained from the area of the curve $R_2(\beta)$, and this may be compared with that to be expected from the theoretical form of the curve $R(\alpha)$.

The theoretical formula to be tested in this work is Prins's modification of the Darwin formula, equation (*2*.90) or (*2*.91) of Chapter II, which takes into account absorption of the radiation in the perfect crystal; for the assumption of negligible absorption underlying Darwin's original formula (*2*.84), or (*6*.2), is not justifiable with the crystals and radiations used in these experiments. In equation (*2*.91), the form of the expression we shall use, η is the parameter determining the angle of incidence. The reflection coefficient $R(\eta)$, which is the same as $R(\alpha)$ except that it is expressed in terms of the parameter η, is the square of the modulus of the complex expression on the right-hand side of equation (*2*.91). If θ is the angle of incidence expressed in radians, the integrated reflection is $\int R(\eta)\,d\theta$, where the integration extends over the range of angles within which $R(\eta)$ is appreciable, in practice a range so small that the quantities A and B in the integrand, which strictly depend on θ, may be taken as constants with the values corresponding to the Bragg angle θ_0. The parameter η is proportional to the deviation of the actual angle of incidence from that corresponding to the middle of the Darwin reflection curve, and it expresses this deviation as a multiple of the half-width of the range of total reflection given by Darwin's theory. This is equal to $\gamma\delta \sec\theta_0 \operatorname{cosec}\theta_0$, where $1-\delta$ is the real part of the refractive index of the crystal for the radiation concerned and $\gamma = PF(2\theta_0)/F(0)$, P being a polarisation factor depending on the state of polarisation of the incident radiation, $F(2\theta_0)$ the structure factor of the crystal unit for the reflection concerned, and $F(0)$ the structure factor for zero angle of scattering. We can therefore write for the integrated reflection, ρ,

$$\rho = \int R(\eta)\,d\theta = \frac{2P\delta}{\sin 2\theta_0}\,\frac{F(2\theta_0)}{F(0)} \int_{-\infty}^{+\infty} \left| \frac{1 + iB/A}{\eta - \dfrac{i\beta}{A} \pm \sqrt{\left(\eta - \dfrac{i\beta}{A}\right)^2 - \left(1 + \dfrac{iB}{A}\right)^2}} \right|^2 d\eta. \tag{6.43}$$

The ambiguity of sign is determined by the consideration that the value of the integrand cannot exceed unity. It is easy to verify that this expression reduces to the Darwin formula (*2*.84) if B and β are zero, corresponding to negligible absorption of radiation in the crystal. The integral in equation (*6*.43) must be evaluated numerically or graphically, and the calculation involves determining the modulus in the integrand for a number of values of η, a somewhat tedious process. In order to perform the calculations it is necessary to make numerical estimates of the values of the quantities A, B, and β, occurring in the expression for $R(\eta)$, and we shall now consider how this may be done.

From equations (*2*.89a) and (*2*.89b) of Chapter II,

$$\frac{A+iB}{\delta+i\beta}=\frac{q}{q_0}. \tag{6.44}$$

If q and q_0 are real, as in the Darwin theory, $q/q_0=PF(2\theta_0)/F(0)$; but the essence of Prins's method of allowing for absorption is the assumption of a complex scattering factor, so this expression must be modified. Suppose for the moment that the unit cell contains only one atom. Then we can write

$$\frac{A+iB}{\delta+i\beta}=P\frac{f'(2\theta_0)+if''(2\theta_0)}{f'(0)+if''(0)}, \tag{6.45}$$

where f' and f'' are the real and imaginary parts respectively of the scattering factor of the atom, as in equation (*4*.6) of Chapter IV. As we have seen in Chapter IV, the ratio of the real to the imaginary part of the scattering factor varies widely for the different electrons in the atom, and so we must consider each separately. If A_s+iB_s is the contribution to $A+iB$ of the electron s in an atom containing more than one electron,

$$A_s+iB_s=P\frac{f_s'(2\theta_0)}{f_s'(0)}\cdot\frac{1+iC_s(2\theta_0)}{1+iC_s(0)}(\delta_s+i\beta_s), \tag{6.46}$$

where $C_s(2\theta)=f_s''(2\theta)/f_s'(2\theta)$, and is the tangent of the angle of phase change of the radiation when it is scattered through an angle 2θ by the electron s. Assuming this phase change to be the same for all angles of scattering for the same electron group, which no doubt is not strictly true, we can write for the whole atom,

$$A+iB=P\sum_s\frac{f_s'(2\theta_0)}{f_s'(0)}(\delta_s+i\beta_s). \tag{6.47}$$

If the frequency of the incident radiation is greater than that of any atomic absorption edge, the absorption, which is proportional to β, will be mainly due to the K electrons. For these, to a fairly good approximation, we may take $f_K'(2\theta)=f_K'(0)$, since their scattering power falls off very slowly with increasing angle of scattering, owing to the small spatial extension of the group in comparison with the wavelength. Equation (*6*.47) may therefore be written in the approximate form

$$A+iB=P\left\{\sum_s\frac{f_s'(2\theta_0)}{f_s'(0)}\delta_s+i\beta\right\}, \tag{6.48}$$

where β is determined from the atomic absorption coefficient; or since δ_s is proportional to $f'_s(0)$

$$A + iB = P\left\{\frac{f'(2\theta_0)}{f'(0)}\delta + i\beta\right\}, \tag{6.49}$$

where $f'(2\theta_0)$ is now the real part of the scattering factor for the whole atom, and $f'(0) = Z$, the number of electrons in the atom. If the scattering factor $f'(2\theta_0)$ is calculated for the atom at rest, we must multiply the right-hand side of equation (*6*.49) by the appropriate temperature factor in order to get the contribution of the atom to $A + iB$ when it is subject to thermal agitation in the crystal lattice. Thus

$$A + iB = Pe^{-M}\left\{\frac{f'(2\theta_0)}{f'(0)}\delta + i\beta\right\}. \tag{6.50}$$

If the crystal unit contains more than one kind of atom, allowance has to be made for the path differences due to the relative positions of the atoms in the unit cell. This allowance may be rather complicated when the crystal is polar, and the radiation has a frequency not very different from that of an absorption edge of one of the atoms in the crystal unit; but if the crystal is non-polar, as in the cases that interest us here, it is easily made. To get the value of $A + iB$ we have simply to use the appropriate value for each atom of the expression on the right-hand side of equation (*6*.50), in place of the value of the scattering factor f in the formula for the structure factor of the unit cell. For example, in the simple case of rock salt, where the structure factor for a spectrum of even indices is just the sum of the scattering factors of the individual atoms,

$$A + iB = P\left\{e^{-M_{Na}}\left(\frac{f_{Na}}{Z_{Na}}\delta_{Na} + i\beta_{Na}\right) + e^{-M_{Cl}}\left(\frac{f_{Cl}}{Z_{Cl}}\delta_{Cl} + i\beta_{Cl}\right)\right\}. \tag{6.51}$$

(i) *Rock Salt.* We shall consider as our first example the case of rock salt, investigated by Renniger.[39] The numerical values of A, B and β are estimated as follows from equation (*6*.51). By equation (*2*.88), p. 63, if μ is the linear absorption coefficient of the crystal, $\beta = \lambda\mu/4\pi$. For rock salt, with Cu Kα radiation ($\lambda = 1{\cdot}54$A), $\mu = 160$; so

$$\beta = 1{\cdot}54 \times 10^{-8} \times 160/4\pi = 0{\cdot}196 \times 10^{-6}.$$

The contributions of sodium and chlorine to β for the whole atom are in the ratio of their respective atomic absorption coefficients. From the International Crystal Tables we find $\mu_{Na} = 117$, $\mu_{Cl} = 604$ for Cu Kα radiation; whence $\beta_{Na} = 0{\cdot}032 \times 10^{-6}$, $\beta_{Cl} = 0{\cdot}164 \times 10^{-6}$. From equation (*2*.60),

$$\frac{\delta_{Na}}{Z_{Na}} = \frac{\delta_{Cl}}{Z_{Cl}} = \frac{\lambda^2 e^2 N}{2\pi mc^2} = 0{\cdot}241 \times 10^{-6},$$

N being the number of atoms of sodium or chlorine in unit volume of the crystal. The values of the scattering factors for the atoms at rest are taken from the tables, and are based on Hartree's calculations, but have been corrected for dispersion by Hönl's method, as described in IV, § 1 (*n*). The values of the temperature factors are those calculated by Waller and James,[15] based on measurements by James and Miss Firth.[5] Using these values, we can at once calculate A and B from equation (*6*.51) by equating real and imaginary parts. The constants deduced by Renninger from these figures for rock salt are shown in Table VI. 5. From these values, the curves reproduced in fig. 27, p. 65, were drawn by Renninger.

TABLE VI. 5

Constants for Prins's reflection curve for rock salt (Renninger).

Spectrum	Polarisation	P	A.10^6	B.10^6	β/α	B/A
200	σ	1	4·83	0·187	0·0406	0·0387
	π	$\cos 2\theta_0 = 0{\cdot}851$	4·11	0·159	0·0477	
400	σ	1	2·96	0·165	0·0663	0·0558
	π	$\cos 2\theta_0 = 0{\cdot}403$	1·19	0·066	0·165	
600	σ	1	1·82	0·135	0·108	0·0745
	π	$\cos 2\theta_0 = 0{\cdot}340$	0·618	0·046	0·317	

The crystals used by Renninger were not natural, but were grown in the laboratory from a melt, and were optically very clear, but their cleavage surfaces gave bad goniometer reflections. They consisted of elements a few square millimetres in area, the largest of which by themselves gave good reflections, but whose surfaces were mutually inclined at angles up to a few minutes. Renninger calls these separate crystals ' crystallites ', to distinguish them from the much smaller ' blocks ' of the mosaic theory. It was found possible to work with single crystallites, and the measurements to be described were made with two corresponding crystallites about 1·4 × 3·0 mm. in area, produced by cleavage from a single one, and isolated by diaphragms of lead foil pasted on the crystal surfaces.

The half-width of the double-reflection curve given by this pair in the order 200 was about 7·1″, much lower than any previously recorded values from rock salt; and it might perhaps have been even lower had it been possible to eliminate entirely reflections from neighbouring crystallites. Determinations of the integrated reflections from the second crystal were made by measuring the area of the double-reflection curve, and also the power that entered the ionisation chamber when the second crystal was removed, and the beam from the first crystal was allowed to enter the chamber directly. Great care was necessary to ensure that the whole of the beam reflected by the first crystal fell on the exposed surface of the crystallite in the second crystal. The measurements were repeated, using an area of 30 to 50 sq. mm. of a fresh cleavage surface as the second reflector, so that a number of different crystallites were irradiated. The reflection curve now showed a number of separate peaks, individually very narrow, but overlapping to a certain extent. The integrated reflections were again measured, and values between 4·47 and $4{\cdot}70 \times 10^{-5}$ were obtained for 200. Measurements were also made from 400 and 600.

As a control, similar measurements were made with natural rock-salt crystals. With these, too, crystallites were observed, and the widths of the reflection curves from some were surprisingly small, as low as 30″ in some cases, although the reflection curve from any considerable area of the surface might extend over a degree or so. The integrated reflections were low for natural crystals, but when the surfaces were polished the values rose considerably.*

Before the values of the integrated reflections from the second crystal are compared with theory, correction must be made for the partial polarisation of the radiation reflected from the first crystal. Let θ_1 and θ_2 be the glancing angles of incidence at the first and second crystals respectively. If each reflects as a perfect crystal the component of the radiation vibrating parallel to the plane of incidence (π) is weakened by the first reflection relative to that vibrating perpendicular to the plane of incidence (σ) in the ratio $|\cos 2\theta_1| : 1$, and reflection at the second crystal reduces it by a further factor $|\cos 2\theta_2|$. The polarisation factor for the double reflection is therefore

$$P' = (1 + |\cos 2\theta_1|\,|\cos 2\theta_2|)/(1 + |\cos 2\theta_1|),$$

instead of

$$P = (1 + |\cos 2\theta_2|)/2,$$

as it would have been had unpolarised radiation fallen on the second crystal. The observed results must therefore be multiplied by a factor P/P', if they are to be compared with values calculated assuming unpolarised incident radiation. Actually this is not quite correct, for the quantities A and B in the Prins formula depend on the state of polarisation also, and the correction should be rather larger. The total effect, however, amounts only to a few per cent. Another correction applied by Renninger is that for the presence of the component of half wave-length in the radiation reflected by the first crystal. This also amounts to a few per cent., but opposes the polarisation correction. The corrected results are shown in Table VI. 6, in which H is the half-width of reflection from a single crystallite, W the width of the reflection curve from a considerable area of the second crystal, and ρ the corrected integrated reflection.

* Compare the results of Bragg, James, and Bosanquet on reflections from freshly cleaved rock salt, quoted in VI, § 1 (*l*).

TABLE VI. 6

Crystal	H	W	$\rho \times 10^5$ 200	400	600
Artificial (untouched cleavage face) - - -	7·1″	Some minutes	4·78	1·05	0·46
Natural (untouched cleavage face) - - -	40″–50″	Some degrees	10·2	2·63	0·98
Natural (polished face) -	900″	Some degrees	27·0	4·50	1·63

Table VI. 7 shows the values of the integrated reflections on the Darwin and Prins theories in comparison with the measured values from the artificial crystal. It will be seen that the agreement as regards order of magnitude is good, and that for the higher orders it is noticeably better for the Prins formula. For 200 the agreement is not so good, and the discrepancy should be too great for experimental error. For small angles of incidence it was, however, very difficult to be sure that the reflection was confined to a single crystallite, and if reflection had taken place to some extent at a second crystallite after the radiation had passed through the first, an increased integrated reflection would have been the result.

TABLE VI. 7

Integrated reflections and half-widths of reflection for the ideal crystal, calculated and observed.

Order	H Darwin	H Prins	H Obs.	Pol.	$\rho \times 10^5$ Darwin	$\rho \times 10^5$ Prins	$\frac{1}{2}(\rho_\sigma + \rho_\pi) \times 10^5$ Darwin	Prins	Obs.
200	4·2″	4·9″	7·1″	σ	4·83	4·43	4·50	4·10	4·78
				π	4·12	3·72			
400				σ	1·72	1·48	1·21	0·99	1·05
				π	0·69	0·49			
600				σ	1·03	0·82	0·69	0·51	0·46
				π	0·35	0·19			

It is plain that the individual crystallites of which Renninger's artificial crystal was built up reflected nearly as perfect crystals, but the crystal as a whole was not truly perfect in the sense that it was regularly built up throughout its entire volume. It consisted of crystallites of linear dimensions from 0·1 to 0·01 cm., each of which was perfect enough and large enough for dynamical reflection to take place within it; but the individual crystallites were inclined to one another, and the whole crystal, examined with a spectrometer in the ordinary way, reflected over a range of several minutes. The intensity of reflection from such a crystal would, however, remain very low, of the order of that to be expected from the dynamical theory. It would not reflect as a mosaic, even although the range of reflection might be considerable.* This large scale distortion is not a mosaic structure, although it must, of course, merge into it as the size of the crystallites decreases, and the individual blocks become so small (10^{-4} to 10^{-6} cm. linear dimensions) that primary extinction cannot take place within them.

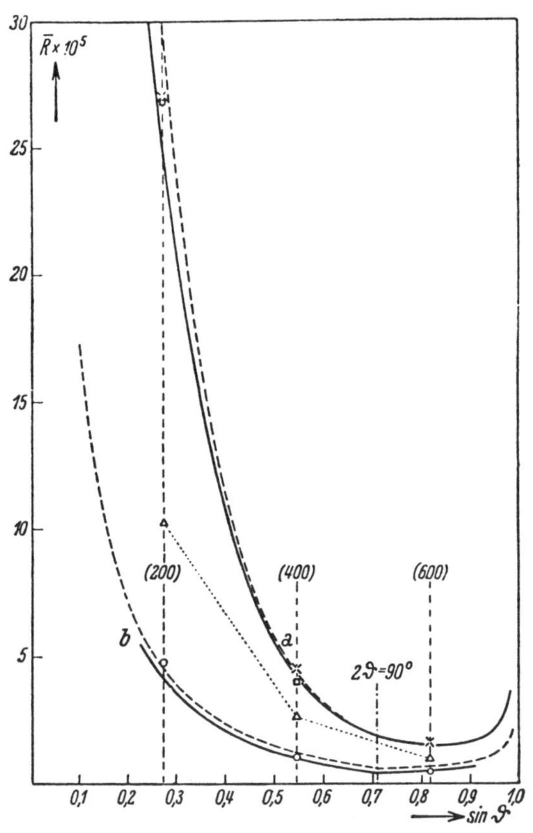

FIG. 118. Integrated reflections obtained from rock-salt crystals with Cu Kα radiation ($\lambda = 1{\cdot}54$A) compared with various theoretical curves. *Curves a*, mosaic crystal without (– – – – – – –) and with (————) allowance for secondary extinction. *Curves b*, perfect crystal, according to Darwin's theory (– – – – – –), with allowance for absorption by Prins's method (————). *Observed values :* ○, from an untouched cleavage of an artificially prepared crystallite; △, from an untouched cleavage surface of a natural crystal: □ ×, from a ground and polished face of a natural crystal (Renninger)

(*Zeit. f. Krist.*, **89**, 344 (1934))

* Compare VI, § 1(*g*), p. 282.

It is important to realise, however, that whether we classify the crystal as a conglomerate of ideally reflecting crystallites or as a mosaic must depend on the wave-length of the radiation used. A truly ideal crystal will reflect dynamically for any wave-length; a crystal made up of perfect crystallites that are optically independent will behave as an ideal crystal in so far as intensity of reflection is concerned, if the crystallites are so large that no appreciable amount of radiation gets through them, even when they have ceased reflecting and the extra absorption due to primary extinction no longer operates. If the crystallites are so small that ordinary absorption allows some radiation to pass through them, and to be reflected in turn from crystallites at a lower level inclined to those in the first layer, the integrated reflection is of course increased; and it is plain that we have here the first step towards the mosaic structure. The limiting size of the crystallites must depend on the wave-length of the radiation used. With Cu Kα, employed by Renninger, the depth of penetration allowed by ordinary absorption is of the order of 0·1 mm. With harder radiation the depth of penetration would have been considerably greater, and similar experiments would have shown the crystal to reflect to some extent as a mosaic.

In fig. 118, the integrated reflection is plotted as a function of sin θ. Curves (*a*) are the theoretical curves for the mosaic crystal, curves (*b*) those for the ideally perfect crystal, the full line being the Prins curve, and the dotted line the Darwin curve. The circles denote Renninger's measurements on the artificial crystal, the triangles those on the freshly cleaved natural crystal, and the crosses and squares those on the same natural crystal, ground and polished. The latter agree very closely with the curve for the mosaic crystal. This work is extremely interesting, for it shows clearly that it is possible to obtain virtually ideal and virtually mosaic crystals of the same substance, and it provides a very striking confirmation of the essential accuracy of the theories of reflection.

(ii) *Calcite*. Renninger's work on rock salt has been discussed first because it is the most detailed quantitative test yet made of the dynamical theory of reflection; but it had already been shown by Allison,[40] Parratt,[41] and others that certain specimens of calcite reflected very nearly in accordance with this theory. The most complete observations are those of Parratt, who measured the reflecting powers of calcite for wave-lengths from 1·5A to 5A, a range including the absorption edge of calcium (3·064A). Conditions for a proper quantitative comparison are rather less favourable with calcite than with rock salt. The unit cell is more complex, and there is a parameter, the distance of oxygen from carbon, that has itself to be fixed by means of intensity measurements, although indeed there is little doubt as to its value. The *f*-curves for calcium, carbon, and oxygen, are less well determined than those of sodium and chlorine, and no data are available for applying the temperature correction, although this is probably not large for the spectrum examined, the first order from the cleavage face (100). With these limitations, the method of calculating the theoretical curves employed by Parratt was substantially the same as that described above for the case of rock salt. The measurements were made with a vacuum double-crystal spectrometer, on account of the high absorption in air of the longest wave-lengths used. In fig. 119, the coefficient of reflection

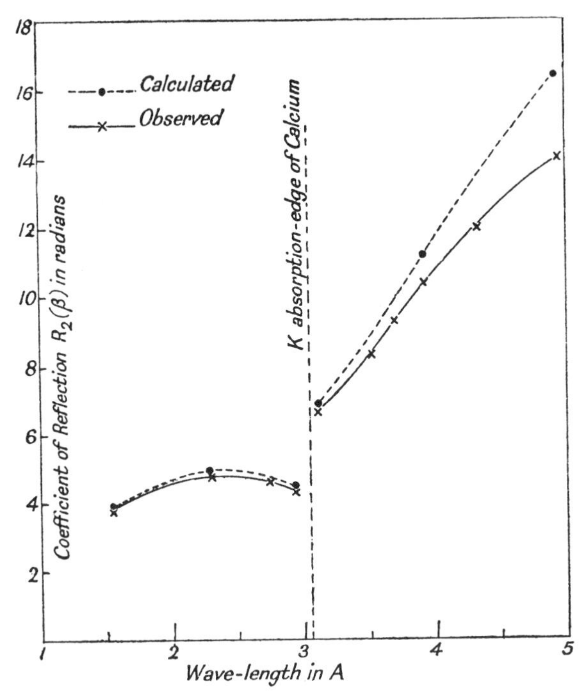

FIG. 119. Observed and calculated coefficients of reflection from a calcite crystal, as a function of the wave-length (Parratt)

from the second crystal is plotted as a function of the wave-length. The circles on the dotted line show the calculated points, and the crosses on the full lines the observed points. The agreement is within the errors of experiment for wave-lengths below about 3·9A. The quantity called the coefficient of reflection by Parratt, and plotted in fig. 119, corresponds to $R_2(\beta)$ of equation (6.38), the polarisation of the radiation being taken into account as in that equation. For further details of the calculation, reference may be made to Parratt's paper. The parameter l used by Parratt corresponds to the parameter η used here, the relation between them being $l=\gamma\eta$, where $\gamma=PF(2\theta_0)/F(0)$; and his angular variable k is expressed in the same units as l. If this is borne in mind, the connection between Parratt's equations and those of the present discussion will be clear.

Parratt also investigated the half-width of the reflection curves $R_2(\beta)$, and found satisfactorily close agreement with theory. The half-widths varied from 5″ for $\lambda = 1{\cdot}54$A to 23·7″ for $\lambda = 4{\cdot}94$A, the calculated values being 4·94″ and 23·5″ respectively. The calcite crystal used by Parratt would appear to reflect as a perfect crystal within the limits imposed by experimental error and the uncertainty of some of the values used in the theoretical calculations. The measurements were made with long-wave-length radiation, and one must remember that this favours agreement with the dynamical formula, although the results of the experiments on line-width indicate that Parratt's calcite crystal was much more nearly a true perfect crystal than Renninger's rock-salt crystal.

4. Intensity Measurements from Powdered Crystals

(*a*) *Introductory :* The results discussed so far in this chapter illustrate the difficulty of interpreting intensity measurements from single crystals owing to the large influence of the state of perfection of the crystal on the value of the integrated reflection. Recognition of this difficulty led to the development of various techniques for measuring absolute integrated reflections from crystal powders. If the crystal particles are small enough, errors due to primary extinction are eliminated; and if they are arranged entirely at random secondary extinction, which depends on a relatively parallel orientation of optically independent crystal fragments, is also enormously reduced, although not in principle entirely eliminated.

Darwin,[4] in discussing the determination of the atomic scattering factors of sodium and chlorine made by Bragg, James, and Bosanquet,[3] suggested that the uncertainties in their results due to extinction might best be overcome by the use of powdered crystals. A. H. Compton and Freeman [42] had independently come to the same conclusion, and had made some preliminary measurements from powdered rock salt, which indicated its general correctness.

The powder method has of course its own limitations. As a method of determining crystal structures it is applicable only to crystals of relatively high symmetry, because of the very great complexity of the system of diffraction haloes given by crystals of low symmetry. Nevertheless, a large number of practically important crystals, notably those of the metals and their alloys, have high symmetry, and the powder method has been of very great importance in dealing with them. The aspect of the matter that will mainly concern us here is, however, the determination of atomic scattering factors, and tests of reflection formulae, by means of absolute measurements on crystals of known structure.

The formulae applicable to determinations of intensities from powdered crystals have been developed in Chapter II, § 2(*i*). The

Debye-Scherrer method, which is that mainly used for crystal analysis, is not very suitable for absolute measurements, although it is of the greatest practical importance, and for this purpose one or other of the methods now to be described has generally been used.

(*b*) *The transmission method:* In one method, illustrated by fig. 23, p. 48, a beam of X-rays passes through a parallel-sided slab formed of the powdered crystal, and the intensities of the diffracted beams that have passed through the slab are compared with that of the direct beam after it has also passed through the slab. Since both diffracted and direct beams have travelled the same distance in the slab, the ratio of their intensities is independent of the absorption coefficient of the crystal for the radiation, and this, in principle, makes the method particularly suitable for absolute determinations.

Suppose the slab to be mounted on the axis of an ionisation spectrometer, and let the slit of the ionisation chamber be of height l and be distant r from the crystal specimen. Then the ratio of P_t, the power in that part of the diffracted beam that enters the chamber, to I_t, the power in the directly transmitted beam, is, by equation (2.49),

$$\frac{P_t}{I_t} = \frac{pltQ}{4\pi r \sin 2\theta_0} \frac{\rho'}{\rho}, \qquad (6.52)$$

in which p is the number of co-operating planes for the spectrum considered, t is the thickness of the powder slab, ρ' and ρ are respectively the average density of the powder specimen, and the density of the crystal in bulk, θ_0 is the Bragg angle, and Q has the significance of equation (2.31). If the ratio P_t/I_t is determined experimentally, Q, and hence $|F|$, can be calculated. Absolute measurements were made by Bearden [20] in this way. The primary beam, before falling on the crystal slab, was reflected from a crystal to make it monochromatic, an essential precaution for absolute determinations. P_t and I_t were measured by allowing first the reflected beam, and then the directly transmitted beam to enter the ionisation chamber for a known time. Bearden determined the structure factors $|F|$ for a series of spectra from rock salt and aluminium with molybdenum $K\alpha$ radiation, and deduced the f-curves for sodium, chlorine and aluminium. He also made careful absolute determinations of the integrated reflections from a single crystal of rock salt, for comparison with the results obtained from the powders. For those spectra for which extinction was small the results obtained by the two methods were in agreement, and they also confirmed fairly well the earlier results of Bragg, James, and Bosanquet; but the intense spectra from the single crystal, which were presumably affected by extinction, were considerably weaker than those from the powders, which appeared to be free from this error. The values of f obtained by Bearden from aluminium powder are indicated in fig. 110 by triangles. For the weaker spectra of high order they agree very closely with the values obtained later by James, Brindley, and Wood [6] from single crystals of aluminium. For the strong spectra of low order, for which extinction was apparently important in the single crystals, the values obtained from the powder are considerably the higher, and agree well with the theoretical curve calculated by Hartree's method. It should perhaps again be emphasised that Bearden's values are independent absolute values, and that the two sets of readings have at no point been brought into agreement.

(c) *The reflection method and the focusing condition:* The power in the diffracted beam is always small in comparison with that in the incident beam, and a direct comparison of the two, involving as it does an initial reflection at a crystal to produce homogeneous radiation, requires a very powerful source of primary radiation. Most determinations have not therefore been absolute in the sense considered above, and it has been usual to compare the intensity of the spectrum to be investigated with that of a spectrum whose absolute intensity is already known as the result of some more or less standard determination. For comparisons of this sort, the spectra diffracted from the surface of a slab of the powder, so thick that none of the incident radiation passes through it, have generally been used. The formula applying in this case is (*2*.50). If I is the total power in the incident beam, and P the power in that part of the diffracted beam entering the ionisation chamber,

$$\frac{\mathrm{P}}{\mathrm{I}}=\frac{\mathrm{Q}pl}{8\pi\mu r\sin\theta_0\left(1+\dfrac{\sin\alpha}{\sin\beta}\right)}, \qquad (6.53)$$

where μ is the linear absorption coefficient of the crystal in bulk, α is the glancing angle of incidence of the primary beam on the surface of the slab, and β the corresponding angle for the diffracted beam, so that $\alpha+\beta=2\theta_0$, θ_0 being the Bragg angle.

One of the advantages of this method, often called the reflection method, to distinguish it from the transmission method discussed above, is that a focusing condition can be found that makes it possible to use an incident beam with a considerable angular divergence, and this of course greatly increases the available intensity. Let A, fig. 120, be a point source of X-rays, and let P be a crystal particle set so as to diffract radiation through an angle $2\theta_0$, giving a spectrum in the direction PB. Let the arc of a circle be drawn through APB. Then, for any other point P′ on this arc, the angle between AP′ and P′B is also $2\theta_0$. Suppose now the arc APB to rotate about the chord AB as axis, forming a surface of revolution—a toroid. Then for any point P on this toroidal surface, the angle between AP and PB is the same. If therefore a crystalline powder is spread with random orientation of the particles over this toroidal surface, any particle which is set so as to produce a spectrum at an angle $2\theta_0$ must diffract this spectrum through the point B, however divergent the primary pencil of radiation from A. In actual practice, if A is a short slit, and if the toroidal surface is replaced by that part of a cylindrical surface tangential to it, that lies

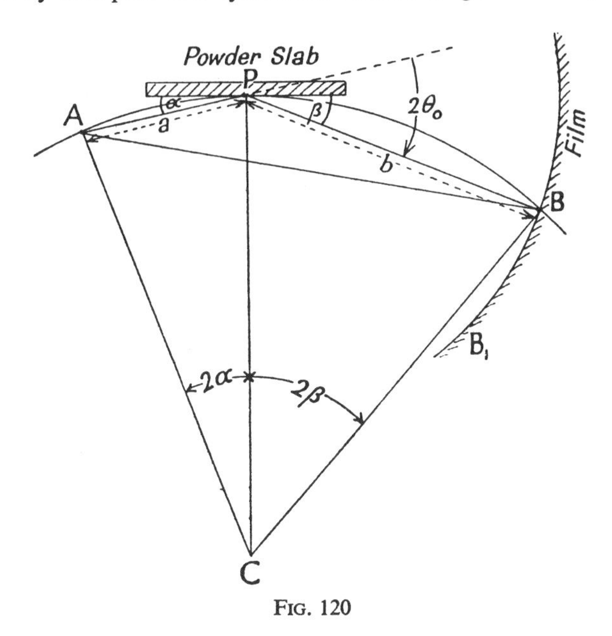

FIG. 120

near the line of contact, a good enough focus for most purposes is obtained. Powder cameras depending on this principle were used by Seemann,[43] and by Bohlin.[44] It will be seen, however, that whether toroid or cylinder is used a different surface will be required to focus each different spectrum. In a method used first by W. H. Bragg,[45] and largely developed by Brentano,[46] the toroidal surface is replaced by a flat surface, tangential to it at the point P. Approximate focus can still be obtained for a primary beam with a divergence of a number of degrees, and it is possible to secure the focus for any spectrum, although of course not for all simultaneously, by a simple rotation of the crystal specimen into the appropriate position. Let C, fig. 120, be the centre of the arc APB, and let the surface of a flat slab of crystal powder be tangential to the arc at P, with its plane perpendicular to the plane of the arc. Let α and β be the glancing angles made with the surface of the slab by the ray AP and the diffracted ray PB respectively, so that $\alpha+\beta=2\theta_0$. It is then plain from the geometry of the figure that $\sin\alpha/\sin\beta=a/b$, where $AP=a$, and $PB=b$; and this is the condition that the surface of the slab shall be tangential to the toroidal surface, and hence that approximate focus shall be obtained at B for rays diffracted from that part of the slab in the immediate neighbourhood of P. The questions of the degree of focusing, and the aberrations produced by the departure of the flat surface from the ideal toroidal surface are treated at length by Brentano, and reference must be made to his papers for details, the discussion of which lies beyond the scope of the present volume.

Suppose now that the powder slab can rotate about an axis through P, which is also the axis of a cylindrical surface BB′, of radius b. A focus will be obtained on this surface for any angle of diffraction $2\theta_0$, provided that the surface of the powder slab makes an angle α with AP such that

$$\sin\alpha/\sin\beta = \sin\alpha/\sin(2\theta_0 - \alpha) = a/b, \tag{6.54}$$

which is constant for a given instrument. If photographic measurements are to be made, BB′ must be the surface of the film, but when the ionisation spectrometer is used, P is the axis of the instrument, and the slit of the ionisation chamber moves along the arc BB′, and is wide enough to take in the whole width of the diffracted beam, which will be narrow because of the focusing effect. When the focusing condition is exactly fulfilled $\sin\alpha/\sin\beta$ is constant for the instrument, whatever the spectrum. It is true that in practice the angle of incidence α varies slightly, because of the finite range of angles in the incident beam, and also because it is usually advisable to oscillate the crystal slab through a small angle during the exposure when a photographic method is used. Nevertheless, the experimental conditions are in general such that no appreciable error is introduced by assuming the factor $(1+\sin\alpha/\sin\beta)$ in equation (*6*.53) to be constant. If then we wish to compare the values of P, say P_1 and P_2, for two spectra from the same crystal powder, we may write

$$\frac{P_1}{P_2} = \frac{p_1}{p_2}\,\frac{Q_1}{Q_2}\,\frac{\sin\theta_2}{\sin\theta_1}, \tag{6.55}$$

provided that the intensity of the incident beam is the same for both spectra. In this way we may readily compare Q_1 and Q_2, and hence $|F_1|$ and $|F_2|$, the values of the structure factors for the two spectra.

Havighurst [47] determined *f*-curves for sodium and chlorine by this method, using powdered rock salt and Mo Kα radiation, and expressed his results in absolute measure by assuming the value of F(220) obtained by Bragg, James, and Bosanquet. He used powders of varying degrees of fineness, and concluded that primary extinction was absent when the linear dimensions of the crystallites were of the order of 10^{-5} cm. Havighurst also used powders of fluorite, lithium fluoride, and sodium fluoride. To standardise his measurements for these substances he compared certain spectra from each with the 220 spectrum of rock salt by the transmission method, and this of course involved a knowledge of the absorption coefficients of the crystals. He determined in this way *f*-curves for lithium, calcium, fluorine, and sodium. The curve for sodium obtained from sodium fluoride differed from that obtained from rock salt in a way that can be adequately explained by the different thermal vibrations of the two substances. Havighurst's results were of importance in the development of the subject, as they confirmed the correctness of the idea of an atomic scattering factor appropriate to each atom, and showed very clearly the effect of particle size on the intensity of reflection.

(*d*) *The mixed-powder method, and the substitution method:* We have seen that the intensities of spectra from the same crystal powder can be compared without a knowledge of the absorption coefficient of the radiation in the powder. It will, however, as a rule be necessary to compare the intensities of the spectra from one powder with the intensity of a standard spectrum from a different substance, such as rock salt or aluminium. Two main methods have been used for this purpose. In one, the crystal powder to be investigated is intimately mixed with a powder of the standard crystal, and it is assumed that the mixture may be treated as a homogeneous powder with a definite absorption coefficient; in the other, which has been called the substitution method, comparison spectra from the standard substance are taken on the same film as those from the substance to be investigated, by interchanging mechanically the powder layers of the two substances during the exposure. The method of mixtures, if the assumption of the homogeneity of the mixed powder as regards absorption is justifiable, requires no knowledge of the absorption coefficients of the crystals; the method of substitution requires a knowledge of the absorption coefficients of both the experimental crystal and the standard crystal to the same order of accuracy as that required in the final comparison of intensities.

The method of mixtures is in principle very attractive but great care is required in using it. It is essential that the powder particles should be so fine that a negligible amount of absorption of the radiation takes place in any one of them, and entirely erroneous results may be obtained if this precaution is neglected. Discussions of the effect of the size of the particles on the effective absorption coefficient of the powder have been given by Brentano [46 (III)], and by Schäfer.[48] The essential point is very simple. Let us suppose radiation to traverse successively two particles of equal size and shape, one of a crystal (*a*), the other of a crystal (*b*), and let each particle be large enough to absorb a considerable fraction of the radiation traversing it. If the radiation passes first through (*a*) and then through (*b*), the scattered radiation will consist almost entirely of the radiation from (*a*), since very little radiation gets through to (*b*); while if the order of the particles is reversed, the scattered radiation will consist almost entirely of radiation from (*b*). It is only if each particle is so small that it absorbs a nearly negligible fraction of the radiation incident on it that both have an equal chance to scatter. Similar considerations must apply to the case of a powder consisting of a very large number of particles, and it is only when all the particles are so small as to satisfy the condition of negligible absorption that one can assume the powder to have an average absorption coefficient.

The considerations put forward in § 1 (*b*) of this chapter, show that when the individual crystallites are small enough for primary extinction to be negligible they are also small enough to fulfil the condition of negligible absorption, except perhaps with crystals containing very heavy elements. The linear dimensions necessary are of the order of 10^{-5} to 10^{-4} cm., and it is exceedingly difficult to produce uniform powders of this kind by purely mechanical means. Various methods of producing them are discussed by Brentano,[49,50] who has employed chemical methods with success in certain cases. It is, moreover, very important that the powder should be fairly uniform. Particles much larger than the optimum size produce extinction errors; and very small particles are also to be avoided, as they cause a broadening of the spectra due to diffraction, which in extreme cases produces wings to the reflection curve, merging into the general background, and making the actual height of the diffraction maximum above the background, and the area of the reflection curve, very difficult to determine with accuracy. Brentano has used aluminium powder, chemically prepared by Kahlbaum, as a standard powder for the method of mixtures. In skilful hands the method is certainly capable of giving excellent results, but neglect of the precautions outlined above may easily lead to very considerable errors.

The substitution method was also suggested by Brentano.[51] It was developed independently by Brindley and Spiers,[52] and in the hands of Brindley and his fellow workers has given very good results. It avoids some of the difficulties of the method of mixtures, particularly those involved in producing particles of appropriate size, but it is subject to other errors, which the method of mixtures avoids. Brindley's powder camera produces, side by side, spectra reflected from the powder to be examined, and from the standard powder. A knowledge of the absorption coefficients of both crystals is therefore required, and this at present involves some uncertainty. The possibility that secondary extinction, as well as ordinary absorption, may take place must be considered; for there may be preferential orientation of the crystal particles, either in the powder itself, or in the larger particles formed by the aggregation of crystallites small enough individually to show no primary extinction. Brentano has found it advisable to introduce a spacing material of some sort, both to bind the powder together, and to help to produce random orientation of the crystallites, particularly when there is reason to suppose that the crystallites tend to break into some special shape owing to cleavage, and this again involves some modification of the normal absorption coefficient of the material. As an example of the use of the substitution method, the work of Brindley and Ridley [53] on potassium chloride may be quoted, while the work of Brentano and Baxter [49] on the L-dispersion of tungsten, discussed in detail in Chapter IV, § 2(*f*), may serve as an example of an important determination made by the method of mixtures. For a further discussion of the methods, reference may be made to a critical account of their difficulties by Brentano.[54]

For determining structures, the method of Debye and Scherrer, as developed by later workers, particularly by Bradley, is the most generally used. For practical work this remains by far the most important of the powder methods, although it is not so suitable for accurate quantitative determinations as the methods described above. One of the difficulties lies in the allowance for the absorption of the radiation in the powder specimen, which is in the form of a rod on the axis of the camera. The absorption correction is a complicated function of the angle of scattering and is not easy to apply with accuracy. Methods of applying the correction have been developed by Claasen,[55] Rusterholz,[56] and Bradley.[57] The latter has given comprehensive tables to facilitate the work.

The discussion of the quantitative aspects of the Debye-Scherrer method belongs more properly to the volume on experimental technique than to the present volume, and we shall not consider it further, except to remark that in the hands of Bradley and his fellow workers it has become an important and accurate method of quantitative measurement, which has given results of the greatest value.

Note on the Intensity of the 222 spectrum from Diamond

The possibility that a forbidden spectrum may appear as a result of multiple reflection has already been considered in Chapter I, § 4, and this has a very important bearing on the measurement of the intensity of diamond 222. Renninger [58] found that the measured intensity of this very weak spectrum varied greatly from specimen to specimen, and, even with the same diamond, depended markedly on the azimuth of the reflecting face with respect to its own normal. He investigated this by rotating a diamond about the zone axis [111], and at the same time recording photographically the intensity of the 222 spectrum. He found that a series of well-marked maxima occurred, the strongest of which had from ten to twenty times the intensity of the true spectrum, which appeared only as a weak background. When Cu $K\alpha$ radiation was used there were in all 36 maxima, of which 12 were particularly strong. The maxima occurred in pairs, which were repeated at intervals of 60° of azimuth, the closest pairs being the strongest maxima.

As Renninger showed, all these maxima can be explained as double reflections of the type referred to as indirect reflections in Chapter I, § 4. The strongest maxima are due to successive reflections 331 and $\bar{1}\bar{1}1$, or $\bar{1}\bar{1}1$ and 331, or reflections corresponding to cyclic interchanges of these indices that keep the index-sum of a pair equal to 222. When the reciprocal-lattice points (2, 2, 2) and (3, 3, 1) lie on the sphere of reflection at the same time, the reflection 331 is in such a direction as to be reflected again by $\bar{1}\bar{1}1$ in the direction of 222 ; and similarly, when (2, 2, 2) and ($\bar{1}$, $\bar{1}$, 1) lie on the sphere at the same time $\bar{1}\bar{1}1$ is reflected again by 331. As the crystal rotates about [111], so that (2, 2, 2) always lies on the sphere, points such as (3, 3, 1) or ($\bar{1}$, $\bar{1}$, 1) pass twice through the sphere for azimuths differing by a little over 10° for Cu$K\alpha_1$(λ=1·537A). The points always lie outside the sphere for $K\alpha_2$(λ=1·541A), so that these maxima are formed by one component of the α doublet only. Altogether 12 maxima of this type can be formed, taking into account the cyclic interchanges of the indices and the order of the reflection. The maxima corresponding to reversal of order differ in azimuth by 180°. Pairs of reflections $11\bar{1}$, 113, and $3\bar{1}\bar{1}$, $\bar{1}33$, also yield twelve possibilities each, and the 36 maxima so far considered are all that are actually observed with copper radiation, although a few others, too weak for observation, are in principle possible. Other geometrically possible pairs give no double reflection, because one of the spectra involved is itself forbidden. It is not difficult to enumerate the possibilities for copper radiation by means of the reciprocal lattice.

With shorter wave-lengths, the number of maxima becomes greater, because many more points lie within the sphere of reflection in any given position, and so there are many more possible intersections of points with the sphere as the lattice rotates. Renninger estimates that with Mo $K\alpha$ (λ=0·710A) a maximum occurs about every degree of azimuth, so that it would be difficult to observe the true spectrum alone in this radiation. Indeed it seems fairly certain that the original observation of the reflection by W. H. Bragg,[22], who used Rh $K\alpha$ (λ=0·615A), was an observation of an indirect reflection, although the true spectrum does in fact exist. As Renninger himself points out (*loc. cit.*), the high values of the observed 222 intensities in comparison with those to be expected theoretically, commented on in § 3 (i) of this chapter, are certainly due to this cause.

References

1. H. G. J. Moseley and C. G. Darwin, *Phil. Mag.*, **26**, 210 (1913).
2. W. H. Bragg, *Phil. Mag.*, **27**, 881 (1914).
3. W. L. Bragg, R. W. James and C. H. Bosanquet, (I) *Phil Mag.*, **41**, 309 (1921); (II) **42**, 1 (1921); (III) **44**, 433 (1922).
4. C. G. Darwin, *Phil. Mag.*, **43**, 800 (1922).
5. R. W. James and E. M. Firth, *Proc. Roy. Soc.*, **A**, **117**, 62 (1927).
6. R. W. James, G. W. Brindley and R. G. Wood, *Proc. Roy. Soc.*, **A**, **125**, 401 (1929).
7. A. H. Compton, *Phys. Rev.*, **9**, 29 (1917).
8. W. L. Bragg and J. West, *Zeit. f. Krist.*, **69**, 118 (1928).
9. Y. Sakisaka, *Jap. Journ. Physics*, **4**, 171 (1927).
10. Y. Sakisaka, *Proc. Math.-Phys. Soc. Japan*, **12**, 189 (1930).
11. R. W. James, *Phil. Mag.*, **49**, 585 (1925).
12. Y. Sakisaka and I. Sumoto, *Proc. Math.-Phys. Soc. Japan*, **13**, 211 (1931).
13. S. Nishikawa, Y. Sakisaka and I. Sumoto, *Phys. Rev.*, **38**, 1078 (1931); **43**, 363 (1933); *Sci. Papers Inst. Phys. Chem. Research*, Tokio, **25**, 20 (1934).
14. E. Fukushima, *Journ. Sci.*, Hiroshima Univ., **3**, 177 (1933).
15. I. Waller and R. W. James, *Proc. Roy. Soc.*, **A**, **117**, 214 (1927).

16. R. W. James, I. Waller and D. R. Hartree, *Proc. Roy. Soc.*, A, **118**, 334 (1928).
17. R. Brill, H. G. Grimm, C. Hermann and C. L. Peters, *Ann. d. Physik*, **34**, 393 (1939).
18. R. W. James and G. W. Brindley, *Proc. Roy. Soc.*, A, **121**, 155 (1928).
19. G. W. Brindley and P. Ridley, *Proc. Phys. Soc.*, **50**, 96 (1938).
20. J. A. Bearden, *Phys. Rev.*, **29**, 20 (1927).
21. W. H. Bragg, *Phil. Trans. Roy. Soc.*, A, **215**, 253 (1915).
22. W. H. Bragg, *Proc. Phys. Soc. Lond.*, **33**, 304 (1921).
23. P. P. Ewald, *Physikal Zeit.*, **26**, 29 (1925); **27**, 182 (1926).
24. W. L. Bragg, C. G. Darwin and R. W. James, *Phil. Mag.*, **1**, 897 (1926).
25. W. Ehrenberg and H. Mark, *Zeit. f. Physik*, **42**, 807 (1927).
26. E. Wagner and H. Kulenkampff, *Ann. d. Physik*, **68**, 369 (1922).
27. B. Davis and W. Stempel, *Phys. Rev.*, **17**, 608 (1921); **19**, 504 (1922).
28. W. Ehrenberg and G. v. Süsich, *Zeit. f. Physik*, **42**, 823 (1927).
29. B. Davis and H. Purks, *Proc. Nat. Acad. Sci.*, **13**, 419 (1927).
30. M. Schwarzschild, *Phys. Rev.*, **32**, 162 (1928).
31. R. C. Spencer, *Phys. Rev.*, **38**, 618 (1931).
32. M. v. Laue, *Zeit. f. Physik*, **72**, 472 (1931).
33. L. P. Smith, *Phys. Rev.*, **46**, 343 (1934).
34. S. K. Allison and J. H. Williams, *Phys. Rev.*, **35**, 149 (1930).
35. W. Ehrenberg, P. P. Ewald and H. Mark, *Zeit. f. Krist.*, **66**, 547 (1928).
36. G. W. Brindley, *Proc. Roy. Soc.*, A, **140**, 301 (1933); *Proc. Leeds Phil. Soc.*, **2**, 271 (1932).
37. M. Renninger, *Zeit. f. techn. Physik*, **16**, 440 (1935).
38. P. P. Ewald and H. Hönl, *Ann. d. Physik*, **25**, 281 (1936).
39. M. Renninger, *Zeit. f. Krist.*, **89**, 344 (1934).
40. S. K. Allison, *Phys. Rev.*, **41**, 1 (1932).
41. L. G. Parratt, *Phys. Rev.*, **41**, 561 (1932).
42. A. H. Compton and N. L. Freeman, *Nature*, **110**, 38 (1922).
43. H. Seemann, *Ann. d. Physik*, **59**, 455 (1919).
44. H. Bohlin, *Ann. d. Physik*, **61**, 421 (1920).
45. W. H. Bragg, *Proc. Phys. Soc. Lond.*, **33**, 222 (1921).
46. J. C. M. Brentano, (I) *Proc. Phys. Soc.*, **37**, 184 (1925); (II) *Phil. Mag.*, **6**, 178 (1928); (III) *Proc. Phys. Soc.*, **47**, 932 (1935); (IV) *Ibid.*, **49**, 61 (1937).
47. R. J. Havighurst, *Proc. Nat. Acad. Sci.*, **12**, 375 (1926); *Phys. Rev.*, **28**, 869 (1926).
48. K. Schäfer, *Zeit. f. Physik*, **86**, 738 (1933).
49. J. C. M. Brentano and A. Baxter, *Zeit. f. Physik*, **89**, 720 (1934).
50. J. C. M. Brentano, *Phil. Mag.*, **4**, 620 (1927); *Zeit. f. Physik*, **70**, 74 (1931).
51. J. C. M. Brentano, *Zeit. f. Physik*, **99**, 65 (1936).
52. G. W. Brindley and F. W. Spiers, *Proc. Phys. Soc.*, **46**, 841 (1934); **50**, 17 (1938).
53. G. W. Brindley and P. Ridley, *Proc. Phys. Soc.*, **50**, 96 (1938).
54. J. C. M. Brentano, *Proc. Phys. Soc.*, **50**, 247 (1938).
55. A. Claasen, *Phil. Mag.*, **9**, 57 (1930).
56. A. Rusterholz, *Helv. Physica Acta*, **4**, 68 (1931).
57. A. J. Bradley, *Proc. Phys. Soc.*, **47**, 879 (1935).
58. M. Renninger *Zeit. f. Physik*, **106**, 141 (1937).

CHAPTER VII

THE USE OF FOURIER SERIES IN CRYSTAL ANALYSIS

1. Triple, Double, and Single Series and their Application

(*a*) *Introductory :* A determination of a crystal structure by means of X-rays is really a determination of the distribution of diffracting matter in the unit cell. At each point in the crystal there is a certain density $\rho(x, y, z)$ of scattering matter, which is a function of the co-ordinates (x, y, z) of the point. The function ρ is to be identified with $|\Psi|^2$, where Ψ is the electronic wave-function at the point (x, y, z), suitably normalised, so that $|\Psi|^2$ is expressed as an electron density. The amplitude of the radiation diffracted from any small element dv in the neighbourhood of the point (x, y, z) is then proportional to $\rho(x, y, z)dv$. It is true that it is often convenient to regard the unit cell as built up of discrete atoms, but the correct idea, and for our present purpose the essential one, is that of a continuous distribution of diffracting matter having maxima in the regions occupied by the atoms.

A crystal structure is essentially a repeating pattern in three dimensions, and the density function $\rho(x, y, z)$ may therefore be expressed as the sum of a suitable Fourier series. A preliminary account of the use of such series has been given in *The Crystalline State*, Vol. I, Chapter IX. In the present chapter we shall describe in greater detail various ways in which the measured intensities of the diffracted beams may be used to define the coefficients of a Fourier series that, when summed, gives direct information about the arrangement of diffracting matter in the unit cell.

There are two main types of such series. (1) If we knew both 'amplitude' and 'phase', the significance of which will be made clear by the following account, for all the diffracted beams, we could build up a complete picture of the distribution of diffracting matter in the crystal. In practice, only the intensities are measured, phase being unknown, and the number of beams is limited; so that any Fourier series is incomplete. Often, however, the phase can be inferred, and enough terms of the series can be got from the experimental measurements to make possible a very fair approximation to the complete picture. It must be realised, however, that we are here supplementing our X-ray measurements by knowledge drawn from other sources. (2) We can base our Fourier representation only on the X-ray measurements of intensity, making no assumptions about phase. The results are then unequivocal, but although the series may give important information about the structure it cannot give a complete representation.

From the consideration of series of the latter type one passes naturally to analytical methods for dealing with scattering from amorphous solids, liquids, and gases. The essential difference between diffraction by crystalline and amorphous matter is that in the former case the diffracted energy is highly concentrated in certain specific directions ('the reflections'), whereas in the latter case it is spread into diffuse haloes. The Fourier series applicable to crystalline diffraction become Fourier integrals in the case of amorphous diffraction. Diffraction by amorphous matter will be dealt with in Chapters IX and X.

The Fourier treatment shows in a very illuminating way the relation of the diffraction of X-rays to the diffraction of light, and to the formation of optical images; and it is at the same time one of the most powerful and widely used methods of utilising the experimental results of X-ray analysis. Use was first made of the Fourier method by W. H. Bragg [1] in 1915. Ewald,[2] in 1921, pointed out the importance of Fourier series in the general theory of crystal lattices, and showed how to express the series for the periodic density distribution in an infinite crystal in terms of the reciprocal-lattice vectors. In a paper written in 1924, in which they discussed the quantum theory of X-ray reflection by crystals, due to Duane [3] and to A. H. Compton,[4] Epstein and Ehrenfest [5] used a triple Fourier series to represent the density of diffracting matter in a crystal, and showed clearly the relationships between the intensities of the X-ray spectra given by the crystal and the coefficients of the terms of this series. Their arguments were based on Bohr's correspondence principle. Duane,[6] in 1925, suggested applying the series given by Epstein and Ehrenfest to the problem of determining the density distribution in the crystal from the observed intensities of the spectra. He realised clearly the ambiguities inherent in the method owing to lack of knowledge of the relative phases of the spectra, and discussed methods of getting over this difficulty. Duane's formulae were applied in the same year by Havighurst [7] to find the distribution of diffracting matter in some simple crystals, and A. H. Compton,[8] in 1926, gave other analyses, and introduced new forms of the series, with the particular aim of determining the electronic distribution in atoms. W. L. Bragg [9] and his collaborators in 1929 extended it to be a standard method of dealing with complex crystals, and Robertson [23] and others have given many examples of the application of Bragg's method to organic crystals. Finally, Warren and Gingrich,[10] and Patterson,[11] in 1934 developed formulae of the second type, which give information about the structure independently of any knowledge of phase. The work of Warren and Gingrich, which is an application of a method proposed by Zernicke and Prins and first practically applied by Debye and Menke to diffraction by liquids, will be discussed in Chapter IX.

(*b*) *The derivation of the triple series:* Let $\rho(x, y, z)$ be the density of scattering matter at the point (x, y, z) of the crystal, the axes being parallel to the primitive translations a, b, and c of the lattice, and not in general rectangular. The structure of the crystal is periodic parallel to x, y, and z, in the distances a, b, and c, respectively. The density at any point in a line parallel to x must therefore be expressible by a Fourier series of the type

$$\sum_h \mathrm{A}_h \cos(2\pi hx/a - \delta_h).$$

If we consider the periodicity parallel to y as well, it is plain that each coefficient A_h may be represented by a series of the type

$$\sum_k \mathrm{A}_{hk} \cos(2\pi ky/b - \delta_k),$$

and similarly for the periodicity in the z direction. For the density at any point (x, y, z) we can therefore write

$$\rho(x, y, z) = \sum_h \sum_k \sum_l \mathrm{A}_{hkl} \cos\left(\frac{2\pi hx}{a} - \delta_h\right) \cos\left(\frac{2\pi ky}{b} - \delta_k\right) \cos\left(\frac{2\pi lz}{c} - \delta_l\right). \tag{7.1}$$

This is the expression given by Epstein and Ehrenfest,[5] and by Duane,[6] but it may be put into another form, which is more generally useful for crystal analysis, and has a more direct physical significance. We use instead of the cosines the complex exponential forms of the periodic functions, including the phase angles δ in the complex amplitudes $\mathrm{A}(hkl)$. We can then write

$$\rho(x, y, z) = \sum_h \sum_{k=-\infty}^{\infty} \sum_l \mathrm{A}(hkl) e^{-2\pi i \left(\frac{hx}{a} + \frac{ky}{b} + \frac{lz}{c}\right)}. \tag{7.2}$$

In this form of the expression it is evident that there is one plane Fourier wave-form for each triplet of whole numbers, h, k, l. Moreover, the planes of constant phase are the planes

$$hx/a + ky/b + lz/c = \text{constant},$$

which are parallel to those lattice-planes of the crystal whose indices are (h, k, l), since h/a, k/b, l/c, are proportional to the direction cosines of the normal to these planes. The value of the expression

$$hx/a + ky/b + lz/c$$

changes by unity in passing from one lattice-plane of the set (hkl) to its neighbour, and the spacing of these planes is therefore equal to the wave-length of the corresponding Fourier wave. Thus to each lattice spectrum there corresponds a set of plane Fourier waves, having wave-fronts parallel to those lattice-planes that may be supposed to produce the spectrum by reflection, and a wave-length equal to the spacing of these planes. To the higher orders from a given set of planes there correspond formally spacings that are submultiples of the true lattice spacings of the planes, and so Fourier waves whose wave-lengths are submultiples of this spacing.

(*c*) *The calculation of the coefficients of the series:* To evaluate the coefficients $A(hkl)$ we proceed in the usual way. Multiply each side of equation (7.2) by the factor

$$e^{2\pi i\left(\frac{h'x}{a}+\frac{k'y}{b}+\frac{l'z}{c}\right)},$$

and integrate each term with respect to x, y, and z, over the ranges 0 to a, 0 to b, and 0 to c, respectively. Every integral on the right-hand side of the equation will then vanish, except that for which $h=h'$, $k=k'$, and $l=l'$, and the value of this remaining integral is $A(h'k'l')abc$.

Let dv be the volume of the elementary parallelepiped whose sides are dx, dy, dz. Then $dv = N\,dx\,dy\,dz$, N being a constant for a given lattice, depending on the angles between the axes of x, y, and z (see Appendix II §§(*b*) and (*d*)). The volume of the unit cell, V, is then $Nabc$, and so we may write the integrated equation (7.2) in the form

$$\int \rho(x, y, z)e^{2\pi i\left(\frac{h'x}{a}+\frac{k'y}{b}+\frac{l'z}{c}\right)}\,dv = VA(h'k'l'), \tag{7.3}$$

where the integration extends throughout the volume of the unit cell of the lattice. The integral on the left-hand side of (7.3) is the structure factor of the unit cell for the spectrum $h'k'l'$, which we have previously denoted by $F(h'k'l')$. This will be plain on reference to Chapter II, § 1(c); for the integral is the form taken by the summation of equation (2.14) when the scattering matter is associated with a continuous distribution of density $\rho(x,y,z)$, instead of with a set of discrete points.

Thus

$$A(hkl) = F(hkl)/V, \tag{7.4}$$

so that the Fourier series (7.2) becomes

$$\rho(x, y, z) = \frac{1}{V}\sum_{h}\sum_{k=-\infty}^{\infty}\sum_{l} F(hkl)e^{-2\pi i\left(\frac{hx}{a}+\frac{ky}{b}+\frac{lz}{c}\right)}. \tag{7.5}$$

To every spectrum that can be given by the crystal there corresponds one term in the Fourier series. Each such term represents a distribution of scattering matter whose density is given by a plane simple-harmonic wave-form with a wave-length equal to the spacing of the planes by which the corresponding spectrum would be produced as a first order. The wave-length is thus equal to one of the lattice spacings of the crystal, or to a submultiple of it. The amplitude of the wave is equal to the structure factor of the unit cell of the crystal for the corresponding spectrum.

(*d*) *The physical interpretation of the terms of the series:* Before proceeding further with the mathematical development it will be well to consider the more physical aspects of the problem. A plane sinusoidal distribution of scattering matter, such as we have considered above, will give two spectra only, the positive and negative first orders corresponding to a spacing equal to the wave-length of the distribution. A sinusoidal distribution will of course include negative densities, and these may be interpreted by supposing the negative part of the distribution to scatter in opposite phase to the positive part. It will then be clear that a sinusoidal distribution, since it consists of equal positive and negative distributions, will give no spectrum of zero order, and it is easy to show that it will give the two first-order spectra, and no others. We can therefore think of each pair of spectra *hkl* and $\bar{h}\bar{k}\bar{l}$ given by the crystal as produced by a certain sinusoidal density distribution, and of the whole array of spectra as produced by a total density distribution that is the sum of the sinusoidal distributions giving the individual pairs of spectra. These distributions must include a uniform distribution, corresponding formally to the spectrum of zero order, which will ensure that the total density is everywhere positive.

The relative phases of the different sinusoidal distributions remain indeterminate, for we can displace any such distribution in a direction perpendicular to its wave-front without altering either the intensity or the angular positions of the pair of spectra that it gives; just as the positions and intensities of the spectra produced by an ordinary diffraction grating, for example, are not in any way changed by displacing the grating in its own plane in a direction perpendicular to its rulings. The relative *phases* of the spectra are, of course, changed by relative shifts of the scattering distributions, but in the case of the spectra produced by a crystal with X-rays we have no means of detecting such phase differences experimentally. In the series (7.5) the phases of the wave are included in the structure factors $F(hkl)$, which are in general complex. The measurements that we are able to make give us $|F(hkl)|$, and not $F(hkl)$, so that the series cannot be evaluated directly in terms of quantities derived from observations of X-ray spectra. Relative shifts of the density distributions affect nothing that we can observe, and the structure of a crystal is not uniquely determined by the positions and intensities of the spectra that can be obtained from it by means of X-rays. To any set of observed spectra there correspond an infinite number of formally possible density distributions. Many of these are physically absurd, and in practice it is often possible to choose the correct one; but such a choice must always involve other than purely optical considerations. The solution of the optical problem is formally indeterminate. We shall return later to this point, but must now continue the discussion of the series.

(*e*) *Reality conditions for the density distribution:* According to (7.5), the density $\rho(x, y, z)$ is in general complex. Physically, a complex scattering factor means a change of phase of the radiation on scattering. If we suppose no such change to occur, $\rho(x, y, z)$ must be real at every point, and this will be so if $F(hkl) = F^*(\bar{h}\bar{k}\bar{l})$, that is to say if the structure factor of the spectrum obtained from one side of a set of planes

(hkl) is the conjugate complex quantity to that of the spectrum obtained by reflection from the opposite side. That this is a condition for the reality of $\rho(x, y, z)$ may be seen as follows. For the sake of brevity, we put

$$2\pi\left(\frac{hx}{a}+\frac{ky}{b}+\frac{lz}{c}\right)=\phi(hkl). \qquad (7.6)$$

Then the terms of the series (7.5) may be written in pairs such as

$$\mathrm{F}(hkl)e^{-i\phi}+\mathrm{F}(\bar{h}\bar{k}\bar{l})e^{+i\phi}.$$

If we put

$$\mathrm{F}(hkl)=|\mathrm{F}(hkl)|e^{i\delta}, \quad \mathrm{F}(\bar{h}\bar{k}\bar{l})=|\mathrm{F}(hkl)|e^{-i\delta}, \qquad (7.7)$$

which is equivalent to the assumption that $\mathrm{F}(hkl)$ and $\mathrm{F}(\bar{h}\bar{k}\bar{l})$ are conjugate complex quantities, the pair of terms reduces to

$$2|\mathrm{F}(hkl)|\cos(\phi-\delta),$$

which is real. The whole series is therefore real if for all values of h, k, and l

$$\mathrm{F}(\bar{h}\bar{k}\bar{l})=\mathrm{F}^*(hkl), \quad \text{or} \quad |\mathrm{F}(hkl)|=|\mathrm{F}(\bar{h}\bar{k}\bar{l})|. \qquad (7.8)$$

This is equivalent to the assumption that the intensity of reflection from one side of a set of planes is equal to that from the other side. We have already discussed this assumption in Chapter II, § 1 (*d*), and have seen that it ceases to be true if the crystal is polar, and if at the same time the phase change on scattering is different for the different atoms in the unit cell, which may be so if the radiation has a frequency very near to one of the absorption frequencies of an atom in the structure, so that dispersion effects are of importance; but this is a very special case, and in practice we can nearly always assume the relation (7.8) to be true. In what follows we shall always do so.

With assumption (7.8), the general form of the Fourier expansion for the density becomes

$$\rho(x, y, z)=\frac{|\mathrm{F}(000)|}{\mathrm{V}}+\frac{2}{\mathrm{V}}\sum_{hkl'}|\mathrm{F}(hkl)|\cos\{\phi(hkl)-\delta(hkl)\}, \qquad (7.9)$$

where the summation is to be taken over all the orders of spectra, except that corresponding positive and negative orders are not to be distinguished. It should be noted that there is only one term $|\mathrm{F}(000)|$, which is equal to the total electron population of the unit cell.

(*f*) *Case of crystal with symmetry centre :* As we have already seen, the phases $\delta(hkl)$ cannot be determined experimentally, and so equation (7.9) cannot be used directly. So long as the values of $|\mathrm{F}(hkl)|$ remain unchanged, the density distribution obtained using any arbitrarily assumed values of $\delta(hkl)$ would always give the same set of spectra, having the same positions and intensities. If we consider only such crystals as have a centre of symmetry in the unit cell, and use such a centre as the origin of co-ordinates, then $\rho(x, y, z)=\rho(\bar{x}, \bar{y}, \bar{z})$, and for this to be true every constituent wave from which ρ is built up must also be symmetrical about the origin. By (7.9), the term representing such a constituent wave may be written

$$2|\mathrm{F}(hkl)|(\cos\phi\cos\delta+\sin\phi\sin\delta). \tag{7.10}$$

Now $\cos\phi$ does not change sign when x, y, and z all change sign together, while $\sin\phi$ does; so that (7.10) can be symmetrical in x, y, z, only if $\sin\delta=0$, that is to say if $\delta=0$ or π. We can therefore write (7.9) in the form

$$\rho(x,y,z)=\frac{|\mathrm{F}(000)|}{\mathrm{V}}+\frac{2}{\mathrm{V}}\sum_{\overline{hkl}}\pm|\mathrm{F}(hkl)|\cos\phi(hkl), \tag{7.11}$$

where each coefficient could, formally, be either positive or negative. Spectra with indices hkl and $\bar{h}\bar{k}\bar{l}$ give identical terms, so that (7.11) can be written as a triple sum over all values of h, k, and l, in the form

$$\rho(x,y,z)=\frac{1}{\mathrm{V}}\sum_{h}\sum_{k=-\infty}^{\infty}\sum_{l}\pm|\mathrm{F}(hkl)|\cos 2\pi\left(\frac{hx}{a}+\frac{ky}{b}+\frac{lz}{c}\right), \tag{7.12}$$

for a crystal with a centre of symmetry.

Formula (7.12) means that if all the possible spectra from a crystal with a centre of symmetry have been measured, densities that would give a set of spectra with the same intensities may be obtained by using the moduli of $\mathrm{F}(hkl)$ calculated from the measured structure factors as coefficients, and taking each as either positive or negative. The degree of indeterminacy thus remains very high, but any density distribution deduced from the formula must satisfy certain conditions if it is to be a physically possible, as distinct from a purely formal solution of the problem, and this in practice limits the possible solutions very greatly. The discussion of this question, which introduces considerations having nothing to do with optical principles, must be postponed until we deal with actual examples.

(*g*) *The Fourier series and the reciprocal lattice :* The results of the preceding paragraphs take a very neat form when expressed in terms of the reciprocal lattice. The vector $\mathbf{r}^*(hkl)$ to the point (h, k, l) of the reciprocal lattice is perpendicular to the planes (hkl) of the crystal lattice, and its magnitude is the reciprocal of the spacing of these planes. It follows therefore from the results of § 1(*b*) of this chapter that the direction of the reciprocal vector $\mathbf{r}^*(hkl)$ is also the direction of the wave-normal of the Fourier wave (hkl), and that the length of the reciprocal vector is numerically equal to the number of waves per unit length in the direction of the wave-normal, that is to say, to the wave-number of the Fourier wave. If, as was suggested by Ewald,[2] we also assign to each point of the reciprocal lattice a weight equal to the structure factor of the corresponding spectrum, this weight will also be equal to the coefficient of the corresponding term in the Fourier series, and the reciprocal lattice will then constitute a complete formal representation of the spectra given by the crystal lattice, and of the set of Fourier waves by the sum of which the density of the scattering matter that produces the spectra can be represented.

Using reciprocal vectors, we may express some of the results already obtained in a rather different way. Let $\mathbf{r}_m^*$ be the reciprocal vector to the point (h, k, l) of the reciprocal lattice, the single symbol m being now used to represent the triplet of numbers h, k, l. If $\mathbf{r}$ is the vector from the origin to the point (x, y, z) in the crystal,

$$\mathbf{r}\cdot\mathbf{r}_m^* = (\mathbf{x}+\mathbf{y}+\mathbf{z})\cdot(h\mathbf{a}^*+k\mathbf{b}^*+l\mathbf{c}^*)$$
$$= h\mathbf{a}^*\cdot\mathbf{x} + k\mathbf{b}^*\cdot\mathbf{y} + l\mathbf{c}^*\cdot\mathbf{z},$$

by the properties of the reciprocal vectors (see Appendix II § (*a*)). Now $\mathbf{a}^*\cdot\mathbf{x} = (\mathbf{a}\cdot\mathbf{a}^*)x/a = x/a$, and similarly $\mathbf{b}^*\cdot\mathbf{y} = y/b$, $\mathbf{c}^*\cdot\mathbf{z} = z/c$, so that

$$\mathbf{r}\cdot\mathbf{r}_m^* = hx/a + ky/b + lz/c. \tag{7.13}$$

The Fourier series (7.5) can therefore be written in the form

$$\rho(\mathbf{r}) = \frac{1}{\mathrm{V}}\sum_m \mathrm{F}(m)e^{-2\pi i(\mathbf{r}\cdot\mathbf{r}_m^*)} \tag{7.14}$$

where
$$\mathrm{F}(m) = \int \rho(\mathbf{r})e^{2\pi i(\mathbf{r}\cdot\mathbf{r}_m^*)}\,dv. \tag{7.15}$$

The value of the phase angle ϕ_m is given by

$$\phi_m = 2\pi(\mathbf{r}\cdot\mathbf{r}_m^*) = 2\pi(\mathbf{r}\cdot\mathbf{n}^*)/d_m, \tag{7.16}$$

where $\mathbf{n}^*$ is the unit vector in the direction of the reciprocal-lattice vector $\mathbf{r}_m^*$, and d_m is the spacing of the planes (hkl). The phase is therefore constant as long as the scalar product $\mathbf{r}\cdot\mathbf{n}^*$ is constant. This scalar product is the projection of the vector $\mathbf{r}$ on the direction of $\mathbf{n}^*$, and is constant so long as the extremity of $\mathbf{r}$ lies on a plane perpendicular to $\mathbf{n}^*$. The wave-fronts of the Fourier waves are therefore planes perpendicular to $\mathbf{n}^*$, and their wave-length is d_m or $1/|\mathbf{r}_m^*|$. The reciprocal lattice thus provides us with a convenient and vivid way of summarising the information given by the X-ray spectra. We shall have occasion to use this device at a later stage.

(*h*) *Examples of the use of triple series:* In the triple series we have so far discussed there is one coefficient for each spectrum given by the crystal. In practice, therefore, a series may have many hundreds of terms, the coefficients of which have all to be determined by observation. To evaluate the series for enough points in the unit cell to give a complete picture of the structure would therefore be a formidable task, the labour of which was for some time considered to be prohibitive, a point of view which led to the development and use of other forms of the series, to be discussed below. Triple series are, however, being increasingly used in crystal analysis, not as a rule to give a complete point-by-point representation of the structure, but rather to obtain the density of scattering matter along lines, or over planes, which may be expected to give information of importance.

Suppose, for example, the electron density along the a edge of the unit cell were required. We should then determine the value of $\rho(x, 0, 0)$ for a series of values of x close enough together to give a picture of the way in which ρ varies along the cell edge. From (7.9), we may write

$$\rho(x, 0\ 0) = \frac{|\mathrm{F}(000)|}{\mathrm{V}} + \frac{2}{\mathrm{V}}\sum_{\overline{hkl}} |\mathrm{F}(hkl)| \cos\left\{\frac{2\pi hx}{a} - \delta(hkl)\right\}. \quad (7.17)$$

If the crystal has a centre of symmetry at the origin, the expression takes the simpler form

$$\rho(x, 0, 0) = \frac{1}{\mathrm{V}} \sum_{h} \sum_{k}^{\infty} \sum_{l} \pm|\mathrm{F}(hkl)| \cos\frac{2\pi hx}{a}. \quad (7.18)$$
$$-\infty$$

Series of this type were first used by Havighurst,[12] who determined the density in some simple crystals, mostly belonging to the cubic system, along lines such as the cell edge [100], the cell diagonal [111], and the face diagonal [110]. All the crystals used had high symmetry, so that certain relationships existed between the values of F that allowed further simplification of the summation, which it is not necessary at this stage to consider in detail.

Curves obtained in this way by Havighurst for the density along the cube edge in rock salt, sodium fluoride, and lithium fluoride are shown in fig. 121. All the crystals are of the rock-salt type, and the alternation of peaks of different height, corresponding to the two types of atom, which occur alternately along the cube edge, is well shown, as well as the very small density between the atoms.

Formally, the sign before any of the coefficients F(hkl) in (7.18) may be either positive or negative, but in crystals of such a simple type it is quite easy to determine the correct sign to use in any particular case; for very simple considerations lead with practical certainty to the structure, and this, together with a knowledge of which atom of the pair has the larger scattering factor, at once determines the signs of the coefficients. For example, if in the case of rock salt the origin is taken at the centre of a chlorine atom, the structure factors are $4(f_{\mathrm{Cl}}+f_{\mathrm{Na}})$ for spectra with even indices, and $4(f_{\mathrm{Cl}}-f_{\mathrm{Na}})$ for spectra with odd indices, and are zero for any other type; and, since $f_{\mathrm{Cl}} > f_{\mathrm{Na}}$ for all angles of scattering, all the coefficients are positive.

It is plain that with a more complicated structure it would have been less easy to decide on the signs of the coefficients. Moreover, the sizes of the unit cells of most crystals are much greater than those of the simple crystals used by Havighurst, and the number of spectra of the general type hkl then becomes very much larger.

Another method of using the triple series is to evaluate the density

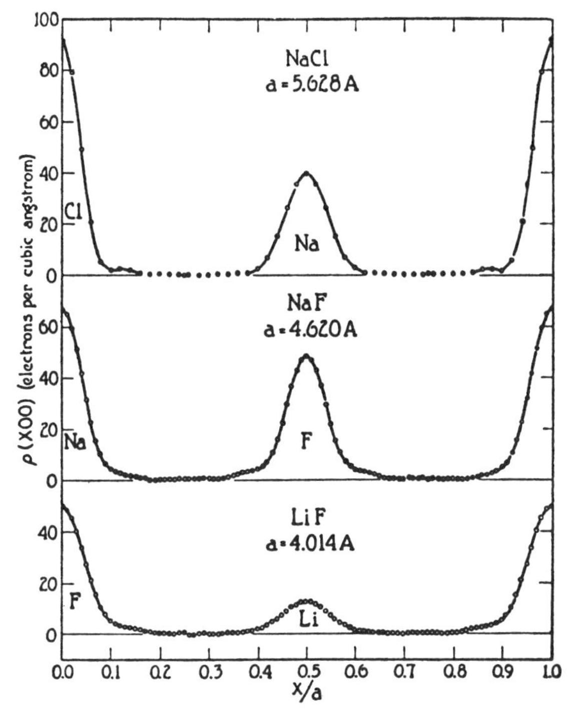

FIG. 121. Electron density along the cube-edge in rock salt, sodium fluoride, and lithium fluoride (Havighurst) (*Phys. Rev.*, **29**, 1 (1927)).

over a plane, so as to obtain a section of the structure at some particular level. This method is being increasingly used to obtain detailed information about structures that have been approximately determined by one of the less laborious methods to be discussed later. We shall consider an example in § 1(*u*). If the density in the plane $z=z_1$ is required, the series to be evaluated is that for $\rho(x, y, z_1)$ or

$$\rho(x, y, z_1)=\frac{|F(000)|}{V}+\frac{2}{V}\sum_{hkl}|F(hkl)|\cos\left\{2\pi\left(\frac{hx}{a}+\frac{ky}{b}+\frac{lz_1}{c}\right)-\delta(hkl)\right\}, \tag{7.19}$$

with z_1 constant. The summation must be carried out for enough values of x and y to give a picture of the distribution of ρ in the plane $z=z_1$. Here again it is possible to simplify the summation if the crystal has a centre of symmetry, or if the symmetry imposes certain relationships between the values of F for different spectra.

(*i*) *The double series:* A method of using Fourier series which has since become a standard method in the analysis of crystals was introduced in 1929 by W. L. Bragg.[9] Instead of the volume density, the surface density of the projection of all the scattering matter contained in the unit cell upon a plane, usually one of the axial planes of the crystal, is determined. If such a projection can be made on more than one plane, an amount of information sufficient to determine the structure can often be obtained, and the work involved in doing this is far less than that involved in an analysis by means of triple series. Moreover, even if it is intended to use the triple series later, it is usual to make a projection of this type as a preliminary measure.

In fig. 122, a unit cell of the structure with edges of length a, b, c, parallel to x, y, z, is shown. Consider an elementary parallelepiped, of length c parallel to z, erected on a base with edges dx and dy in the

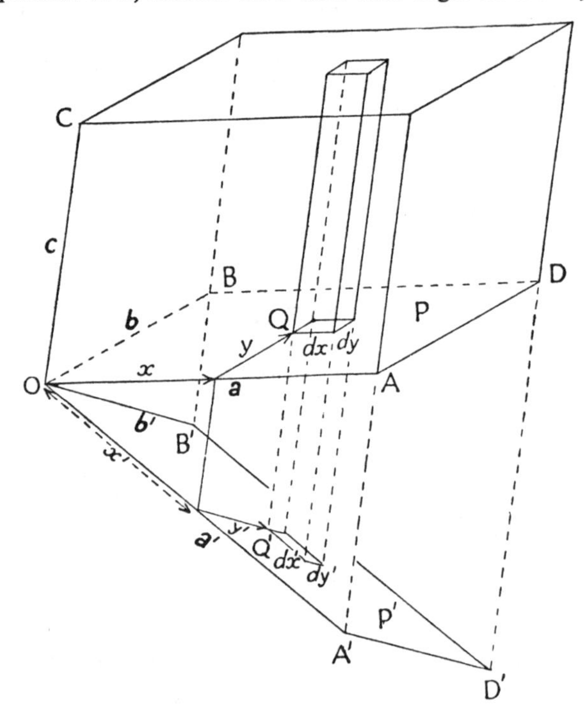

FIG. 122

neighbourhood of the point Q(x, y), in the plane $z=0$. The total scattering matter in this parallelepiped, divided by the area of its base, we shall denote by $\sigma(x, y)$, and this we shall call the surface density at the point (x, y) when the structure is projected parallel to c on to the ab plane. This defines the sense in which the word 'projection' is used in this connection.

Let Nabc be the volume of the unit cell, and A the area of the face ab. An elementary parallelepiped with edges dx, dy, dz has a volume

N $dx\,dy\,dz$, and the element of area with edges dx, dy has an area $(\mathrm{A}/ab)\,dx\,dy$. Thus

$$\sigma(x, y) = \mathrm{N}\,dx\,dy \int_0^c \rho(x, y, z)\,dz / (\mathrm{A}/ab)\,dx\,dy$$

$$= \mathrm{N}(ab/\mathrm{A}) \int_0^c \rho(x, y, z)\,dz. \tag{7.20}$$

We now substitute the series (7.5) for $\rho(x, y, z)$ in the integrand of (7.20), and perform the integration with respect to z. All the terms vanish on integration, except those for which $l=0$, and for these the integration simply introduces a factor c. Thus, since $\mathrm{V}=\mathrm{N}abc$, we can write for the density of the projection

$$\sigma(x, y) = \frac{1}{\mathrm{A}} \sum_{h}\sum_{k}{}_{-\infty}^{+\infty} \mathrm{F}(hk0)\, e^{-2\pi i\left(\frac{hx}{a}+\frac{ky}{b}\right)}. \tag{7.21}$$

For the projection in a direction parallel to c onto the ab plane, only the structure factors F$(hk0)$, belonging to those crystal planes for which c is the zone axis, that is to say to the planes of the zone [001], are required. It may be necessary to measure a hundred or more spectra, but this is quite practicable, and the summation of the series involves no excessive amount of work, particularly when certain shortened methods, some of which we shall discuss later, are used.

The reality conditions discussed in § 1 (e) of this chapter apply here also. The series is real if $|\mathrm{F}(hk0)| = |\mathrm{F}(\bar{h}\bar{k}0)|$, and the corresponding form for $\sigma(x, y)$ is

$$\sigma(x, y) = \frac{\mathrm{F}(000)}{\mathrm{A}} + \frac{2}{\mathrm{A}} \sum_{h,k} |\mathrm{F}(hk0)| \cos\{\phi(hk0) - \delta(hk0)\}, \tag{7.22}$$

where again the sum is to be taken over all orders of spectra belonging to the zone [001], without distinction between corresponding positive and negative orders. If the projection has a centre of symmetry, as it sometimes may, even when the structure as a whole has none, and if this centre is taken as the origin of the co-ordinates x, y in the projection, the series reduces to a form corresponding to (7.12) :

$$\sigma(x, y) = \frac{1}{\mathrm{A}} \sum_{h}\sum_{k}{}_{-\infty}^{+\infty} \pm|\mathrm{F}(hk0)| \cos 2\pi\left(\frac{hx}{a}+\frac{ky}{b}\right), \tag{7.23}$$

where the indeterminacy of the phases of the constituent waves is reduced to an ambiguity in the signs of the coefficients.

In practice, it is usual as we have done here, to project the structure onto one of the axial planes of the crystal, but it is evident that any plane may be used. Let us, for brevity, refer to the axial plane as the plane P, and let P′(A′OB′D′) be any other plane, making any angle with the direction of c. Let Q′(x', y') be the projection on the plane P′ of Q(x, y) in the plane P (fig. 122), the co-ordinates x' and y' being referred to the axes a', b', which are the projections of the axes a and b on the plane P′. Then

$$x'/a' = x/a, \quad y'/b' = y/b.$$

If $\sigma'(x', y')$ is the density of the projection on the plane P′ in the neighbourhood of Q′, it is clear that

$$A'\sigma'(x', y') = A\sigma(x, y),$$

where A′ is the area of the projection of the face ab of the unit cell, whose area is A, onto the plane P′. Formula (7.21) therefore also applies to the density of the projection at any point in the plane P′ if x', y', a', b', and A′ are substituted for x, y, a, b, and A, respectively.

Formally, there is some advantage in projecting the structure onto a plane perpendicular to the direction of projection. When this direction is that of the c axis, the corresponding plane of projection is that containing the a^* and b^* axes of the reciprocal lattice, which we may conveniently call the plane P*. In fig. 123 let $OA_1D_1B_1$ be the projection of the ab face of the unit cell (OADB), fig. 122, on the plane P*. The distances between the pairs of lines OB_1, A_1D_1, and OA_1, B_1D_1, are respectively the a and b spacings of the lattice, $d(100)$ and $d(010)$. $OA_1(=a_1)$ and $OB_1(=b_1)$ form the primitive translations of a net in two dimensions, lying in the same plane as the net a^*b^* of the reciprocal lattice. These two nets are reciprocal to one another; for the direction of the a^* axis is perpendicular to b_1, which lies in the bc plane of the crystal lattice, and $1/a^*$ is equal to $d(100)$, which is also the a_1 spacing of the a_1b_1 net; and similarly b^* is perpendicular to a_1, and $1/b^*$ is equal to $d(010)$, and so to the b_1 spacing of the a_1b_1 net.

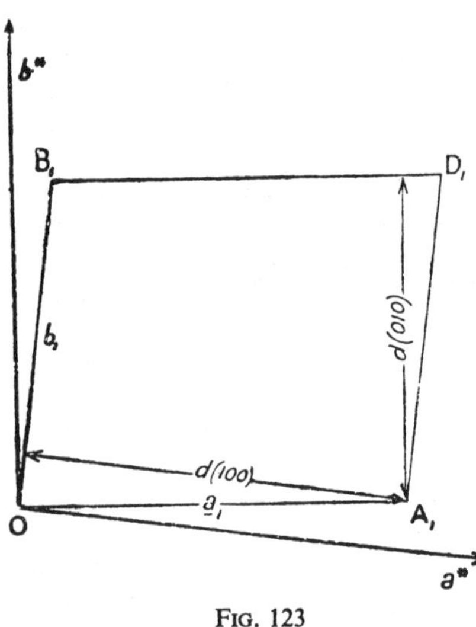

FIG. 123

The density $\sigma_1(x_1, y_1)$ of the projection at the point (x_1, y_1) in the plane P* is given by

$$\sigma_1(x_1, y_1) = \frac{1}{A_1} \sum_{h}^{+\infty}\sum_{k} F(hk0) e^{-2\pi i \left(\frac{hx_1}{a_1} + \frac{ky_1}{b_1}\right)}. \tag{7.24}$$

(lower summation limit: $-\infty$)

One Fourier wave corresponds to each spectrum of the type $hk0$, and to each of these there corresponds a point of the reciprocal net a^*b^*. The vector $h\mathbf{a}^* + k\mathbf{b}^*$ from the origin to the point (h, k) of this net lies in the plane P*, and is normal to the corresponding Fourier wave-fronts. The points of the reciprocal lattice associated with the c-direction of projection, or the direction of the zone axis [001], are those of the net $(h, k, 0)$. The projection of the structure onto the plane of this net gives a doubly periodic density distribution based on the net with translations $a_1 b_1$, which is reciprocal to the net $(h, k, 0)$. We shall return to this correspondence later.

The projection need not of course be made along one of the principal axes. Any zone axis [*uvw*] may be used, and the spectra contributing to the sum would be those belonging to the zone [*uvw*], whose indices satisfy the relation $hu + kv + lw = 0$ (see Appendix II, § (*f*)).

W. L. Bragg [9] first applied the double series to find the structure of diopside, and an account of this determination will be found in *The Crystalline State*, Vol. I, Chapter IX. A general account of the method, and of some more recent examples of its application, will be found in a report by Robertson [14] published by the Physical Society.

(*j*) *The projection of a slice of the crystal cell*: It is possible to project a portion of the unit cell instead of the whole of it. For example, we might project all the scattering matter lying between the planes $z = z_1$ and $z = z_2$ onto the c plane. To do this, all that is necessary is to take the integration with respect to z in (7.20) from z_1 to z_2, instead of from 0 to c. Let us call the projection so obtained $S_{12}(x, y)$. Then, by (7.20),

$$S_{12}(x, y) = N(ab/A) \int_{z_1}^{z_2} \rho(x, y, z)\, dz. \tag{7.25}$$

The actual integral that occurs in (7.25) is, by (7.5),

$$\int_{z_1}^{z_2} e^{-2\pi i l z/c}\, dz = \frac{c}{2\pi i l}\left\{ e^{-\frac{2\pi i l z_1}{c}} - e^{-\frac{2\pi i l z_2}{c}} \right\}$$

$$= \frac{c}{\pi l} e^{\pi i l (z_1 + z_2)/c} \sin\{\pi l (z_2 - z_1)/c\}. \tag{7.26}$$

From (7.5) and (7.25), therefore,

$$S_{12}(x, y) = \frac{z_2 - z_1}{cA} \sum_h \sum_k \sum_l{}_{-\infty}^{+\infty} F(hkl) e^{-2\pi i \left(\frac{hx}{a} + \frac{ky}{b} + \frac{l(z_1+z_2)}{2c}\right)} \frac{\sin\{\pi l (z_2 - z_1)/c\}}{\pi l (z_2 - z_1)/c} \tag{7.27}$$

$$= \frac{z_2 - z_1}{cA} \sum_h \sum_k \sum_l{}_{-\infty}^{+\infty} |F(hkl)| \cos\left\{ 2\pi\left(\frac{hx}{a} + \frac{ky}{b} + \frac{l(z_1 + z_2)}{2c}\right) - \delta(hkl)\right\} \times \frac{\sin\{\pi l (z_2 - z_1)/c\}}{\pi l (z_2 - z_1)/c}. \tag{7.28}$$

The last factor in (7.27) and (7.28) approaches unity as $z_2 - z_1$ approaches zero, and in this case $S_{12}(x, y)$ becomes in the limit equal to the density in the plane $z = z_1$, and is equivalent to (7.19). If, on the other hand,

$z_2 = z_1 + c$, the slice includes the whole unit cell. The last factor in (7.27) is then zero unless $l = 0$, in which case it is unity, and the expression reduces at once to (7.21). An expression equivalent to (7.27) has been given by Booth.[13] The series for $S_{12}(x, y)$ is essentially a three-dimensional series, and requires a knowledge of all the coefficients $F(hkl)$.

(*k*) *Methods of handling the double series :* The practical application of the methods of Fourier synthesis is a subject belonging more properly to a volume on experimental technique, but a few points of some general importance may be dealt with here. The summation even of a double series involves a considerable amount of numerical work, and it is important to reduce this as much as possible by using shortened methods such as those devised by Beevers and Lipson,[15] and by Robertson.[16] We shall here give a short outline of one such method. For details reference may be made to a paper by Lipson and Beevers.[17]

In considering how to sum the series it is of great assistance to think of the Fourier components as represented by the points of the reciprocal lattice, to each of which a weight equal to the structure factor of the corresponding spectrum has been given. For the two-dimensional series (7.22) the points concerned are those lying in the reciprocal net $(h, k, 0)$, which contains the origin, and is perpendicular to the zone axis [001] of the crystal lattice. The structure factor $F(hk0)$ associated with the point (h, k) of this net is either equal to, or is the conjugate complex quantity of $F(\bar{h}\bar{k}0)$ associated with the point $(\bar{h}, \bar{k})$, and the two points together give the term

$$2|F(hk0)| \cos \{hu + kv - \delta(hk)\}$$

in the series for $A\sigma(x, y)$. We have here introduced for the sake of brevity the notation

$$u = 2\pi x/a, \qquad v = 2\pi y/b, \qquad w = 2\pi z/c, \tag{7.29}$$

and have also written $\delta(hk)$, instead of $\delta(hk0)$, for the phase angle. All the terms of the series are now included by considering only those points in the reciprocal net for which h is positive or zero or, of course, alternatively, those for which k is positive or zero. We shall here consider the former case.

To the origin, there corresponds the term $|F(000)|$. The points along the a^* axis, excluding the origin, correspond to a summation with respect to h only, and give a contribution

$$2\sum_{1}^{H} |F(h00)| \cos \{hu - \delta(h0)\}, \tag{7.30}$$

and, similarly, the points along the b^* axis give a contribution

$$2\sum_{1}^{K} |F(0k0)| \cos \{kv - \delta(0k)\}. \tag{7.31}$$

The remaining terms are associated with points lying on neither axis, for which neither h nor k is zero. The points with positive k give a contribution

$$2\sum_{1}^{H}\sum_{1}^{K}|F(hk0)| \cos\{hu+kv-\delta(hk)\}, \tag{7.32}$$

and those with negative k a contribution

$$2\sum_{1}^{H}\sum_{1}^{K}|F(h\bar{k}0)| \cos\{hu-kv-\delta(h\bar{k})\}. \tag{7.33}$$

The whole series is given by the sum of these contributions. In these expressions, H and K have been used to denote the largest values of h and k occurring in the summations. In practice, these are seldom greater than 15 or 20, instead of being infinite as in the theoretical series. It will be necessary to consider at a later stage what is the effect of this limitation.

We now introduce the abbreviations

$$C(hk)=|F(hk0)| \cos\delta(hk), \qquad S(hk)=|F(hk0)| \sin\delta(hk), \tag{7.34}$$

and expand the cosines in the expressions (7.29) to (7.33). After a certain amount of reduction, the sum of the series can be written in the form

$$A\sigma(u,v)=A(0,u)+\sum_{1}^{K}A(k,u)\cos kv+\sum_{1}^{K}B(k,u)\sin kv, \tag{7.35}$$

where

$$A(0,u)=F(000)+2\sum_{1}^{H}\{C(h0)\cos hu+S(h0)\sin hu\},$$

$$A(k,u)=2C(0k)+2\sum_{1}^{H}\{[C(hk)+C(h\bar{k})]\cos hu+[S(hk)+S(h\bar{k})]\sin hu\},$$

$$B(k,u)=2S(0k)+2\sum_{1}^{H}\{[S(hk)-S(h\bar{k})]\cos hu-[C(hk)-C(h\bar{k})]\sin hu\}. \tag{7.36}$$

The coefficients $A(0,u)$, $A(k,u)$, and $B(k,u)$ have first to be determined by summations carried out over h for each value of k, and finally the summation of (7.35) with respect to k must be made.* This has to be done for each point (u, v) (or (x, y)) for which the value of σ is required. A division of the cell edge into 60 parts, so that u and v vary in steps of 6 degrees is suitable if the maximum values of the indices h and k lie between 10 and 20, but other intervals may be necessary in some cases.

Lipson and Beevers have greatly facilitated the work of summing these series by preparing sets of cardboard strips upon which values of

* The series can of course equally well be written in such a form that the summation with respect to h is the final operation. The choice will depend on the circumstances of the particular case. Reference may be made to the paper by Lipson and Beevers (*loc. cit.*) for practical details.

A sin hu and A cos hu are printed. Such a strip is represented in fig. 124.

99 S 3	0	31	58	80	94	99	80	58	31	0	$\overline{31}$	$\overline{58}$	$\overline{80}$	$\overline{94}$	$\overline{99}$

FIG. 124. Values of 99 sin $3u$ for values of u at 6° intervals from $u=0°$ to $u=90°$

The strip shown as an example in the figure gives the values of 99 sin $3u$ for values of u at intervals of 6°. Each strip extends from $u=0°$ to $u=90°$, covering one quarter of the side of the unit cell. The index values h of the strips go up to 20, and the amplitudes A range from -99 to $+99$ at intervals of unity. Sets of strips are prepared for both sines and cosines, and they fit into specially constructed boxes, so that the strip for any amplitude and any index can be instantly found. In summing any series, say of the type $\sum_h \text{A}(h) \sin hu$, all that has to be done is to take from the sine box those strips with the appropriate values of A(h) and h, place them one below the other so that the vertical columns correspond, and add up the columns of figures. This gives at once the values of the sum for a set of 16 successive values of u. The same strips can, of course, be used to carry out summations with respect to k. It is not necessary to have strips with values of u or v higher than 90°, since the values of the sines and cosines repeat themselves in known ways for larger angles, and can be written down directly once the values up to 90° are known. For example, cos hu is symmetrical about $u=90°$ if h is even, and antisymmetrical if h is odd, and there are analogous rules for the sines; so that if odd and even values of h and k are kept separate during the summations, there is no difficulty in extending the computation to cover the whole area of the cell-face, although the symmetry of the projection will usually make it unnecessary to carry out the work for more than a fraction of the area.

The final result of the calculations is to give a set of numbers representing the values of the density $\sigma(x, y)$ at points spaced regularly and fairly closely over the whole area of the projection. These numbers are plotted at the appropriate points in a scale diagram, and contour lines based upon them are drawn for values of the density differing by suitable intervals. Such a diagram resembles a contour map of a mountainous country, the 'mountains' corresponding to concentrations of density of the scattering matter, that is to say to the presence of atoms. Even although we are dealing with the projection of the whole of the matter in the unit cell upon one plane, so that overlapping of the atoms is to be expected, it is surprising how often the individual atoms are resolved in such a way that the projection gives a vivid picture of the molecule. An example of this is given by the projection on the ac plane of the unit cell of p-dinitrobenzene, fig. 125, due to

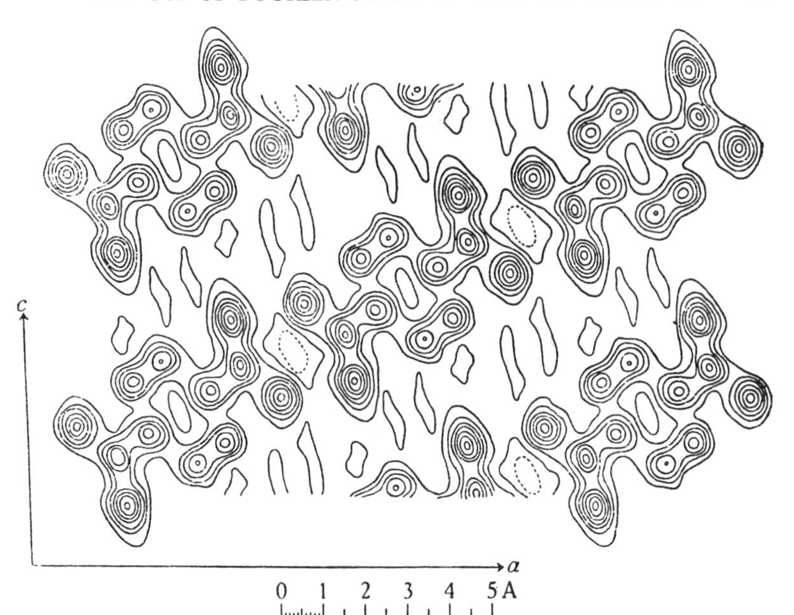

FIG. 125. Projection of the structure of *p*-dinitrobenzene in the direction of the *b* axis. The crystal is monoclinic (James, King & Horrocks)
(*Proc. Roy. Soc.*, **A**, **153**, 225 (1935))

James, King and Horrocks,[18] in which the benzene rings and the nitro groups are very clearly shown with no overlapping. The rings are actually very nearly regular hexagons, but they are inclined to the direction of projection, and so appear distorted. In this case, a fairly clear and well-resolved set of contours could also be obtained by projection on the *ab* plane, so that the co-ordinates of the atoms in the unit cell could be determined from the two projections taken together.

Very often the projection of the density on a plane has a centre of symmetry, and if this is taken as the origin of the co-ordinates x, y, or u, v, the expressions (7.35) and (7.36) may be considerably simplified; for in this case $\delta(hk)=0$ or π, $\mathrm{F}(hk0)$ is real, $\mathrm{C}(hk)=\mathrm{F}(hk0)$, and $\mathrm{S}(hk)=0$, so that the equations (7.36) become

$$\begin{aligned} \mathrm{A}(0, u) &= \mathrm{F}(000) + 2\sum_{1}^{\mathrm{H}} \mathrm{F}(h00) \cos hu, \\ \mathrm{A}(k, u) &= 2\mathrm{F}(0k0) + 2\sum_{1}^{\mathrm{H}} \{\mathrm{F}(hk0) + \mathrm{F}(h\bar{k}0)\} \cos hu, \\ -\mathrm{B}(k, u) &= 2\sum_{1}^{\mathrm{H}} \{\mathrm{F}(hk0) - \mathrm{F}(h\bar{k}0)\} \sin hu. \end{aligned} \tag{7.37}$$

The projection may have other symmetry elements, involving relations between the structure factors that still further simplify the summations. For example, if $\mathrm{F}(hk0)=\mathrm{F}(h\bar{k}0)$, as it will if the projection has a symmetry plane perpendicular to b, for which a two-fold rotation axis in the crystal parallel to a is a sufficient condition, $B(k, u) = 0$, and the sum becomes

$$A\sigma(u, v) = F(000) + 2\sum_{1}^{H} F(h00)\cos hu + 2\sum_{1}^{K} F(0k0)\cos kv$$
$$+ 4\sum_{1}^{H}\sum_{1}^{K} F(hk0)\cos hu \cos kv. \qquad (7.38)$$

(*l*) *Practical application of the double series when the projection is centrosymmetrical:* In the expressions (7.37) or (7.38) the coefficients F are all real, but may be either positive or negative; and which sign to use in any given case can be determined only by considerations other than those derived from observations of the diffraction of X-rays by the crystal. A structure has in effect to be guessed, and the structure factors corresponding to this tentative arrangement of the atoms must then be worked out from the assumed atomic co-ordinates, with values of the atomic scattering factors taken from tables such as those given by James and Brindley.[19] The algebraic signs of the structure factors so determined are then given to the corresponding Fourier coefficients, but for the numerical values of the coefficients the *measured* values of $|F(hk0)|$ are used. The final solution when the projection has a centre of symmetry is thus independent of any theoretical values of f; for these are used merely as an aid to determining the correct signs of the coefficients. The assumed values of the co-ordinates will usually be close enough to the true ones to give the signs of the important coefficients correctly, but the preliminary Fourier projection made using these will generally indicate atomic positions different enough from those assumed to change the signs of some of the smaller coefficients. A revised projection must now be made with the consequent corrections to the signs of the coefficients and, if necessary as a result of this, still another. Although the structure cannot be worked out directly, and processes of trial and error must still be used, the method of Fourier synthesis is nevertheless greatly superior to the method of merely comparing the values of observed and calculated intensities, since it uses in a very direct way all the information at our disposal as a result of the observations of the spectra; and the superiority of the method is particularly marked where the structure has a number of parameters. Once a structure near enough to the truth to give the signs of the important coefficients has been guessed, the variation of the parameters to give the best fit with observation is done automatically by means of the series, at all events in the case of a crystal with a centre of symmetry, in which only observed numerical data are used in the final solution.

The type of information used to obtain a preliminary assumed structure varies from crystal to crystal. There is no routine method of procedure, and each crystal presents a separate problem, with its own points of interest and difficulty, often affording considerable scope for the experience and ingenuity of the investigator. The size and probable shape of the molecule may be known from previous work on substances of similar type, and the possibilities of packing together such molecules into the space available in the unit cell, with due regard to any restrictions that may be imposed by the symmetry of the space-group to which the structure belongs, must be investigated. Such considerations sometimes set very close limits to the possible positions of the atoms. A standard procedure is to examine the crystal optically, and if possible to measure its three principal refractive indices. If a characteristic of the structure is the packing of flat molecules, such for example as those of the organic ring compounds, into layers, the crystal is likely to show strong negative double refraction, with the small refractive index associated with a direction of the electric vector of the light waves perpendicular to the sheets of molecules. Valuable information as to the orientation of flat or chain-like molecules may sometimes be obtained by examining the diffuse reflections due to thermal movement. The methods used are in fact those used in any structure determination by the method of trial and error, and their systematic discussion belongs to a volume on experimental technique, and not to the present chapter.

(*m*) *Some examples of the use of the double series when the projection has a centre of symmetry:* We shall now consider more closely two actual examples of the application of the double Fourier series, with the object rather of illustrating the differences in the treatment of individual cases than of indicating any general method of procedure.

The first case we shall consider is that of potassium di-hydrogen phosphate, KH_2PO_4, a relatively simple structure, to which the Fourier method was applied by West.[20] The unit cell of the crystal is body-centred tetragonal, and the positions of the K and P atoms are fixed by the symmetry of the crystal at definite points in the structure. The oxygen atoms surround the phosphorus atoms in groups of four, arranged nearly at the corners of a regular tetrahedron, each group having a phosphorus atom at its centre. We shall not here consider the nature of the crystal unit in detail. The crystal had already been analysed by Hendricks,[21] and the aim of West's work was to apply the Fourier method to the accurate determination of the shape and size of the PO_4 group.

To illustrate the particular method used by West to determine the signs of the Fourier coefficients we need only consider the projection along the tetragonal axis onto the plane (001), which is shown diagrammatically in fig. 126. The black circles in fig. 126 *a* represent the atoms of K and P, which in this projection fall on top of one another, their positions being fixed by the symmetry of the cell. The O atoms in the projection are arranged in squares about the K + P positions, and are denoted by white circles in fig. 126 *a*. It will be seen that the projection has a centre of symmetry, but no planes of symmetry. The structure factors F($hk0$), which are those required for this projection, fall into two classes, (i) those for which h is even, to which all the atoms contribute, and (ii) those for which h is odd, to which K and P make no contribution, the spectra of this type being due to oxygen alone, and therefore all weak in comparison with those of type (i). The contributions of K and P to the relatively strong spectra of type (i) are all in phase, and we may conclude that they govern the signs of the structure factors for these spectra. We may therefore take the signs of the coefficients for the terms with h even as being those due to the K and P atoms alone, and these are very easily calculated. A preliminary projection was therefore made using terms in the sum with h even only. The type of projection obtained is illustrated in fig. 126 *b*, which gives only the neighbourhood of one of the K + P peaks, that enclosed in dotted lines in fig. 126 *a*. It will be seen that around this peak there are eight smaller peaks, symmetrically disposed about the diagonals of the cell, instead of four. This is because the structure that would give no spectra with h odd, which is what this projection corresponds to, would have this symmetry about the diagonals. The diffracting matter of the four oxygen atoms would have to be disposed symmetrically in eight maxima about the K + P peak, instead of in four. But the co-ordinates of the O atoms can be read off from this projection, and using them we can calculate the signs of the coefficients with h odd. A projection made using these alone is shown in fig. 126 *c*. This gives four positive maxima in the positions of four of the peaks of fig. 126 *b*. The peaks symmetrical with these about the cell diagonals have, however, become hollows of negative density. The true distribution is got by superposing figs. 126 *b* and 126 *c*, when the corresponding positive and negative densities neutralise each other, and we are left with four oxygen peaks only around each K + P peak, each with double the density of the eight peaks obtained from the spectra with h even. This is an ingenious solution of a particular problem, which might well have applications to

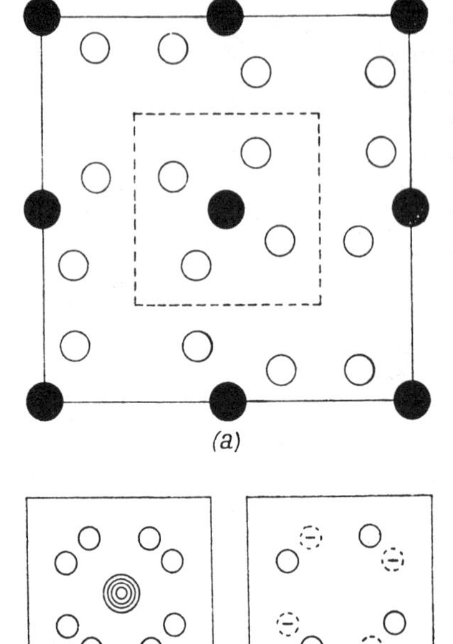

FIG. 126

other fairly simple structures. The attack is relatively direct, and the superiority of the Fourier method to that of detailed comparison of observed and calculated spectra is evident.

A particularly beautiful example of the method is the analysis of the phthalocyanine structure by Robertson.[22] Although this molecule is very complicated, its analysis is perhaps the most direct yet carried out. There are 32 carbon atoms and 8 nitrogen atoms in the molecule, but it was possible to obtain the electron-density map directly by means of the double Fourier series without making any initial assumptions as to the form of the molecule, and indeed, as Robertson points out, without assuming the existence of separate atoms at all.

The molecule of free phthalocyanine has a centre of symmetry at which no atom is situated, but it is possible to prepare a series of compounds isomorphous with one another and with the free compound, which contain a metal atom in this position. It can be proved quite definitely, by X-ray methods, that the metal atom enters the compound at the centre of symmetry. Because the series of compounds is isomorphous, and the dimensions of the unit cell remain closely similar, we may infer a general similarity of structure. Since the metal atom lies at the centre of symmetry, it is easy to calculate the magnitude and the sign of its contribution to the structure factor of any spectrum, and it is plain that for all spectra, except perhaps for some of quite high order, the corresponding Fourier coefficients will be positive, that is to say the Fourier waves will have crests and not troughs at the centre of symmetry. Suppose now we compare a spectrum given by a metallic compound with the corresponding spectrum given by the free compound. If the intensity of the spectrum from the metallic compound is greater than that from the free compound we know that the contributions to the structure factor from the metal, and from the non-metallic part of the molecule, have the same sign. If, on the other hand, the intensity of the spectrum from the metallic compound is less than the intensity of that from the free compound we know that the contribution from the non-metallic part, that is to say in effect from the free molecule, must be opposite in sign to that from the metal. Thus it is possible to determine quite directly the signs of the coefficients for the free compound, and hence to carry out a direct summation of the Fourier series. The projection parallel to the b axis of the crystal is shown in fig. 127, together with a key diagram. The peaks corresponding to the forty atoms of carbon and nitrogen show out very clearly. It must be remembered that this is a perfectly direct optical determination, with no initial assumptions as to the nature or the number of atoms in the molecule, and it forms a striking confirmation of the general correctness of the Fourier method of representation. Unfortunately, the method used to analyse the structure is not one of general application, since it depends on the existence of a series of isomorphous compounds ; but it is nevertheless a good example of

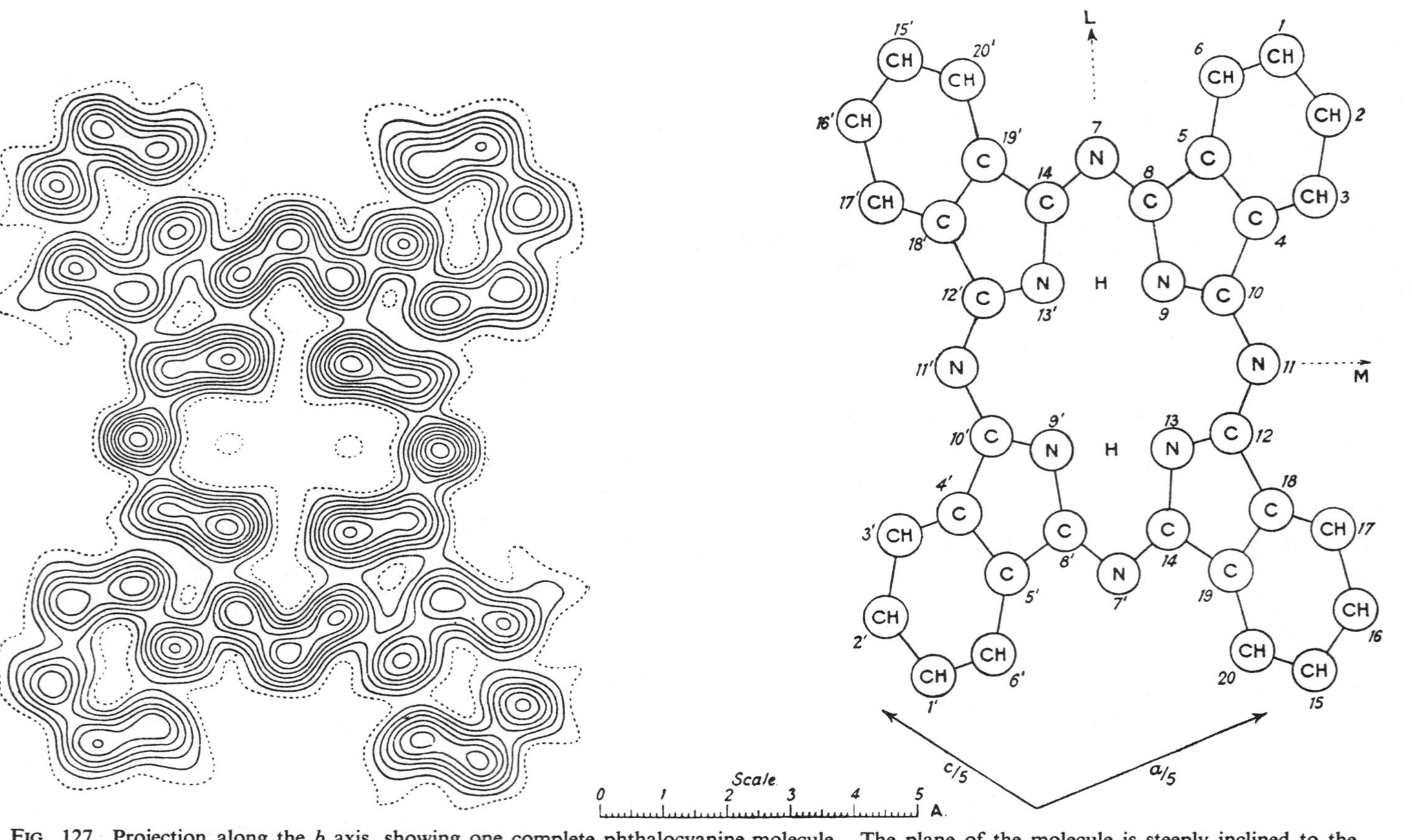

FIG. 127. Projection along the *b* axis, showing one complete phthalocyanine molecule. The plane of the molecule is steeply inclined to the plane of the projection, the M direction making an angle of 46° with the *b* axis, and the L direction 2·3°. Each contour line represents a density increment of one electron per A.², the one electron line being dotted.

the ingenious exploitation of a particular circumstance to effect an analysis—a method which in some form or other must always be used. It will, however, rarely happen that so beautifully direct a solution is possible, and methods of trial and error will as a rule have to be used. Reference to a number of typical determinations by this method will be found in the bibliography at the end of this chapter [23].

(*n*) *The use of the double series when the projection has no centre of symmetry:* The general expression for the structure factor $F(hkl)$ is, by equation (2.14) p. 31,

$$F(hkl) = \sum_s f_s e^{2\pi i(hu_s + kv_s + lw_s)}$$

$$= A + iB,$$

where

$$A = \sum_s f_s \cos 2\pi(hu_s + kv_s + lw_s)$$

$$B = \sum_s f_s \sin 2\pi(hu_s + kv_s + lw_s), \qquad (7.39)$$

f_s being the scattering factor for the atom whose co-ordinates, expressed as fractions of the corresponding sides of the unit cell, are u_s, v_s, and w_s, and the summation being over all the atoms in the unit cell. If the crystal has a centre of symmetry, which is also the origin of co-ordinates, B, the sum of the sine terms, vanishes, and $F(hkl)$ is real. If there is no centre of symmetry, B does not vanish, and $F(hkl)$ is complex ; but it is plain from (7.39) that so long as the f's are real $F(\bar{h}\bar{k}\bar{l}) = A - iB = F^*(hkl)$, the result obtained in VII, § 1(*c*), equation (7.8). The phase angle $\delta(hkl)$ of (7.8) is evidently given by

$$\tan \delta(hkl) = B/A$$

$$\cos \delta(hkl) = A/|F(hkl)|, \quad \sin \delta(hkl) = B/|F(hkl)|. \qquad (7.40)$$

To calculate the phase angles, we therefore need to know the atomic scattering factors, f, for the atoms in the unit cell, and also the atomic co-ordinates. It may often happen that although the structure as a whole has no centre of symmetry, its projection onto one or more of the axial planes has. For example, structures belonging to the orthorhombic holoaxial class, which is characterised by the possession of three two-fold axes, have no centre of symmetry, but their projections on all three axial planes are centro-symmetrical. Sometimes, however, the required projection has no centre of symmetry, and it will then be necessary to use the series (7.21), or (7.35), that form of it suitable for summation. The quantities $C(hk)$ and $S(hk)$, defined in (7.34), which appear in the coefficients of (7.35), involve the phase angle $\delta(hk0)$.

The lines on which it is necessary to proceed are as follows. By some means, an approximation to the structure must be guessed. Using the assumed co-ordinates of the atoms, and the values of f from the tables, we calculate $\sin \delta(hk0)$ and $\cos \delta(hk0)$ from (7.39) and (7.40) for all spectra of the type $(hk0)$ lying within the range of observation. Then, using the *observed* values of $|F(hk0)|$, and the calculated values of $\sin\delta$ and $\cos\delta$, we calculate the quantities $C(hk)$ and $S(hk)$, and form the coefficients $C(hk)+C(h\bar{k})$, $S(hk)+S(h\bar{k})$, $C(hk)-C(h\bar{k})$, and $S(hk)-S(h\bar{k})$, that occur in the summations (7.36). With these values, the summations (7.36) and (7.35) are carried out. If the initial guess at the structure has been fairly accurate, a good enough projection should be obtained to make it possible to correct the assumed co-ordinates. A re-calculation of the phase angles must then be carried out with the corrected co-ordinates, and with the revised phase angles a second projection must be made. The process must be repeated until no important change is introduced by further corrections.

The amount of work involved in carrying out the summation for the unsymmetrical projection is not seriously greater than that for the symmetrical one, the increase being largely in the preliminary calculations of the phases. Robertson [24] has carried out a successful analysis of the structure of resorcinol by this method. Another example is the analysis of 4:4′-dinitrodiphenyl by van Niekerk.[25]

When the projection has no centre of symmetry, the values of the phase angles used in the summation depend directly on the assumed atomic co-ordinates, and also on the theoretical values of the atomic scattering factors. The method is therefore less satisfactory in principle than it is when the projection has a centre of symmetry; for then the final summation depends on no numerical values other than observed ones, theoretical values of the scattering factors being used only to help to decide between the two possibilities, 0 and π, for any given phase angle. It is important to make some test to see whether the process of successive approximations described above does tend stably towards the correct solution. A projection will not in general be attempted until there is fair agreement between observed and calculated intensities for the more important spectra, and we may perhaps expect the projection obtained to depend a good deal on the assumed co-ordinates from which the relative phases of the Fourier waves are calculated.

The following simple example may serve to illustrate the difficulty under consideration. A projection was made of a fictitious crystal structure with a rectangular unit cell containing two identical atoms. The separation of these atoms was assumed to be parallel to the *a* edge of the unit cell, and equal to one third of the cell edge. The *a* and *b* spacings were assumed each to be equal to 5A, and the projection was made in a direction parallel to the *c* axis. The scattering factors of the atoms were supposed to be equal to those of K^+, as given in the tables of James and Brindley, and all structure factors of the type $F(hk0)$ for spacings greater than 0·5A were calculated, and included in the summation.

With such a crystal, the obvious course would be to take the origin mid-way between the two atoms, which would have co-ordinates

$\pm(u, 0, 0)$, where $u=60°$. A projection was made in this way, assuming as a tentative value of the parameter $u=54°$. The 'observed' values of $|F|$, that is to say those calculated assuming $u=60°$, were used as coefficients. The assumed parameter, $u=54°$, gives the wrong sign to all coefficients for which $h=8$, but correct signs to all the others. The values of $\sigma(x, 0)$ at points along the cell edge, which are very easily calculated, are shown in fig. 128 *b*. There are well-marked peaks

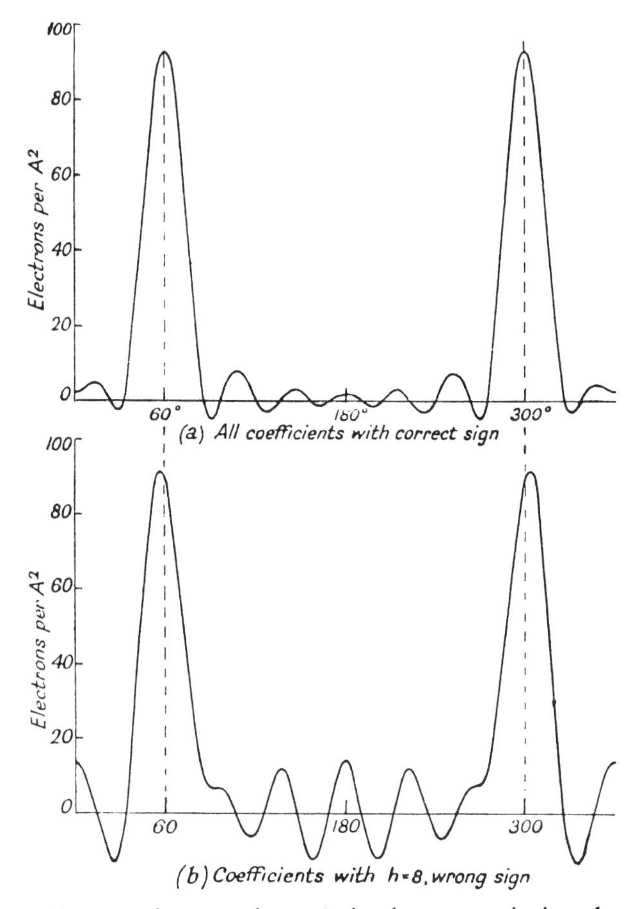

FIG. 128. Illustrating successive approximations to a projection where the crystal has a centre of symmetry

at about $u=\pm58°$, and this correction to the assumed values of $\pm54°$ is enough to give the correct signs of the coefficients with $h=8$, so that the next approximation is in this case the final one. It is shown in fig. 128 *a*. The large oscillations in the background of fig. 128 *b* are due to the reversal of phase of the waves with $h=8$, and the smaller oscillations in the corrected figure, 128 *a*, are due to the systematic rejection of all Fourier waves with lengths shorter than 0·5A. They are analogous to diffraction rings about the main maxima, and will be discussed in the second section of this chapter.

This example is of course trivial in itself, but it is introduced for comparison with the next one, in which the projection of the same structure is treated as one with no centre of symmetry, by choosing the origin at the centre of one of the atoms. The atomic co-ordinates are then 0, and $(2u, 0, 0)$, and the structure factor is given by

$$F(hk0) = f(1 + 2\cos 2hu) + if \sin 2hu.$$

The values of $\cos\delta$ and $\sin\delta$ were calculated for $u = 108°$, and the values of $C(hk)$ and $S(hk)$ were determined using these values and the

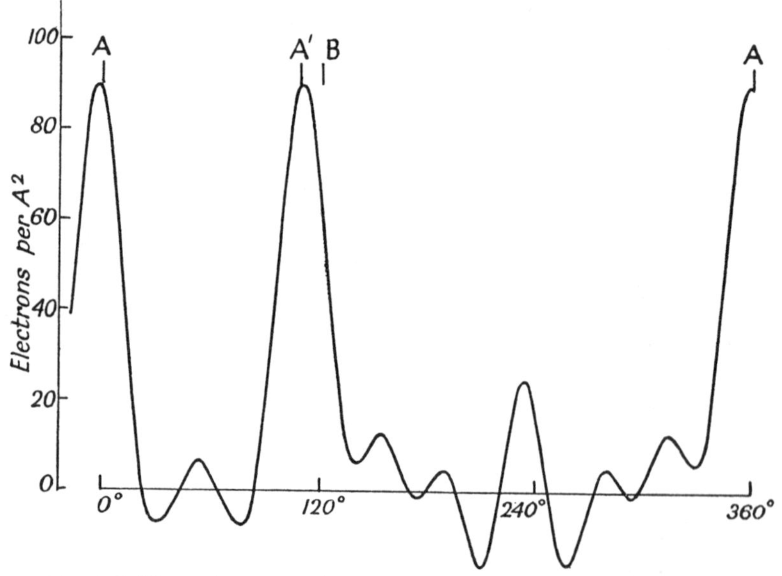

FIG. 129. Illustrating successive approximations to a projection where structure has no centre of symmetry. A and B are the correct positions of the maxima; A and A′, those assumed in calculating the phases

'observed' values of $|F(hk0)|$ for $u = 120°$. The resulting values of $\sigma(x, 0)$ are shown in fig. 129 for rather more than the whole cell edge. The lines A and A′ show the assumed values of the co-ordinates, and A and B the actual values on which the 'observed' values of $|F|$ are based. It will be seen that the first approximation yields a value of u differing very little from the assumed one, about 111°, instead of 108°, and a further approximation would presumably yield a further slow approach to the correct value. The irregularity of the background shows that the phases have not been guessed correctly, but this criterion might be difficult to apply in practice in the case of a more complicated crystal. The general indications in this simple case are that the projection tends to give values of the parameters rather largely influenced by the assumed values, and that the approach to the correct values in successive approximations may be rather slow. It is clear that caution must be used in interpreting the results obtained from projections having no centres of symmetry. If the projection shows clean definite maxima near the positions to be expected from the assumed parameters one may infer that the assumed structure cannot be far wrong, and that there is good general agreement between observed and calculated structure factors, but it is probable that the exact values of the parameters cannot be so reliably inferred as in the case of the structure with a symmetry centre, in which the corrections introduced by the successive approximations are changes of sign of certain coefficients in the series.

In the simple example taken above, the values of f used in calculating the phase angles are the same as those giving the observed structure factors. In practice this will not generally be true, since theoretical values of f, to which no reliable corrections for the thermal motions of the atoms can be made, must be used in calculating the phases, while the observed intensities depend on the actual effective values of the scattering factors for the atoms in the crystal.

(*o*) *One-dimensional series:* Let ABCD, EFGH, fig. 130, be two successive lattice-planes of a crystal structure, and let the parallelepiped between them enclose a space of volume equal to that of the unit cell. Consider a thin slab of this cell of thickness dx, parallel to the lattice-planes, and at a distance x from ABCD.

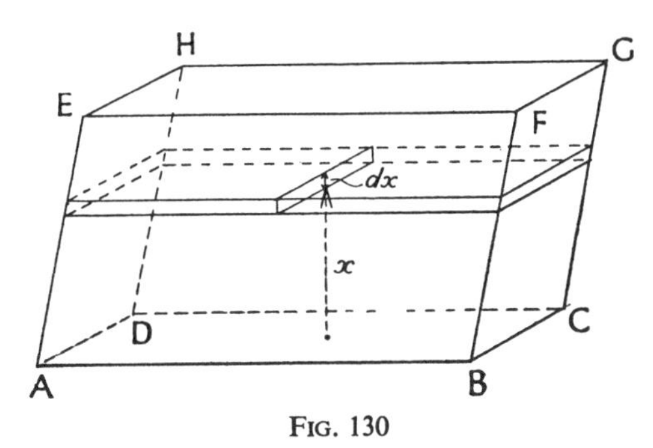

FIG. 130

Let $S(x)\,dx$ be the total amount of diffracting matter lying within this slab. The function $S(x)$ is periodic in the distance d, the spacing of the planes, and can therefore be expressed as the sum of a Fourier series of the simple or one-dimensional type,

$$S(x) = \sum_{-\infty}^{+\infty} A(n)\, e^{-2\pi i n x/d}. \tag{7.41}$$

The coefficients $A(n)$ are determined in the usual way, and are given by

$$d.A(n) = \int_0^d S(x)\, e^{2\pi i n x/d}\, dx. \tag{7.42}$$

If the phase of the radiation scattered from the plane ABCD is taken as zero, the phase of the contribution from the slab at a distance x from ABCD to the nth order spectrum from the planes is $2\pi nx/d$; so that the integral on the right-hand side of (7.42) is equal to F(n), the structure factor for the nth order spectrum from the planes of spacing d. Thus the series for S(x) may be written

$$S(x) = \frac{1}{d}\sum_{n=-\infty}^{+\infty} F(n)e^{-2\pi inx/d}. \tag{7.43}$$

If the scattering density S(x) is everywhere real, F(n) and F$(\bar{n})$ are conjugate quantities, and, just as in VII, § 1 (e), the series (7.43) may be expressed in the form

$$S(x) = \frac{F(0)}{d} + \frac{2}{d}\sum_{n=1}^{n=\infty} |F(n)| \cos\left\{\frac{2\pi nx}{d} - \delta(n)\right\}. \tag{7.44}$$

If the structure has a symmetry centre, or a plane of symmetry, or a rotation or screw axis parallel to the lattice-planes, the origin for x may always be so chosen that S(x) is symmetrical in x; and in this case $\delta(n)$ must be either 0 or π, so that a distribution consistent with an observed set of values of $|F(n)|$ is given by the series

$$S(x) = \frac{F(0)}{d} + \frac{2}{d}\sum_{n=1}^{n=\infty} \pm |F(n)| \cos\frac{2\pi nx}{d}, \tag{7.45}$$

the argument being exactly as in VII, § 1 (f).

This is the simplest of all the types of Fourier series that are applied to crystal analysis. It gives information about the distances of atoms from the reflecting planes, without regard to their positions parallel to these planes. The use of such series was first suggested by W. H. Bragg [1] in 1915. The derivation of the series, and some examples of its application, were given by A. H. Compton in 1926, in his book *X-Rays and Electrons.*

In applying the series it is of course necessary to assume tentative structures, or to use information other than that given purely by the use of X-rays, to estimate the phases of the Fourier components. The form (7.45) of the series has usually been employed, in which only the algebraic signs of the coefficients have to be determined. Some examples have been given in *The Crystalline State*, Vol. I, Chapter IX, an interesting one being that of the alums, due to J. M. Cork,[26] in which the signs of the coefficients were determined by examining a series of isomorphous compounds.

The coefficients $|F(n)|$ are the structure factors of the successive orders from one set of lattice-planes, and unless the spacing of these planes is fairly wide the number of terms in the series will be small and the resolution consequently low. In fig. 131 we give as an example a curve calculated by means of the one-dimensional series showing the distribution of diffracting matter between the (001) planes of β-alumina (Na_2O, $11Al_2O_3$). The planes have a spacing of 22·45A, and 16 terms of the series have been used. Actually spectra up to 0, 0, 32 were included, but the spectra of odd order are absent for this set of planes. We are indebted to Mr. C. A. Beevers for the results upon which fig. 131 is based. Rather more than one quarter of the distance between successive planes is shown in the figure, in which S(x) is plotted as a function of the angle $2\pi x/d$ expressed in degrees, so that 360° represents the full spacing d (=22·45A). The distribution is symmetrical about x=0, d/4, and 3d/4 ; or 0°, 90°, and 270° ; so that the complete

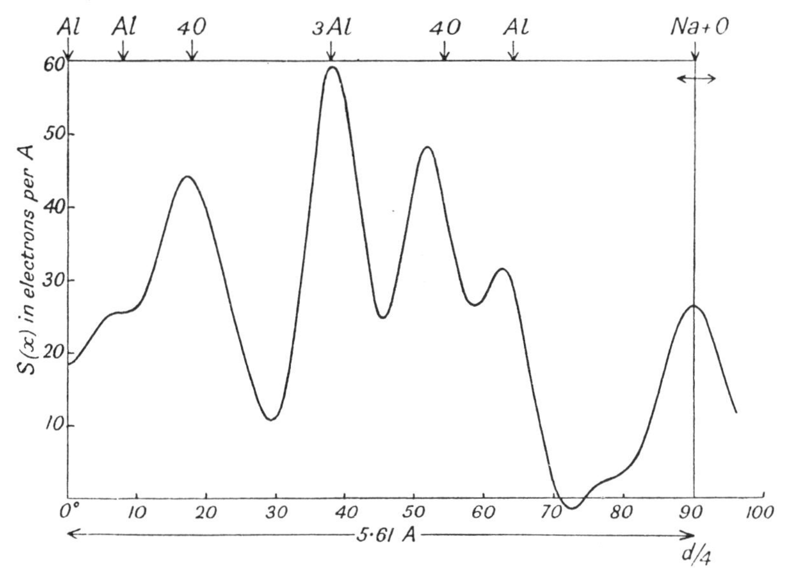

FIG. 131. The distribution of diffracting matter between the (001) planes of β-alumina (Na_2O, 11Al_2O_3), calculated by means of a one-dimensional Fourier series

distribution is easily obtained from the portion of the curve shown. The positions of the atoms between the planes as determined by Bragg, Gottfried and West [27] are indicated by arrows at the top of the figure.

(*p*) *Fourier methods requiring no knowledge of the phases of the spectra : Patterson's series :* In 1934, A. L. Patterson [11] developed a method of obtaining information about interatomic distances in crystals by using a Fourier series the coefficients of which are the values of $|F(hkl)|^2$, instead of those of F(hkl), as in the series so far discussed. Such a series, which takes no account of the relative phases of the spectra, must evidently give less information than could be obtained in an ideal case from the ordinary series if full knowledge of the relative phases were available. On the other hand, the information derived from it depends directly on observation, and not in any way on assumptions based on tentative ideas of the structure of the crystal. Patterson's method arose as a result of discussions of another direct method introduced by Warren and Gingrich,[10] the consideration of which we shall postpone until the scattering of X-rays by amorphous matter has been dealt with. Patterson's work is, however, a direct extension of the ordinary Fourier methods, and is useful in obtaining preliminary estimates of the structure before applying the ordinary series. It must therefore be considered here.

Suppose the density of scattering matter at a point (x, y, z) in a crystal to be $\rho(x, y, z)$, which is given by the Fourier series (7.5). This density will have very small values except in the neighbourhood of certain points, the positions of the atoms. Now consider the function $\rho(x+u, y+v, z+w)$, where u, v, w are parameters supposed constant in any given case while x, y, z vary. This latter function is the same function of the variables $x+u, y+v, z+w$ as $\rho(x, y, z)$ is of x, y, z, and it represents a distribution exactly similar to that corresponding to $\rho(x, y, z)$, but displaced with reference to it, without rotation, through a distance whose components parallel to x, y, z are $-u, -v, -w$.

Consider now the product $\rho(x, y, z)\,\rho(x+u, y+v, z+w)$. Its value at any point (x, y, z) will in general be small, but it may reach appreciable values at certain points if u, v, w are the components of the distance between two of the atoms in the structure, that is to say between two of the density maxima in the distribution $\rho(x, y, z)$. For in this case one maximum of the distribution $\rho(x, y, z)$ coincides with another maximum of the distribution $\rho(x+u, y+v, z+w)$. The coincidence of the two distributions will be complete if u, v, and w are zero, or any multiples of the lattice translations of the crystal, but there will be some degree of coincidence, as we have just seen, if they are the components of any interatomic distance.

As a measure of the degree of coincidence, we may take the function $P(u, v, w)$, given by

$$P(u, v, w) = N\int_0^a\int_0^b\int_0^c \rho(x, y, z)\rho(x+u, y+v, z+w)\,dx\,dy\,dz, \quad (7.46)$$

where a, b, c are the lattice translations of the unit cell, whose volume V is $Nabc$. The integrand is the product of the densities of the two distributions for the same point (x, y, z), which is small unless there is a coincidence of maxima at the point concerned. The greater the general coincidence of maxima in the two distributions, the greater the value of the integral. $P(u, v, w)$ is a function of u, v, and w only. Let us consider it as a density distribution in the co-ordinate space of u, v, w. This distribution is periodic in the distances a, b, c parallel to u, v, w, and it has maxima whenever the vector from the origin to the point (u, v, w) is equal to the vector distance between two atoms in the original crystal structure. The main maxima occur whenever this vector distance is a lattice vector of the crystal, that is to say whenever u, v, and w are any multiples, positive, negative, or zero, of a, b and c respectively.

Since $P(u, v, w)$ is periodic, it may be expressed as a Fourier series, and we must now find the coefficients of this series. By equations (7.46) and (7.5),

$$P(u, v, w) = \frac{N}{V^2} \sum_{hkl} \sum_{h'k'l'} \int_0^a \int_0^b \int_0^c \left[F(hkl)\, F(h'k'l')\, e^{-2\pi i \left(\frac{hx}{a}+\frac{ky}{b}+\frac{lz}{c}\right)} \right.$$
$$\left. \times\, e^{-2\pi i \left(\frac{h'(x+u)}{a}+\frac{k'(y+v)}{b}+\frac{l'(z+w)}{c}\right)} \right] dx\, dy\, dz. \quad (7.47)$$

This integral vanishes unless $h = -h'$, $k = -k'$, $l = -l'$. In this case, $F(hkl) = F^*(h'k'l')$, and (7.47) becomes

$$P(u, v, w) = \frac{N}{V^2} \sum_{hkl} abc\, |F(hkl)|^2\, e^{-2\pi i \left(\frac{hu}{a}+\frac{kv}{b}+\frac{lw}{c}\right)}$$
$$= \frac{1}{V} \sum_h \sum_{k=-\infty}^{+\infty} \sum_l |F(hkl)|^2 \cos 2\pi \left(\frac{hu}{a}+\frac{kv}{b}+\frac{lw}{c}\right) \cdot \quad (7.48)$$

The series for $P(u, v, w)$ is thus of the same type as series (7.12) for the density distribution for a crystal with a centre of symmetry, except that the coefficients are now necessarily positive, and proportional to the *intensities* of the spectra. This is Patterson's series. Its evaluation yields a density distribution which is periodic with the periodicity of the crystal, has maxima at the origin and at points derived from the origin by the lattice translations of the crystal, and subsidiary maxima at vector distances from the origin equal to the vector distances between every pair of atoms in the crystal.

(*q*) *The properties of Patterson's series :* We may discuss the essential properties of the series by considering the one-dimensional case, dealt with in VII, § 1 (*o*), in which $S(x)dx$ is the amount of diffracting matter lying within a slab of the unit cell of thickness dx at a distance x from the yz plane. Patterson's series in this case is given by

$$P(u) = \int_0^a S(x) S(x+u)\, dx$$
$$= \frac{1}{a} |F(0)|^2 + \frac{2}{a} \sum_1^h |F(h00)|^2 \cos 2\pi \frac{hu}{a} \cdot \quad (7.49)$$

$P(u)$ has principal maxima when u is equal to any multiple of a, and subsidiary maxima when u is equal to the distance between any two of the maxima of the original distribution $S(x)$. We may discuss the properties of the series analytically in the following way. Let there be atoms in the structure at points whose x co-ordinates are $x_1, x_2 \ldots, x_n, x_m \ldots$, and let $f_1, f_2 \ldots, f_n, f_m \ldots$, be their scattering factors. Then

$$F(h00) = \Sigma_m f_m\, e^{2\pi \frac{ihx_m}{a}} \quad (7.50)$$

and

$$|\mathrm{F}(h00)|^2 = \sum_m \sum_n f_m f_n^* e^{\frac{2\pi i h}{a}(x_m - x_n)}$$

$$= \sum_m |f_m|^2 + \sum_m \sum_n^{m \neq n} 2 f_m f_n^* \cos \frac{2\pi h}{a}(x_m - x_n). \tag{7.51}$$

Assuming the f's to be real, and substituting (7.51) in (7.49), we obtain

$$\mathrm{P}(u) = \frac{1}{a} |\mathrm{F}(000)|^2 + \frac{2}{a} \sum \mathrm{B}(h) \cos 2\pi \frac{hu}{a}$$

$$+ \frac{2}{a} \sum_m \sum_n^{m \neq n} \left[\sum_h f_m(h) f_n(h) \left\{ \cos \frac{2\pi h}{a}(u + x_m - x_n) \right.\right.$$

$$\left.\left. + \cos \frac{2\pi h}{a}(u - x_m + x_n) \right\} \right], \tag{7.52}$$

where $\mathrm{B}(h) = \sum_m f_m^2(h)$.

It must be remembered that f_m and f_n are functions of h, and we have written them $f_m(h)$ and $f_n(h)$ to emphasise this.

In equation (7.52), each summation with respect to h is a Fourier series with positive and steadily decreasing coefficients. The first such series, the second term in (7.52), gives a distribution with maxima at $u = 0$, $\pm a$, $\pm 2a$, and so on. These maxima will be the sharper the more slowly $\mathrm{B}(h)$ dies away with increasing h, although if the series terminates while $\mathrm{B}(h)$ is still appreciable subsidiary maxima and minima will appear. In practice, the coefficients die away rather rapidly, since they depend on f^2, and the maxima are rather broad. These maxima give us no useful information. They are due simply to the necessary coincidence of $\mathrm{S}(x)$ and $\mathrm{S}(x+u)$ when u is any multiple of the identity period a.

The third term in (7.52) shows that corresponding to any pair of atoms m and n there are two Fourier series, each with a coefficient $f_m(h) f_n(h)$ for the hth term. These series yield distributions with maxima at $u = na \pm (x_m - x_n)$, symmetrically disposed therefore about each main maximum. The heights of these maxima are proportional to $\sum_h f_m(h) f_n(h)$. It is easy to generalise the results just obtained to two and three dimensions.

A full three-dimensional analysis by this method is impracticable on account of the amount of work involved, although, as we shall see later, very valuable use has been made of a three-dimensional series for certain special cases. Patterson's early applications of the method were to projections in two dimensions. If $\sigma(x, y)$ is the density of the projection of the diffracting matter on the xy plane, the corresponding Patterson series is

$$\mathrm{P}(u, v) = (\mathrm{A}/ab) \int_0^a \int_0^b \sigma(x, y)\, \sigma(x+u, y+v)\, dx\, dy$$

$$= \frac{1}{\mathrm{A}} \sum_{h}^{+\infty}\sum_{k}^{} |\mathrm{F}(hk0)|^2 \cos 2\pi \left(\frac{hu}{a} + \frac{kv}{b} \right), \tag{7.53}$$

$-\infty$

A being the area of the face *ab*. This distribution will have main maxima, corresponding to coincidences of the two distributions $\sigma(x, y)$ and $\sigma(x+u, y+v)$, at the corners of the network *ab*, Symmetrically disposed about each maximum will be pairs of subsidiary maxima, corresponding to each interatomic distance in the projection $\sigma(x, y)$. A simple example will make the relation between the two types of distribution clear. In fig. 132*a*, is shown the projection of the positions of the atoms in the unit cell of iron pyrites, FeS_2, on one of the cube faces. The black circles indicate the iron atoms, and the white circles the sulphur atoms. Fig. 132*b* shows the positions of the maxima in the corresponding Patterson projection $P(u, v)$.

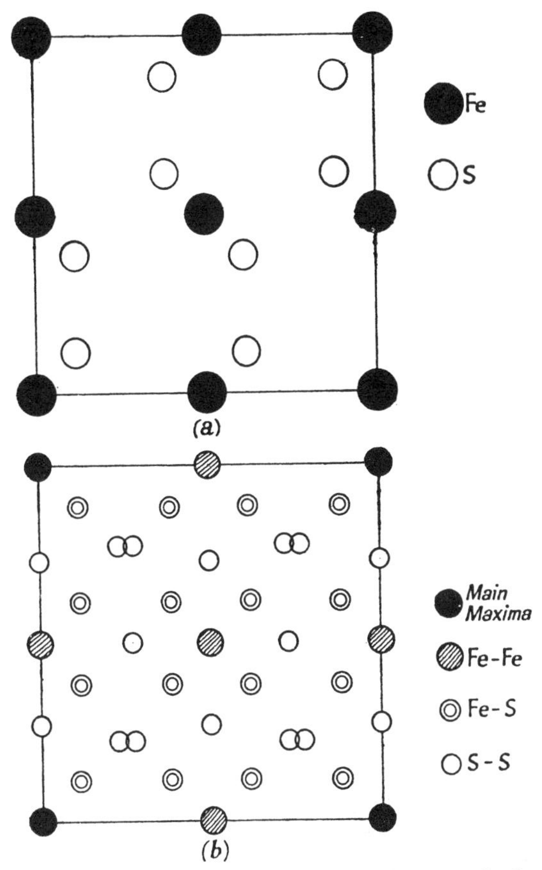

FIG. 132. (*a*) Projection of a unit cell of FeS_2 on a cube face
(*b*) Patterson projection of the same unit

Even for such a comparatively simple structure, the $|F|^2$ projection is not easy to interpret, and it would be more difficult still if the true projection were not already known. In the figure, all the positions are shown at which maxima should occur, but in the actual projection, the maxima will be rather diffuse, and the fainter or closer ones may not be resolved. In the $|F|^2$ projection, the vector corresponding to a given interatomic distance is always drawn from the origin, no matter where the two atoms may lie in the true projection. If there are several equal and parallel interatomic distances in the true projection, all give the same peak in the $|F|^2$ projection, which will be correspondingly strong. Interatomic distances between two heavy atoms will also show up clearly in the $|F|^2$ projection, since the height of a corresponding maximum is proportional to $\sum_h f_m f_n$, the sum being taken over all the orders of spectra concerned. Very valuable information may sometimes be obtained in such cases, and the method may be a useful adjunct to the process of trial and error used as a preliminary to a full crystal analysis ; but it is unlikely to provide information leading to the direct analysis of an unknown and complicated structure, because of the many interatomic distances in the true projection, and the number of overlapping and unresolved maxima in the $|F|^2$ projection in such a case. Many of the difficulties due to lack of resolution are avoided by a method employing a three-dimensional series, which we shall consider in detail in a later paragraph.

(*r*) *Methods of increasing the resolution of the Patterson distribution :* As we have seen in § 1(*q*) of this chapter, the maxima in the $|F|^2$ distribution are rather diffuse, because the coefficients $|F|^2$ of the series diminish rapidly. Again it is easiest to consider the one-dimensional case, although the results can be applied quite directly to the two-dimensional one. The maxima corresponding to any particular pair of atoms m and n are those given by the Fourier series in the third term of equation (7.52). The coefficients of these series are $f_m(h)\, f_n(h)$, which diminish rapidly with increasing h. If we can in any systematic way decrease the rate at which the coefficients fall off, we shall sharpen the maxima in the distribution given by the series. A number of devices suggested by Patterson [11] for increasing the sharpness of the maxima amount to such a systematic modification of the coefficients, and although useful, have in reality a mathematical rather than a physical significance.

A set of equal coefficients, not decreasing at all, would correspond physically to point atoms, without temperature motion, and the structure factor for this case could be obtained by writing Z_m in place of f_m in (7.50), Z_m being the number of electrons in an atom of type m. If all the atoms in the unit cell were alike, each with a scattering factor f_m, we could in effect reduce their action to that of point atoms by writing instead of the actual structure factor $F(h)$ the quantity $F(h)Z_m/f_m$, and the series whose coefficients were the squares of these quantities would correspond to a distribution with considerably sharper maxima than those given by the series whose coefficients are the values of $|F(h)|^2$, although the termination of the series at a finite number of terms would necessarily introduce false detail. In practice, the values of $F(h)$ are those appropriate to a crystal whose atoms are in thermal movement, while the values of f_m used in the modified coefficients would presumably be tabulated values corresponding to atoms at rest. The modified coefficients would not therefore correspond exactly to point atoms, and the false detail would accordingly be reduced.

The atoms in the unit cell of a crystal are not in general all of one kind. Patterson therefore suggests using in the coefficients instead of $F(h)$, $F(h)/\bar{f}$, where

$$\bar{f} = \sum_m f_m / \sum_m Z_m = \sum_m f_m / F(0), \tag{7.54}$$

the summation being over all the atoms in the unit cell. The general effect of this is to produce a scattering factor corresponding more closely to one due to point atoms, and to sharpen the maxima in the resulting distribution. Patterson gives in his paper a number of examples, an interesting one being the case of copper sulphate. Here the maxima in the distribution made using the modified coefficients $|F(h)|^2/|\bar{f}|^2$ are much more clearly defined than those in the ordinary $|F|^2$ distribution.

Patterson makes the further suggestion that the main maxima at the corners of the lattice-cell, corresponding in the one-dimensional case we are here considering to the second term in (7.52), should be removed, since they give no useful information, and prevent resolution of other maxima. This removal may be carried out in principle by using instead of $|F(h)|^2$ a modified coefficient $|F_0(h)|^2$, where

$$|F_0(h)|^2 = |F(h)|^2 - \sum_m |f_m|^2. \tag{7.55}$$

The second term in (7.52) then vanishes, and we are left only with the peaks of the third term, corresponding to interatomic distances within the unit cell, but not to the lattice translations. In practice, however, there might be some danger of introducing false maxima by this process, since the values of $\sum_m |f_m|^2$ would have to depend on theoretical f-curves, and might be considerably in error.

(*s*) *Harker's application of the Patterson method:* As we have seen above, the two-dimensional $|F|^2$ series is not very satisfactory in practice, because the projection of every interatomic distance in the unit cell onto the plane considered has its corresponding maximum in the $P(u, v)$ distribution, so that there is nearly always overlapping of maxima. Moreover, the individual maxima are rather diffuse because of the comparatively small number of effective terms in the series, whose coefficients diminish rapidly with increasing order. The series for the three-dimensional Patterson distribution, $P(u, v, w)$, cannot be evaluated completely without excessive labour, but it has been shown by Harker [28] that very useful results can nevertheless be obtained by applying it to special cases of crystals with certain symmetry elements. We shall consider some of the more important examples: (i) Let the crystal have a *twofold rotation axis*, and suppose this to be chosen as the b axis of co-ordinates. Then for any atom with co-ordinates (x, y, z) there is another with co-ordinates $(\bar{x}, y, \bar{z})$. The vector between these two atoms has components $(2x, 0, 2y)$, and the positions of the corresponding maxima in the distribution $P(u, v, w)$ can be found by evaluating $P(u, 0, w)$ only; for the corresponding vector must lie in the uw plane, since all vectors in the distribution start from the origin. Thus

$$\begin{aligned} VP(u, 0, w) &= \sum_{h}\sum_{k}\sum_{l}{}_{-\infty}^{+\infty} |F(hkl)|^2 \cos 2\pi\left(\frac{hu}{a}+\frac{lw}{c}\right) \\ &= \sum_{h}\sum_{l}\left[\cos 2\pi\left(\frac{hu}{a}+\frac{lw}{c}\right)\sum_{k}|F(hkl)|^2\right] \\ &= \sum_{h}\sum_{l} C(hl)\cos 2\pi\left(\frac{hu}{a}+\frac{lw}{c}\right), \end{aligned} \tag{7.56}$$

where
$$C(hl) = \sum_{k}|F(hkl)|^2. \tag{7.57}$$

All the interatomic vectors not actually in the plane xz are eliminated in this way, although they would, of course, appear in the two-dimensional distribution $P(u, w)$. It will be seen that the evaluation of the series involves the measurement of the intensities of all spectra of the general type hkl, of which there may be a large number. On the other hand, a very accurate estimate of the intensities is not necessary, and the work involved is by no means prohibitive. The preliminary summations to form the coefficients $C(hl)$ may be thought of in the following way. Suppose a rotation photograph to be taken with the crystal rotating about the b axis. Layer lines and row lines will be formed, and each row line corresponds to a given pair of values of h and l, while each layer line corresponds to a given value of k. One coefficient $C(hl)$ would therefore be given by the sum of the squares of the structure factors of all the spectra in one row line. This is a convenient geometrical way of visualising the summations to be carried out; it is not necessarily the most convenient way to do them in practice. By this method, information about the distances of atoms from the two-fold axis may be obtained, which may be of considerable value in deciding upon a possible structure. The same method can be used for three-, four-, or six-fold axes; for the vectors between equivalent pairs of atoms are in all these cases parallel to the plane $y=0$.

(ii) If the axis parallel to b is a *two-fold screw axis*, an atom with co-ordinates (x, y, z) gives rise to a second with co-ordinates $(\bar{x}, y+b/2, \bar{z})$, and the vector between them has components $(2x, -b/2, 2z)$. The corresponding maxima in the $P(u, v, w)$ distribution have co-ordinates $\pm(2x, -b/2, 2z)$ and $\pm(2x, +b/2, 2z)$, since all is symmetrical about the origin. In this case we evaluate the series for the plane $v=b/2$, and obtain

$$VP(u, b/2, w) = \sum_{h}\sum_{k}\sum_{l}{}_{-\infty}^{+\infty} |F(hkl)|^2 \cos 2\pi\left(\frac{hu}{a}+\frac{k}{2}+\frac{lw}{c}\right)$$

$$= \sum_{h}\sum_{l} C(hl) \cos 2\pi\left(\frac{hu}{a}+\frac{lw}{c}\right),$$

where $$C(hl) = \sum_{k}(-1)^k |F(hkl)|^2. \qquad (7.58)$$

The components parallel to the plane $y=0$ of the vectors from the origin to the maxima in $P(u, b/2, w)$ give the directions, and double the distances, of the corresponding atoms from the two-fold screw axis. In the summations for $C(hl)$ in this case, the intensities of the spectra on the even layer lines are taken as positive and those of the spectra on the odd layer lines as negative.

(iii) Suppose the crystal has a *reflection-plane of symmetry*, perpendicular to the b axis, and passing through the origin of co-ordinates. To any atom with co-ordinates (x, y, z) there corresponds a second of the same kind with co-ordinates $(x, \bar{y}, z)$, and the vector between them has components $(0, 2y, 0)$. The corresponding maxima in the $P(u, v, w)$ distribution lie on the b axis at distances $\pm 2y$ from the origin. We therefore evaluate $P(u, v, w)$ for points on the axis of v. Thus

$$VP(0, v, 0) = \sum_{h}\sum_{k}\sum_{l}{}_{-\infty}^{+\infty} |F(hkl)|^2 \cos 2\pi\frac{kv}{b}$$

$$= \sum_{k} B(k) \cos\frac{2\pi kv}{b},$$

where $$B(k) = \sum_{h}\sum_{l} |F(hkl)|^2. \qquad (7.59)$$

The summation for $B(k)$ now includes squares of the structure factors of all spectra on the kth layer line. The distances of the maxima in $P(0, v, 0)$ from the origin are equal to twice the distances of the corresponding atoms from the mirror plane.

(iv) If the crystal has a *glide-plane of symmetry* perpendicular to the b axis, with a glide $\frac{1}{2}c$, equivalent points have co-ordinates (x, y, z) and $(x, \bar{y}, z+\frac{1}{2}c)$. The maxima in the distribution $P(u, v, w)$ corresponding to the distances between these atoms have co-ordinates $\pm(0, 2y, -\frac{1}{2}c)$

and $\pm(0, 2y, +\frac{1}{2}c)$. We evaluate $P(u, v, w)$ for the line $u=0$, $z=\frac{1}{2}c$, and obtain

$$VP(0, v, \tfrac{1}{2}c) = \sum_{h}\sum_{k}\sum_{l}{}_{-\infty}^{+\infty} |F(hkl)|^2 \cos 2\pi\left(\frac{kv}{b}+\frac{l}{2}\right)$$

$$= \sum_{k} B(k) \cos \frac{2\pi kv}{b}, \tag{7.60}$$

where $$B(k) = \sum_{h}\sum_{l} (-1)^l |F(hkl)|^2.$$

Here again, the information given is the distance of the atom from the glide-plane, which is half the value of v corresponding to a maximum. Other types of glide-plane may be dealt with in a similar way.

In Table VII. 2, we give a list, taken from Harker's paper, of the appropriate forms of $P(u, v, w)$ for different types of symmetry element. The notation is that of the *International Crystal Tables*, which is also that used in Vol. I of *The Crystalline State*.

TABLE VII. 2

Symmetry Element	Form of $P(u, v, w)$
(a) Axes parallel to b axis :	
(i) $2, 4, 4_2, \bar{4}, 6, 6_2, 3, \bar{3}, 6_3$	$P(u, 0, w)$
(ii) $2_1, 4_1, 4_3, 6_1, 6_5$	$P(u, b/2, w)$
(iii) $3_1, 6_2, 6_4$	$P(u, b/3, w)$
(b) Planes perpendicular to b axis :	
(i) Reflection plane	$P(0, v, 0)$
(ii) Glide $\frac{1}{2}a$	$P(\frac{1}{2}a, v, 0)$
(iii) Glide $\frac{1}{2}a+\frac{1}{2}c$	$P(\frac{1}{2}a, v, \frac{1}{2}c)$
(iv) Glide $\frac{1}{2}c$	$P(0, v, \frac{1}{2}c)$
(v) Glide $(a+c)/4$	$P(a/4, v, c/4)$
(vi) Glide $(3a+c)/4$	$P(3a/4, v, c/4)$

If some atoms are known to be in special positions, other simplifications may be possible.

Harker's method is likely to be of considerable value in the analysis of the more complicated structures. It will in general be used in conjunction with the ordinary two-dimensional F-series, and its main function is to fix some co-ordinates of important atoms in the structure in order to determine the appropriate phases of the spectra for use in the F-series.

(t) *An example of the use of the Patterson-Harker method:* As a good example of the use of this method we may take the determination of the structure of penta-erythritol, $C(CH_2.OH)_4$, by Llewellyn, Cox and Goodwin.[29] The structure is based on a body-centred tetragonal lattice, and the symmetry is that of the space-group $I\bar{4}$. There are two molecules in each unit cell, and the central carbon atoms of these molecules lie at the cell corners and cell centres. Through them pass axes of four-fold rotatory inversion. The carbon atoms of the CH_2 groups, and the oxygen atoms of the OH groups lie in general positions, and there are eight of each in the unit cell, four associated with each molecule, and derivable from one another by the action of the axis of rotatory inversion.

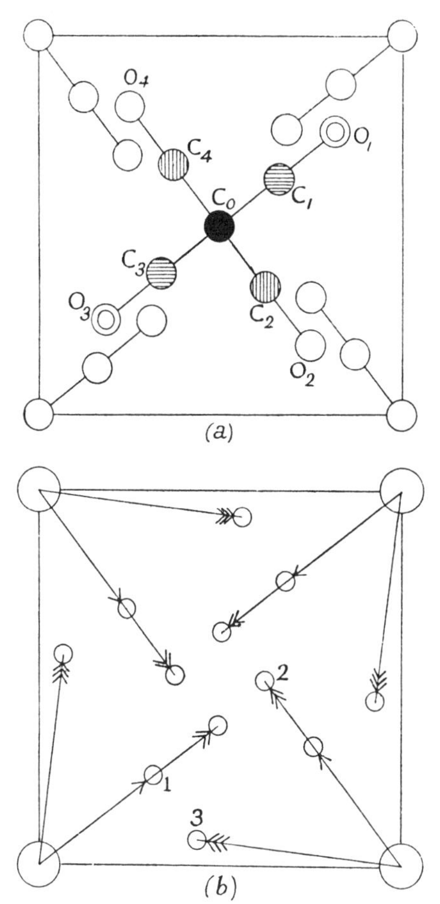

FIG. 133. (*a*) The projection of the positions of the atoms in the unit cell of penta-erythritol on the *c* face; (*b*) The Patterson-Harker section P(*u*, *v*, 0) for the same crystal

Fig. 133*a* shows the projection of the structure on the *c* face of the unit cell. Similarly shaded atoms in the molecule at the cell centre lie in the same plane for space-group reasons. The plane containing the carbon atoms C_1 and C_3 lies as far above the central carbon atom C_0 as the plane containing the atoms C_2 and C_4 lies below it; and similarly with O_1 and O_3, and O_2 and O_4. Atoms in the molecules at the cell corners lie in planes half a unit above and below those containing the corresponding atoms in the molecule at the cell centre.

Fig. 133*b* shows the Patterson-Harker distribution P$(u, v, 0)$, and is a section of the three-dimensional distribution P(u, v, w) by the plane $w=0$. For the sake of clearness, only the positions of the maxima in the distribution re shown, but it must be remembered that the original diagram is a contour diagram, similar to those shown in figs. 125 and 127, from which the positions of the maxima have to be estimated. Apart from the main maxima at the cell corners, there should be only *two* independent maxima in this distribution—those corresponding to the spacings between C_1 and C_3, and between O_1 and O_3. Each of these pairs of atoms lies in a plane parallel to the plane $w=0$, because of the fourfold axis of inversion; and owing to the action of this axis each of these maxima will be repeated symmetrically four times about the cell centres and cell corners. Such maxima have been called 'space-group maxima'. They are the maxima denoted by (1) and (2) in fig. 133*b*, together with those derived from (1) and (2) by the action of the four-fold axes. In addition to these maxima there are, however, maxima of a third type, marked (3) in fig. 133*b*. These cannot be space-group maxima, and must be due to atoms which happen for reasons not directly connected with the symmetry of the crystal to have very nearly the same z co-ordinates.

An elimination of the various possibilities, for the details of which reference must be made to the original paper, shows that the oxygen atoms $O_1O_2O_3O_4$ lie very nearly in the same plane as the central carbon atom, and that the maxima of type (3) correspond to spacings such as O_1O_2 and O_2O_3. If this interpretation is correct, other maxima must occur corresponding to the spacings between the central carbon atom C_0 and the oxygen atoms. No such maxima are found in the present instance, and it is necessary to account for their absence. In fig. 133*b* the vectors from the cell corners to the maxima, which are equal and parallel to the corresponding interatomic distances, are indicated by arrows. Those with single heads correspond to vector distances of the type C_1C_3 or C_2C_4, and those with double heads to distances of the type O_1O_3 and O_2O_4. These vectors correspond to the true space-group maxima. The three-headed arrows correspond to vector distances of the type O_1O_2 and O_2O_3, which happen for reasons unconnected with the essential symmetry of the structure to lie parallel to the plane $w=0$. It will be noticed that the distance O_1O_3 is very nearly double the distance C_1C_3, and that the corresponding vectors are very nearly parallel. The vector corresponding to the distance C_0O_1 will thus be very nearly equal to that corresponding to the distance C_1C_3, and the maxima corresponding to it will thus coincide with the maxima of type (1), which accounts for their non-appearance as independent maxima in the distribution.

From the information contained in fig. 133*b* it was possible to fix the general dimensions and orientation of the molecules, and so to calculate the phases of the terms ($hk0$) of the ordinary two-dimensional Fourier series from which the structure was finally determined. It is important to remember that fig. 133*b* was obtained from the results of observation alone, without any assumption as to the nature of the structure. In its evaluation, Llewellyn, Cox and Goodwin used only visually estimated intensities of the spectra of the general type hkl; accurate measurements were made of the intensities of the spectra of type $hk0$ alone, which were used in the final projection.

It is instructive to compare the Patterson-Harker distribution $P(u, v, 0)$ shown in fig. 133*b* with the two-dimensional distribution $P(u, v)$. In the latter there would have been in principle one maximum for each type of interatomic distance shown in the projection on the *c* face, fig. 133*a*. It is plain that the resulting distribution would have been far more complex than that shown in fig. 133*b*, and that to interpret it would in all probability have been impossible.

(*u*) *Summary:* In the preceding paragraphs we have discussed a variety of methods by means of which information can be obtained about the structure of crystals by summing appropriate Fourier series. We have seen that there is no method that gives the structure directly. It is always necessary to infer by means other than those depending on observations of an optical character the phases of the spectra, and so of the corresponding terms of the Fourier series. If the crystal has a centre of symmetry, it is possible, by choosing this as origin of co-ordinates, to reduce the indeterminacy of phase to an indeterminacy of algebraic sign; but the phases, or the signs, have always to be inferred in someway before the series can be summed. For this purpose, it is necessary to make a preliminary guess at the structure near enough to allow the phases of the spectra to be calculated with such accuracy that a recognisable structure appears as a result of the first summation of the series.

To do this, we may use our knowledge of the sizes and shapes of the atoms and molecules of which the crystal is built up, and of the way in which the symmetry elements of the structure limit the possibilities of packing them together; or we may use information about the orientation of the molecules obtained by measuring the refractive indices, or by observing the diffuse spectra due to thermal movement, or the magnetic anisotropy.[30] Another method is to employ Patterson series, for which it is not necessary to know the phases of the spectra, to obtain information about the distances between important atoms in the structure, or, in some cases, their distances from planes or axes of symmetry. By some or all of these means it may be possible to obtain a fairly accurate preliminary idea of the structure.

Once this preliminary guess at the structure has been made, it is possible to calculate from it the phases of the spectra for use in the series, or the signs of the coefficients, if the crystal has a symmetry centre. This is probably the greatest labour in the whole process of analysis, and it is usual therefore to make first of all two-dimensional projections of the structure on one or more of the axial planes, using the double series, since this involves the calculation of the phases of only the comparatively small number of spectra of the types $hk0$, $0kl$, and $h0l$. The projections so obtained will in some cases give the structure with all the accuracy that is immediately needed, and indeed, until comparatively recently, any further analysis was rarely attempted; but it will often happen that overlapping of the molecules in one or more of the projections prevents a really clear picture of the molecule from being obtained.

From the two-dimensional projections, together with the results of appropriate Patterson analyses, it may now be possible to calculate the phases of the spectra of type *hkl* with enough certainty to make the use of a triple series practicable, and so to investigate the finer points of the structure. A method that has been used with success for this purpose is to determine the density of the scattering matter over planes at a series of heights in the unit cell, so chosen as to give the information required, by means of the series (7.19). In the summation of such triple series, methods exactly analogous to those described in some detail in § 1 (*j*) in the case of the double series may be used. A striking example of an analysis of a complicated structure made by taking sections of the structure in this way is that of geranylamine hydrochloride by Jeffrey,[31] illustrated in fig. 134, which shows a series of sections of the structure

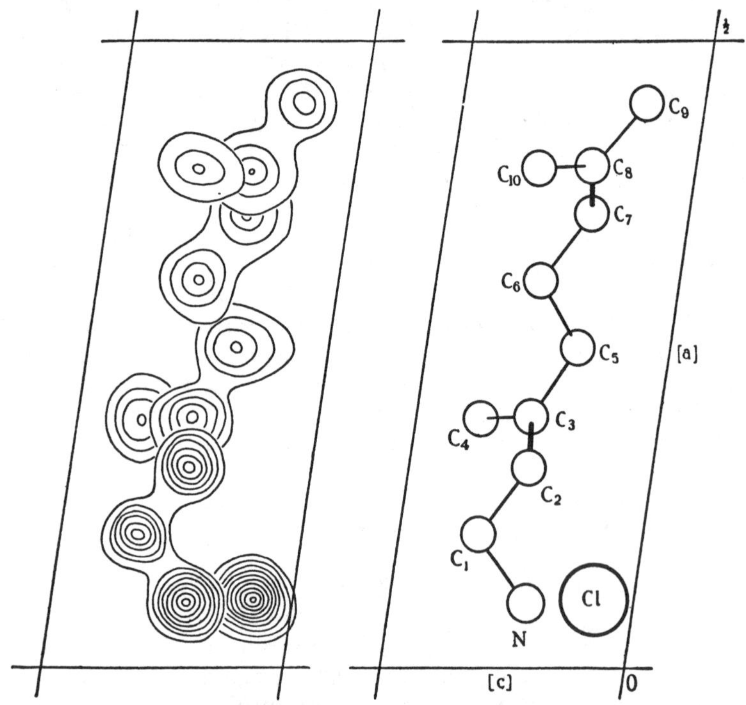

FIG. 134. Projection of the molecule, and Fourier sections, of geranylamine hydrochloride on (010) (Jeffrey)
(*Proc. Roy. Soc.*, **A**, **183**, 388 (1945))

projected on the plane (010). Such determinations do involve a considerable amount of labour, as an examination of the list of spectra* used by Jeffrey in analysing of this crystal will show, but with present-day methods of analysis, applied step by step in the manner indicated, there is now every chance that the labour will produce results.

It may at times happen that the presence of a heavy atom in the

* Reproduced in Publication No. 61 of the British Rubber Producers' Research Association.

structure has a determining effect on the phases of the spectra, so that if it is possible to fix the position of this atom the signs of the coefficients in the Fourier series follow with some degree of certainty. An example of this has already been given in § 1(*m*) in considering Robertson's analysis of the phthalocyanines, and the point is of equal importance in using triple series.

2. The Fourier Projection considered as an Optical Image

(*a*) *Introductory :* In this section we shall consider how the projection of a crystal structure onto a plane, obtained by summing the double Fourier series, is related to an optical image. We shall see that although calculation, and not a lens, is used in the final stages of its formation, and although certain essential data, used automatically by a lens, namely the relative phases of the spectra, have to be inferred by means other than optical before the calculation can be carried out, the projection is nevertheless formally an optical image.

No optical image can ever be a completely faithful reproduction of the object; for the practical necessity of using lenses of finite aperture imposes limits to the fineness of the detail that can be resolved. We shall find that analogous effects due to lack of optical resolution manifest themselves in the projection of the crystal structure. It is a fundamental principle of physical optics that detail on a scale appreciably smaller than that of the wave-length of the light used to observe it cannot be distinguished by any optical arrangement. The wavelengths of the X-rays used in examining crystal structures are of the order of the atomic dimensions, and while therefore we may expect to get the positions of the centres of the atoms with some accuracy from the positions of the maxima in the Fourier projection, we shall not expect the distribution of density around those maxima to represent in any detail the electronic distribution in the atoms, even in those cases where each maximum in the projection stands out clearly from its neighbours, without overlapping. The regions round the density maxima in the projection are in fact analogous to the diffraction discs seen when particles whose dimensions are smaller than the wave-length of light are viewed in the ultramicroscope. We must therefore beware of interpreting rings of subsidiary maxima and minima such as may sometimes surround the main maxima in the Fourier projection as being due to details of the atomic structure. They will in general no more correspond to anything in the atom itself than the rings surrounding the image of a bright star seen in a telescope of small aperture correspond to anything in the structure of the star. In both cases the rings are diffraction effects, and are due to the limited resolving power of the optical system used.

This much we can say without any formal analysis, using only the principles of physical optics. In the succeeding paragraphs, we shall follow in some detail the parallel between the methods of X-ray analysis and microscopic vision. W. L. Bragg [32] has considered this question in a number of papers, and has given an introductory account of it in Vol. I of *The Crystalline State*, Chapter IX. The present treatment is based on this, but is more detailed.

(*b*) *The reciprocal relation between the net of a two-dimensional grating and the array of spectra produced by it*: We shall first of all consider some properties of a two-dimensional plane grating used in the ordinary way to produce spectra with monochromatic light. Let the grating consist of small holes in a plane opaque screen, arranged at the points of a regular network having primitive translations represented by the vectors **a** and **b**. Suppose a train of plane waves w_0, of wave-length λ, to fall at normal incidence on the grating, so that each opening becomes the source of coherent waves in the same phase. In fig. 135, let G represent the grating, and L a lens upon which the diffracted radiation falls. The

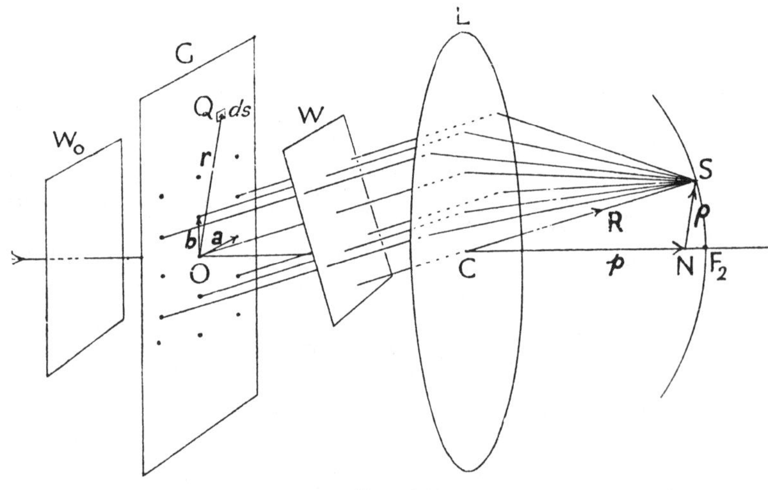

FIG. 135

wavelets leaving the separate grating openings combine at a distance from the grating to form a number of separate trains of plane waves, travelling in different directions, and each such wave-train is brought to a focus at a point in the second focal surface of the lens. The points so produced constitute the array of spectra formed by the grating. We shall suppose the second focal surface of the lens to be spherical, of radius f, with its centre at C, the second nodal point of the lens.

Let S, fig. 135, be a spectrum, and let CS be represented by the vector **R**, so that $|\mathbf{R}| = f$. The vector **R** is in the direction of the wave-normal of the diffracted plane wave-train w that produces the spectrum S. The conditions for the formation of a spectrum are that the projections of both the primitive translations, a and b, of the grating onto the direction of **R**, that is to say the scalar products of both the vectors **a** and **b** with a unit vector in the direction of **R**, should be an integral multiple of the wave-length.

The unit vector in the direction of **R** is $\mathbf{R}/|\mathbf{R}|$, or $\mathbf{R}/f$, so that the condition for the formation of a spectrum can be written

$$\frac{1}{f}(\mathbf{R}\cdot\mathbf{a}) = h\lambda, \quad \frac{1}{f}(\mathbf{R}\cdot\mathbf{b}) = k\lambda, \tag{7.61}$$

h and k being any integers, including zero.

Let SN, fig. 135, be the perpendicular from S on to the axis of the lens, and let NS be the vector $\boldsymbol{\rho}$, and CN the vector $\mathbf{p}$. Then

$$\mathbf{R} = \mathbf{p} + \boldsymbol{\rho}, \tag{7.62}$$

and since $\mathbf{p}$ is perpendicular both to **a** and to **b**, (7.61) can be written in the form

$$\frac{1}{f\lambda}(\boldsymbol{\rho}\cdot\mathbf{a}) = h, \quad \frac{1}{f\lambda}(\boldsymbol{\rho}\cdot\mathbf{b}) = k. \tag{7.63}$$

The values of $\boldsymbol{\rho}$ corresponding to the possible lattice spectra are obtained by giving to h and k all possible integral values.

Let $\mathbf{a}^*$ and $\mathbf{b}^*$ be the primitive translation vectors of the net reciprocal to the grating net. The two nets are coplanar, $\mathbf{a}^*$ is perpendicular to **b**, and $\mathbf{b}^*$ to **a**, while the axial lengths $|\mathbf{a}^*|$ and $|\mathbf{b}^*|$ are equal respectively to the reciprocals of the a and b spacings of the net (a, b), so that

$$\mathbf{a}\cdot\mathbf{a}^* = \mathbf{b}\cdot\mathbf{b}^* = 1, \quad \mathbf{a}\cdot\mathbf{b}^* = \mathbf{b}\cdot\mathbf{a}^* = 0. \tag{7.64}$$

Let $\mathbf{r}^*$ be the vector to the point (h, k) of the reciprocal net, so that

$$\mathbf{r}^* = h\mathbf{a}^* + k\mathbf{b}^*. \tag{7.65}$$

Then $\mathbf{a}\cdot\mathbf{r}^* = h$, and $\mathbf{b}\cdot\mathbf{r}^* = k$, and it is plain that if

$$\frac{1}{f\lambda}\boldsymbol{\rho} = \mathbf{r}^*, \tag{7.66}$$

the conditions (7.63) are fulfilled. The vector distance $\boldsymbol{\rho}$ from the axis of the lens to a spectrum $S(h, k)$ is thus parallel to and proportional to the vector from the origin to the point (h, k) in the net reciprocal to the grating net.

To get this relationship accurately fulfilled we have taken the focal surface of the lens as a sphere of radius f. Let a plane be drawn tangent to this sphere at the point where it cuts the axis of the lens, that is to say at the second focal point F_2, and suppose all the possible spectrum points projected onto this plane in a direction parallel to the axis. The resulting array of points on the plane will form a net reciprocal to that of the grating.

The various sets of parallel and equally spaced rows in which the points of the grating net can be considered as lying can be assigned

Miller indices (h, k), and from the properties of the reciprocal net it is plain that the vector $\boldsymbol{\rho}$ to the spectrum $S(h, k)$ is perpendicular to the rows (h, k) of the grating, and also that the length of the vector $\mathbf{r}^*$ $(=|\mathbf{r}^*(hk)|)$ is the reciprocal of the spacing $d(hk)$ of the rows (h, k) of the grating. The term 'spacing' is here used in the extended sense of § 1(*a*) of Chapter I, or of Appendix II, § (*e*).

The spectra will not, of course, all have the same intensity, and to get a complete representation of the behaviour of the grating we must associate with each reciprocal-lattice point (h, k) the *amplitude* $A(hk)$ of the corresponding spectrum. $A(hk)$ may be represented conveniently by a complex quantity, so as to include a phase factor. It depends on the angle of diffraction, and on the form of an individual opening of the grating, and is the analogue of the structure factor of an X-ray spectrum.

The significance of the *phase* of the spectrum can be seen in the following way. In fig. 135 the grating may be moved in any direction through any distance, so long as its orientation remains the same, without altering in any way either the positions or the intensities of the spectra ; for nothing in the above argument depends on the actual positions of the grating points so long as their relative positions, and the orientation of the grating as a whole, remain unchanged. The phase of the wave arriving at S will, however, be changed. If each grating point is moved through a distance represented by a vector $\mathbf{u}$ lying in the plane of the grating, the distance travelled by the light from each grating opening before reaching S will be altered by $(\mathbf{u}\cdot\mathbf{R})/f$, and the phase of the wave arriving at S by $2\pi(\mathbf{u}\cdot\mathbf{R})/\lambda f$. If each grating unit consists of an exactly similar set of points, instead of being a single point, we can consider points of each type as forming a separate grating. Each such grating contributes its own amplitude and phase to the wave arriving at S, and the resultant amplitude and phase thus depend on the arrangement of the points in each unit. A knowledge of the relative phases of the spectra is thus an essential part of the information required if we are to deduce from the array of spectra given by the grating the structure of a grating element. It is, of course, just this that we attempt to do in analysing a crystal, and we must now consider the bearing of the results obtained in this paragraph on the corresponding problem in crystal analysis.

(*c*) *The relationship between the two-dimensional grating and the projection of a crystal structure on a plane :* The correspondence discussed in the last paragraph between the array of spectra produced by a two-dimensional grating and the net reciprocal to the grating net is the physical analogue of the formal representation of the spectra given by a crystal by means of the reciprocal lattice, which was discussed in § 1(*g*) of this chapter. The analogy is particularly close if we reduce the crystal problem to a problem in two dimensions by considering only the projection on a plane.

As before, suppose the crystal structure to be projected parallel to the c axis onto the plane (001). The density of the projection $\sigma(x, y)$ is periodic in the directions x and y in the distances a and b respectively. For the sake of simplicity, we shall assume the c axis to be perpendicular to the plane (001), so that the latter contains not only the axes a and b, but also the reciprocal axes a^* and b^*, which simplifies the discussion without affecting the general validity of the results obtained.

Consider now a two-dimensional *optical* grating with primitive translations a and b. Let a plane wave fall normally on this grating so that each point of its surface scatters in the same phase, and let the amplitude scattered from the element of surface ds in the neighbourhood of the point (x, y) be $\sigma(x, y)ds$, $\sigma(x, y)$ being the density at the corresponding point in the crystal projection considered above. Suppose now this grating to be used instead of the point grating discussed in the last paragraph. Let the element ds be situated at Q (fig. 135), and let $\mathbf{r}$ be the vector from the origin to this point, so that $\mathbf{x}+\mathbf{y}=\mathbf{r}$. The phase difference between the radiation scattered in the direction R from the origin O and from the element ds is $2\pi(\mathbf{r}\cdot\mathbf{R})/\lambda f=2\pi(\mathbf{r}\cdot\boldsymbol{\rho})/\lambda f$ by (7.62), and so the amplitude scattered into the spectrum $S(h, k)$ by each unit mesh of the lattice will be

$$A(hk)=\int \sigma(x, y)\, e^{\frac{2\pi i}{\lambda f}(\mathbf{r}\cdot\boldsymbol{\rho})}\, ds, \tag{7.67}$$

where the integration is taken over the whole area, A, of the mesh; and, since $\boldsymbol{\rho}/f\lambda=\mathbf{r}^*$, this can be written

$$\begin{aligned} A(hk)&=\frac{A}{ab}\int_0^a\int_0^b \sigma(x, y)\, e^{2\pi i(\mathbf{r}\cdot\mathbf{r}^*)}\, dx\, dy \\ &=\frac{A}{ab}\int_0^a\int_0^b \sigma(x, y)\, e^{2\pi i\left(\frac{hx}{a}+\frac{ky}{b}\right)}\, dx\, dy, \end{aligned} \tag{7.68}$$

by (7.13).

The right-hand side of (7.68) is simply the structure factor $F(hk0)$ for the corresponding crystal unit, the projection of which on the plane (001) has the density $\sigma(x, y)$. This can be seen as follows. By (7.3),

$$\begin{aligned} F(hkl)&=\int \rho(x, y, z)\, e^{2\pi i\left(\frac{hx}{a}+\frac{ky}{b}+\frac{lz}{c}\right)}\, dv \\ &=N\int_0^a\int_0^b\int_0^c \rho(x, y, z)\, e^{2\pi i\left(\frac{hx}{a}+\frac{ky}{b}+\frac{lz}{c}\right)}\, dx\, dy\, dz. \end{aligned}$$

Then
$$\begin{aligned} F(hk0)&=N\int_0^a\int_0^b e^{2\pi i\left(\frac{hx}{a}+\frac{ky}{b}\right)}\, dx\, dy\int_0^c \rho(x, y, z)\, dz \\ &=\frac{A}{ab}\int_0^a\int_0^b \sigma(x, y)\, e^{2\pi i\left(\frac{hx}{a}+\frac{ky}{b}\right)}\, dx\, dy, \end{aligned} \tag{7.69}$$

by (7.20).

Thus the amplitudes $A(hk)$ of the spectra due to the plane grating of density $\sigma(x, y)$ correspond exactly to the structure factors $F(hk0)$ from a crystal that would produce a density distribution $\sigma(x, y)$ on the plane (001). From the reciprocal relationship between the plane grating and its spectra, derived in the last paragraph, it is clear that the points of the net $(hk0)$ of the reciprocal lattice of the crystal, weighted with the corresponding structure factors $F(hk0)$, correspond formally exactly to the array of spectra produced by a plane grating having primitive translations a and b, and a scattering density $\sigma(x, y)$ at the point (x, y) equal to the density of the projection of the unit cell of the crystal onto the plane ab. In the general case, the plane of the projection should be the plane of the reciprocal axes a^*, b^*, and not the plane ab, as will be evident from the discussion in § 1(i) of this chapter, p. 351 ; but in the case we have considered, in which the c axis is assumed to be perpendicular to the a and b axes, the two planes coincide. It is easy to modify the argument to suit the general case, but no difference in principle is introduced by so doing.

It will now be easy to carry over formally to the crystal problem any results we may deduce by considering the more familiar optical problem of the grating. That we can by no physical process produce simultaneously the set of spectra $(hk0)$ from the crystal, arranged on the reciprocal net, does not in the least impair the usefulness of the formal analogy.

(d) *Abbe's theory of image formation :* Suppose the grating in fig. 135 to lie outside the first principal focus of the lens. Then a real image of it will be formed somewhere outside the second principal focus. The question arises of the relationship of this image to the array of spectra formed in the second focal surface of the lens. For simplicity, consider the case of a linear grating consisting of a row of small, equally spaced, holes, ... O_1, O_0, O_1', fig. 136. These form a row of spectra

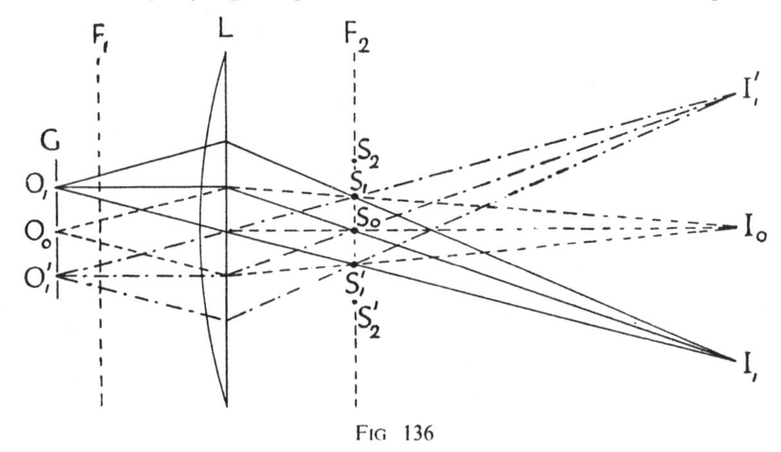

Fig 136

... S_2', S_1', S_0, S_1, S_2, ... , in the second focal plane, F_2, of the lens L, each of which may be considered as a source of light having an amplitude proportional to that of the corresponding spectrum. The phases of these sources are not independent, but the relation between them is determined by the form of the grating elements. The image may be considered as formed by the interaction of the waves from these sources. If I_1 is the image of the hole at O_1, then the optical path from O_1 to I_1 is the same by all routes through the lens, so that waves from all the spectral sources reinforce at I_1 ; and the same is true for the images of the other holes. The waves from the two corresponding spectra S_1 and S_1', the first orders on either side, produce by their interaction a set of interference fringes, and the amplitude of the disturbance due to them in the image plane I_1I_2 ... can be represented by a set of sinusoidal stationary waves. The single wave from the spectrum of zero order, S_0, produces a uniform amplitude in the image plane, and this together with the sinusoidal amplitude distribution due to the first-order spectra gives rise to a waxing and waning of the amplitude with a linear period equal to that of the grating itself, multiplied by the appropriate magnification factor for the image. The distribution of amplitude bears, however, no necessary resemblance to that transmitted by the grating itself. The two second order spectra, S_2 and S_2', produce a similar set of stationary waves with half the linear period of those given by the first-order spectra, and if these are added to the contributions from the first and zero orders, a modification of the amplitude in the image surface, bringing it in general closer to that in the object surface, is the result. The third-order spectra will produce another set of stationary waves, with a period equal to one third that of the fundamental, and so on. These successive sets of stationary waves represent in fact the Fourier components, by the sum of which the periodic distribution of amplitude transmitted by the grating can be represented. According to Abbe,[33] in order that there should be anything approaching good representation of the object by the image it is necessary for a considerable number of spectra to give their contributions to the Fourier series. If spectra are excluded, by reducing the aperture of the lens for example, there will be a distribution of amplitude in the image plane corresponding to that transmitted by a grating that would have produced the spectra actually observed, and no others. As more and more orders are excluded, the representation of the object by the image gets more and more faulty, until when only the first and zero orders are used the distribution of amplitude in the image is of the type

$$A_0 + 2A_1 \cos 2\pi x/g',$$

where g' is the grating constant multiplied by the appropriate magnification factor, whatever the actual nature of the individual grating elements. If all orders except the zero order are excluded, there will be uniform illumination in the image plane, and all trace of the periodicity of the object will have vanished in the image. If, on the other hand, the odd-order spectra are excluded by appropriate means, while the even orders are retained, the image will show a periodicity half that of the true image of the grating. A very good acount of Abbe's theory is given by A. B. Porter,[34] who also describes experiments in which various components of the array of spectra produced by a cross-grating were cut off, and it was verified that the corresponding variation of the image was in general accordance with the predictions of Abbe's theory.

The extension of the theory to the case of the cross-grating will be evident. Each pair of spectra (hk) and $(\bar{h}\bar{k})$ gives rise to a set of stationary waves in the image plane, the lines of constant phase being parallel to the rows (hk) of the grating. The image is formed by the superposition of the waves produced by each pair of corresponding spectra, which cross one another in various directions.

In practice, only a limited number of spectra can ever be used. In the case of the linear grating, with central illumination, the first-order spectrum is produced at an angle θ_1 given by $\sin\theta_1 = \lambda/g$, g being the grating constant, so that if u is the angle subtended by the radius of the lens in the plane of the grating, no periodicity will appear in the image if $\sin u < \lambda/g$. Even if the aperture of the lens is large enough to include all the spectra the grating gives, and in the practical design of microscopes it is possible to approach this condition, there is still a limit to the resolution. The number of spectra that the grating gives is limited by the wave-length; for if θ_m is the angle of diffraction of the mth-order spectrum, $\sin\theta_m = m\lambda/g$, and it follows that $m \ngtr g/\lambda$. The shorter the wave-length, therefore, the greater the number of spectra. Now the amplitude of any given grating spectrum relative to that of the spectrum of zero order is a function of $(\sin\theta)/\lambda$, and so is independent of the wave-length for a given order. If then the theory outlined above is correct it follows that each additional spectrum must give more information about the grating. In principle, therefore, an infinite number of spectra, corresponding to an infinitely short wave-length, are required to give complete representation of the object by the image, while if $\lambda > g$, no trace of periodicity can appear at all.

Principles similar to those discussed above apply if the structure of the object is not periodic, but the image must then be represented by a Fourier integral, instead of by the sum of a Fourier series.

(*e*) *The relationship between the spectra and the Fourier components of the image:* It is not difficult to calculate the wave-length and the amplitude of the standing waves produced in the image surface by the waves coming from the pairs of spectra, and to show their correspondence to the terms of the Fourier series representing the form of the grating, but it is perhaps instructive to proceed rather differently. Let us suppose the light, when it has reached the spectrum surface, to be reversed, so that each spectrum may now be considered as a source giving rise to waves that pass out again through the lens. If we consider the spectra to be point sources, each gives rise to a train of plane waves, and the interaction of the plane waves from all the spectra should build up again in the plane of the grating a distribution of amplitude equal to that which left it; or, in other words, they should combine to give an image of the grating coinciding with the grating itself. This image may be an imperfect representation of the grating for two reasons. First, owing to the finite aperture of the lens, not all the diffracted waves given by the grating with the wave-length actually used are able to form spectra in the focal surface. Not all the trains of waves given by the grating are therefore present in the image built up by reversing the waves from the spectra. Secondly, the number of spectra given by the grating depends on the wave-length of the light used, or, in other words, the modulation of the amplitude of the primary wave-train by the grating depends on the wave-length of the light. Periodicities in the grating on a scale appreciably smaller than the wave-length of the light give rise to no corresponding periodic variations in the amplitude of the wave transmitted by the grating. The spectra produced relate in fact to periodicities on a scale greater than the wave-length, and detail on a scale finer than this does not affect the spectra, and so can have no counterpart in the image.

Let us consider the spectra formed by the plane grating discussed in §§ 2(*b*) and 2(*c*) of this chapter. We take the spectra in pairs, fixed by the equal and opposite vectors $\boldsymbol{\rho}(hk)$ and $\boldsymbol{\rho}(\bar{h}\bar{k})$. In fig. 137, let

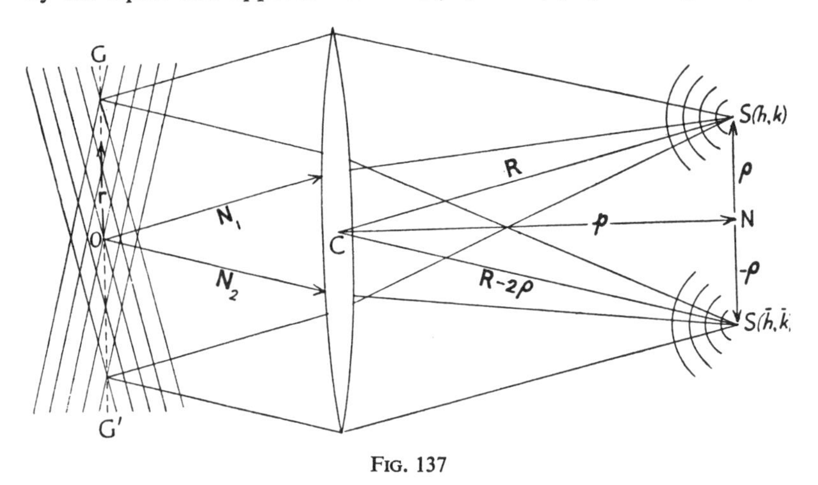

FIG. 137

S(hk) and S$(\bar{h}\bar{k})$ be the two spectra, and let A(hk) and A$(\bar{h}\bar{k})$ be the corresponding amplitudes, which may be complex and include a phase factor, but which will be conjugate complex quantities, so long as the actual process of scattering by the grating introduces no relative phase differences. The two trains of waves to which these spectra give rise are indicated in fig. 137, and what is required is their resultant in GG′, the plane of the grating.

We choose as origin the point O in which the plane GG′ intersects the axis of the lens. The displacement at a point determined by a vector $\mathbf{r}$ from the origin due to a plane wave of amplitude A is $Ae^{2\pi i(\nu t-\mathbf{N}\cdot\mathbf{r})}$, where $\mathbf{N}$ is the wave-vector, a vector of magnitude $1/\lambda$ in the direction of the wave-normal. Let $\mathbf{N}_1$ and $\mathbf{N}_2$ be the wave-vectors of the two plane waves that give rise to the spectra $S(hk)$ and $S(\bar{h}\bar{k})$ respectively. When the whole wave-system is reversed, the signs of these vectors are reversed, and the displacement $D(\mathbf{r})$ due to the two waves at the point fixed by the vector $\mathbf{r}$ is given by

$$D(\mathbf{r}) = e^{2\pi i\nu t}\{A(hk)e^{2\pi i\mathbf{N}_1\cdot\mathbf{r}} + A(\bar{h}\bar{k})e^{2\pi i\mathbf{N}_2\cdot\mathbf{r}}\}. \qquad (7.70)$$

Now $\mathbf{N}_1$ is in the direction of $\mathbf{R}$, and has magnitude $1/\lambda$, and since $|\mathbf{R}| = f$,

$$\mathbf{N}_1 = \frac{1}{\lambda f}\mathbf{R} = \frac{1}{\lambda f}(\mathbf{p}+\boldsymbol{\rho}), \qquad (7.71)$$

where $\mathbf{p}$ is the vector CN ; and similarly,

$$\mathbf{N}_2 = \frac{1}{\lambda f}(\mathbf{R}-2\boldsymbol{\rho}) = \frac{1}{\lambda f}(\mathbf{p}-\boldsymbol{\rho}). \qquad (7.72)$$

We are interested in the disturbance only in the plane of the grating, so that $\mathbf{r}$ is perpendicular to $\mathbf{p}$ and the scalar product $\mathbf{r}\cdot\mathbf{p} = 0$. Thus,

$$\mathbf{N}_1\cdot\mathbf{r} = \frac{1}{\lambda f}(\boldsymbol{\rho}\cdot\mathbf{r}), \quad \mathbf{N}_2\cdot\mathbf{r} = -\frac{1}{\lambda f}(\boldsymbol{\rho}\cdot\mathbf{r}). \qquad (7.73)$$

Since $A(hk)$ and $A(\bar{h}\bar{k})$ are conjugate, if

$$A(hk) = |A(hk)|e^{-i\delta(hk)}, \text{ then } A(\bar{h}\bar{k}) = |A(hk)|e^{i\delta(hk)}. \qquad (7.74)$$

Substitution of (7.73) and (7.74) in (7.70) then gives at once for $D(\mathbf{r})$, the disturbance due to the two waves at a point in the plane of the grating fixed by the vector $\mathbf{r}$,

$$D(\mathbf{r}) = 2|A(hk)| \cos\left\{\frac{2\pi}{\lambda f}(\boldsymbol{\rho}\cdot\mathbf{r}) - \delta(hk)\right\} e^{2\pi i\nu t}. \qquad (7.75)$$

By equation (7.66) of VII, § 2(*b*), $\boldsymbol{\rho}/\lambda f = \mathbf{r}^*(hk)$, the vector to the point (h, k) of the net reciprocal to the grating net; and if (x, y) are the components of the vector $\mathbf{r}$ referred to axes parallel to the translations a and b of the grating, then by (7.13)

$$\mathbf{r}\cdot\mathbf{r}^* = hx/a + ky/b,$$

since $z = 0$. Thus (7.75) may be written

$$D(x, y) = 2|A(hk)| \cos\left\{2\pi\left(\frac{hx}{a}+\frac{ky}{b}\right) - \delta(hk)\right\} e^{2\pi i\nu t}. \qquad (7.76)$$

The total amplitude $\sigma(x, y)$ at the point (x, y) in the plane of the grating is therefore

$$\sigma(x, y) = \mathrm{A}(00) + \sum_{h, k} 2|\mathrm{A}(hk)| \cos\left\{2\pi\left(\frac{hx}{a} + \frac{ky}{b}\right) - \delta(hk)\right\}, \quad (7.77)$$

which is a Fourier series of exactly the same form as the double series (7.22) derived for the projection of the crystal structure on a plane. Each term represents the amplitude of a standing wave such as (7.76). The nodal and antinodal lines of such a wave are parallel to the rows (h, k) of the grating, and the wave-length is equal to the spacing of the rows in the extended sense. To each pair of spectra $\mathrm{S}(hk)$ and $\mathrm{S}(\bar{h}\bar{k})$ there corresponds one such stationary wave.

The sum of the series (7.77) gives the amplitude at each point of the optical image of the grating. In so far as the series is incomplete, owing to the omission of terms of finite magnitude, the representation of the object is imperfect. The series may be incomplete because the aperture of the lens is not large enough to receive all the diffracted wave-trains, and will in any case be incomplete because the number of wave-trains is limited by the wave-length of the radiation used.

We have seen in VII, § 2(*c*), how the two-dimensional projection $\sigma(x, y)$ of the crystal structure corresponds formally exactly with the distribution of transmitted amplitude from a certain two-dimensional grating having the same translations. The array of spectra produced by the grating forms a network reciprocal to the grating net, and corresponds to the array of lattice spectra on the net $(hk0)$ of the reciprocal lattice of the crystal. We now see how the optical image of the grating is represented by a Fourier series of exactly the same type as that giving the density $\sigma(x, y)$ of the projection of the crystal structure. This projection must therefore be regarded formally as an optical image, and it will be subject to the imperfections of optical images. In practice, the number of spectra we can obtain from a crystal is always limited by the wave-length of the radiation used. Limiting the number of spectra we now see to be equivalent to reducing the aperture of the optical device used to examine the structure. We know that optically this gives rise to false detail in the image, and to lack of exact correspondence between image and object, and we must therefore be on the look-out for similar imperfections in the Fourier projection of the crystal structure. In the next paragraph we shall consider this point in more detail.

The discussion given above assumes the spectral sources in the second focal plane of the lens to be points, which give rise on reversal of the waves through the lens to plane waves. A set of point spectra corresponds to a grating of infinite extent, and their actual formation requires a lens of unlimited aperture, as does the formation of truly plane waves on their reversal. We have tacitly assumed the existence of such a lens, but have used only a limited number of the point spectra to produce the optical image in the plane GG′. The number of diffracted wave-trains actually produced, even by an infinite grating, depends on the wave-length used, and the number of these that can form spectra in the second focal plane of the lens on the aperture of the lens. Using a limited number of spectra as sources, and so a limited number of plane waves to form the image, we may therefore conveniently speak of by analogy as limiting the aperture of the equivalent optical system ; but it must be remembered that the analogy is not completely accurate.

If the grating is of limited extent, the spectra will not be points, but will be formed by diffracted wave-trains, each of which spreads over a range of angles inversely proportional to the linear dimensions of the grating, and a limitation of the aperture of the lens used to form the spectra will have a similar effect. Let us, however, assume the aperture of the lens to be of unlimited extent, so that any point source in the second focal plane corresponds to a truly plane wave on reversal. If the grating is of limited extent, each spectrum extends over a finite area of the focal plane, and the reversed wave-train corresponding to it will not be a single plane wave-train but a slightly divergent one containing a continuously variable range of wave-vectors. The addition of a number of such slightly divergent wave-trains produces in the plane GG′ a set of maxima in the regions of the actual grating points, but effectively zero displacement in regions outside that occupied by the actual grating. The Fourier series has become a Fourier integral, representing a structure that is no longer truly periodic, because no longer of infinite extent.

Effects of this kind need only be considered if the linear dimensions of the grating are comparable with the wave-length of the radiation used, a condition not met with in ordinary crystal analysis, to which the analogies worked out in the first part of this paragraph apply with all necessary accuracy. The discussion of diffraction by very small crystals will be taken up in Chapter X.

(*f*) *Diffraction effects in the Fourier projection :* We have now to consider how far the individual maxima in a two-dimensional Fourier projection, in which the atomic centres are well separated, correspond to the true projections of the scattering matter in the atoms onto a plane. To do this, we shall consider the projection of a crystal composed of atoms all of one kind, situated at the corners of a primitive space-lattice, and scattering as points. We shall see that by considering this simple case we can derive quite useful results, which may be applied with a fair degree of approximation to the case of scattering by an actual atom. The scattering factor for a point atom is independent of the direction of scattering, and so we may suppose the structure factors to be equal to F(0) for all spectra given by the crystal. The component terms of the Fourier series representing the projection will therefore all have the same amplitude.

As before, let the direction of projection be that of the *c* axis, which we shall further suppose perpendicular to the *ab* plane of the crystal lattice. This plane will then also be the a^*b^* plane of the reciprocal lattice. The reciprocal-lattice points corresponding to the spectra $hk0$ concerned in the projection lie in this plane. Let us suppose that in making the projection all spectra are included whose reciprocal-lattice points lie on the plane within a distance R* of the origin. If then θ_0 is the greatest glancing angle for any spectrum used,

$$2 \sin \theta_0 \lesseqgtr \lambda R^* \tag{7.78}$$

and $R^* \lesseqgtr 1/d_0$, where d_0 is the corresponding smallest spacing for which a spectrum is measured.

If A is the area of the mesh *ab* of the crystal lattice, the area of the reciprocal-lattice mesh is 1/A, and a number of spectra approximately equal to $\pi R^{*2}A$ will be used in making the projection.

The corresponding two-dimensional optical grating will consist of scattering points arranged on a network with translations *a* and *b*. Let u_0 (fig. 138) be the greatest angle that a wave-train diffracted by the

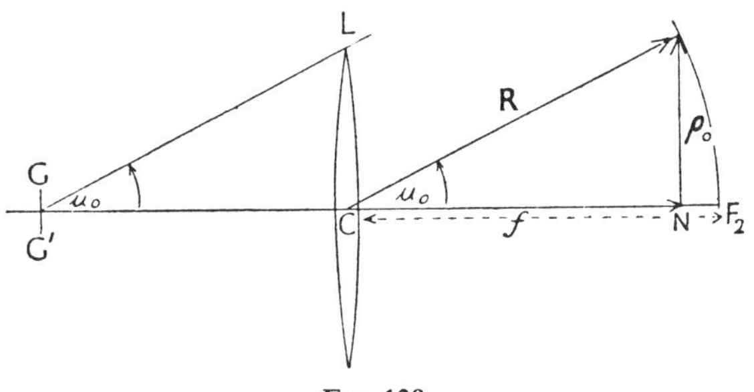

FIG. 138

grating may make with the axis if it is to enter the lens, so that $\sin u_0$ is the numerical aperture of the equivalent optical system. All the spectra lie within a circle of radius ρ_0 on the focal surface, and from fig. 138, in which the notation is the same as in figs. 135 and 137, we see that $\sin u_0 = \rho_0/f$. But, by (7.66), $\rho_0/f\lambda = R^*$, the corresponding greatest value of the reciprocal vector, and so

$$\sin u_0 = \lambda R^*. \tag{7.79}$$

On comparing (7.78) and (7.79), we see that if all spectra up to a glancing angle θ_0 are included in making the projection of the crystal structure, the numerical aperture of the equivalent optical system is $2 \sin \theta_0$.

Suppose now the spacing of the lattice is large in comparison with λ. Then a very large number of spectra will be included, and we may suppose the corresponding points to be spread uniformly in the reciprocal-lattice plane with a surface density equal to A points per unit area. As A becomes larger and larger, the equivalent optical case tends to the image of a single point, and the series $\sigma(x, y)$ for the density of the projection tends to an integral as a limit, whose value we shall now determine.

For the simple lattice we are considering, if one of the lattice-points is taken at the origin the values of $\delta(hk)$ are all zero, and the coefficients $F(hk0)$ are all positive; and since they are also all equal to $F(0)$, the series (7.23) can be written

$$\sigma(x, y) = \frac{F(0)}{A} \sum_{-h}^{+h} \sum_{-k}^{+k} \cos 2\pi \left(\frac{hx}{a} + \frac{ky}{b}\right),$$

or

$$\sigma(\mathbf{r}) = \frac{F(0)}{A} \sum_{-h}^{+h} \sum_{-k}^{+k} \cos 2\pi (\mathbf{r} \cdot \mathbf{r}^*), \tag{7.80}$$

where $\mathbf{r}$ is the vector to the point (x, y) in the plane of the projection, and $\mathbf{r}^*$ is the reciprocal-lattice vector to the point (h, k).

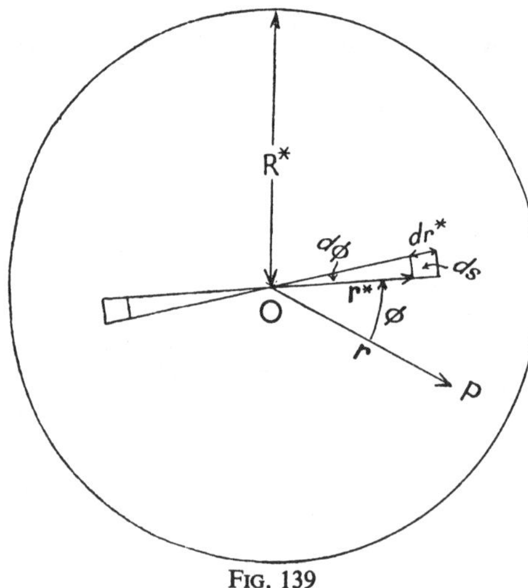

FIG. 139

Let the circle in fig. 139 be the circle of radius R^* drawn on the reciprocal-lattice plane, on which we now suppose the points to be closely spread with a surface density A, so that we can suppose the vector $\mathbf{r}^*$ to a reciprocal-lattice point to be continuously variable. The element of area ds at the point (r^*, ϕ) contains $Ar^*dr^*d\phi$ points, and together with the corresponding element of area at $(-r^*, \pi + \phi)$ contributes an amount

$$\frac{2F(0)}{A} Ar^* \cos 2\pi (\mathbf{r} \cdot \mathbf{r}^*) dr^* d\phi$$

to the sum (7.80). In fig. 139, the plane of the projection and the reciprocal-lattice plane are shown superposed with a common origin O. P is the point $\mathbf{r}$ in the plane of projection for which the value of the density $\sigma(\mathbf{r})$ is to be calculated. The scalar product $\mathbf{r} \cdot \mathbf{r}^*$ can be written $rr^* \cos \phi$, where ϕ is the angle between the directions of $\mathbf{r}$ and $\mathbf{r}^*$, and the whole sum can be replaced by the integral

$$\sigma(r) = 2F(0) \int_0^{R^*} \int_0^{\pi} r^* \cos (2\pi rr^* \cos \phi) dr^* d\phi. \tag{7.81}$$

This integral is one that occurs in the theory of diffraction by circular apertures, and contains a Bessel function of unit order. If $J_0(t)$ and $J_1(t)$ are Bessel functions of zero and unit order respectively, then

$$\int_0^{\pi} \cos(t\cos\phi)\,d\phi = \pi J_0(t) \tag{7.82†}$$

and

$$\int_0^{z} tJ_0(t)\,dt = zJ_1(z). \tag{7.83†}$$

Using (7.82), we may write (7.81) in the form

$$\sigma(r) = 2\pi F(0)\int_0^{R^*} r^* J_0(2\pi r r^*)\,dr^*$$

or, by (7.83),

$$\sigma(r) = 2F(0)\pi R^{*2}\frac{J_1(2\pi r R^*)}{2\pi r R^*}. \tag{7.84}$$

Equation (7.84) gives the value of the density $\sigma(r)$ at a distance r from one of the lattice-points. The factor $F(0)\pi R^{*2}$ can be left out of account in discussing the distribution of density, which is determined by the factor $2J_1(m)/m$, where

$$m = 2\pi r R^*. \tag{7.85}$$

The variation of this factor with m is not unlike that of the familiar diffraction factor for a slit, $(\sin m)/m$. It has a principal maximum at $m=0$, equal to unity, and subsidiary lateral maxima and minima. Its zero values are given by the roots of the equation $J_1(m)=0$,‡ with the exception of $m=0$. The first four roots, other than $m=0$, occur at

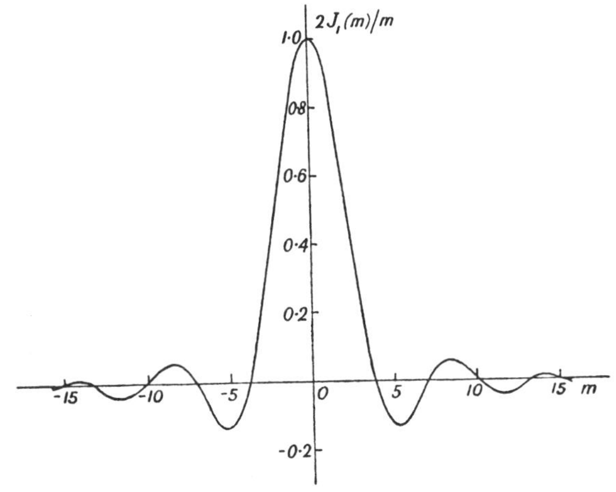

FIG. 140. Graph of the function $2J_1(m)/m$

† See, for example, G. N. Watson, *The Theory of Bessel Functions*, Camb., 1922, pp. 21, 45.

‡ See Table X. 3, p. 575.

values of m equal to 3·832, 7·016, 10·173, and 13·323. In fig. 140, the curve of $2J_1(m)/m$ is shown plotted as a function of m. The curve falls to zero for the first time when $m=3{\cdot}832$, or when

$$r=\frac{3{\cdot}832}{2\pi R^*}=\frac{0{\cdot}61}{R^*}=\frac{0{\cdot}61\lambda}{2\sin\theta_0}=\frac{0{\cdot}61\lambda}{a} \tag{7.86}$$

by (7.78) and (7.79), a being the numerical aperture of the equivalent optical system.

The projections of the point-like atoms made by using a limited number of spectra are therefore not points. If all the measurable spectra up to a given value of the glancing angle are used, the result is equivalent optically to that of using a lens of finite circular aperture to form an image of a point source. An appreciable diffraction disc extends over a radius given by (7.86), and is surrounded by alternately negative and positive minima and maxima. In an actual optical image, the negative regions do not occur, since the intensity is proportional to the square of the amplitude; but in the case of the Fourier projection we are actually plotting resultant amplitudes and not intensities. Since every point source will produce in the projection a diffraction disc of radius $0{\cdot}61/R^*$ it follows that no detail of the structure can be distinguished in the projection unless it is on a coarser scale than this. Suppose, for example, we measure all spectra up to a glancing angle 30°, so that $2\sin\theta_0=1$, then no detail on a scale finer than $0{\cdot}61\lambda$ could be distinguished, and if the observations were made with CuKα radiation ($\lambda=1{\cdot}54$A) the limit of resolution would be about 0·94A, which is too coarse to show any of the finer details of the atomic structure. In practice, greater glancing angles and shorter wave-lengths are sometimes used, but it is scarcely possible to increase this degree of resolution by a factor of four. The greatest possible value of R^* that can ever be used is the radius of the limiting sphere (Chapter I, § 2(a)), which is $2/\lambda$, and this gives a theoretical limit of $0{\cdot}3\lambda$, which, however, can never quite be reached in practice.

We may express this result rather differently. If d_0 is the smallest spacing of the reflecting planes used in the projection, the term ‘ spacing ’ being used in its extended sense, then d_0 is the wave-length of the shortest Fourier wave that goes to build up the projection. If R^* is the length of the corresponding reciprocal-lattice vector, $R^*=1/d_0$, and by (7.86)

$$r=0{\cdot}61d_0. \tag{7.87}$$

Detail on a smaller scale than this cannot therefore be resolved.

The integral form (7.81) of the series for $\sigma(r)$ is not of course periodic, and is the limiting form when the spacing becomes indefinitely great, and when the atoms scatter as points ; but nevertheless, the analysis given is of fairly direct application to the case of an actual crystal. As we have seen in Chapter III, § 4(e), much of the diffracting matter in an atom lies within a distance of the atomic centre considerably less than the wave-lengths of the radiations used in crystal analysis. For these inner groups the value of the scattering factor f falls off very slowly with increasing θ, and is not greatly reduced even at the largest glancing angles that can be used. The condition of the point-like atom is thus fairly well satisfied for such groups. Moreover, if the spacing of the crystal is not too small, well over 100 spectra may be included in the projection, and the error made by treating the summation as an integration is not very great. We may therefore expect the inner electron groups of the atom to produce a distribution of density in the projection not unlike that given by (7.84), with rings of alternate positive and negative density surrounding the central maximum. These will be superposed on the distribution due to the outer electrons, which produce no appreciable diffraction rings, since the value of f for them falls off rapidly, and has as a rule become negligible at the largest glancing angles used. The series must, however, be terminated while the values of its coefficients are appreciable, owing to the slow decrease of f for the inner electron groups, and it is this that gives rise to the diffraction rings. This point was tested in a very direct way by W. L. Bragg and J. West,[35] from whose work the treatment given above is taken, with certain inessential differences in the method of presentation.

(*g*) *Some examples of false detail due to diffraction:* Bragg and West considered the projection of an artificial rock-salt crystal, in which the radial distribution of the electrons in the atoms was that calculated by Hartree by means of the method of the self-consistent field. It has been shown by James, Waller, and Hartree [36] that the observed X-ray diffraction from a rock-salt crystal corresponds very closely with that to be expected from the Hartree distribution if an appropriate temperature factor is applied. Bragg and West considered two cases, one in which the atoms, or rather ions, were supposed at rest, without zero-point energy, at the lattice-points, and the other in which they were supposed subject to the proper thermal motions at room temperature, 290° Abs.

The projection was made on the plane (110). The atoms are not entirely separated, even in this projection, but it is possible to disentangle with some certainty the effects due to the partial overlap of their projections. In the case of the crystal with the atoms at rest, the diffraction rings show up very clearly round the projections of both atoms. In fig. 141, taken from the paper by Bragg and West, the density $\sigma(r)$ in electrons per A^2 for the case in which the atoms are at rest is shown plotted as a function of r, the distance from the atomic centre, for the projections of Cl^- and Na^+, and on the same figure the positions of the maxima and minima of the diffraction pattern obtained on the assumption that the atoms scatter as points are indicated by means of arrows. The projection was made assuming RhKα radiation ($\lambda = 0{\cdot}615$A) to be used, and all spectra up to a maximum value of θ_0

equal to 30° are included, so that the numerical aperture $2 \sin \theta_0$ is equal to unity, and $d_0 = 0{\cdot}615$A. The greatest negative value of the

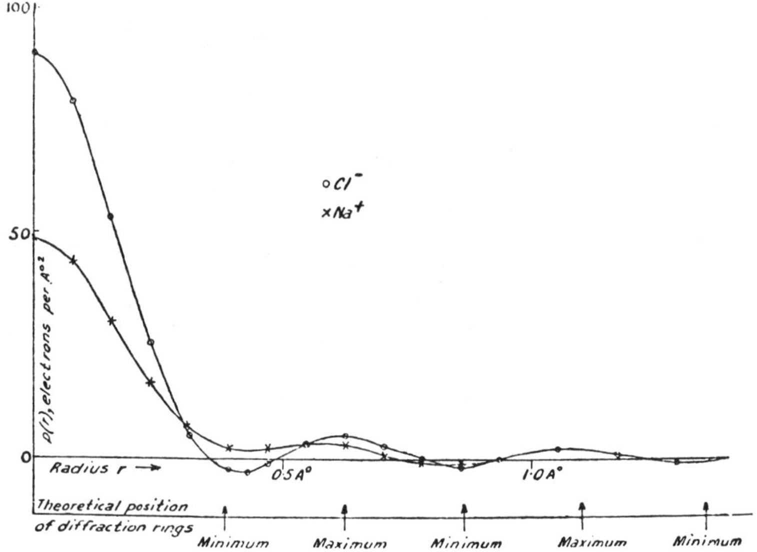

FIG. 141. Diffraction rings of Cl^- and Na^+ compared with optical rings (Bragg and West)
(*Phil. Mag.*, **10**, 823 (1930))

first diffraction ring should occur at $0{\cdot}81\, d_0$, and the greatest positive value of the next at $1{\cdot}33\, d_0$. The general agreement between the theoretical and observed sizes of the central diffraction maximum, which is about the same for both Na and Cl, as it should be, is very striking. The agreement is better for Cl^- than for Na^+, but this again is as it should be, since there is a greater concentration of electrons towards the centre of the heavier atoms. The maxima and minima have quite evidently nothing whatever to do with electronic rings in the atoms ; indeed the projection of the actual atomic electron-cloud onto a plane shows no maxima and minima, and $d\sigma/dr$ should be everywhere negative.

The projection made using the values of f appropriate to the atoms in thermal motion at room temperature shows only vestiges of the diffraction rings, and no negative densities. The spreading of the average electron distribution near the centre of the atom is enough to do away with most of the false detail, although of course no more information about the actual distribution of the electrons in the atom can be obtained from this projection than from the other. Bragg and West point out, however, that if it is desired to count the number of electrons in each atom by estimating from the projection the value of the integral $\int \sigma(r) ds$, ds being an element of area of the projection, for each peak, the projection in which the atoms are subjected to thermal motions is the better one for the purpose. The structure factors of the spectra have fallen effectively to zero at the largest angle of diffraction employed, and this has the effect of trimming up the projection, and of removing the spurious rings, while leaving the total value of $\int \sigma(r)ds$ for the unit cell unchanged; for this is always equal to F(000). It will still be impossible to assign with certainty an exact number of electrons to each atom, for the outer regions of the projections of the different atoms overlap considerably. Great care must therefore be used in drawing any conclusions about the state of ionisation of the atoms in the crystal from such electron counts.

In many crystals, particularly those composed of the lighter atoms, such as carbon, oxygen, or nitrogen, the natural thermal movements are enough to do away with diffraction rings in the projection. If the crystal contains heavier atoms the rings may still be quite evident, even in the actual crystal subjected to thermal motion. An interesting case of this was discussed by Parker and Whitehouse,[37] who made careful projections of the structure of iron pyrites, FeS_2, in an attempt to determine the electron distribution in the homopolar bond in the S_2 group. They found definite diffraction rings round both the iron and the sulphur atoms, which completely masked the effect looked for. By applying an artificial temperature factor, as suggested by Bragg and West, to make the Fourier series converge within the number of terms they were able to employ, they got rid of the false detail due to the rings. The electron count of 25·9 for each iron atom indicated that the iron was not ionised; on the other hand, when the diffraction effects were removed, the details of the S_2 group were so smeared out by the necessary thermal movement that no indication of the probable electron distribution between the two atoms could be obtained. If heavy atoms such as iodine are present, the diffraction rings due to them may completely mask the positions of the lighter atoms, especially if long waves, such as those of CuKα, are used.

3. Applications of the Methods of Fourier Analysis to the Determination of the Electron Distribution in Atoms.

(*a*) *Fourier integrals:* Let $f(x)$ be a function periodic in x with a period a, so that it may be expressed by a one-dimensional Fourier series of the type

$$f(x) = \sum_{n=0}^{n=\infty} A(n) e^{-2\pi i \frac{nx}{a}}, \qquad (7.88)$$

where

$$A(n) = \frac{1}{a}\int_0^a f(x) e^{2\pi i \frac{nx}{a}}. \qquad (7.89)$$

Now let the identity period a increase, until it finally becomes infinite. The values of $2\pi nx/a$ for successive values of n become more and more nearly equal.

Put
$$\frac{2\pi n}{a} = p. \tag{7.90}$$

Then as a increases to infinity, p becomes a continuously variable quantity, and the series (7.88) tends to a certain integral. Let $A(p)\,dp$ be the total amplitude contributed to (7.88) in the short range dp of the variable p. Then the series (7.88) can be written

$$f(x) = \int_{-\infty}^{+\infty} A(p) e^{-ipx}\, dp. \tag{7.91}$$

The integral theorem of Fourier (see Appendix IV) enables us to write down the coefficient $A(p)$ as an integral with respect to x, the analogue of equation (7.89). The theorem states that if $f(x)$ is given by equation (7.91), then

$$A(p) = \frac{1}{2\pi} \int_{-\infty}^{+\infty} f(x) e^{ipx}\, dx. \tag{7.92}$$

Equations (7.91) and (7.92) are the generalisations of Fourier's series to functions that are single-valued but no longer periodic.

If the function $f(x)$ is symmetrical about $x=0$, it will be expressible in the form

$$f(x) = \int_{0}^{\infty} A(p) \cos px\, dp, \tag{7.93}$$

and in this case it may be shown that

$$A(p) = \frac{1}{\pi} \int_{-\infty}^{+\infty} f(x) \cos px\, dx; \tag{7.94}$$

while for a function antisymmetrical in x the corresponding forms are:

if
$$f(x) = \int_{0}^{\infty} A(p) \sin px\, dp, \tag{7.95}$$

then
$$A(p) = \frac{1}{\pi} \int_{-\infty}^{+\infty} f(x) \sin px\, dx. \tag{7.96}$$

A derivation of this Fourier integral will be found in Appendix IV.

(*b*) *The determination of the radial electron density*, $U(r)$, *from the atomic scattering factor*, f: In Chapter III, § 2(*b*), equation (3.4), we obtained for the atomic scattering factor f the expression

$$f = \int_{0}^{\infty} U(r) \frac{\sin \mu r}{\mu r}\, dr, \tag{7.97}$$

where $\mu = 4\pi(\sin\theta)/\lambda$, and $U(r)$ is the radial electron density, the number of electrons lying between radial distances r and $r+dr$ from the centre of the atom being $U(r)dr$. If $U(r)$ is known as a function of r it is easy to calculate f as a function of μ or of $(\sin\theta)/\lambda$ from equation (7.97); and the values of the f-curves given in the tables, and used in crystal analysis, are calculated in this way from various assumed forms of $U(r)$, such as those of Hartree, Thomas, or Pauling and Sherman. This aspect of the matter was discussed fully in Chapter III, Section 4. We must now consider the converse problem.

Suppose we have determined f as a function of μ, by experiment. Then equation (7.97) can be regarded as an integral equation for $U(r)$. If we can solve it, we can determine the radial electron distribution in the atom from the atomic scattering factor.

Equation (7.97) can be written

$$\mu f(\mu) = \int_0^\infty \frac{U(r)}{r} \sin(\mu r)\, dr, \tag{7.98}$$

and has then exactly the same form as (7.95). By (7.96) therefore

$$\frac{U(r)}{r} = \frac{1}{\pi}\int_{-\infty}^{+\infty} \mu f(\mu) \sin(\mu r)\, d\mu \tag{7.99}$$

$$= \frac{2}{\pi}\int_0^\infty \mu f(\mu) \sin(\mu r)\, d\mu. \tag{7.100}$$

The integral (7.100) can be evaluated graphically or numerically for any desired value of r if $f(\mu)$ is known as a function of μ. In principle then the problem is solved, but in actual cases there are serious practical difficulties, connected with the rate of convergence of the integral; for the factor $\mu f(\mu)$ decreases very slowly with increasing μ, owing to the form of the atomic f-curves.

A. H. Compton,[38] and Wollan,[39] have applied equation (7.100) to determine the radial electron density for the monatomic gases helium, neon, and argon, using values of the scattering factors determined by direct measurement of the scattering of X-rays by the gases. Such measurements have been made by Barett,[40] Herzog,[41] Wollan,[42] and others, and will be discussed in more detail in Chapter IX. The radiation scattered from a gas contains incoherent as well as coherent radiation, and this must be corrected for, since equation (7.97) applies essentially to the coherent radiation. The method of applying the correction is discussed in Chapter IX, § 1(*b*). We shall here suppose it to have been made already to the scattering factor $f(\mu)$.

In evaluating the integral we first plot $\mu f(\mu)$ as a function of μ. Some examples of such curves are shown in fig. 142. They are equivalent to the $B(x)$ curves of Compton and Wollan. Compton's $B(x)$ is equal to $(2/Z)\mu f(\mu)$, Z being the total number of electrons in the atom, and his x is μ/π. We shall denote $\mu f(\mu)$ by $B(\mu)$. Suppose now we wish to determine $U(r)/r$ for a given value of r, say r_1. For any convenient set of values of μ we multiply the corresponding ordinates of the $B(\mu)$ curve by the values of $\sin(\mu r_1)$, and plot the resulting products against μ. The area between this curve and the axis of μ gives the value of $U(r_1)/r_1$. The difficulty of carrying out this process in practice is due to the incomplete knowledge we have of the form of the curve for $\mu f(\mu)$ for large values of μ. At the largest values of μ

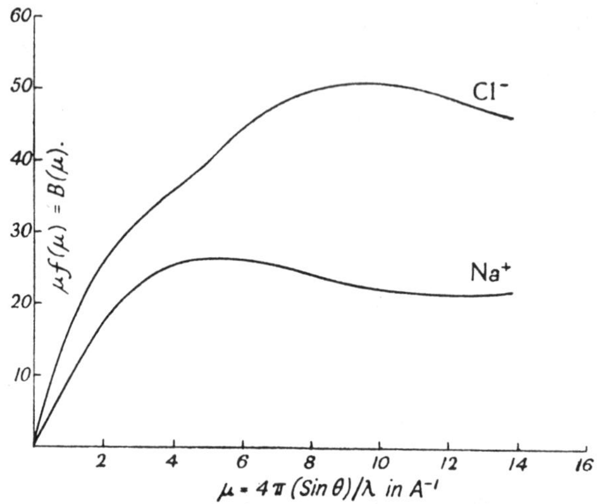

FIG. 142. Curves showing $\mu f(\mu)$ as a function of μ, for Cl^- and Na^+

for which f can be measured in any practical case $\mu f(\mu)$ is still large, and it is very difficult to estimate in what way it will tend to zero. To illustrate this, we have drawn in fig. 142 the $B(\mu)$ curve for the Hartree distributions Cl^- and Na^+ up to values of $(\sin\theta)/\lambda = 1{\cdot}1$, or $\mu = 13{\cdot}83$. It will be seen how little information these curves give as to the way in which they tend to zero for large values of μ. In practice, measurements of the scattering cannot be made for values of μ very much greater than this, because it becomes necessary to use inconveniently hard radiation. Nor can we easily measure the scattering for very small values of μ; but here we may get some help from the slope of the curve at the origin. If $B(\mu) = \mu f(\mu)$, then $dB/d\mu = f(\mu) + \mu df/d\mu$. When $\mu = 0$, $f(0) = Z$, the total number of electrons in the atom, and $(df/d\mu)_0 = 0$ from the general form of the diffraction curves, so that $dB/d\mu = Z$, when $\mu = 0$, which gives the slope of the curve at the origin.

For large values of r, the function $B(\mu)\sin\mu r$ oscillates rapidly, and the exact form of the extrapolated curve for $B(\mu)$ matters very little. This is only another way of saying that the electronic distribution at large radii affects the form of the f-curve only for small values of $(\sin\theta)/\lambda$. For small values of r, however, when $\sin(\mu r)$ oscillates only slowly, the extrapolated part of the $B(\mu)$ curve becomes of the first importance. The way in which the f-curve falls off for large values of $(\sin\theta)/\lambda$ depends almost entirely on the electronic distribution $U(r)$ for small values of r; and conversely, if we are trying to determine $U(r)$ from f, it is the values of f for large $(\sin \theta)/\lambda$ that are significant. All errors in measuring f, or in estimating it by extrapolation, are magnified by multiplication by μ, and no reliable information about the inner

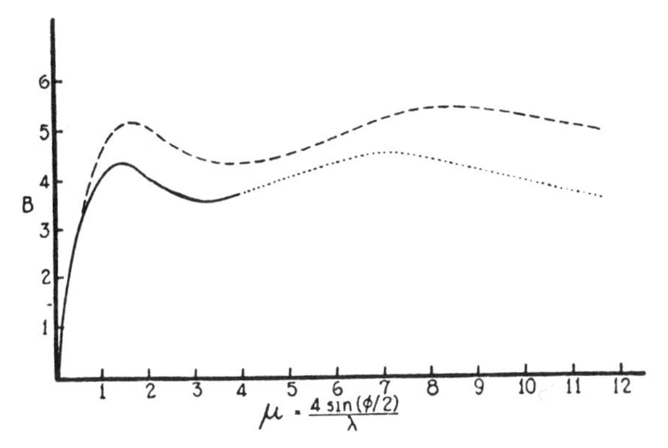

FIG. 143 *a*. Broken line is $B(\mu)$-curve for Na^{+} from the Hartree atom. The solid line is the experimental part of B-curve for Ne. The dotted line shows how this curve was extrapolated

electrons can be obtained by this method. This is of course just what the principles of optics would lead us to expect. We cannot resolve detail on a scale smaller than that of the wave-length of the radiation used whatever mathematical means we may employ to discuss the observations. We can get the outer part of the B curve in practice only by using short wave-lengths; for $\sin \theta$ cannot be greater than unity, and μ is always less than $4\pi/\lambda$. In any extensive extrapolation of the B curve, we are putting into it the results of preconceived ideas, based on theoretical knowledge of the electronic distribution in the atom. We naturally get out of our calculations something not unlike the distribution upon which the extrapolation was founded; but the process is in no sense a direct determination of the details of the electronic distribution. In figs. 143 *a* and 143 *b* the B curve, and the resulting calculated distribution

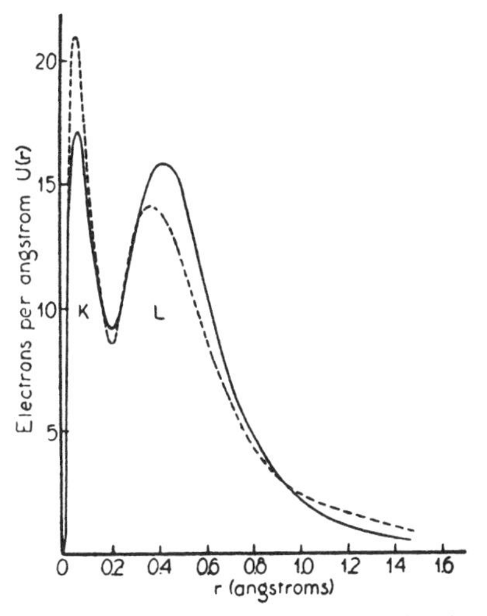

FIG. 143 *b*. Radial electron distribution curves for Ne. The solid line calculated from the extrapolated $B(\mu)$ curve of fig. 143(*a*). The dotted line is the Hartree distribution

for neon, taken from a paper by Wollan,[39] are shown. The dotted part of the B curve is due to extrapolation, based on the known variation of B for the Hartree distribution Na^+. It is plain that without any knowledge of this distribution it would have been impossible to predict the position of the second maximum in the B curve. The inner part of the distribution in fig. 143 *b* is thus experimental only in a very limited sense. The whole K peak lies within a distance of 0·2 A of the atomic centre, and a wave-length of 0·7 A was used in the experiments. The detail in the curve is thus well below the limits of possible experimental resolution.

(*c*) *Series for the radial electron distribution:* An expression for the radial distribution $U(r)$ was first given by A. H. Compton [8] in the form of a series. He considered the atoms as forming a crystal lattice of spacing D, wide enough for there to be no appreciable overlapping of the electronic distributions of neighbouring atoms. For such a crystal, the one-dimensional series (7.45) gives $S(x)$, where $S(x)dx$ is the number of electrons lying in a slice of the crystal of thickness dx at a distance x from the atomic centre. Compton showed that $U(r)$, the radial density for a radius r, and $S(r)$, the value of $S(x)$ when $x=r$, are connected by the relation

$$\frac{dS(r)}{dr} = -\frac{U(r)}{2r}, \tag{7.101}$$

from which the series for $U(r)$ can be obtained by differentiation of the series for $S(r)$. Thus

$$U(r) = \frac{8\pi r}{D^2} \sum_1^\infty nf(n) \sin\left(\frac{2\pi nr}{D}\right). \tag{7.102}$$

In this expression, $f(n)$ is the value of the structure factor, here the atomic scattering factor of the atom, for reflection in the nth order from the planes of spacing D.

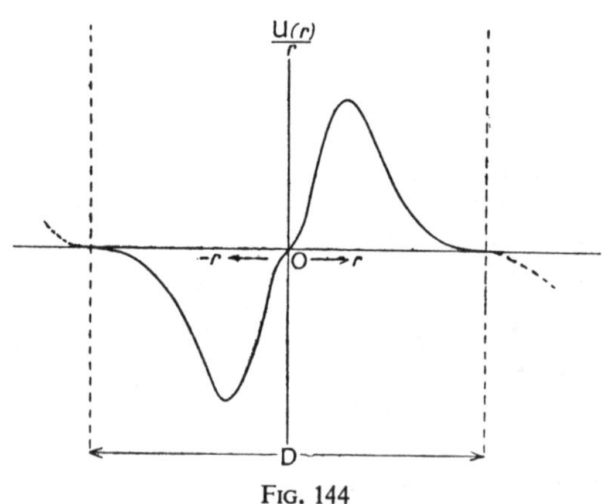

FIG. 144

We may derive the series directly in the following way. $U(r)$, the radial density function for a single atom, is essentially a symmetrical function of r. $U(r)/r$ must be zero for $r=0$; for $\rho(r)$, the actual charge-density, is finite everywhere, and

$$U(r)=4\pi r^2\rho(r),$$

or

$$U(r)/r=4\pi r\rho(r),$$

and must vanish for $r=0$, since $\rho(0)$ is finite. The function $U(r)/r$ must therefore be antisymmetrical in r, about the origin $r=0$. If it becomes effectively zero at distances $\pm D/2$ from 0, we may consider the portion of the curve $U(r)/r$, as shown in fig. 144, as one repeat of a periodic function having a period D, and can expand the *whole function* as a Fourier series with a fundamental period D. Because the function is antisymmetrical about $r=0$, we must use a sine series, and so

$$\frac{U(r)}{r}=\sum_1^{\infty} A(n)\sin\frac{2\pi nr}{D}. \tag{7.103}$$

The coefficient $A(n)$ is given in the usual way by

$$A(n)=\frac{2}{D}\int_{-D/2}^{+D/2}\frac{U(r)}{r}\sin\left(\frac{2\pi nr}{D}\right)dr. \tag{7.104}$$

Now $f(n)$, the scattering factor for the nth order spectrum from planes of spacing D, is given by

$$f(n)=\int_0^{D/2}\frac{U(r)}{r}\left\{\sin\left(\frac{2\pi nr}{D}\right)\Big/\frac{2\pi n}{D}\right\}dr,$$

and so $A(n)=8\pi nf(n)/D^2$, and the Fourier series (7.103) becomes

$$\frac{U(r)}{r}=\frac{8\pi}{D^2}\sum_1^{\infty} nf(n)\sin\left(\frac{2\pi nr}{D}\right). \tag{7.105}$$

This series reduces to the Fourier integral (7.99) if D becomes indefinitely great.

This series has been used by Compton,[8] Havighurst,[12] and Bearden [44] to determine the radial distribution in a number of atoms, from values of the scattering factor derived from measurements of the intensity of reflection of X-rays by crystals. Compton carried out the analysis for sodium and chlorine, basing his calculations on the results obtained by Bragg, James and Bosanquet; Havighurst used his own measurements on NaF, NaCl, and CaF_2 to determine radial distributions for Na, Cl, and Ca; and Bearden again used his own measurements from NaCl and aluminium. All the distributions so obtained show well-marked maxima and minima. A typical example is Havighurst's [12] curve for Ca^{++}, which is reproduced in fig. 145.

It is very tempting to interpret the maxima as being due to the different electron groups in the atom; they are, however, almost certainly diffraction effects, due to the termination of the Fourier series while its coefficients are still appreciable. It will be seen that the coefficients in the series (7.105) are not $f(n)$, but $nf(n)$, so that the convergence of the series is very slow. Measurements of f for values of $(\sin\theta)/\lambda$ greater than about 1·2 A^{-1} were not available. Increasing the value of the fictitious spacing D involves the extrapolation of the f-curve beyond

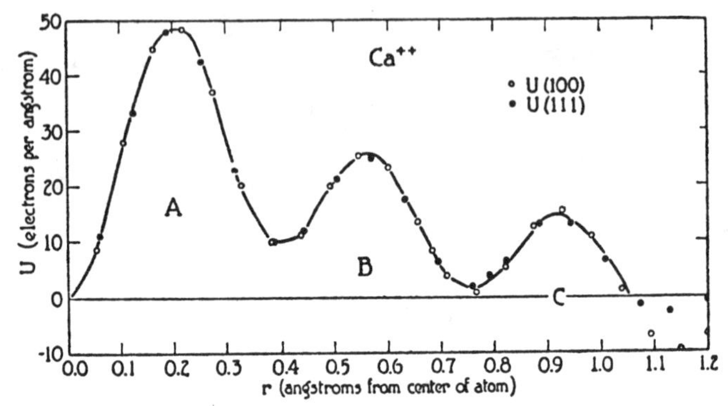

FIG. 145. Radial electron density in Ca^{++} as calculated by Havighurst from the radial Fourier series. The maxima are due to diffraction effects and not to rings of electrons (Havighurst)
(*Phy. Rev.*, **29**, 1 (1927))

the range of measurement for both large and small values of $(\sin\theta)/\lambda$, and the method ceases to be one employing only experimental results. The radial series, because of the nature of its coefficients, is peculiarly liable to give false detail, having no counterpart in the atomic structure. Here again, the application of an artificial temperature factor will remove this false detail, by making the series converge more quickly, and so lessening the importance of the higher harmonics of the series, to the omission of which the false detail is due. This will make it easier to estimate the total number of electrons in the atom from the area beneath the $U(r)$ curve, but it will not of course give any more information about the details of the electron distribution, for these are in effect smeared out by the assumed thermal motion.

It will be noticed that the curve of fig. 145 has the same form whether the series is calculated using the spectra from the (100) planes or from the (111) planes. This, however, is no proof that the peaks correspond to anything real, since if the greatest value of $(\sin\theta)/\lambda$ is the same for both sets of spectra the effective aperture of the equivalent optical system is the same in both cases, and will produce similar diffraction rings.

References

1. W. H. Bragg, *Phil. Trans. Roy. Soc.*, A, **215**, 253 (1915).
2. P. P. Ewald, *Zeit. f. Krist.*, **56**, 129 (1921).
3. W. Duane, *Proc. Nat. Acad. Sci.*, **9**, 158 (1923).
4. A. H. Compton, *Proc. Nat. Acad. Sci.*, **9**, 359 (1923).
5. P. Epstein and P. S. Ehrenfest, *Proc. Nat. Acad. Sci.*, **10**, 133 (1924).
6. W. Duane, *Proc. Nat. Acad. Sci.*, **11**, 489 (1925).
7. R. J. Havighurst, *Proc. Nat. Acad. Sci.*, **11**, 502, 507 (1925).
8. A. H. Compton, *X-Rays and Electrons*, Van Nostrand (1926), p. 151 *et seq.*
9. W. L. Bragg, *Proc. Roy. Soc.*, A, **123**, 537 (1929); *Zeit f. Krist.*, **70**, 475 (1929).
10. B. E. Warren and N. S. Gingrich, *Phys. Rev.*, **46**, 368 (1934).
11. A. L. Patterson, (*a*) *Phys. Rev.*, **46**, 372 (1934); (*b*) *Zeit. f. Krist.*, **90**, 517 (1935).
12. R. J. Havighurst, *Phys. Rev.*, **29**, 1 (1927); *Jour. Amer. Chem. Soc.*, **48**, 2113 (1926).
13. A. D. Booth, *Trans. Faraday Soc.*, **41**, 434 (1945).
14. J. M. Robertson, Physical Society, *Reports on Progress in Physics*, **4**, 332 (1937).
15. C. A. Beevers and H. Lipson, *Phil. Mag.*, **17**, 855 (1934).
16. J. M. Robertson, *Phil. Mag.*, **21**, 176 (1936).
17. H. Lipson and C. A. Beevers, *Proc. Phys. Soc.*, **48**, 772 (1936).
18. R. W. James, H. King and H. Horrocks, *Proc. Roy. Soc.*, A, **153**, 225 (1935).
19. R. W. James and G. W. Brindley, *Phil. Mag.*, **12**, 81 (1931).
20. J. West, *Zeit. f. Krist.*, **74**, 306 (1930).
21. S. B. Hendricks, *Amer. Jour. Sci.*, **14**, 269 (1927).
22. J. M. Robertson, *Jour. Chem. Soc.*, 615 (1935), 1195 (1936).
23. The following list contains references to some structure determinations of organic compounds made by the method of the double Fourier series, and not specifically quoted in the text. The list is not intended to be exhaustive, but merely illustrative:
    Hexachlorbenzene, by K. Lonsdale, *Proc. Roy. Soc.*, A, **133**, 536 (1931); Anthracene, Naphthalene and Durene, by J. M. Robertson, *Proc. Roy. Soc.*, A, **140**, 79 (1933), A, **142**, 674 (1933), A, **142**, 659 (1933): Chrysene, by J. Iball, *Proc. Roy. Soc.*, A, **146**, 140 (1934): Dibenzyl, by J. M. Robertson, *Proc. Roy. Soc.*, A, **150**, 348 (1935): Stilbene, by J. M. Robertson and I. Woodward, *Proc. Roy. Soc.*, A, **162**, 568 (1937): Cyanuric-triazide, by I. E. Knaggs, *Proc. Roy. Soc.*, A, **150**, 576 (1935): The carboxyl group and oxalic acid dihydrate, by J. M. Robertson and I. Woodward, *J.C.S.*, 1817 (1936).
24. J. M. Robertson, *Proc. Roy. Soc.*, A, **157**, 79 (1936).
25. J. N. van Niekerk, *Proc. Roy. Soc.*, A, **181**, 314 (1943).

26. J. M. Cork, *Phil. Mag.*, **4**, 688 (1927).
27. W. L. Bragg, C. Gottfried and J. West, *Zeit. f. Krist.*, **77**, 255 (1931).
28. D. Harker, *Jour. Chem. Physics*, **4**, 381 (1936).
29. F. J. Llewellyn, E. G. Cox and T. H. Goodwin, *J.C.S.*, 883 (1937).
30. K. S. Krishnan, B. C. Guha and S. Banerjee, *Phil. Trans. Roy. Soc.*, **A**, **231**, 235 (1933); **A**, **234**, 265 (1935); K. Lonsdale and K. S. Krishnan, *Proc. Roy. Soc.*, **A**, **156**, 597 (1936); K. Lonsdale, *Reports on Progress of Physics*, **4**, 368 (1938); *Science Progress*, **32**, 677 (1938).
31. G. A. Jeffrey, *Proc. Roy. Soc.*, **A**, **183**, 388 (1945).
32. W. L. Bragg, *British Jour. of Radiology*, **2**, 65 (1929). . .
33. E. Abbe, *Archiv. f. mikroskopische. Anat.*, **9**, 413 (1887); *Collected Works*, **1**, 45 (1904).
34. A. B. Porter, *Phil. Mag.*, **11**, 154 (1906).
35. W. L. Bragg and J. West, *Phil. Mag.*, **10**, 823 (1930).
36. R. W. James, I. Waller and D. R. Hartree, *Proc. Roy. Soc.*, **A**, **118**, 334 (1928).
37. H. M. Parker and W. J. Whitehouse, *Phil. Mag.*, **14**, 939 (1932).
38. A. H. Compton, *Phys. Rev.*, **35**, 925 (1930).
39. E. O. Wollan, *Phys. Rev.*, **38**, 15 (1931); *Reviews of Modern Physics*, **4**, 205 (1932).
40. C. S. Barrett, *Phys. Rev.*, **32**, 22 (1928).
41. G. Herzog, *Zeit. f. Physik*, **69**, 207 (1931), **70**, 583 (1931).
42. E. O. Wollan, *Phys. Rev.*, **37**, 862 (1931).
43. R. J. Havighurst, *Phys. Rev.*, **31**, 16 (1928).
44. J. A. Bearden, *Phys. Rev.*, **29**, 20 (1927).

CHAPTER VIII

LAUE'S DEVELOPMENT OF THE DYNAMICAL THEORY—KOSSEL LINES

1. The Dynamical Theory in Terms of a Continuous Charge Distribution

(*a*) *Introductory:* In Chapter VII, we thought of a crystal as a continuous, three-dimensionally periodic distribution of scattering matter, which could be represented by means of a suitable triple Fourier series. We must now see how the dynamical theory of X-ray reflection, developed by Ewald, of which a preliminary account was given in Section 4 of Chapter II, can be made to fit into such a picture. In the dynamical theory as originally given by Ewald, the crystal was assumed to consist of electric dipoles, situated exactly at the lattice-points. The wave-field within the crystal was considered as built up of the waves scattered by the dipoles, and the conditions were sought that the wave-field produced by the oscillations of all the dipoles should be just that necessary to maintain those oscillations. Because the dipoles are situated only at the lattice-points, only the fields at those points are relevant; but if the scattering is to be regarded as due to matter continuously distributed throughout the crystal, it is clear that dynamical equilibrium must be set up between the wave-field and the scattering matter at every point of the continuum.

A rigid treatment of the problem, based on the quantum theory, would require a knowledge of the wave-functions corresponding to the electrons in the crystal, and the perturbations of these wave-functions by the incident electromagnetic field would then have to be calculated by methods analogous to those outlined in Chapter III, § 3(*c*), for the case of scattering by a single atom. A start along these lines has been made by Kohler,[1] who shows that the Schrödinger current-density, from which the wave-field due to the scattered radiation is to be calculated, is at every point in the crystal proportional to the rate of change of the electric field at that point. It appears that the continuous electron density ρ in the crystal, now of course the Schrödinger charge-density, may be taken into account by ascribing to the crystal at each point a dielectric constant η, given by

$$\eta = 1 - \frac{e}{\pi m \nu^2} \rho, \tag{8.1}$$

ν being the frequency of the electromagnetic waves concerned. The result agrees with the classical dispersion theory, if it is assumed, as is in fact done in deriving it, that ν is far from any natural absorption frequency of the system.

A method of treating the problem, formally very similar to that suggested by Kohler's analysis, was given by Laue [2] as early as 1931. Laue's method is based on classical ideas, and is physically easy to understand, and it is to a large extent justified formally by Kohler's results. We shall therefore give some account of his theory here.

The crystal is thought of as a continuous distribution of negative charge, with the positive charges concentrated at the atomic centres. An electromagnetic wave passing through it is scattered by the negative charges, but scarcely at all by the massive positive charges. In calculating the resultant wave-field, we may therefore suppose the positive charges to be distributed in any way that is convenient, and Laue assumes them to be so distributed that in the absence of any disturbing field positive and negative charges exactly neutralise each other at every point. An electric field applied to the system will cause a relative displacement of the positive and negative distributions, and thus an electric polarisation at every point proportional to the field strength at that point, and so to be expressed in terms of an appropriate dielectric constant. This dielectric constant is a three-dimensionally periodic function of the co-ordinates of the crystal. It must not be confused with the ordinary dielectric constant applicable to electrostatic problems, which of course depends on the distribution of the positive charges. It is a purely fictitious dielectric constant, applicable only to the calculation of the scattered wave-field, for radiation within the frequency limits assumed in the problem, which do not include any absorption frequencies of the system, so that dispersion effects need not be considered.

(*b*) *The fictitious dielectric constant and polarisation expressed as a Fourier series:* Let η be the fictitious dielectric constant, and let **E** and **D** be the vectors representing the electric intensity and electric induction in the wave-field, so that

$$\mathbf{D} = \eta \mathbf{E}. \tag{8.2}$$

Then if **P** is the dielectric polarisation, or the electric moment per unit volume, at any point,

$$4\pi \mathbf{P} = (\eta - 1)\mathbf{E} = (1 - 1/\eta)\mathbf{D} = \phi \mathbf{D}. \tag{8.3}$$

The quantity ϕ might be called the polarisability of the medium. It too is a three-dimensionally periodic function of the co-ordinates, and so can be expressed as a triple Fourier series. The most convenient form of the series for our purpose is that of equation (7.14) of Chapter VII, § 1(*g*), in which the phases of the Fourier waves, in so far as they are functions of the co-ordinates, are expressed in terms of the reciprocal-lattice vectors $\mathbf{r}_m^*$. We write

$$\phi = \sum \phi_m e^{-j(\mathbf{r}\cdot\mathbf{r}_m^*)} \tag{8.4}$$

$$j = 2\pi\sqrt{-1},$$

with the coefficients ϕ_m given by

$$\phi_m = \frac{1}{V}\int \phi e^{j(\mathbf{r}\cdot\mathbf{r}_m^*)}\,dv. \tag{8.5}$$

The index m denotes the triplet of numbers (h, k, l) specifying a reciprocal-lattice point. The vector $\mathbf{r}_m^*$ is that from the origin to this point, and the summation is to be taken over all reciprocal-lattice points, and so is really a triple summation over h, k, and l. The integration in (*8*.5) is to be taken throughout the unit cell of the structure, the volume of which is V.

In the absence of absorption in the medium, the dielectric constant η and the polarisation are real, and so ϕ_m and ϕ_{-m} are either equal if real, or conjugate quantities if complex, just as for the corresponding quantities $F(m)$ and $F(\overline{m})$ in § 1(*e*) of Chapter VII.

Equations (*8*.4) and (*8*.5) are closely related, not only formally, but physically, to the corresponding equations (*7*.14) and (*7*.15) of Chapter VII. To discuss the relationship, we consider the radiation scattered by a unit cell of the structure. The electric moment of a volume dv of the crystal is $\phi\mathbf{D}dv/4\pi$. Let a plane wave of wave-length λ and frequency ν pass through the crystal in the direction defined by the unit vector $\mathbf{s}_0$. Then, by the classical theory of scattering by dipoles, the scattered amplitude at unit distance in a direction defined by the unit vector $\mathbf{s}$, due to a unit cell of the structure is

$$\eta|E|\frac{\nu^2}{c^2}\int \phi e^{i\kappa(\mathbf{S}\cdot\mathbf{r})}\,dv, \tag{8.6}$$

where $\mathbf{S} = \mathbf{s} - \mathbf{s}_0$, and $\kappa = 2\pi/\lambda$ in the usual notation. It is assumed that the radiation is so polarised that the direction of the electric vector is perpendicular to the plane containing $\mathbf{s}$ and $\mathbf{s}_0$.

The corresponding expression for the scattering by a unit cell in terms of the charge density, when the frequency of the radiation is much greater than any natural frequency of the system, is

$$-\frac{e^2}{mc^2}|\mathbf{E}|\int \rho e^{i\kappa(\mathbf{S}\cdot\mathbf{r})}\,dv. \tag{8.7}$$

Thus, comparing (*8*.6) and (*8*.7), we see that

$$\phi/\rho = -e^2/\pi\eta m\nu^2 \simeq -e^2/\pi m\nu^2, \tag{8.8}$$

since, in all cases to which the results are applied, η differs from unity by only a few parts in a million. When the scattering takes place under such conditions as to produce the spectrum m, $S/\lambda = \mathbf{r}_m^*$, and so by (*8*.7) the structure factor $F(m)$ is given by

$$F(m) = \int \rho e^{j(\mathbf{r}\cdot\mathbf{r}_m^*)}\,dv = -\frac{\pi m\nu^2}{e^2}\int \phi e^{j(\mathbf{r}\cdot\mathbf{r}_m^*)}\,dv = -\frac{\pi m\nu^2 V}{e^2}\phi_m,$$

by (8.5), so that

$$\phi_m = -\frac{e^2}{\pi m \nu^2}\frac{\mathrm{F}(m)}{\mathrm{V}} = -\frac{e^2\lambda^2}{mc^2\pi}\frac{\mathrm{F}(m)}{\mathrm{V}}. \tag{8.9}$$

We have thus expressed the fictitious dielectric constant in terms of the structure factors of the unit cell for the various possible lattice spectra.

It will be convenient to have an expansion for $1/\eta$, which is also a three-dimensionally periodic function of the co-ordinates. By (8.3) and (8.4),

$$-\frac{1}{\eta} = \phi - 1 = -1 + \sum_m \phi_m e^{-j(\mathbf{r}\cdot\mathbf{r}_m^*)}$$

$$= \sum_m \phi'_m e^{-j(\mathbf{r}\cdot\mathbf{r}_m^*)}, \tag{8.10}$$

where $\phi'_0 = \phi_0 - 1$, and $\phi'_m = \phi_m$ if $m \neq 0$. (8.10a)

In Ewald's theory, $\eta = 1$, and so $\phi = 0$, everywhere except at the lattice points themselves; but at these points $e^{j(\mathbf{r}\cdot\mathbf{r}_m^*)} = 1$, and so, by (8.5), $\phi_m = \phi_0$ for all indices. The formulae to be derived in terms of the continuous distribution must therefore reduce to Ewald's formulae if these substitutions are made, and if $\mathbf{r}$ corresponds always to a lattice-point.

(c) *The fundamental equations of the wave-field:* The field vectors must everywhere obey Maxwell's equations for an electromagnetic field in an insulating medium:

$$d\mathbf{D}/dt = c \operatorname{curl} \mathbf{H}, \quad d\mathbf{H}/dt = -c \operatorname{curl} \mathbf{E} = -c \operatorname{curl} (\mathbf{D}/\eta). \tag{8.11}$$

According to Ewald's theory, the wave-field consists of a large number of plane waves, the wave-vectors of any two of which differ by a vector in the reciprocal lattice (Chapter II, § 4(c)), and we therefore try a substitution of the following type,

$$\mathbf{D} = e^{j\nu t} \sum_m \mathbf{D}_m e^{-j(\mathbf{K}_m\cdot\mathbf{r})},$$

$$\mathbf{H} = e^{j\nu t} \sum_m \mathbf{H}_m e^{-j(\mathbf{K}_m\cdot\mathbf{r})}, \tag{8.12}$$

where $\mathbf{K}_m = \mathbf{K}_0 + \mathbf{r}_m^*$, (8.12a)

which represents a field built up of a triply infinite series of plane waves whose wave-vectors differ from each other in the required way. We have to determine $\mathbf{K}_0$, and hence the other wave-vectors, in such a way that (8.12) satisfies Maxwell's equations, and any other conditions that may be imposed by the nature of the particular problem considered. We shall deal first with the condition that Maxwell's equations must be satisfied.

From the second equation of (8.12),

$$\operatorname{curl} \mathbf{H} = je^{j\nu t} \sum_m (\mathbf{H}_m \times \mathbf{K}_m) e^{-j(\mathbf{K}_m\cdot\mathbf{r})}, \tag{8.13}$$

and from the first equation,

$$\frac{1}{c}\frac{d\mathbf{D}}{dt}=j\frac{\nu}{c}e^{j\nu t}\sum_{m}\mathbf{D}_m e^{-j(\mathbf{K}_m\cdot\mathbf{r})}. \tag{8.14}$$

By (8.11), the right-hand sides of the last two equations must be equal at all times at every point of the wave-field, a condition that can be fulfilled only if corresponding terms in the sums are equal. Thus

$$k\mathbf{D}_m=\mathbf{H}_m\times\mathbf{K}_m, \tag{8.15}$$

where $k=1/\lambda=\nu/c$, λ being the wave-length in empty space corresponding to frequency ν. The physical meaning of this equation is that the electric induction in each constituent wave is perpendicular to the corresponding wave-normal and magnetic field.

Again, from (8.10) and (8.12),

$$\mathbf{D}/\eta=-e^{j\nu t}\sum_{p}\sum_{q}\phi_p'\mathbf{D}_q e^{-j(\mathbf{K}_q+\mathbf{r}_p^*\cdot\mathbf{r})}, \tag{8.16}$$

or, since $\mathbf{K}_q+\mathbf{r}_p^*=\mathbf{K}_{p+q}$, by (8.12a),

$$\mathbf{D}/\eta=-e^{j\nu t}\sum_{p}\sum_{q}\phi_p'\mathbf{D}_q e^{-j(\mathbf{K}_{p+q}\cdot\mathbf{r})}.$$

We now take $p+q$ as an index of summation, instead of p, as we may, since the sums are to infinity, putting $m=p+q$. Then

$$\mathbf{D}/\eta=-e^{j\nu t}\sum_{m}\sum_{q}\phi_{m-q}'\mathbf{D}_q e^{-j(\mathbf{K}_m\cdot\mathbf{r})}. \tag{8.17}$$

From (8.17),

$$\operatorname{curl}(\mathbf{D}/\eta)=je^{j\nu t}\sum_{m}\left\{\mathbf{K}_m\times\sum_{q}\phi_{m-q}'\mathbf{D}_q\right\}e^{-j(\mathbf{K}_m\cdot\mathbf{r})}, \tag{8.18}$$

and from (8.12),

$$-\frac{1}{c}\frac{d\mathbf{H}}{dt}=-\frac{\nu j}{c}\sum_{m}\mathbf{H}_m e^{-j(\mathbf{K}_m\cdot\mathbf{r})}, \tag{8.19}$$

whence, using Maxwell's equations as before, and equating corresponding terms on the right-hand sides of (8.18) and (8.19), we find

$$k\mathbf{H}_m=-\sum_{q}\phi_{m-q}'(\mathbf{K}_m\times\mathbf{D}_q), \tag{8.20}$$

and on substituting this in (8.15), obtain

$$k^2\mathbf{D}_m=\sum_{q}\phi_{m-q}'\{\mathbf{K}_m\times(\mathbf{K}_m\times\mathbf{D}_q)\}. \tag{8.21}$$

By the rule for the double vector product,

$$\mathbf{A}\times(\mathbf{B}\times\mathbf{C})=\mathbf{B}(\mathbf{A}\cdot\mathbf{C})-\mathbf{C}(\mathbf{A}\cdot\mathbf{B}),$$

$$\begin{aligned}\mathbf{K}_m\times(\mathbf{K}_m\times\mathbf{D}_q)&=\mathbf{K}_m(\mathbf{K}_m\cdot\mathbf{D}_q)-\mathbf{K}_m^2\mathbf{D}_q\\&=-\mathbf{K}_m^2\{\mathbf{D}_q-\mathbf{n}_m(\mathbf{n}_m\cdot\mathbf{D}_q)\},\end{aligned}$$

where $\mathbf{n}_m=\mathbf{K}_m/|\mathbf{K}_m|$, and is the unit vector in the direction of $\mathbf{K}_m$.

The quantity between the curly brackets is therefore the vector difference between $\mathbf{D}_q$ and the projection of $\mathbf{D}_q$ on the direction of $\mathbf{K}_m$; that is to say, it is the component of $\mathbf{D}_q$ in a direction perpendicular to $\mathbf{K}_m$, and this component we denote by $\mathbf{D}_{q[m]}$.

Equation (*8*.21) therefore becomes

$$k^2\mathbf{D}_m = -\mathbf{K}_m^2 \sum_q \phi'_{m-q}\mathbf{D}_{q[m]}$$

$$= -\mathbf{K}_m^2 \phi'_0 \mathbf{D}_{m[m]} - \mathbf{K}_m^2 \sum_{q \neq m} \phi'_{m-q}\mathbf{D}_{q[m]}. \qquad (8.22)$$

Now $\mathbf{D}_{m[m]} = \mathbf{D}_m$, and hence, using (*8*.10a), we obtain finally from (*8*.22)

$$\frac{\mathrm{K}_m^2 - k^2}{\mathrm{K}_m^2}\mathbf{D}_m = \sum_q \phi_{m-q}\mathbf{D}_{q[m]}, \qquad (8.23)$$

which is the fundamental equation of the wave-field in the crystal, giving the amplitude of the wave m. It is of course only one of an infinite number of linear equations connecting the values of D for the different waves, and they cannot be solved as a general problem. It will be seen, however, from the form of (*8*.23), that those waves for which $|\mathbf{K}_m|$ is very nearly equal to k, that is to say, whose velocity of propagation within the crystal is very nearly equal to that of light, will be far more intense than any other waves. This is the result obtained by Ewald, which we have already discussed in Section 4 of Chapter II. As we saw there, there will be in general only a few waves, in some of the most important cases only two, that need be considered at all; and since the wave-vectors of the constituent waves of the wave-field bear exactly the same relationship to one another as the dipole waves of the Ewald theory, the practical method of deciding which waves are of importance by means of the sphere of reflection in the reciprocal lattice can again be used, and the argument from this point on runs almost precisely as in the original theory, and need not be repeated in detail.

As in II, § 4(*b*), we put $|\mathbf{K}_m| = k(1+\epsilon_m)$, ϵ_m being always small for waves that are of appreciable amplitude. If the square and higher powers of ϵ_m are neglected, equation (*8*.23) becomes

$$2\epsilon_m \mathbf{D}_m = \sum_q \phi_{m-q}\mathbf{D}_{q[m]}. \qquad (8.24)$$

By (*8*.12), (*8*.12a), and (*8*.24), the induction due to the whole wave-field at any point is

$$\mathbf{D} = e^{j\nu t} \sum_m \left(\frac{1}{2\epsilon_m} \sum_q \phi_{m-q}\mathbf{D}_{q[m]}\right) e^{-j(\mathbf{K}_m \cdot \mathbf{r})}$$

$$= e^{j\{\nu t - \mathbf{K}_0 \cdot \mathbf{r}\}} \sum_m \left(\frac{1}{2\epsilon_m} \sum_q \phi_{m-q}\mathbf{D}_{q[m]}\right) e^{-j(\mathbf{r}^*_m \cdot \mathbf{r})}. \qquad (8.25)$$

Equation (*8*.25), as it stands, represents a plane wave of wave-vector $\mathbf{K}_0$ travelling through the crystal. The amplitude of this wave is given by the summation, and varies in a three-dimensionally periodic manner from point to point, but is constant at any given point. The amplitude function is a triple Fourier series, and the different Fourier waves are parallel to the lattice-planes, and have wave-lengths equal to the lattice spacings, and to submultiples of them. The set of waves is in fact just that used to represent the density function $\rho(x, y, z)$ in Chapter VII, but the coefficient and phase of the wave (m) are now given by the quantity in the brackets in (*8*.25). The sums concerned are of course vector sums. The electromagnetic field in the crystal thus consists of a series of nodes and loops of disturbance, having the periodicity of the lattice.

To compare equation (*8*.25) with Ewald's result, we put $\phi_{m-q}=\phi_0$ for all terms, and remembering that we are now concerned only with the lattice-points, at which the phase factors $e^{-j(\mathbf{K}_m\cdot\mathbf{r})}$ are all equal, we may write (*8*.25) in the form

$$\mathbf{D}=e^{j\nu t}\phi_0\sum_m\frac{1}{2\epsilon_m}\sum_q\mathbf{D}_{q[m]}e^{-j(\mathbf{K}_q\cdot\mathbf{r})}$$

$$=\phi_0\sum_m\frac{1}{2\epsilon_m}\mathbf{D}_{[m]},\tag{8.26}$$

where $\mathbf{D}_{[m]}$ is the component perpendicular to $\mathbf{K}_m$ of $\mathbf{D}$, the total induction. This corresponds exactly to Ewald's equation

$$\mathbf{b}=\frac{1}{\Omega}\sum\frac{1}{\epsilon_m}\mathbf{b}_{[m]},\tag{8.27}$$

equation (*2*.104) of Chapter II, with $\phi_0=2/\Omega$, since $\mathbf{b}$, the amplitude of the dipole oscillation, is proportional to $\mathbf{D}$; but while (*8*.27) is a statement only about the dipole oscillators at the lattice-points, (*8*.25) gives the electromagnetic field at any point in the crystal.

(*d*) *The case in which two waves only are appreciable:* The most important case in practice is that in which only two waves make appreciable contributions to the wave-field. We have already discussed this in Chapter II, § 4(*g*), but shall here obtain some additional results. We denote the waves of appreciable amplitude by the indices (0) and (m), and consider two cases, (*a*), that in which the induction vector in both waves is perpendicular to the vectors $\mathbf{K}_0$ and $\mathbf{K}_m$, and (*b*), that in which it lies in the plane containing them.

In case (*a*), by (*8*.24),

$$(\phi_0-2\epsilon_0)\mathbf{D}_0+\phi_{-m}\mathbf{D}_m=0,$$
$$\phi_m\mathbf{D}_0+(\phi_0-2\epsilon_m)\mathbf{D}_m=0,\tag{8.28}$$

and in case (*b*),

$$(\phi_0-2\epsilon_0)\mathbf{D}_0+\phi_{-m}\cos 2\theta\mathbf{D}_m=0,$$
$$\phi_m\cos 2\theta\mathbf{D}_0+(\phi_0-2\epsilon_m)\mathbf{D}_m=0,\tag{8.29}$$

where 2θ is the angle between the directions of the vectors $\mathbf{K}_0$ and $\mathbf{K}_m$. In every practical case, θ may be treated as constant and equal to the Bragg angle θ_0. The conditions that these equations should have a solution are, in case (*a*)

$$(\phi_0 - 2\epsilon_0)(\phi_0 - 2\epsilon_m) = \phi_m\phi_{-m}, \tag{8.30}$$

and in case (*b*),

$$(\phi_0 - 2\epsilon_0)(\phi_0 - 2\epsilon_m) = \phi_m\phi_{-m}\cos^2 2\theta_0. \tag{8.31}$$

We put

$$\xi_0 = k(\epsilon_0 - \tfrac{1}{2}\phi_0), \quad \xi_m = k(\epsilon_m - \tfrac{1}{2}\phi_0), \tag{8.32}$$

so that in case (*a*)

$$\xi_0\xi_m = \tfrac{1}{4}k^2\phi_m\phi_{-m}, \tag{8.33}$$

and in case (*b*)

$$\xi_0\xi_m = \tfrac{1}{4}k^2\phi_m\phi_{-m}\cos^2 2\theta_0. \tag{8.34}$$

Since $\phi_0 = 2/\Omega$, ξ_0 and ξ_m have the same significance as in equation (2.110), which may be written

$$\xi_0\xi_m = \tfrac{1}{4}k^2\phi_0^2. \tag{8.35}$$

The real axis of the hyperbola forming the dispersion surface is thus $k \sec\theta_0\sqrt{\phi_m\phi_{-m}}$, instead of $k\phi_0\sec\theta_0$, in case (*a*), and

$$k|\cos 2\theta_0| \sec\theta_0\sqrt{\phi_m\phi_{-m}}$$

in case (*b*). For a crystal without absorption, ϕ_m and ϕ_{-m} are conjugate complex quantities, or are equal if real, so that the expressions for the diameters can be written $k|\phi_m|\sec\theta_0$ and $k|\phi_m|\sec\theta_0|\cos 2\theta_0|$, and by (8.9) are therefore proportional to the structure factors of the spectra concerned.

The width of the range of total reflection for a crystal without absorption, when reflection is symmetrical, which is $(2/\Omega)\sec\theta_0 \operatorname{cosec}\theta_0$ according to the dipole theory (§ 4(*k*), Chapter II), now becomes $|\phi_m|\sec\theta_0 \operatorname{cosec}\theta_0$, in agreement with Darwin's theory.

Darwin's theory gives as the angular width of total reflection $2\delta\sec\theta_0 \operatorname{cosec}\theta_0$, where $1-\delta$ is the refractive index of the crystal. For rays passing through the crystal without interference, there is only one wave, and the refractive index is then given by

$$n = \frac{|\mathbf{K}_0|}{k} = 1 + \epsilon_0 = 1 + \tfrac{1}{2}\phi_0, \tag{8.36}$$

by (8.29), since $\mathbf{D}_m$ is now zero. By (8.9),

$$\phi_0 = -\frac{e^2}{\pi m v^2}\frac{\mathrm{F}(0)}{\mathrm{V}} = -\frac{\mathrm{N}e^2}{\pi m v^2}, \tag{8.37}$$

N being the number of electrons per unit volume, so that

$$n = 1 - (\mathrm{N}e^2/2\pi m v^2) = 1 - \delta, \tag{8.38}$$

in agreement with our previous result. Comparing (*8*.37) and (*8*.38), we find, using (*8*.9),

$$|\phi_m| = 2\delta|F(m)|/|F(0)|,$$

which gives for the range of total reflection in case (*a*)

$$2\delta \sec\theta_0 \operatorname{cosec}\theta_0 |F(m)|/|F(0)|,$$

in agreement with the results of Darwin's theory. (Chapter II, § 3(*e*)).

Fig. 146 shows the trace of the dispersion surface for two waves (0) and (*m*), drawn for case (*a*). This may be compared with fig. 32, p. 77, from which it differs in one respect only: the upper branch of the hyperbola no longer passes through the Laue point L, the position of

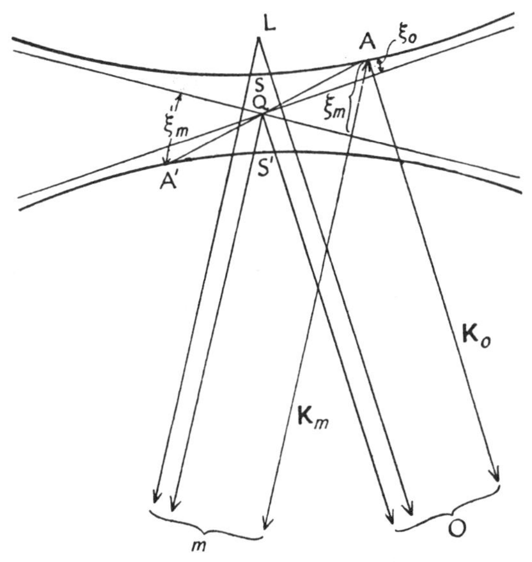

FIG. 146. Trace of the dispersion surface for two waves

which is fixed by the condition $L0 = Lm = k = 1/\lambda$. The lines 0Q, *m*Q have length $k(1 + \frac{1}{2}\phi_0)$, and the asymptotes of the hyperbolas are the tangents at their point of intersection, Q, to the circles of this radius having centres at the reciprocal-lattice points (0) and (*m*). It is to be remembered that in a figure drawn to scale the lengths *m*Q and 0Q should be about 10^5 times as great as that of SS′, the real axis of the hyperbola, so that within the bounds of the figure the circles are indistinguishable from their tangents. The dispersion *surface* is obtained by supposing the whole figure to rotate through a small angle about the line 0*m*.

We may here recapitulate briefly the significance of the dispersion surface. Consider any point A on it. Then A0 and A*m* are the wave-vectors $\mathbf{K}_0$ and $\mathbf{K}_m$ of waves that satisfy the fundamental equations, and which, by their superposition, produce a dynamically self-consistent wave-field in the crystal. In an unbounded crystal, any number of such wave-fields, having any relative amplitudes, can co-exist, each being associated with some point on the dispersion surface. Electric inductions perpendicular to the plane Am0 are associated with wave-points on the hyperbola shown, those lying *in* this plane with wave-points on a second hyperbola, having the same asymptotes, but a real axis of length SS′ cos $2\theta_0$.

Even in the infinite crystal, there are certain relationships between the amplitudes and phases of the wave-fields corresponding to different wave-points. For case (*a*), by (*8*.28),

$$\frac{\mathbf{D}_0}{\mathbf{D}_m}=\frac{2\epsilon_m-\phi_0}{\phi_m}=\frac{\phi_{-m}}{2\epsilon_0-\phi_0},$$

or

$$\frac{\mathbf{D}_0}{\mathbf{D}_m}=\frac{2\xi_m}{k\phi_m}=\frac{k\phi_{-m}}{2\xi_0}. \tag{8.39}$$

We note first that ξ_m has the same sign so long as A remains on the one branch of the hyperbola, and that ϕ_m and ϕ_{-m} are constants. The phase difference between $\mathbf{D}_0$ and $\mathbf{D}_m$ is thus constant for all wave-points on the same branch of the hyperbola. If A and A′ are points at the opposite ends of the same diameter, it is plain from the figure that the corresponding values of ξ_m are numerically equal but opposite in sign ; and the same is true for ξ_0. The ratio $\mathbf{D}_0/\mathbf{D}_m$ thus has the same value but the opposite sign for wave-points at the opposite ends of a diameter. In other words, the *phase difference* between $\mathbf{D}_0$ and $\mathbf{D}_m$, which is constant for all wave-points on the same branch of the hyperbola, differs by π for points on different branches, a result we shall require later.

(*e*) *Determination of the wave-points when the crystal is bounded by a plane surface and a primary wave enters it from outside :* This problem has already been considered in §§ 4(*i*), (*j*), and (*k*) of Chapter II, and it will be assumed that the reader is already familiar with the discussion given there, although a certain amount of repetition is unavoidable. Let a plane wave represented by the vector $\mathbf{k}_0$ fall on the surface of the crystal, which we shall suppose to be a plane-parallel slab, and give rise to the wave $\mathbf{K}_0$ inside it. The second wave $\mathbf{K}_m$ set up within the crystal as a result of the interaction with the lattice gives rise to a wave $\mathbf{k}_m$ on leaving it. Two distinct cases arise: Case I, in which $\mathbf{K}_m$ is directed into the crystal and leaves it again at the lower surface; and Case II, in which the wave $\mathbf{k}_m$ leaves the crystal again at the surface at which the primary wave entered. We shall refer to Cases I and II in what follows as those of transmission and reflection respectively. The difference between them will be clear from fig. 147, in which *mm* denotes in each case the direction of the crystal planes concerned in the production of the spectrum, which are inclined at an angle ϕ to the surface. The vectors $\mathbf{k}_0$ and $\mathbf{k}_m$ have the same absolute value $k(=1/\lambda)$, but $|\mathbf{K}_0|$ and $|\mathbf{K}_m|$ differ slightly from it and have the values $k(1+\epsilon_0)$ and $k(1+\epsilon_m)$ respectively. In any particular case, the directions of $\mathbf{K}_m$ and $\mathbf{k}_m$, and of $\mathbf{K}_0$ and $\mathbf{k}_0$ will differ by angles of the order of seconds, and $\mathbf{K}_m$ will be very nearly the reflection of $\mathbf{K}_0$ in the planes mm at the Bragg

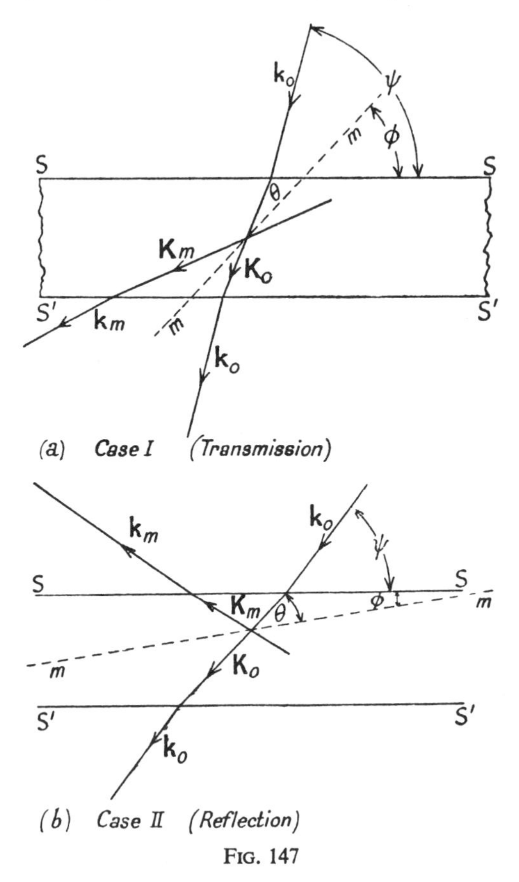

FIG. 147

angle θ_0. If ψ is the glancing angle of incidence of $\mathbf{k}_0$ on the crystal face, $\psi=\theta_0+\phi$, very nearly, while the glancing angle at which $\mathbf{k}_m$ will leave the crystal face in Case II (reflection) will be very nearly indeed $\theta_0-\phi$. The condition for Case I is therefore nearly enough, $\phi>\theta_0$, and for Case II, $\phi<\theta_0$.

In fig. 148, which is drawn so as to correspond to the reflection problem (fig. 147, II), let O be the origin of the reciprocal lattice, and (m) the second reciprocal-lattice point concerned, so that $\overrightarrow{Om}=\mathbf{r}_m^*$. Let PO be of length k, and parallel to $\mathbf{k}_0$, representing therefore the wave-vector of the incident radiation. The line PZ is normal to the surface SS of the crystal. A necessary boundary condition is that the components of $\mathbf{K}_0$ and $\mathbf{k}_0$ parallel to the surface should be equal, and so the wave-point must lie somewhere on PZ. Its position is therefore fixed, since it must also lie on the dispersion surface. Let A be a point of intersection of PZ with the dispersion surface, and let it lie at a distance PA from P, which we shall denote by gk†. AO and Am

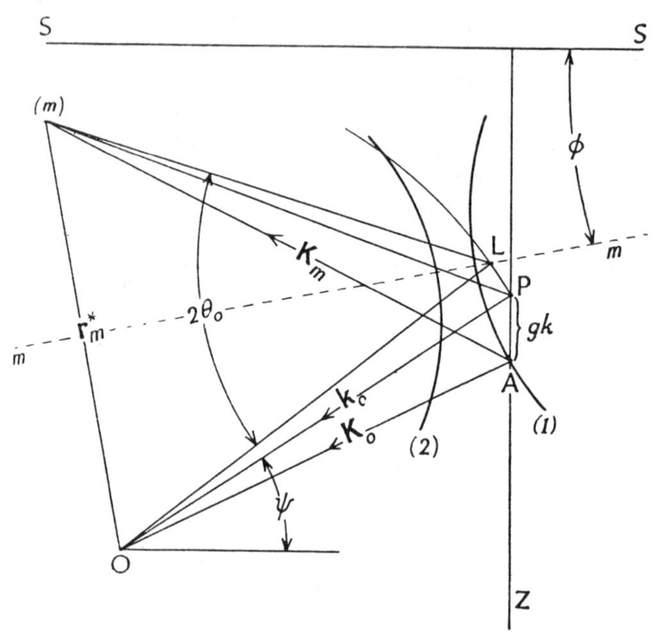

FIG. 148. The dispersion surface and wave-vectors for two waves in the reflection problem

now represent in magnitude and direction a pair of wave-vectors $\mathbf{K}_0$ and $\mathbf{K}_m$ (the figure of course is not even approximately to scale). PO makes an angle ψ with the surface. Let L be the Laue point, equidistant k from both O and (m). The angle OLm will then be $2\theta_0$, and the angle POL, which by the figure we see to be equal to $\theta_0+\phi-\psi$, is the angle between the actual direction of incidence and that corresponding to reflection strictly according to Bragg's law. We shall write

$$\theta_0+\phi-\psi = -\Delta\psi_0. \tag{8.40}$$

It will always be a small angle in the cases that interest us, at the most of the order of minutes of arc.

Put $$Pm = k(1+\alpha_m);$$

† g is the quantity called by Ewald the 'accommodation' (*Anpassung*), and is denoted by δ in his papers. We have not used δ here, as there may be some risk of confusion with the δ used in the expression for the refractive index, here and elsewhere in the book.

then $\qquad k\alpha_m = \mathrm{P}m - k = \mathrm{P}m - \mathrm{L}m = \mathrm{PL}\sin 2\theta_0,$

so that $\qquad k\alpha_m = k(\mathrm{P\hat{O}L})\sin 2\theta_0,$

and
$$\alpha_m = -\Delta\psi_0 \sin 2\theta_0. \tag{8.41}$$

Since A is the wave-point,

$$\mathrm{A}m = k(1+\epsilon_m), \quad \text{and} \quad \mathrm{AO} = k(1+\epsilon_0).$$

Now $\mathrm{A}m = \mathrm{P}m$ + component of PA parallel to $\mathrm{A}m$

$$= k(1+\alpha_m) - kg\cos m\hat{\mathrm{P}}\mathrm{A}$$
$$= k(1+\alpha_m) - kg\sin(\psi - 2\theta_0), \text{ very nearly.}$$

Whence
$$\epsilon_m = \alpha_m - g\gamma_m, \tag{8.42}$$

where
$$\gamma_m = \sin(\psi - 2\theta_0) = \sin(\phi - \theta_0) \quad \text{very nearly,} \tag{8.42a}$$

since, in the cases that concern us, ψ is always very nearly equal to $\phi + \theta_0$; and in an exactly similar way,

$$\epsilon_0 = -g\gamma_0 \tag{8.43}$$

where
$$\gamma_0 = \sin\psi = \sin(\phi + \theta_0) \quad \text{very nearly.} \tag{8.43a}$$

We have thus expressed ϵ_0 and ϵ_m in terms of the conditions of incidence.

(*f*) *The calculation of the ratio* $\mathbf{D}_m/\mathbf{D}_0$ *:*

Put
$$\mathbf{D}_m = x\mathbf{D}_0. \tag{8.44}$$

Then, using equations (8.28), (8.42), and (8.43), and eliminating g, we obtain for the state of polarisation (*a*), in which $\mathbf{D}_0$ and $\mathbf{D}_m$ are both perpendicular to the plane containing $\mathbf{K}_0$ and $\mathbf{K}_m$,

$$\frac{\gamma_m}{\gamma_0}\phi_{-m}x^2 + \eta x - \phi_m = 0, \tag{8.45}$$

where
$$\eta = 2\alpha_m - \phi_0\left(1 - \frac{\gamma_m}{\gamma_0}\right)$$
$$= -2\Delta\psi_0 \sin 2\theta_0 - \phi_0\left(1 - \frac{\gamma_m}{\gamma_0}\right), \tag{8.46}$$

by (8.41). Equation (8.45) is a quadratic in x, corresponding to the fact that there will in general be two intersections of PZ with the hyperboloidal dispersion surface for a given state of polarisation. Its solution is

$$x = \left(-\eta \pm \sqrt{\eta^2 + 4\frac{\gamma_m}{\gamma_0}\phi_m\phi_{-m}}\right) \Big/ 2\frac{\gamma_m}{\gamma_0}\phi_{-m}. \tag{8.47}$$

(*g*) *Total reflection at a crystal surface* (*Case* II) *:* In this case, since $\theta_0 > \phi$, γ_m/γ_0 is essentially negative. If we also assume the crystal to be non-absorbing, ϕ_m and ϕ_{-m} are either equal, or conjugate complex quantities, and $\phi_m\phi_{-m}=|\phi_m|^2$, and is positive, so that x will be complex if

$$-2|\phi_m|\sqrt{\left|\frac{\gamma_m}{\gamma_0}\right|}<\eta<2|\phi_m|\sqrt{\left|\frac{\gamma_m}{\gamma_0}\right|}. \tag{8.48}$$

Within this range, which corresponds to the range of total reflection in the Darwin-Ewald theory, there is no intersection of the normal PZ with the dispersion surface (cf. § 4(k), Chapter II). By (8.46),

$$\eta=-2\Delta\psi_0\sin 2\theta_0-\phi_0\left(1+\left|\frac{\gamma_m}{\gamma_0}\right|\right), \tag{8.49}$$

and the middle of the range of total reflection occurs when $\eta=0$, or when

$$\Delta\psi_0=\Delta\theta_0=-\frac{\phi_0}{2\sin 2\theta_0}\left(1+\left|\frac{\gamma_m}{\gamma_0}\right|\right), \tag{8.50}$$

which gives the correction to Bragg's law for the case of unsymmetrical reflection. The quantity ϕ_0 is negative, and is equal to -2δ, where $1-\delta$ is the refractive index of the crystal. The middle of the range of total reflection therefore occurs at a greater glancing angle than θ_0. For symmetrical reflection, $|\gamma_m/\gamma_0|=1$, and the correction reduces to $2\delta/\sin 2\theta_0$, the value previously obtained.

We may write (8.49) in the form

$$-\frac{\eta}{2\sin 2\theta_0}=\Delta\psi_0-\frac{\delta}{\sin 2\theta_0}\left(1+\left|\frac{\gamma_m}{\gamma_0}\right|\right)=\Delta\psi_1, \tag{8.51}$$

where $\Delta\psi_1$ is the value of the angle of incidence measured from the middle of the range of total reflection, which gives a physical interpretation to η. The relationship of the various quantities concerned may be seen from fig. 149.

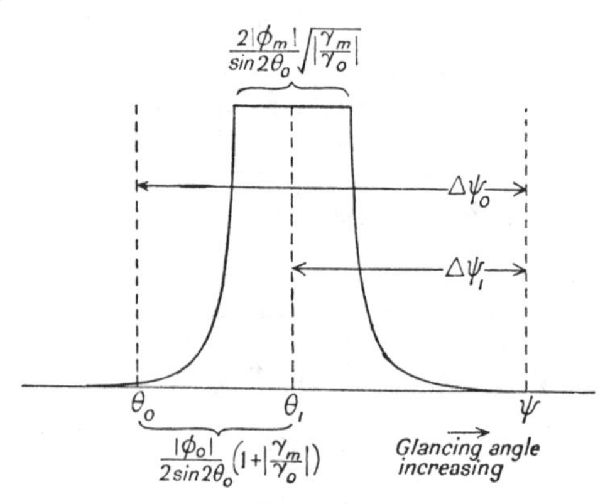

FIG. 149

The range of total reflection, by (8.48) and (8.51), is

$$2|\phi_m|\sqrt{\left|\frac{\gamma_m}{\gamma_0}\right|}\operatorname{cosec} 2\theta_0,$$

which agrees with Darwin's result for $|\gamma_m/\gamma_0| = 1$, the case of symmetrical reflection.

If x is real and positive, the corresponding values of $\mathbf{D}_0$ and $\mathbf{D}_m$ will have the same phase; if x is real and negative, they will differ in phase by π; but if x is complex, as in the range of total reflection, they will in general differ in phase by some amount between 0 and π.

(*h*) *The application of the boundary conditions in Case* II (*reflection*): We now use the conditions that must hold good between the values of the electric induction at two adjacent points on opposite sides of the boundary between two media, which here, since the fictitious dielectric constant with which we are concerned differs very little from unity, can be applied in the form that the electric induction is continuous in crossing the boundary. This method was first used by Kohler [3].

Let the crystal consist of a parallel-sided slab of thickness d, and let $\mathbf{D}_0^e$ and $\mathbf{D}_m^e$ be the induction in the incident and emergent waves at the upper surface of the crystal. At the lower surface, only the primary wave will emerge in Case II, and we denote its electric vector by $\mathbf{D}_e^0$. We still consider the state of polarisation (*a*), in which the induction vectors in all the waves are parallel. The boundary conditions then give the following relations: at the upper surface,

$$\mathbf{D}_0^{(1)} + \mathbf{D}_0^{(2)} = \mathbf{D}_0^e\,; \quad x_1\mathbf{D}_0^{(1)} + x_2\mathbf{D}_0^{(2)} = \mathbf{D}_m^e, \tag{8.52}$$

and at the lower surface,

$$c_1\mathbf{D}_0^{(1)} + c_2\mathbf{D}_0^{(2)} = \mathbf{D}_e^0\,; \quad c_1x_1\mathbf{D}_0^{(1)} + c_2x_2\mathbf{D}_0^{(2)} = 0, \tag{8.53}$$

c_1 and c_2 being phase factors. The indices (1) and (2) throughout refer to the wave-fields corresponding to the two possible wave-points A_1 and A_2.

If $\mathbf{D}_0^{(1)}$ refers to a point just below the upper surface of the crystal, the corresponding value at a point just above the lower surface is $\mathbf{D}_0^{(1)}e^{-j(\mathbf{K}_0^{(1)}\cdot\mathbf{r})}$, with the condition that $\mathbf{z}\cdot\mathbf{r} = d$, $\mathbf{z}$ being a unit vector in the direction of the normal to the surface. By fig. 148,

$$\mathbf{K}_0 = \mathbf{k}_0 + kg\mathbf{z},$$

and so
$$\mathbf{K}_0\cdot\mathbf{r} = \mathbf{k}_0\cdot\mathbf{r} + kgd,$$

whence
$$\mathrm{D}_0^{(1)}e^{-j(\mathbf{K}_0^{(1)}\cdot\mathbf{r})} = \mathrm{D}_0^{(1)}e^{-j\mathbf{k}_0\cdot\mathbf{r}}\,e^{-jkg_1d}.$$

We can therefore put

$$c_1 = e^{-jkg_1d}, \quad c_2 = e^{-jkg_2d}. \tag{8.54}$$

The second equation of (8.53) now gives

$$\mathbf{D}_0^{(1)}/\mathbf{D}_0^{(2)} = -c_2x_2/c_1x_1 = -(x_2/x_1)e^{-jkd(g_2-g_1)}. \tag{8.55}$$

If the crystal has absorption, the scattering functions ϕ_m and ϕ_{-m} are complex, so that x, and consequently g, will be complex, even outside the range of total reflection. Let us put

$$g_1 = g_1' + ig_1'', \quad g_2 = g_2' + ig_2''.$$

Then (8.55) becomes

$$\mathbf{D}_0^{(1)}/\mathbf{D}_0^{(2)} = -(x_2/x_1)e^{-jkd(g_2'-g_1')}e^{-2\pi kd(g_2''-g_1'')}. \tag{8.56}$$

The value of x_2/x_1 does not depend on the thickness of the crystal, and in general $g_1'' \neq g_2''$, so that, provided $e^{2\pi kd|g_2''-g_1''|} \gg 1$, a condition that can always be satisfied if the crystal is thick enough, and the absorption not actually zero, either $\mathbf{D}_0^{(1)}$ or $\mathbf{D}_0^{(2)}$ will be negligibly small, according to whether $g_2'' - g_1'' \gtrless 0$. There will consequently be only one primary and one secondary wave inside the crystal in this case. This result was first obtained by Kohler,[3] and it greatly simplifies the reflection problem (Case II); for there is in nature no such thing as a perfectly non-absorbing crystal, and we may always assume the existence of only one effective wave-point, even when we are dealing with a crystal so transparent that the forms of the expression applicable to a crystal without absorption may otherwise be used.

(*i*) *The reflection coefficient in Case II:* Applying the results of the last paragraph to (8.52), with either $\mathbf{D}_0^{(1)}$ or $\mathbf{D}_0^{(2)}$ zero, we obtain at once

$$\mathbf{D}_m^e/\mathbf{D}_0^e = x \tag{8.57}$$

or, by (8.47),

$$\left|\frac{\mathbf{D}_m^e}{\mathbf{D}_0^e}\right|^2 = \left|\frac{\gamma_0}{\gamma_m}\right|^2 \left|\frac{\eta \mp \sqrt{\eta^2 - 4|\gamma_m/\gamma_0|\phi_m\phi_{-m}}}{2\phi_{-m}}\right|^2, \tag{8.58}$$

since in the reflection problem γ_m/γ_0 is negative.

This expression gives the ratio of the intensities in the reflected and incident wave-trains, and for symmetrical reflection, for which $|\gamma_m/\gamma_0| = 1$, and for a non-absorbing crystal, for which ϕ_m and ϕ_{-m} are conjugate quantities, it reduces at once to Darwin's formula. For unsymmetrical reflection, (8.58) gives for the range in which $\eta^2 < 4|\gamma_m/\gamma_0||\phi_m|^2$

$$\gamma_m|\mathbf{D}_m^e|^2 = \gamma_0|\mathbf{D}_0^e|^2, \tag{8.59}$$

which corresponds to total reflection of energy from the surface of the crystal, since the energy falling on it per unit area is proportional to $\gamma_0|\mathbf{D}_0^e|^2$, while that leaving it per unit area is proportional to $\gamma_m|\mathbf{D}_m^e|^2$. The energy falling on a given area of the crystal surface is therefore exactly balanced by the amount leaving it.

For symmetrical reflection, (8.58) can be written in the form

$$\left|\frac{\mathbf{D}_m^e}{\mathbf{D}_0^e}\right|^2 = \left|\frac{2\phi_m}{\eta \pm \sqrt{\eta^2 - 4\phi_m\phi_{-m}}}\right|^2. \tag{8.60}$$

As in § 3(*h*) of Chapter II, we can represent absorption in the crystal by giving the scattering functions ϕ_m and ϕ_{-m} complex values, and at the same time making the refractive index, and so ϕ_0, complex. If we put

$$-\tfrac{1}{2}\phi_0=\delta+i\beta\ ;\quad -\tfrac{1}{2}\phi_m=A_1+iB_1\ ;\quad -\tfrac{1}{2}\phi_{-m}=A_2+iB_2, \qquad (8.61)$$

relations which correspond to equations (2.89a) and (2.89b) of Chapter II, and use (8.49), equation (8.60) reduces immediately to Prins's formula for the reflection coefficient from an absorbing crystal, equation (2.90) of p. 63. We have discussed the application of this equation in some detail in Chapter II, §§ 3(*h*) and 3(*i*), and also in Chapter VI, § 3(*j*). An example of its application to the polar crystal, zinc sulphide, is given by Kohler (loc. cit.).

(j) *Detailed discussion of the values of x for a non-absorbing crystal in Case II:* In our further discussion of the field inside the crystal we follow closely the treatment given by Laue.[4(i)] When the crystal is non-absorbing, ϕ_m and ϕ_{-m} are conjugate, and can be expressed in the form

$$\phi_m=|\phi_m|e^{i\theta_m},\quad \phi_{-m}=|\phi_m|e^{-i\theta_m}, \qquad (8.62)$$

and since in Case II γ_m/γ_0 is essentially negative, equation (8.47) may then be written

$$x=\sqrt{\left|\frac{\gamma_0}{\gamma_m}\right|}\,e^{i\theta_m}\{p\mp\sqrt{p^2-1}\}, \qquad (8.63)$$

where

$$p=\eta\Big/2|\phi_m|\sqrt{\left|\frac{\gamma_m}{\gamma_0}\right|}\,. \qquad (8.64)$$

Three distinct cases must be considered:

(i) Let the glancing angle be less than that corresponding to the lower limit of the range of total reflection, so that $p>1$. We make the substitution

$$p=\cosh u,\quad \text{with } u \text{ positive}. \qquad (8.65)$$

Then

$$x=\sqrt{\left|\frac{\gamma_0}{\gamma_m}\right|}\,e^{i\theta_m}(\cosh u\mp\sinh u),$$

or

$$\left.\begin{aligned} x_1&=\sqrt{\left|\frac{\gamma_0}{\gamma_m}\right|}\,e^{-u+i\theta_m}\\ x_2&=\sqrt{\left|\frac{\gamma_0}{\gamma_m}\right|}\,e^{u+i\theta_m}\end{aligned}\right\}. \qquad (8.66)$$

(ii) Let the glancing angle belong to the range corresponding to total reflection, so that $-1<p<+1$. The appropriate substitution is now

$$p=\sin v\quad (-\pi/2<v<+\pi/2), \qquad (8.67)$$

giving

$$x = \sqrt{\left|\frac{\gamma_0}{\gamma_m}\right|}\, e^{i\theta_m}(\sin v \mp i \cos v),$$

or

$$\left.\begin{aligned} x_1 &= -i\sqrt{\left|\frac{\gamma_0}{\gamma_m}\right|}\, e^{iv+i\theta_m} \\ x_2 &= +i\sqrt{\left|\frac{\gamma_0}{\gamma_m}\right|}\, e^{-iv+i\theta_m} \end{aligned}\right\}. \tag{8.68}$$

(iii) Let the glancing angle be greater than that corresponding to the upper limit of total reflection. Then $p < -1$, and the substitution is

$$-p = \cosh u, \text{ with } u \text{ positive,} \tag{8.69}$$

which leads to

$$\left.\begin{aligned} x_1 &= -\sqrt{\left|\frac{\gamma_0}{\gamma_m}\right|}\, e^{u+i\theta_m} \\ x_2 &= -\sqrt{\left|\frac{\gamma_0}{\gamma_m}\right|}\, e^{-u+i\theta_m} \end{aligned}\right\}. \tag{8.70}$$

(*k*) *Extinction within the range of total reflection:* In ranges (i) and (iii) of the angle of incidence, the value of g corresponding to the complex values of x determined in the last paragraph are real. For, by (*8*.43) and (*8*.39),

$$g = -\frac{1}{2\gamma_0}(\phi_0 + x\phi_{-m}) = -\frac{1}{2\gamma_0}(\phi_0 + x|\phi_m|e^{-i\theta_m}), \tag{8.71}$$

so that g is real in ranges (i) and (iii), but complex in range (ii) corresponding to total reflection. For this case, by (*8*.68) and (*8*.71), we can write for g_2

$$g_2 = -\frac{1}{2\gamma_0}\left[\phi_0 + \sqrt{\left|\frac{\gamma_0}{\gamma_m}\right|}\,|\phi_m|(\sin v + i\cos v)\right]. \tag{8.72}$$

The field in the crystal due to $\mathbf{D}_0^{(2)}$ is $\mathbf{D}_0^{(2)}\, e^{-2\pi i(\mathbf{K}_0^{(2)}\cdot\mathbf{r})}$, and at a depth d below the surface of the crystal

$$\mathbf{K}_0^{(2)}\cdot\mathbf{r} = \mathbf{k}_0\cdot\mathbf{r} + kdg_2. \tag{8.73}$$

If then we substitute for g_2 in (*8*.73) from (*8*.72), we see that the expression for the field at a depth d below the surface of the crystal contains, as well as a complex phase factor, a real damping factor $e^{-\frac{1}{2}\zeta d}$, where

$$\tfrac{1}{2}\zeta = \frac{2\pi k}{\sqrt{|\gamma_0\gamma_m|}}\,|\phi_m| \cos v. \tag{8.74}$$

The other value of x would give a value of $\mathbf{D}_0^{(1)}$ increasing without limit with increasing depth below the surface of the crystal, and so is inadmissible.

Equation (*8*.74) gives the primary extinction in the range of total reflection, which we have already discussed in Chapter II, § 3(*g*). For the case of symmetrical reflection, $\gamma_0 = |\gamma_m| = \sin\theta_0$, quite nearly enough, since the whole angular range of total reflection is of the order of seconds of arc. In the middle of the range, $\cos v = 1$, so that

$$\tfrac{1}{2}\zeta = \frac{2\pi k|\phi_m|}{\sin\theta_0} = \frac{2\lambda}{\sin\theta_0}\,\frac{e^2}{mc^2}\,\frac{\mathrm{F}(m)}{\mathrm{V}}, \tag{8.75}$$

by (8.9). This expression may be compared with the value of the primary extinction given in equation (2.85) of Chapter II, which represents the mean extinction coefficient per atomic plane. We must therefore multiply (8.75) by a, the spacing of the atomic planes, before comparing the two expressions. At either end of the range of total reflection the extinction vanishes, and it is zero everywhere outside it.

We have already seen that in ranges (i) and (iii) also, only one value of x is admissible, and we can easily see in each case which it is. For example, in range (iii), u is zero at the limit of the range of total reflection, and is positive and increasing as the angle of incidence increases beyond this limit. By (8.70) therefore, only x_2 is admissible, since x_1, and therefore $\mathbf{D}_m^{(1)}$, will increase without limit as u increases, that is to say, as the angle of incidence departs more and more from that corresponding to reflection. A similar argument shows that in range (i) only x_1 is admissible.

(*l*) *The intensity of the wave-field in the crystal in the reflection problem (Case II), as a function of the angle of incidence :* When only one wave-point is effective, the total field in the crystal at a point at a vector distance $\mathbf{r}$ from the origin can be written

$$\mathbf{D} = e^{j\nu t}\{\mathbf{D}_0 e^{-j(\mathbf{K}_0\cdot\mathbf{r})} + \mathbf{D}_m e^{-j(\mathbf{K}_m\cdot\mathbf{r})}\}, \quad \text{by (8.12).}$$

By (8.52), $\quad \mathbf{D}_m = x\mathbf{D}_0^e \quad \text{and} \quad \mathbf{D}_0 = \mathbf{D}_0^e,$

so that

$$\mathbf{D} = \mathbf{D}_0^e\, e^{j\{\nu t - \mathbf{K}_0\cdot\mathbf{r}\}}\{1 + xe^{-j(\mathbf{r}\cdot\mathbf{r}_m^*)}\}, \tag{8.76}$$

by (8.12a).

The ratio $|\mathbf{D}|^2/|\mathbf{D}_0^e|^2$ of the intensity within the crystal at the point $\mathbf{r}$ to that at the same point had only the primary wave been effective we denote by R. Then

$$\mathrm{R} = |1 + xe^{-j(\mathbf{r}\cdot\mathbf{r}_m^*)}|^2. \tag{8.77}$$

Since x is constant for any given angle of incidence, and $\mathbf{r}\cdot\mathbf{r}_m^*$ is constant in any plane parallel to the lattice-planes m, we see that the intensity of the field within the crystal is constant in sheets parallel to the reflecting planes, and that apart from effects due to extinction, it will vary with a periodicity equal to that of the spacing of the planes if the reflection takes place in a first order, and equal to a submultiple of that spacing if the reflection takes place in a higher order.

For ranges (i) and (iii), with glancing angles smaller and larger respectively than those corresponding to the range of total reflection, x is real, x_1 being the appropriate value in range (i), and x_2 in range (iii). Using (8.66) and (8.70), we then find

$$R = \left| 1 \pm \sqrt{\left|\frac{\gamma_0}{\gamma_m}\right|} e^{-u} e^{i(2\pi \mathbf{r}\cdot\mathbf{r}_m^* - \theta_m)} \right|^2$$

$$= 1 + \left|\frac{\gamma_0}{\gamma_m}\right| e^{-2u} \pm 2e^{-u} \sqrt{\left|\frac{\gamma_0}{\gamma_m}\right|} \cos(2\pi \mathbf{r}\cdot\mathbf{r}_m^* - \theta_m), \qquad (8.78)$$

the positive sign applying in range (i) and the negative in range (iii).

For the range of total reflection x_2 applies, and is complex, but in addition $\mathbf{K}_0$ itself is complex, as we saw when discussing the extinction coefficient. By (8.72), (8.73) and (8.74), we see that the value of R obtained by substituting the value of x_2 from (8.68) in (8.77) must be multiplied by the extinction factor $e^{-\zeta d}$, d being the depth of the point considered below the surface of the crystal. Thus

$$R = \left| 1 + i \sqrt{\left|\frac{\gamma_0}{\gamma_m}\right|} e^{-i\{2\pi \mathbf{r}\cdot\mathbf{r}_m^* + v - \theta_m\}} \right|^2 e^{-\zeta d}$$

$$= \left[1 + \left|\frac{\gamma_0}{\gamma_m}\right| + 2\sqrt{\left|\frac{\gamma_0}{\gamma_m}\right|} \sin\{2\pi \mathbf{r}\cdot\mathbf{r}_m^* + v - \theta_m\} \right] e^{-\zeta d}. \qquad (8.79)$$

We consider now the case of a primitive point-lattice, and calculate the field actually in the atomic planes. In this case, $\mathbf{r}\cdot\mathbf{r}_m^*$ is always integral. For such a lattice, ϕ_m and ϕ_{-m} are real, in the absence of absorption, so that θ_m is either 0 or π; and the fact that the refractive indices of most materials for X-rays are less than unity indicates that π is the correct choice, since it corresponds to scattering with change of phase of this amount (cf. Chapter IV, §§ 1(*b*) and 1(*c*)). We therefore put $\theta_m = \pi$, and obtain for the intensity of the field in the positions occupied by the atoms the following values of R.

For ranges (i) and (iii),

$$R = 1 + \left|\frac{\gamma_0}{\gamma_m}\right| e^{-2u} \mp 2e^{-u} \sqrt{\left|\frac{\gamma_0}{\gamma_m}\right|}, \qquad (8.80)$$

the negative sign applying for range (i), and the positive for range (iii). For range (ii), total reflection,

$$R = \left\{ 1 + \left|\frac{\gamma_0}{\gamma_m}\right| - 2\sqrt{\left|\frac{\gamma_0}{\gamma_m}\right|} \sin v \right\} e^{-\zeta d}. \qquad (8.81)$$

In (8.80), u is positive, and equal to zero at each limit of the range of total reflection, increasing as that range is left in either direction. In (8.81), $\sin v = 1$ for the lower limit and -1 for the upper limit of the range of total reflection. The solutions therefore fit at these limits, giving

$$R = 1 + \left|\frac{\gamma_0}{\gamma_m}\right| - 2\sqrt{\left|\frac{\gamma_0}{\gamma_m}\right|}$$

for the lower limit, and

$$R = 1 + \left|\frac{\gamma_0}{\gamma_m}\right| + 2\sqrt{\left|\frac{\gamma_0}{\gamma_m}\right|}$$

for the upper limit.

We can now draw curves showing the way in which the intensity at the position occupied by an atom varies as the angle of incidence of the radiation varies. We take for discussion the special case of symmetrical reflection, in which $|\gamma_0/\gamma_m| = 1$. Curves for this case are shown in fig. 150. The abscissae are values of $-p$, or $-\eta/2|\phi_m|$ in the case

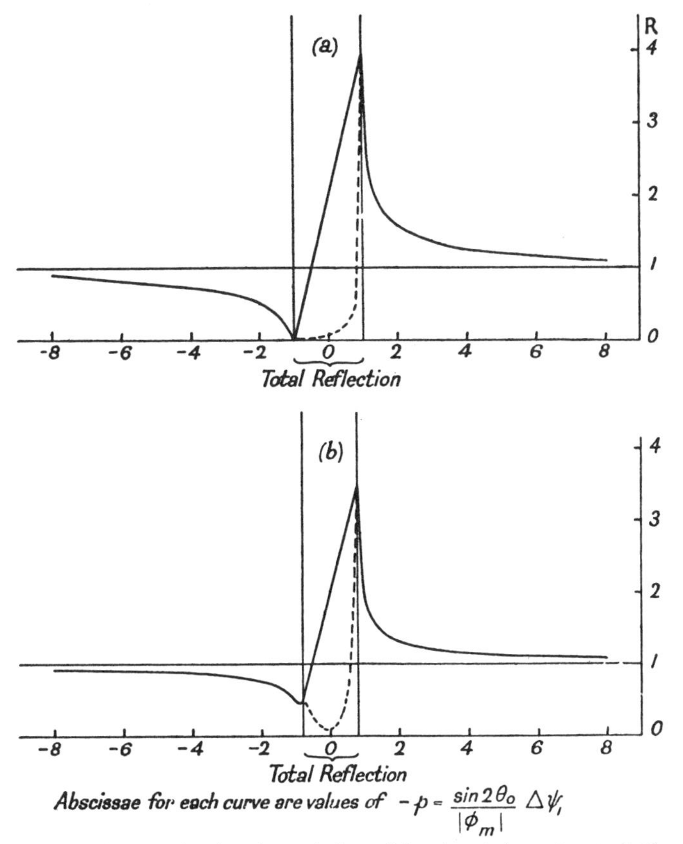

FIG. 150. Curves showing the variation of the electric intensity at a lattice-point in dynamical reflection with the angle of incidence of the radiation, (*a*) when the electric vector is perpendicular to the plane containing the directions of incidence and reflection, (*b*) when it lies in the plane

considered, by (8.64). By (8.51) therefore, we must multiply the abscissae by $|\phi_m|/\sin 2\theta_0$ to convert the units into angles in radians measured from the middle of the range of total reflection. It is easy to draw the curves if we remember that $p = \cosh u$ in range (i), and

$-\cosh u$ in range (iii), while $p = \sin v$ in range (ii), with v varying from $+\pi/2$ to $-\pi/2$ as p decreases and $\Delta\psi_1$ increases.

Let us follow out in detail what happens as the glancing angle varies, starting from values considerably less than those corresponding to total reflection and increasing. The variation in the intensity of the field at the positions occupied by the atoms is shown in fig. 150, in which the ordinates give values of R. The corresponding variations in the positions of the wave-point may be understood from fig. 148. To begin with, p is positive and $\Delta\psi_1$ negative. The relevant wave-point A lies on the branch (1) of the hyperbola, and $|\mathbf{K}_m|$ differs considerably from k. The wave D_m is hardly excited at all, and the field is very nearly that given by D_0^e alone, so that R is nearly unity. As the glancing angle ψ increases, P in fig. 148 moves towards L, and the single effective wave-point A moves along branch (1) of the hyperbola. The difference of phase between $\mathbf{D}_0$ and $\mathbf{D}_m$ remains constant, and is such that the two waves oppose each other. As the angle of incidence increases, and $\mathbf{D}_m$ grows, the resultant field in the atomic planes decreases, and becomes zero in the particular case we are considering at the lower limit of the range of total reflection. At this point, the planes of *minimum* disturbance in the crystal coincide with the atomic planes that produce the reflection, of order m. For a first-order reflection, the planes of maximum field lie halfway between the atomic planes; for a second order there are two such maxima between successive atomic planes, and so on.

As the angle of incidence increases still further, we enter the range of total reflection. There is now no real position of the wave-point; $\mathbf{D}_0$ and $\mathbf{D}_m$ are both fully excited, and the phase difference between them no longer remains constant, but alters progressively, until the two fields agree in phase in the atomic planes at the upper limit of total reflection.

If we neglect extinction for the moment, we see that the resultant field intensity rises in a linear way with increasing glancing angle, and R becomes 4 in the case of symmetrical reflection at the upper limit of the range of total reflection. As the angle of incidence increases beyond this point, the effective wave-point makes its appearance on the other branch of the hyperbola. The phase difference between $\mathbf{D}_0$ and $\mathbf{D}_m$ again remains constant, but $\mathbf{D}_m$ becomes less important as $\Delta\psi_1$ increases, rapidly at first, and then more slowly, so that the field once more approaches that due to the wave D_0^e alone, and R tends to unity from above. We must think of the whole set of nodes and loops of intensity in the crystal as being displaced while the angle of incidence passes through the range corresponding to total reflection. At the lower limit of this range the planes of minimum intensity, and at the upper limit the planes of maximum intensity, coincide with the atomic planes. Outside the range of total reflection, on either side, the maxima and minima occupy the same positions as at the corresponding limits, but the difference between the maximum and minimum intensity gets smaller as the excitation of the wave $\mathbf{D}_m$ diminishes, and the interference effects die away.

The full-line curve in fig. 150(*a*) illustrates the variation of R at the position occupied by an atom when extinction is neglected, and refers therefore to atoms lying only a little way below the surface of the crystal. For atoms situated more deeply within the crystal extinction becomes very important inside the range of total reflection, but is zero at its limits, and beyond. The dotted line shows the course of the curve within the range of total reflection, drawn for the case in which the intensity of the field is reduced to 1/20 of its value by extinction at the middle of the range.

The whole discussion has hitherto been confined to case (*a*), in which $\mathbf{D}_0$ and $\mathbf{D}_m$ are perpendicular to the plane containing $\mathbf{K}_0$ and $\mathbf{K}_m$. If they lie *in* this plane (Case (*b*)), expressions (*8*.80) and (*8*.81) must be modified by multiplying the *last term* in each by a factor $|\cos 2\theta_0|$. The range of total reflection is now given by

$$-2|\cos 2\theta_0|\,|\phi_m|\sqrt{\left|\frac{\gamma_m}{\gamma_0}\right|} < \eta < +2|\cos 2\theta_0|\,|\phi_m|\sqrt{\left|\frac{\gamma_m}{\gamma_0}\right|}, \qquad (8.82)$$

and a parameter $p' = p/|\cos 2\theta_0|$ replaces the parameter p used in the discussion of case (*a*). The variation of R for the position of an atom in case (*b*) is shown by curve (*b*) in fig. 150, which is drawn for $|\cos 2\theta_0| = 0{\cdot}75$. The maxima and minima in the curve without extinction are less well marked than in case (*a*), because $\mathbf{D}_0$ and $\mathbf{D}_m$ are no longer parallel, but inclined to one another at an angle $2\theta_0$, so that complete interference and reinforcement can no longer take place. The dotted part of the curve again shows the effects of extinction.

Curves (*a*) and (*b*) of fig. 150 are drawn for symmetrical reflection. If the reflection is unsymmetrical, R may still become zero in case (*a*). Equation (*8*.80) shows that the position of the zero value is given by

$$e^{-u} = \sqrt{|\gamma_m/\gamma_0|}, \text{ or } p = \frac{1}{2}\left(\sqrt{\left|\frac{\gamma_0}{\gamma_m}\right|} + \sqrt{\left|\frac{\gamma_m}{\gamma_0}\right|}\right), \text{ provided that } |\gamma_0| \lesseqgtr |\gamma_m|.$$

The zero value will in this case lie a little below the lower limit of total reflection.

(*m*) *The transmission coefficient in Case I:* In this case, $\mathbf{k}_m$ leaves the lower surface of the crystal slab, $\theta_0 < \phi$, γ_m is positive, and γ_0/γ_m is an essentially positive quantity. For a non-absorbing crystal, by (*8*.47), $x\phi_{-m}$, and hence the accommodation g, is always real. Two wave-points are always effective, and the field inside the crystal is made up of the two wave-fields corresponding to them, superposed. If we consider transmission through a plane-parallel slab of crystal, the boundary conditions for the state of polarisation (*a*) are, at the upper surface,

$$\mathbf{D}_0^{(1)} + \mathbf{D}_0^{(2)} = \mathbf{D}_0^e\ ; \quad x_1\mathbf{D}_0^{(1)} + x_2\mathbf{D}_0^{(2)} = 0, \qquad (8.83)$$

and at the lower surface,

$$c_1\mathbf{D}_0^{(1)}+c_2\mathbf{D}_0^{(2)}=\mathbf{D}_e^0;\quad c_1x_1\mathbf{D}_0^{(1)}+c_2x_2\mathbf{D}_0^{(2)}=c_3\mathbf{D}_e^m. \tag{8.84}$$

c_1, c_2, c_3 are phase factors, c_1 and c_2 having the same significance as in (8.54), while the condition that the tangential component of $\mathbf{k}_m$ must be the same as that of $\mathbf{K}_m$ leads to the result that $|c_3|=1$, if the glancing angle made by $\mathbf{K}_m$ with the surface is not too small. For a discussion of this point, reference may be made to Kohler's paper. Equations (8.83) and (8.84) lead to the following result for the ratio of the intensity in the incident beam to that in the transmitted interference beam:

$$\mathrm{T}_m=\left|\frac{\mathbf{D}_e^m}{\mathbf{D}_0^e}\right|^2=\frac{x_1^2x_2^2}{|x_1-x_2|^2}|c_1-c_2|^2. \tag{8.85}$$

The factor $|c_1-c_2|^2$ is equal to $2\{1-\cos 2\pi k(g_1-g_2)d\}$, and so depends on the thickness of the crystal. It represents interference in a plane-parallel plate, and in any actual case the cosine term would be averaged out by unavoidable small variations in the thickness of the plate, and would vanish. In this case, the transmission coefficient for the wave m becomes

$$\mathrm{T}_m=\frac{2x_1^2x_2^2}{|x_1-x_2|^2}=\frac{2|\phi_m|^2}{\eta^2+4|\gamma_m/\gamma_0|\,|\phi_m|^2}, \tag{8.86}$$

for a non-absorbing crystal, by (8.47). T_m has a maximum value of $\frac{1}{2}|\gamma_0/\gamma_m|$ when $\eta=0$, and is therefore equal to $\frac{1}{2}$ when $|\gamma_0/\gamma_m|=1$, as it is if the reflecting planes within the crystal are perpendicular to the surface of the crystal plate, an arrangement that is frequently used in practice, and which has been discussed for the case of the mosaic crystal in Chapter II, § 2(h). By (8.46), we see that the maximum occurs when

$$\Delta\psi_0=-\frac{\phi_0}{2\sin 2\theta_0}\left(1-\frac{\gamma_m}{\gamma_0}\right),$$

which gives the deviation from Bragg's law. Since ϕ_0 is essentially negative, and γ_m/γ_0 positive, we may put

$$\Delta\psi_0=\frac{|\phi_0|}{2\sin 2\theta_0}\left(1-\left|\frac{\gamma_m}{\gamma_0}\right|\right). \tag{8.87}$$

When the reflecting planes are normal to the surface of the plate, $\Delta\psi_0=0$, that is to say, reflection occurs at the Bragg angle. The angular width of the transmitted beam in this case is suitably expressed by the width at half-maximum, which, by (8.86), is evidently $4\sqrt{\dfrac{\gamma_m}{\gamma_0}}|\phi_m|$ in terms of η, or $2\sqrt{\gamma_m/\gamma_0}|\phi_m|\operatorname{cosec}2\theta_0$ in angular measure, which is the same as the width of the range of total reflection in Case II.

(n) *The intensity of the wave-field in the crystal in Case I:* Since the total wave-field is now made up of two superimposed wave-fields, corresponding to the two wave-points, the state of vibration within the crystal is much less easily expressed than it was in the reflection problem. For a full discussion, we must refer the reader to Laue's paper [4 (i)]. The two wave-vectors of each of the pairs $\mathbf{K}_0^{(1)}$, $\mathbf{K}_0^{(2)}$ and $\mathbf{K}_m^{(1)}$, $\mathbf{K}_m^{(2)}$ differ slightly, both in magnitude and direction, and the effect of this is to cause fluctuations in the intensity in the crystal, of the same nature as beats. The intensity is still constant, or nearly so, in planes parallel to the lattice-planes (m), but for any given direction of incidence the surfaces of maximum disturbance lie at different distances from the atomic planes at different depths below the surface of the crystal, and structurally equivalent atoms at different depths are not excited in the same way. Upon a short-period fluctuation of intensity, similar to that discussed in the reflection problem, are superimposed 'beat' fluctuations of longer period, due to the interaction of two slightly differing wave-fields. The existence of this effect was shown by Ewald in his original treatment of the dynamical theory, and we have already referred to it briefly in § 4(k) of Chapter II.

One result deduced by v. Laue, which we shall use later, may be stated here. We have considered an incident wave represented by the vector $\mathbf{k}_0$, giving rise to waves $\mathbf{K}_0$ and $\mathbf{K}_m$ inside the crystal, which in their turn correspond respectively to waves $\mathbf{k}_0$ and $\mathbf{k}_m$ outside the crystal on the other side of the plate. These conditions are represented diagrammatically in fig. 151, in which the dotted line shows the direction of the atomic planes m. Now let a wave be incident on the *upper* surface of the crystal in the direction $\mathbf{k}_m$. Inside the crystal, the corresponding wave-vector is $\mathbf{K}_m$, and this gives rise to an interference wave $\mathbf{K}_0$, which we can picture as reflected from the lower surface of the planes m, and which corresponds to an emergent wave $\mathbf{k}_0$ below the crystal. Laue shows that if the two waves $\mathbf{k}_0$ and $\mathbf{k}_m$ are incident together on the upper side of the crystal, and if their intensities are so adjusted that the same energy falls per unit area on the surface in both wave-trains, then the total field intensity in the crystal is the sum of those due to the two primary waves, and is independent of the depth below the surface. The two wave-fields are complementary. If one gives rise at any point to an intensity in excess of the primary field, the other gives rise to an intensity as much below it, and this holds for every angle of incidence.

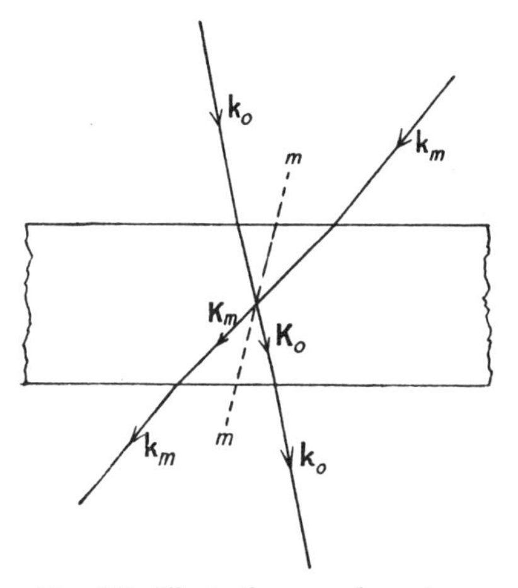

FIG. 151. Illustrating complementary reflection

For a detailed proof of this property of the field, reference must be made to Laue's paper (§§ 3. and 4.).

2. Diffraction Phenomena when the Source of Radiation lies within the Crystal

(*a*) *Introductory:* Suppose now that the atoms of the crystal lattice themselves become independent sources of monochromatic radiation. The waves spreading out from any atom must be diffracted by the other atoms of the lattice, and must ultimately give rise to disturbances travelling out into the space surrounding the crystal. We wish to investigate the nature of the external wave-field so produced. Because the sources within the crystal are independent, the total intensity distribution outside the crystal will be the sum of those due to the individual atoms. It is clear that a direct discussion of this type of crystal interference would be difficult, for we should have to start with a spherical wave spreading out from a point inside the crystal, and to consider its interaction with all the atoms of the lattice. It is true that we can make a plausible prediction of the type of disturbance to be expected by applying the elementary theory of reflection of waves at the lattice-planes; although we are not entitled to do this without further justification when the sources of radiation are close to the planes themselves. Neglecting this question for the moment, let us assume that the waves emanating from any point within the lattice are reflected at the lattice-planes to an appreciable degree only at angles that correspond to Bragg's law. The directions of appreciable reflection from any set of planes $m(=hkl)$ will then lie on the surface of a cone, whose axis is normal to the planes, and so in the direction of the reciprocal-lattice vector $\mathbf{r}_m^*$, and whose semi-vertical angle α is given by the equation

$$2d_m \cos\alpha = \lambda$$

or

$$\cos\alpha = \tfrac{1}{2}\lambda|\mathbf{r}_m^*|. \qquad (8.88)$$

If all the atoms in the crystal are emitting, we should expect reflections from both sides of the planes, and so the cone corresponding to the planes m is to be thought of as a complete double cone. To every reciprocal-lattice vector there corresponds such a cone, and the whole set of cones must be thought of as rigidly attached to the crystal. For any position of a photographic plate relative to the crystal, we may therefore expect an appropriate set of diffraction maxima, which will be ellipses, parabolas, or hyperbolas—the intersections of the cones with the plane of the plate.

Interference effects due to monochromatic radiation excited within the crystal were first announced by Clark and Duane [5] in 1922, and further investigated by them in a series of papers. They reflected white radiation from a tungsten anticathode from the faces of a number of

Plate X — See other side

PLATE X

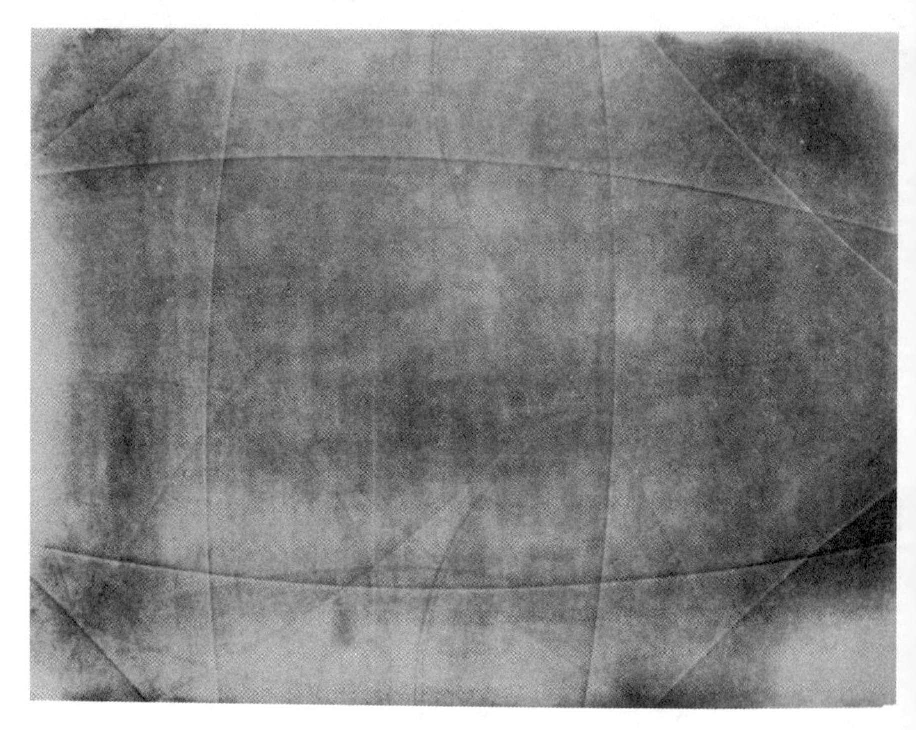

FIG. 152. Kossel lines from copper crystal used as anticathode. Plate parallel to cube face (Voges)
(*Ann. der Physik*, **27**, 694 (1936))

crystals, mostly alkaline halides or polyhalides, and examined the reflected radiation by means of an ionisation spectrometer. When the crystals contained elements such as iodine, or caesium, they observed certain peaks in the reflection curve which they ascribed to the reflection of the characteristic K radiations of the elements concerned at the crystal planes. The conditions for producing interference of this type were certainly present, but the effect would be extremely hard to observe with the experimental arrangement used, and it is certain that the main peaks observed were due to other causes, probably to the effect of the K absorption edge of the heavy element on the form of the reflection curve*; and although some small residuum of the phenomena observed by Clark and Duane may have been due to a true interference of the fluorescent radiation, the first to obtain undoubted effects of this kind were Kossel, Loeck, and Voges,[6] in 1934. They used a single crystal of copper as the anticathode in an X-ray tube. The characteristic radiation was therefore excited by the electron beam in the crystal itself, and interference effects, the general nature of which was in accord with the elementary theory outlined above, were observed on a photographic plate placed near the crystal, in the first experiments, actually within the tube. We shall refer to the interference lines so obtained as *Kossel lines*. In subsequent work,[7, 8] the technique was developed, and reference should be made to the beautiful photographs published by Voges,[8] one of which is shown in fig. 152 (Pl. X). Similar effects, when the emission of the radiation is stimulated by X-rays, have been obtained by Borrmann.[9] Although the simple theory is able to predict the general geometry of the effect it cannot give us much information about the actual distribution of intensity, and for this, more powerful methods are needed.

(*b*) *The reciprocity theorem in optics:* A beautifully neat solution of the problem of interference when the source lies inside the crystal has been given by v. Laue,[4] who uses the reciprocity theorem in optics, an old theorem, the general formulation of which in terms of the electromagnetic theory was given by H. A. Lorentz [10] in 1905. In the form in which we shall need it, the theorem may be stated as follows. If a source of radiation and a point of observation are interchanged, the intensity, measured in terms of electric displacement, is the same at the new place of observation as at the old.

Laue quotes an experiment performed by Selenyi [11] in 1913, which serves as a good example of an application of the theorem to an optical problem. Let a pencil of light radiating from a point P_1 (fig. 153) inside a block of glass, and limited by a screen S, fall on the boundary between the glass block and the air beneath it, the angles of

* It has been suggested by Mrs. Lonsdale that some of the effects observed by Clark and Duane were diffuse maxima, due to thermal movement. See Chapter V p. 239.

incidence being so arranged that according to the geometrical theory all the light should be totally reflected. There will be no appreciable disturbance below the glass block, provided that the point of observation lies at a distance from the surface greater than a few times the wave-length of the radiation used. If, however, we consider a point P_2 within a distance of the surface comparable with the wave-length of light, we know that the field in which it lies is not zero.

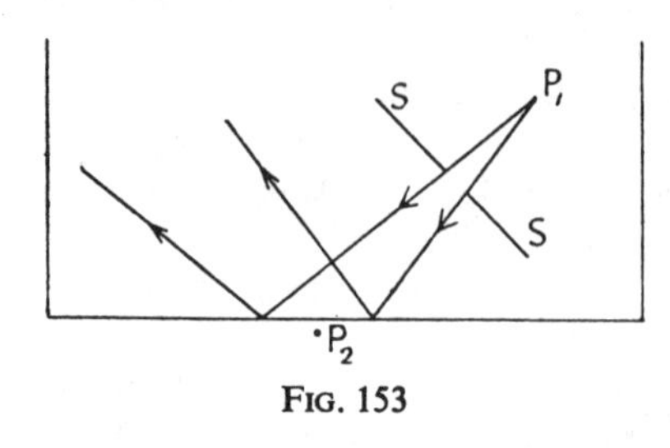

FIG. 153

Inhomogeneous waves travel parallel to the surface of the glass in the air, their amplitude dying away very rapidly as the distance from the surface increases. According to the reciprocity theorem, therefore, a source of light of strength equal to P_1 placed at P_2, very close to the glass surface, should send light to the position occupied by P_1, a point which no light should reach according to the geometrical theory; and the intensity at P_1 should be equal to that at P_2 with the original arrangement of the source. That this actually occurs was demonstrated by Selenyi.

(c) *The application of the reciprocity theorem to the problem of the diffraction of radiation excited within the crystal:* Suppose a certain atom A inside the crystal to be emitting radiation, and let the resultant intensity be observed at a point P outside the crystal and at a considerable distance from it. We wish to calculate the intensity at this point. Suppose the source A removed to P. Then it will produce a certain intensity at its original position inside the crystal, which will be just the same as the original intensity at P. Now when the source is removed to P outside the crystal it will send out spherical waves, which by the time they reach the crystal may be treated as plane, since we have supposed P to lie at a considerable distance. The intensity of the field at the position of the atom A within the crystal is just that of the wave-field set up inside the crystal when a plane wave-train enters it from without, a problem that we have already solved for the ideal crystal in the first section of this chapter.

We have seen that the intensity of the field at a point occupied by an atom reaches its maximum in the case of reflection, for example, when the angle of incidence is such as to correspond to one limit of the range of total reflection. Let the direction of incidence then be that of the vector $\mathbf{k}_0$. If the crystal is now stimulated so that the atom A emits radiation, it will radiate a maximum amount outside the crystal in the direction $-\mathbf{k}_0$, and the variation with direction of the intensity radiated will be exactly the same as the variation of intensity at the position of the atom A as the direction of the radiation incident from without varies in a corresponding way. The curves in fig. 150 were drawn to show the stimulation of a plane of atoms due to an external source of radiation as the direction of the incident wave-vector $\mathbf{k}_0$ varied, but they may equally well be taken as showing the variation of the radiation in the direction $-\mathbf{k}_0$ outside the crystal when an atom belonging to the plane becomes a source of radiation of the same frequency.

For interference of a given order m, the directions of $\mathbf{k}_0$ corresponding to maximum excitation of the atoms lie on a cone having the normal to the lattice-planes (m), or the reciprocal vector $\mathbf{r}^*_m$, as axis, and a semi-vertical angle equal to within a few seconds to that given by Bragg's law. Thus, when the atom itself becomes a source, the radiation outside the crystal will show a maximum around the same cone. The predictions of the elementary theory as to the geometry of the cones are therefore closely confirmed, but the reciprocity theorem also allows us to consider questions of intensity and fine structure in detail. Since the emission of fluorescent radiation from atoms is essentially an incoherent process, all the atomic sources are independent, and the total intensity is the sum of the intensities due to the individual atoms.

(*d*) *Detailed consideration of the diffraction cones corresponding to the reflection problem* (*Case II*)*:* The geometrical conditions are illustrated by fig. 154. The planes SS and *mm* are parallel respectively to the crystal surface and to the reflecting lattice-planes, and the angle between them is ϕ. OR is the direction of the normal to these planes, and so of the reciprocal-lattice vector $\mathbf{r}^*_m$. The cone is drawn corresponding to the Bragg angle θ_0. In the case considered, $\theta_0 > \phi$, and the cone does not intersect the plane SS. The intensity scattered into the cones is to be derived from the wave-field in the crystal in the reflection problem, and figs. 150(*a*) and (*b*) apply. We consider first radiation scattered along directions adjacent to OP, and polarised with the induction vector perpendicular to the plane ROP, to which fig. 150(*a*) applies. Beginning with a direction of scattering OQ, lying within the Bragg cone, we suppose the angle ROQ to increase, which corresponds to a

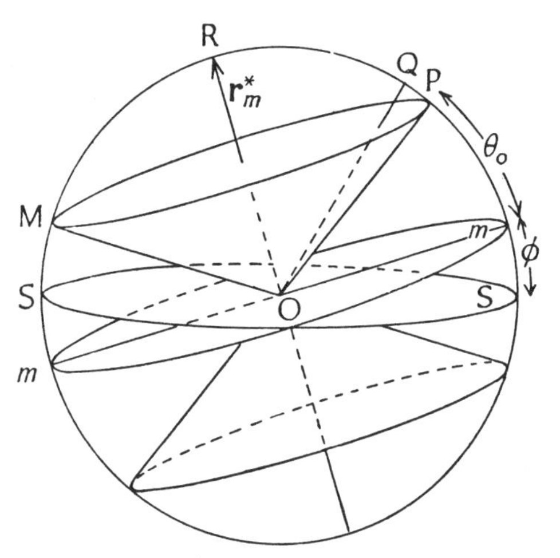

FIG. 154. Diffraction cones for Case II (Reflection): $\theta_0 > \phi$

decrease in the glancing angle in fig. 150(*a*). When OQ lies well within the Bragg cone, there will be a uniform scattering corresponding to R = 1 in fig. 150. As the angle ROQ increases, the intensity of the scattered radiation rises to a maximum, some four or five times that of the background, in a direction corresponding to the upper limit of the range of total reflection in the dynamical problem. Thereafter, as the angle of the cone increases still more, the intensity of the scattered radiation falls off very rapidly, at a rate depending, however, on the depth of the atomic source below the crystal surface. The fall observed in any actual case must be an average effect, due to the radiation from a very large number of atoms at different distances below the surface ; but at the lower limit of the range of total reflection the intensity scattered should fall to zero, and so be *less* than that of the general background. For still greater angles of the cone, the intensity rises once more, tending again to that of the background. The cones corresponding to the diffraction maxima thus have a fine structure, according to the dynamical theory. The angular widths involved are small, a matter of a minute or so at the most, and a nearly perfect crystal will be required if the effect is to be observed. If greater scattered intensity corresponds to greater blackening, as on a photographic film, each curved line corresponding to the intersection of a diffraction cone with the film should show a border lighter than the general background on its *convex* side.

If the electric field-vector lies in the plane ROP, curve (*b*) of fig. 150 applies, and the same general conclusions hold good, except that the maxima and minima will be less well marked. It may be noted that if the radiation from the atoms is unpolarised, as is to be expected in the case of fluorescent radiation, there must be some degree of polarisation in the radiation scattered into the diffraction cones, because of the different intensities corresponding to cases (*a*) and (*b*).

Our theoretical discussion throughout this chapter has related only to the case in which the intersection of the surfaces SS and *mm* is perpendicular to the plane ROP, which of course is true for only two directions on the cone. The more general case is discussed by Laue, but we shall not consider it here, since it alters nothing in principle, although it affects the variation of intensity around the cone.

(*e*) *The total excess or defect of intensity in the Kossel lines :* In the last paragraph, we considered the fine structure of the Kossel lines in Case II (Reflection). We may, however, wish to know whether the total intensity in a line is greater or less at any part of it than that of the general background. As a criterion of this, Laue calculates a quantity S, defined by

$$\mathrm{S} = \int \frac{|\mathbf{D}|^2 - |\mathbf{D}_0^e|^2}{|\mathbf{D}_0^e|^2}\, d\psi. \tag{8.89}$$

If S is positive, a Kossel line produced under conditions of resolution not good enough to show up its fine structure will appear darker than the background; if S is negative, it will appear lighter. The quantity S may be thought of as made up of contributions from the three ranges (i), (ii), and (iii), which are denoted by S_1, S_2, and S_3 respectively. For the non-absorbing crystal, ranges (i) and (iii) give

$$S_1 + S_3 = \frac{2C|\phi_m|}{3 \sin 2\theta_0} \sqrt{\left|\frac{\gamma_0}{\gamma_m}\right|}, \qquad (8.90)$$

where $C = 1$ or $|\cos 2\theta_0|$, according to the state of polarisation.

The contribution S_2, corresponding to the range of total reflection in the dynamical problem, can easily be calculated for two limiting cases, (1) when the atom emitting radiation lies near the surface of the crystal, so that extinction can be neglected, and (2) when it lies so far below the surface that owing to the effects of extinction the field in its neighbourhood would be zero if the radiation were incident from outside. Case (1) gives,

$$S = S_1 + S_2 + S_3 = \frac{8C|\phi_m|}{3 \sin 2\theta_0} \sqrt{\left|\frac{\gamma_0}{\gamma_m}\right|}, \qquad (8.91)$$

an expression that is essentially positive, and denotes an excess of intensity in the Kossel line.

In a later paper, Laue [12] shows that S may be negative in case (2), when the atom lies deep in the crystal; for then $|D|^2 = 0$, and the integrand in (8.89) becomes -1. Taking into account the width of the range of total reflection, we obtain at once

$$S_2 = -\frac{2C|\phi_m|}{\sin 2\theta_0} \sqrt{\left|\frac{\gamma_m}{\gamma_0}\right|}, \qquad (8.92)$$

giving

$$S = S_1 + S_2 + S_3 = \frac{2C|\phi_m|}{3 \sin 2\theta_0} \sqrt{\left|\frac{\gamma_0}{\gamma_m}\right|} \left(1 - 3\left|\frac{\gamma_m}{\gamma_0}\right|\right). \qquad (8.93)$$

The quantity in brackets in (8.93) may become negative if the ratio $|\gamma_m/\gamma_0|$ is large enough, so that under suitable conditions of incidence the Kossel line may appear as a lighter line on a dark background. What will actually be observed in any case is of course an average effect, due to atoms lying at a variety of depths in the crystal, but it must be remembered that extinction is often much more important than ordinary absorption, so that certainly when the atoms within the crystal are excited by X-rays we may expect the majority of emitters to be in situations in which (8.93) would be valid. We shall return to this question in a later paragraph.

(*f*) *The nature of the diffraction cones corresponding to Case I* (*transmission*) *in the dynamical problem:* The geometrical conditions are represented by fig. 155. Since $\phi > \theta_0$, the planes SS now intersect the cone. An incident wave travelling in the direction PO will give rise to an interference wave travelling in the crystal in the direction OM. To obtain the intensity of the radiation scattered in the direction OP when the atoms of the lattice become sources of radiation we must therefore use Case I of the dynamical problem. As we have already

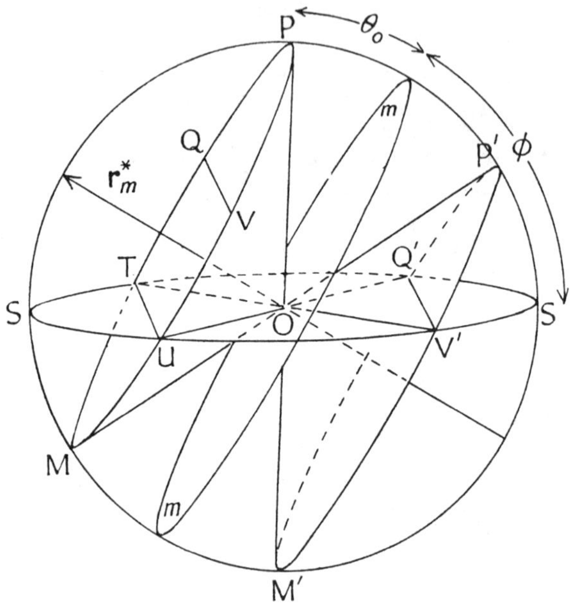

FIG. 155. Composite diffraction cones: $\phi > \theta_0$

seen in § 1 (*m*) of this chapter, the field in the region of a given atom now depends on its depth below the surface, so that the conditions in the interference cone cannot be calculated so easily as in Case II. It can, however, be stated that the radiation in the directions OP and OP′, complementary directions on the two cones m and $\bar{m}$, will be complementary, so that if there is an excess of intensity over that of the background in the direction OP, there will be a corresponding defect in the direction of OP′. This follows at once from the result quoted in § 1 (*n*) that the sum of the fields due to waves incident in the directions PO and P′O, and carrying equal energies to the crystal per unit area, is constant, and independent of the depth in the crystal. Therefore the sum of the intensities emitted by the crystal in the two complementary directions, when an atom within it becomes a source of radiation, must be constant, so that if the excess over the background is positive in one direction, it must be negative in the complementary direction. Not only the total radiation within the diffraction line, but also the fine structure, should be complementary in the two directions. A photographic plate, parallel for example to the crystal surface, should show hyperbolic traces of the two halves of the cone. If the resolution is not enough to show fine detail, one branch of the hyperbola should give a trace darker than its surroundings, and the other a trace lighter ; and if the fine detail is resolved, it should be complementary in the two traces. In general, the two branches of the hyperbola, corresponding to the diffraction cones hkl and $\bar{h}\bar{k}\bar{l}$ must be complementary in intensity. The analysis in terms of the dynamical theory shows that in the simple case considered the cone lying nearest to the normal to the crystal surface should contain the excess of radiation, and so should give the darker trace.

If the planes m lie normal to the surface, the excess radiation in both cones must be zero, but the fine structure is not symmetrical, as a consideration of the geometry of the reflection as illustrated by fig. 148 will show; so that if the resolution is good enough to show fine detail, a complementary light and dark structure may still appear. Examples of this are to be found in the paper by Voges (fig. 4), in which the complementary fine structures of 002 and 00$\bar{2}$, and of 020 and 0$\bar{2}$0 show up very well; whereas, in the corresponding photograph taken by Borrmann with X-rays (fig. 5), in which the resolution is not great enough to show the fine structure, the corresponding traces do not occur, although those for 111 and $\bar{1}\bar{1}\bar{1}$, and 1$\bar{1}\bar{1}$ and $\bar{1}$11, when the planes are no longer perpendicular to the crystal face, show up excellently as light and dark pairs, the dark branch being in each case the nearer of the pair to the normal to the surface.

(*g*) *Composite cones :* Referring again to fig. 155, we see that when radiation falls on the crystal from outside, all rays directed towards O from points in the arc QPV of the cone give rise to diffracted rays in directions from O to points on the arc UMT, which are therefore directed into the crystal. In calculating the intensity of the portion QPV of the diffraction cone when the atoms of the crystal become the sources of radiation, we must therefore use Case I of the dynamical theory, and it is to this portion of the cone that the discussion of the last paragraph applies. On the other hand, rays directed towards O from points between Q and T give rise to reflections in directions from O towards points on the arc UV. The intensities of the parts QT and UV of the diffraction cone must therefore be calculated from the formulae for the reflection problem (Case II) of the dynamical theory, with appropriate modifications for the obliquity of the surface. Such composite cones must always occur unless the planes m are normal to the crystal surface, when the whole of the exposed cone corresponds to Case I. The composite arc is always that one of a pair closest to the normal to the crystal face.

It is interesting to consider whether we may expect a change of intensity from light to dark in the Kossel lines corresponding to such composite cones. From fig. 155, we see that radiation incident along QO in the corresponding problem of dynamical reflection would be reflected along OU, and so would emerge very nearly indeed parallel to the surface. This means that $|\gamma_m|$ would be very small, and so, by (*8*.93), the total intensity excess in the Kossel line would be positive. On the other hand, the point T on the arc would correspond to radiation incident along TO at very nearly grazing incidence to the crystal surface, and so to a very small value of $|\gamma_0|$. The value of S given by (*8*.93) would in this case be negative, and the intensity of the Kossel line, in so far as it depends on atoms lying not too close to the surface, will be less than that of the background. We should therefore expect a not very well resolved line to change from dark to light between those points on it corresponding to Q and T, the light part of the line corresponding to radiation leaving the surface at nearly grazing incidence. Examples of such changes from dark to light lines are well shown in fig. 5 of Borrmann's paper, for the lines $1\bar{1}1$ and $11\bar{1}$, and the same effect may be seen in some of the photographs obtained by Voges, for example, in fig. 5.

(*h*) *Experimental details :* As we have already mentioned, undoubted diffraction effects due to the excitation of atomic sources inside a crystal were first obtained in 1934 by Kossel, Loeck, and Voges,[6] from a single crystal of copper, used as the anticathode in an X-ray tube. The beam of electrons impinging on the crystal caused the atoms forming the lattice to emit the characteristic K radiation of copper. The experimental difficulties are considerable, as the intensities to be expected are not large. The theory shows that there will be a strong background of radiation scattered in all directions, and that the excess scattering in the directions of the interference cones is not very large. In the earlier experiments, the film was placed inside the vacuum tube, at a distance of about 2·2 cm. from the crystal, but in later experiments it was placed outside the tube, at a distance of 50 or 60 cm. In this way it was possible to reduce the strength of the background relative to that of the lines.

Although copper is an easily deformable crystal, which might be expected to show little primary extinction, and to reflect as a mosaic, it appears that in suitably prepared specimens portions at any rate are fairly perfect.* In the experiments of Voges, the anticathode spot is sharply focussed, and can be arranged to fall on some perfect region of the crystal. Thus the angular width of the lines is very small, not greatly in excess of that indicated by the dynamical theory, and up to a considerable distance there is no appreciable spreading of the lines, the breadth of which is determined almost entirely by that of the anticathode spot. The falling off of intensity within the interference cones is thus proportional to $1/r$, and that in the background to $1/r^2$,

* Compare the experiments of Renninger with rock salt. Chapter VI, § 3 (*j*).

r being the distance of the plate from the crystal; so that the contrast between lines and background increases as the distance increases, an advantage that offsets the disadvantage of the longer exposure that is necessary. Voges' photographs show up very beautifully the fine structure of the lines, and it is clear from their width that the fragment of crystal concerned was acting as if it were very nearly perfect, so that Laue's theory should apply fairly closely, since a copper crystal, with appropriate choice of axes, may be considered as based on a simple lattice.

Fig. 156 shows a photometer curve obtained by Voges of a 200 reflection, with $\phi=0$ (Case II, symmetrical reflection). This may be

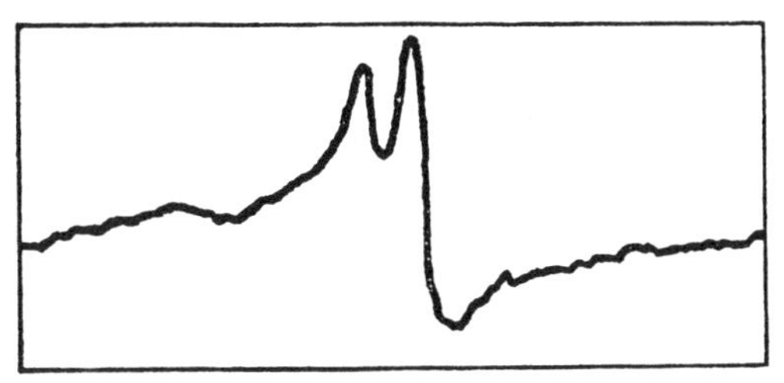

FIG. 156. Trace of a photometer curve by Voges of the 200 reflection from copper showing light-dark effect in Case II (Reflection) with $\phi=0$ (Symmetrical reflection)

compared with fig. 150. It must be remembered that the Cu Kα line is a doublet, which accounts for the double peak. The α_1 peak is also reduced by the minimum to the right of the α_2 peak, so that the peaks appear more nearly equal in strength than they should. Qualitatively at least, the predictions of the theory are borne out by these experiments, which provide a remarkable confirmation of the essential correctness of the dynamical theory. It is of course not to be expected that completely detailed agreement between theory and experiment will be obtained, for the theory still relates to a somewhat idealised state of affairs. For example, atoms have size, and in estimating the effective field acting on them we are not concerned solely with the field strength exactly in the atomic planes. Moreover, the effects of extinction are very difficult to estimate quantitatively in any particular case.

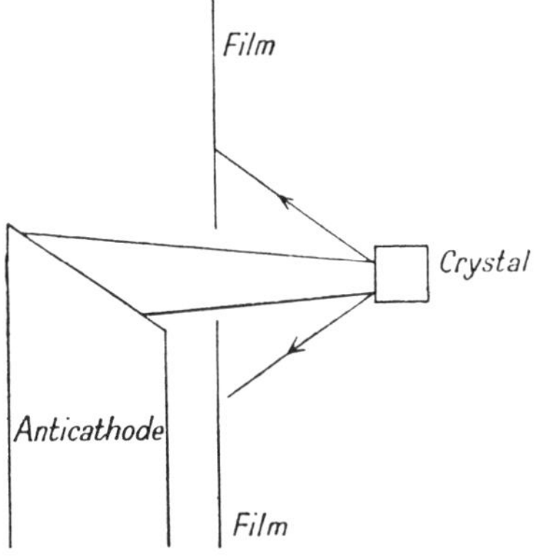

FIG. 157. Illustrating the excitation of Kossel lines with X-rays

Borrmann [9] has used X-rays to excite the fluorescent radiation. The general arrangement is that shown in fig. 157, and one of his photographs, taken with the cube axis of the crystal perpendicular to the plane of the film is shown in fig. 158 (Pl. XI). The spots are those of the ordinary Laue pattern, obtained by reflection backwards from the crystal. The fine light and dark curves are the traces of the diffraction cones due to the interference of the fluorescent radiation from within the crystal. They show no fine structure, for the crystals used had been artificially roughened, and then afterwards etched until the contrast between the lines and the background was as great as possible, a condition that was reached when the crystal was still too imperfect to show fine structure. So long as we are interested only in the total excess or defect of the radiation above the background, the maximum will be reached when primary extinction is not too large. Borrmann's photographs record this integrated excess, and for the same reason, as we have already mentioned, the traces corresponding to the planes that are normal to the surface do not appear, although they do so in Voges' photographs in virtue of their fine structure.

(*i*) *The geometry of the cones :* We are not primarily concerned here with the detailed geometry of interference patterns of the Kossel type, but only with the optical principles underlying their production, and we must be content therefore with calling attention to a few of their more salient features. If plane photographic plates or films are used, all the curves must of course be conic sections. In principle, every reciprocal-lattice vector of the crystal is the axis of a diffraction cone of semi-vertical angle α given by $\cos\alpha = \frac{1}{2}\lambda|\mathbf{r}_m^*|$. Bragg's law has been assumed as an amply good enough approximation in discussing the geometry of the cones, since departures from it are of the order of seconds only.

An inspection of the photographs obtained shows that intersections of three and sometimes of more curves at a single point are not uncommon, and it is not difficult to see how these may come about. Let us suppose we are dealing with radiation leaving the cube face of a face-centred cubic crystal. The reciprocal-lattice points in the plane $c^* = 0$ are shown in fig. 159, together with the reciprocal-lattice vectors that lie parallel to this plane. The circle drawn with the origin as centre has radius $2/\lambda$, and is the trace in the plane $c^* = 0$ of a sphere of the same radius. The plane (020) of the reciprocal lattice cuts this sphere in a small circle, which defines the diffraction cone 020. In the same way, a plane through the point (220) of the reciprocal lattice, perpendicular to the lattice vector to this point, the trace of which in the plane of the figure is indicated by the symbol 220, defines by its intersection with the sphere the cone 220. The traces of a number of such planes are shown in the figure, appropriately numbered. Each reciprocal-lattice vector drawn corresponds to a double cone, and so to two traces. Only such planes as intersect the circle can give interference cones. It

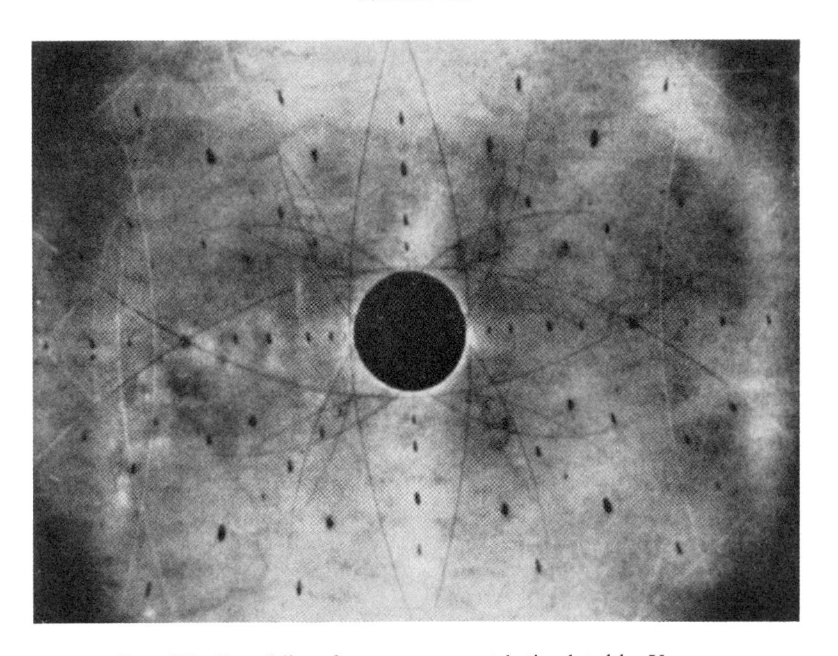

FIG. 158. Kossel lines from copper crystal stimulated by X-rays.
Plate parallel to [100] (Bormann)
(*Ann. der Physik*, **27**, 669 (1936))

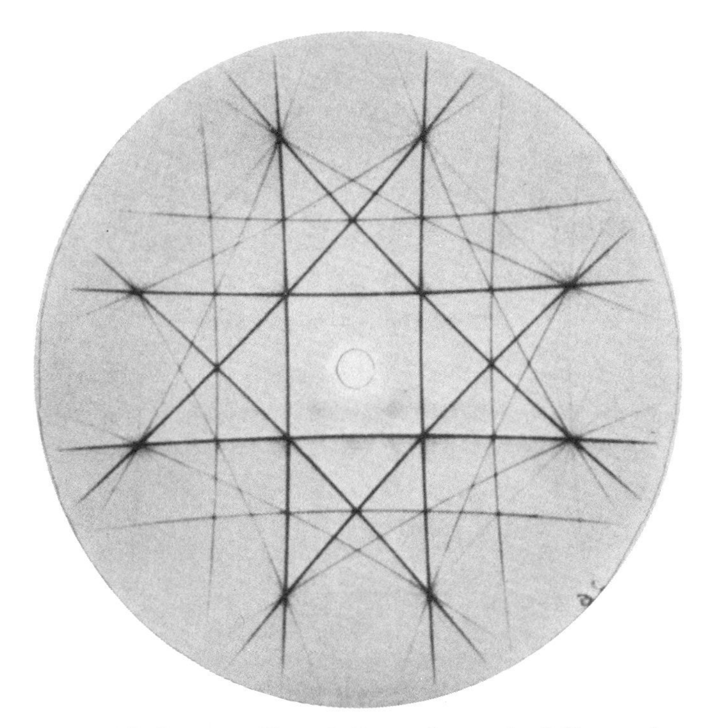

FIG. 162. Seemann wide-angle diagram from rock salt (Seemann)
(*Physik, Mitteilungen*, (41))

(P. 448)

Plate XI — See other side

will be noticed that a number of intersections of three or more of these traces occur. For example, 020, 420, and 400 intersect at A, and 200, 020, and 220 at B. Each point of intersection of three lines represents

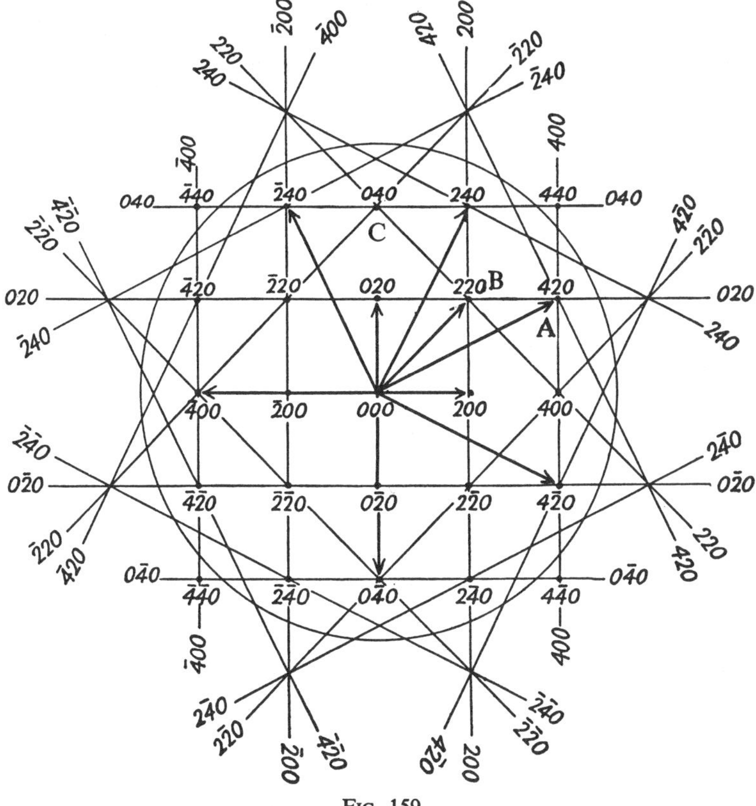

Fig. 159

a line of intersection of three planes, and the radius to the point where such a line of intersection meets the sphere must define a direction that is common to three diffraction cones. Only such points as lie within the circle can give rise to observable intersections of interference lines. The traces of the cones defined by these sets of planes on a photographic plate parallel to the cube face must of course be hyperbolas. Many more cones are possible than those indicated in the figure, which shows only those corresponding to the zone [001]. Those corresponding to other zones can, however, be dealt with in the same way, by drawing the appropriate reciprocal-lattice net. For further details, reference may be made to a paper by Kossel.[13]

In our discussion of the geometry of the cones we have so far assumed the Bragg law to be valid, according to which the intersections just considered should be exact. Kossel and Voges have, however, pointed out that in some actual cases they are not quite exact. In particular, when an intersection occurs involving three lines, two of which show little fine structure, but the third of which shows a well-marked light border, a careful examination reveals that the intersection of the two simple dark lines takes place not at the darkest maximum of the third line, but at the boundary between the light and dark region, or even sometimes within the light region itself. This confirms in a very satisfactory way the predictions of the dynamical theory. Fig. 150 shows that the maximum of a line showing fine structure lies at a glancing angle definitely *above* that corresponding even to the corrected Bragg law, which would place it at the middle of the range of total reflection. The diffraction cone of maximum intensity corresponding to such a line would therefore have an angle smaller than that predicted by Bragg's law. The study of such intersections should give interesting results in connection with the fine structure of the cones.

In addition to intersections of the type we have so far discussed, which depend only on the geometry of the lattice, and occur whatever wave-length is used, provided that it lies within the range necessary to give lattice spectra at all, there are other intersections and coincidences that depend on the ratio λ/a. Suppose the wave-length used to be initially too large to give any lattice spectra, and let it gradually decrease. The sphere of radius $2/\lambda$, the trace of which is shown in fig. 159, will expand as λ decreases. When its radius becomes greater than $|\mathbf{r}^*(111)|$ in the case we are here considering, that of the face-centred lattice, the first diffraction cones will appear, and as the sphere grows with diminishing λ the angles of these cones get greater. At a certain value of λ/a contacts between pairs of these cones occur, and as λ diminishes still further, they will intersect. The relation of any pair of diffraction cones to one another thus depends on λ/a. We may take a simple example, which may be followed easily from fig. 159.

Suppose the sphere to pass exactly through the point C in the figure, the point (040) of the reciprocal net. The diffraction cones 220 and $\bar{2}20$ then touch along their common generator OC, which is the zone

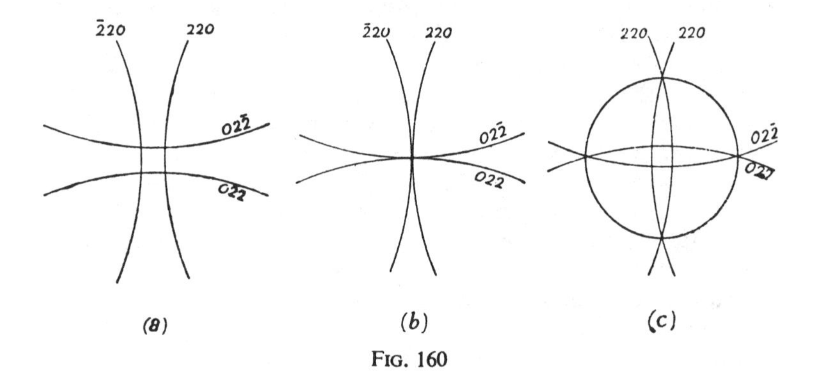

Fig. 160

axis [010], and the same will be true for the cones 022 and 02$\bar{2}$. The traces of these cones on a photographic plate with its plane perpendicular to this zone axis will be as shown in fig. 160(*b*), which, by fig. 159, corresponds to $2/\lambda = 4/a$, or to $\lambda/a = 1/2$. The corresponding traces when λ/a is a little greater or a little less than 1/2 are shown in fig. 160(*a*) and (*c*) respectively. In fig. 160(*c*), the trace of the cone 040 also appears, and once λ/a is small enough for it to appear at all it will always pass through the intersections of the hyperbolas, forming an intersection of the first type, depending only on the lattice geometry.

It is plain that if we observe traces of the types shown in fig. 160(*a*) and (*c*) we shall know that λ/a differs very little from 1/2, and that we shall be able to estimate its departure from this value by measuring the distances between the apses of the two hyperbolas. Let this distance be x. Then the angle of the cone corresponding to one of the hyperbolas in case (*a*) is $45° - x/2D$, D being the distance of the plate from the crystal. The angle 45° is that between [010] and [220], which is exactly known from the geometry of the lattice. A determination of λ/a thus depends only on the measurement of the small correction. Kossel,[14] to whom this method is due, points out that it is not as a rule necessary to measure D. Suppose, for example, we find traces of pairs of cones of the types m and n in the region of the pole of the zone axis $[p]$, as shown in fig. 161. The actual value of the ratio λ/a is greater than that corresponding to a contact of the hyperbolas m, but smaller than that corresponding to the contact of the hyperbolas n, and its value for both these cases is readily calculated from the lattice geometry. For example, when the hyperbolas m are in contact

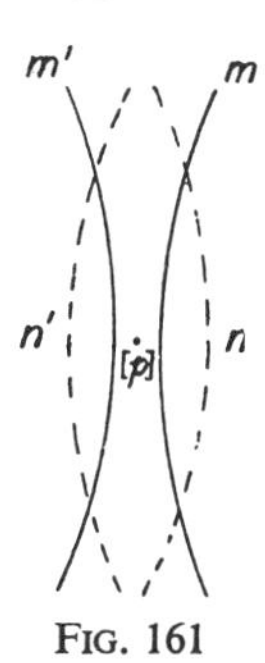

FIG. 161

$$\cos \alpha_{mp} = \tfrac{1}{2}\lambda|\mathbf{r}_m^*| = \frac{1}{2}\frac{\lambda}{a}\sqrt{\Sigma m^2}, \tag{8.94}$$

α_{mp} being the angle between the zone axes $[m]$ and $[p]$, which is easily calculated from the lattice geometry. It is to be remembered that m, n, and p are single symbols standing for triplets of numbers such as hkl, so that Σm^2 stands for $h^2 + k^2 + l^2$. Once the value of λ/a for the two types of contact have been calculated, the determination of the actual value, which lies between them, is a matter of interpolation, depending only on the ratio of x_m to x_n, the distances between the apses of the two types of hyperbola. These measurements are made directly on the film, and there is no need to know the distance of the crystal. By such methods, which appear to have considerable possibilities, Kossel has made an accurate determination of the ratio λ/a for the copper lattice, using Cu Kα radiation, arriving at a value in very close agreement with the best precision measurements by other methods. One of the most interesting developments of the method should be its application to the study of the fine structure of the lines.

(*j*) *Seemann's wide-angle diagrams:* Interference patterns of the same geometrical form, but of different physical origin, have long been known, the so-called wide-angle diagrams of Seemann.[15] A cone of rays with an angle of 60° or more falls on a slice of crystal. Because of the great variety of directions in the primary beam, a large number of the lattice-planes in the crystal are in the correct positions to give reflections, and the reflected maxima must lie on the surfaces of the set of cones that we have discussed above. With this arrangement, it is possible to screen the plate from the effects of the primary beam, and to photograph only the radiation reflected from the planes. It is in fact a case of ordinary reflection of radiation from an external primary source, with angles of incidence continuously variable over a considerable range. Fig. 162 (Pl. XI) is a good example of such a pattern obtained by Seemann from a rock-salt crystal, with a cube face as surface. The correspondence between the hyperbolic traces of the cones of radiation, and the network of lines in fig. 159 will at once be apparent.

When the sources of radiation lie inside the crystal, every direction of incidence is possible, and if all the lines in principle possible actually appeared there would be far more lines present on an interference pattern due to fluorescent radiation from within the crystal than appear even on the Seemann diagram, in which the range of possible angles of incidence is still limited. That comparatively few lines do in fact appear is due to the faintness of all but the strongest reflections. We are here dealing essentially with an excess or defect of intensity relative to a background that must always be present and must always be fairly strong; and the intensities concerned are altogether of a different order of magnitude from those used in producing the Seemann patterns.

(*k*) *Divergent-beam photography:* An interesting method of photography, in which very divergent beams are used, and which produces diagrams geometrically closely similar to those of Kossel and Seemann, has been developed by Mrs. Lonsdale.[16] In this method, a slice of crystal one or two millimetres in thickness is placed close against the anticathode of a tube of a type designed by Dr. A. Müller, which gives a beam of X-rays diverging externally, with a spread of nearly 180°, from what is virtually a point source. For the purpose of this discussion, we may assume a point source of monochromatic radiation to lie in one face of the crystal. The photographic plate or film is placed at a distance of about 10 cm. on the side of the crystal remote from the source.

There is of course a general blackening of the film produced by radiation diverging at all angles from the source, but if the thickness and nature of the crystal are suitable, a pattern of lines in the form of a series of conic sections appears on the film, having the same geometrical arrangement, apart from fine structure, as the Kossel lines already discussed. The lines are of two kinds. Most of them are lighter than the background on the original film, and are deficiency lines, due to the reflection within the crystal of radiation that would otherwise have travelled in the directions concerned. They are, however, sometimes accompanied by lines darker than the background, parallel to, but not in general coincident with the deficiency lines, which are due to reflection of characteristic radiation directly to the film. The formation of the two types of line can be understood from fig. 163.

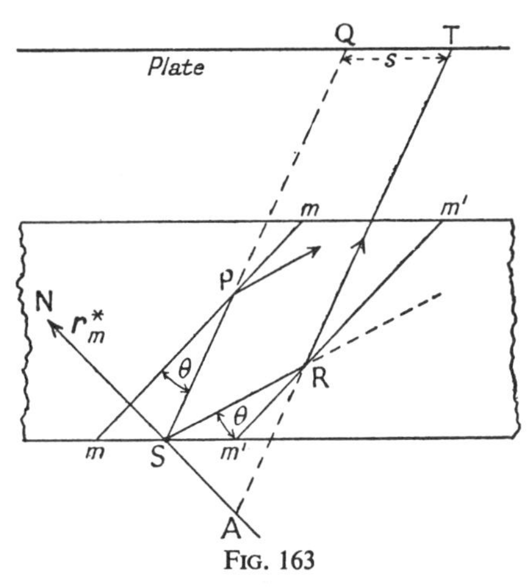

FIG. 163

Let *mm*, *m'm'* be crystal planes, and let S be the point source of radiation. Consider a reflection formed at a glancing angle θ. A ray such as SP is reflected from the plane *mm* at this angle, and of course from any other planes of the same set that lie in its path, so that the radiation in the direction SQ is deficient in intensity after it has left the crystal. Considering the matter in three dimensions, we see that all such directions as SQ lie on a cone with axis SN, normal to the reflecting planes, and so in the direction of the reciprocal-lattice vector $\mathbf{r}_m^*$. The vertex of the cone is at S, and its semi-vertical angle α is given by

$$\cos\alpha = \sin\theta = \tfrac{1}{2}\lambda|\mathbf{r}_m^*| = \tfrac{1}{2}\lambda/d_m. \qquad (8.95)$$

The cone will be very sharply defined if the crystal is ordinarily perfect, and its trace on the photographic plate is a conic section, which will appear lighter than the background (darker, of course, on a print from the original plate).

A ray such as SR will be reflected from the plane *m'm'*, and all possible directions of reflection from this plane will lie on a cone, having the same angle and axis as the deficiency cone, but with its vertex at A. The intersection of this cone with the plate gives a reflection line, darker than the background, which follows closely the corresponding deficiency line but lies just on the convex side of it. Reflection can take place from a large number of planes parallel to *m'm'*, and to each such plane corresponds a cone with the same angle but a slightly different vertex, so that the reflection lines are not in general as sharp as the deficiency lines. It will be seen from the diagram that a reflection line cannot be formed at all if the angle made by the plane *m'm'* with the lower crystal surface is less than θ, and that they will be most sharply defined when the angle is not very different from 90°. The deficiency lines will therefore be commoner than the reflection lines.

To give good deficiency lines, the crystal must be so thick that a considerable amount of reflection takes place during the transmission of the radiation through it, and yet not so thick that the characteristic radiation is heavily absorbed. Reflection lines from thick crystals on the other hand are weak, and tend to be diffuse. A thin crystal gives stronger reflection lines but weak deficiency lines. The distance s between a reflection line and the corresponding deficiency line depends in a simple geometrical way on the distance of the point of reflection from the source, and is independent of the distance of the plate from the source. By measuring the distance between the two kinds of line on actual films, it is found that the layer of crystal farthest from the source is responsible for most of the reflection lines.

Divergent-beam photographs can be used to determine the ratio λ/a for a cubic crystal with great precision from coincidences between intersections of the lines. The method used is exactly analogous to that described by Kossel, and discussed in § 2(i) of this chapter. Mrs. Lonsdale has made a determination of this kind for diamond, and has also pointed out that divergent-beam photographs, in principle, allow the space-group and cell-dimensions of any crystal that will give such a photograph to be determined. Since the exposure times are very short (30 seconds, for example, was found to be enoughwith a diamond 1·5 mm. in thickness) the method has considerable possibilities in the preliminary stages of a crystal analysis.

An interesting point is the relationship of the nature of the photograph obtained to the state of perfection of the crystal. It appears that if the crystal is highly perfect, showing a large amount of primary extinction, good photographs are not obtained. The deficiency lines obtained from such a crystal, although very sharp if they appear at all, are often absent or exceedingly faint. It is clear from fig. 163 that if the crystal is very perfect, so that the atomic planes are exactly parallel throughout its whole volume, the deficiency cone will be very sharply defined, and may be so narrow that its effect on the photograph is difficult to see. Mrs. Lonsdale has found that although many organic crystals show the effect very well, yet some do not. Dipping such crystals in liquid air for a few seconds often caused a pattern to appear when a photograph was again taken after the crystal had regained room temperature. Presumably, the sudden fall of temperature on immersion in liquid air had destroyed the perfection of the crystal, breaking it up into a mosaic. Some organic crystals, so soft that one would naturally assume their structure to be mosaic in character, show this effect, and so presumably, in spite of their softness, show considerable primary extinction, a point of importance in crystal analysis, when intensity measurements have to be made. The method has been found useful in distinguishing between Type I and Type II diamonds.

The first divergent-beam photographs showing deficiency lines were obtained as early as 1914 by Rutherford and Andrade,[17] who used the method to determine the wave-lengths of γ-rays from radium, by passing them through slices of rock salt. It is noteworthy that they found the photographs obtained with very perfect slices of crystal, especially selected for the purpose, to be much inferior to those obtained with material of a more usual degree of imperfection.

(*l*) *Laue's explanation of Kikuchi lines:* Laue [4,19] has suggested that the so-called Kikuchi lines in electron diffraction, the formation of which is not easily accounted for satisfactorily by elementary considerations, owe their origin to processes analogous to those discussed in this chapter. Electron diffraction patterns obtained from single crystals often show, in addition to the characteristic cross-grating pattern, pairs of parallel straight lines stretching across the plate. The lines occur in pairs, of which one line is usually darker and the other lighter than the background, and each pair is related to some set of atomic planes in the crystal in such a way that an extension of a plane parallel to this set and passing through the crystal would cut the photographic plate in a line midway between them. The phenomenon was first observed by Kikuchi.[21] Examples of Kikuchi patterns are shown in fig. 160, Plate XXIX, of Vol. I. In electron diffraction patterns, the angle of deviation is always small, and the pairs of apparently straight lines on the Kikuchi patterns are really small portions of the two branches of a hyperbola in the neighbourhood of the apses.

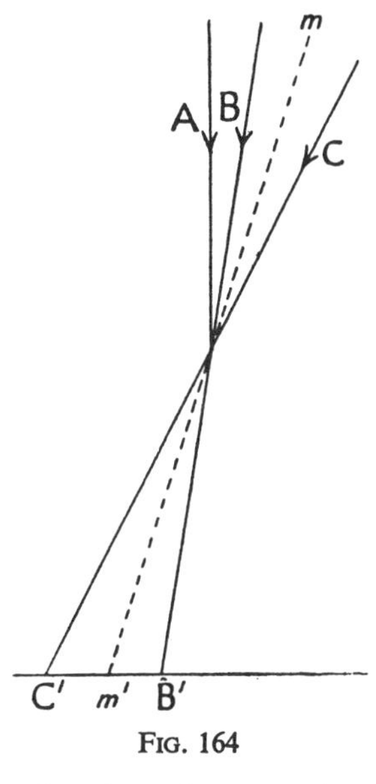

FIG. 164

All the explanations that have been given of these lines ascribe them to reflection by atomic planes of radiation travelling within the crystal. It has been supposed that a general radiation of the same wave-length as the incident radiation is set up inside the crystal, which although travelling in all directions is strongest in the direction of the primary beam. In fig. 164, let A be the direction of the incident electron beam, and *m* the trace of a lattice-plane. The general electron radiation in the direction B is partially reflected towards C′ and partly transmitted towards B′; while that in the direction C is partly reflected towards B′ and partly transmitted towards C′. If now the radiation in the direction B is stronger than that in the direction C, less intensity will be restored to the direction B′ by the reflection of C than will be taken from it by the reflection of

B to C′ ; while more radiation from the direction of B will be reflected into the direction of C′ than is taken away from it by the reflection of C towards B′. Thus C′ will be darker, and B′ lighter than the background, if we think in terms of the blackening of a photographic plate. This theory gives no satisfactory explanation of the origin of the general radiation, and it also indicates that if the plane *m* contains the direction of incidence everything should be symmetrical about the plane, and the lines should disappear, which, however, is by no means so in practice.

The theory of diffraction of electrons in a crystal lattice, as dealt with by Bethe,[18] is essentially a dynamical theory, although it is complicated by the fact that the periodic variations of the potential energy of the lattice relative to the energy in the incident beam, which correspond to the distribution of the scattering function in the case of diffraction of X-rays, are large, so that quantities that can safely be neglected in one theory may be of importance in the other. We may, however, assume with some confidence that the general nature of the wave-function within the crystal when dynamical self-consistency is attained is similar to that of the distribution of the wave-field in the corresponding X-ray problem.

Laue shows that a reciprocity theorem holds good in the wave-mechanical problem, in the form that if with a given set of conditions a source of electrons at a point P_1 causes a certain value of the wave-function u at a point P_2, then the same electron source displaced to P_2 will cause the same value of the wave-function at P_1. Thus, if an atom in the crystal becomes in any way the source of electron waves, we may calculate the corresponding wave-intensity in any direction outside the crystal from that at the position of the atom in question when an electron wave enters the crystal from the opposite direction, just as we did in the X-ray problem.

We have now to consider the way in which the sources of electron waves in the crystal may be produced. These must be essentially incoherent, for the coherent radiation scattered from the atoms is the origin of the ordinary diffraction pattern. Laue [19] considers the source of the necessary electron waves to be the inelastic scattering of the incident electrons by electrons bound in the atoms of the crystal, with losses of energy amounting perhaps to a few electron-volts. Since, in order to produce Kikuchi lines, electrons of 30,000 to 40,000 electron-volts are generally used, these energy losses will correspond to a negligibly small change in wave-length, and the inelastically-scattered electrons could without error be treated as having the same wave-length as the incident electrons. Born's theory of collisions indicates that electron waves scattered in this way with loss of energy from different atoms will be incoherent, and Laue therefore supposes the Kikuchi lines to owe their origin to this inelastic scattering, in virtue of which the atoms become independent sources of radiation.

In applying the theory developed for X-rays to explain Kikuchi lines, we must not expect more than a very general qualitative agreement. While any such theory will give complementary cones, accounting for the dark and light pairs, the decision as to which line of the pair should be dark and which light depends on questions of phase not easily answered in the case of electrons; and indeed the arrangement of the pairs does appear as a rule to be the opposite to that predicted by the simple theory in the case of X-rays. There are also other features in the detailed structure that are not accounted for by the simple theory. For example, the general background between pairs of lines is often quite different from that outside them. Such effects are plainly to be seen in photographs obtained by Boersch [20] and Voges' photographs suggest that a similar effect occurs with X-rays. Laue considers that the inelastic scattering responsible for the Kikuchi lines must be that from electrons bound in atoms, and not from the conductivity electrons. Although the theory is still in a relatively incomplete state, it appears to offer a very promising field for future investigation.

References

1. M. Kohler, *Berl. Sitzungsber.*, 334 (1935).
2. M. v. Laue, *Ergeb. der exact. Naturwiss.*, **10**, 133 (1931).
3. M. Kohler, *Ann. der Physik*, **18**, 265 (1933).
4. M. v. Laue, (i) *Ann. der Physik*, **23**, 705 (1935); (ii) *Die Interferenzen von Röntgen- und Elektronenstrahlen, Fünf Vorträge*, (Berlin, Springer, 1935).
5. G. L. Clark and W. Duane, *Proc. Nat. Acad. Sci.*, Washington, **8**, 90 (1922), **9**, 126 (1923), 131 (1923), **10**, 48 (1924).
6. W. Kossel, V. Loeck and H. Voges, *Zeit. f. Physik*, **94**, 139 (1935).
7. W. Kossel and H. Voges, *Ann. der Physik*, **23**, 677 (1935).
8. H. Voges, *Ann. der Physik*, **27**, 694 (1936).
9. G. Borrmann, *Ann. der Physik*, **27**, 669 (1936).
10. H. A. Lorentz, *Proc. Amsterdam Acad.*, **8**, 401 (1905).
11. P. Selenyi, *Comptes Rendus*, **157**, 1408 (1913).
12. M. v. Laue, *Ann. der Physik*, **28**, 528 (1937).
13. W. Kossel, *Ann. der Physik*, **25**, 512 (1936).
14. W. Kossel, *Ann. der Physik*, **26**, 533 (1936).
15. H. Seemann, *Physikal. Zeit.*, **20**, 169 (1919); *Naturwiss.*, **23**, 735 (1926).
16. K. Lonsdale, *Nature*, **151**, 52 (1943), **153**, 22, 433 (1944).
17. E. Rutherford and E. N. da C. Andrade, *Phil. Mag.*, **28**, 263 (1914).
18. H. Bethe, *Ann. der Physik*, **87**, 55 (1928).
19. M. v. Laue, *Ann. der Physik*, **25**, 569 (1936); *Physikal. Zeit.*, **37**, 544 (1936).
20. H. Boersch, *Physikal. Zeit.*, **38**, 1000 (1937).
21. S. Kikuchi, *Proc. Jap. Acad. Sci.*, **4**, 271, 275, 354, 475 (1928); *Jap. Jour. Phys.*, **5**, 83 (1928-9); *Physikal. Zeit.*, **31**, 777 (1930).

CHAPTER IX

THE SCATTERING OF X-RAYS BY GASES, LIQUIDS AND AMORPHOUS SOLIDS

1. Introductory

(*a*) *General survey of the subject:* In this chapter, and in the succeeding one, we shall consider the scattering of X-rays by amorphous matter, such as gases, liquids, and vitreous and non-crystalline solids. The distinction between the amorphous and crystalline states is by no means clear-cut from the point of view of the phenomena to be discussed. There is in reality no such thing as a completely random assemblage of atoms or molecules; for molecules have a finite size, and the mere fact that a molecule A occupies a certain portion of space at a given instant means that that particular portion of space is for the time being forbidden to another molecule B. No arrangement of molecules of finite size can therefore be a random one in the sense that all positions in space are equally probable at any instant for any molecule of the assemblage. Even in a gas, the most nearly random assemblage of molecules that we can think of, there must be a rudimentary degree of arrangement. No two molecules can approach one another to within a distance smaller than the sum of their molecular radii, and this, as we shall see below, is in principle enough to give rise to diffraction phenomena when radiation of appropriate wave-length passes through the gas, although at ordinary pressures the effects are too slight to be readily observed.

In a very important paper, published in 1915, Debye [1] showed that the collection of electrons into groups to form atoms must give rise to diffraction phenomena, even in the case of a monatomic gas; and he calculated the distribution of scattered intensity about the primary beam that would be produced by certain definite arrangements of electrons in rings about the atomic nuclei. This aspect of the subject we have already discussed in detail in Chapter III, under the heading of the *Atomic Scattering Factor*. In this chapter, we shall discuss first the effects that must be produced by the unavoidable degree of molecular arrangement called into being by the mere fact that the molecules occupy spaçe. Such arrangement, as we have already said, exists in principle even in a gas, but as a cause of diffraction effects becomes important only for liquids and non-crystalline solids. There must be every gradation between this irregular or amorphous arrangement and the regular geometrical patterns characteristic of crystals; and one of our tasks will be to attempt to follow the effects on the diffraction pattern of a transition from one to the other. Secondly, we shall have to consider the diffraction effects that must be produced by the arrangement of the atoms within each molecule into a regular configuration.

The repetition of this configuration in each molecule, with random position and random orientation, as in a gas, will give rise to well-marked diffraction phenomena. Debye has called such effects 'Internal Interference Effects', and those due to the mutual configuration of the molecules, conditioned by their packing and dimensions, 'External Interference Effects'. It is plain that we shall not in general be able to separate them. In principle, both must always occur together, although sometimes the one effect and sometimes the other will have predominant importance. In the case of a gas at ordinary pressures, for example, the observed diffraction effects are almost entirely those due to internal interference. We shall see too that we may sometimes regard diffraction by crystals as a limiting case of the external effect, and sometimes of the internal effect, according to whether we consider the crystal in bulk, or in the form of a powder.

A third type of structure, to be considered in the next chapter, is that of the fibre, or hair-like substance, in which there is an arrangement of atoms that may at times verge on the crystalline, so that the internal interference effects are similar to those produced by very elongated crystals with a very small extension at right angles to the direction of the fibres. The orientation of the different crystal-like individuals in the collection may, however, be very nearly at random about the direction of their lengths, while their elongated shape causes them to lie very nearly parallel to each other. This type of arrangement gives diffraction effects showing affinity to those produced by assemblages of particles having strong preferential orientation, and small dimensions in certain directions, and for the discussion of such problems it is necessary to consider the effect of crystal size on the diffraction pattern.

It will be seen too that a gas may be regarded, at least formally, as the limiting case of a random arrangement of crystalline particles, with the number of crystal units in each particle reduced to one, or, in the case of diatomic gases such as O_2 and N_2, to two. It is interesting to trace the passage from one type of aggregate to the other in the light of the diffraction effects they produce, and it is clear that this again is really a problem of crystal size.

The first to show clearly that a regular crystalline arrangement is not essential for the production of diffraction effects were Debye,[1] in the paper already mentioned, and Ehrenfest,[2] both in 1915 ; and one result of this work was the discovery by Debye and Scherrer [3] of the diffraction haloes produced by powdered crystals. The same method was developed independently a little later by Hull,[4] in the United States. Diffraction haloes produced by paraffin wax had been observed as early as 1913 by Friedrich.[5]

Debye and Scherrer,[3] in 1916, investigated the diffraction patterns produced by benzene, in the hope of being able to determine from them the structure of the benzene molecule. They found definite interference rings, and showed that their angular radii depended on the wave-length of the radiation used in the manner to be expected if the rings were due to diffraction ; but they were not able to interpret the diffraction pattern in terms of the molecular structure. Their failure to do so, as Debye himself points out, was due to the masking of the internal interference effects, due to the regular arrangement of the atoms in the molecule, by the external effects, due to the fact that because of the finite size of the molecules, intermolecular distances lying within a certain range occur in the liquid with greater frequency than any other distances. The existence of such effects was very definitely shown by Keesom and de Smedt,[6] who obtained diffraction rings even with the monatomic liquid argon.

In the years following, a very large amount of experimental work was done on diffraction by liquids in various parts of the world, notably in India, as a result of the theoretical work of Raman and Ramanathan,[7] and in the United States, by Stewart and his fellow-workers. Stewart[8] in particular upheld the view that each molecule in a liquid, because of its form and size, produced a certain order, or regularity of orientation, among the molecules in its neighbourhood. For this condition, he introduced the name ' cybotaxis ', and he ascribed to its existence the diffraction haloes given by liquids. It is probable that the earlier tendency was to overestimate the importance of such regularities in producing diffraction effects, which can often be accounted for by that departure from completely random arrangement which is a consequence of the finite size of the molecules, without its being necessary to assume any further local regularity of structure. Nevertheless, effects due to local arrangement are doubtless often of considerable importance in determining the details of the diffraction patterns. Recent work on liquid crystals has shown that a very high degree of order may sometimes exist in a liquid, and in this connection reference should be made to the report of a discussion on liquid crystals published in the *Transactions of the Faraday Society*, Vol. *29*, 1933, p. 881.

The importance of the mere departure from random arrangement in producing diffraction effects was very clearly brought out by Debye[9] in 1925, in a very important paper on the diffraction of X-rays by gases, in which the idea of a probability function expressing the probability of the occurrence of any given intermolecular distance was introduced. Similar ideas were put forward by Zernicke and Prins[10] in a stimulating paper, which has formed the starting-point of much of the recent work on the interpretation of liquid diffraction patterns. They discussed the form of the probability function in the case of a one-dimensional arrangement of linear molecules having a finite length, and showed how it was related to the diffraction pattern that such an arrangement would produce; and they also showed how to apply Fourier's integral theorem to the determination of the probability function from the observed diffraction pattern. The first quantitative application of this idea was made by Debye and Menke[11] to the case of liquid mercury;

but more recently, Warren and his fellow-workers at the Massachusetts Institute of Technology have used the Zernicke and Prins formula as the basis of very important work on the structure of vitreous solids, some of which we shall discuss in detail below.

No attempt will be made to deal exhaustively with the very large amount of work that has been done on diffraction by amorphous materials, the detailed discussion of which lies beyond the scope of this volume. We shall deal with the optical principles according to which the diffraction patterns must be interpreted, and with specific experimental work only where it appears to illustrate the application of these principles. For a more detailed discussion of the experimental work, reference may be made to Randall's book, *The Diffraction of X-Rays and Electrons by Amorphous Solids, Liquids, and Gases.* A general introduction to the subject will also be found in Chapter VIII of Volume I of the present work.

It has been found convenient to divide the treatment of the topics referred to in this introduction into two chapters. In the present chapter, we deal with diffraction by gases, and liquids, and by solids in which questions of particle size do not arise. In the next chapter, we deal with those problems involving the idea of particle size and orientation, and, as a particular case of this, with diffraction by fibrous materials.

(*b*) *Formulae for incoherent scattering:* The formulae that we shall develop in the following sections are those for the coherent scattered radiation, which alone can produce interference effects; but in any experimental measurement of the scattered radiation, the total scattering, coherent or unmodified, and incoherent or modified, must be included. In dealing with the radiation scattered by a crystal in the direction of a spectrum it is not as a rule necessary to consider the incoherent radiation, which forms a very small fraction of the scattered intensity, and is moreover allowed for in practice in estimating the background. In the case of amorphous bodies, however, the incoherent radiation may form an important fraction of the total radiation scattered, particularly for large values of $(\sin\theta)/\lambda$, and it must therefore always be allowed for in any comparison of theory with experiment.

The general method of correcting for incoherent radiation is to calculate its amount theoretically, from one of the several formulae that have been deduced for this purpose. We have discussed some of these in Chapter III, §§ 2(*d*), 3(*e*) and 3(*h*), and in Chapter V, § 3(*k*), but it will be convenient to recapitulate them briefly for reference. The most accurate formula which is based on a non-relativistic wave-equation, and on a total atomic wave-function expressed as the product of the individual electronic wave-functions is probably the Hartree-Waller equation (*3*.56) of Chapter III:

$$I_{inc} = \sum_k (1 - |f_k|^2) - \sum_j \sum_k |f_{jk}|^2. \qquad (9.1)$$

The summations are here to be taken over all the electrons in the atom. For the definition of the term $|f_{jk}|^2$, which has its origin in electron exchange, reference may be made to Chapter III, § 3(*h*). It can often be neglected, and (*9*.1) then reduces to the Compton-Raman formula.

$$I_{inc} = \sum_k (1 - |f_k|^2). \tag{9.2}$$

For some atoms, the values of f_k for the individual electronic wave-functions are available, or may be calculated from data such as those given by James and Brindley ; but the assumption is often made that the electronic scattering factor may be replaced by f/Z, f being the total scattering factor, and Z the total number of electrons in the atom. Equation (*9*.2) then becomes

$$I_{inc} = Z - f^2/Z, \tag{9.3}$$

a form which, although very approximate, is often accurate enough, and has the advantage of being very easy to evaluate numerically. The total intensity of the incoherent radiation is the sum of the individual contributions from all the atoms in the assemblage.

A formula for I_{inc} based on the Thomas-Fermi model of the atom (Chapter III, § 4(*d*)) has been given by Heisenberg,[18] who obtains the following result,

$$I_{inc} = Z\left[1 - \int_0^{\xi_0} \xi^2 \, d\xi \left\{\left(\frac{\phi(\xi)}{\xi}\right)^{\frac{1}{2}} - v\right\}\left\{\left(\frac{\phi(\xi)}{\xi}\right)^{\frac{1}{2}} + \tfrac{1}{2}v\right\}\right], \tag{9.4}$$

where $\xi = rZ^{1/3}/a$, r being the distance measured from the centre of the atom, and a a distance characteristic of an atom of atomic number Z, equal to $0{\cdot}47/Z^{1/3}$A. The function $\phi(\xi)$ is the potential at the radius r, expressed in units Ze/a; and $v = \mu a/(6\pi Z)^{1/3} = 0{\cdot}176\mu/Z^{2/3}$, where $\mu = 4\pi(\sin\theta)/\lambda$. The upper limit of integration, ξ_0, is determined by the equation

$$\left(\frac{\phi(\xi_0)}{\xi_0}\right)^{\frac{1}{2}} - v = 0. \tag{9.5}$$

Bewilogua [19] has given figures, reproduced in Table IX. 1, which allow the value of I_{inc} to be calculated as a function of μ for any atomic number.

TABLE IX. 1

Values of I_{inc}/Z as a function of $v = 0{\cdot}176\mu/Z^{2/3}$ (Bewilogua).

v	I_{inc}/Z	v	I_{inc}/Z
0·05	0·319	0·6	0·909
0·1	0·486	0·7	0·929
0·2	0·674	0·8	0·944
0·3	0·776	0·9	0·954
0·4	0·839	1·0	0·963
0·5	0·880		

A further correction, not hitherto considered, is necessary in the case of incoherent radiation. It was shown by Breit,[12] that when radiation is scattered by a free electron, which recoils, and therefore produces, according to the principles of the Compton effect, radiation of a modified frequency, the ratio of the intensity of the radiation scattered by the recoiling electron to that scattered by an electron at rest is equal to $(\omega'/\omega)^3$, ω' and ω being respectively the frequencies of the modified and unmodified radiation.

Now according to the theory of the Compton effect,

$$(\omega'/\omega)^3 = 1 \Big/ \left\{ 1 + \frac{2h\lambda}{mc}(\sin^2\theta)/\lambda^2 \right\}^3 = 1/\mathrm{B}^3. \tag{9.6}$$

The same formula has been derived by Dirac,[13] Gordon,[14] and Schrödinger,[15] using the principles of quantum mechanics. If the relativistic form of the wave-equation is used, the correcting factor, which has been derived by Klein and Nishina,[16] and by Waller,[17] is more complicated, but it reduces to the Breit-Dirac formula (*9.6*) for the wave-lengths employed in experiments on scattering of X-rays. We may therefore write

$$\mathrm{I}_{\text{total}} = \mathrm{I}_{\text{coh}} + \mathrm{I}_{\text{inc}}/\mathrm{B}^3, \tag{9.7}$$

where I_{inc} is to be calculated from one of the formulae discussed in this section. The discussion of I(coh) for the different cases that arise will occupy us for the remainder of this chapter. It will be noticed that B, and hence the total scattered radiation, is no longer a function of $(\sin\theta)/\lambda$ only, but for a given value of $(\sin\theta)/\lambda$ depends on the wave-length. The coherent scattering, on the other hand, is always a function of $(\sin\theta)/\lambda$.

If Cu Kα radiation is used, ($\lambda = 1{\cdot}54$ A), the value of B may be written $1 + 0{\cdot}0744(\sin^2\theta)/\lambda^2$, in which $(\sin\theta)/\lambda$ is to be expressed in reciprocal Ångström units. The correction is thus by no means negligible for the larger angles of scattering.

(*c*) *General formulae for the diffraction of coherent radiation by assemblages of atoms:* Let a parallel beam of X-rays, whose direction is defined by the unit vector $\mathbf{s}_0$ fall on an assemblage of atoms, and consider the radiation scattered in the direction defined by the unit vector $\mathbf{s}$. The conditions of scattering are conveniently defined in the usual way by the vector S, where

$$\mathbf{S} = \mathbf{s} - \mathbf{s}_0; \quad |\mathbf{S}| = 2\sin\theta, \tag{9.8}$$

2θ being the angle of scatttering.

Let the positions of the atoms relative to any convenient origin be defined at some instant t by the vectors $\mathbf{r}_1, \mathbf{r}_2, \ldots \mathbf{r}_p, \mathbf{r}_q \ldots$ and let their atomic scattering factors be $f_1, f_2, \ldots f_p, f_q \ldots$ respectively.

We consider the intensity, I, of the coherent radiation scattered by this particular configuration of atoms, as observed at a point at a distance R, large in comparison with the linear dimensions of the assemblage. Using the results of Chapter I, §§ 1(*a*) and 3(*b*), and Chapter II, § 1(*b*), we may write for the amplitude of the scattered radiation at the point concerned

$$A = \frac{e^2}{mc^2} \frac{P}{R} \sum_p f_p \, e^{i\omega t + i\kappa \mathbf{r}_p \cdot \mathbf{S}}, \tag{9.9}$$

where P is a factor depending on the state of polarisation of the incident radiation. The corresponding intensity is given by the value of $|A|^2$. If the incident radiation is unpolarised, we may write the intensity $|A|^2$ in the form

$$|A|^2 = \frac{1}{R^2} \left(\frac{e^2}{mc^2}\right)^2 \frac{1+\cos^2 2\theta}{2} \sum_p \sum_q f_p f_q^* \, e^{i\kappa(\mathbf{r}_p - \mathbf{r}_q)\cdot \mathbf{S}}. \tag{9.10}$$

This is a general expression for the intensity scattered under the conditions defined by the vector S by any assemblage of atoms. Each sum in the double summation is to be taken from 1 to N, N being the total number of atoms in the scattering assemblage. There will be N terms for which $p=q$ and $\mathbf{r}_p = \mathbf{r}_q$. For each of these the exponential factor is unity. There remain $N(N-1)$ terms for which $p \neq q$. For such a term, $\mathbf{r}_p - \mathbf{r}_q = \mathbf{r}_{pq}$, the vector distance between the atoms p and q. Equation (9.10) can thus be written

$$|A|^2 = C \left[\sum_{p=1}^{p=N} |f_p|^2 + \sum_p^{p\neq q} \sum_q f_p f_q^* \, e^{i\kappa \mathbf{S}\cdot \mathbf{r}_{pq}} \right], \tag{9.11}$$

where

$$C = \frac{1}{R^2} \left(\frac{e^2}{mc^2}\right)^2 \frac{1+\cos^2 2\theta}{2}. \tag{9.11a}$$

If, as will usually be the case, the atomic scattering factors are all real, we need not distinguish between f and its conjugate f^*. The two terms in the double sum involving the pair of atoms p and q then together become $2f_p f_q \cos(\kappa \mathbf{S}\cdot \mathbf{r}_{pq})$, and the expression for the intensity scattered by the given configuration of atoms becomes

$$I = \sum_p f_p^2 + 2 \sum_{pq} f_p f_q \cos(\kappa \mathbf{S}\cdot \mathbf{r}_{pq}) \tag{9.12}$$

where

$$I = |A|^2 / C, \tag{9.12a}$$

and is the intensity scattered by the assemblage expressed in terms of that scattered by a single classical electron under the same conditions. In what follows, intensities will always be expressed this way unless the contrary is stated, and will be denoted by the letter I. In (9.12) the first summation contains N terms, and extends over every atom in the assemblage; the second contains $\frac{1}{2}N(N-1)$ terms, one for every pair of *different* atoms in the assemblage.

Equations (9.11) and (9.12) are the fundamental expressions upon which the treatment of all problems of scattering by assemblages of particles is based, and our remaining work consists in applying them to the different cases that arise. In Chapter I, we have already in effect applied (*9*.12) to the case of scattering by a lattice composed of identical points. In (*1*.9) the summation is expressed in a form convenient for dealing with the rigid lattice, and leads to equation (*1*.13); but equation (*9*.12), which for this particular problem would take the form

$$I = f^2\left[N + 2\sum_{pq}\cos(\kappa\mathbf{S}\cdot\mathbf{r}_{pq})\right], \tag{9.13}$$

when summed for the distance between every pair of different atoms in the lattice, is exactly equivalent to (*1*.13). We shall refer again to this equivalence in a later section.

In gases and liquids, the atoms are constantly moving, and what is observed is an average intensity, and not the intensity corresponding to a given fixed configuration of atoms. If we could follow the movements of any particular pair of atoms p and q, we should find the vector distance $\mathbf{r}_{pq}$ continually varying in magnitude and direction. The problem of determining the scattered intensity observed in such a case is one of averaging. It will be assumed in what follows that the movements of the atoms are so slow that the electric field in the incident radiation makes a very large number of complete vibrations before any appreciable change in configuration occurs, an assumption fully justified in the cases dealt with. On the other hand, during the time required to make an observation, a system such as a gas or a liquid will assume an enormous number of configurations if the positions of individual atoms are considered. In averaging the intensity, we have to consider that scattered by each possible configuration, weight this with the probability of the occurrence of that configuration, and average the intensity so weighted over all possible configurations. The methods used will differ for different types of assemblage, and the most important will be considered in turn.

2. The Scattering of Coherent Radiation by Assemblages of Monatomic Molecules

(*a*) *Scattering by a gas consisting of point atoms :* It will be instructive to consider first of all an ideal case, that of a gas consisting of identical atoms which are mathematical points, exercising no action on each other, so that any atom is equally likely to occupy any position in the volume occupied by the assemblage. Suppose the volume of the gas irradiated to be a sphere of radius R, which contains N atoms. The most convenient form of the expression for the scattered intensity corresponding to a given instantaneous configuration of the atoms is here (*9*.10), which, since the atoms are all identical, now becomes

$$I = |f|^2\left[N + \sum_{p}^{p\neq q}\sum_{q} e^{i\kappa\mathbf{S}\cdot\mathbf{r}_p - \mathbf{r}_q}\right]. \tag{9.14}$$

This expression has to be averaged over all possible configurations. We consider first a given pair of atoms p and q, and determine the average value of the term in the double summation referring to these. The probability that p shall lie in an element of volume dv_p at a vector distance $\mathbf{r}_p$ from the origin, while at the same time q shall lie in an element of volume dv_q at a vector distance $\mathbf{r}_q$ is $dv_p\, dv_q/\mathrm{V}^2$, V being the volume of the sphere of radius R. The average value of the contribution of the atoms p and q to the scattered intensity is obtained by multiplying the term in (9.14) referring to p and q by this probability and integrating the result over all possible elements of volume, that is to say, over the whole sphere, for both dv_p and dv_q. The double sum has $\mathrm{N}(\mathrm{N}-1)$ terms, and the average value will be the same for each, so that the average value of (9.14) is given by

$$\bar{\mathrm{I}}=|f|^2\left[\mathrm{N}+\mathrm{N}(\mathrm{N}-1)\iint e^{i\kappa\mathbf{S}\cdot(\mathbf{r}_p-\mathbf{r}_q)}\frac{dv_p}{\mathrm{V}}\frac{dv_q}{\mathrm{V}}\right], \tag{9.15}$$

each integration being extended over the whole volume of the sphere. In the case we are considering, the positions of the atoms p and q are entirely independent of each other, so that the double integral may be written as the product of two independent integrals. If X_p is the value of the integral referring to p,

$$\mathrm{X}_p=\int e^{i\kappa\mathbf{S}\cdot\mathbf{r}}\,dv/\mathrm{V}. \tag{9.16}$$

We have here omitted the suffix p, $\mathbf{r}$ being the vector distance of dv from the origin. It is convenient to take the origin at the centre of the sphere of gas considered. As a direction of reference we take the direction of the vector S. Let $\mathbf{r}$ make an angle α with S. Then

$$\kappa\mathbf{S}\cdot\mathbf{r}=4\pi r\cos\alpha(\sin\theta)/\lambda=\mu r\cos\alpha$$

where

$$\mu=4\pi(\sin\theta)/\lambda. \tag{9.17}$$

As the element of volume dv, we take the volume lying between the spheres of radii r and $r+dr$ and the cones having S as axis and semi-vertical angles α and $\alpha+d\alpha$. Then

$$\mathrm{X}_p=\frac{2\pi}{\mathrm{V}}\int_0^{\pi}\int_0^{R} r^2\, e^{i\mu r\cos\alpha}\sin\alpha\, dr\, d\alpha. \tag{9.18}$$

Integrating first with respect to α, we obtain

$$\mathrm{X}_p=\frac{4\pi}{\mathrm{V}}\int_0^{R} r^2\frac{\sin\mu r}{\mu r}\,dr. \tag{9.19}$$

Further integration, with respect to r, gives

$$\mathrm{X}_p=\frac{4\pi}{\mathrm{V}}\frac{\mathrm{R}^3}{3}\Phi(\mu\mathrm{R}), \tag{9.20}$$

where

$$\Phi(x)=3(\sin x-x\cos x)/x^3. \tag{9.21}$$

The integral X_q in (9.15), referring to atom q, has the same value as X_p, so that finally

$$\overline{I}(\mu) = |f|^2[N + N(N-1)\{\Phi(\mu R)\}^2]. \qquad (9.22)$$

The function $\Phi(x)$ is of some importance in the theory of diffraction. $\Phi(\mu R)$ is the amplitude scattered in the direction defined by μ (scattering angle 2θ) of a quantity of diffracting matter of unit strength uniformly distributed throughout a sphere of radius R, just as $(\sin \mu R)/\mu R$ is the amplitude scattered under the same conditions by unit diffracting matter uniformly spread over the surface of a sphere of radius R. The general properties of the two functions are similar; both are unity when $x=0$, but $\Phi(x)$ has zero values when $x = \tan x$, which are the values of x corresponding to the subsidiary maxima and minima of $(\sin x)/x$. They occur when $x = 1{\cdot}430\pi$, $2{\cdot}459\pi$, $3{\cdot}471\pi$, $4{\cdot}477\pi$, $4{\cdot}482\pi$ ... approaching $(n+\frac{1}{2})\pi$ for large n. Graphs of the two functions are shown in fig. 165. It will be seen that while $\Phi(x)$

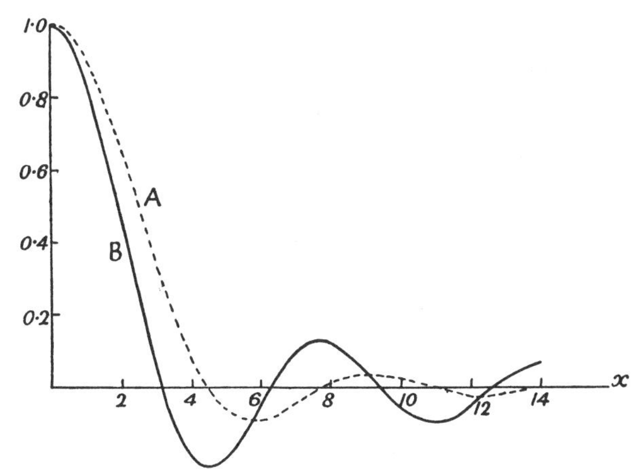

FIG. 165. Curves of $\Phi(x) = 3(\sin x - x \cos x)/x^3$, A, and $(\sin x)/x$, B

(curve A) falls to zero less rapidly with increasing x than $(\sin x)/x$ (curve B), its subsequent oscillations are less well marked, and die out more rapidly, the value of the function becoming unimportant if $x > 1{\cdot}430\pi$.

In equation (9.22), the second term is thus of very little importance if $\mu R > 1{\cdot}430\pi$, that is to say if $\sin \theta > 0{\cdot}37\lambda/R$, R being the radius of the scattering volume of gas. If, for example, copper Kα radiation is used ($\lambda = 1{\cdot}54$ A), and the radius of the specimen of gas is 1 mm., the second term in (9.22) becomes zero for the first time when θ is about 0·02 seconds of arc, and at all angles at which scattering can be observed, we should have simply

$$\overline{I} = N|f|^2. \qquad (9.23)$$

The average scattered intensity would thus be just the sum of the intensities scattered by the individual atoms, a result of course to be expected for a random assemblage of scatterers, at all angles at which the phase differences between the contributions scattered by the different members of the assemblage are appreciable. At zero angle of scattering, $\Phi(\mu R) = 1$, whatever the value of R. The scattered intensity in the direction of the primary beam is thus, by (*9.22*), $N^2|f|^2$. Physically, this means that in the direction of the primary beam there is no path difference between the contributions to the scattered beam from the different atoms. The total *amplitude* is thus the sum of the individual amplitudes, and therefore proportional to Nf, so that the resultant intensity is proportional to $N^2|f|^2$. If, however, the scattering atoms are distributed over a region whose linear dimensions are very much greater than the wave-length of the scattered radiation, even a very small angle of scattering will introduce large phase differences, which cause destructive interference of the radiation scattered by atoms in different parts of the assemblage.

This strong maximum value of the scattered intensity at zero angle of scattering must occur for every assemblage of scattering particles. It corresponds to the zero-order spectrum from a crystal lattice. In this case, however, the regular periodic arrangement of the atoms allows agreement of phase to take place, under suitable conditions, in directions other than that of the primary beam. With amorphous matter such agreement can occur only at zero angle of scattering, but then it is always present. The angular breadth of this zero-order maximum is excessively small in any practical case, and the term corresponding to it is omitted in the ordinary formulae for scattering by liquids and gases, which do not therefore apply for $\mu = 0$, or in the immediate neighbourhood of $\mu = 0$.

Equation (*9.22*) may be written

$$\bar{I}(\mu) = |f|^2[N\{1 - \Phi^2(\mu R)\} + N^2\Phi^2(\mu R)], \qquad (9.24)$$

which is identical in form with the Compton-Raman expression for the total scattering by an atom, discussed in § 2(*c*) of Chapter III; and indeed, mathematically, the derivation of each expression follows the same lines. $Nf\Phi(\mu R)$ is in effect the scattering factor of a sphere of radius R, and uniform scattering density fN/V per unit volume, but because R is very large in comparison with λ, its value is appreciable only in the immediate neighbourhood of $\mu = 0$. The exact form of the zero-angle maximum depends, of course, on the shape of the volume scattering the radiation. It will have the form discussed above only if this is spherical, but similar principles apply whatever the shape. The zero-angle scattering is appreciable only over a range of angles comparable with the ratio of λ to the linear dimensions of the scattering assemblage. It should be remembered too that a primary beam having one definite direction is in any case an abstraction, and is not in practice attainable. Any observable zero-order scattering would be the result of integration over a small range of directions of the primary beam.

(*b*) *Scattering by a gas consisting of monatomic molecules of finite size:* Suppose now that the atoms of the gas discussed in the last paragraph are no longer points, but spheres with an effective diameter, or distance of closest approach, s. The distance between two given atoms p and q can no longer assume all values with equal probability; for values smaller than s are clearly impossible, and all other values may not be equally probable. The treatment given in the last paragraph must therefore be modified. We must now write for the probability that atom p lies in the element of volume dv_p, while at the same time the atom q lies in dv_q, $\mathrm{W}(\mathbf{r}_{pq})\,dv_p\,dv_q/\mathrm{V}^2$. The probability function $\mathrm{W}(\mathbf{r}_{pq})$ will be a spherically symmetrical function of the distance between the two atoms, and will be the same for any pair of atoms in the assemblage. In what follows we shall denote it simply by W. The integral in equation (9.15) is now to be replaced by Y, where

$$\mathrm{Y}=\iint \mathrm{W} e^{i\kappa \mathbf{S}\cdot\mathbf{r}(p-\mathbf{r}_q)} \frac{dv_p}{\mathrm{V}}\frac{dv_q}{\mathrm{V}}. \tag{9.25}$$

For a completely random distribution of atoms, W is unity. It will also approach unity for large distances between the atoms. It is therefore convenient to write instead of W, $1-(1-\mathrm{W})$. The integral Y may then be written as $\mathrm{Y}_1-\mathrm{Y}_2$. Y_1 is obtained by putting $\mathrm{W}=1$ in (9.25). It is the double integral $\mathrm{X}_p\mathrm{X}_q$ discussed in the last paragraph where its value was found to be exceedingly small except in the immediate neighbourhood of $\mu=0$. In what follows we shall neglect it.

The second integral, Y_2, is given by

$$\mathrm{Y}_2=\iint (1-\mathrm{W}) e^{i\kappa \mathbf{S}\cdot\mathbf{r}_{pq}} \frac{dv_p}{\mathrm{V}}\frac{dv_q}{\mathrm{V}}. \tag{9.26}$$

To evaluate it, we first suppose p to lie anywhere within the volume V with equal probability. Integration with respect to dv_p then yields the factor unity. If now r is the radial distance of q from p, the position of p now being taken as origin, and if α is the angle between $\mathbf{r}$ and $\mathbf{S}$, then

$$\mathrm{Y}_2=\frac{1}{\mathrm{V}}\iint 2\pi(1-\mathrm{W})r^2\, e^{i\mu r\cos\alpha}\sin\alpha\, d\alpha\, dr, \tag{9.27}$$

as in (9.18). The limits of integration for α are from 0 to π. For r, we may take limits 0 to infinity, since W approaches unity, and the integrand therefore zero, for large values of r. This method of dealing with the integration avoids difficulties connected with convergence. Integrating with respect to α, we obtain

$$\mathrm{Y}_2=\frac{4\pi}{\mathrm{V}}\int_0^\infty (1-\mathrm{W})r^2\frac{\sin\mu r}{\mu r}\,dr, \tag{9.28}$$

and for the average scattered intensity, neglecting Y_1 as inappreciable,

$$\bar{I}(\mu) = |f|^2 \left[N - \frac{4\pi N(N-1)}{V} \int_0^\infty (1-W) r^2 \frac{\sin \mu r}{\mu r} dr \right]. \tag{9.29}$$

Equation (9.29) applies generally to any assemblage of similar atoms for which the average probability function $W(r)$ is spherically symmetrical. It would, for example, apply to a liquid composed of monatomic molecules. We shall, however, defer the detailed consideration of scattering by liquids until a later paragraph, and shall apply (9.29) to the case of a gas.

If the gas molecules are treated as spheres with a definite radius s, and if the pressure is not too high, $W(r) = 0$ if $r < s$, and has a uniform value, very nearly indeed unity, if $r > s$. By (9.19), (9.20), and (9.29), therefore,

$$\bar{I}(\mu) = |f|^2 \left[N - \frac{N(N-1)}{V} \frac{4\pi s^3}{3} \Phi(\mu s) \right]. \tag{9.30}$$

Now $4\pi N s^3/3$ is the total volume occupied by the spheres of action of all the molecules in the volume V, which we shall denote by Ω. Since N is a very large number, we may also replace $N-1$ by N without appreciable error, so that (9.30) becomes

$$\bar{I}(\mu) = N|f|^2 \left\{ 1 - \frac{\Omega}{V} \Phi(\mu s) \right\}, \tag{9.31}$$

an expression first obtained by Debye.[9] The ratio Ω/V is a very small fraction for a gas at ordinary pressures, so that the second term in (9.31) may be neglected, and, as for the gas with point atoms, we find the average scattered intensity to be the sum of the intensities scattered by the individual atoms.

At very high pressures the importance of the second term increases, and although at pressures at which Ω/V is any appreciable fraction of unity the assumptions made above as to the nature of the probability function $W(r)$ are certainly too simple, it is instructive to consider this case, since we can follow from it very easily the general effect of increasing the closeness of the packing of the atoms.

The value of the function $\Phi(\mu s)$, which is unity for $\mu = 0$, becomes zero for the first time when $(\sin \theta)/\lambda = 0{\cdot}37/s$. If $s = 2$A, which corresponds approximately to the case of argon, $\Phi(\mu s)$ becomes zero when $(\sin \theta)/\lambda = 0{\cdot}179\text{A}^{-1}$, or, with Cu K$\alpha$ radiation, when 2θ, the angle of scattering, is about 32°. If at the same time Ω/V is appreciable, there will be a marked maximum of scattering at about that angle, the scattering at smaller angles being reduced because of the increasing importance of the second term. A diffraction halo will thus be produced, whose angular width is determined mainly by the distance of closest approach of the atoms in the assemblage.

In fig. 166, the values of $f^2\{1 - \Omega\Phi(\mu s)/V\}$ are shown, plotted as a function of $(\sin \theta)/\lambda$. In drawing the curve, the values of f^2 for argon have been used, with $s = 2$A. Curves are drawn for $\Omega/V = 0, 0{\cdot}1, 0{\cdot}5$, and $0{\cdot}8$. The curve for $\Omega/V = 0$ is simply the f^2 curve for argon. When $\Omega/V = 0{\cdot}1$ this curve is reduced at small angles, but only refined

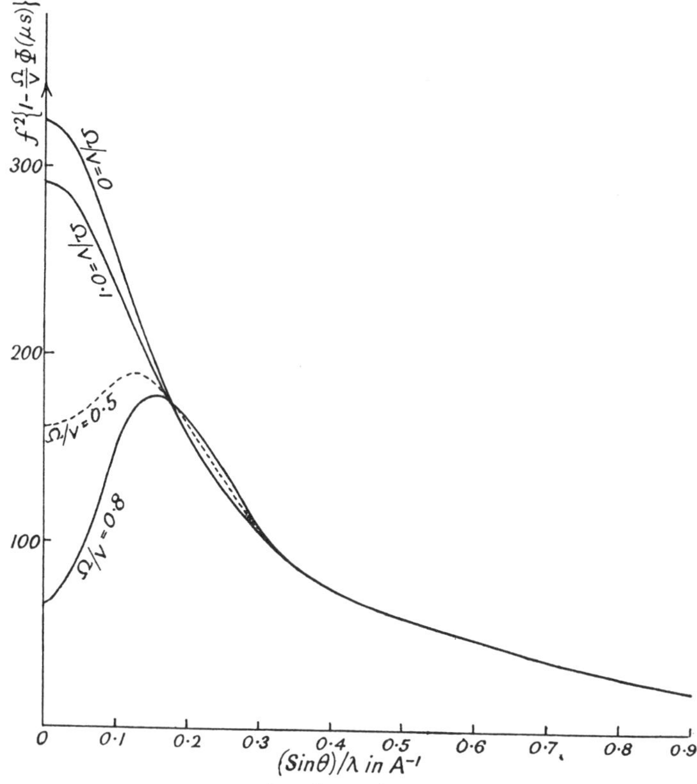

FIG. 166. Curves of $f^2\left\{1 - \frac{\Omega}{V}\Phi(\mu s)\right\}$ for argon. $(\mu = 4\pi(\sin\theta)/\lambda)$

quantitative measurements could detect the existence of any diffraction effect. At standard temperature and pressure, for a gas having $s = 2$A, Ω/V is rather less than 10^{-3}, so that pressures of some hundreds of atmospheres would be necessary to produce any appreciable effect; and no large effect could be produced until the average interatomic distance was so small that the simple assumption of a constant probability function for $r > s$ was no longer justifiable. The curves in fig. 166 are thus a little artificial, but they do indicate the important general principle, that while the scattering from a gas under ordinary conditions tends to a maximum at the smallest angles of scattering, as the compression gets greater the scattering at small angles falls off; so that at very high pressures, or in the liquid state, a maximum of scattering develops at a moderate angle, determined mainly by the distance of closest approach of the atoms. The development of the halo characteristic of liquid scattering is thus in principle accounted for.

(*c*) *Experimental tests of the scattering formula for monatomic gases:* Measurements of the scattering of X-rays by monatomic gases have now been made by a number of workers; by Scherrer and Stäger,[20] for mercury vapour; by Barrett,[21] for helium and argon; by Herzog,[22] for argon; and by Wollan,[23] for helium, neon, and argon. Herzog's most recent determination is a careful absolute measurement of the total scattering, coherent and incoherent, from argon, for one particular angle of scattering. The other determinations are not absolute, and give the relative scattering at different angles; but the results obtained by various workers agree well among themselves, and Herzog's work allows them to be expressed in absolute measure.

The experimental conditions were in all cases such that the coherent radiation should be given by the first term of (9.31), but the measured intensity includes that of the incoherent radiation. Equations (9.31), (9.1), and (9.6) then lead to the following expression for the total scattered intensity per atom, expressed in terms of the scattering from a single classical electron, and corrected for polarisation:

$$I_{tot} = |f|^2 + \left\{ \sum_k (1 - |f_k|^2) - \sum_j{}' \sum_k{}' f_{jk} \right\} \Big/ B^3,$$

where f_k is the scattering factor for the electron k of the atom, and f is the total atomic scattering factor, which is equal to Σf_k.

A comparison of Wollan's experimental results for helium with theory, made by Herzog,[24] is illustrated in fig. 167. The full curves show the values of the coherent, incoherent, and total scattering calculated on the basis of the Waller-Hartree theory, and including the Breit-Dirac correction factor $1/B^3$ for $\lambda = 0{\cdot}71$A, the wave-length for which Wollan's results are given. Hartree's values for f_k for the electron groups of helium, determined by the method of the self-consistent field were used. The exchange term in the incoherent scattering here makes no contribution, since the two electrons in normal helium have opposite spins. The circles show Wollan's experimental values. The dotted curve shows the theoretical total scattering when the Breit-Dirac correction is neglected. It will be seen that the agreement of theory with experiment is materially improved by the correction, which is quite considerable for the larger angles of scattering. The curves for helium illustrate the importance of the incoherent radiation in scattering by gases and amorphous bodies. The incoherent scattering for helium is, of course, relatively large; for heavier atoms it will form a smaller fraction of the total scattering at any given value of $(\sin\theta)/\lambda$.

Herzog shows also that in the case of neon the Waller-Hartree formula for the total scattering gives appreciably better agreement with experiment than Compton's approximation (9.3), which, however, gives fairly good agreement for heavier atoms, such as argon. In general, moreover, data will not be available for the evaluation of the Waller-Hartree formula, and it will be necessary to use either Compton's

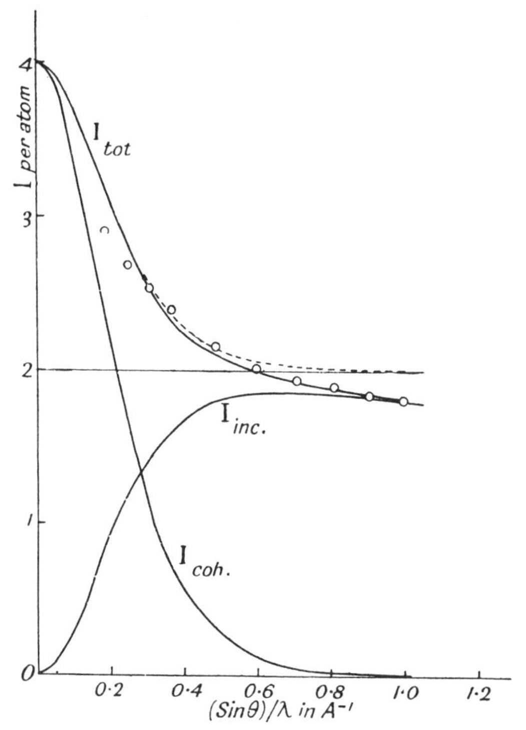

FIG. 167. Coherent, incoherent and total scattering from helium. Theoretical curves according to Herzog.[24] The circles indicate experimental values obtained by Wollan [23]

approximation or the Heisenberg formula (9.4), which is based on the Thomas-Fermi atomic model.

In his absolute test of the scattering formula, mentioned above, Herzog used argon, with Cu $K\alpha$ radiation ($\lambda = 1{\cdot}54$A) as the primary radiation, and measured the intensities with an ionisation chamber and valve electrometer. He found the ratio of the intensity scattered through an angle of 40° to the intensity of the primary beam to be $2{\cdot}88 \times 10^{-7}$ for argon at 1575 mm. pressure and at a temperature of 291·3 abs. The calculated value was $2{\cdot}87 \times 10^{-7}$ for the same conditions, the probable error in both experimental and theoretical determinations being estimated as of the order of 6 per cent. With Cu $K\alpha$ radiation, the incoherent scattering from argon at 40° forms only about 1·5 per cent. of the total scattering, and the determination must therefore be regarded as confirming the formula for the coherent radiation, and as giving little beyond the order of magnitude of the incoherent radiation.

This work on scattering from monatomic gases furnishes one of the most direct proofs of the essential correctness of the theory of the atomic scattering factor. Tests made by using crystals are all subject to the limitation that measurements of the scattering can only be made in those directions in which a spectrum occurs. The atoms moreover are not free, and disturbance of the outer electron shells must result from this, whereas in the case of the monatomic gas the atoms are for all experimental purposes entirely free. On the other hand, the figures quoted above for argon show that gases scatter only a very small fraction of the primary radiation, so that the actual measurements are exceedingly difficult if absolute values are required. Reference has already been made in § 3(*b*) of Chapter VII to attempts to determine the electronic distribution in the atoms of monatomic gases from their scattering curves for coherent radiation.

(*d*) *The probability function* $W(r)$, *and the density function* $\rho(r)$: In considering the problem of scattering by a gas, we have used a function $W(\mathbf{r})$ to express the probability that the centres of two specified atoms p and q should lie at a vector distance $\mathbf{r}$ apart. We might imagine the two atoms to be marked in some way, so that their positions in the assemblage could be identified. If we could then make a very large number of independent determinations of the distance between them at a series of instants chosen at random, the frequency with which distances whose magnitudes, independently of their directions, lay between r and $r+dr$ would occur would be proportional to $4\pi r^2 W(r)\,dr$. It is here assumed that the function $W(r)$ is spherically symmetrical, depending only on the magnitude of r and not on its direction, an assumption justified in all the cases we shall consider. The numerical value of $W(r)$ is so chosen as to be unity when all distances of separation are equally probable, that is to say, for a truly random assemblage.

This is the significance of $W(r)$ as we have already used it; but for many purposes it is convenient to consider another function $\rho(r)$, so defined that $4\pi r^2 \rho(r)\,dr$ is the average number of atoms lying at any instant at distances between r and $r+dr$ from the centre of a specified atom p. We consider for a moment an infinitely extended assemblage of atoms whose average properties are the same throughout. Then $\rho(r)$ will be the same function of r whichever atom is chosen as origin. It is a kind of average atomic density at a given radius from *any one atom*, which is not of course the same as the average density of the assemblage. We shall refer to it in what follows as the atomic density function.

The two functions $\rho(r)$ and $W(r)$ have the same dependence on r, and differ only by a constant factor; for the probability that any atom q will lie at a distance r from p must be proportional to the average number of atoms to be found at any instant at a distance r from p. The relation between the two functions may be found as follows. Suppose the distribution to be a completely random one. Then $\rho(r)$ is plainly the average number of atoms per unit volume, which is uniform, and equal say to ρ_0. It is also equal to N/V, where N is the number of atoms in volume V. For a completely random distribution, however, $W(r) = 1$, and so

$$\rho(r) = \rho_0 W(r) = \frac{N}{V} W(r). \tag{9.32}$$

(*e*) *Formulae for scattering by monatomic liquids:* As we have seen already, there is no sharply defined boundary separating the problems of diffraction by matter in the gaseous and liquid states. Equation

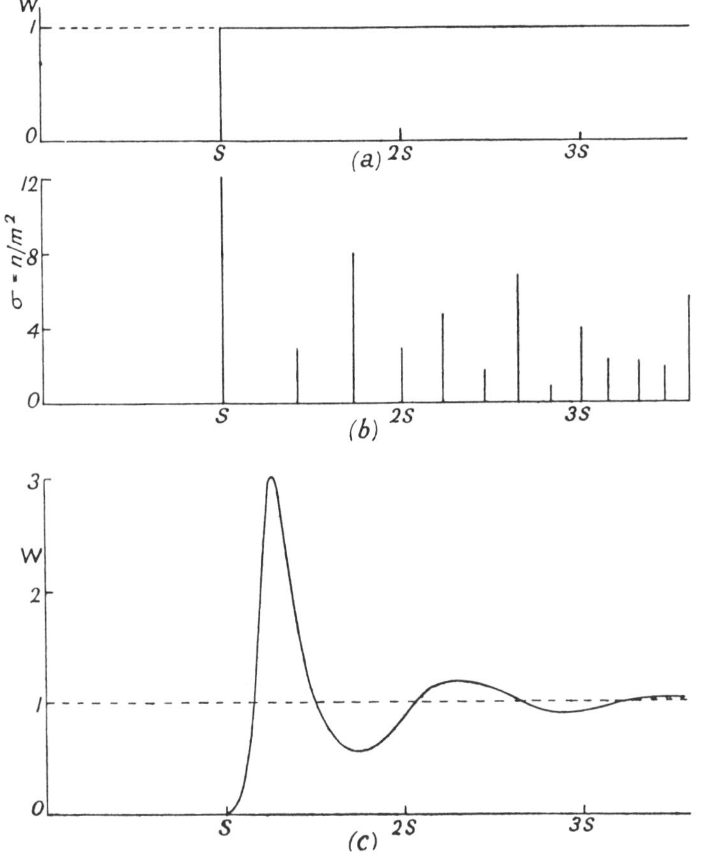

FIG. 168. Distribution functions: (*a*) gas consisting of monatomic molecules at low pressures, (*b*) close-packed spheres, (*c*) irregular close packing of spheres

(*9*.29) applies directly to the monatomic liquid, if the appropriate probability function $W(r)$, or the related atomic density function $\rho(r)$, defined by (*9*.32), is used. $W(r)$ will, however, be a function of a much more complicated type than that applicable to the case of a gas at moderate pressures, for which we assumed $W(r)=0$ if $r<s$, $W(r)=1$ if $r>s$. This simple probability function is illustrated in fig. 168 (*a*). In the monatomic liquid, the atoms will be in a condition which might be called irregular close packing, and all interatomic distances are by no means equally probable for $r>s$.

Consider for a moment a strictly close-packed assemblage of hard spheres of diameter s. Here only certain interatomic distances can occur at all, the smallest of which is s, which also occurs most frequently. The next possible interatomic distance is $\sqrt{2}s$, the next $\sqrt{3}s$, the next $2s$ (or $\sqrt{4}s$); and so on. For all these distances $W(r)$, or $\rho(r)$, has a value different from zero: its value for any other distance is zero. If we imagine the close-packed assemblage to be rotated about one of the atoms as centre into all possible positions, the centres of all the other atoms lie on a set of spheres of radii s, $\sqrt{2}s$, $\sqrt{3}s$, ... , and none will lie between these spheres. The radial distribution of the atomic centres is discontinuous, and is illustrated in fig. 168 (*b*). The ordinates in this curve are proportional to the number of atomic centres per unit area lying on spheres of definite radii, measured from the centre of any one atom. If n is the number of atomic centres at distance ms from any given atom, the ordinates show the values of n/m^2, which is zero except at those radii for which $m=\sqrt{\frac{1}{2}(p^2+q^2+r^2)}$, p, q, r, being any integers, positive, negative or zero, the sum of which is even.

If now the rigid close-packed structure is somewhat loosened, and a certain degree of randomness is introduced into the arrangement, the concentration of the possible atomic positions onto spheres of definite radii will be somewhat smeared out, and $\rho(r)$ will become a continuous function of r; but maxima will persist for radii corresponding to those of the spheres, or rather to radii a little greater, since the whole assemblage is somewhat loosened, and there will be minima with values well below the average density between the maxima. We shall expect the first of the maxima to be particularly well marked, and the second and third appreciable, but $\rho(r)$ will approximate to the uniform average density ρ_0 very rapidly as r increases. Even in the relatively loose association of atoms characteristic of the liquid state, the demands on space made by the atoms necessarily introduces a certain degree of order, and if only small regions of the liquid are considered, the state of affairs in any one of them approaches geometrically to that in a crystalline array; but there will be no regular relationship between the atomic arrangement in two such regions that are widely separated, as there would be in a true crystal, and the value of $\rho(r)$ will approximate to that for a random array at distances that are quite small multiples of the atomic diameter. Fig. 168(c) shows schematically the nature of the probability function, or the density function in this case.

This way of considering the matter was first clearly put forward by Zernicke and Prins,[10] who explained thus the occurrence of the well-known diffraction haloes characteristic of scattering by liquids. They discussed the distribution function for a number of rods of equal length, arranged along a line of length rather greater than the sum of the lengths of the rods themselves, and found it to show the general characteristics indicated above for the three-dimensional case. They then discussed diffraction of radiation by an irregular linear grating, the positions of whose openings were governed by the distribution function determined for the rods, and found it to exhibit the general characteristics of diffraction by liquids.

Zernicke and Prins [10] also pointed out that by using a formula of the type of (9.29) it is possible from observations of the scattering of X-rays by a liquid to determine experimentally the distribution functions $W(r)$, or $\rho(r)$. At the time no suitable experimental material was available for testing the formula, but Debye and Menke [11] later made careful measurements of the scattering of X-rays by mercury, a monatomic liquid, and by applying the theory of Zernicke and Prins obtained a distribution function exactly of the predicted form.

We shall first consider the theory of the method. To do this, we start from equation (9.29), in which it is convenient to replace $W(r)$ by the density function $\rho(r)$. Using (9.32), and writing N instead of $N-1$, as we may, since N is a very large number, we obtain

$$\bar{I}_a(\mu) = \frac{\bar{I}(\mu)}{N} = |f|^2\left\{1 - 4\pi\int_0^\infty \{\rho_0 - \rho(r)\}\, r^2 \frac{\sin \mu r}{\mu r}\, dr\right\}, \quad (9.33)$$

or

$$\mu\left(\frac{\bar{I}_a(\mu)}{|f|^2} - 1\right) = 4\pi\int_0^\infty \{\rho(r) - \rho_0\}\, r \sin \mu r\, dr, \quad (9.34)$$

where $\bar{I}_a(\mu)$ is the average scattering per atom, corrected for the polarisation factor and expressed in terms of the intensity scattered by a classical electron at unit distance.

Equation (9.34) is of the form

$$f(\mu) = \int_0^\infty \phi(r) \sin \mu r\, dr. \quad (9.35)$$

By Fourier's integral theorem (Appendix IV; and Chapter VII, § 3(a)), therefore,

$$\phi(r) = \frac{2}{\pi}\int_0^\infty f(\mu) \sin \mu r\, d\mu. \quad (9.36)$$

Following Zernicke and Prins, we put

$$\frac{\bar{I}_a}{|f|^2} - 1 = i(\mu), \quad (9.37)$$

and, comparing (*9*.34) and (*9*.35) and using (*9*.36), obtain

$$4\pi r^2\{\rho(r)-\rho_0\}=\frac{2r}{\pi}\int_0^\infty \mu i(\mu)\sin\mu r\,d\mu. \qquad (9.38)$$

If $i(\mu)$ can be determined as the result of experiment, the integral in (*9*.38) can be evaluated numerically for each of a series of values of r, and so $\rho(r)-\rho_0$, the difference between the actual radial atomic density and that corresponding to a uniform distribution, or, alternatively, $W(r)-1$, may be determined as a function of r.

(*f*) *Examples of scattering by monatomic liquids:* The first real experimental test of the theory of Zernicke and Prins was made by Debye and Menke,[11] who measured the intensity distribution of copper and molybdenum Kα radiation scattered from liquid mercury. Reference may be made to the original papers for experimental details. The results, corrected for absorption, and for various geometrical factors characteristic of the apparatus used, and divided by the polarisation factor $(1+\cos^2 2\theta)/2$, give the sum of the coherent scattering $\bar{I}_a(\mu)$, and the incoherent scattering I_{inc}, although not in absolute measure. To express the intensity in absolute measure, and to correct for the incoherent radiation, use is made of the fact that, for large values of μ, $\bar{I}_a(\mu)$ approaches f^2, the scattering per atom characteristic of an entirely random assemblage of atoms; for the integrand in (*9*.33) is the product of two factors, the first of which approaches zero for very small r, and the second of which approaches zero for large μ unless r is very small, so that, for large values of μ, the integral has always very small values. We must therefore draw the curve for the total observed intensity, corrected for polarisation, on such a scale that at large values of μ it coincides as nearly as possible with the sum of the curves for f^2 and I_{inc}. The values of f^2 can be taken from one of the existing tables of scattering factors, and I_{inc} can be calculated from one of the formulae given in § 1(*b*) of this chapter. Debye and Menke used Heisenberg's formula (*9*.4) for the incoherent scattering, which is based on the Thomas-Fermi model of the atom, and values of f^2 calculated according to the same model. When the curve for the total intensity has been standardised in this way, the curve for the coherent scattering per atom, $\bar{I}_a$, may be drawn by subtracting from the values of the total scattering at a series of values of μ, those of the incoherent scattering at the corresponding values.*

The curve for $W(r)$ obtained by Debye and Menke for mercury is reproduced in fig. 117, p. 195 of Vol. I of *The Crystalline State*. It will be seen that it shows exactly the type of distribution predicted above from general considerations, a distance of approach of a little over 3A being much more probable than any other.

* Another method of standardising the observed intensity curve has been suggested and used by Hultgren, Gingrich, and Warren. Cf. *Jour. Chem. Phys.* **3**, 351, (1935).

Tarasov and Warren [25] have made similar observations using liquid sodium, also a monatomic liquid, and one better suited to scattering observations than mercury, owing to the much smaller correction for absorption of X-rays in the specimen. Fig. 169 (i) curve (*a*) shows the

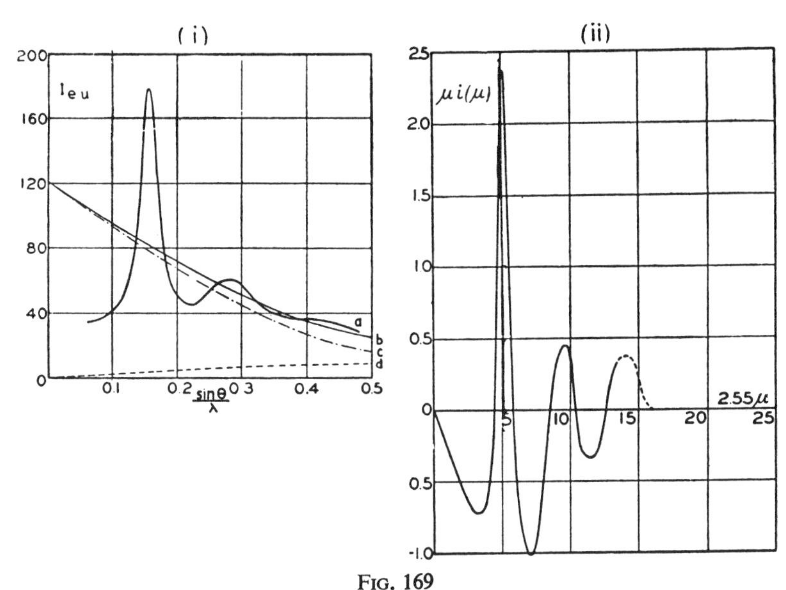

FIG. 169

(i) (*a*) Corrected experimental intensity curve for liquid sodium in electron units per atom. (*b*) Total independent scattering per atom. (*c*) Independent coherent scattering per atom. (*d*) Incoherent scattering per atom

(ii) The curve $\mu i(\mu)$ for liquid sodium. The abscissae $2{\cdot}55\mu$ were convenient for the harmonic analyser used

(Tarasov and Warren, *Jour. Chem. Phys.*, **4**, 236 (1936))

experimental intensity curve as a function of $(\sin\theta)/\lambda$. The total independent scattering, assuming a completely random arrangement of atoms, is shown by curve (*b*), and curve (*a*) is drawn to such a scale as to approximate to the independent scattering at large angles. Curves (*c*) and (*d*) show the correction for incoherent radiation. Fig. 169 (ii) shows the curve for $\mu i(\mu)$, the scale of the abscissae being chosen to suit the harmonic analyser used in determining the value of the integral (9.38).

Fig. 170 curve (*a*) shows the values of $4\pi r^2 \rho(r)$ in electrons per A, plotted against *r*, the distance from any given atom in A units. The average radial density curve $4\pi r^2 \rho_0$ (curve *b*), and the distribution of neighbours in crystalline sodium are also shown. There is a well-marked maximum in $\rho(r)$ at about 4A, a distance a little greater than he first possible interatomic distance in crystalline sodium, while distances appreciably less than this do not occur at all.

A sufficient explanation of the type of distribution obtained appears to be the space occupied by the atoms themselves, and it does not seem to be necessary to assume any physical connection between the pairs of atoms. In support of this view, Debye [11] quotes an experiment

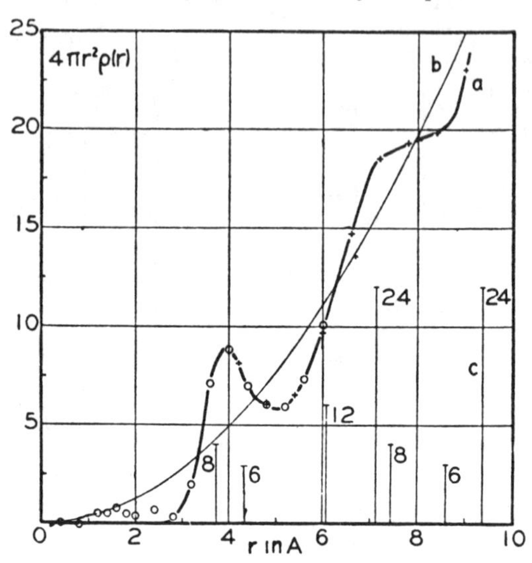

FIG. 170. (*a*) Radial distribution curve for liquid sodium, (*b*) average density curve, (*c*) distribution of neighbours in crystalline sodium (Tarasov and Warren) (*Jour. Chem. Phys.*, **4**, 236 (1936))

made with a number of similar steel balls, two of which were marked, loosely packed in a flat box with a glass lid. The box was shaken, and the distance between the marked pair was recorded. Records were kept over a period of weeks, the box being left in a convenient place in the laboratory, so that anyone entering the room could shake it and record the result. A distribution curve showing the frequency of occurrence of distances lying within certain ranges was plotted, which showed the general characteristics of the distribution curve for mercury. The type of distribution must here have been determined purely by spatial considerations.

Equation (9.38) has been applied by Warren and his fellow-workers to scattering by crystalline and vitreous solids with very interesting results. We shall consider this work in Section 4 of this chapter.

3. The Scattering of Coherent Radiation by Assemblages of Polyatomic Molecules

(*a*) *Introductory:* In Section 2 of this chapter we have followed through from the gaseous state to the liquid state the problem of the scattering of X-rays by assemblages of molecules consisting of single atoms. In the present section, we shall suppose the individual members of the assemblage to be complex molecules, all alike, but each made up of a number of atoms, which preserve fixed relative positions. Leaving out of account the zero-order scattering, we found the average intensity scattered from an assemblage of monatomic molecules in the gaseous state at moderate pressure to be the sum of the intensities scattered by the individual atoms. The only interference effects that are appreciable are those due to the distribution of scattering matter in the individual atoms, which are taken into account by ascribing to each atom a characteristic scattering factor, or f-factor. It is not until the packing of the atoms becomes so close as to approximate to that in the liquid state that interference effects due to their relative positions in the assemblage become important. Following Debye, we may thus distinguish between (*a*) internal interference effects, depending upon the distribution of scattering matter within the individual members of the assemblage, and (*b*) external interference effects, depending upon the relative positions of these individual members. In the case of the monatomic liquid, the characteristic diffraction haloes are due entirely to the external effects.

In the case of assemblages of polyatomic molecules, the internal effects depend not only on the scattering factors of the component atoms, but also on the relative positions of the atoms in the molecules.

Again, the external interference effects are entirely negligible if the molecules form a gas at ordinary pressures. At very high pressures, and in the liquid state, a diffraction halo due to the necessary preponderence of certain intermolecular distances develops. Conditions are, however, more complicated than in the monatomic liquid; for if the individual molecules are not spherically symmetrical, if, for example, they are elongated, or electrically polar, then there may well be a preferential relative orientation of neighbouring molecules, or a local structure more highly organised than that due merely to considerations of geometrical packing. This type of effect has been called by Stewart [8] 'cybotaxis'. It is probable that its importance in producing diffraction effects has often been overestimated, and that even in polyatomic liquids, if the temperature is not too low, the diffraction effects are largely governed by simple packing effects which we have seen to be quite adequate to produce diffraction haloes; but local structure in the liquid, dependent on the molecular structure, is undoubtedly important in many cases, and we shall consider some examples below. It should be noted that even if local structure does occur, the probability function $W(r)$ and the density function $\rho(r)$ will still be spherically symmetrical; for they are average effects, and in the true liquid we must suppose all possible orientations of any local structures to occur in any volume of the liquid large enough to scatter a measurable intensity of X-rays.

(*b*) *Scattering by gases consisting of polyatomic molecules:* The fundamental expression for the scattered intensity is again (9.12). Unless the pressure of the gas is very high, external interferences may be neglected, and the total scattered intensity is the sum of the intensities due to the individual molecules. In terms of equation (9.12), this means that in summing up over all the pairs of atoms in the assemblage, we need not take into account those terms in which the atoms p and q lie in different molecules. Vectors such as $\mathbf{r}_{pq}$ will always refer to the distance between two atoms in the same molecule.

In a gas under the assumed conditions, all orientations of the molecules will be equally likely, and in any finite time all orientations will be equally represented by molecules in the assemblage. Let N now be the number of *molecules* in the assemblage, and let each contain n atoms, all of which may be different. The instantaneous intensity I_m scattered by a molecule may now be written, according to (9.12),

$$\mathrm{I}_m = \sum_p f_p^2 + 2\sum_{pq} f_p f_q \cos(\kappa\mathbf{S}\cdot\mathbf{r}_{pq}), \tag{9.39}$$

where the summation is to extend only over the n atoms in the molecule, and will therefore have n^2 terms in all. To get the total observed intensity scattered by the gas, we now suppose the molecule to assume all orientations in space with equal probability, determine the average value of I_m for all these orientations, and multiply by N, the total number of scattering molecules.

To determine the average, consider the pair of atoms p and q in the molecule. We have to find the average value of $\cos(\kappa\mathbf{S}\cdot\mathbf{r}_{pq})$ when $\mathbf{r}_{pq}$ can take any orientation with equal probability. Let l_{pq} be the magnitude of the vector $\mathbf{r}_{pq}$. Then $\kappa\mathbf{S}\cdot\mathbf{r}_{pq} = \mu l_{pq}\cos\alpha$, where α is the angle between the directions of S and $\mathbf{r}_{pq}$. The probability that $\mathbf{r}_{pq}$ lies in a direction making an angle between α and $\alpha + d\alpha$ with S is $\frac{1}{2}\sin\alpha\, d\alpha$ and therefore

$$\overline{\cos(\kappa\mathbf{S}\cdot\mathbf{r}_{pq})} = \int_0^\pi \tfrac{1}{2}\cos(\mu l_{pq}\cos\alpha)\sin\alpha\, d\alpha = \frac{\sin \mu l_{pq}}{\mu l_{pq}}. \tag{9.40}$$

The observed intensity from the assemblage of N molecules is therefore

$$\bar{\mathrm{I}}(\mu) = \mathrm{N}\left[\sum_p f_p^2 + 2\sum_{pq} f_p f_q \frac{\sin\mu l_{pq}}{\mu l_{pq}}\right], \tag{9.41}$$

where the first summation extends over all the atoms in a single molecule and the second over every pair of different atoms in the molecule, l_{pq} being the distance between the atoms p and q.

It must again be remarked that this formula does not apply to angles of scattering in the immediate neighbourhood of zero. Its value as it stands, for $\mu = 0$, is $\mathrm{N}(\Sigma f_p)^2$, whereas the true value should be $\mathrm{N}^2(\Sigma f_p)^2$. But in deriving (9.41) from (9.12), all terms referring to atoms in different molecules have been neglected. If, in (9.12), we consider for the moment only terms in which atom p lies in a molecule A, and atom q in a second molecule B we obtain a contribution $2\sum_p \sum_q f_p f_q \cos(\kappa \mathbf{S}\cdot\mathbf{r}_{pq})$, where p is any atom in A and q any atom in B. When $\mathbf{S}=0$, this sum is equal to $2(\Sigma f_p)^2$. There are $\frac{1}{2}N(N-1)$ distinct pairs of molecules, so that for zero angle of scattering an extra term $N(N-1)(\Sigma f_p)^2$ has to be added to the value given by (9.41), giving a total $N^2(\Sigma f_p)^2$. The extra term due to the external interferences rapidly becomes negligible as μ assumes appreciable values, and the formula applicable to the scattering at finite angles as usually observed is (9.41).

(c) *Scattering by gases consisting of diatomic molecules:* If each molecule has two atoms p and q at a distance l apart, the scattered intensity is given by

$$\bar{I}(\mu)=N\left\{f_p^2+f_q^2+2f_pf_q\frac{\sin\mu l}{\mu l}\right\}, \tag{9.42}$$

which reduces to the simple form

$$\bar{I}(\mu)=2Nf^2\left\{1+\frac{\sin\mu l}{\mu l}\right\}, \tag{9.43}$$

if the two atoms are identical. The first term of (9.43), or the first two of (9.42), represent simply the sum of the intensities scattered by the two atoms independently. The last term is the interference term and is due to the arrangement of the atoms in each molecule in groups of two, at a definite distance apart. The expression (9.41) for the scattering by a gas composed of more complicated molecules is built up in the same way of the sum of the intensities due to the individual molecules, and a series of interference terms of the same type as that for the diatomic molecule, one for each pair of different atoms in the molecule.

We shall consider the case of the diatomic molecule in a little more detail. It is best to consider the quantity $\bar{I}(\mu)/N$, which is the average scattered intensity per molecule, corrected for polarisation and expressed in terms of the scattering by a classical electron as unity. We shall denote this quantity by $\bar{I}_m$. Then, for a diatomic molecule,

$$\bar{I}_m=2f^2\left\{1+\frac{\sin\mu l}{\mu l}\right\}. \tag{9.44}$$

The value of $\bar{I}_m$ approaches $4f^2$ for small values of μ, leaving out of account zero-angle scattering, and $2f^2$ for large μ. For intermediate values it exhibits maxima and minima, which in the actual diffraction pattern correspond to diffuse haloes surrounding the direction of the primary beam. In contrast with what is found in the case of liquids, the intensity of scattering at small angles increases instead of falling away.

The right-hand side of (9.44) is the product of two factors, each of which is a function of μ. We shall consider for the moment only the second, which contains the effect of the internal interference due to the molecular configuration. The value of the interference term $(\sin \mu l)/\mu l$ is unity for $\mu = 0$, and becomes zero for $\mu l = n\pi$ $(n = 1, 2, 3, \ldots)$. The subsidiary maxima occur when $\mu l = 2{\cdot}459\pi, 4{\cdot}477\pi, \ldots$. The first maximum of the factor $1 + (\sin \mu l)/\mu l$, apart from that at $\mu = 0$, will occur for $\mu l = 2{\cdot}459\pi$, or $\sin\theta = 0{\cdot}615\lambda/l$, the second will occur for $\sin\theta = 1{\cdot}119\lambda/l$, and the third for $\sin\theta = 1{\cdot}621\lambda/l$, the position of the mth maximum approaching more and more closely to

$$\sin\theta = (2m + \tfrac{1}{2})\lambda/4l$$

as m becomes large. The maxima beyond the third are however weak and ill-defined.

If the atoms of the gas scattered as points, the factor f^2 would be independent of the angle of scattering, and the factor $1 + (\sin \mu l)/\mu l$ would give the variation of the scattering with angle. Curve (*a*) in fig. 171 shows the scattered intensity $\bar{I}_m$ as a function of $(\sin\theta)/\lambda$ for

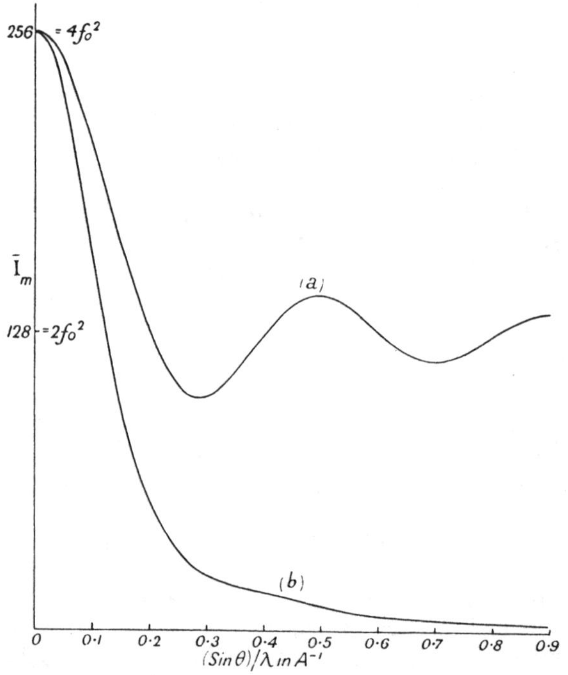

FIG. 171. The effect of the atomic scattering factor on the scattering curve for diatomic molecules. (*a*) Theoretical curve for $f = 8$, $l = 1{\cdot}25$A, when the atoms scatter as points. (*b*) Theoretical curve when the value of f appropriate to oxygen is used

two point atoms with $f = 8$, at a distance of 1·25A apart. This corresponds approximately to the interatomic distance in the oxygen molecule. In an actual gas such as oxygen, the atoms do not scatter as points, and the factor f^2 decreases very rapidly as μ increases. This has the effect of making the maxima and minima in the scattering curve much less well defined, a conclusion first reached by Debye.[26] In drawing curve (*b*) in fig. 171, the angular variation of f for oxygen, as given in the tables of James and Brindley, has been used. It will be seen that hardly a vestige remains of the diffraction maxima that are so prominent in curve (*a*). It would appear that it is difficult if not impossible to determine optically from the positions of the maxima in the scattering curve the distances between the two atoms in molecules such as oxygen or nitrogen. This theoretical prediction is entirely confirmed by experiment. The results of Gajewski [27] for nitrogen (N_2) and oxygen (O_2) show that in the case of nitrogen the scattering curve falls away quite smoothly with increasing angle of scattering with no trace of a diffraction maximum. In the case of oxygen, a vestige of the first maximum is observable, but it is not possible to determine its position with any accuracy. By assuming likely distances between the pairs of atoms, and using theoretical values of the scattering factors, Gajewski finds that the observed results are entirely consistent with interatomic distances of 1·15A and 1·20A for nitrogen and oxygen respectively.

With chlorine, also a diatomic molecule, we might expect the diffraction maxima to be more clearly marked, since in the heavier atoms there will be a greater concentration of electrons near the atomic centre, and consequently a less rapid drop of the factor f^2 with increasing μ. Some observations by Richter [28] confirm this expectation. His curve for chlorine is shown in fig. 172. The upper curve is that which would have been obtained had the chlorine atoms scattered as points: the lower curve is that calculated on the assumption that the distance between the chlorine

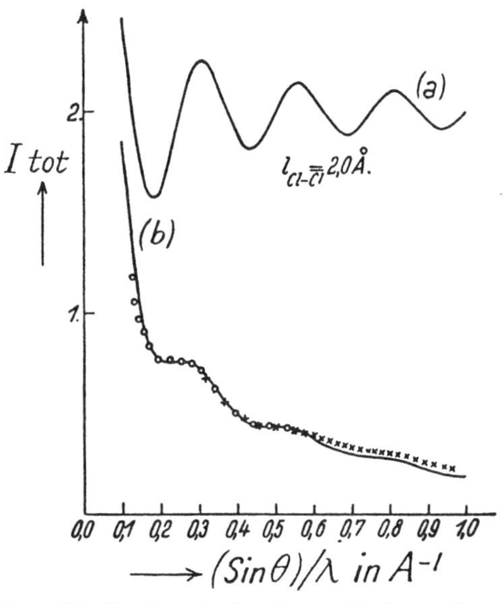

FIG. 172. Total scattering from chlorine. Curve (*a*) calculated assuming atoms to scatter as points. Curve (*b*) calculated assuming Hartree distribution (Richter),

o = experimental points with Cu Kα radiation
x = " " " Mo Kα "

(*Physikal. Zeit.*, **33**, 587 (1932))

atoms is 2·8A, and that the atomic scattering factor of each is given by the Hartree distribution. The circles and crosses show the observed values obtained with Cu Kα and Mo Kα radiation respectively. It will be seen that even here there are no real maxima in the scattering curve; they are represented merely by roughly horizontal portions of the curve, and any direct determination of their positions is impossible.

(*d*) *Scattering by molecules with more than two atoms:* Equation (9.41) shows that there is one interference term for each pair of atoms in the molecule, and that the contribution of a given pair does not depend on the direction of the line joining them, but only on its length. For a more complicated molecule, the contribution of the interference terms to the scattering is the sum of a series of terms of the same type as that for the diatomic molecule, one for each interatomic distance in the molecule. If the molecule is of irregular structure, with many different interatomic distances, the result of this superposition is to produce confusion, since the maxima corresponding to the different interatomic distances are diffuse; but if the molecule is regular, so that the same interatomic distance occurs in it a number of times, the corresponding interference terms will all give identical distributions, and well-marked maxima and minima may be produced.

Good examples of this are molecules such as carbon tetrachloride,[29] and silicon tetrachloride (CCl_4 and $SiCl_4$), in which a central carbon or silicon atom is surrounded symmetrically by four chlorine atoms arranged at the corners of a regular tetrahedron. The Cl–Cl distance, the edge of the tetrahedron, occurs six times, and the C–Cl distance, the radius of the tetrahedron, four times. The formula for the molecular scattering is thus

$$\bar{I}_m = f_C^2 + 4f_{Cl}^2 + 12f_{Cl}^2 \frac{\sin \mu l}{\mu l} + 8f_C f_{Cl} \frac{\sin \mu r}{\mu r}, \qquad (9.45)$$

where l is the edge of the tetrahedron and r its radius. Observations of the scattering from CCl_4 by Bewilogua,[30] and by van der Grinten,[31] and of that from $SiCl_4$ by James,[32] are fully in accord with this formula. A curve obtained by van der Grinten, who used monochromatised Cu Kα radiation, is shown in fig. 173, together with the calculated curve based on the tetrahedral molecule and the Hartree atomic model. The measurements of scattering are not absolute, and the scale of the two curves is made to fit artificially, but the general agreement in the positions and relative magnitudes of the observed and calculated maxima is excellent. A Cl–Cl distance of 2·99A was assumed in calculating the theoretical curve. James obtained similar agreement in general features between the observed and calculated curves for $SiCl_4$, assuming a tetrahedral molecule with an edge of 3·35A. It can in general be said that the characteristic features of the scattering of X-rays by gases are fully accounted for by the theory given above. Such discrepancies as do occur between theory and experiment may easily be explained by errors of observation, and by an imperfect knowledge of the atomic scattering factors.

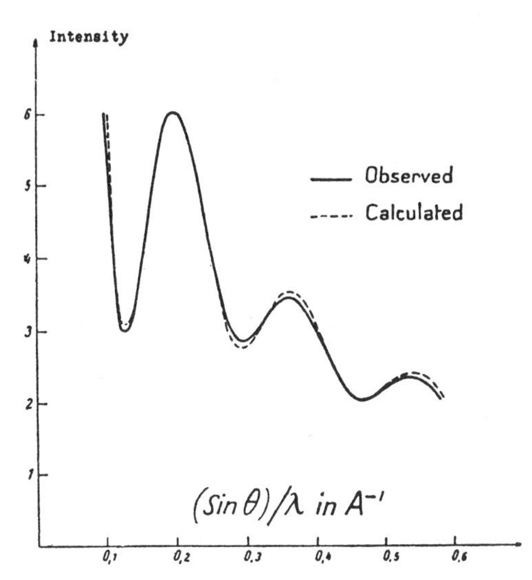

FIG. 173. Comparison between observed and calculated scattering curves for CCl_4 (van der Grinten) (*Physikal. Zeit.*, **34**, 609 (1933))

The nature of the scattering formula is such that it is not possible to determine molecular structures directly and uniquely from the diffraction patterns they produce. The only practicable method is to assume a structure, and then to see whether the scattering deduced from it by means of the formula agrees with that observed. In simple cases, such as that of CCl_4, this process amounts in effect to a unique determination, and suffices to show quite definitely that the four chlorine atoms do not, for example, lie at the corners of a square; but when the molecule is less symmetrical, the determination must be less certain. Nevertheless, the method has proved to be very valuable in investigating molecular structure. As an interesting example of the way in which the scattering varies in a series of chemically closely related compounds we may mention Bewilogua's [30] determination of the four chlorine derivatives of methane, CCl_4, $CHCl_3$, CH_2Cl_2, and CH_3Cl, and Ehrhardt's [33] investigation of the various forms of dichlorethylene and dichlorethane, $C_2H_2Cl_2$ and $C_2H_4Cl_2$.

(*e*) *Experimental methods:* Only a very brief space can be devoted to experimental methods, the discussion of which really lies outside the scope of this volume. Only such points as will help in the theoretical consideration of the results will be dealt with. The methods in general use for studying the scattering of X-rays by gases were largely developed by Debye and his school at Leipzig. A good account of the methods of making the observations and of applying the necessary corrections will be found in the paper by Bewilogua [30] already quoted.

The methods used are photographic, and the principles of the apparatus are illustrated in fig. 174. A narrow and nearly parallel beam of X-rays, XX, passes through the gas or vapour, contained in a suitable cell, which can, if necessary, be heated electrically. The cell has a thin flat window, W, made usually of aluminium foil, parallel to which the primary beam passes, as close as possible to it without touching it. The window is limited by a slit, SS, of two or three millimetres width, bounded by parallel knife-edges. The film is held in a semi-cylindrical camera, the axis of which is parallel to the slit, and lies in the plane of the window. The scattered radiation reaching any given point of the film comes from an approximately equal volume, whatever the angle of scattering. Corrections have to be made for the absorption of the scattered radiation by the window, which depends considerably on the angle of scattering, and it is therefore important to have the primary beam as monochromatic as possible. Usually, the direct beam from a copper target, filtered through nickel foil to remove most of the β radiation, has been used, and this is adequate for most purposes. The exposure times even with this method may be several hours, and to make the primary beam monochromatic by reflecting it from a suitable crystal would lengthen them unduly for most work, although van der Grinten [31] has successfully used this method in his work on CCl_4.

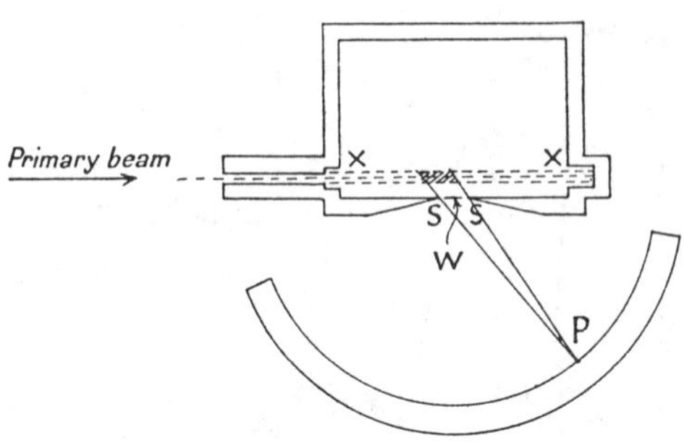

FIG. 174. Diagrammatic representation of the method of observing scattering of X-rays by gases

In all discussions of the experimental results given here it is assumed that the necessary corrections have been made to the observed intensities, and that the scattering formulae apply to a definite volume of gas, uniformly irradiated, scattering without absorption in all directions.

(*f*) *The effect of temperature on scattering by gases:* In deducing the formulae for the internal interference effects when radiation is scattered from an assemblage of gas molecules, it has been assumed that the atoms in the individual molecules occupy fixed relative positions. This is not quite true, since intra-molecular oscillations due to thermal movement must occur; this must have the effect of reducing the importance of the interference effects for the larger values of $(\sin\theta)/\lambda$. Since this effect must be present in principle, it is now necessary to enquire whether it is appreciable in practice. The binding of atom to atom within the molecule is in general much stronger than the binding of molecule to molecule in the crystalline state. We are dealing in fact with the 'optical frequencies' discussed in § 1(*p*) of Chapter V, rather than with the 'acoustic frequencies' which are mainly responsible for the temperature effect in the case of X-ray reflection from a crystal; so that we must be prepared to find the temperature effect in the case of scattering by a gas correspondingly smaller.

The question has been examined theoretically and practically by James,[32] who has applied his results to the case of silicon tetrachloride, for which it is possible to calculate the magnitude of the effect to be expected, from the known optical and Raman frequencies of the molecule. It will be enough here to give an outline of the theory, for details of which the original papers may be consulted.

The position of the atom p is assumed to be given at any instant by the vector $\mathbf{r}_p + \mathbf{u}_p$, where $\mathbf{u}_p$ represents the vector displacement of the atom p from its equilibrium position at the instant considered, because of the thermal movement. The instantaneous intensity of equation (9.39) may then be written, using (9.10), in the form

$$\mathrm{I}_m = \sum_p f_p^2 + \sum_p^{p \neq q} \sum_q f_p f_q \, e^{i\kappa \mathbf{S}\cdot\mathbf{r}_{pq}} \, e^{i\kappa \mathbf{S}\cdot(\mathbf{u}_p - \mathbf{u}_q)}, \tag{9.46}$$

and in averaging the terms referring to pairs of different atoms it is now necessary to consider not only all the possible orientations of the molecule, but all the possible displacements $\mathbf{u}_p$ and $\mathbf{u}_q$.

The oscillations of the n atoms composing the molecule are not all independent: they form a coupled system having $3n$ degrees of freedom. Now any small oscillation of such a system can be expressed as the sum of $3n$ normal co-ordinates corresponding to $3n$ normal modes of vibration. If the molecule is free, six of these are in general needed to define its motions of translation and rotation, leaving $3n - 6$ for the intra-molecular vibrations. The $SiCl_4$ molecule, for example, contains 5 atoms, and there are 9 independent normal modes, not all of which, however, have different frequencies. We have dealt with the representation of the thermal vibrations of a lattice in terms of a set of normal co-ordinates in Chapter V, and the line of thought here is similar; but because of the different nature of the problem, a rather different notation is convenient.

Let $p_1 \ldots p_k \ldots p_{3n}$ be the $3n$ normal co-ordinates of the system. These are so defined that if T and V are respectively the kinetic and potential energies of the system,

$$\mathrm{T} = \tfrac{1}{2} \sum_{k=3n}^{k=1} \dot{p}_k^2, \qquad \mathrm{V} = \tfrac{1}{2} \sum_{k=3n}^{k=1} \omega_k^2 p_k^2. \tag{9.47}$$

By Lagrange's equations, the normal co-ordinates so defined satisfy $3n$ independent equations of the type

$$\ddot{p}_k + \omega_k^2 p_k = 0,$$

the solution of which is

$$p_k = \mathrm{A}_k \, e^{i\omega_k t}, \tag{9.48}$$

A_k being in general a complex quantity giving both amplitude and phase. The full treatment of the problem shows that it is only the displacements of the atoms parallel to the line joining p and q that appreciably modify the interference term associated with this pair of atoms. Let us call these displacements ξ_p and ξ_q. By the theory of normal co-ordinates, each of these can be expressed as a linear combination of the normal co-ordinates, in the form

$$\xi_q = \sum_k \alpha_{qk}\, p_k = \sum_k A_k \alpha_{qk}\, e^{i\omega_k t}. \tag{9.49}$$

The coefficients α_{pk}, α_{qk}, ... determine the relative amplitudes of the vibrations of the different atoms as a result of the normal mode k, and their values depend on the nature of the vibrating system. The relative amplitudes, and the phases, of the different normal modes are determined by the coefficients A_k, and are initially arbitrary; but in a problem such as the present one, where we are concerned with a very large number of molecules in thermal equilibrium, the mean relative values of the coefficients A_k are determined by statistical principles. There are thus two processes, a geometrical averaging of (9.46), and the application of statistical principles to obtain the dependence of the average result on temperature.

The geometrical averaging, for the details of which reference may be made to the original paper, leads to the result

$$\bar{I}_m(\mu) = \sum f_p^2 + 2\sum_{pq} f_p f_q \frac{\sin \mu l_{pq}}{\mu l_{pq}} e^{-A}, \tag{9.50}$$

where

$$A = \tfrac{1}{2}\mu^2 \sum_k (\alpha_{pk} - \alpha_{qk})^2 \overline{p_k^2}. \tag{9.51}$$

By (9.49),

$$(\alpha_{pk} - \alpha_{qk})^2 \overline{p_k^2} = \overline{(\xi_p - \xi_q)_k^2} = \overline{(\delta l_{pq}^2)_k}; \tag{9.52}$$

that is to say, it is the mean square of the variation in the distance between the atoms p and q due to the normal mode k. Moreover, since the phases of the different normal co-ordinates are all independent, if $\overline{\delta l_{pq}^2}$ is the mean-square variation of the same distance under the influence of all the normal modes of vibration,

$$\overline{\delta l_{pq}^2} = \sum_k \overline{(\delta l_{pq}^2)_k}, \tag{9.53}$$

so that (9.51) may be written in the form

$$A = \tfrac{1}{2}\mu^2 \overline{\delta l_{pq}^2} = 8\pi^2 \frac{\sin^2\theta}{\lambda^2} \overline{\delta l_{pq}^2}, \tag{9.54}$$

a result analogous to equation (1.33), or (5.1), for the vibrations of the crystal lattice.

Equation (9.50) is an approximation. There are further terms, which we have not considered here, since they are always very small numerically, and correspond to changes of the scattered intensity well beyond the limits of observation.

As a simple example of the application of the formulae, we shall consider the case of a gas whose molecules consist of two atoms of mass m_1 and m_2 at a distance l apart. The vibrations along the line joining the two atoms can be expressed in terms of one normal co-ordinate p. Let the displacements of the atoms in this line be ξ_1 and ξ_2. Because of the conservation of centre of mass of the system $m_1\xi_1 + m_2\xi_2 = 0$, and by (9.47) and (9.49)

$$\mathrm{T} = \tfrac{1}{2}(m_1\dot{\xi}_1^2 + m_2\dot{\xi}_2^2) = \tfrac{1}{2}\dot{p}^2; \quad \xi_1 = \alpha_1 p, \quad \xi_2 = \alpha_2 p.$$

These equations lead at once to

$$\alpha_1 = \sqrt{\mathrm{M}}/m_1, \quad \alpha_2 = -\sqrt{\mathrm{M}}/m_2; \quad \alpha_1 - \alpha_2 = \frac{1}{\sqrt{\mathrm{M}}}, \tag{9.55}$$

where

$$\frac{1}{\mathrm{M}} = \frac{1}{m_1} + \frac{1}{m_2}.$$

By (9.51) and (9.55) therefore,

$$\mathrm{A} = \tfrac{1}{2}\mu^2\,\overline{p^2}/\mathrm{M} = 8\pi^2\frac{\sin^2\theta}{\lambda^2}\frac{\overline{p^2}}{\mathrm{M}}. \tag{9.56}$$

Bloch [34] has shown that for a simple harmonic oscillator

$$\overline{p^2} = \frac{h}{8\pi^2\nu}\coth\left(\frac{h\nu}{2k\mathrm{T}}\right), \tag{9.57}$$

h being Planck's constant, k, Boltzmann's constant, ν the frequency of the oscillator, and T the absolute temperature. Thus

$$\mathrm{A} = \frac{h}{\mathrm{M}\nu}\coth\left(\frac{h\nu}{2k\mathrm{T}}\right)\frac{\sin^2\theta}{\lambda^2}. \tag{9.58}$$

For chlorine, for example, the Raman effect shows ν to be

$$1{\cdot}665 \times 10^{13}\ \mathrm{sec}^{-1}.$$

At 17° C. (290° abs.), $h\nu/2k\mathrm{T} = 1{\cdot}37$, and $\coth(h\nu/2k\mathrm{T}) = 1{\cdot}14$. The mass of the chlorine atom is $58{\cdot}6 \times 10^{-24}$ gm. These figures give $\mathrm{A} = 0{\cdot}046$ for Cu Kα radiation at a scattering angle of 120°. The temperature correction thus amounts to about 5 per cent of the interference term alone. The most important part of the scattered intensity, Σf_p^2, is not affected by the temperature at all. At the larger angles of scattering, at which alone A will be appreciable, the interference term produces only slight oscillations of the total intensity. The temperature effect produces only relatively small variations of these oscillations, too small to be observed.

James [32] has made detailed calculations of the magnitude of the temperature correction for CCl_4 and $SiCl_4$, using Trumpy's [35] identification of the normal vibrations of the molecule with those responsible for the Raman and infra-red optical frequencies, and calculating the mean-square change in length of the edge and radius of the tetrahedron. In Table IX. 2, the total value of $\sqrt{\overline{\delta l_{pq}^2}}$ in A units for the edges and radii of the molecules at 0°, and 373°, and 573° absolute are shown.

TABLE IX. 2

Values of $\sqrt{\overline{\delta l_{pq}^2}}$ in A units

Temp. abs.	Radius		Edge	
	CCl_4	$SiCl_4$	CCl_4	$SiCl_4$
0°	0·051	0·043	0·054	0·061
373°	0·056	0·049	0·076	0·098
573°	0·062	0·056	0·092	0·120

In fig. 175, the curve X represents the theoretical value of the scattering from $SiCl_4$ when the atoms are assumed to be without vibration, the

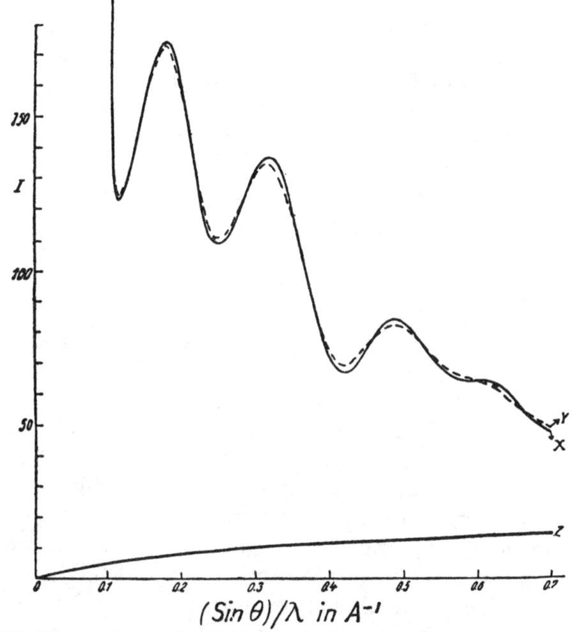

FIG. 175. Theoretical scattering curves for $SiCl_4$ (James)
x Coherent scattering when the atoms are assumed to be without vibration
y Curve with temperature correction applied
z Incoherent scattering
(*Physikal. Zeit.*, **33**, 737 (1932))

dotted curve Y shows the scattering at 573° abs. (300° C.), calculated on the basis of the figures in Table IX. 2. It will be seen that the differences between the two curves are within the errors of experiment, and that such discrepancies as still exist between theory and observation are not to be ascribed to neglect of the thermal movement.

Table IX. 2 shows that at absolute zero the values of $\sqrt{\overline{\delta l_{pq}^2}}$ due to the zero-point vibrations are quite large fractions of their values at 300° C. The change in the scattering curve between 100° C. and 300° C. should be quite inappreciable, a result confirmed by James, who measured the scattered intensity from $SiCl_4$ at these two temperatures. Except at quite small angles of scattering, where the difference observed can hardly be ascribed to a temperature effect, the scattering curves obtained at the two temperatures were identical within the errors of experiment. Similar results were obtained by van der Grinten [31] who used CCl_4 vapour, and monochromatised Cu $K\alpha$ radiation.

(*g*) *Scattering of electrons by gas molecules:* Formula (9.41), with appropriate interpretation of the scattering factor, and the factor C, also applies to the scattering of electrons by gases, and much work on the investigation of molecular structure has been done in this way, of which the pioneer work of Mark and Wierl [36] may be mentioned here. Some account of this work has been given in Vol. I, but its detailed consideration does not lie within the scope of this volume, and space allows no more than a brief reference for purposes of comparison. Electrons have certain advantages over X-rays in scattering experiments of this kind. Very small quantities of the gas or vapour can be used, and the necessary exposure times are to be measured in seconds rather than in hours. The wave-lengths of the electrons easily obtainable with voltages such as 30 or 40 kilovolts, are considerably shorter than those of the X-rays suitable for scattering experiments. Consequently, greater values of $(\sin\theta)/\lambda$ can be used, and correspondingly more maxima are obtained. On the other hand, there are certain difficulties* in interpreting quantitatively the diffraction haloes produced by the electrons. For one thing, the atomic scattering factors for electrons are less well understood, and less controlled by experiment than those for X-rays, and they fall off even more rapidly with increasing $(\sin\theta)/\lambda$. Diffraction haloes obtained with electrons appear visually to show well-marked maxima, but when a photometer curve is made it is often found that there are no real maxima in the photographic blackening, but merely a series of steps, the apparent maximum being a subjective effect; and the relation between the position of this subjective maximum and that of the true maximum is by no means clear. A comparison of observed and calculated scattering curves, taking into account the atomic scattering factors, is as necessary in the case of electrons as

* See for example: L. Bewilogua, *Physikal Zeit.*, **32**, 114 (1931); **33**, 688 (1932).

in that of X-rays. If due allowance is made for the difficulties, the method is one of great elegance and power, and is capable of great development.

(*h*) *Scattering by liquids with complex molecules:* If the distances between the molecules in the assemblage are reduced until they become comparable with those in the liquid state, the external interference effects become important. Debye [9] has extended the treatment given in IX, § 2(*b*) to assemblages of diatomic molecules, and Menke [37] has considered the general case of molecules of any form. Suppose, as before, that there are N scattering molecules, each containing n atoms, We shall denote the different *molecules* by the Greek letters μ, ν, ... and the different atoms in the individual molecules by p, q, ... as before. Let corresponding points C in each molecule be chosen as convenient molecular centres. The vector distance from the main origin O for the assemblage to C_μ, the centre of molecule μ, is denoted by $\mathbf{r}_\mu$, and the distance from C_μ to atom p in molecule μ by $\mathbf{v}_{\mu p}$, and the vector from O to this atom by $\mathbf{r}_{\mu p}$. Then, for the atom p in molecule μ,

$$\mathbf{r}_{\mu p} = \mathbf{r}_\mu + \mathbf{v}_{\mu p}. \tag{9.59}$$

This notation is illustrated by fig. 176.

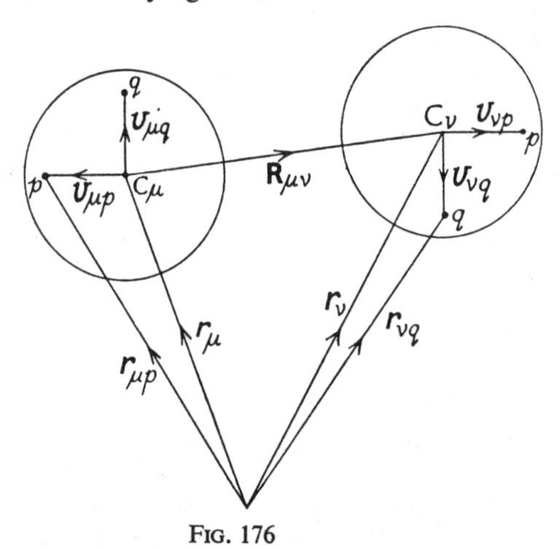

FIG. 176

The amplitude scattered by molecule μ is proportional to $\sum_p f_p e^{i\kappa \mathbf{S}\cdot\mathbf{r}_{\mu p}}$ for the given configuration, where $\sum_p$ denotes summation over the n atoms in the individual molecule. The total amplitude scattered by the assemblage is proportional to

$$\underset{\mu}{\mathrm{S}} \sum_p f_p e^{i\kappa \mathbf{S}\cdot\mathbf{r}_{\mu p}},$$

where S denotes summation over the different molecules. The total scattered intensity for the given configuration is proportional to the square of the modulus of this expression, so that, as in IX, § 1(*c*),

$$\mathrm{I}(\mu)=\underset{\mu}{\mathrm{S}}\,\underset{\nu}{\mathrm{S}}\,\sum_{p}\sum_{q} f_p f_q\, e^{i\kappa\mathbf{S}\cdot\mathbf{r}_{\mu p}-\mathbf{r}_{\nu q}}. \tag{9.60}$$

To get the observed intensity, we have to average (9.60) over all possible orientations of the individual molecules, and all possible intermolecular distances. Using (9.59), we may write (9.60) in the form

$$\mathrm{I}(\mu)=\underset{\mu}{\mathrm{S}}\,\underset{\nu}{\mathrm{S}}\,\sum_{p}\sum_{q} f_p f_q\, e^{i\kappa\mathbf{S}\cdot\mathbf{r}_{\mu}-\mathbf{r}_{\nu}}\, e^{i\kappa\mathbf{S}\cdot\mathbf{v}_{\mu p}-\mathbf{v}_{\nu q}}. \tag{9.61}$$

There are now two types of term to consider, (*a*) those for which $\mu=\nu$, corresponding to the internal interference effects, and (*b*) those for which $\mu\neq\nu$, corresponding to the external effects. There are N terms of type (*a*), and for each of these the first exponential factor in (9.61) is unity. The vector $\mathbf{v}_{\mu p}-\mathbf{v}_{\nu q}$ occurring in the second factor is l_{pq}, the vector distance between the atoms p and q in the same molecule; and when the molecules are allowed to take all possible orientations in space, the terms of type (*a*) lead simply to the expression

$$\bar{\mathrm{I}}_i=\mathrm{N}\left\{\sum_p f_p^2+2\sum_{pq} f_p f_q \frac{\sin \mu l_{pq}}{\mu l_{pq}}\right\}=\mathrm{N}\bar{\mathrm{I}}_m, \tag{9.62}$$

for the intensity contribution of the internal terms, which is, of course, simply equation (9.41), the intensity scattered by the molecules in the gaseous state.

The terms of type (*b*) refer to atoms lying in different molecules. They contribute an average intensity $\bar{\mathrm{I}}_e$, which we must now calculate. The vector difference $\mathbf{r}_\mu-\mathbf{r}_\nu$ in (9.61) is now the vector $\mathbf{R}_{\mu\nu}$ between the centres of the two atoms μ and ν. Just as in IX, § 2(*b*), we now write for the probability that the centre of the atom μ should lie in the volume element $d\mathrm{V}_\mu$ at a vector distance $\mathbf{r}_\mu$ from O, while at the same time the centre of molecule ν lies in $d\mathrm{V}_\nu$ at a vector distance $\mathbf{r}_\nu$,

$$\mathrm{W}(\mathbf{R}_{\mu\nu})\, d\mathrm{V}_\mu\, d\mathrm{V}_\nu/\mathrm{V}^2,$$

V being the total volume of the scattering gas. We may suppose $\mathrm{W}(\mathbf{R})$ to be radially symmetrical on the average, and to be the same for any pair of different molecules. There are N(N − 1) such pairs altogether, and by an argument exactly analogous to that used in deriving equation (9.29), we obtain for the contribution of the external terms to (9.61), the zero-order maximum being neglected,

$$\bar{\mathrm{I}}_e=-\sum_p\sum_q \frac{4\pi\mathrm{N}(\mathrm{N}-1)}{\mathrm{V}}\,\overline{[f_p f_q\, e^{i\kappa\mathbf{S}\cdot\mathbf{v}_{\mu p}-\mathbf{v}_{\nu q}}]}\int_0^\infty (1-\mathrm{W})\,\mathrm{R}^2\frac{\sin\mu\mathrm{R}}{\mu\mathrm{R}}\,d\mathrm{R}. \tag{9.63}$$

The exponential term in (9.63) may be written as the product of two factors, and since p and q lie in different molecules, $\mathbf{v}_{\mu p}$ and $\mathbf{v}_{\nu q}$ are entirely independent of one another. The two factors may therefore be averaged over all directions in space independently, and the average value for any atom of type p will be the same for each molecule, and equal to $(\sin \mu l_{cp})/\mu l_{cp}$, l_{cp} being the distance of the atom of type p from the molecular centre. Thus

$$\bar{I}_e = -F_e \cdot \frac{4\pi N(N-1)}{V} \int_0^\infty (1-W) R^2 \frac{\sin \mu R}{\mu R} dR, \tag{9.64}$$

where

$$F_e = \sum_p \sum_q f_p f_q \frac{\sin \mu l_{cp}}{\mu l_{cp}} \cdot \frac{\sin \mu l_{cq}}{\mu l_{cq}}. \tag{9.65}$$

The total scattered intensity $\bar{I}(\mu)$ is the sum of the expressions (9.62) for the internal terms and (9.64) for the external terms. Finally then

$$\bar{I}_M(\mu) = \frac{I(\mu)}{N} = \bar{I}_m - \frac{4\pi(N-1)}{V} F_e \int_0^\infty (1-W) R^2 \frac{\sin \mu R}{\mu R} dR, \tag{9.66}$$

a result first given by Menke.

The above argument assumes a spherically symmetrical probability function W(R), and random molecular orientation. We must inquire a little more closely into what this means physically. It must be remembered that the smallest portion of liquid from which it is possible to obtain a measurable intensity of scattered radiation contains a vast number of molecules. Random orientation of the molecules implies that in the whole irradiated region no one direction of a molecule shall be more likely than any other. It says nothing about the local orientations in the immediate neighbourhood of a given molecule which may indeed be non-random. But so long as such regions of local arrangement in different parts of the assemblage occur with all possible orientations, and this will always be so in a true liquid, the average molecular orientation over the whole assemblage may be considered as a random one for the purpose of calculating the intensity; and from similar considerations we also see that the probability function W(R), or the radial atomic density function with respect to any particular type of atom, will be spherically symmetrical.

Another point to be noted is that (9.66), as it stands, appears to depend on the quite arbitrarily chosen position of the molecular centre C, and the scattered intensity must clearly be independent of any such choice. It must, however, be remembered that the form of the function W(R) will depend on the position of C in the molecule, so that the product of F_e with the integral in (9.66) may still be independent of the choice of C.

Equation (9.66) may be transformed by Fourier's integral theorem. As in (9.33), we introduce the density function $\rho(R) = NW(R)/V$, and replace $N-1$ by N. Then (9.66) becomes

$$\frac{\mu \bar{I}_m}{F_e} \left\{ 1 - \frac{\bar{I}_M}{\bar{I}_m} \right\} = 4\pi \int_0^\infty \{\rho_0 - \rho(R)\} R \sin \mu R \, dR, \tag{9.67}$$

Plate XII — See other side

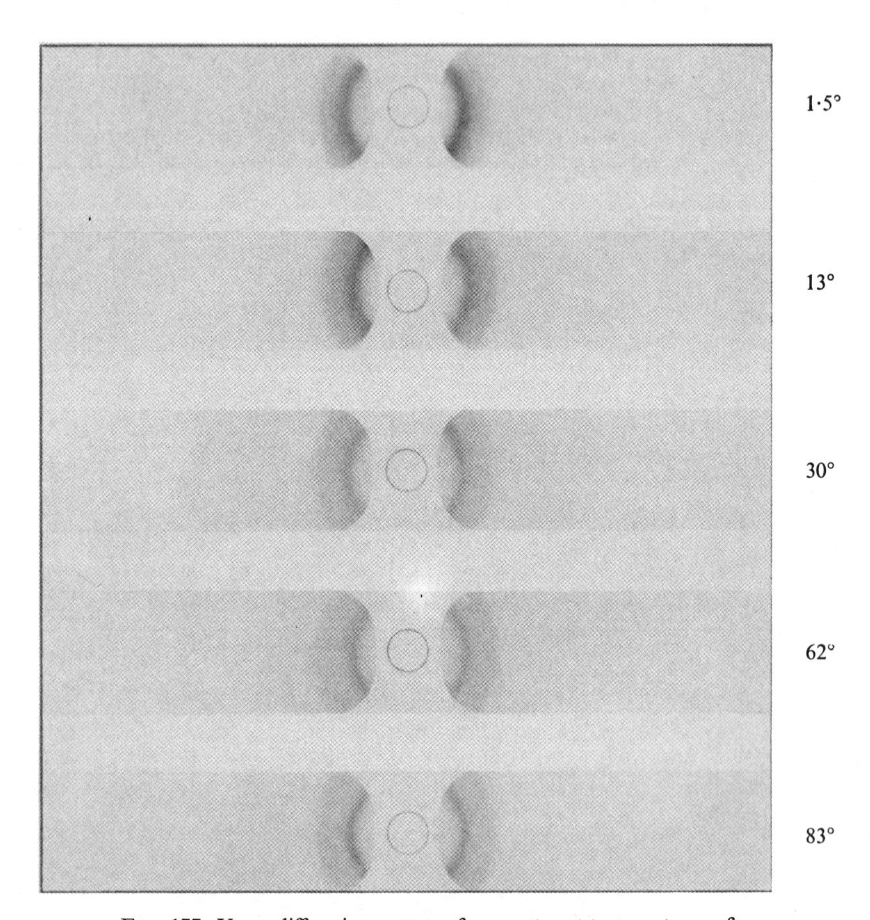

FIG. 177. X-ray diffraction patterns from water at temperatures of 1·5°, 13° 30°, 62° and 83° C. (Morgan and Warren)
(*Jour. Chem. Phys.*, **6**, 666 (1938))

and applying Fourier's theorem, we obtain, as in IX, § 2(*e*),

$$4\pi R^2\{\rho(R) - \rho_0\} = \frac{2R}{\pi}\int_0^\infty \mu\phi(\mu)\sin\mu R\, d\mu, \qquad (9.68)$$

where

$$\phi(\mu) = \frac{\bar{I}_m}{F_e}\left\{\frac{\bar{I}_M}{\bar{I}_m} - 1\right\}. \qquad (9.69)$$

Evaluation of this integral would lead to a determination of the radial density function $\rho(R)$. In determining the function $\phi(\mu)$, it would be necessary to determine $\bar{I}_M$ as a function of μ by observation. $\bar{I}_m$ could be determined either by observation of the molecular scattering in the gaseous state, or by calculation if the molecular structure were known. F_e could also be determined as a function of μ, given the structure of the molecule. In normalising the observed values of $\bar{I}_M$, it would again have to be assumed that the ratio $\bar{I}_M/\bar{I}_m$ approaches unity for large values of μ, or in other words, that the large-angle scattering approaches the independent scattering. Menke applied his theory to the case of liquid CCl_4, a molecule of a very symmetrical type. He did not actually obtain the expression for the distribution function in the form (9.68), which appears to have first been given by Zachariasen,[38] although it is in fact a perfectly obvious extension of (9.66), and he employed a more indirect method of comparing theory with experiment. The results obtained are of the type to be expected, but no direct determination of the probability function was made. As an example of the application of the formula to a liquid in which local parallel orientation probably occurs, we may mention the work of Warren [39] on long-chain liquid paraffins.

(*i*) *The structure of water:* As an example of the methods outlined in the last section, we shall consider in more detail the case of water. The diffraction of X-rays by water has been studied by Stewart,[40] Meyer,[41] Amaldi,[42] Katzoff,[43] and more recently by Morgan and Warren.[44] All these workers agree as to the general features of the diffraction curve. Katzoff was the first to apply the Fourier integral formula to determine the radial distribution, but the most complete experimental study is that of Morgan and Warren, who worked at a number of different temperatures ranging from 1·5° C. to 83° C. The X-ray diffraction patterns they obtained are illustrated in fig. 177 (Pl. XII), and the intensity curves, corrected for absorption and polarisation, in fig. 178. So far as molecular dimensions are concerned, water can be regarded as a monatomic liquid, the molecular diameter being that of the oxygen atom, about 2·8A. It is, however, clear from the diffraction curves that the arrangement in the liquid state is not that of irregular close packing such as is shown, for example, by liquid sodium. There is a second maximum, not quite clearly resolved from the main maximum, at an angle smaller than that characteristic of the close-packed arrangement. It will be noticed that this peak becomes much less marked as the temperature rises.

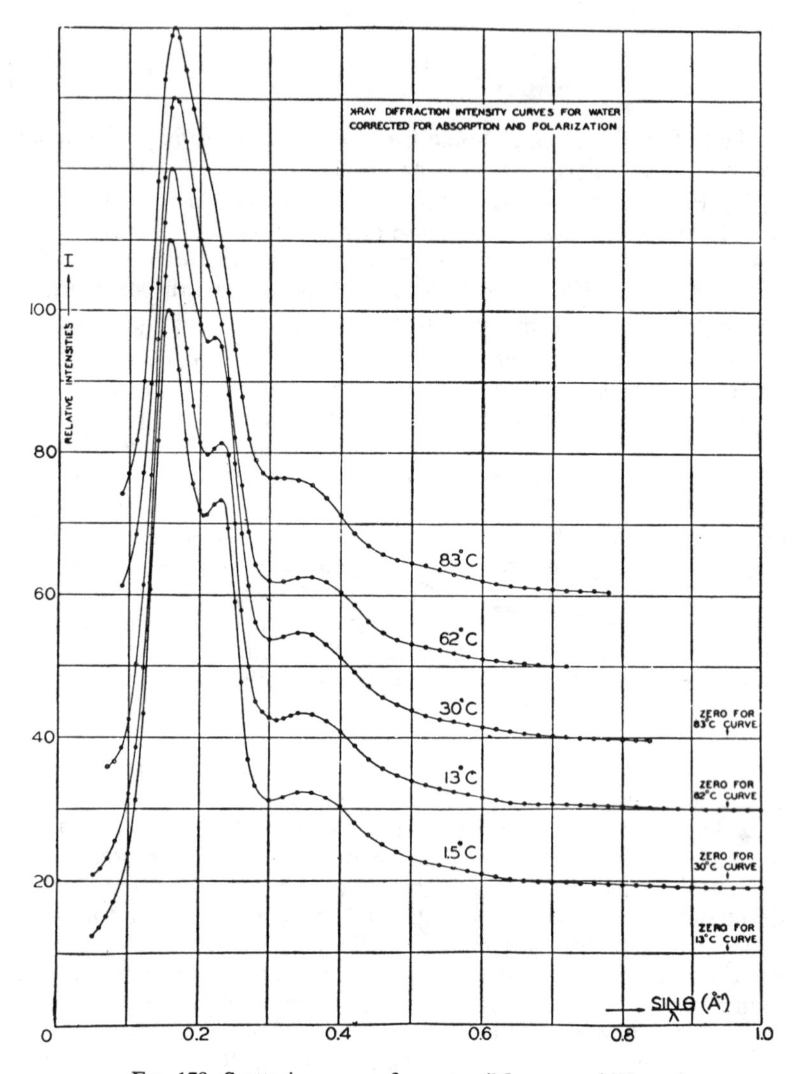

FIG. 178. Scattering curves for water (Morgan and Warren)
(*Jour. Chem. Phys.*, **6**, 666 (1938))

The corresponding radial distribution curves are shown in fig. 179. The main maximum occurs at a radius of about 2·8A, and a second and broader maximum at about 4·5A, which is only very faintly marked at the higher temperatures. The number of nearest neigh-

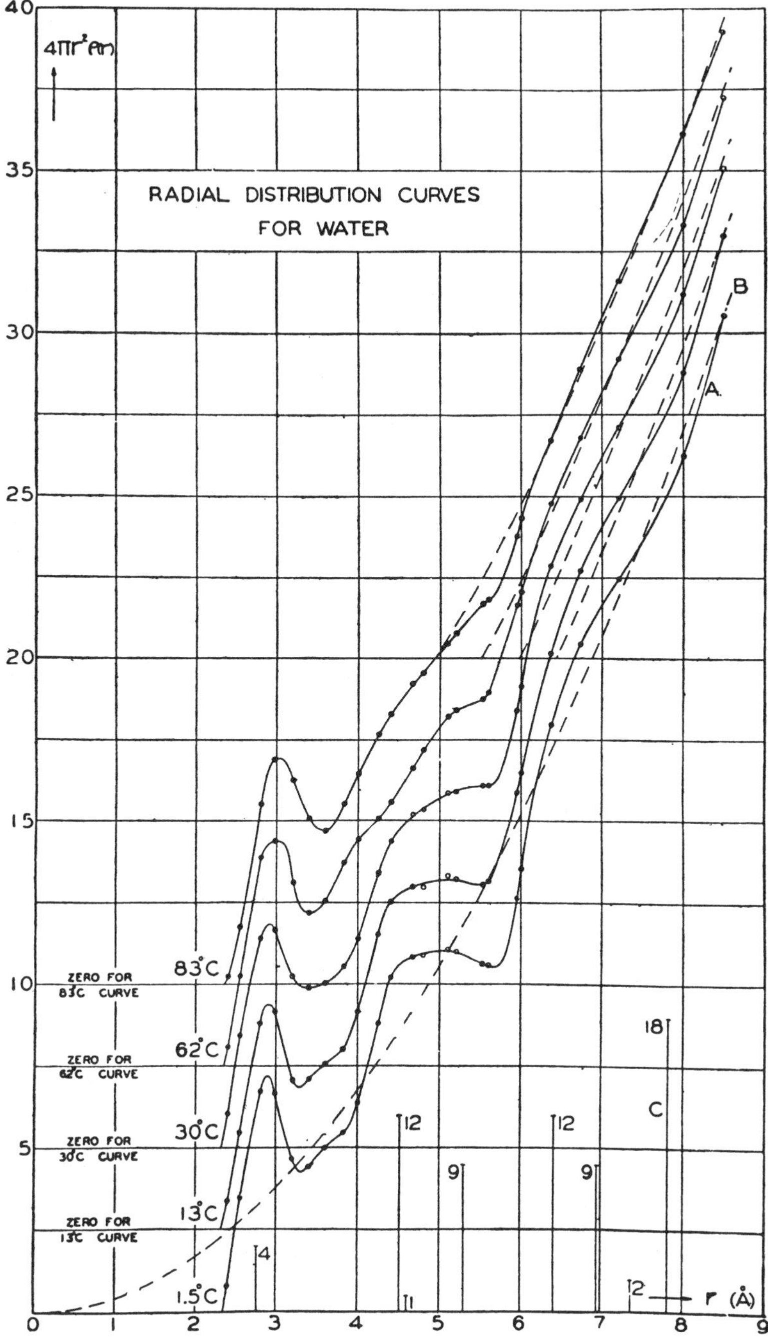

FIG. 179. Radial distribution curves for water (Morgan and Warren, *loc. cit.*)

bours appears to lie between 3 and 4 on the average, and to diminish slightly for higher temperatures.

The earlier diffraction curves obtained for water were discussed by Bernal and Fowler [45] in a very important paper on the theory of water. According to optical evidence, the H_2O molecule is V-shaped, with the O–H distance about 1·0A, and the $H\widehat{O}H$ angle not far from the tetrahedral angle, 109°. The scattering of X-rays is due almost exclusively to the electrons, so that, effectively, the water molecules will scatter as oxygen atoms, except perhaps at very small angles, when the distortion of the outer electron shells by the hydrogen nuclei will affect the scattering factor. Geometrically, we may, as we have seen, consider the water molecule to be a sphere of 2·8A diameter; but electrically, it is strongly polar, and we may expect association to occur more readily in certain relative positions of the molecules than in others. According to Bernal and Fowler, a close-packed assemblage of such spheres would have a density 1·84 times as great as that of water, so that the arrangement in water cannot be an irregular close packing, and this view, as we have seen above, is supported by the diffraction curves.

The structure and polar properties of the water molecule suggest that there will be a tendency to a fourfold co-ordination, with each molecule surrounded tetrahedrally by four others, a type of arrangement actually found in the ice crystal. A similar type of co-ordination is found in the crystalline forms of silica, in which the links from Si to Si are made through oxygen atoms tetrahedrally arranged about silicon atoms.[46] Three main types of crystalline silica are known, quartz, cristobalite, and tridymite, all of which are variants of this type of co-ordination, and ordinary ice corresponds to the tridymite structure.

Bernal and Fowler considered that it was not possible to explain water as an irregular structure of the tridymite type, because of the very great diminution in volume when ice melts, and suggested that it corresponds rather to an irregular quartz structure, which is 17 per cent. denser than the tridymite structure. They considered also that the diffration pattern given by water was better explained on the basis of an irregular quartz-like structure than on that of a tridymite-like structure. They suggested the existence of three chief types of arrangement of the H_2O molecules in water: Water I, tridymite ice-like, and present to a certain degree at low temperatures below 4°C.; Water II, quartz-like, predominating at ordinary temperatures; and Water III, close packed an ideal liquid, predominating at high temperatures for some distance below the critical temperature. These forms they suppose to pass continuously into each other with change of temperature, the liquid at any particular temperature being homogeneous, but the average arrangement of the molecules resembling, I, II, or III, in greater or less degree.

Morgan and Warren, however, as a result of their radial analysis consider that the quartz-like arrangement of the hypothetical Water II probably does not occur, since in it the next nearest neighbours should lie at 4·2A instead of at 4·5A, as they find experimentally. They conclude that water is best described as a broken-down ice structure, in which each molecule is striving to bind itself to four neighbours, just as in ice, but in which bonds are continually breaking and reforming, so that at any instant a molecule will be bonded on the average to fewer than four neighbours, and will have other neighbours at a continuous variety of distances.

It will be clear from this discussion of what is after all a comparatively simple case, that the interpretation of the diffraction patterns given by liquids composed of complex molecules is a matter of considerable difficulty, and is likely to yield little direct information about unknown structures. Even an ideally accurate determination of W(R) leads only to a spherical average of the distribution of diffracting matter about any given molecule. This distribution will be determined mainly by the fact that the molecules themselves occupy space, and relatively little by their internal structure; and in general the most that it will be possible to say will be that a certain assumed structure and arrangement is not inconsistent with the observed diffraction effects. It will rarely if ever be possible to make a direct determination of a molecular structure.

4. Analysis of Diffraction Patterns due to Powdered Crystals and Vitreous Solids

(*a*) *Introductory :* Formulae of the type developed in §§ 1(*c*) and 2(*e*) of this chapter may be used to give the average distribution of diffracting matter around any one atom in a solid structure, and have been applied both to powdered crystals and to vitreous and amorphous solids. It might at first sight appear that a method giving only an average radial distribution of density about the average atom in a crystal structure could have little to offer in comparison with the usual methods of crystal analysis, but this is not necessarily so. The ordinary methods are methods of trial and error, in which likely structures are first assumed, and then tested by comparison with observation. The methods discussed in the previous section for finding the average radial distribution about a member of an assemblage of atoms are, on the other hand, direct, and assume nothing about the structure; and although the information given by them is necessarily limited, it is definite. The knowledge that certain interatomic distances occur in the crystal may provide just the clue needed, and may lead to a correct guess at the structure. It may be compared with the type of information given by the Patterson Fourier series, discussed in Chapter VII, to which indeed the radial distribution formula is analogous, that certain interatomic distances occur in certain directions; and as we saw in Chapter. VII, such information may be of great value in structure analysis.

In dealing with scattering by solids we have to distinguish in principle between two cases: (*a*) The volume of solid producing the scattering may form a continuous structure, analogous, for example, to a liquid, in which there are no holes of more than atomic dimensions. (*b*) The scattering specimen may be made up of discrete particles, like a specimen of powder. Formulae of the type we have discussed above apply only to scattering by assemblages of atoms in which the average radial distribution is spherically symmetrical, and will therefore apply to solids of type (*a*) only if these are amorphous or vitreous. To apply them to crystalline solids, it is necessary to use powdered specimens, so that all orientations occur with equal probability in the irradiated sample; and here, of course, the question of crystal size is important. Each particle must be so large that effects due to its limited size do not come into consideration. We shall then in effect again be dealing with the average scattering from a single piece of crystal of appreciable size, which is allowed to assume all possible orientations in space. This condition may be assumed to apply in practice if the dimensions of the powder particle exceed about 10^{-4} cm. in all directions. We shall discuss the effect of very small particles in a later section.

(*b*) *Powdered crystals containing only one kind of atom:* Let a crystal powder containing a very large number of particles, having every possible orientation, scatter monochromatic X-rays. The diffraction pattern produced is the ordinary Debye-Scherrer ring system. It may also be considered as the average distribution of scattered radiation that would be obtained from a single crystal particle if that particle were allowed to take all possible orientations in space. We shall here assume that all the particles are alike, and that each contains a very large number of crystal units, and shall consider the average intensity scattered per crystal particle.

In such an assemblage, the distance between corresponding points of neighbouring particles will be large compared with the wave-length of the radiation, and the different particles will have random distribution. External interferences depending on the distances from particle to particle may therefore be neglected. Formally, the interference pattern we are interested in may be regarded as the internal interference pattern, or molecular scattering $\bar{I}_m$, from a gas whose molecules are the crystal particles. Each of these 'molecules' has a very large number of identical interatomic distances, regularly and periodically arranged, and so particularly well-defined maxima occur in the scattered radiation—the Debye-Scherrer rings.

Suppose for simplicity that the crystal lattice is simple, and that each particle contains N atoms of one kind only, of scattering factor *f*. Any two given atoms, *p* and *q*, remain at a fixed distance apart in the crystal. The process of averaging is therefore in principle rather different from that used in the case of the gas or liquid, where any pair of atoms was supposed to take all possible relative positions, although the result is formally the same. We consider any atom p, and suppose the crystal to assume all possible orientations about it. We may then define a spherically symmetrical distribution function $\rho(r)$, such that $4\pi r^2\rho(r)\,dr$ atoms lie at distances between r and $r+dr$ from the centre of the atom p. If each crystal particle contains a very large number of atoms, we may suppose $\rho(r)$ to be the same for all the atoms, since those at the edges of the particles form a very small fraction of the whole assemblage; and the summation over all the atoms may thus be replaced by multiplication by N. Treating the crystal particles as large molecules, we may use equation (9.41), and replacing the summation over q by an integration over r, we obtain

$$\bar{\mathrm{I}}(\mu)=\mathrm{N}f^2\left[1+4\pi\int r^2\rho(r)\frac{\sin\mu r}{\mu r}\,dr\right], \tag{9.70}$$

where the integration is to extend over the whole volume of the average crystal particle.

Equation (9.70) may be written in the form

$$\bar{\mathrm{I}}(\mu)=\mathrm{N}f^2\left[1+4\pi\int r^2\{\rho(r)-\rho_0\}\frac{\sin\mu r}{\mu r}\,dr+4\pi\rho_0\int r^2\frac{\sin\mu r}{\mu r}\,dr\right], \tag{9.71}$$

ρ_0 being the average atomic density in the crystal. As in the case of the liquid, $\rho(r)-\rho_0$ will approach zero as r becomes large, and we may suppose the crystal fragment to be so large that this condition is satisfied. The limits of integration of the first integral may then be taken from 0 to infinity. The second integral, as we have already seen, may be neglected except for angles of scattering in the immediate neighbourhood of the primary beam, so long as the dimensions of the crystal particles are large in comparison with λ, a condition that is always fulfilled in ordinary powder-photograph work. The expression for the scattered intensity then becomes

$$\bar{\mathrm{I}}(\mu)=\mathrm{N}f^2\left[1+4\pi\int_0^\infty r^2\{\rho(r)-\rho_0\}\frac{\sin\mu r}{\mu r}\,dr\right], \tag{9.72}$$

which is identical with (9.33), and gives, as before, on applying Fourier's integral theorem,

$$4\pi r^2\rho(r)=4\pi r^2\rho_0+\frac{2r}{\pi}\int_0^\infty \mu i(\mu)\sin(\mu r)\,d\mu, \tag{9.73}$$

where

$$i(\mu)=\frac{\bar{\mathrm{I}}(\mu)}{\mathrm{N}f^2}-1. \tag{9.74}$$

Equation (9.72) gives the intensity distribution in the Debye-Scherrer ring system, as a function of the angle of scattering and the average density distribution relative to any one atom. It does not include the intensity of the zero order, since, considering a single particle, this must be proportional to N^2. Equation (9.70) includes the zero order. When $\mu=0$, the value of the integral is simply the total number of atoms surrounding any one atom, which is of course N − 1; so that $\bar{I}(\mu)$ becomes $N^2 f^2$. The absence of the zero order must be remembered in using (9.73).

By determining $\bar{I}(\mu)$, and thence $i(\mu)$, experimentally, and evaluating the Fourier integral of (9.73) for a series of values of r, we may determine $\rho(r)$. Formula (9.72) was first applied in this way by Warren and Gingrich.[47]

(c) *The crystal powder or solid containing more than one kind of atom* : Before considering the application of this method to practical cases we must modify (9.72) and (9.73) so that they can be used when the crystal is not a simple lattice composed of one kind of atom. We shall find that a simple Fourier inversion, of the type used to transform (9.72) to (9.73), is no longer possible, and that approximations have to be used. The following method is used by Warren, Krutter and Morningstar.[48]

Let a certain group containing atoms of types m, n, ... be chosen as a unit of the structure, and let there be N such groups in each crystal particle. The average scattering per particle can now be written in the form

$$\bar{I}(\mu) = N\sum_m f_m^2 + \sum_p^{p \neq q}\sum_q f_p f_q \frac{\sin \mu r_{pq}}{\mu r_{pq}}, \tag{9.75}$$

the first sum being taken over all the atoms in a group, and the others over every pair of different atoms in the crystal particle, irrespective of which group they may belong to.

We now introduce a weighted density function $\rho_m(r)$, defined in the following way. Let spheres of radii r and $r + dr$ be described about the centre of an atom of type m, and suppose the average number of atoms of types m, n, ... lying between these spheres to be a_m, a_n, ... Then

$$4\pi r^2 \rho_m(r)\, dr = \sum_m a_m f_m, \tag{9.76}$$

the sum being taken over all the atoms in the group chosen as unit. Then, since there are N unit groups, (9.75) may be written in the form

$$\bar{I}(\mu) = N\left[\sum_m f_m^2 + \sum_m f_m \int 4\pi r^2 \rho_m(r) \frac{\sin \mu r}{\mu r}\, dr\right], \tag{9.77}$$

the summations again extending over all the atoms of the unit group.

In (9.77), both f_m and $\rho_m(r)$ are functions of μ, so that it is not possible to carry out a direct inversion by Fourier's integral theorem. There is thus no direct way of determining $\rho_m(r)$ from the observed scattering, as there is in the case of the monatomic assemblage. The following approximation, however, allows useful results to be obtained. Let us assume that the scattering factor of an atom of type m can be written

$$f_m = K_m f_e, \tag{9.78}$$

K_m being a constant for a given atom, and f_e a kind of average scattering factor per electron for the group of atoms under consideration, defined by the equation

$$f_e = \sum_m f_m / \sum_m Z_m; \tag{9.79}$$

that is to say, it is the sum of the scattering factors for the atoms of the group divided by the total number of electrons in the group, and is, of course, a function of μ. This assumes that the factor giving the angular dependence of f is the same for all atoms, an assumption which, while certainly not true, nevertheless gives an approximation good enough to yield useful results with suitably chosen values of K_m. The value of f_m defined by (9.78) and (9.79) is also used in the equation for the density function, (9.76). We may then express $\rho_m(r)$ in terms of the scattering factor per electron, f_e, writing

$$\rho_m(r) = f_e\, g_m(r), \tag{9.80}$$

and (9.77) becomes

$$\bar{I}(\mu) = N\left[\sum_m f_m^2 + 4\pi f_e^2 \int \left\{\sum_m K_m\, g_m(r)\right\} r^2 \frac{\sin \mu r}{\mu r}\, dr\right], \tag{9.81}$$

K_m and $g_m(r)$ being now no longer functions of μ. We now write $g_m(r) - g_0$ for the deviation of the electron density function from the mean electron density g_0, and, by the same argument as that used in deriving (9.72), obtain from (9.81)

$$\bar{I}(\mu)/N = \sum_m f_m^2 - 4\pi f_e^2 \int_0^\infty \sum_m K_m\{g_0 - g_m(r)\}\, r^2 \frac{\sin \mu r}{\mu r}\, dr. \tag{9.82}$$

Fourier's theorem may now be applied, and gives

$$4\pi r^2 \sum_m K_m\, g_m(r) = 4\pi r^2 g_0 \sum_m K_m + \frac{2r}{\pi} \int_0^\infty \mu i(\mu) \sin \mu r\, d\mu, \tag{9.83}$$

with

$$i(\mu) = \left(\frac{\bar{I}(\mu)}{N} - \Sigma_m f_m^2\right) \Big/ f_e^2 = \sum_m K_m^2 \left(\frac{I(\mu)}{N\, \Sigma_m f_m^2} - 1\right). \tag{9.84}$$

It should be noted once more that equations of the type (9.83) and (9.73) are valid only if the crystals are so large that the region over which the integration extends is great enough for the integrand to become sensibly zero. Only then can the integration be extended to infinity without the form and size of the particles being taken into account. We can get a general idea of the effect of very small particles by considering equations (9.71) and (9.81), which still apply. The density function $g_m(r)$ will fluctuate with increasing r in the usual way, but instead of converging to a steady value, equal to the average density for large r, it will fall to quite small values. Now the low value of scattered intensity at small angles, apart from the zero-angle scattering, in the case of liquids is due to the destructive interference of the scattered components from the larger values of r. If the radii of the particles are too small, these contributions are absent, and the intensity tends to rise at the smaller angles of scattering. We get in fact the molecular type of scattering, since these very small crystals are in effect large molecules, which, if the packing of the particles is of the looseness of ordinary powder, will be far enough apart to be optically independent.

(*d*) *Some examples :* Warren and Gingrich [47] applied the method outlined in the last paragraph to the determination of the number of near neighbours to any sulphur atom in rhombic sulphur, and this will serve as a good example of a crystal containing only one kind of atom. The experimental scattering curve is shown in fig. 180(*a*), corrected for

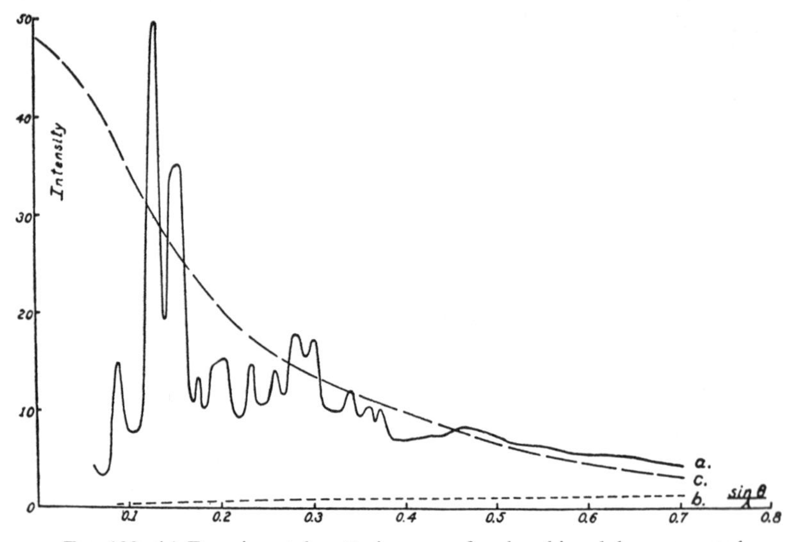

FIG. 180. (*a*) Experimental scattering curve for rhombic sulphur, corrected for absorption; (*b*) Incoherent scattering; (*c*) Independent coherent scattering (Warren and Gingrich)
(*Phys. Rev.*, **46**, 368 (1934))

polarisation and absorption in the specimen. The curve Nf^2, drawn on the assumption that all the atoms scatter independently, is shown in fig. 180(*c*), and curve 180(*b*) shows the incoherent scattering. The assumption made in standardising the curves is that the observed scattering for large values of $(\sin\theta)/\lambda$ is the sum of the independent and incoherent scattering, and this fixes the absolute scale of the scattering curve. The peaks in the scattering curve are, of course, the interference maxima in the Debye-Scherrer ring system. Fig. 181 shows the corresponding $\mu i(\mu)$ curve, and fig. 182 the radial density $4\pi r^2\rho(r)$ as a function of r, the distance from any atom. A well-marked peak, having an area of 2·3 atoms, occurs for $r = 2{\cdot}35$A. The figures are absolute, or at all events approximately so, and the result may be interpreted by supposing each sulphur atom to have two close neighbours at this distance. This suggests either closed rings or long chains of sulphur atoms, and since on chemical grounds a ring molecule S_8 has been suggested for rhombic sulphur, this provided a basis for attacking the structure, which was in fact determined completely, shortly afterwards, by Warren and Burwell.[49]

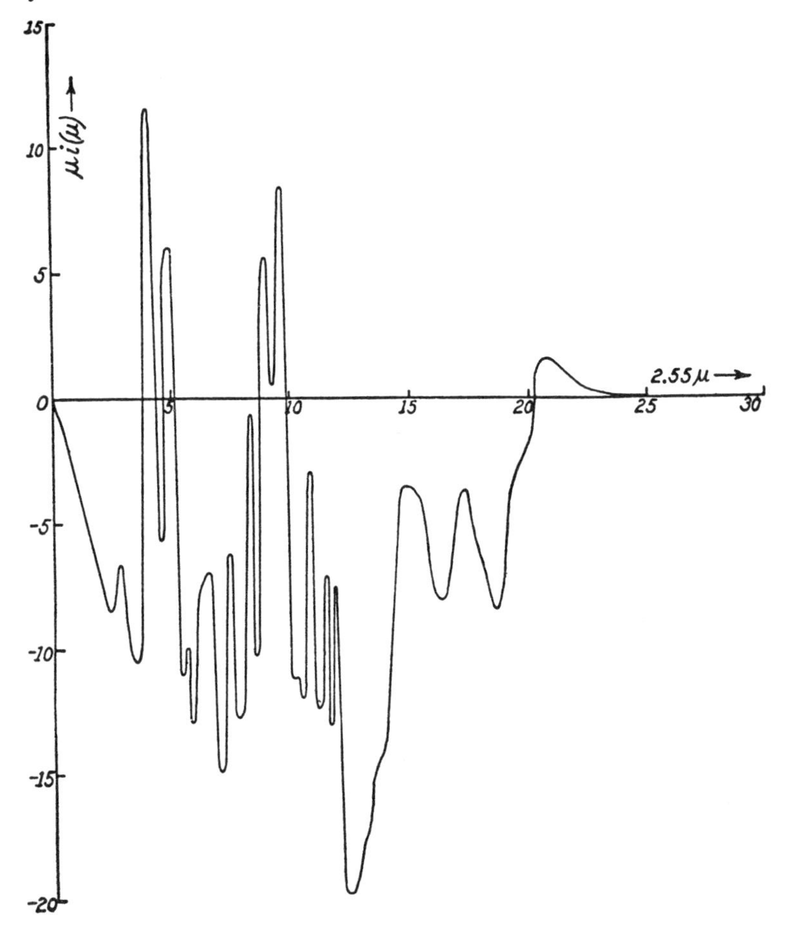

FIG. 181. Curve of $\mu i(\mu)$ for rhombic sulphur, as a function of μ (Warren and Gingrich, *loc. cit.*)

The distribution curve in fig. 182 may be compared with that for liquid sodium (fig. 170). It will be noticed that the first peak is almost completely isolated in the case of the crystalline solid sulphur, but not in the case of the liquid sodium. In the crystal, the distance between closest neighbours is much more definite than in a liquid, and distances a little greater than the least distance of approach of two atoms do not

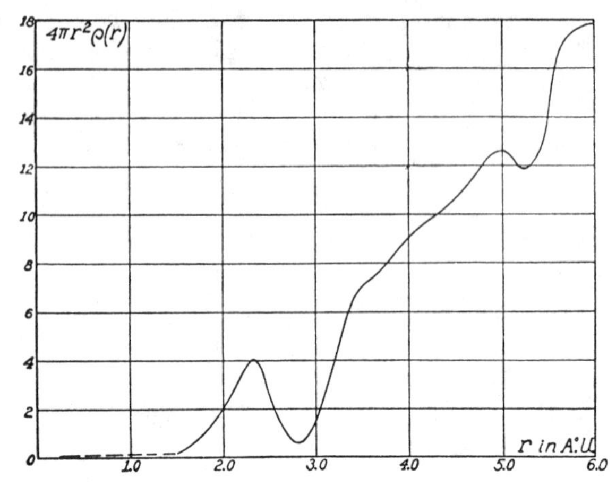

FIG. 182. Radial density distribution in rhombic sulphur, giving the number of atoms per A unit at any radial distance from a given atom (Warren and Gingrich, *loc. cit.*)

occur at all, but there is no such absolute prohibition in the case of a liquid.

Similar methods have been used to determine the structure of the different crystalline forms of phosphorus,[50] and of carbon black.[51]

(*e*) *The analysis of vitreous solids :* Some of the most interesting applications of the method of radial Fourier analysis have been made in studying the vitreous state by Warren and his fellow-workers.[52] Space does not allow more than a relatively brief referencet o this work, and we shall take as a typical example the case of vitreous silica, investigated by Warren, Krutter, and Morningstar.[48] In Plate XV, p. 195, of Vol. I of *The Crystalline State*, the diffraction pattern of vitreous silica is shown, together with that of cristobalite, one of the crystalline forms of silica. In fig. 183 (Pl. XIII), microphotometer records of the diffraction patterns of vitreous silica, cristobalite, and silica gel, are shown for comparison. It will be noticed that the main maxima of all three patterns occur very nearly at the same value of $(\sin \theta)/\lambda$, but that vitreous silica shows only the one very broad maximum (within the range of angles shown), the intensity decreasing to small values at small angles of scattering, while the crystalline form, cristobalite, shows a number of sharp maxima. Silica gel, on the other hand, shows a broad maximum similar to that for vitreous silica, but shows also the increasing intensity towards small angles of scattering that is characteristic of scattering by assemblages of independent molecules.

The result of applying the radial analysis to the scattering curves for vitreous silica is shown in fig. 184. The ordinates here are the values of $4\pi r^2 \sum_m g_m(r) K_m$, and the curve is really two curves superposed, one

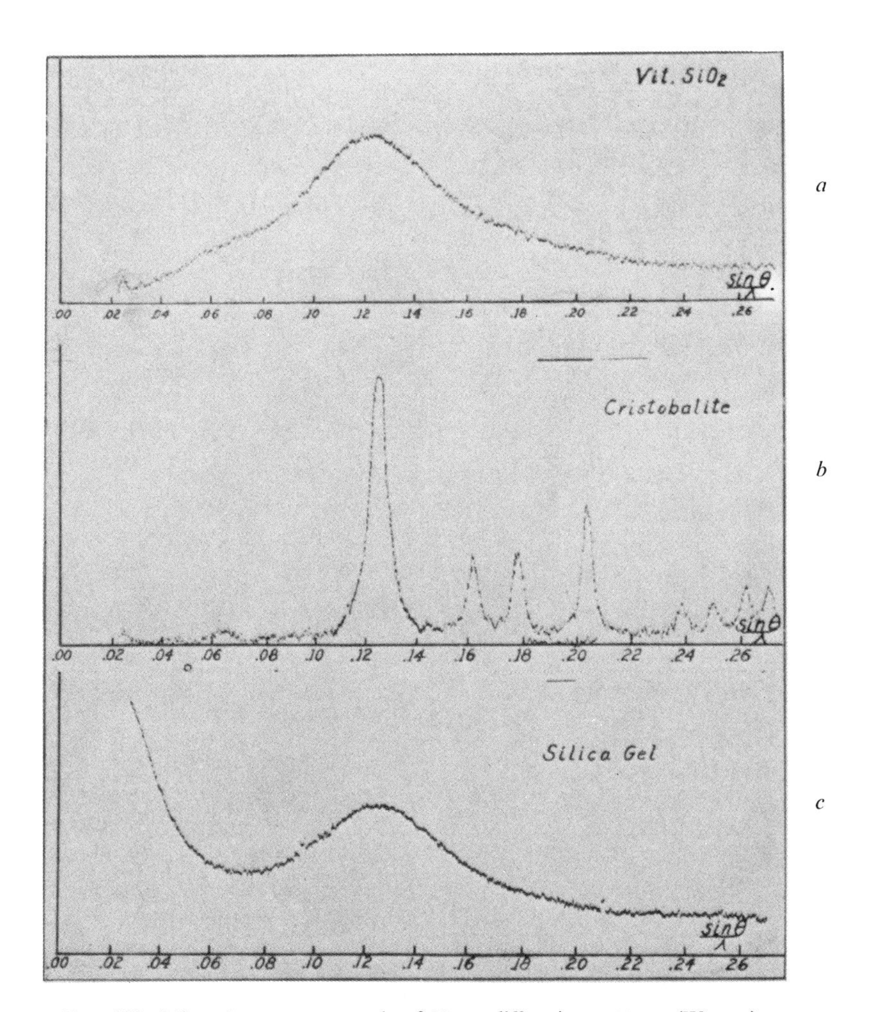

FIG. 183. Microphotometer records of X-ray diffraction patterns (Warren). (*a*) Vitreous silica, (*b*) cristobalite, (*c*) dried commercial silica gel
(*Jour. of App. Phys.*, **8,** 645 (1937))

Plate XIII — See other side

for the distribution about a silicon atom, and one for that about an oxygen atom. The first peak occurs at $r = 1{\cdot}62$A, and the average silicon-oxygen distance in the crystalline silicates is 1·60A. This suggests that the type of local co-ordination characteristic of the crystalline forms persists in vitreous silica. If there are n oxygen atoms around each silicon atom as nearest neighbours, there must be $n/2$

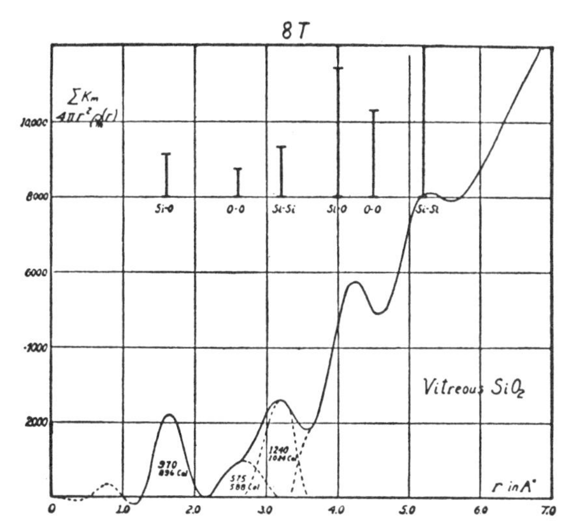

FIG. 184. Radial distribution curve for vitreous silica (Warren, Krutter and Morningstar)
(*Jour. Amer. Ceramic Soc.*, **19**, 202 (1936))

silicon atoms around each oxygen atom. By (9.76) and (9.78), the area under the first peak will be $K_{Si}\, nK_0 + 2K_0(n/2)\, K_{Si}$, and this may be equated to the observed area, 970 elec.². On inserting the values of K_{Si} and K_0 chosen as most suitable to give an approximation to the f-curves over the range of the experiment, it was found that n was approximately 4, as in crystalline silica. The position of the second peak in fig. 184 at 2·65A supports the view that the arrangement is still tetrahedral in the vitreous state. The local scheme of co-ordination about any one silicon atom appears thus to be the same in the crystalline and vitreous states. To quote Warren 'The X-ray results are completely explained by picturing glassy silica as a random network in which each silicon is surrounded tetrahedrally by four oxygens, and each oxygen bonded to two silicons, the two bonds to an oxygen being roughly diametrically opposite. The orientation of one tetrahedral group with respect to a neighbouring group can be practically at random. This is the simplest picture of silica glass, free from all specialised assumptions, which will completely explain the X-ray diffraction pattern. There is a definite scheme of structure involved, each atom has a definite number of nearest neighbours at a definite distance, but no unit of structure repeats itself identically at regular intervals in three dimensions, and hence the material is not crystalline.' Fig. 185, taken from a paper by Zachariasen,[53] represents in a schematic way in two dimensions the difference between a crystal and a glass.

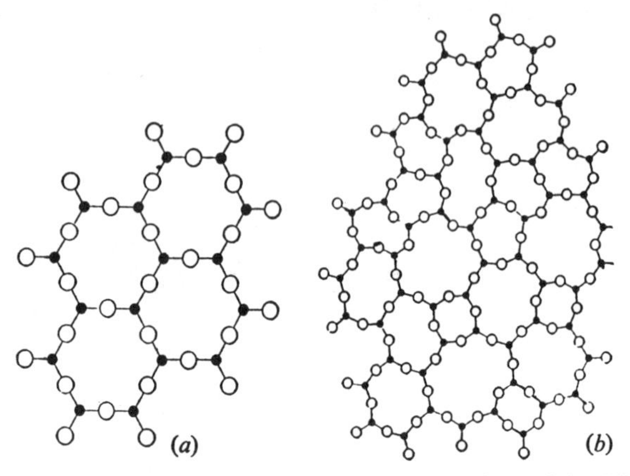

FIG. 185. Diagrammatic representation in two dimensions of the difference between a crystal (*a*) and a glass (*b*) (Zachariasen)
(*Jour. Amer. Chem. Soc.*, **54**, 3841 (1932))

The diffraction pattern of vitreous silica has been interpreted by some workers as due to extremely small crystals of cristobalite. The effect of decreasing the size of the crystals is, of course, to broaden the lines in the diffraction pattern, and it is argued that the difference between figs. 183(*a*) and (*b*) is simply that (*a*) may be considered as produced by progressive broadening of the lines in (*b*) as the individual crystals get smaller and smaller. Formally perhaps, this is a permissible interpretation; but if we investigate, as Warren has done, what reduction in size of the crystallites is necessary, we see that each would contain only about a single unit of the cristobalite structure. Such particles can hardly be called crystals, and the question also arises of their connection with one another, whether there are breaks at the boundaries, or whether there is a continuous scheme of bònding analogous to that shown in fig. 185. The type of diffraction pattern given by vitreous silica, with its small intensity at small angles of scattering, is similar to that given by a liquid, and indicates that the structure is essentially continuous, or at all events that there are no large-scale inhomogeneities comparable in size with the particles themselves. Such discontinuities would give rise to small-angle scattering, as we have seen in § 4(*c*). Small-angle scattering does occur in the case of silica gel (fig. 183 (*c*)), which presumably consists of very small discrete particles (10 to 100A) with voids between them. It is the particular nature of the silicon-oxygen link, which is strong yet flexible, that allows glass-like structures of the kind discussed above to be formed. The molten substance on solidification rapidly forms a number of such bonds, and thereby becomes so stiff and viscous as to prevent any reorganisation and crystallisation from taking place. The random network is nearly as stable as the crystal, and though not, in principle, in the most stable state, can continue to exist almost indefinitely.

REFERENCES

1. P. Debye, *Ann. Physik*, **46**, 809 (1915).
2. P. Ehrenfest, *Proc. Amsterdam Acad.*, **17**, 1132, 1184 (1915).
3. P. Debye and P. Scherrer, *Nach. der Gött. Ges.*, 1 and 16 (1916).
4. A. W. Hull, *Phys. Rev.*, **10**, 661 (1917).
5. W. Friedrich, *Physikal. Zeit.*, **14**, 317 (1913).
6. W. H. Keesom and J. de Smedt, *Proc. Amsterdam Acad.*, **25**, 118 (1922); *Physica*, **2**, 191 (1922).
7. C. V. Raman and K. R. Ramanathan, *Nature*, **111**, 185 (1923); *Phil. Mag.*, **47**, 671 (1924).
8. G. W. Stewart, *Reviews of Modern Physics*, **2**, 116 (1930).
9. P. Debye, (i) *Jour. Math. and Phys.*, **4**, 153 (1925); (ii) *Physikal. Zeit.*, **28**, 135 (1927).
10. F. Zernicke and J. A. Prins, *Zeit. Physik*, **41**, 184 (1927).
11. P. Debye and H. Menke, (i) *Physikal. Zeit.*, **31**, 797 (1930); (ii) *Ergebnis. d. techn. Röntgenkunde*, **2**, 1 (1931).
12. G. Breit, *Phys. Rev.*, **27**, 362 (1926).
13. P. A. M. Dirac, *Proc. Roy. Soc.*, **111** A, 405 (1926).
14. W. Gordon, *Zeit. f. Physik*, **40**, 117 (1926).
15. E. Schrödinger, *Ann. der Physik*, **82**, 257 (1927).
16. O. Klein and Y. Nishina, *Zeit. f. Physik*, **52**, 853 (1929).
17. I. Waller, *Zeit. f. Phys.*, **61**, 837 (1930).
18. W. Heisenberg, *Physikal. Zeit.*, **32**, 737 (1931).
19. L. Bewilogua, *Physikal. Zeit.*, **32**, 740 (1931).
20. P. Scherrer and A. Stäger, *Helv. Phys. Acta.*, **1**, 518 (1928).
21. C. S. Barrett, *Phys. Rev.*, **32**, 22 (1928).
22. G. Herzog, *Zeit. f. Physik*, **69**, 207 (1931); *Helv. Phys. Acta.*, **6**, 508 (1933).
23. E. O. Wollan, *Phys. Rev.*, **37**, 862 (1931).
24. G. Herzog, *Zeit. f. Physik*, **70**, 583, 590 (1931).
25. L. P. Tarasov and B. E. Warren, *Jour. Chem. Phys.*, **4**, 236 (1936).
26. P. Debye, *Physikal. Zeit.*, **31**, 419 (1930); *Proc. Phys. Soc.*, **42**, 340 (1930).
27. H. Gajewski, *Physikal. Zeit.*, **33**, 122 (1932).
28. H. Richter, *Physikal. Zeit.*, **33**, 587 (1932).
29. P. Debye, L. Bewilogua and F. Ehrhardt, *Physikal. Zeit.*, **30**, 84 (1929); P. Debye, *Physikal. Zeit.*, **30**, 524 (1929); *Zeit. f. Elektrochemie*, **36**, 612 (1930).

30. L. Bewilogua, *Physikal. Zeit.*, **32**, 265 (1932).
31. W. van der Grinten, *Physikal. Zeit.*, **34**, 609 (1933).
32. R. W. James, *Physikal. Zeit.*, **33**, 737 (1932).
33. F. Ehrhardt, *Physikal. Zeit.*, **33**, 605 (1932).
34. F. Bloch, *Zeit. f. Physik*, **74**, 295 (1932).
35. B. Trumpy, *Zeit. f. Physik*, **66**, 790 (1930).
36. H. Mark and R. Wierl, *Zeit. f. Physik*, **60**, 741 (1930); R. Wierl, *Ann. der Physik*, **8**, 521 (1931).
37. H. Menke, *Physikal. Zeit.*, **33**, 593 (1932).
38. W. Zachariasen, *Phys. Rev.*, **47**, 277 (1935).
39. B. E. Warren, *Phys. Rev.*, **44**, 969 (1933).
40. G. W. Stewart, *Phys. Rev.*, **35**, 1426 (1930); **37**, 9 (1931).
41. H. H. Meyer, *Ann. der Physik*, **5**, 701 (1930).
42. E. Amaldi, *Physikal. Zeit.*, **32**, 914 (1931).
43. S. Katzoff, *Jour. Chem. Phys.*, **2**, 841 (1934).
44. J. Morgan and B. E. Warren, *Jour. Chem. Phys.*, **6**, 666 (1938).
45. J. D. Bernal and R. H. Fowler, *Jour. Chem. Phys.*, **1**, 515 (1933).
46. W. L. Bragg, *Zeit. f. Krist.*, **74**, 237 (1930).
47. B. E. Warren and N. S. Gingrich, *Phys. Rev.*, **46**, 368 (1934).
48. B. E. Warren, H. Krutter and O. Morningstar, *Jour. Amer. Ceramic Soc.*, **19**, 202 (1936).
49. B. E. Warren and J. T. Burwell, *Jour. Chem. Phys.*, **3**, 6 (1934).
50. R. Hultgren, N. S. Gingrich and B. E. Warren, *Jour. Chem. Phys.*, **3**, 351 (1935).
51. B. E. Warren, *Jour. Chem. Phys.*, **2**, 551 (1934).
52. B. E. Warren, *Zeit. f. Krist.*, **86**, 349 (1933); *Jour. Amer. Ceramic Soc.*, **17**, 249 (1934); B. E. Warren and A. D. Loring, *ibid.*, **18**, 269 (1935); B. E. Warren, *Jour. Appl. Phys.*, **8**, 645 (1937); B. E. Warren and J. Biscoe, *Jour. Amer. Ceramic Soc.*, **21**, 259 and 287 (1938); see also J. T. Randall, H. P. Rooksby and B. S. Cooper, *Zeit. f. Krist.*, **75**, 196 (1930).
53. W. Zachariasen, *Jour. Amer. Chem. Soc.*, **54**, 3841 (1932).

CHAPTER X

DIFFRACTION BY SMALL CRYSTALS, AND ITS RELATIONSHIP TO DIFFRACTION BY AMORPHOUS MATERIAL

1. A Comparison between Diffraction by Crystal Powders and by Polyatomic Molecules

(*a*) *Introductory:* In this chapter, an attempt will first be made to trace the relationship between the formulae for scattering by individual molecules, which were developed in the last chapter, and those for scattering by crystals. We have already seen, in § 4(*b*) of the last chapter, that there is no essential difference, from the point of view of the optical principles involved, between scattering by a polyatomic gas and scattering by a powder composed of identical crystallites with completely random orientation, provided that the average distance between the crystallites is so great in comparison with the wave-length of the radiation used that external interference effects can be neglected. Each crystallite may be considered as a large molecule, and equation (9.41) must apply to the scattered intensity, which is now expressed in terms of the distances between each pair of atoms in an individual particle.

In Chapter I, we developed equations for the intensity scattered by a crystal lattice composed of identical points forming a parallelepipedal block. Taking as a starting-point scattering by a powder composed of such small blocks with a random orientation, it ought to be possible to derive, for example, the formula for scattering by a diatomic gas, by supposing the number of particles in each crystallite reduced to two. We shall find that it is indeed possible to deduce in this way equation (9.43) of Chapter IX, as a special case of the intensity distribution in the Debye-Scherrer ring system for a crystal powder.

The discussion of such problems leads naturally to the consideration of the effect of crystal size on the distribution of intensity in the diffracted radiation, of scattering by lattices with imperfections, and by such substances as fibres and hairs. The second, third, and fourth sections of this chapter will be devoted to these topics.

(*b*) *The interference function and the structure factor in terms of the reciprocal lattice:* The problems to be discussed in this chapter are most easily handled by the methods introduced in Chapter I, §§ 1(*b*) and 1(*c*), and employed in Chapter V, which involve the use of the reciprocal lattice. Let the triplet of numbers (ξ, η, ζ) define a point in the reciprocal-lattice space at a vector distance $\xi\mathbf{a}^* + \eta\mathbf{b}^* + \zeta\mathbf{c}^*$ from the origin. Then the scattering function $\mathrm{I}(\xi, \eta, \zeta)$ gives the intensity of the radiation scattered in the direction defined by the unit vector **s** when the direction of the incident radiation is defined by the unit vector $\mathbf{s}_0$, if

$$(\mathbf{s}-\mathbf{s}_0)/\lambda=\mathbf{S}/\lambda=\xi\mathbf{a}^*+\eta\mathbf{b}^*+\zeta\mathbf{c}^*. \qquad (10.1)$$

$I(\xi, \eta, \zeta)$ can thus be thought of as a distribution in the reciprocal-lattice space, and its value at any point gives the scattered intensity when the conditions of incidence and scattering are such that the extremity of the vector $\mathbf{S}/\lambda$, as defined by (*10*.1), lies at that point.

As before, $I(\xi, \eta, \zeta)$ gives the intensity as a multiple of that scattered by a single classical electron under the same conditions, and it is expressible as the product of two factors, one of which is the interference function $I_0(\xi, \eta, \zeta)$, as defined in Chapter I, § 1 (*c*), which is the intensity function when each lattice-unit is a point of unit scattering power. The other factor, which we shall denote by $|F(\xi, \eta, \zeta)|^2$, is the square of the modulus of the structure factor of unit cell of the lattice, and this depends on the details of the distribution of the scattering matter in the unit cell.

The significance of these two factors may be understood by considering the simple case of the optical diffraction grating. The formula for the intensity of the light diffracted by such a grating, expressed as a function of the angle of scattering, consists of two factors, one of which, depending only on the spacing and number of lines in the grating, determines the positions and sharpness of the spectra, but indicates the same intensity for all of them, independent of order. This factor is the interference function. It is obtained by supposing the grating openings to be infinitely narrow in comparison with the wave-length of the light used. The second factor, corresponding to the structure factor, depends on the form and structure of the individual grating lines, and determines the way in which the intensities of the spectra vary with their order.

The total intensity function $I(\xi, \eta, \zeta)$ for any point in the reciprocal-lattice space may thus be written

$$I(\xi, \eta, \zeta)=|F(\xi, \eta, \zeta)|^2\, I_0(\xi, \eta, \zeta). \qquad (10.2)$$

By equation (*2*.13) of Chapter II, the structure factor may be written

$$F=\Sigma f e^{i\kappa\mathbf{S}\cdot\boldsymbol{\rho}}, \qquad (10.3)$$

where $\boldsymbol{\rho}$ is the vector distance of any atom in the unit cell from the origin of that cell. If we write

$$\boldsymbol{\rho}=u\mathbf{a}+v\mathbf{b}+w\mathbf{c},$$

as in (*2*.6), and use (*10*.1), and the properties of reciprocal vectors, the expression for the structure factor becomes

$$F(\xi, \eta, \zeta)=\Sigma f e^{2\pi i(\xi u+\eta v+\zeta w)}, \qquad (10.4)$$

the summation being over all the atoms in one unit cell. In using the structure factor hitherto, we have always been concerned with scattering in the direction of one of the lattice spectra, for which ξ, η, and ζ have the integral values of h, k, and l. Equation (*10*.4) is more general, and applies to scattering in any direction, ξ, η, and ζ being treated as continuously variable. So long as the lattice contains a large number of units, the interference function, I_0, has appreciable values only if ξ, η, and ζ are very nearly indeed integral, so that in such a case we may simply assign to each reciprocal-lattice point a weight $|F(hkl)|^2$ in order to represent in the reciprocal-lattice space the intensities of the spectra that can be given by the crystal; and this is, in fact, the procedure adopted in structure analysis. It must be remembered, however, that if we are considering exceedingly minute crystals, I_0 may have values different from zero throughout a volume of appreciable magnitude about each reciprocal-lattice point. The values of $|F|^2$ for other than integral values of ξ, η, and ζ then become significant, and may appreciably modify the intensity function.

We may consider as a simple example a face-centred cubic lattice, with cube edge a. This we may regard as based on a simple cubic lattice of edge a, each unit cell of which contains four points with values of (u, v, w) equal to $(0, 0, 0)$, $(\frac{1}{2}, \frac{1}{2}, 0)$, $(0, \frac{1}{2}, \frac{1}{2})$, and $(\frac{1}{2}, 0, \frac{1}{2})$. If each atom has a scattering factor f, we find for this case, from (*10*.4),

$$|F(\xi, \eta, \zeta)|^2 = 4|f|^2\{1 + \cos \pi\xi \cos \pi\eta + \cos \pi\eta \cos \pi\zeta + \cos \pi\zeta \cos \pi\xi\}. \qquad (10.5)$$

If we confine our attention to integral values of the parameters, that is to say, to the reciprocal-lattice points (h, k, l), we find that $|F|^2$ vanishes unless h, k, l are either all even or all odd; so that the effective reciprocal-lattice points form a body-centred cubic lattice with edge $2a^*$. A section of the distribution $|F(\xi, \eta, \zeta)|^2$ by the $(1\bar{1}0)$ plane of this lattice passing through the origin is shown in fig. 186, which is

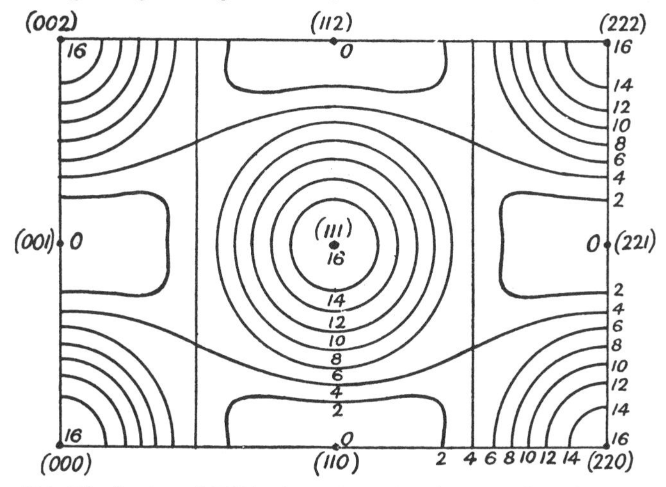

FIG. 186. Distribution of $|F|^2$ in the reciprocal-lattice space for a face-centred lattice, and section by plane $(1\bar{1}0)$

drawn on the assumption that $f=1$, that is to say, that the atoms scatter as points with unit scattering power. The values of $|F|^2$ at the reciprocal-lattice points are either 16, as at the points (000), (111), (222), (002), (220), or zero, as at the points (001), (110), (112), (221). Contours are drawn at intervals of two units in the values of $|F|^2$.

It is clear from this diagram that there is an appropriate value of $|F|^2$ for every point in the reciprocal-lattice space, and so for every direction of scattering, since to every point (ξ, η, ζ) there corresponds a value of S/λ. It is also plain that the maxima in $|F|^2$ are relatively flat, while, as we have seen, those in I_0 are exceedingly sharp if the crystal contains more than very few unit cells; so that only if the crystal is very minute will it be necessary to consider the values of $|F|^2$ at points other than the reciprocal-lattice points themselves. To obtain the true value of $|F|^2$, we must multiply each value shown in the figure by the appropriate value of $|f|^2$, itself a function of ξ, η, and ζ.

The atomic scattering factor is spherically symmetrical in the reciprocal-lattice space, since it is a function of $|S|/\lambda$, and it diminishes rapidly with increasing distance from the origin.

We have considered a very special case, which gives a distribution of $|F|^2$ itself periodic in a simple way in the reciprocal-lattice space, a result not to be expected unless the atoms occupy special positions in the unit cell; and even in the case we have considered the periodicity is destroyed if we take the factor $|f|^2$ into account. The interference factor I_0, on the other hand, is always periodic, the distribution corresponding to it repeating itself exactly about each reciprocal-lattice point.*

One more point should be considered in connection with the particular case we have discussed. The face-centred cubic lattice is a true space-lattice, and can be based on a rhombohedral primitive unit cell whose translations are the three vectors from the origin to the points $(\frac{1}{2}, 0, 0)$, $(0, \frac{1}{2}, \frac{1}{2})$, and $(\frac{1}{2}, 0, \frac{1}{2})$. The lattice reciprocal to this can easily be shown to be the body-centred cubic lattice, itself a particular case of a rhombohedral lattice, formed by the whole-number points for which $|F(\xi, \eta, \zeta)|^2$, as given in (*10.5*), has values different from zero. This, of course, must be so. To any given lattice there corresponds only one reciprocal

* It is sometimes advantageous to plot as a distribution in the reciprocal space the structure factor F itself, either of the unit cell, or of some well-defined group of atoms, such for example as the benzene ring. Such a plot may be called the Fourier transform of the group of atoms. By plotting the reciprocal net of a crystal of which the group is a constituent, in the correct orientation relative to the transform, it is possible to estimate rapidly the contribution of the group to the structure factors of the spectra represented by the points of the net, and to obtain an idea of the way in which these contributions are changed by changes in orientation of the group. Space does not permit of a detailed discussion of this method, which belongs properly to a volume on technique, although it is also of considerable theoretical importance. For a discussion of Fourier transforms reference may be made to a paper by Ewald (*Proc. Phys. Soc.*, **52,** 167 (1940)); and for accounts of the application of the method to the scattering by anthracene of X-rays and electrons respectively to papers by Knott (*Proc. Phys. Soc.*, **52,** 229 (1940)), and Charlesby, Wilman, and Finch (*ibid.*, **51,** 479 (1939)).

lattice, the nature of which cannot depend on the way in which we choose to regard it as a geometrical convenience. An apparent difficulty arises, however, since $|F|^2$ will be constant throughout the reciprocal-lattice space for a primitive lattice, assuming the atoms to scatter as points. Thus, in determining the intensity distribution, we have now to consider only the variation of the interference function I_0, instead of that of the product of $|F|^2$ and I_0. It must be remembered, however, that the distribution of the interference function about the reciprocal-lattice points will differ according to whether we consider the crystal as based on a cubic lattice, and bounded by cube faces, or as based on a rhombohedral lattice and bounded by rhombohedral faces. If we considered one case, say that of the rectangular block of crystal, from both points of view, we should get the same result, although the summations involved in terms of the rhombohedral lattice would be of considerable complexity. This question has been raised in order to emphasise that the choice of axes, and so the splitting of the intensity function into the factors I_0 and $|F|^2$, is to some extent arbitrary, although a definite choice will often be dictated by the mathematical possibilities of the problem under consideration. We shall consider the manner in which the distribution of the interference function about each reciprocal-lattice point depends on the external form of the crystal fragment in §§ 2(*h*) *et seq*. below.

(*c*) *Scattering by irregular assemblages of crystallites:* Consider now a powder composed of crystal fragments, all of uniform size, but with entirely random orientation. Each particle is in effect optically independent of all the others, and so the total intensity is the sum of the intensities scattered by the individual crystal fragments. We suppose, as before, that the lattice is primitive, and that the atoms scatter as points. An average intensity distribution $I_0(|S|/\lambda)$ may then be obtained by supposing the distribution corresponding to the interference function $I_0(\xi, \eta, \zeta)$ to assume all possible orientations about the origin with equal probability, and finding the average density at each point. In this average distribution, each reciprocal-lattice point gives rise to a sphere about the origin as centre, whose radius is the corresponding value of $|S|/\lambda$. The regions of finite density about the lattice points give rise to spherically symmetrical shells of finite density, with maxima on the spheres described by the lattice-points themselves; and these shells will be sharply defined or diffuse according to whether the number of scattering points in the individual crystallites is large or small. The intersections of these spheres of maximum density with the sphere of reflection drawn for any particular direction of incidence determines the angular distribution of the Debye-Scherrer rings given by the powder, and the sharpness or diffuseness of the maxima determines the sharpness or diffuseness of the corresponding haloes.

For a truly random orientation of the crystal particles, the haloes will show an intensity and breadth symmetrical about the direction of incidence of the radiation; but if there is a tendency for a particular crystal axis to lie parallel to a certain direction, the average distribution $\bar{I}_0(S/\lambda)$ will not be spherically symmetrical, but will have values greater in certain regions of the spherical shells than in others. To take an extreme case, suppose the *a* axes of all the crystallites to lie parallel to the same direction, any orientation of a fragment about this direction being, however, equally probable. Each reciprocal-lattice point then describes a ring, a circle of latitude on the sphere corresponding to random orientation, having the *a* direction as axis. The intersection of these rings with the sphere of reflection now gives not a continuous halo but a series of points. All these points lie in those planes of the reciprocal lattice that are parallel to the axes *b** and *c**, and if the direction of the incident beam is parallel to those planes the diffraction pattern is just that produced by a complete rotation of a single crystal about the *a* axis. Diffraction occurs only in the directions defined by the intersection of the Debye-Scherrer rings with the layer lines. Between the layer lines the ring no longer develops. The sharpness radially of the directions so defined depends on the number of particles in the individual crystallites. Small deviations of the crystal axes from the one definite direction will cause a broadening of the spots along the circumferences of the rings. It is plain that we may have every possible gradation between the case discussed and that of a completely random orientation, and that the method we have just considered is suitable for dealing with problems of preferential orientation as well as with that of the effect of crystal size.

(*d*) *The linear crystallite:* If each crystal fragment is a parallelepiped having N_1, N_2, N_3 points in rows parallel to the edges *a, b, c*, respectively, the interference function is given by

$$I_0(\xi, \eta, \zeta) = \frac{\sin^2 N_1\pi\xi}{\sin^2 \pi\xi} \frac{\sin^2 N_2\pi\eta}{\sin^2 \pi\eta} \frac{\sin^2 N_3\pi\zeta}{\sin^2 \pi\zeta}, \qquad (10.6)$$

as in § 1(*c*) of Chapter I.

Suppose now that the fragments become smaller and smaller in the directions of **b** and **c**, while the number of points parallel to **a** remains large. The regions of finite density in the distribution $I_0(\xi, \eta, \zeta)$ for any one crystallite now become wider in the *b*c** planes, while remaining narrow parallel to **a***. If, finally, each crystallite becomes simply a row of N equally spaced points along the *a* axis, the flat density distributions about each reciprocal-lattice point unite, forming a series of flat sheets coincident with the *b*c** planes of the reciprocal lattice, and perpendicular to the *a* axis. Since now $N_2 = N_3 = 1$, equation (*10*.6) becomes

$$I_0 = \frac{\sin^2 N_1\pi\xi}{\sin^2 \pi\xi}, \qquad (10.7)$$

the interference function being a function of ξ only. The larger N_1, the narrower the density maxima in the neighbourhood of the planes $\xi = h$ (h integral). This density distribution is represented graphically in fig. 187, and is that corresponding to the case of the linear point-diffraction-grating. The parallel lines are the traces of the planes of

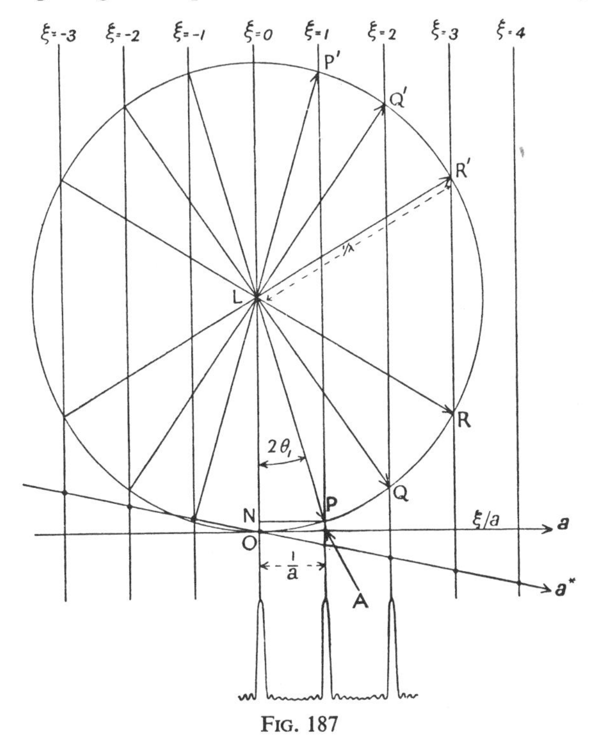

FIG. 187

maximum density, which are perpendicular to the plane of the paper, and the graph at the bottom of the figure illustrates the variation of I_0 as a function of ξ/a when there are only ten points in the row.

If the radiation is incident in a direction perpendicular to **a**, the centre of the sphere of reflection lies at L, at a distance $1/\lambda$ from O, OL being perpendicular to OA. The spacing of the sheets of maximum density is $1/a$. The loci of maximum scattering for the first, second, and third order spectra are the cones formed by joining L to the circles of intersection of the planes of maximum density with the sphere of reflection; and LP, LQ, and LR, lying in the plane containing LO and OA, are generators of these cones. Corresponding cones lie to the left of OL. The linear grating must be supposed to lie at L, and to be parallel to OA.

From the diagram, it is clear that $2\theta_1$, the angle of diffraction for the first order spectrum, is given by

$$\sin 2\theta_1 = NP/LP = OA/OL = 1/a \div 1/\lambda = \lambda/a,$$

which is the ordinary formula for the linear diffraction grating at perpendicular incidence. If distances measured from O parallel to OA are denoted by x, then $x=\xi/a$, and the interference function may be written

$$I_0(x)=\sin^2(N\pi xa)/\sin^2(\pi xa). \qquad (10.8)$$

(*e*) *The transition from the case of the linear crystallite to that of the diatomic molecule :* Let the number of grating elements now be reduced to two. Then

$$I_0(x)=\sin^2(2\pi xa)/\sin^2(\pi xa)=4\cos^2(\pi xa)=2+2\cos(2\pi xa). \qquad (10.9)$$

The maxima in $I_0(x)$ are broad and diffuse, and zero is reached only midway between the reciprocal-lattice planes. The diffraction maxima now correspond to ordinary interference fringes from two point sources, observed at a distance large compared with the distance between them.

Let us now suppose the scattering sample to consist of a very large number of such diatomic scatterers, with every possible orientation equally likely. We have then, formally, the case of the diatomic gas. The appropriate average intensity distribution is obtained by allowing the distribution $I_0(x)$ of equation (*10*.9) to rotate about O into all possible orientations, and then finding the average value at any point.

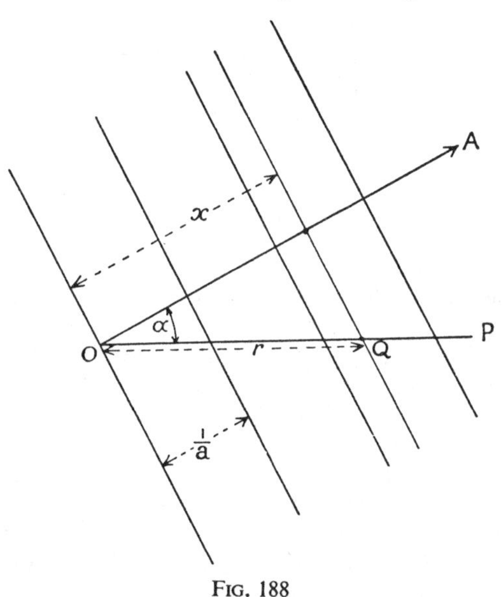

FIG. 188

Let OA, fig. 188, be the a axis of the distribution $I_0(x)$ corresponding to one of the molecules, and let it make an angle α with some line OP, chosen as a direction of reference. If there are a very large number of molecules, n, and if all orientations are equally likely, the number whose x axes lie within a solid angle $d\Omega$ in the neighbourhood of OA is $n\, d\Omega/4\pi$, and the contribution of these to the density at the point Q in OP, distant r from O, is $n\mathrm{I}_0(x)\, d\Omega/4\pi$, where $x = r\cos\alpha$. For a given value of r, $\mathrm{I}_0(x)$ depends only on α, so that in integrating over all solid angles to get the average density at Q we may put $d\Omega = 2\pi\sin\alpha\, d\alpha$, and integrate with respect to α from 0 to π. Thus, by (*10*.9),

$$\bar{\mathrm{I}}_0(r) = \tfrac{1}{2}n\int_0^\pi \{2 + 2\cos(2\pi ar\cos\alpha)\}\sin\alpha\, d\alpha = 2n\left\{1 + \frac{\sin 2\pi ar}{2\pi ar}\right\}. \quad (10.10)$$

The average intensity function $\bar{\mathrm{I}}_0(r)$ is thus spherically symmetrical about O, and the angular distribution of the scattering is axially symmetrical about the direction of the primary beam, and for a given angle of scattering is independent of that direction.

To get the scattered intensity for any value of $(\sin\theta)/\lambda$, all that is necessary is to put $r = |\mathrm{S}|/\lambda = 2(\sin\theta)/\lambda$ in (*10*.10). This gives

$$\bar{\mathrm{I}}_0(\sin\theta/\lambda) = 2n\left\{1 + \frac{\sin\mu a}{\mu a}\right\}, \quad (10.11)$$

where $\mu = 4\pi(\sin\theta)/\lambda$, which is the ordinary formula for scattering by a diatomic molecule consisting of point atoms with unit scattering power at a distance a apart.

In fig. 189, the plain and shaded rings represent diagrammatically a section through the centre of the spherically symmetrical distribution $\bar{\mathrm{I}}_0(r)$. The circle ROR′ is the trace of the sphere of reflection drawn for the direction of incidence LO. The value of $\bar{\mathrm{I}}_0(r)$ at Q gives the intensity scattered in the direction LQ.

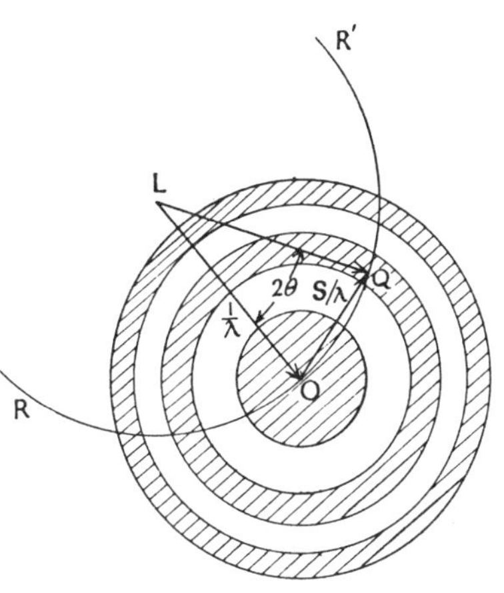

FIG. 189

If the atoms do not scatter as points, (*10*.11) must be multiplied by $|f|^2$, where f is the scattering power of each atom, which is itself a function of $(\sin\theta)/\lambda$, and so of $|\mathrm{S}|/\lambda$. The factor $|f|^2$ can thus also be represented by a spherically symmetrical density distribution in the reciprocal-lattice space, and the actual value of the scattered intensity may therefore be obtained by multiplying together the values of the two distributions at the corresponding point, or, since each distribution is spherically symmetrical, at the corresponding radius. If $|f|^2$ had not been spherically symmetrical, it would have been necessary first to multiply it by I_0 for each point of the distribution, and then to average the product.

We have here considered the diatomic molecule as the limiting case of the crystal lattice. If we carry the process one step further, the distribution I_0 for a single atom becomes simply a uniform distribution in the reciprocal-lattice space, and the controlling distribution is that for $|f|^2$ alone.

(f) The interference function $I_0(\xi, \eta, \zeta)$ expressed as a Fourier series: The interference function $I_0(\xi, \eta, \zeta)$ for a parallelepipedal lattice of scattering points is represented by a certain density distribution in the reciprocal-lattice space, which repeats itself identically about each reciprocal-lattice point. This distribution, like the reciprocal lattice itself, must be supposed to extend to infinity, whatever the number of elements in the crystal lattice, and must therefore be expressible in terms of a suitable triple Fourier series, periodic in ξ, η, and ζ, with period unity. The number of elements in the crystal lattice determines the form of the density distribution about each reciprocal-lattice point, and determines, therefore, the coefficients of the terms in the Fourier series and, as we shall see, their number.

Let us consider first of all the case of the linear grating consisting of a single row of N scattering points, discussed in § 1(*d*) above. The interference function is a function of ξ only, and the corresponding distribution has constant density over any plane ξ = const. By (*10*.7) and (*9*.13), it can be expressed in the equivalent forms

$$I_0(\xi) = \frac{\sin^2 N\pi\xi}{\sin^2 \pi\xi} = N + 2\sum_{n=1}^{n=N-1} (N-n)\cos 2\pi n\xi, \qquad (10.12)$$

which were derived from the same starting-point by different methods, and which are trigonometrically identical, as it is not difficult to show. In the second form of the expression, there is one term in the summation for each possible interatomic distance in the row of points, the corresponding coefficient being the number of times that distance occurs. The expression is thus a Fourier series for $I_0(\xi)$, with a limited number of terms.

As a simple example, we may consider a row of three points, for which (*10*.12) becomes

$$I_0(\xi) = 3 + 4\cos 2\pi\xi + 2\cos 4\pi\xi, \qquad (10.13)$$

a series with three terms only, a constant term, the fundamental, and the first harmonic. Fig. 190 shows graphically the three terms plotted as functions of ξ, together with their sum. The sharpening of the main maxima by the addition of the first harmonic to the fundamental is clearly shown, and also the origin of the subsidiary maxima. Addition of further harmonics, corresponding to an increase in the number of points in the row, will still further sharpen the main maxima, and will increase the number of the subsidiary maxima, and diminish their relative importance, until, when the row contains a very large number of points, appreciable intensity is confined to the immediate neighbourhood of the integral values of ξ.

The intensity distribution corresponding to a row of points may therefore be considered as the sum of a set of plane sinusoidal distributions with wave-lengths equal to the reciprocal of the spacing of the

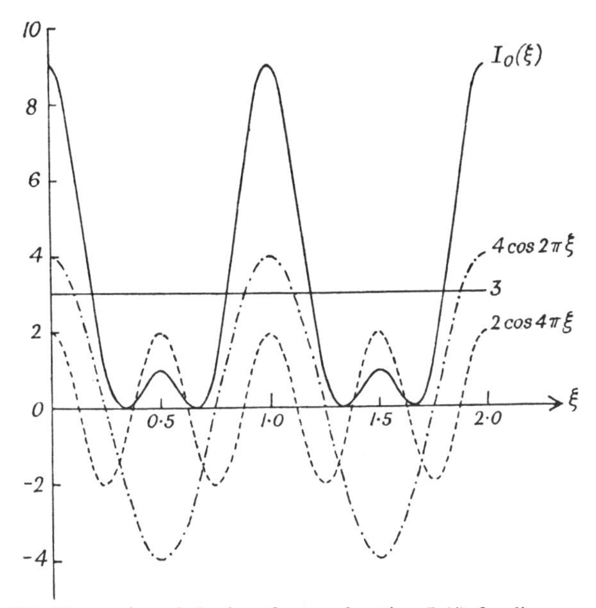

FIG. 190. Harmonics of the interference function $I_0(\xi)$ for linear grating with three elements

points, and to submultiples of this, plus a certain constant positive distribution. The number of waves, including the fundamental, is one less than the number of points in the row. The coefficient of the fundamental is $2(N-1)$, and those of the succeeding harmonics $2(N-2)$, $2(N-3)$ ..., the series terminating when the coefficients become zero.

This line of thought may readily be extended to three dimensions. For a lattice having N_1 points parallel to a, N_2 parallel to b, and N_3 parallel to c, $I_0(\xi, \eta, \zeta)$ is the product of three factors of the type of (*10*.12) and can be written

$$I_0(\xi, \eta, \zeta) = \sum_{-(N-1)}^{N-1}\sum\sum (N_1 - |p|)(N_2 - |q|)(N_3 - |s|) \cos 2\pi p\xi \cos 2\pi q\eta \cos 2\pi s\zeta, \tag{10.14}$$

p, q, and s being integers over which the summations are to be taken, with $N = N_1$, N_2, and N_3 for the summations over p, q, and s respectively.

Making use of the relation $2\cos A \cos B = \cos(A+B) + \cos(A-B)$, we can easily express (*10*.14) in the form

$$I_0(\xi, \eta, \zeta) = \sum_{-(N-1)}^{N-1} \sum \sum (N_1 - |p|)(N_2 - |q|)(N_3 - |s|) \cos 2\pi(p\xi + q\eta + s\zeta), \tag{10.15}$$

From the definition of ξ, η, and ζ, in § 1(*b*) above, we may write

$$\cos 2\pi(p\xi + q\eta + s\zeta) = \cos 2\pi\left(\frac{px^*}{a^*} + \frac{qy^*}{b^*} + \frac{sz^*}{c^*}\right), \tag{10.16}$$

where x^*, y^*, and z^* are now the co-ordinates of any point in the reciprocal-lattice space referred to the reciprocal-lattice axes. The expression (*10*.16) represents a plane sinusoidal distribution, the planes of constant phase being parallel to the planes (p, q, s) of the reciprocal lattice, and the wave-length of the distribution being the spacing of these planes, the term spacing being used in the general sense. Because of the reciprocal relation of the two lattices, the planes (p, q, s) of the reciprocal lattice are perpendicular to the zone axis $[p, q, s]$ of the *crystal* lattice, which is parallel to the line joining the origin to the point (p, q, s) of that lattice.

The lattice may be thought of as consisting of a series of rows of points, all parallel to this zone axis. The spacing of the points in any row is the distance of the point (p, q, s) from the origin, when p, q, and s have no common factor. Successive points in the same row are those obtained by multiplying p, q, and s by 1, 2, 3, ... , and so on. All the points in the lattice lie on such rows, and the distance between any two points is equal to the distance between two points in some such row. To every such distance there corresponds one wave of the type (*10*.16).

Suppose, for example, $p = q = s = 1$. The term corresponding to this in the summation is

$$(N_1 - 1)(N_2 - 1)(N_3 - 1) \cos 2\pi(\xi + \eta + \zeta).$$

The point (1, 1, 1) of the crystal lattice lies in the zone [1, 1, 1], and the distance between two points in this or any parallel row is the length of the diagonal of one of the elementary lattice-cells. There are

$$(N_1 - 1)(N_2 - 1)(N_3 - 1)$$

such cells, and so this particular distance is represented this number of times in the lattice. This number is the coefficient of the wave form $\cos 2\pi(\xi + \eta + \zeta)$, whose planes of constant phase are perpendicular to the diagonals [1, 1, 1] of the cells. The wave-length is the reciprocal of the cell diagonal, that is to say, of the distance between successive points in the diagonal rows. The values $p = q = s = -1$ give the same coefficient, and the same wave form. We may ascribe them to the distances between the same points measured in the opposite direction. Thus the shortest distances parallel to the cell diagonal [1, 1, 1] contribute in all

$$2(N_1 - 1)(N_2 - 1)(N_3 - 1) \cos 2\pi(\xi + \eta + \zeta)$$

to the sum. To $p=q=s=\pm 2$ there corresponds a wave distribution parallel to that for $p=q=s=\pm 1$, but having half the wave-length, and a coefficient $2(N_1-2)(N_2-2)(N_3-2)$, which is the number of times the double spacing occurs in rows parallel to [1, 1, 1] in the lattice, when each occurrence of the spacing is counted once in each direction.

The Fourier waves corresponding to the interatomic distances belonging to any one zone $[p, q, s]$ give rise to an intensity distribution in parallel sheets, whose successive and identical planes of maximum density coincide with the planes (p, q, s) of the reciprocal lattice, and the maxima are the sharper the greater the number of possible interatomic distances parallel to this zone axis. The density in any such distribution is of the type characteristic of the linear grating, although the coefficients decrease according to a different rule. By the criss-crossing of all possible density distributions of this kind, corresponding to all possible zone axes of the crystal lattice, the intensity distribution $I_0(\xi, \eta, \zeta)$ in the reciprocal-lattice space is built up. All the distributions have maxima at the origin.

Formally, there is much similarity between this method of representing the density distribution $I_0(\xi, \eta, \zeta)$ in the reciprocal-lattice space, and that of representing the scattering density, $\rho(x, y, z)$, in the crystal-lattice space, which we discussed in Chapter VII. We saw there that each term in the triple Fourier series for the density corresponds to a point in the reciprocal lattice, the wave-normal of the Fourier wave being in each case in the direction of the vector from the origin to that point, and the wave-length the reciprocal of the magnitude of that vector. The smaller the number of spectra observed, that is to say, the smaller the number of effective reciprocal-lattice points, the more diffuse the maxima representing the lattice-points in the density distribution $\rho(x, y, z)$. In the triple Fourier series representing $I_0(\xi, \eta, \zeta)$, there is one term for each vector in the *crystal* lattice, the wave-normal of the Fourier wave being the direction of that vector, and the wave-length equal to the reciprocal of its magnitude. Sharpness of the maxima in the distribution $I_0(\xi, \eta, \zeta)$ corresponds to sharpness in the directions of the diffraction maxima; and the maxima become more diffuse, and the directions of the scattering maxima less well defined as the number of crystal-lattice points becomes smaller and smaller.

The parallelism between the two cases is not, however, quite complete. If the crystal lattice consists of atoms scattering as points, the coefficients of all the terms in the series for $\rho(x, y, z)$ are equal, that is to say, there is one term of the same weight for each type of distance between points in the reciprocal lattice. In the series for $I_0(\xi, \eta, \zeta)$, each Fourier term is weighted with a coefficient equal to the number of times the corresponding atomic spacing occurs in the crystal lattice.

(g) *Scattering from crystallites in random orientation:* The average intensity distribution corresponding to a random orientation of parallelepipedal crystallites may be obtained directly from equation (*10*.15) by the method of averaging that led to (*10*.10). The term

$$\cos 2\pi(p\xi + q\eta + s\zeta)$$

may be written $\cos 2\pi x l_{pqs}$, where x is a distance measured from the origin along a line perpendicular to the planes $p\xi + q\eta + s\zeta = \text{const.}$, and l_{pqs} is the distance from the origin to the point (p, q, s) of the crystal lattice. On averaging over all possible orientations of the crystal lattice, we find for the average value of the interference function at a distance r from the origin, just as in § 1 (*e*),

$$\bar{\mathrm{I}}_0(r) = \sum_p^{\mathrm{N}-1} \sum_q \sum_s_{-(\mathrm{N}-1)} (\mathrm{N}_1 - |p|)(\mathrm{N}_2 - |q|)(\mathrm{N}_3 - |s|) \frac{\sin 2\pi r l_{pqs}}{2\pi r l_{pqs}}. \quad (10.17)$$

Since $r = |\mathrm{S}|/\lambda, = 2(\sin\theta)/\lambda$, this expression gives the coherent intensity scattered by a random array of crystallites as a function of $(\sin\theta)/\lambda$ in terms of the distances between every possible pair of points in the lattice. The coefficient of the term involving l_{pqs} is equal to the number of times the corresponding interatomic distance occurs in the crystal block. Equation (*10*.17) gives the distribution of intensity in a Debye-Scherrer ring system for a powder composed of uniform parallelepipedal crystallites. It may be compared with equation (*9*.41) for a polyatomic gas, remembering that f has been taken equal to unity for all the atoms of the lattice.

It is interesting to consider an actual case, that of a simple cubic crystallite having $\mathrm{N}_1 = \mathrm{N}_2 = \mathrm{N}_3 = 3$, and therefore only 27 points in all. If a is the cube spacing, and if we write

$$x = 4\pi a(\sin\theta)/\lambda = \mu a, \quad (10.18)$$

we find from (*10*.17) for this case

$$\bar{\mathrm{I}}_0(x) = 27 + 108\frac{\sin x}{x} + 144\frac{\sin\sqrt{2}x}{\sqrt{2}x} + 64\frac{\sin\sqrt{3}x}{\sqrt{3}x} + 54\frac{\sin 2x}{2x}$$
$$+ 144\frac{\sin\sqrt{5}x}{\sqrt{5}x} + 96\frac{\sin\sqrt{6}x}{\sqrt{6}x} + 36\frac{\sin 2\sqrt{2}x}{2\sqrt{2}x} + 48\frac{\sin 3x}{3x}$$
$$+ 8\frac{\sin 2\sqrt{3}x}{2\sqrt{3}x}. \quad (10.19)$$

The behaviour of this expression as a function of x is shown in fig. 191 for values of x up to 17. In the same figure, the positions of the spectra given by an infinite crystal lattice with the same spacing are indicated. It will be seen that although the maxima are broad and in some cases unresolved, the main features of the ring system are already developing, even with these very minute crystals. The relatively large sizes of the maxima 100, 120, 110, and 112, correspond to the relatively large number of occurrences of the smaller interatomic spacings. The broad

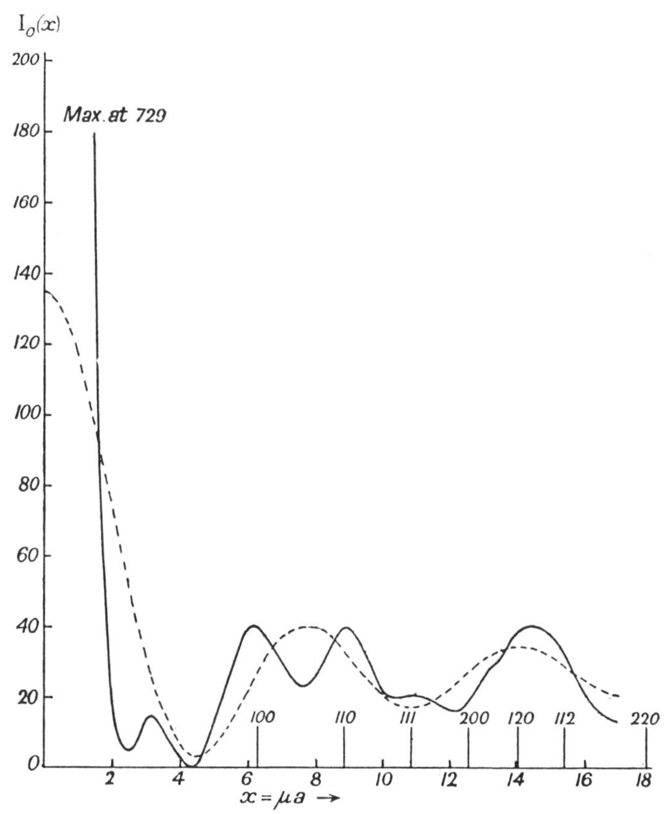

FIG. 191. Radial intensity distribution in a ring system due to random arrangement of simple cubic lattices, each with 27 point atoms. The positions of the spectra for a powder composed of large crystallites based on the same lattice are indicated. The dotted curve shows the contribution to $I_0(x)$ of the constant term, together with that due to the smallest interatomic distance

group of maxima 100, 110, occupies a position approximately the same as that of the main halo due to a liquid composed of atoms in irregular close packing with a closest distance of approach a. In the case of a crystal, the broad halo becomes broken up by interference into a few fine lines, because of the regular arrangement of the atoms, but the powder spectra of simple crystals, such as those of the metals and alloys, tend to show their stronger lines in this same region, because in the crystal, just as in the amorphous liquid, certain small interatomic spacings occur more frequently than any others. In fig. 191, the dotted curve shows the contribution to $\bar{I}_0(x)$ of the constant term together with that due to the smallest interatomic distance a.

In fig. 191, a small maximum will be noticed at a value of x about half that corresponding to 100, indicating at first sight a doubling of the (100) spacing. This peak is due to the subsidiary maxima in the interference function, which lie half-way between the reciprocal-lattice points. It will be seen that as the intensity distribution rotates about the origin, the subsidiary maxima nearest to the origin must produce fairly well-defined spheres in the average distribution. The corresponding spheres produced by the subsidiary maxima at greater distance from the origin will be masked by those due to the main maxima.

As the crystals get larger, the main maxima get sharper and sharper, corresponding to the addition of higher and higher 'harmonics' to the series (*10*.17); but the labour of calculating the intensity curves in this way soon becomes prohibitive, owing to the very rapid growth of the number of terms.

2. The Effect of Crystal Size on the Widths of the Diffraction Spectra

(*a*) *Introductory:* In the last paragraph, we discussed diffraction by exceedingly minute crystals, consisting of only a few units. We have now to deal with the mathematical methods that have been devised for estimating the broadening of the diffraction maxima when the number of units in the crystal is small, but not so small that the methods of the last paragraph are applicable. In Chapter I, § 1 (*c*), we have already discussed briefly the effect of crystal size on the breadth of the interference maxima, and have seen that the angular breadth is of the order of the wave-length used divided by the linear dimensions of the crystal. Broadening of lines in powder photographs due to this cause was first observed by Scherrer,[1] in the case of colloidal gold, and some of his results are illustrated in Plate XII, Vol. I. Other examples of such broadening are shown in Plates XIII and XIV of the same volume. In practice, particles whose linear dimensions are greater than 10^{-4} cm. may be treated as of infinite size for this purpose, the broadening of the lines due to them on account of diffraction being much smaller than the unavoidable breadth due to the finite size of the specimen and the aperture of the camera.

With particles small enough to produce appreciable broadening only average effects can be obtained, and it will be necessary to use a powder in which the crystal particles may assume every possible orientation, but in which the individual particles will in general be uniform neither in size nor in shape, so that although it is possible to derive mathematical expressions for the line-breadth in certain cases, their application remains a matter of some difficulty, and cannot be expected to yield very precise values for the sizes of the particles.

The methods outlined in Section 1 of this chapter are very suitable for dealing with the problem, and indeed reference has already been made to it at several points. The first fairly complete discussion was given by Laue,[2] on whose work subsequent treatments have been based. Scherrer [1] and Seljakov [3] have given formulae for line-breadth for the case of cubic crystals, and a simplified derivation of Laue's expression has been given by Warren.[4] Formulae have also been derived by Patterson.[5] More recently, Stokes and Wilson [6] have shown how to calculate the line-breadth in the case of a powder composed of identical particles without the use of approximation functions to represent the interference function, and future experimental work is likely to be based on their formula, which is of a more general type than the earlier ones.

(*b*) *The breadths of Debye-Scherrer rings:* The case of a crystal powder composed of a large number of identical particles was first discussed in detail by Laue,[2] who considered the general problem in which the incident radiation is neither exactly parallel nor exactly homogeneous, and comes from a finite aperture. We shall here consider only the effect of crystal size—the true diffraction width of the line—and shall assume the incident radiation to be parallel and homogeneous.

Laue defines the breadth of the line as that of a line of uniform intensity, equal to the maximum intensity of the actual line, that would give the same integrated intensity as the actual line. Thus, if $I(\epsilon)$ is the intensity at an angular distance ϵ from the direction of maximum scattering, and $I(m)$ is the maximum intensity of the line, the breadth β is given by

$$\beta = \int I(\epsilon)\, d\epsilon / I(m). \qquad (10.20)$$

The breadth so defined is called the integral line-breadth.

In order to calculate the line-breadth, we use the methods employing the reciprocal lattice, developed in the first section of this chapter. In fig. 192, let R be the reciprocal-lattice point (h, k, l) corresponding to one of the crystal particles. The vector from the origin to the point R is given by

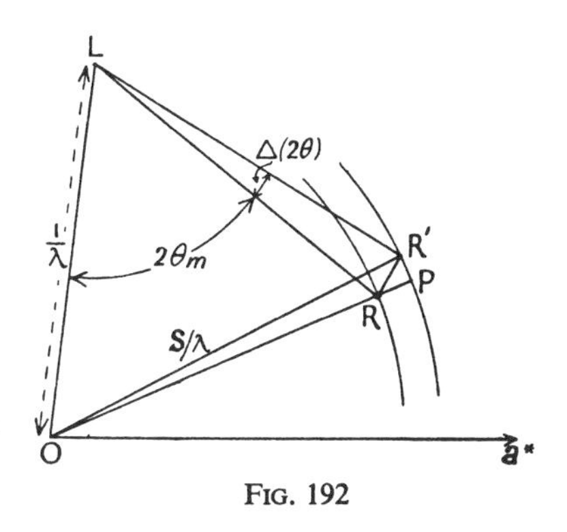

FIG. 192

$$\overrightarrow{OR} = h\mathbf{a}^* + k\mathbf{b}^* + l\mathbf{c}^*. \qquad (10.21)$$

Let R′, in he neighbourhood of R, be the point $(h+\xi, k+\eta, l+\zeta)$ referred to the origin O. Then (ξ, η, ζ) are its co-ordinates referred to R as origin, and expressed as fractions of the reciprocal-lattice translations. The value of the interference function at R′ may then be written $I_0(\xi, \eta, \zeta)$, its distribution about each reciprocal-lattice point being the same.

Let $\overrightarrow{LO}$ be a vector of length $1/\lambda$ in the direction of the incident radiation, L being the centre of the sphere of reflection. If this sphere passes through R, an interference maximum is formed in the direction LR, at an angle of scattering $2\theta_m$, and $\overrightarrow{OR}$ is the vector $\mathbf{S}_m/\lambda$, the magnitude of which is $2(\sin\theta_m)/\lambda$. For a neighbouring direction LR′ the intensity has fallen to a value determined by that of the interference function I_0 at R′, where R′ lies on the sphere of reflection.

We have here to consider the case of a large number of identical crystallites with random orientation. We must therefore suppose the distribution corresponding to the interference function $I_0(\xi, \eta, \zeta)$ to rotate into all possible orientations about O as centre with equal probability, and find its average value at each point. This will give the average interference function $\bar{I}_0$ appropriate to diffraction by the assemblage of crystallites with random orientation. It is a function of $|\mathbf{S}|/\lambda$ only, and will have the same value all over the surface of any sphere having O as centre. The maxima of the average interference function lie on the spheres traced out by the reciprocal-lattice points as the lattice rotates about O, and any such sphere defines by its intersection with the sphere of reflection a cone of maximum scattering, one of the Debye-Scherrer rings given by the powder.

The value of $\bar{I}_0(\rho)$ at a given distance ρ from the origin determines the intensity of radiation diffracted through an angle 2θ, where

$$\rho = 2(\sin\theta)/\lambda = |\mathbf{S}|/\lambda. \qquad (10.22)$$

In the mathematical process of determining the average it is immaterial whether we suppose the distribution I_0 corresponding to a single crystal to rotate about O and determine the average at a fixed point R′, or whether we suppose the vector $\overrightarrow{OR'}$ to assume with equal probability all possible positions in the fixed distribution and determine the average value of I_0 at its extremity. We shall here use the latter method. If ds is a small element of area of the surface of a sphere of radius OR′ in the neighbourhood of R′ in the fixed distribution, the probability that when OR′ rotates at random about O into all possible positions it passes through ds is $ds/4\pi \mathrm{OR}'^2$, and $\bar{I}_0(\rho)$, the average value of the interference function at any point on a sphere of radius $\rho(=\mathrm{OR}')$ is given by

$$\bar{I}_0(\rho) = \frac{1}{4\pi\rho^2}\int I_0(\xi, \eta, \zeta)\,ds. \qquad (10.23)$$

The integrated intensity across a line is then obtained by integrating (10.23) with respect to ρ, and the integral breadth in ρ of the line is given by

$$\beta(\rho) = \int \bar{I}_0(\rho)\,d\rho/\bar{I}_0(m), \qquad (10.24)$$

$\underline{I}_0(m)$ being the maximum value of $\bar{I}_0$ on the sphere of radius OR, the distance from the origin to the reciprocal-lattice point considered.

We have here assumed the lattice-units to scatter as points, so that the intensity is given by the interference function alone. If each unit consists of a number of atoms, it is necessary to multiply the value of I_0 at each point by the corresponding value of the square of the structure factor $|F|^2$, itself a function of ξ, η, ζ. So long as the crystal contains a considerable number of lattice-units, $|F|^2$ varies with the co-ordinates much more slowly than I_0, and may be taken as a factor constant for any given spectrum. The value of F does not then affect the line-breadth, since $|F|^2$ will occur as a factor both in the numerator and denominator of (*10*.24). For very minute crystals this statement would need modification, and the variation of $|F|^2$ with the co-ordinates would have to be taken into account.

In principle, the integration in (*10*.24) is to be taken over the whole surface of the sphere of radius ρ, but in any actual case the integrand has appreciable values only in the immediate neighbourhood of the reciprocal-lattice point. We need therefore consider only positions of R′ very near to R, and to a close enough approximation may suppose R′ to lie not on the surface of the sphere, but on the plane tangent to it at the point P where it is cut by OR. We may therefore assign to R′ the rectangular co-ordinates (x, y, z), x being the distance RP, and y and z rectangular co-ordinates in the tangent plane, referred to P as origin. The element ds then becomes $dy\,dz$, and the value of $\bar{I}_0(\rho)$ for any value of ρ is obtained by integrating (*10*.23) with respect to y and z, x meanwhile remaining constant. In particular, $\bar{I}_0(m)$, the value corresponding to the maximum of the interference line, is obtained by carrying out the integration with x equal to zero.

Up to this point, the argument followed is substantially that used by Laue. In his further treatment, however, he considers the case of a crystal in the form of a rectangular parallelepiped, for which I_0 is known, and in carrying out the integration he uses not the actual interference function, but functions that approximate to it, and are more easily handled mathematically.

Recently, however, it has been shown by Stokes and Wilson [6] that a more exact solution of the problem than that given by Laue is possible. They consider the case of a crystal belonging to the cubic system, and determine the breadth of the line *hkl*. The mathematical work is simplified by referring the crystal to orthorhombic axes such that the spectrum whose indices are *hkl* when referred to the cubic axes has indices *h*′00 referred to the orthorhombic axes. This can always be done for a cubic lattice, and it has the evident advantage that the rectangular axes, x, y, z, relative to which the integration in the reciprocal-lattice space is to be carried out, are now parallel to the orthogonal axes ξ, η, ζ, if these refer to the transformed crystal axes. If **a, b, c,** and $\mathbf{a}^*$, $\mathbf{b}^*$, $\mathbf{c}^*$, are respectively the primitive translations of the orthorhombic lattice and its reciprocal lattice,

$$x = |\mathbf{a}^*|\xi, \quad y = |\mathbf{b}^*|\eta, \quad z = |\mathbf{c}^*|\zeta. \tag{10.25}$$

If $\mathbf{r}_n$ and $\mathbf{r}_{n'}$ are vectors from the origin to corresponding points in the cells n and n' of one of the crystal particles, so that

$$\begin{aligned} \mathbf{r}_n &= n_1\mathbf{a} + n_2\mathbf{b} + n_3\mathbf{c}, \\ \mathbf{r}_{n'} &= n_1'\mathbf{a} + n_2'\mathbf{b} + n_3'\mathbf{c}, \end{aligned} \tag{10.26}$$

the interference function I_0 for the particle is given by

$$\mathrm{I}_0(\mathbf{S}/\lambda) = \sum_n \sum_{n'} e^{-i\kappa \mathbf{S}\cdot\mathbf{r}_n - \mathbf{r}_{n'}}. \tag{10.27}$$

If we put

$$n_1' = n_1 + m_1, \quad n_2' = n_2 + m_2, \quad n_3' = n_3 + m_3, \tag{10.28}$$

and use

$$\mathbf{S}/\lambda = \xi\mathbf{a}^* + \eta\mathbf{b}^* + \zeta\mathbf{c}^*,$$

we obtain

$$\mathrm{I}_0(\xi, \eta, \zeta) = \sum_{n_1} \sum_{n_2} \sum_{n_3} \sum_{m_1} \sum_{m_2} \sum_{m_3} \exp\{2\pi i(m_1\xi + m_2\eta + m_3\zeta)\}. \tag{10.29}$$

The maximum value of the average interference function $\bar{\mathrm{I}}_0(m)$ is obtained by putting $\xi = 0$ in (*10*.29), and then carrying out the integrations with respect to y and z, or, by (*10*.25), with respect to η and ζ, if we first multiply by b^* and c^*. The limits of integration for both η and ζ we take as $\pm\frac{1}{2}$, that is to say, we integrate through a complete reciprocal-lattice cell. The justification for using rectangular axes at all is that I_0 is appreciable only in the immediate vicinity of the reciprocal-lattice points, and the precise limits are therefore unimportant, so long as we confine the integration to one cell. With the limits $\pm\frac{1}{2}$, the integrals with respect to η and ζ vanish unless m_2 and m_3 are zero, in which case both integrations lead to a factor $+1$; so that, by (*10*.23),

$$4\pi\rho_m^2 \bar{\mathrm{I}}_0(m) = b^*c^* \sum_{n_1} \sum_{n_2} \sum_{n_3} \sum_{m_1} 1. \tag{10.30}$$

To obtain the integrated intensity across a line we must integrate (*10*.29) with respect to ξ also, at the same time multiplying by a^*, since $x = \xi a^*$. The same remarks apply as to the limits, and in this case the integral vanishes unless $m_1 = 0$ also, so that

$$4\pi\rho_m^2 \int \bar{\mathrm{I}}_0(\rho)\, d\rho = a^*b^*c^* \sum_{n_1} \sum_{n_2} \sum_{n_3} 1. \tag{10.31}$$

From (*10*.23) it will be seen that, strictly, the integration with respect to ρ involves the factor $1/4\pi\rho^2$ also; but since the range of values of ρ on either side of ρ_m over which the integrand is appreciable is extremely small, this factor has here been taken as constant in the integration.

The triple summation in (*10*.31) is the number of unit cells in a crystal particle, or V/*abc*, V being the volume of the particle. Thus, by (*10*.24), (*10*.30), and (*10*.31), the integral breadth in ρ of the line is given by

$$\beta(\rho) = \mathrm{V}/a^2bc \sum_{n_1}\sum_{n_2}\sum_{n_3}\sum_{m_1} 1 = \mathrm{V/S}, \text{ say.} \qquad (10.32)$$

Consider first the summation with respect to n_1, with n_2, n_3 and m_1 remaining constant. We are then considering a single row of lattice-cells parallel to **a**, and so perpendicular to the planes (h, k, l) referred to the cubic axes. There is one term in the sum over n_1 for every point in this row that is distant m_1a from another point in the same row. If there are in all N points in this row, the lattice distance $|m_1|a$ occurs $\mathrm{N} - |m_1|$ times in it, so that the denominator of (*10*.32) may be written

$$\mathrm{S} = abc \sum_{m_1}\sum_{n_2}\sum_{n_3} (\mathrm{N} - |m_1|)\,a. \qquad (10.33)$$

Let the full line in fig. 193 represent the crystal particle, and AB the row of cells considered, of length $\mathrm{T} = \mathrm{N}a$, and let the dotted line represent the same crystal particle, displaced without rotation through a distance $t = m_1a$ parallel to AB. The length $\mathrm{A'B} = \mathrm{T} - t = (\mathrm{N} - |m_1|)\,a$ of the row is common to the crystal and its displaced 'ghost'. If the volume of the crystal fragment is very much greater than the volume of a unit cell, we may replace the summations in (*10*.33) by integrations. Passing through an area *ds* perpendicular to *a* there are *ds*/*bc* rows of cells parallel to the direction considered, and to a change *dt* in the length of *t* there correspond *dt*/*a* possible values of m_1. Thus, (*10*.33) becomes

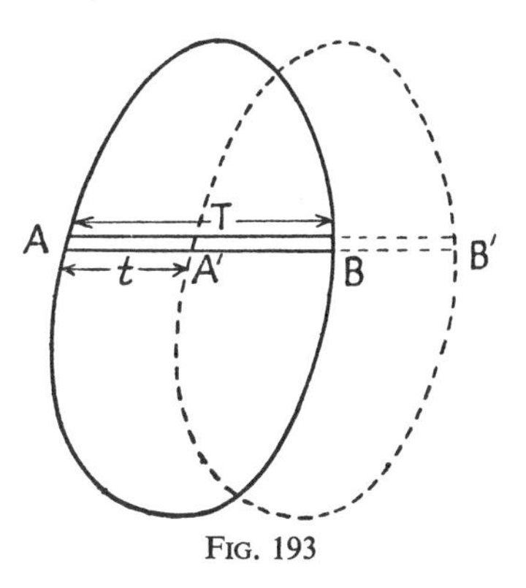

FIG. 193

$$\mathrm{S} = \iint (\mathrm{T} - t)\, ds\, dt, \qquad (10.34)$$

where the integrations extend over the values of the variables for which the crystal and its ghost have any volume in common. In the integrand of (*10*.34), $(\mathrm{T} - t)\,ds$ is that volume of a tube of cross-section *ds*, whose total length is T, which is common to the crystal and its ghost when the displacement is t; and the integral of this with respect to s is the total volume common to the two with this displacement. This common volume we denote by V_t. Then

$$\mathrm{S} = \int \mathrm{V}_t\, dt, \qquad (10.35)$$

the integration being over all values of t, positive or negative, for which crystal and ghost have any volume in common. By (*10*.32), the value of the integral line-breadth in ρ is now

$$\beta(\rho) = \mathrm{V} \Big/ \int \mathrm{V}_t \, dt. \tag{10.36}$$

The result may be expressed somewhat differently. In (*10*.34), let us consider a single tube of cross-section ds, and carry out the integration with respect to t. The limits are $\pm\mathrm{T}$, since outside these values of t there is no overlap between the crystal and its ghost for the given tube. The integration with respect to t yields

$$\mathrm{S} = \int \mathrm{T}^2 \, ds = \int \mathrm{T} \, dv, \tag{10.37}$$

dv being the volume of an elementary tube of the crystal of total length T in the direction of a, perpendicular to the reflecting planes (h, k, l). The corresponding expression for the integral line-breadth is

$$\beta(\rho) = \mathrm{V} \Big/ \int \mathrm{T} \, dv, \tag{10.38}$$

which is the reciprocal of the *volume average* of the thickness of the crystal in the direction considered.

In practice, we measure the angular breadth of the line, $\beta(\epsilon)$, or $\beta(2\theta)$, expressed as an angular range of scattering. By (*10*.22),

$$d\rho = 2 \cos \theta \, d\theta / \lambda,$$

so that $d(2\theta) = (\lambda/\cos\theta) d\rho$, and $\beta(2\theta) = (\lambda/\cos\theta)\beta(\rho)$. We have thus the alternative expressions

$$\beta(2\theta) = \lambda \mathrm{V} / \cos\theta \int \mathrm{T} \, dv, \tag{10.39}$$

or

$$\beta(2\theta) = \lambda \mathrm{V} / \cos\theta \int \mathrm{V}_t \, dt, \tag{10.40}$$

the second of which is probably the more generally useful. These important results are due to Stokes and Wilson.[6]

The quantity $(\int \mathrm{T} \, dv)/\mathrm{V}$, or $(\int \mathrm{V}_t \, dt)/\mathrm{V}$, is the average thickness that applies for the purpose of calculating the line-breadth in the case of a random assemblage of similar particles, all of volume V. It is a weighted mean of the thickness of the crystal perpendicular to the reflecting planes producing the spectrum under consideration, and the first form of the expression shows that the weight to be assigned to any thickness is proportional to the volume of the crystal having this thickness in the direction considered.

(c) *The generality of the expression for the integral line-breadth:* Equations (*10*.39) and (*10*.40) were deduced on the assumption that the crystal lattice was cubic. Their form suggests, however, that they may apply to crystals of any symmetry, as Stokes and Wilson remark in their paper. It would seem that this is true if the crystal particles contain so many units that integration throughout their volume can replace summation over cells, and if the region about a reciprocal-lattice point within which the interference function is appreciable is a small fraction of the reciprocal-lattice cell, conditions assumed to hold in deducing the expressions, and nearly always fulfilled in practice in work of this kind. Laue [13] has shown that, within the limits of approximation stated, the distribution of the interference function about the reciprocal-lattice points depends almost entirely upon the external form of the crystal, and not to any appreciable extent upon the nature of the underlying lattice. The distribution can in fact be obtained from a volume integration taken throughout the space occupied by the crystal, or by suitable transformation, from an integration over the surface of the crystal, or, in the case of a crystal bounded by plane faces, from integrals taken along its edges.

Since the formulae for line-breadth are obtained by suitable integrations of the interference function in the reciprocal-lattice space they depend only on the distribution of this function, and if this does not depend on the type of lattice, neither does the line-breadth. A mathematical discussion of the matter is given in §§ 2(*h*), (*i*), (*j*), and (*k*) below, but a simple example may be considered here by way of illustration.

The distribution of the interference function due to a parallelepiped of an orthorhombic crystal is given by

$$I_0(\xi, \eta, \zeta) = \frac{\sin^2(\pi N_1 \xi)}{\sin^2(\pi \xi)} \frac{\sin^2(\pi N_2 \eta)}{\sin^2(\pi \eta)} \frac{\sin^2(\pi N_3 \zeta)}{\sin^2(\pi \zeta)},$$

or $I_0(X, Y, Z)$

$$= (N_1 N_2 N_3)^2 \frac{\sin^2(\pi N_1 aX)}{N_1^2 \sin^2(\pi aX)} \frac{\sin^2(\pi N_2 bY)}{N_2^2 \sin^2(\pi bY)} \frac{\sin^2(\pi N_3 cZ)}{N_3^2 \sin^2(\pi cZ)}, \qquad (10.41)$$

if X, Y, Z are the actual rectangular co-ordinates of a point in the reciprocal-lattice space, since $\xi = aX$, $\eta = bY$, $\zeta = cZ$. If πaX, πbY, πcZ are so small that the arguments of the sines can be used in the denominator, instead of the sines themselves, (*10*.41) may be written

$$I_0(X, Y, Z) = (N_1 N_2 N_3)^2 \frac{\sin^2(\pi AX)}{(\pi AX)^2} \frac{\sin^2(\pi BY)}{(\pi BY)^2} \frac{\sin^2(\pi CZ)}{(\pi CZ)^2}, \qquad (10.42)$$

A, B, and C being the three edges of the parallelepiped; and this expression, apart from a numerical factor, is the same as that for scattering by a uniform distribution of diffracting matter having the same boundaries as the crystal. The expression (*10*.41) is periodic, and repeats identically about each reciprocal-lattice point, while (*10*.42) is not a periodic function, and represents a distribution about the origin only; but within the limits of approximation stated, the interference function of the crystal may be represented by distributions such as (*10*.42) about each reciprocal-lattice point, which depend only on the dimensions of the crystal and not on the lattice constants *a*, *b*, and *c*.

The regions around the maxima within which (*10*.41) has appreciable values have dimensions of the order $1/N_1a$, $1/N_2b$, $1/N_3c$ in X, Y, and Z respectively. A crystal in which N_1, N_2 and N_3 are as small as 30 will thus amply satisfy the conditions that the sines may be replaced by their arguments in the denominator of (*10*.41). It may be remarked that the planes $I_0=0$ coincide for the two expressions (*10*.41) and (*10*.42).

(*d*) *Calculation of the integral line-breadth in some special cases : the Scherrer constant :* We may write the expression for the line-breadth in the form

$$\beta = K\lambda/L\cos\theta, \tag{10.43}$$

L being some linear dimension of the crystal particle, and K a numerical constant. Let us assume L to be the cube-root of the volume of a single crystal particle; then K, which is sometimes called the Scherrer constant, will in general depend both on the external form of the crystal particles and on the order of the spectrum. By comparing (*10*.40) and (*10*.43), we see that

$$K = VL\Big/\int V_t\,dt = V^{4/3}\Big/\int V_t\,dt = V^{4/3}\Big/\int T\,dv, \tag{10.44}$$

V_t being the volume common to the crystal and its ghost when their displacement is *t*.

If the crystals are spheres, K, by symmetry, will be the same for all orders. Its value is readily calculated in this case by either method. The thickness of a sphere of radius R parallel to any diameter at a distance *r* from its centre is $2\sqrt{R^2-r^2}$. The tubular element of volume to which this thickness corresponds is $2\sqrt{R^2-r^2}\times 2\pi r\,dr$. Thus

$$\int T\,dv = 8\pi\int_0^R r(R^2-r^2)\,dr = 2\pi R^4 = \pi D^4/8,$$

D being the diameter of the sphere. By (*10*.39) therefore,

$$\beta = \lambda V/2\pi R^4\cos\theta = 2\lambda/3R\cos\theta = \lambda/(3D/4)\cos\theta. \tag{10.45}$$

The Scherrer constant K in this case is $(4/3)(\pi/6)^{1/3}$, or 1·0747.

Space will allow us to consider only one more case, that of a cube of crystal with its edges parallel to the axes, the crystal itself having cubic symmetry. Let A be the length of the edge of the cube. When the ghost cube is displaced a distance *t* perpendicular to the planes (*h, k, l*), the volume common to the crystal and its ghost is a parallelepiped with edges $A-|h|t/N$, $A-|k|t/N$, $A-|l|t/N$, where $N^2=h^2+k^2+l^2$.

The common volume vanishes when $t = t' = \mathrm{NA}/|h|$, h being the largest index, so that the limits of integration are $\pm t'$. Thus,

$$\int \mathrm{V}_t\, dt = \int_t^{t'} (\mathrm{A} - |h|t/\mathrm{N})(\mathrm{A} - |k|t/\mathrm{N})(\mathrm{A} - |l|t/\mathrm{N})\, dt$$

$$= \mathrm{A}^4\mathrm{N}(6h^2 - 2|hk| + |kl| - 2|lh|)/6|h|^3. \qquad (10.46)$$

Since in this case $\mathrm{V} = \mathrm{A}^3$,

$$\beta = \mathrm{K}\lambda/\mathrm{A}\cos\theta$$

with $$\mathrm{K} = 6|h|^3/(h^2 + k^2 + l^2)^{1/2}(6h^2 - 2|hk| + |kl| - 2|lh|). \qquad (10.47)$$

In applying this formula, it is to be remembered that h is the largest index numerically. For some calculations for other regular solids, reference may be made to the work of Stokes and Wilson.[6] In Table X. 1, taken from their paper, values of K for several solids are shown for different crystallographic indices.

TABLE X. 1

Calculated values of the Scherrer constant K (*Cubic crystal lattice*)

Reflection	Cube	Tetrahedron	Octahedron	Sphere
100	1·0000	1·3867	1·1006	1·0747
110	1·0607	0·9806	1·0376	1·0747
111	1·1547	1·2009	1·1438	1·0747
210	1·0733	1·2403	1·1075	1·0747
211	1·1527	1·1323	1·1061	1·0747
221	1·1429	1·1556	1·1185	1·0747
310	1·0672	1·3156	1·1138	1·0747

(*e*) *The use of approximation functions : Laue's formula :* Much of the experimental work on line-breadths has been based on a formula due to Laue, who considered the case of a rectangular parallelepiped of crystal. The interference function $\mathrm{I}_0(\xi, \eta, \zeta)$ for such a crystal has been considered in detail in § 1(*c*) of Chapter I, and has the form given in (*10*.41). For small variations of ξ, we may write as before

$$\mathrm{N}_1^2 \sin^2(\pi\mathrm{N}_1\xi)/(\pi\mathrm{N}_1\xi)^2,$$

instead of $\sin^2(\pi\mathrm{N}_1\xi)/\sin^2(\pi\xi)$. Either fraction has appreciable values only when $\xi < 1/\mathrm{N}_1$, so that the conditions for using this approximation are fulfilled for all but the most minute crystals; but even with this approximation the integrations involved in (*10*.23) cannot readily be carried out for a general reciprocal-lattice point (h, k, l), and Laue therefore replaces this approximate expression by others more amenable to mathematical treatment. One such function is $\mathrm{N}^2 e^{-\pi\mathrm{N}^2\xi^2}$, which has the same maximum value, and the same integral with respect to

ξ as $N^2 \sin^2(\pi N\xi)/(\pi N\xi)^2$. The degree of approximation of the two functions is shown in fig. 194. Treating the factors of I_0 containing η and ζ in the same way, we obtain from (*10*.41) and (*10*.23)

$$\bar{I}_0(\rho) = B \int \exp\{-\pi(N_1^2 \xi^2 + N_2^2 \eta^2 + N_3^2 \zeta^2)\}\, ds, \qquad (10.48)$$

where B includes factors the exact nature of which is not relevant to the present discussion.

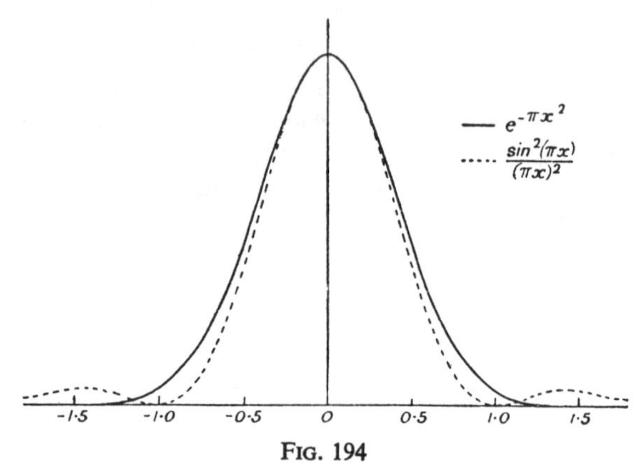

FIG. 194

In order to see the type of error introduced by an approximation of this kind, we consider the case of a cubic crystal, the results for which we may compare with those obtained by Stokes and Wilson, which were discussed in the last paragraph. Suppose the crystal particles to be themselves cubes of edge A $(=Na)$. Equation (*10*.48) then reduces to

$$\bar{I}_0(\rho) = B \int \exp\{-\pi N^2(\xi^2 + \eta^2 + \zeta^2)\}\, ds. \qquad (10.49)$$

If X, Y, Z are the actual co-ordinates of a point in the reciprocal-lattice space, referred to the reciprocal-lattice point R as origin, $aX = \xi$, $aY = \eta$, $aZ = \zeta$ in the case of the cubic crystal, so that

$$\bar{I}_0(\rho) = B \int \exp\{-\pi N^2 a^2(X^2 + Y^2 + Z^2)\}\, ds, \qquad (10.50)$$

It is now plain that the approximation used gives a distribution of I_0 having spherical symmetry about the reciprocal-lattice point R. In performing the integration, we use axes x, y, z with R as origin, such that x coincides with RP (fig. 192), and y and z are parallel to the tangent plane to the sphere of radius OR at the point R. Then, since both sets of axes are rectangular, $X^2 + Y^2 + Z^2 = x^2 + y^2 + z^2$, and $ds = dy\,dz$ to the necessary degree of approximation, just as in § 2(*b*). Taking the limits of integration for y and z as $\pm\infty$, as we may, since the value of the integrand is appreciable only in the immediate neighbourhood of R, and remembering that $\int_{-\infty}^{+\infty} \exp(-\pi u^2)\,du = 1$, we obtain on integration

$$\bar{I}_0(\rho) = (B/N^2 a^2)\exp(-\pi N^2 a^2 x^2). \qquad (10.51)$$

From (*10*.51) we see that $\bar{I}_0(m) = B/N^2 a^2$, and the integrated intensity across a line is given by

$$\int \bar{I}_0(\rho)\,d\rho = \int \bar{I}_0(\rho)\,dx = B/N^3 a^3.$$

Thus

$$\beta(\rho) = \int \bar{I}_0(\rho)\,d\rho / \bar{I}_0(m) = 1/Na = 1/A,$$

and

$$\beta(2\theta) = (\lambda/\cos\theta)\beta(\rho) = \lambda/A\cos\theta, \qquad (10.52)$$

giving a value unity for the Scherrer constant K, the same for all spectra. The result given in equation (*10*.52) was first obtained by Scherrer.[1] It is not in agreement with that obtained in X, § 2(*d*), for we there saw that the Scherrer constant depends on the order of the spectrum, even when the crystal particles have the form of cubes.

It is not difficult to see in what respect the approximation is inadequate, for it assumes at the outset a spherically symmetrical distribution of I_0 about the reciprocal-lattice point. The actual distribution corresponding to a crystal of cubic form is, however, by no means spherically symmetrical. It has been discussed fully in § 1(*c*) of the first chapter, where it was shown that the distribution has its maximum extensions in directions perpendicular to the cube faces, that is to say, along the axes ξ, η, ζ. A distribution of this type is represented diagrammatically in fig. 195. Let OR represent the direction of the reciprocal-lattice vector to the point R. The distribution of the intensity across the corresponding line in the powder photograph is obtained by integrating I_0 across a succession of parallel planes perpendicular to OR. Let SS′ be the trace of such a plane. Evidently the distribution will vary according to the direction of OR relative to the axes of extension of the interference function, that is to say, according to the spectrum. In the case of 100, for example, four out of the six directions of maximum extension lie in the plane

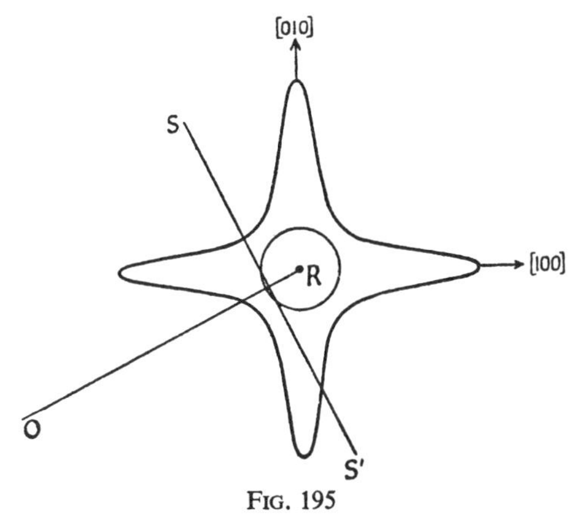

FIG. 195

$x=0$, passing through R, and so contribute mainly to the region of maximum intensity of the line. In the case of 111, on the other hand, the axis of x is equally inclined to all six directions of extension, and these make a greater contribution to the outer parts of the line, which is therefore broadened. We should infer from these simple considerations that for cubic particles the Scherrer constant would be greatest for 111 and least for 100, which is in agreement with the calculations of Stokes and Wilson. If the distribution of I_0 is spherically symmetrical about the reciprocal-lattice point, its intersection by the sphere of integration, or rather by the plane used to represent a limited portion of this sphere, depends only on the distance of the plane from the reciprocal-lattice point, and not on its direction, so that the distribution of intensity in the lines, expressed in terms of ρ, is the same for all spectra.

Laue applies the same method to the case of a powder composed of parallelepipedal blocks of an orthorhombic crystal, and obtains for the average effective thickness, ϵ, defined by the equation

$$\beta(2\theta)=\lambda/\epsilon \cos\theta, \tag{10.53}$$

$$\epsilon=\sqrt{\frac{(h/a)^2+(k/b)^2+(l/c)^2}{(h/N_1 a^2)^2+(k/N_2 b^2)^2+(l/N_3 c^2)^2}} \tag{10.54}$$

for the spectrum *hkl*. In this more general case again, the characteristic extensions of the interference function perpendicular to the faces of the crystal block have been removed by the approximation used, for the surfaces of constant I_0 are now ellipsoids. It is plain that such approximations will always tend to even out the differences between the breadths of lines of different orders, and they can in any case be applied only to crystals having simple shapes. Equations (*10*.39) and (*10*.40), on the other hand, apply to crystals of any shape, provided only that all the crystal particles are identical, and that their orientation is random.

(*f*) *Experimental application of the formulae for line-breadth:* Powders consisting of crystalline particles of identical shape and size, and of regular external form, such as were assumed in deriving the formulae of the few last paragraphs, do not exist, and it is not therefore possible to obtain more than an approximate idea of the average magnitude of the particles, or of their shape, from measurements of line-breadth. Before any such estimates can be made, it is necessary to determine the true diffraction width of the lines, which is not, of course, the width as measured on the photographic film. A powder consisting of particles of 'infinite' size ($t>10^{-4}$ cm.) will always give lines of finite breadth, much greater than the true diffraction breadth. The observed breadth of the lines given by such particles depends on the geometry of the apparatus used—the width of the slit, the divergence of the incident beam, and the amount of absorption of the radiation in the specimen, all of which were neglected in deriving the formulae of the last paragraph. In general too, the radiation will not be truly homogeneous. If, for example, Kα radiation is used, the two components of the doublet, α_1 and α_2, will generally be imperfectly resolved, which must, of course, increase the apparent breadth of the lines. The first step in any experimental determination is therefore to deduce from the observed width of the lines the true broadening due to diffraction.

Some workers have attempted to design their cameras in such a way as to minimise the effect of absorption and to take it and the other geometrical factors into account by calculation. Brill and Pelzer,[7,8] for example, used a powder specimen in the form of a very thin hollow cylinder. Such a cylinder will give lines with a double maximum, whose degree of resolution depends on the particle size. Assumptions have to be made about the shape of the lines given by an infinite particle, which, of course, depends on the geometry of the apparatus. The details of the experimental arrangements belong more properly to a volume on technique, but reference may be made to a review by Cameron and Patterson,[9] in which the relative merits of the different methods are discussed.

A more recent application of the Laue formula has been made by F. W. Jones,[10] whose method appears to have considerable advantages. He allows experimentally for the unavoidable breadth of the lines produced by the geometrical conditions by mixing with the powder to be investigated a second powder composed of particles large enough to be considered as infinite. The lines produced by these particles have, of course, a finite breadth, which serves as a measure of the broadening produced by causes other than pure diffraction, and which may be compared with the width of the lines produced by the very small particles. We shall call such lines *s*-lines, and the lines whose breadth is to be investigated *m*-lines, in accordance with the notation used by Jones.

Let the angular distribution in an *s*-line be given by $I_s(\max)f(x)$, x being the actual distance measured along the film from the middle of the line. By (*10*.20), b, the breadth of the line, is given by

$$b = \int f(x)\,dx. \qquad (10.55)$$

The distribution of intensity due to broadening by diffraction alone is $I_d(x) = I_d(\max)F(kx)$, k being a parameter depending on the size of the particles. The true diffraction width, β, of a line is then given by

$$\beta = \int F(kx)\,dx. \qquad (10.56)$$

In fig. 196, let the continuous curve (c) be the curve $I_s(x)$, supposed unaffected by diffraction. To produce an m-line at the same angle, we must suppose each element dx of the curve (c) to be broadened by diffraction to produce a curve such as (d), shown by the dotted line.

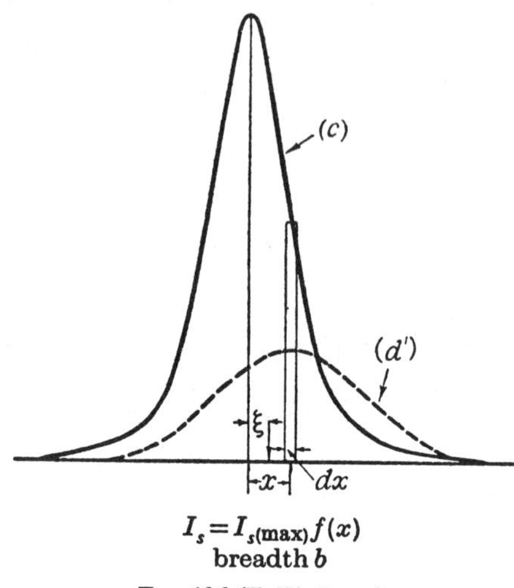

FIG. 196 (F. W. Jones)
(*Proc. Roy. Soc.*, **A 166**, 16 (1938))

The breadth of this curve is equal to β, and its area is the same as that of the element dx which is $I_s(\max)f(x)\,dx$. Thus, if I is the maximum value of the curve (d) corresponding to the element dx,

$$\beta I = I_s(\max) f(x)\,dx. \qquad (10.57)$$

The resultant observed m-line is built up by superposing a large number of such curves, one for each element of the s-line. $I(\xi)$, the intensity contributed by the element dx to the resultant m-line at a point distant ξ from its mid-point is $IF(kx - k\xi)$, and so $I_m(\xi)$, the total intensity at the point ξ in the m-line is given by

$$I_m(\xi) = \int IF(kx - k\xi)\,dx = I_s(\max)\int F(kx - k\xi) f(x)\,dx/\beta, \qquad (10.58)$$

by (*10*.57).

The maximum intensity of the m-line occurs at $\xi = 0$, and is obtained by putting $\xi = 0$ in (*10*.58), while the integrated intensity of the line is obtained by integrating (*10*.58) with respect to ξ. The observed breadth of the m-line, denoted by B, which is equal to the ratio of the integrated intensity to the maximum intensity, is therefore given by

$$B = \iint F(kx - k\xi) f(x)\,dx\,d\xi \Big/ \int F(kx) f(x)\,dx$$

$$= \int F(kx)\,dx \int f(x)\,dx \Big/ \int F(kx) f(x)\,dx, \qquad (10.59)$$

since $$\int_{-\infty}^{+\infty} F(kx-k\xi)\,d\xi = \int_{-\infty}^{+\infty} F(k\xi)\,d\xi = \int_{-\infty}^{+\infty} F(kx)\,dx.$$

By (*10*.55), (*10*.56), and (*10*.59), therefore,

$$\beta/B = \int F(kx) f(x)\,dx \Big/ \int f(x)\,dx, \tag{10.60}$$

$$b/B = \int F(kx) f(x)\,dx \Big/ \int F(kx)\,dx. \tag{10.61}$$

In applying these results practically, the curve $f(x)$ is determined experimentally from a microphotometer curve of an s-line at an angle at which the α-doublet is resolved, and a hypothetical distribution, such as $e^{-k^2x^2}$, $1/(1+k^2x^2)$, or $\sin^2(kx)/(kx)^2$, is assumed for $F(kx)$. The integrations of (*10*.60) and (*10*.61) are carried out numerically for a number of assumed values of k, and a curve is drawn connecting β/B and b/B. Curves of this type obtained by Jones are shown in fig. 197. The fraction β/B approaches zero for 'infinite' particles (observed breadth $B \gg$ diffraction breadth β), and unity for vanishingly small particles (whole effective broadening due to diffraction); while b/B approaches zero for vanishingly small particles, and unity for large particles.

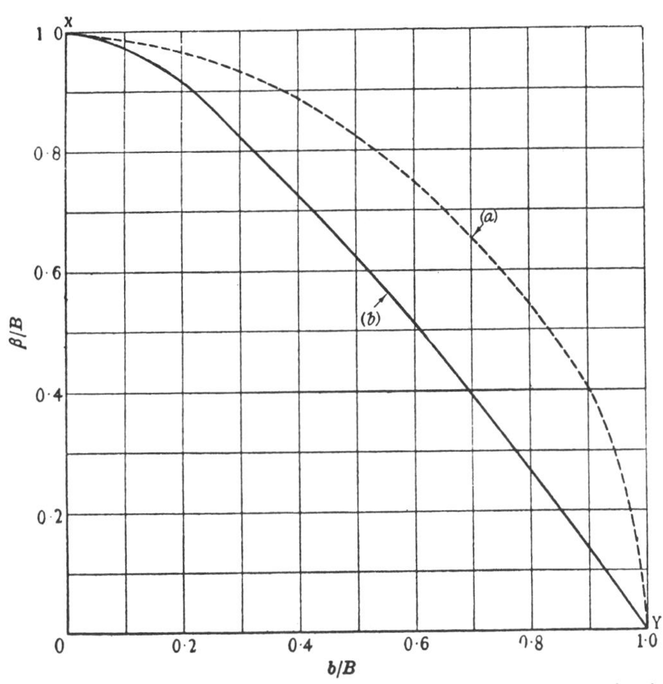

FIG. 197. Correction curve for experimental conditions in determination of particle sizes (F. W. Jones)
(*Proc. Roy. Soc.*, **A 166**, 16 (1938))

The fraction b/B, the ratio of the observed breadth of an s-line to that of an m-line, can be determined experimentally, and from the curve the corresponding value of β/B, and so of β, can be found. The values of b and B must, of course, be determined for the same angle of scattering, but this can easily be done by interpolating between values of b for s-lines on either side of the m-line under investigation. Once the value of β has been determined, Laue's theory may be applied.

Curve (a), fig. 197, was obtained using $\mathrm{F}(kx)=e^{-k^2x^2}$; and

$$\mathrm{F}(kx)=\sin^2(kx)/(kx)^2$$

gives a curve that is almost identical. Curve (b) was obtained with $\mathrm{F}(kx)=1/(1+k^2x^2)$. The form of the correction curve thus depends considerably on the type of distribution assumed. If the particles are of regular form and identical in size, curve (a) has the best theoretical foundation, but if the particles have a range in size, a distribution such as $1/(1+k^2x^2)$, which gives a sharper peak and a broader base, may well be in better accord with reality. So far as possible, the experimental arrangements should be such that small values of b/B can be used, since the form assigned to $\mathrm{F}(kx)$ then matters least.

Correction has also to be made for the presence of the α-doublet, but for an account of this, reference may be made to the original paper.

Some values obtained by Jones for colloidal gold are shown in Table X. 2. The breadths are given in millimetres, as measured on the film, the radius of the camera being 45·3 mm. The values of L are in A-units.

TABLE X. 2. COLLOIDAL GOLD

Indices	B	b/B	β	L(A)	L calc.
111	0·45	0·58	0·24	310	315
200	0·59	0·42	0·42	180	212
220	0·54	0·48	0·35	240	232
311	0·56	0·48	0·36	250	229
222	0·49	0·55	0·28	340	315
331	0·75	0·41	0·54	230	—
422	1·22	0·28	1·02	180	—

The large variation in L with index shows that the particles of colloidal gold were not cubes. By applying Laue's formula, Jones found that his results could be explained by supposing them to be elongated in a direction perpendicular to the (111) planes, their length in that direction being about 700A, and their dimensions in perpendicular directions about 180A. The values of L calculated on this assumption are shown in the last column of the table.

The effect of a distribution in size of the particles is difficult to estimate. It has been discussed by Patterson,[11] and also by Jones (*loc. cit.*) who considers that in general the observed mean size will be greater than the true mean size, but that the difference between them is unlikely to exceed 30 per cent of the observed mean value.

The formulae of the last two paragraphs have been derived on the assumption that each lattice-unit scatters as a single point, so that only the interference function I_0 and not the structure factor of the unit cell need be considered in determining the intensity distribution $I(\xi, \eta, \zeta)$. The formulae apply equally well if the structure factor may be assumed to have a definite value characteristic of each spectrum, constant so long as that spectrum alone is considered, the assumption always made in structure analysis. We have already seen in § 1(*b*) of this chapter that while this assumption is, in principle, incorrect, it is nevertheless fully justified in practice, unless the crystal is exceedingly minute. For crystals containing very few unit cells, it might, however, be necessary to take into account the variation of the structure factor as a function of ξ, η, and ζ in the region of the reciprocal-lattice point in deciding on the right type of distribution function $F(kx)$ to assume. For cryttals of simple structures, the general result of such a correction would be so sharpen the interference maxima, and neglect of it would therefore tend to lead to an overestimate of the size of the crystal particles. The effect would, however, certainly be inappreciable except for crystals containing very few unit cells.

(*g*) *Some examples of the diffraction of electrons by very small single crystals:* It is not possible to observe the effects characteristic of the diffraction of X-rays by very small single crystals, for the intensities involved are far too small. A very small number of atomic planes are, however, enough to reflect a considerable fraction of an incident electron beam, and effects due to electron diffraction by small single crystals are therefore not uncommon. We shall consider briefly some of the possibilities.

(i) *Two-dimensional patterns:* An array of spots on a network, which has the appearance of a diffraction pattern produced by a two-dimensional grating, is frequently observed. Some examples are shown in Vol. I, Plates XXVII, and XXVIII (B), and possible explanations are there discussed (p. 261). The reciprocal-lattice methods we have considered in this chapter are very suitable for dealing with this, and with other problems in electron diffraction.

We consider, for simplicity, the case of a very thin sheet of a crystal, based on a cubic lattice, upon which monochromatic radiation falls normally, parallel to the *a* axis. Fig. 198 represents a section of the reciprocal lattice, through the origin and perpendicular to [010]. Let L be the centre of the sphere of reflection for the case considered. For 30,000-volt electrons the wave-length is about 1/10 of that for Mo Kα radiation, so that the sphere of reflection has a radius 10 times greater than that for the corresponding problem with X-rays. The angles of diffraction for corresponding spectra are therefore much smaller, and within the region of the reciprocal lattice corresponding to quite a number of spectra the sphere of reflection departs only slightly from the reciprocal-lattice plane through the origin and perpendicular to [100], of which the trace in fig. 198 is the line OB. Since the crystal is very thin in the *a* direction, the region of appreciable intensity distribution

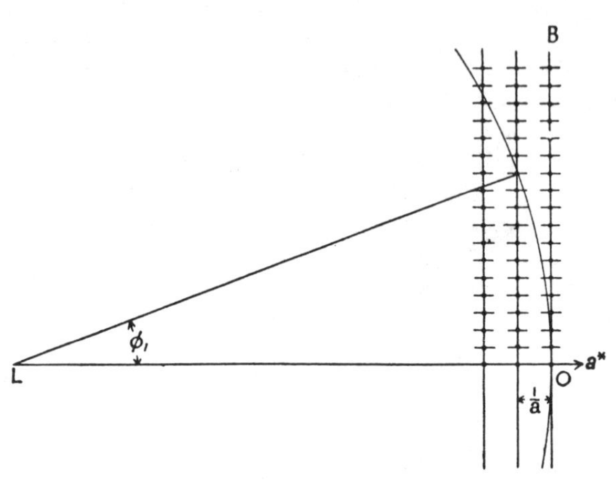

FIG. 198

surrounding each reciprocal-lattice point will be elongated parallel to **a***. We represent these regions by short straight lines in the diagram. It is clear that the sphere of reflection intersects a number of these lines in the neighbourhood of O, and that a network of spectra, corresponding to a network of points on the reciprocal-lattice plane (100) will be produced.

As the angle of diffraction ϕ increases, the sphere will no longer intersect the short lines, and for such angles no spectra will be observed. It is evident from the diagram that they may, however, reappear at angles near those at which the sphere cuts the successive reciprocal-lattice planes. The sphere cuts these planes in circles, so that one might expect to see, surrounding the central region in which the two-dimensional pattern appears, a series of circular regions in which spots may also appear, the angular breadth of each successive ring being narrower than that of its predecessor. Such rings are well shown in Plate XXVIII (B), Vol. I.

If ϕ_1 is the angle at which the first ring should appear outside the central region, it is clear from the figure that

$$\cos\phi_1 = (1/\lambda - a^*) \div 1/\lambda = 1 - \lambda a^* = 1 - \lambda/a. \qquad (10.62)$$

If all the angles considered are small, we may write $\cos\phi_1 = 1 - \frac{1}{2}\phi_1^2$, so that $\phi_1 = \sqrt{2\lambda/a}$; and for the nth ring, $\phi_n = \sqrt{2n\lambda/a}$.

The intensity distribution becomes zero at a distance a^*/N, or $1/t$ from the plane OB, N being the number of planes in the thickness of the crystal and t the thickness itself. The angular width of the central maximum will therefore be $2\sqrt{2\lambda/t}$, or $\sqrt{8\lambda/t}$, a result given by W. L. Bragg.[12] (cf. Vol. I, p. 261). For the rings to be visible, the crystal must be thin, but not too thin, so that there is a definite region in the reciprocal-lattice space in which no intersections of the sphere of reflection with a region of appreciable intensity distribution can occur.

The rings occur in those directions in which there is agreement in phase between the waves scattered by successive points in rows parallel to the direction in which the primary beam is travelling, and they are closely analogous to the Fabry-Pérot rings produced with light in parallel half-silvered plates.

As was pointed out in Vol. I, there are probably several causes conspiring to produce the two-dimensional pattern. The possibility of its occurrence will be increased by slight variations of the angle of incidence, by lack of exact homogeneity in the wave-length of the radiation, or by distortion of the lattice-planes, the effects of all of which may be followed by considering the reciprocal-lattice diagram.

(ii) *Effects due to subsidiary maxima:* The possibility that some of the forbidden spots and rings observed in diffraction patterns obtained with electrons may be due to the subsidiary maxima in the interference function ought not to be lost sight of. Let us consider, for example, the case of a rectangular slab of crystal containing only three planes perpendicular to [100], and a considerable number in other directions. The relevant factor in the interference function will be

$$\sin^2(3\pi\xi)/\sin^2(\pi\xi),$$

which will have the value 9 for the reciprocal-lattice points, and will reach zero for $\xi = \pm 1/3$: but it will have subsidiary maxima equal to unity for $\xi = \pm 1/2, 3/2, \ldots$, that is to say, half-way between the

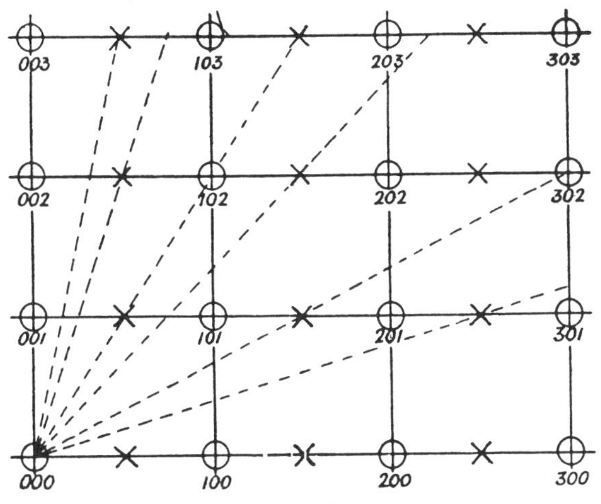

FIG. 199

reciprocal-lattice points in the a^* direction. These subsidiary maxima have therefore an intensity 1/9 that of the main maxima, and are by no means negligible. In fig. 199, a section of the reciprocal lattice through the origin and perpendicular to [010] is shown. The circles represent the positions of the main maxima, and the crosses those of the subsidiary maxima. It is plain from the diagram that apparent doublings of spacings (spectra at half angles) occur for 100, 102, ... 302, ... 104, 304, ... and so on; and in general all planes of type $2m+1$, $2n$, $2p$ will show apparently doubled spacings if the subsidiary maxima are mistaken for main maxima. Abnormal spectra must not, of course, be ascribed to this cause unless the general appearance of the diffraction pattern is consistent with the existence of very small crystals, but the possibility of this and of similar effects with crystals of other dimensions must not be forgotten.

(*h*) *The effect of the external form of the crystal on the interference function:* The shape of the region about the reciprocal-lattice points within which the interference function has appreciable values depends on the external form of the crystal. We have already studied this dependence in the case of the parallelepipedal crystal, for which it can be worked out exactly, but we shall now consider the general case, since, apart from its interest as an optical problem, it has some application to electron diffraction, and to the problem of the dependence of line-breadth on crystal size.

We consider the amplitude scattered from a certain volume of a crystal lattice, bounded by a closed surface of any form. Let $\rho(\mathbf{r})$ be the density of scattering material in the crystal at a point at a vector distance $\mathbf{r}$ from the origin O of the lattice. Then the amplitude scattered by the volume considered, under the conditions defined by the vector $\mathbf{S}$, may be written

$$A = \int \rho(\mathbf{r})\, e^{i\kappa \mathbf{S}\cdot\mathbf{r}}\, d\tau, \qquad (10.63)$$

$d\tau$ being an element of volume of the crystal, and the integration being throughout the whole volume considered. The amplitude A is a function of S, and so of the co-ordinates (ξ, η, ζ) of a point in the reciprocal-lattice space. $|A|^2$ is indeed closely related to the interference function I_0 for the piece of crystal, and is equal to it for a primitive point lattice. Evaluation of the integral will give the required distribution.

By equation (7.14) of Chapter VII, the density $\rho(\mathbf{r})$ at any point in the crystal may be expressed as a Fourier series in the form

$$\rho(\mathbf{r}) = \frac{1}{V} \sum_m F(m)\, e^{-2\pi i(\mathbf{r}\cdot\mathbf{r}_m^*)}, \qquad (10.64)$$

$\mathbf{r}_m^*$ being the reciprocal-lattice vector to the point m (hkl), $F(m)$ the structure factor for the corresponding spectrum, and V the volume of the unit cell of the crystal lattice. Now S/λ is the vector from the origin to the point (ξ, η, ζ) of the reciprocal-lattice space, corresponding to which the interference function is required, and so the vector $\boldsymbol{\rho}_m$, defined by

$$\boldsymbol{\rho}_m = \mathrm{S}/\lambda - \mathbf{r}^*_m, \tag{10.65}$$

is the vector distance of the point (ξ, η, ζ) from the reciprocal-lattice point (m). Substituting (*10*.64) in (*10*.63), and using (*10*.65), we obtain

$$\mathrm{A} = \frac{1}{\mathrm{V}} \sum_m \mathrm{F}(m) \int e^{2\pi i(\boldsymbol{\rho}_m \cdot \mathbf{r})}\, d\tau. \tag{10.66}$$

Unless $|\boldsymbol{\rho}_m|$ is quite small, the integrand in (*10*.66) is a rapidly oscillating function, and the value of the integral is exceedingly small. We may therefore, to a good approximation, regard each term of (*10*.66) as giving the *amplitude* distribution about one point of the reciprocal lattice, the distribution about any one such point having a negligible value in the region of any of the other points. The error made by doing this is small and can be estimated for certain cases. Thus, if $\mathrm{E}(\boldsymbol{\rho})$ is the amplitude distribution about any one lattice-point, (m),

$$\mathrm{E}(\boldsymbol{\rho}) = \frac{\mathrm{F}(m)}{\mathrm{V}} \int e^{2\pi i(\boldsymbol{\rho} \cdot \mathbf{r})}\, d\tau. \tag{10.67}$$

For a point lattice, the true amplitude distribution, represented by the whole series (*10*.66) with $\mathrm{F}(m)$ constant for all spectra, is periodic, having the same distribution about each reciprocal-lattice point. $\mathrm{E}(\boldsymbol{\rho})$, given by (*10*.67), is not periodic ; its value is appreciable only about *one* lattice-point.

We can easily integrate (*10*.67) directly if the crystal has the parallelepipedal form discussed previously. If (u, v, w) are the co-ordinates of the extremity of the vector $\mathbf{r}$ in terms of the lattice translations,

$$\boldsymbol{\rho} \cdot \mathbf{r} = \xi u + \eta v + \zeta w.$$

The element $d\tau$ may be written $\mathrm{V}\, du\, dv\, dw$. We put $\mathrm{F} = 1$ for the point lattice. Then

$$\mathrm{E}(\boldsymbol{\rho}) = \int_0^{\mathrm{N}_1} \int_0^{\mathrm{N}_2} \int_0^{\mathrm{N}_3} e^{2\pi i(\xi u + \eta v + \zeta w)}\, du\, dv\, dw. \tag{10.68}$$

The integral may be written as the product of three independent integrals, the value of the first of which is $(e^{2\pi i \mathrm{N}_1 \xi} - 1)/2\pi i \xi$, and the modulus of this is $\sin(\pi \mathrm{N}_1 \xi)/\pi\xi$. The value of $|\mathrm{E}|^2$ is thus the same approximation to the true interference function that we have already used in § 2(*c*), in discussing problems of line-breadth. We may therefore write

$$\mathrm{I}_0(\xi, \eta, \zeta) = \left(\frac{\pi\xi}{\sin \pi\xi} \cdot \frac{\pi\eta}{\sin \pi\eta} \cdot \frac{\pi\zeta}{\sin \pi\zeta}\right)^2 |\mathrm{E}(\boldsymbol{\rho})|^2. \tag{10.69}$$

$|E|^2$ and I_0 differ by a factor which is unity at the lattice-points themselves, and which varies with increasing $|\boldsymbol{\rho}|$ much more slowly than $E(\boldsymbol{\rho})$ itself. For most purposes, therefore, $|E(\boldsymbol{\rho})|^2$ can be used instead of the interference function in discussing the intensity distributions about a reciprocal-lattice point. Following Laue,[13] we shall call $E(\boldsymbol{\rho})$ the crystal-form factor. It will be seen that, apart from a numerical factor, it represents the amplitude scattered under the given conditions by a *uniform* volume distribution of scattering matter, having the same boundaries as the crystal, and does not therefore depend on the type of lattice upon which the crystal is built.

(*i*) *The crystal-form factor expressed as a surface integral :* Laue [13] has shown that the integral (*10*.67) may be transformed into an integral taken over the external surface of the crystal.

Let (x, y, z) and (X, Y, Z) be the components of $\mathbf{r}$ and $\boldsymbol{\rho}$ respectively, referred to a set of rectangular co-ordinates. Then

$$\boldsymbol{\rho}\cdot\mathbf{r} = Xx + Yy + Zz$$

and (*10*.67) becomes

$$E(\boldsymbol{\rho}) = \frac{1}{V}\int e^{2\pi i(Xx+Yy+Zz)}\,dx\,dy\,dz. \qquad (10.70)$$

We now put

$$\phi = e^{2\pi i(Xx+Yy+Zz)}, \qquad (10.71)$$

which satisfies the differential equation

$$\nabla^2\phi + 4\pi^2\rho^2\phi = 0. \qquad (10.72)$$

Substituting the value of ϕ given by (*10*.72) in (*10*.70), we obtain

$$E(\boldsymbol{\rho}) = -\frac{1}{4\pi^2 V\rho^2}\int \nabla^2\phi\,dx\,dy\,dz,$$

which, by Green's theorem, transforms at once to the surface integral

$$E(\boldsymbol{\rho}) = \frac{1}{4\pi^2 V\rho^2}\int \frac{\partial\phi}{\partial n}\,d\sigma, \qquad (10.73)$$

in which $\partial/\partial n$ represents differentiation along the normal to the surface element $d\sigma$, drawn *inward* into the volume occupied by the crystal, the integration being taken over the whole external surface. If l, m, n are the direction cosines of this normal, we find from (*10*.71)

$$\frac{\partial\phi}{\partial n} = 2\pi i(lX + mY + nZ)\phi$$
$$= 2\pi i\rho_n\phi,$$

ρ_n being the component of $\boldsymbol{\rho}$ parallel to the inward drawn normal. Thus

$$E(\boldsymbol{\rho}) = \frac{i}{2\pi V\rho^2}\int \rho_n\, e^{2\pi i(\boldsymbol{\rho}\cdot\mathbf{r})}\,d\sigma. \qquad (10.74)$$

The integral takes a simpler form if the crystal is bounded by plane faces. For the integration over such a face, F, fig. 200, ρ_n is constant, and the extremity of $\mathbf{r}$ moves in the face. Let $\mathbf{r}_p$ be the vector from the

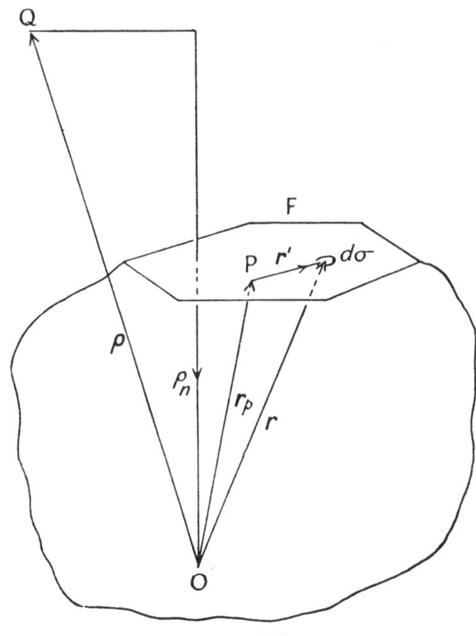

FIG. 200

origin O to any fixed point P in the face F, and $\mathbf{r}'$ the vector in the face from P to the extremity of $\mathbf{r}$, so that $\mathbf{r}=\mathbf{r}_p+\mathbf{r}'$. Then the contribution of this face to $E(\boldsymbol{\rho})$ may be written

$$E_F(\boldsymbol{\rho})=\frac{i\rho_n}{2\pi V\rho^2}e^{2\pi i(\mathbf{r}_p\cdot\boldsymbol{\rho})}\int_F e^{2\pi i(\boldsymbol{\rho}\cdot\mathbf{r}')}\,d\sigma. \tag{10.75}$$

Each face by which the crystal is bounded gives such a contribution.

We can infer certain properties of $E(\boldsymbol{\rho})$ without actually evaluating the integral for any particular case.

(i) The *intensity* distribution is symmetrical about each reciprocal-lattice point, whatever the geometrical symmetry of the crystal fragment. For changing $\boldsymbol{\rho}$ into $-\boldsymbol{\rho}$ turns ρ_n into $-\rho_n$; and thus, from (*10*.74), $E(\boldsymbol{\rho})$ is the conjugate complex quantity to $E(-\boldsymbol{\rho})$, so that

$$|E(\boldsymbol{\rho})|=|E(-\boldsymbol{\rho})|. \tag{10.76}$$

(ii) For a given value of $|\boldsymbol{\rho}|$, $|E_F(\boldsymbol{\rho})|$ has its greatest value when $\boldsymbol{\rho}$ is perpendicular to F. In this case $\boldsymbol{\rho}_n=\boldsymbol{\rho}$, and $\boldsymbol{\rho}\cdot\mathbf{r}'$ vanishes, since $\mathbf{r}'$ is perpendicular to $\boldsymbol{\rho}$. The integral in (*10*.75) is thus equal to F, the area of the face, and

$$|E_F(\boldsymbol{\rho})|=\frac{F}{2\pi V|\boldsymbol{\rho}|}, \tag{10.77}$$

and falls off inversely as $|\boldsymbol{\rho}|$. If we keep $|\boldsymbol{\rho}|$ constant, and turn the vector $\boldsymbol{\rho}$ away from the normal, the integral in (*10*.75) becomes less than F, and diminishes the more rapidly with increasing deviation of $\boldsymbol{\rho}$ from the normal the greater the area of the face. Thus the intensity distribution shows a projection or 'spine', as Laue terms it, perpendicular to any well-marked flat face of the crystal, which is sharper and more strongly marked the larger the face. Because of the symmetry condition (*10*.76), there must be two identical spines in opposite directions corresponding to each face.

(*j*) *The crystal-form factor expressed as the sum of a set of integrals along the crystal edges:* By a method exactly analogous to that used to transform the volume integral to the surface integral, we may transform the integral in (*10*.75) over the face into a line integral round the contour of the face. To do this, we take the z co-ordinate axis perpendicular to the face, in the direction of the inward drawn normal, and the axes of x and y in the face itself. Over the face F, z is constant, and (*10*.75) becomes

$$\mathrm{E_F}(\boldsymbol{\rho}) = \frac{i\rho_n}{2\pi \mathrm{V}\rho^2}\, e^{2\pi i z\mathrm{Z}} \int_{\mathrm{F}} e^{2\pi i(\mathrm{X}x+\mathrm{Y}y)}\, dx\, dy. \qquad (10.78)$$

The integrand, denoted by ϕ, now satisfies the equation

$$\frac{\partial^2\phi}{\partial x^2} + \frac{\partial^2\phi}{\partial y^2} + 4\pi^2(\mathrm{X}^2+\mathrm{Y}^2)\phi = 0.$$

If ρ_t is the component of $\boldsymbol{\rho}$ parallel to the face F, $\rho_t^2 = \mathrm{X}^2+\mathrm{Y}^2$, and proceeding as in the transformation from the volume to the surface integral, we obtain

$$\mathrm{E_F}(\boldsymbol{\rho}) = \frac{-\rho_n}{4\pi^2 \mathrm{V}\rho^2\rho_t^2} \int \rho_\nu\, e^{2\pi i(\boldsymbol{\rho}\cdot\mathbf{r})}\, ds, \qquad (10.79)$$

the integral being taken round the boundary of the face F, and ρ_ν being the component of $\boldsymbol{\rho}$ that lies in the surface and is normal to the element ds of the boundary. The positive direction of the normal is that towards the interior of the boundary curve.

If the crystal is bounded by plane faces, these meet in straight lines, and the total value of $\mathrm{E}(\boldsymbol{\rho})$ can be expressed as a sum of integrals along each *edge* ; and in this form it can be evaluated, although the algebra becomes complicated except in quite simple cases. Let $\mathbf{r} = \mathbf{R} + \mathbf{s}$, where $\mathbf{R}$ is the vector from O to the mid-point of the edge L, so that $\mathbf{s}$ is in the edge itself; and let ρ_s be the component of $\boldsymbol{\rho}$ parallel to the edge L. Then the contribution of the edge L to (*10*.79) may be written

$$\begin{aligned} \mathrm{E_{FL}}(\boldsymbol{\rho}) &= \frac{-\rho_n\rho_\nu}{4\pi^2 \mathrm{V}\rho^2\rho_t^2}\, e^{2\pi i(\mathbf{R}\cdot\boldsymbol{\rho})} \int_{-\frac{1}{2}\mathrm{L}}^{+\frac{1}{2}\mathrm{L}} e^{2\pi i\rho_s s}\, ds \\ &= \frac{-\rho_n\rho_\nu \mathrm{L}}{4\pi^2 \mathrm{V}\rho^2\rho_t^2}\, e^{2\pi i(\mathbf{R}\cdot\boldsymbol{\rho})}\, \frac{\sin \pi \mathrm{L}\rho_s}{\pi \mathrm{L}\rho_s}. \end{aligned} \qquad (10.80)$$

In applying the formula we may proceed as follows. In fig. 201 let M be the origin of the vector $\boldsymbol{\rho}$, that is to say, the reciprocal-lattice point. Through M draw a line LL′ parallel to L, the edge whose contribution is to be considered. From M, draw rectangular axes, n, ν, s, n being perpendicular to the face F, ν perpendicular to L and parallel to F, and s along the edge L; n and ν are both positive when drawn inwards. The quantities ρ_n, ρ_ν, and ρ_s in (*10*.80) are MP, MQ, and MR, the components of $\boldsymbol{\rho}$ parallel to these three axes, and $\rho_t^2 = \rho_\nu^2 + \rho_s^2$.

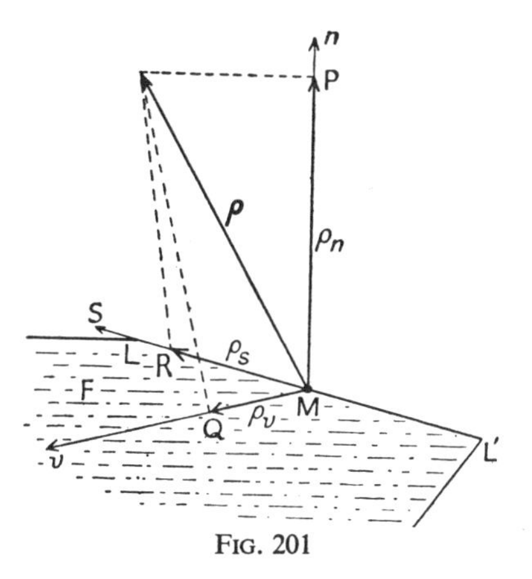

FIG. 201

Each edge makes two contributions, for it forms part of the boundaries of two faces. The two contributions will differ in the factor $-\rho_n\rho_\nu/\rho_t^2$, for the directions of the axes of n and ν, and so the values of ρ_n, ρ_ν, and ρ_t depend on the direction of the face of the boundary of which the edge is considered as forming part. For each edge of the crystal a corresponding line must be drawn through M, and $\boldsymbol{\rho}$ resolved accordingly. The exponential factor in (*10*.80) is a phase factor which depends on the distance of the middle of the edge considered from the origin, as well as on $\boldsymbol{\rho}$, and these factors must, of course, be taken into account when forming the sum for $E(\boldsymbol{\rho})$.

If the crystal, considered as a geometrical figure, has a centre of symmetry, pairs of edges, or pairs of faces, derived from one another by the operation of the symmetry centre, may be considered together, and the work of summation may be considerably reduced. For example, it is easy to see from (*10*.80) that a pair of parallel edges, related by a symmetry centre, which is also chosen as origin, give together the contribution

$$E'_{FL} = -\frac{\rho_n\rho_\nu L}{2\pi^2 V\rho^2\rho_t^2}\cos(\mathbf{R}\cdot\boldsymbol{\rho})\frac{\sin \pi L\rho_s}{\pi L\rho_s}, \qquad (10.81)$$

while, by (*10*.74), a pair of opposite faces gives

$$E'_F = -\frac{\rho_n}{\pi V\rho^2}\int_F \sin 2\pi(\mathbf{r}\cdot\boldsymbol{\rho})\,d\sigma. \qquad (10.82)$$

For an actual example of the calculation of $E(\boldsymbol{\rho})$ reference may be made to a paper by Laue and Riewe,[14] who discuss the octahedron in detail. Without detailed calculation, however, it is possible to draw certain general conclusions for crystals bounded by a relatively small number of well-marked faces and edges. We have already seen that the intensity distribution has two extensions, or 'spines', in opposite directions perpendicular to each face, and that in these directions $E(\boldsymbol{\rho})$ falls off inversely as $|\boldsymbol{\rho}|$. It is clear, however, from (*10*.80) that for any given direction $|E_{FL}|$, and so in general $|E(\boldsymbol{\rho})|$ itself, will fall off inversely as $|\boldsymbol{\rho}|^3$. But if $\boldsymbol{\rho}$ is perpendicular to an edge L, ρ_s vanishes, and $|E_{FL}|$ falls off inversely as $|\boldsymbol{\rho}|^2$. In such directions, the value of $|E_{FL}|$ has subsidiary maxima, which fall away as the direction of $\boldsymbol{\rho}$ departs from that of the normal, the more rapidly the longer L. The intensity distribution will thus have subsidiary spines in directions perpendicular to well-marked edges.

We may take as an example the case of a crystal in the form of a regular octahedron. It is convenient to think of the octahedron as inscribed in a cube (fig. 202). The main spines will be perpendicular to the octahedral faces, and so along the cube diagonals. There will

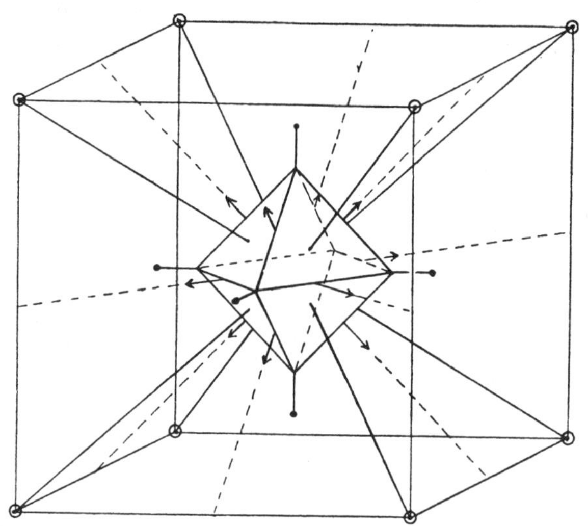

FIG. 202. Diagrammatic representation of the interference function for an octahedral crystallite

be subsidiary spines along the four-fold axes perpendicular to the cube faces, since each of these directions is perpendicular to four edges of the octahedron; and yet others along the two-fold axes through the cube centre and perpendicular to the cube edges, since each of these directions is perpendicular to two edges of the octahedron. The distribution is illustrated diagrammatically in fig. 202.

It must be noted that we have considered only the amplitude factor, that is to say, the maximum possible value of $|E(\boldsymbol{\rho})|$. If the variation of the different factors is taken into account, it will be found that the intensity distribution is intersected by certain surfaces of zero intensity, which cut the distribution up into regions of subsidiary maxima and minima. We have considered these in detail in Chapter I for the simple case of the parallelepipedal lattice.

(*a*)

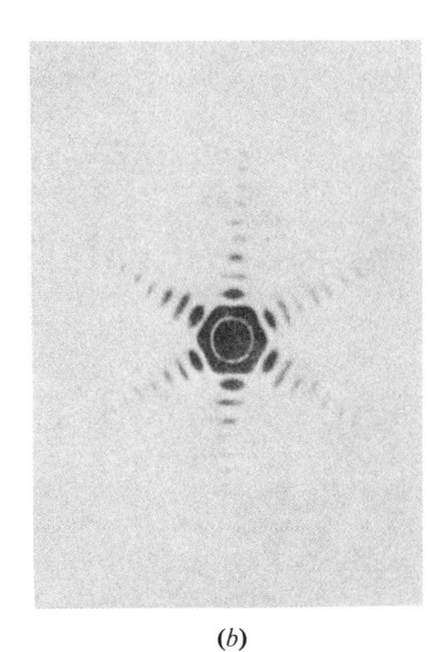

(*b*)

FIG. 203. Diffraction of light by (*a*) a triangular, (*b*) a hexagonal aperture (Scheiner and Hirayama)

(Reproduced in Laue and Riewe, *Zeit. f. Krist.*, **A, 95,** 408 (1936))

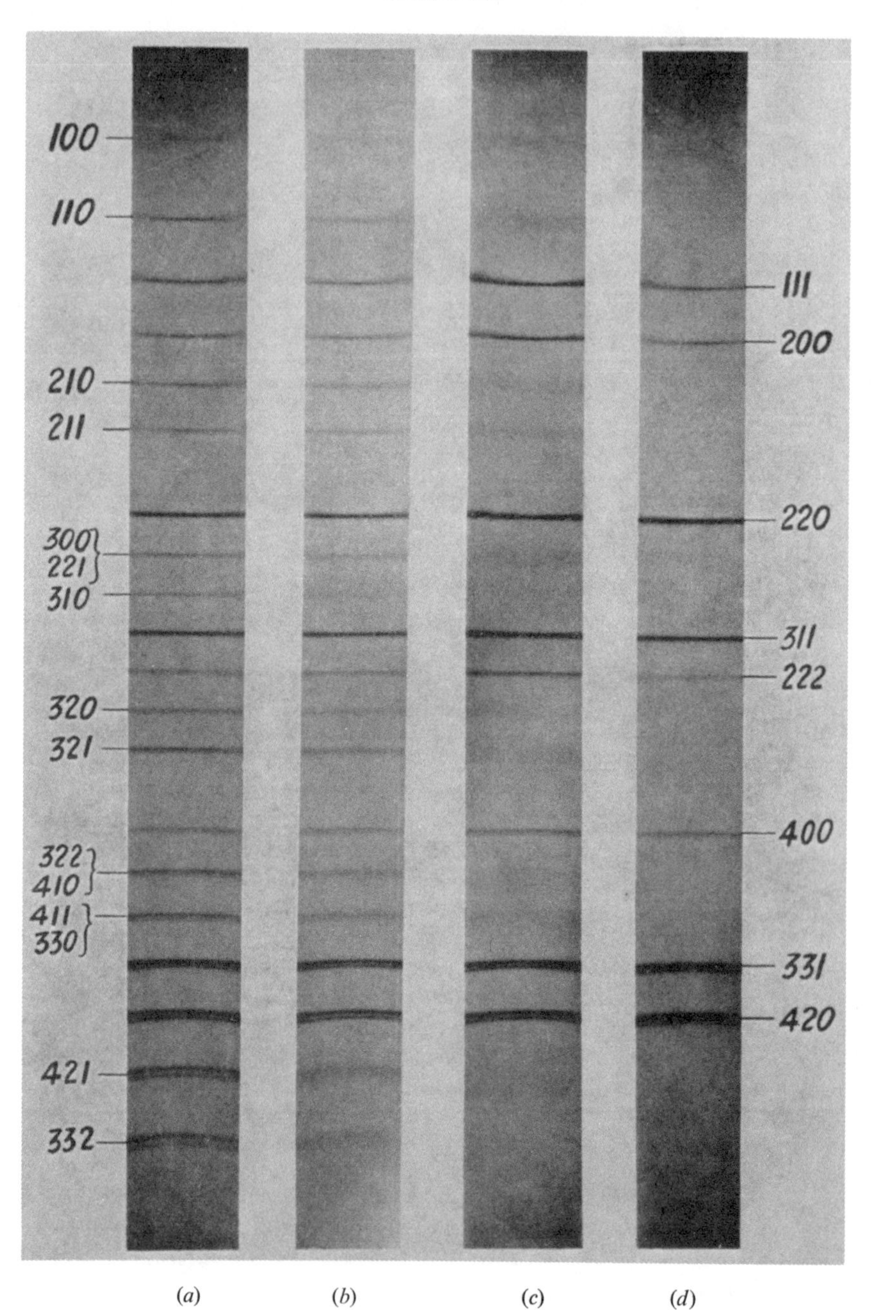

(*a*) (*b*) (*c*) (*d*)

FIG. 204. Photographs illustrating the development of a super-lattice in the alloy Cu_3Au (Jones and Sykes)
(*Proc. Roy. Soc.*, **A, 157,** 213 (1936))

Kirchner and Lassen [15, 16] have obtained certain star-shaped interference figures by electron diffraction in thin films of silver and gold, which have been explained by Laue in terms of the spines on the intensity distributions of small octahedral crystallites.

(*k*) *The optical analogy to the interference function :* The theorems we have just considered have close analogies in ordinary optical diffraction. We saw in Chapter VII, Section 2, that the array of points in the reciprocal lattice is the analogue of the array of spectra produced by an optical grating, the spectra in the case of a two-dimensional grating being actually arranged on a net reciprocal to that of the grating. In the same way, the distribution of intensity about the lattice-points, or the crystal-form factor, is analogous to the diffraction pattern produced by an aperture in ordinary Fraunhofer diffraction.

It is a well-known result that if (x, y) are the rectangular co-ordinates of a point in the plane of an aperture limiting a lens, and (X, Y) the co-ordinates of a point in the focal plane of the lens, referred to parallel axes, and expressed in suitable units, the amplitude at the point (X, Y) is given by

$$A(X, Y) = \int e^{2\pi i(xX+yY)}\, dx\, dy, \qquad (10.83)$$

the integration being over the whole aperture. Equation (*10*.83) has the same form as (*10*.78), and it was pointed out by Laue [17] that it could therefore be expressed as an integral taken round the edge of the aperture, and that if the aperture has straight sides, this way of regarding the matter leads directly to the result that the diffraction pattern will show extensions of large intensity in directions perpendicular to these sides. Diffraction patterns from triangular and hexagonal apertures are shown in fig. 203 (Pl. XIV), which illustrate the point excellently. The 'spines' in the two-dimensional patterns are very well marked, and it should be noticed that the triangular aperture gives six of them, and not three, just as theory predicts. The diffraction pattern has a centre of symmetry, although the aperture has not. The regions of zero intensity intersecting the intensity distribution are also well shown.

This theorem in Fraunhofer diffraction, although rediscovered by Laue, seems to have been arrived at by Abbe, and appears in a dissertation by Straubel [18] dating from 1888.

3. The Effect of certain Types of Fault and Imperfection of the Lattice on the Diffraction Spectra

(*a*) *Scattering by crystals with non-identical unit cells based on a regular lattice :* A crystal may be based on a regular lattice, but the structure factor may not be the same for every unit cell. Examples of this are given by certain metallic alloys, capable of showing a super-structure.

In figs 204(*a*), (*b*), (*c*) and (*d*) (Pl. XV) are reproduced some photographs from a paper by Jones and Sykes [19 (ii)] that illustrate the growth of a super-lattice structure in the alloy of copper and gold, Cu_3Au. When this alloy is quenched from a high enough temperature the powder photographs show the lines characteristic of a face-centred cubic lattice (fig. 204(*d*)), which indicates that the atoms of the two kinds fill the available sites on such a lattice in a random way. When, however, the alloy is annealed at suitable temperatures, segregation, more or less complete, of the copper and gold atoms takes place, so that the gold lies on one of the four interpenetrating simple cubic lattices of which the face-centred lattice is built up, and the copper atoms on the other three. This may, of course, happen in four different ways. The result of such ordering, starting at a number of unrelated points throughout the crystal, without definite phase relationship, is to produce in the powder photograph the lines characteristic of the simple cubic lattice. The lines belonging to this lattice that do not also belong to the face-centred lattice may be seen in fig. 204(*c*) as faint and very broad lines lying between the sharper lines due to the face-centred lattice. The broadness of the lines indicates that the growth of the super-lattice has started from a number of very minute nuclei. In fig. 204(*b*), certain of these nuclei have grown at the expense of others, and the super-lattice lines have become much sharper, while in fig. 204(*a*) the nuclei have grown until they are comparable in size with the crystal particles themselves, and all the lines in the photograph are now equally sharp. While this explanation of the diffuseness of the super-lattice lines is broadly correct, it does not follow without further investigation that the sizes of independent regions can be determined simply by applying the formulae for particle size, and the sizes so calculated were in fact found by Jones and Sykes to depend in a marked way on the indices of the lines.

A similar mixture of sharp and diffuse lines has been observed by Edwards and Lipson [20] in powder photographs from hexagonal cobalt, which they have explained by assuming the structure, a close-packed one, to consist of layers not uniformly in the sequence characteristic of hexagonal close-packing. When one layer of close-packed atoms has been formed, there are two possible ways for the second layer to be placed on it (Vol. I, p. 144). Once the second layer has formed, however, the siting of all further layers is determined if the structure is to be truly hexagonal. It will consist of pairs of layers following one another in the relation already started. It may be supposed, however, that after a number of layers have been formed in this way a fault occurs, the next layer forming not in the correct position for hexagonal symmetry, but in the alternative position. From this point, the sequence of pairs characteristic of hexagonal symmetry may again go on until another fault occurs, and so on. A structure formed in this way is still a close-packed structure, but it is not a perfectly regular structure, and contains faults occurring at random intervals.

The problem of diffraction by lattices containing faults of this kind has been treated by Wilson [21] by a method that is a simple extension of that used by Stokes and Wilson [6] for calculating the integral line-breadths due to small crystal particles. He shows that the line-breadths for powders composed of crystals with faults depend on the probability that two non-adjacent unit cells of the structure are alike.

It will be assumed in what follows that the lattice of the crystal is regular, so that each unit cell comes in its true position, but that the structure factors of the cells are not all identical. Let $\mathbf{r}_n$ and $\mathbf{r}_{n'}$ be the vectors to the nth and n'th cells of a crystal particle, and let the structure factors of these cells be F_n and $F_{n'}$ respectively. We may then write for the intensity scattered by the crystal, under conditions specified by the vector S, in the usual notation,

$$I = \sum_n \sum_{n'} F_n F^*_{n'} \exp\{-i\kappa \mathbf{S}\cdot\mathbf{r}_n - \mathbf{r}_{n'}\}, \tag{10.84}$$

the summation being taken over all the values of n and n' occurring in the piece of crystal. Equation (*10*.84) is formally the same as (*9*.10), and is true for any assemblage of scattering units. Putting

$$\mathbf{r}_{n'} - \mathbf{r}_n = m_1\mathbf{a} + m_2\mathbf{b} + m_3\mathbf{c} = \mathbf{r}_m, \tag{10.85}$$

we may write (*10*.84) in the form

$$I(\xi, \eta, \zeta) = \sum_n \sum_{m_1} \sum_{m_2} \sum_{m_3} F_n F^*_{n+m} \exp\{2\pi i(m_1\xi + m_2\eta + m_3\zeta)\}, \tag{10.86}$$

just as in deriving (*10*.29) from (*10*.27). Here, for simplicity, we have used the suffix n to represent the three indices n_1, n_2, n_3, of (*10*.29), and $n+m$ the three indices n_1+m_1, n_2+m_2, n_3+m_3.

If M is the number of times the lattice vector $\mathbf{r}_m$ occurs in the crystal, and J_m is the average value of $F_n F^*_{n+m}$ for two cells separated by this vector, we may write (*10*.86) in the form

$$I(\xi, \eta, \zeta) = \sum_m M J_m \exp\{2\pi i(m_1\xi + m_2\eta + m_3\zeta)\}. \tag{10.87}$$

If, for example, the crystal is a parallelepiped containing N_1, N_2, N_3 cells respectively parallel to a, b, c,

$$I(\xi, \eta, \zeta) = \sum_{m_1 m_2 m_3} (N_1 - |m_1|)(N_2 - |m_2|)(N_3 - |m_3|) J_m \exp\{(2\pi i(m_1\xi + m_2\eta + m_3\zeta)\}. \tag{10.88}$$

In summing, negative values of m must also be taken into account, that is to say $\mathbf{r}_m$ and $\mathbf{r}_{-m}$ are to be counted as separate vectors.

If J_m is constant, it can be taken outside the summation, as a factor. The remaining factor, involving the summations, is then the interference function, which in this case is equivalent to (*10*.41). If the crystal has a less regular shape, M cannot be expressed so easily, but the principle is the same. When all the cells are alike, J_m is constant, and equal to $|F|^2$, which is itself a function of ξ, η, ζ, and the summation is the interference function, which is always a periodic function of the reciprocal-lattice co-ordinates, having its maxima at the reciprocal-lattice points, which, as we have seen in X, § 1(*f*), can be expressed as a triple Fourier series. If J_m is not constant, the intensity function cannot be divided into the two factors $|F|^2$ and I_0; for J_m is itself a function of m and at the same time of ξ, η, and ζ, and the summation no longer has the form of a Fourier series. The type of distribution may differ from point to point of the reciprocal lattice, and there is no longer any reason to expect the integral breadth in ρ to be independent of the spectrum, even for a powder composed of spherical particles.

The form of the intensity function for the case of hexagonal cobalt discussed above has been worked out in detail by Wilson, who finds a variation of the sharpness from spectrum to spectrum of the same general nature as that found experimentally by Edwards and Lipson. It is not possible to give details here, and reference must be made to the original papers.

In dealing with powder photographs it is the integral line-breadth that is required. To calculate this, we use the methods discussed in § 2(*b*) of this chapter, starting from (*10*.86) instead of from (*10*.29). We consider as before a cubic lattice, referred to orthorhombic axes such that the spectrum considered becomes *h*00 when referred to them. The co-ordinates ξ, η, ζ in (*10*.86) refer to the axes reciprocal to the orthorhombic axes. Integrating (*10*.86) with respect to η and ζ, with $\xi = 0$, we obtain for the maximum intensity

$$4\pi\rho_m^2 \bar{I}(m) = b^*c^* \sum_{n_1}\sum_{n_2}\sum_{n_3}\sum_{m_1} F_{n_1, n_2, n_3} F^*_{n_1+m_1, n_2, n_3}, \qquad (10.89)$$

and for the integrated intensity

$$4\pi\rho_m^2 \int \bar{I}(\rho)\,d\rho = a^*b^*c^* \sum_{n_1}\sum_{n_2}\sum_{n_3} |F_{n_1 n_2 n_3}|^2. \qquad (10.90)$$

If J_t is the average value of the product $F_n F^*_{n+m_1}$ for two cells at a distance $t = m_1 a$ apart, in a row of cells parallel to a, we may write the right-hand side of (*10*.89) in the form

$$a^*b^*c^* \sum_{n_2}\sum_{n_3}\sum_{m_1} (N - |m_1|)\, a J_t,$$

N being the total number of cells in the row (n_2, n_3) parallel to a, and transforming the summations into integrations, exactly as in X, § 2(*b*), obtain for the line-breadth in ρ

$$\beta(\rho) = \int (\rho)\,d\rho / \bar{I}(m = VJ_0 \Big/ \int V_t J_t\,dt, \qquad (10.91)$$

or

$$\beta(2\theta) = \lambda V J_0 / \cos\theta \int V_t J_t\,dt. \qquad (10.92)$$

V_t is the volume common to the crystal and its ghost, displaced through a distance t, and J_t is the average value of the product $F_n F_{n'}^*$ for two cells at a distance t apart in the direction of displacement, which is perpendicular to the crystal planes producing the reflection considered. Equation (*10*.92) reduces to the usual formula for the integral line-breadth when J_t is independent of t, as it is in a crystal with no faults.

To calculate J_t it is first necessary to determine the probability that two cells at a given distance t apart in a given direction should have the same structure factor. The following method has been used by Wilson.[21] Suppose there are n possible structure factors, all equally likely, and consider a cell (2) at a distance $t+dt$ from a cell (1). Cell (2) may be like cell (1) either because the cell at a distance t is like cell (1) and no further fault occurs in the distance dt, or because the cell at t is unlike cell (1) and a fault occurs in the distance dt such as to make cell (2) like cell (1). The fault may produce $(n-1)$ possible types of cell, but only one of these will be like the cell (1), so that if $w\,dt$ is the probability that a fault of some kind occurs in the distance dt, the probability that it will make cell (2) like cell (1) is $w\,dt/(n-1)$. Let $P(t)$ be the probability that two cells at a distance t apart in the direction considered are alike, so that the probability that they are unlike is $1-P(t)$. Then the probability that the cell at $t+dt$ is like cell (1) is the sum of the probabilities of the two types of occurrence considered above, or

$$P(t+dt) = P(t)(1-w\,dt) + \{1-P(t)\}\,w\,dt/(n-1). \qquad (10.93)$$

Since dt is small,

$$P(t+dt) = P(t) + (dP(t)/dt)\,dt, \qquad (10.94)$$

and (*10*.93) and (*10*.94) together lead to

$$(n-1)\,dP/dt = -nwP(t) + w, \qquad (10.95)$$

which, on integration, gives

$$P(t) = [1 + (n-1)\exp\{-nwt/(n-1)\}]/n, \qquad (10.96)$$

the constant of integration being determined by the condition $P(0)=1$.

Equation (*10*.96) reduces to $P(t)=1/n$ for large t, as it should, since, when the distance between the cells (1) and (2) is large, the nature of cell (1) can have very little influence on that of cell (2), so that the chance of their being identical is simply the chance that of n equally probable choices the single favourable choice may be made.

If $F_1, F_2, \ldots F_n$ are the n possible structure factors, the average value of $F_n F_{n'}^*$, or J_t, is given by

$$J_t = \sum_{p=1}^{p=n} F_p[P(t)F_p^* + \{1-P(t)\}(\sum_{k}^{k\neq p} F_k^*)/(n-1)]/n. \qquad (10.97)$$

To proceed further it is necessary to know the values of F for the different possible unit cells. In the case of the alloy Cu_3Au, investigated experimentally by Jones and Sykes,[19] and theoretically by Wilson,[21] to whose papers reference must be made for details, there are four possible values of F. The available atomic sites lie on a face-centred lattice, and when the indices of the spectra *hkl* are all even or all odd, the condition corresponding to such a lattice, the structure factors are all equal, and J_t reduces to $(F_{Au}+3F_{Cu})^2$, which is independent of *t*. The sharpness of these lines is therefore limited only by the size of the crystal fragments; they are the sharp lines of the main lattice. The super-lattice lines are those for which the indices are neither all even nor all odd, and for these the value of F is either $F_{Au}-F_{Cu}$, or $F_{Cu}-F_{Au}$. In all cases, J_t reduces to

$$J_t=(F_{Au}-F_{Cu})^2 \exp(-4wt/3). \qquad (10.98)$$

The effective value of $\overline{|F|^2}$ in equation (*10*.84) then falls off with increasing distance *t* between the two cells, which means that even a crystal of infinite extent will behave as regards the line-breadths of the super-lattice lines as if it had a finite size, so long as *w*, the probability of a fault occurring in unit distance, is finite. Moreover, since *w* will as a rule differ in different directions in the crystal, the diffuseness of the lines may differ for different spectra, for *t* is perpendicular to the lattice-planes producing the spectrum considered.

Equation (*10*.98) may be written in the form

$$J_t=J_0 \exp(-\delta|t|), \qquad (10.99)$$

δ being small in comparison with the reciprocal dimensions of the crystal. The integrand of the integral in (*10*.92) will then become small before *t*, the displacement of the ghost from the crystal, has become more than a small fraction of the possible displacement. To a close approximation therefore we may put $V_t=V$ in (*10*.92), and write

$$\begin{aligned}\beta(2\theta) &= \lambda/\cos\theta \int_{-\infty}^{+\infty} \exp(-\delta|t|)\,dt \\ &= \delta\lambda/2\cos\theta. \qquad (10.100)\end{aligned}$$

The apparent crystal size, ϵ, which is given by $\beta=\lambda/\epsilon\cos\theta$, is therefore $2/\delta$, or $3/2w$ in the case considered. Thus, the greater the probability of a fault occurring in a given distance, the smaller the effective particle size.

(*b*) *Diffraction by a distorted lattice*: Consider now a lattice, composed of identical cells, that has become distorted owing to displacements of the cells from their ideal positions. Such distortion might, for example, be produced by the cold working of a metal. In Chapter V, we considered in some detail lattice distortions set up by thermal vibrations. These could be considered as made up of displacements due to the various normal modes of vibration possible to the crystal, and the average to be considered was therefore a time average over the displacements due to these independent modes. We have here to consider a space average over displacements that are non-periodic in character. A preliminary discussion of this problem has been given by Stokes and Wilson.[22]

Suppose the unit cell fixed by the vector $\mathbf{r}$ from the origin to be displaced bodily by a vector distance $\mathbf{u}(\mathbf{r})$. Such a system of displacements will in general involve a rotation as well as a translation, which is given at any point by $\frac{1}{2}\nabla\times\mathbf{u}$; but unless the rotation of a cell is large it will have relatively little effect on the phase of the radiation scattered by the cell, and we shall neglect it in calculating the structure factors. If F′ is the structure factor of the displaced cell,

$$\mathrm{F}' = \mathrm{F}\exp\{i\kappa\mathbf{S}\cdot\mathbf{u}(\mathbf{r})\} \qquad (10.101)$$

in the usual notation.

Consider a second cell at a distance t from the first in the direction of the unit vector $\mathbf{n}$. Its structure factor F″ will be

$$\mathrm{F}'' = \mathrm{F}\exp\{i\kappa\mathbf{S}\cdot\mathbf{u}(\mathbf{r}+\mathbf{n}t)\}. \qquad (10.102)$$

If the displacements are non-periodic, the relative displacements of the two cells may be assumed large and random for large values of t, so that the mean value of $\mathrm{F}'\mathrm{F}''^*$ will then vanish. For small values of t, $\mathbf{u}(\mathbf{r}+\mathbf{n}t)$ may be expanded in powers of t, and if terms in t^2, and higher powers, are neglected

$$\mathbf{u}(\mathbf{r}+\mathbf{n}t) = \mathbf{u}(\mathbf{r}) + t\mathbf{n}\cdot\nabla\mathbf{u}. \qquad (10.103)$$

Here $\nabla\mathbf{u}$ is a tensor, the symmetrical part of which gives the strains in the crystal at the point considered. With this approximation, using (10.101) and (10.102), we may write

$$\mathrm{J}_t = \overline{\mathrm{F}'\,\mathrm{F}''^*} = |\mathrm{F}|^2\,\overline{\exp\{i\kappa\mathbf{S}\cdot t\mathbf{n}\cdot\nabla\mathbf{u}\}}. \qquad (10.104)$$

If we consider the *hkl* spectrum, we take $\mathbf{n}$ normal to the (*hkl*) planes, since it is only displacements perpendicular to these planes that affect the phase of F for this spectrum. In this case,

$$\mathbf{n} = \mathrm{S}/|\mathrm{S}|, \quad \text{or} \quad \mathrm{S} = \mathbf{n}|\mathrm{S}| = 2\mathbf{n}(\sin\theta),$$

and $\kappa|\mathrm{S}| = 4\pi(\sin\theta)/\lambda = \mu$ in the notation previously used. Thus, for the *hkl* spectrum,

$$\mathrm{J}_t = \overline{\mathrm{F}'\,\mathrm{F}''^*} = |\mathrm{F}|^2\,\overline{\exp\{i\mu t\mathbf{n}\cdot\nabla\mathbf{u}\cdot\mathbf{n}\}}. \qquad (10.105)$$

From the properties of the tensor $\nabla\mathbf{u}$ it is easy to show that $\mathbf{n}\cdot\nabla\mathbf{u}\cdot\mathbf{n}$ is the tensile strain e_{hh} in the direction of $\mathbf{n}$, and perpendicular to the reflecting planes, so that

$$\mathrm{J}_t = |\mathrm{F}|^2\,\overline{\exp(i\mu e_{hh}t)}, \qquad (10.106)$$

an equation obtained by Stokes and Wilson, which should be a good approximation for any system of strains that makes J_t approach zero rapidly for large values of t.

Suppose now that $\phi(e)\,de$ is the fraction of a crystal for which e_{hh}, the tensile strain in the direction considered, lies between e and $e+de$. Then

$$\mathrm{J}_t = |\mathrm{F}|^2\int_{-\infty}^{+\infty}\phi(e)\exp(i\mu et)\,de. \qquad (10.107)$$

By (*10*.92), if J_t vanishes except for small values of t, so that V_t is always very nearly equal to V when the integrand is appreciable, we may write for the integral line-breadth for a powder composed of distorted particles such as we are considering

$$\beta(2\theta) = |F|^2 \lambda / \cos\theta \int J_t \, dt, \tag{10.108}$$

and the apparent particle size ϵ is therefore given by

$$\epsilon = \int J_t \, dt / |F|^2 = \int_{-\infty}^{+\infty} \int_{-\infty}^{+\infty} \phi(e) \exp(i\mu te) \, de \, dt. \tag{10.109}$$

To evaluate the integrals in (*10*.109) we first notice that the integrand is a rapidly oscillating function for large e. Large numerical values of e therefore contribute nothing to the integral, and we may give to $\phi(e)$ its value when $e=0$, and take it outside the sign of integration, and integrate with respect to e between the finite limits X and $-X$, which can afterwards be made to tend to infinity. Thus

$$\epsilon = \frac{2\phi(0)}{\mu} \int_{-\infty}^{+\infty} \frac{\sin(X\mu t)}{\mu t} \, d(\mu t). \tag{10.110}$$

The value of the integral is π, whatever the value of X, so that

$$\epsilon = 2\pi\phi(0)/\mu = (\lambda/2 \sin\theta)\, \phi(0), \tag{10.111}$$

and

$$\beta = 2 \tan\theta / \phi(0). \tag{10.112}$$

We can thus define a quantity

$$\eta = \beta \cot\theta = 2/\phi(0), \tag{10.113}$$

which should be a measure of the apparent strain in the crystal, and independent of the wave-length or order of reflection from a given set of planes.

Measurements of the integral breadth alone will give that fraction of the crystal in which the strain is negligible. Plainly, the larger this is, the smaller the integral breadth. According to Stokes and Wilson, it is in principle possible to obtain the distribution of strain $\phi(e)$ by analysing the photometric curves of broadened and unbroadened lines. Space does not, however, allow this point to be discussed.

Stokes and Wilson consider in detail two simple types of distortion. The first of these, which assumes all values of $|e_{hh}|$ between zero and a certain maximum to be equally likely, and $|e_{hh}|$ to be given by the direct stress divided by Young's modulus, leads to $\phi(0) = 1/2(e_{hh})_{\max}$, and so to

$$\eta = \beta \cot\theta = 4(e_{hh})_{\max}. \tag{10.114}$$

If the further assumption that the maximum stress is independent of direction is made, e_{hh} = maximum stress/Young's modulus, and the equation

$$\eta = \beta \cot\theta = A + BH \tag{10.115}$$

is found to apply, where A and B are constants for the given spectrum, and

$$\mathrm{H} \equiv (k^2 l^2 + l^2 h^2 + h^2 k^2)/(h^2 + k^2 + l^2). \qquad (10.116)$$

The second assumption, that $\overline{\exp(ix)} \equiv \exp(-\overline{x^2}/2)$, the stress being on the average isotropic, which is accurately true for a Gaussian distribution, leads to

$$\mathrm{J}_t = |\mathrm{F}|^2 \exp(-\tfrac{1}{2}\mu^2 \overline{e_{hh}^2}\, t^2),$$

and so to

$$\epsilon = \int_{-\infty}^{+\infty} \exp(-\tfrac{1}{2}\mu^2 \overline{e_{hh}^2}\, t^2)\, dt = (\lambda/2 \sin\theta)(2\pi\, \overline{e_{hh}^2})^{-\frac{1}{2}}, \qquad (10.117)$$

$$\beta = 2(2\pi\, \overline{e_{hh}^2})^{\frac{1}{2}} \tan\theta, \qquad (10.118)$$

and

$$\eta = \beta \cot\theta = 2(2\pi\, \overline{e_{hh}^2})^{\frac{1}{2}}. \qquad (10.119)$$

In this case,

$$\eta^2 = \mathrm{A} + \mathrm{BH}, \qquad (10.120)$$

H having the value of (*10*.116).

Experimental work on metal filings and wires gives results agreeing fairly well with both (*10*.115) and (*10*.120), but is not accurate enough to decide between them. It is indeed likely that more accurate measurements would agree with neither formula; for the assumptions upon which each is based are unavoidably oversimplified. For details of the calculations, reference must be made to the original papers.

(c) *Diffraction by lattices with periodic distortions:* The distortions discussed in the last paragraph were assumed to be non-periodic and random in character. There is, however, some evidence that periodic errors occur in certain crystals, producing effects analogous to the optical ghosts that accompany the spectra given by a diffraction grating with a periodic error of ruling. The effect on the spectra of periodic distortions due to the normal modes of vibration of the crystal were discussed in V, § 1 (*i*), in dealing with the temperature factor. We have now to consider a stationary disturbance, periodic in space, and not a time average; but very similar mathematical methods may with advantage be used. For simplicity, we consider only the reflections from a single set of planes, the $h00$ planes, but the generalisation of the method to three dimensions will be obvious.

Let $\mathrm{S}(x)\,dx$ be the amount of scattering matter lying in the unit cell between planes distant x and $x+dx$ from the origin. By (7.43), p. 370, we may express $\mathrm{S}(x)$ as a Fourier series in the form

$$a\,\mathrm{S}(x) = \sum_{h=-\infty}^{h=\infty} \mathrm{F}(h) \exp(-2\pi i h \mathrm{G} x), \qquad (10.121)$$

G being the reciprocal of a, the spacing of the planes, and so the number of repetitions of the undistorted lattice in unit length in the direction of x. The spectra given by these planes are represented by the points $(x^*, 0, 0)$, with $x^* = \pm h\mathrm{G}$, in the reciprocal-lattice space, or the points $(\xi, 0, 0)$, with $\xi = \pm h$ in the notation previously used.

Let this distribution be now subjected to a distortion such that a plane at a distance x from the origin is displaced in the direction of x by a distance $u(x)$, itself a periodic function of x, and so expressible as a Fourier series

$$u(x) = \sum_{m=-\infty}^{m=\infty} \mathrm{B}(m) \exp(-2\pi i m g x), \qquad (10.122)$$

g being the number of repetitions of the disturbance in unit length. If the displacement is small, so that u^2 may always be neglected in comparison with u, we may write for the disturbed density $\mathrm{S}'(x)$, as in V, § 1 (*i*)

$$\mathrm{S}'(x) = \mathrm{S}(x) - \frac{\partial}{\partial x}(u\mathrm{S}), \qquad (10.123)$$

or, using (*10*.121) and (*10*.122),

$$a\,\mathrm{S}'(x) = \sum_h \mathrm{F}(h) \exp(-2\pi\, ih\mathrm{G}x) + 2\pi i \sum_h \sum_m \mathrm{F}(h)\mathrm{B}(m)(h\mathrm{G} + mg) \exp\{-2\pi i(h\mathrm{G} + mg)x\}. \qquad (10.124)$$

Equation (*10*.124) shows that the disturbed density $\mathrm{S}'(x)$ will contain in addition to the original sinusoidal components of frequency $h\mathrm{G}$, which give the spectra from the undisturbed planes, components with frequencies $|h|\mathrm{G} \pm |m|g$. To each such sinusoidal component there corresponds a possible spectrum, and these new frequencies, which have appeared as a result of the periodic disturbance, correspond to the ghost spectra. These may be thought of as satellite spectra accompanying the main lattice-spectra, their representative points in the reciprocal-lattice space corresponding to $x^* = \pm(|h|\mathrm{G} \pm |m|g)$, or $\xi = \pm(|h| \pm |m|g/\mathrm{G})$. If the periodic disturbance repeats itself in a distance $\mathrm{Q}a$, $\mathrm{G} = \mathrm{Q}g$, and the values of ξ for the satellite spectra are $h \pm m/\mathrm{Q}$. Here, Q may or may not be a whole number, for it is not necessary that the disturbance should repeat itself in an integral number of lattice-units.

Equation (*10*.124) shows that the amplitude of any ghost spectrum of order m is proportional to the distance of its representative point from the reciprocal-lattice origin, and also to the amplitude $\mathrm{B}(m)$ of the mth Fourier component of the periodic disturbance.

The treatment given here assumes a continuous scattering density $\mathrm{S}(x)$, and a displacement $u(x)$ that is also a continuous function of the co-ordinates x, and the resulting displacements have been considered as small enough for their squares to be neglected. We might, however, have supposed each lattice-plane to be displaced as a whole, without distortion, by an amount that is a periodic function of its undisplaced distance from the origin. If the displacement is also sinusoidal, it is not difficult to treat this case without the restriction to small displacements, as Kochendörfer [23] has shown.

Let a be the spacing of the lattice-planes, and pa the undisplaced distance from the origin of the pth plane. If this plane is displaced from its true position through a distance B cos $2\pi pag$, we may write for the amplitude scattered by a set of N planes, each of unit scattering power,

$$A = \sum_{p=0}^{p=N-1} \exp\left[i\mu(pa + B\cos 2\pi\, pag)\right], \qquad (10.125)$$

where $\mu = 4\pi(\sin\theta)/\lambda = 2\pi\xi G$.

To proceed further, we use the expansion

$$\exp(iv\cos x) = J_0(v) + 2\sum_{m=1}^{m=\infty} i^m J_m(v)\cos mx, \qquad (10.126)$$

$J_m(v)$ being the Bessel function of order m. Putting $v=\mu B$, and $x = 2\pi pag$, we substitute this in (10.125), at the same time expressing the cosines in the exponential form, and obtain for A the expansion

$$A = \sum_{m=-\infty}^{m=\infty} A_m = \sum_{m=-\infty}^{m=\infty} i^{|m|} J_{|m|}(\mu B) \sum_{p=0}^{p=N-1} \exp\{ipa(\mu + 2\pi mg)\}. \qquad (10.127)$$

The mth term of this expansion is a geometrical series of the same nature as that met with in the theory of the linear diffraction grating, which was discussed in detail in Chapter I, § 1(a). It has appreciable values only when $\frac{1}{2}a(\mu + 2\pi mg)$ is an integral multiple, h, of π, that is to say, when

$$\mu = 2\pi(hG - mg). \qquad (10.128)$$

For these values of μ, the term has maxima of value $NJ_{|m|}(\mu B)$, and the larger N, the number of planes, the sharper are these maxima.

The intensity is given as a function of μ by $|A|^2$. Because of the sharpness of the maxima in A, the only terms in $|A|^2$ that have appreciable value are those of the type $|A_m|^2$, and we can write, with an accuracy that becomes greater the larger the value of N,

$$|A|^2 = N^2 \sum_{m=-\infty}^{m=\infty} J^2_{|m|}(\mu B)\, \frac{\sin^2\{\frac{1}{2}Na(\mu + 2\pi mg)\}}{\sin^2\{\frac{1}{2}a(\mu + 2\pi mg)\}}. \qquad (10.129)$$

The term $m=0$ gives the main spectra, which occur at the same positions as those from the undisturbed lattice, given by $\mu = 2\pi hG$, corresponding to $\xi = h$ in the reciprocal lattice row perpendicular to the planes. The intensities of the main spectra are, however, no longer all equal, that of the hth order being proportional to $J_0^2(2\pi hGB)$, or $J_0^2(2\pi hB/a)$. The terms $\pm m$ represent the ghosts of mth order accompanying the main spectra. They occur at angles given by

$$\mu = 2\pi(hG \mp mg), \quad \text{or} \quad \xi = n \mp mg/G = h \mp m/Q. \qquad (10.130)$$

The intensities of these ghosts are proportional to

$$J^2_{|m|}(\mu B), \quad \text{or} \quad J^2_{|m|}(2\pi\xi B/a),$$

where μ or ξ have the values corresponding to (*10*.130). In principle, therefore, ghosts of all orders m accompany each main spectrum of order h, including that of zero order, even although the disturbance is represented by a purely sinusoidal function, with no higher Fourier components. The ghosts accompanying the zero order will as a rule be very faint, since the values of the Bessel functions that determine their intensities will be small.

Daniel and Lipson [24] have given an approximate treatment of the problem, neglecting the squares of the displacements. According to this, only the first-order ghosts should appear with a sinusoidal displacement, and their intensities should be proportional to the squares of the orders of the corresponding main spectra. The relation between this approximate solution and that given by (*10*.129) is not difficult to see. For small values of μB, $J_m(\mu B)$ is proportional to $(\mu B)^m$. Thus, if μB is small enough throughout the relevant range of μ, the main spectra will be of nearly constant intensity, and the first-order ghosts will have intensities proportional to μ^2, μ being that value corresponding to the position of the ghost, so that ghosts of order $\pm m$ accompanying a single main spectrum will not be equal in intensity. The intensities of the second-and third-order ghosts will be proportional to μ^4 and μ^6 respectively, and will be very weak if μ is small, but will increase with increasing order. If μB becomes large, the intensity relationships differ considerably from those given by the approximate treatment.

In fig. 205(*a*) and (*b*) are shown the *amplitudes* of the main spectra and the corresponding ghosts for $B/a = 1/20\pi$, or $\mu B = \xi/10$, and $B/a = 1/10\pi$, or $\mu B = \xi/5$, plotted as a function of ξ, with $g = G/10$, or $Q = 10$. The maximum displacements are here approximately 1/63 and 1/31 of the main lattice spacing respectively. In the first case, the conditions are very nearly those given by the approximate treatment, which we may consider as closely applicable when $B/a < 1/100$. In the second case, however, the departures from the predictions of the approximate theory are already quite considerable. For large enough values of B/a the ghosts of the first order accompanying spectra of higher order may become more intense than the main spectra.

In some respects, the assumption of a bodily displacement of discrete lattice-planes, without distortion, would appear to be more physical than that of a continuous distortion of a continuous density distribution, although it is not in fact easy to decide which is the better representation in the case of a complex lattice-cell. The assumption of a continuous density distribution allows us to deal without difficulty with displacements that are periodic but not sinusoidal, but is not easily applicable to other than small displacements. The essential difference between the two types of assumption may be seen from the following example.

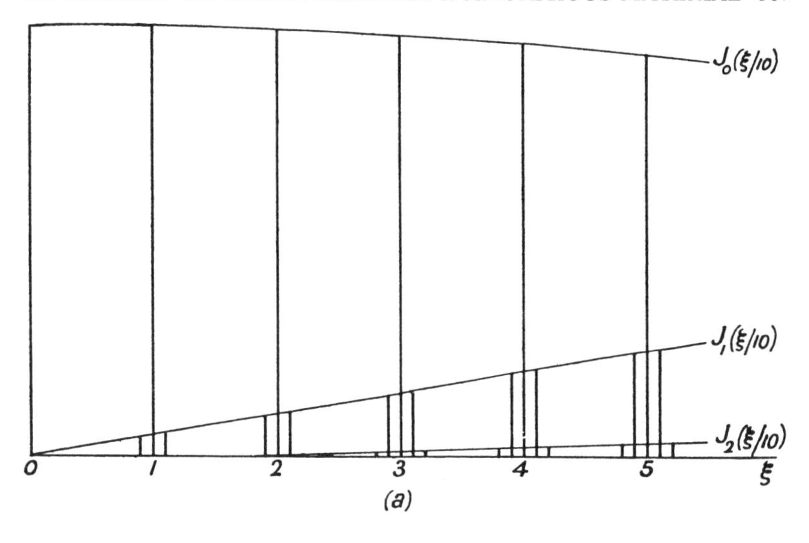

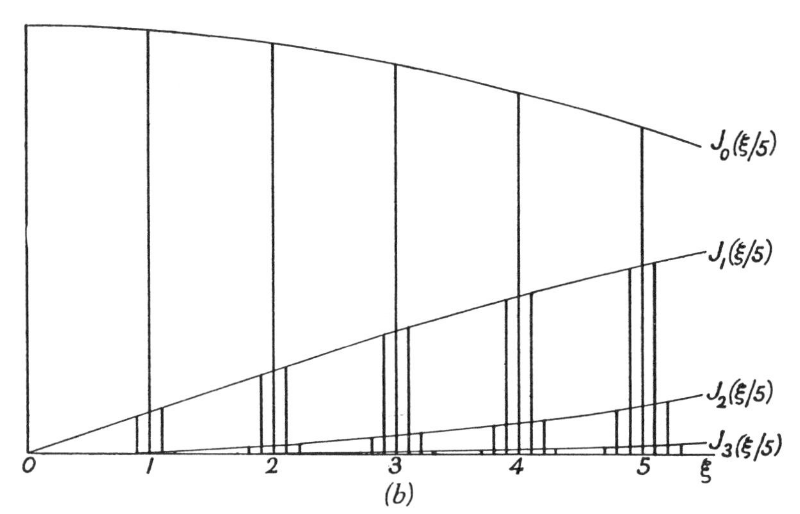

FIG. 205. Main spectra and ghost spectra for a set of planes with a sinusoidal distortion of spacing, the period of the distortion being 10 times that of the plane ($Q=10$). In 205(*a*) the maximum displacement is $1/20\pi$ of the spacing of the planes, in 205(*b*) the ratio is $1/10\pi$. Abscissae are $\xi=20(\sin\theta)/\lambda$

Suppose a grating to consist of equal transparent strips separated by opaque intervals. We might suppose it to be distorted by movement of the transparent strips as a whole, without alteration of their widths, which corresponds to the displacement of discrete lattice-planes; or we might suppose the displacement at any point to be a continuous function of the distance measured across the grating. Any given strip will then in general be altered in width as well as in position. If we suppose the total amplitude transmitted by each line nevertheless to remain the same, this second case corresponds to the assumption of the distortion of a continuous scattering density. It is plain that the two cases are not the same.

Another possible type of periodic disturbance can be represented by a modulation of the periodic density function $S(x)$ by a factor itself periodic in x with a different frequency, a case very easily treated by a method closely similar to that used above for the continuous distortion. A distribution so modulated will give satellite spectra in the same positions as those given by a displacement with the same periodicity. The intensities of the groups of satellites are now, however, the same for all orders of the main spectra, including the zero order.

(*d*) *Experimental evidence of structures with periodic faults :* In 1940, Chao and Taylor [25] described oscillation photographs taken about the *c* axes of some plagioclase felspars, some of which showed subsidiary layer lines between the prominent layer lines corresponding to the spacing of about 7A characteristic of the albite structure. In some cases, subsidiary lines appeared midway between the main layer lines, as if the true length of the cell in the *c* direction were really double that of the fundamental spacing; in other cases, pairs of layer lines, symmetrically disposed about a position midway between the main layer lines, and corresponding closely to the positions of the 2nd and 3rd, and 7th and 8th lines for a spacing 5×7A were observed. Very much weaker reflections, corresponding possibly to some of the other layer lines for this spacing, were also sometimes visible.

The explanation of the phenomena suggested by Chao and Taylor is that these crystals are not homogeneous throughout, but consist of regular alternations of the albite and anorthite structures, two closely related types of felspar structure, in layers approximately parallel to (001). The *c* axis of the albite structure has a length of about 7A, and that of the anorthite structure is about twice as great, so that a regular alternation of layers of the two structures is, in its optical effect, equivalent to a structure with a regular periodic spacing of about 7A having a periodically varying structure factor; and from this point of view, the extra layer lines could be considered as optical ghosts corresponding to this type of modulation. By suitable arrangements of layers it was found possible to obtain agreement between the observed and calculated intensities of the layer lines, but space does not allow us to consider the details here, for which reference must be made to the original work.

Another very interesting case has been discussed by Daniel and Lipson,[24, 26] that of the alloy Cu_4FeNi_3, which has a single-phase, face-centred cubic structure above about 800° C., but at lower temperatures consists of two phases, each a face-centred cubic structure. It was discovered by Bradley that when this alloy was heated at 650° C. for one hour the diffraction lines remained sharp, but each was flanked by two slightly diffuse, but quite strong side-bands, or satellite spectra. The appearance of the lines, and the way in which their intensities varied with the order of the spectra, suggested that they were due to a periodic variation of the lattice spacing rather than to a variation of the structure factor superposed on a definite spacing. This variation of spacing is presumably set up during the dissociation of the alloy, and the appearance of the lines depends on the temperature and on the length of the heat treatment.

Daniel and Lipson [26] found that it was possible to explain the results quantitatively fairly well in terms of Kochendörfer's formula, discussed in the last paragraph, on the assumption that a similar periodic variation of spacing takes place along all three cubic axes, although it was necessary to make allowance for extinction, a considerable amount of which had been introduced by the heat treatment. From the spacing of the satellites, Q, the number of unit cells over which one complete period of the variation extends, was found to vary from 50 to as much as 100, when the heat treatment had been lengthy. The diffuseness of the side-bands was taken as an indication that there was a considerable variation of Q in any given specimen, although it must also be remembered in this connection that a limited length of the regions of coherence of the periodic modification would have the same effect, while leaving the main spectra unbroadened.

From the variations of the intensity of the satellites with order it was possible to make estimates of the ratio B/a, the ratio of the maximum displacement of a plane to the undisturbed spacing. Daniel and Lipson actually tabulate Qb/a, where b is the maximum variation of the distance between successive planes from a, and this is equal to $2\pi B/Q$. The tabulated quantity is therefore $2\pi B/a$, and is actually independent of Q. The values of B/a found necessary to explain the results range from 0·05 to 0·11, according to the length and the temperature of the heat treatment, smaller values being on the whole associated with short times and high temperatures. The values of b range from 0·018A to 0·049A, in a spacing of 3·58A and are thus relatively small, in spite of the large values of B/a. It should be noted, however, that the intensities of the satellites accompanying any given order are determined primarily by B/a, and not by b, as (*10*.129) shows, since $J_m(\mu B)$ is equal to $J_m(2\pi\xi B/a)$.

Quite recently, Mr. D. H. Saunder[27], working at the University of Cape Town, has obtained what appears to be definite evidence of a periodic variation of structure in some crystals of complexes of 4 : 4′ dinitrodiphenyl with 4-bromo and 4-iododiphenyl. Oscillation photographs about one of the axes of these crystals show what at first sight appear to be a set of layer lines of which the lines of order 0, 7, and 14 are sharp and normal in appearance, the remainder being fainter and considerably more diffuse. Closer inspection, in the case of the bromo compound, shows the spacing of the diffuse layer lines to be abnormal, and careful measurement shows the arrangement to be that illustrated diagrammatically in fig. 206, in which, however, the abnormality of the spacing has been somewhat exaggerated.* The 2nd, 4th, and 6th layer lines appear in fact to be in the precise positions of ghosts $0+g$, $0+2g$, and $0+3g$, accompanying the zero-order layer line, and the 1st, 3rd, and 5th to be in those of the ghosts $G-3g$, $G-2g$, and $G-g$ accompanying the 7th, or 1st sharp layer line, g being not an exact submultiple of G, the frequency corresponding to the sharp layer lines. If the arrangement of lines is so interpreted, it is found that the distribution of diffuseness with order m is the same in each group of ghosts.

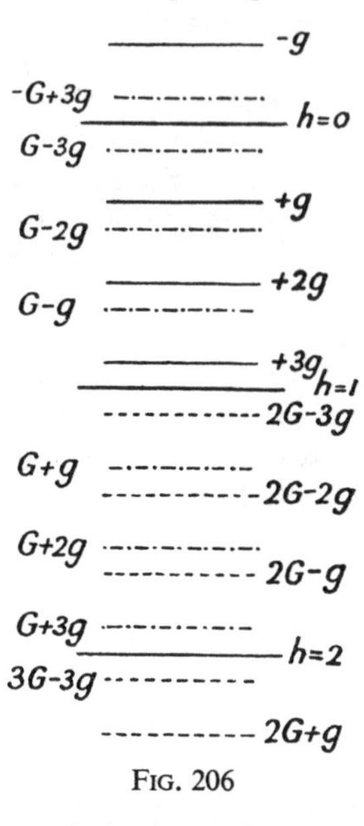

FIG. 206

The structure of the crystal is not fully known, although its type can be inferred with some certainty from work on similar complexes.[27] The dinitrodiphenyl molecules lie parallel to one another in face-centred array, nearly in the c planes, and at a distance of 3·69A apart in the direction of the c axis, about which the oscillation photographs under discussion were taken. The long axes of these molecules are parallel to the a axis of the crystal. The molecules of the halogen compound run in a direction nearly parallel to the c axis, and nearly perpendicular to the planes of the dinitrodiphenyl molecules, through tubular cavities formed by the face-centred array in which these lie.

The dinitrodiphenyl molecules alone would produce the sharp layer lines 0, 7, and 14; and the arrangement of the subsidiary layer lines is exactly that to be expected of a series of optical ghosts, such as would occur if the regular sequence of the planes of spacing 3·69A had been modified, in the case of the bromo complex, by a periodic error repeating itself in distance 12·7A, a distance a little less than that corresponding to seven layers of the dinitrodiphenyl molecules. Since 12·7A is the length of a molecule of 4-bromodiphenyl, it is natural to suppose that the presence of the halogen molecules running through the structure imposes a periodic variation of the fundamental spacing.

The diffuseness of the ghost spectra varies from order to order, but the elongation of the spots is in all cases parallel to b. If the crystal were composed of optically independent fragments with a small extension parallel to b, all the spots would be equally diffuse; and it would

* A similar, but much more marked, abnormality occurs in the case of the complex with 4-chlorodiphenyl.

thus appear that the explanation of the effect must be sought in random faults occurring in the *b* direction, superposed on a fundamentally regular lattice spacing. The following simplified model has been found to give an explanation of the main features of the pattern observed.* The crystal is assumed to consist of layers of dinitrodiphenyl molecules, all exactly alike, placed one above the other with their planes parallel to (001), and their lengths parallel to the *a* axis. Through this structure the chains of halogen molecules pass, with their lengths parallel to the *c* axis. The halogen molecules lying in any one chain are assumed to maintain an unchanged regular sequence throughout the crystal. In any one (020) plane, the structure is supposed to be without faults, corresponding atoms in different halogen chains all having the same *c* co-ordinates, so that the consequent displacements of the dinitrodiphenyl molecules are all in phase. It is supposed, however, that in different (020) planes there may be relative displacements of the halogen molecules, so that in passing from one (020) plane to the next there may be a sudden change in the *c* co-ordinates of corresponding halogen molecules. It is assumed that there is a certain probability that within any distance measured parallel to *b* such a change, or fault, may occur, and the simplest possible assumption, that there are two possible sets of co-ordinates for the halogen molecules, is made. It is shown that the diffuseness of the ghost maxima is governed by a factor which is a function of *m*, the order of the ghost, the form of which suggests that the spots on the ghost layer lines might show sharp nuclei within relatively diffuse wings; and on the photographs such spots are in fact to be seen. It can also be shown that the scattering by the halogen molecules alone, assuming the same probability of faults in the structure, would produce a set of layer lines having the positions and diffuseness of the ghosts accompanying the zero-order spectrum, the strength of which on the photographs can be ascribed to this cause, the true ghosts being those accompanying the higher orders.

4. Diffraction by Fibrous Materials

(*a*) *Introductory:* We shall conclude this chapter by discussing briefly the diffraction of X-rays by fibrous materials, a subject that is becoming of increasing importance because of the scientific interest and practical importance of many of the substances concerned. The typical fibrous substance is composed of long chain-like structures, each with some element of pattern repeating itself at regular intervals along the length of the chain. Cellulose, for example, consists of chains of β-glucose residues; and the fibrous proteins, such as wool, hair, silk, horn, quills, muscle fibres, and many other naturally occurring bodies, of polypeptide chains. The work on the structure of cellulose was

* James and Saunder, *Proc. Roy. Soc.*, **A 190**, 518 (1947).

initiated by Herzog, Jancke, and Polanyi [29] as early as 1920, and was continued by Mark and Andress.[28] In 1925, Katz [30] discovered that while unstretched rubber gave X-ray patterns showing diffuse haloes, such as are given by liquids and amorphous solids, the pattern changed when the rubber was stretched to one consisting of spots arranged on layer lines, resembling a rotation photograph produced by a crystal. More recently, the work of Astbury [31] and his associates on wool and hair, and the related fibrous proteins, has added much to our knowledge; and the biological importance of the substances investigated make the results of unusual interest.

The present volume is not the place to deal with the subject in detail. It can here be treated only as a problem in diffraction, closely related to those already considered in the first section of this chapter. The appearance of the spots on the fibre photographs shows that they are formed by crystallites, or crystal-like bodies, containing a considerable number of repetitions of the pattern along the direction of the fibres, and a comparatively small number of repetitions in directions transverse to it, the directions of the fibre axes of the crystallites in the irradiated specimen being approximately parallel to one another and perpendicular to the incident beam. It is interesting to trace the development of such diffraction patterns from those produced by fibres consisting of single rows of units, to those consisting of a number of parallel rows in crystalline array, in order to see at what stage a typically crystalline pattern makes its appearance. The actual structures themselves, and their arrangement and mutual interaction, are too complex for exact theoretical treatment, and simplification will be necessary; but even so, interesting results may be obtained.

A brief account of some of the experimental work on fibres has already been given in Volume I, Chapter VIII, p. 203, where a typical fibre photograph is shown in Plate XVI. For fuller accounts, and detailed bibliographies, reference may be made to the following works: Meyer and Mark, *Der Aufbau der hochpolymeren organischen Naturstoffe* (Leipzig, 1930); Astbury, *The Fundamentals of Fibre Structure* (Oxford, 1933); Randall, *The Diffraction of X-rays and Electrons by Amorphous Solids, Liquids and Gases* (Chapman and Hall, 1933), Chapter VII; Astbury, *Textile Fibres under the X-rays* (Imperial Chemical Industries, 1943), which gives an excellent series of reproductions of fibre photographs.

(*b*) *Diffraction by elongated crystallites with parallel arrangement and random orientation about the direction of their lengths:* The diffraction pattern produced by a single straight row of regularly spaced points, which is the ideally simple fibre, has already been discussed in § 1 (*d*) of this chapter, where it was considered as a degenerate case of diffraction by a parallelepipedal lattice in which the number of points in the rows parallel to two of the axes had been reduced to one. We shall again treat the problem in the same way, but shall now investigate the manner in which the complexity of the diffraction pattern grows as the number of points in these rows again increases, remaining, however, quite small, while the number of points parallel to the long axis remains large.

We shall suppose for simplicity that the lattice is orthogonal, although it is quite easy to generalise the treatment to lattices of any type, if necessary. Let the *c* axis be the fibre axis, and let the number of points, N_3, parallel to this be large and equal to N. Let N_1 and N_2 be respectively the number of points in rows parallel to *a* and *b*.

Since N is large, the distribution of the interference function corresponding to a crystallite is such that it has appreciable values only in a set of planes perpendicular to *c*, having a spacing $1/c$. These are the planes $\zeta = \text{integer}$, in the notation already used. If each unit scatters as a point, and if $N_1 = N_2 = 1$, the interference function I_0 is constant and equal to N^2 anywhere in these planes, the thickness of the region of appreciable I_0 on either side of these planes being negligible in practice if N is more than a few hundred. The diffraction pattern produced when radiation is incident in a direction perpendicular to the line of points has been fully discussed in § 1(*d*), and is illustrated by fig. 187. The diffracted beams form cones, which intersect any plane perpendicular to the direction of the incident radiation, such for example as a photographic plate, in a series of hyperbolas, the well-known layer lines. If the specimen consists of a very large number of exactly similar chains, all parallel to each other, but having no ordered arrangement except their parallelism, the distribution of intensity in the pattern is the same as for a single row, assuming always that the individual rows are so far apart that external interference, depending upon the positions of chain relative to chain, can be neglected. We shall for the time being make this assumption, which is certainly a great oversimplification, but we shall try to make some allowance for the effect later.

For chains longer than a few hundred units, the width of a layer line is determined by factors having nothing to do with diffraction, such as the divergence of the incident beam and the geometry of the camera. The distribution of the intensity along the layer lines will show no maxima and minima, and the well-marked maxima and minima that actually occur on the typical fibre photograph indicate that the structures producing them have considerably greater complexity than the simple one considered, as indeed one would expect.

In any actual chain, each repeating unit will itself have a structure, and will consist of a group of atoms of definite composition and configuration. Such a structure must cause maxima and minima of intensity along the layer lines, even with a single chain, but will not in general give rise to such sharply defined maxima as are actually observed. We accordingly consider first the effect of increasing the number of rows parallel to *c* in each crystallite, when each unit scatters as a point, leaving the consideration of the structure factor of the unit until later.

First, suppose $N_1=2$, $N_2=1$, $N_3=N$. Each crystallite is then a grid consisting of two parallel rows. Again we need only consider the values of I_0 in the plane ζ=integer, in which it is a function of ξ only, and by (*10*.6) is given by

$$I_0(\xi)=N^2\frac{\sin^2 2\pi\xi}{\sin^2 \pi\xi}=4N^2\cos^2\pi\xi. \qquad (10.131)$$

This type of distribution is illustrated in fig. 207, in which portions of the infinite planes $\zeta=0$ and $\zeta=1$ are shown. $I_0(\xi)$ is constant in any such plane along lines perpendicular to a and c and the variation parallel to a^* is given by (*10*.131).

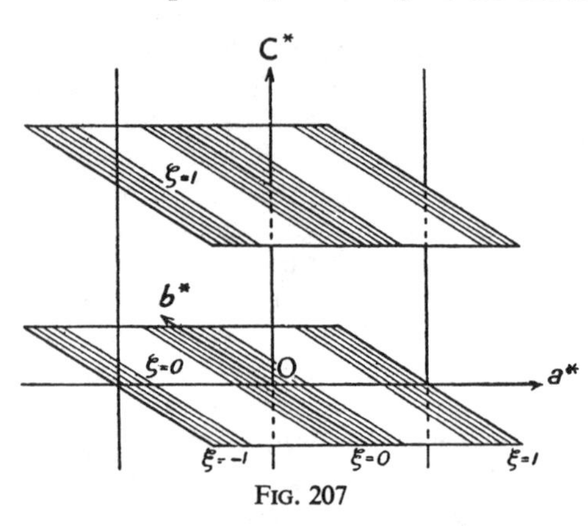

FIG. 207

Suppose now that the specimen irradiated contains a very large number of such crystallites, with their c axes all parallel to a common direction, but with all orientations about the c axis equally probable. If external interferences can be neglected, and if the positions of the crystallites parallel to their lengths are random, the average interference function $\bar{I}_0(r)$ will be axially symmetrical about c, and may be obtained by finding the average value at any radius when the distribution rotates about the c axis into all orientations with equal probability. We need consider only the planes ζ=integer, in each of which the average distribution will be the same.

Let OQ, fig. 208, be a line through the origin in the equatorial plane $\zeta=0$, and parallel to the a axis of one of the crystallites. If x is the actual distance OQ, $x=\xi/a$, or $\xi=ax$. The value of $I_0(\xi)$ is constant along a line such as QP perpendicular to OQ, and so the contribution of the crystallite considered to the total value of I_0 at a point P distant r from O on a line OP making an angle α with OQ is

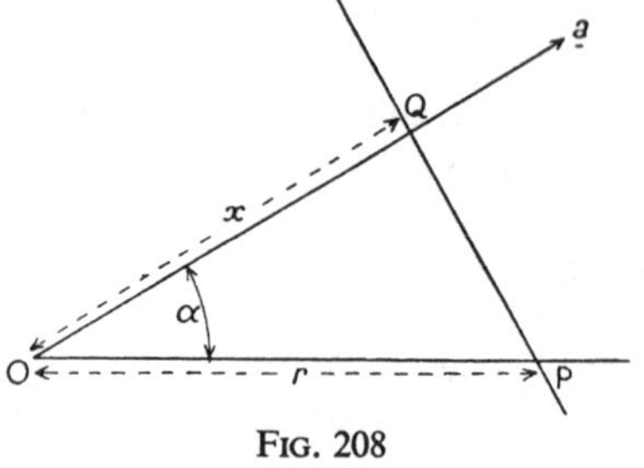

FIG. 208

$$I_0(xa)=I_0(ra\cos\alpha)=4N^2\cos^2(\pi ar\cos\alpha),$$

by (*10*.131).

The probability that the a axis of a given crystallite lies in a direction making an angle between α and $\alpha + d\alpha$ with OP is $d\alpha/2\pi$, and so the average value of I_0 for random orientation about c is

$$\bar{I}_0(r) = \frac{1}{2\pi}\int_0^{2\pi} 4N^2 \cos^2(\pi ar \cos\alpha)\, d\alpha$$

$$= \frac{N^2}{\pi}\int_0^{2\pi} \{1 + \cos(2\pi ar \cos\alpha)\}\, d\alpha$$

$$= 2N^2\{1 + J_0(2\pi ar)\}, \qquad (10.132)$$

where $J_0(x)$ is the Bessel function of zero order, defined by

$$J_0(x) = \frac{1}{2\pi}\int_0^{2\pi} \cos(x\cos\theta)\, d\theta.$$

The variation of the Bessel function $J_0(x)$ with x is similar to that of the more familiar diffraction function $(\sin x)/x$, except that the subsidiary maxima and minima are rather more strongly marked. Table X. 3, shows the maxima, minima, and roots of the functions $J_0(x)$ and $J_1(x)$ up to $x = 40$.

TABLE X. 3

Maxima, Minima, and roots of $J_0(x)$ and $J_1(x)$. (Jahnke-Emde).

x	$J_0(x)$	$J_1(x)$	x	$J_0(x)$	$J_1(x)$
0	1	0	19·6195	+0·1801	0
2·4048	0	+0·5191	21·2116	0	+0·1733
3·8317	−0·4028	0	22·7601	−0·1672	0
5·5201	0	−0·3403	24·3525	0	−0·1617
7·0156	+0·3001	0	25·9037	+0·1567	0
8·6537	0	+0·2715	27·4935	0	+0·1522
10·1735	−0·2497	0	29·0468	−0·1480	0
11·7915	0	−0·2325	30·6346	0	−0·1442
13·3237	+0·2184	0	32·1897	+0·1406	0
14·9309	0	+0·2065	33·7758	0	+0·1373
16·4706	−0·1965	0	35·3323	−0·1342	0
18·0711	0	−0·1877	36·9171	0	−0·1313
			38·4748	+0·1286	0

NOTE. The values of x for the maxima and minima of $J_0(x)$ are those of the roots of $J_1(x)$, the function of unit order; and *vice versa*.

Detailed tables of the functions are to be found in G. N. Watson's *Theory of Bessel Functions* (Cambridge, 1922), or in Jahnke-Emde, *Funktionen-tafeln* (Leipzig, 1933). Curve A, in fig. 213, shows the variation of $4\{1 + J_0(x)\}$ with x.

Equation (*10*.132) applies only to the planes ζ = integer. The value of $\bar{I}_0$ for any value of ζ is given by

$$\bar{I}_0(r, \zeta) = 2\{1 + J_0(2\pi ar)\}\frac{\sin^2 N\pi\zeta}{\sin^2 \pi\zeta}. \qquad (10.133)$$

Since N is large, we may assume $\bar{I}_0(r, \zeta)$ to be sensibly equal to zero except in the planes ζ = integer. The average distribution corresponding to (*10*.133) is then illustrated diagrammatically in fig. 209. The shaded

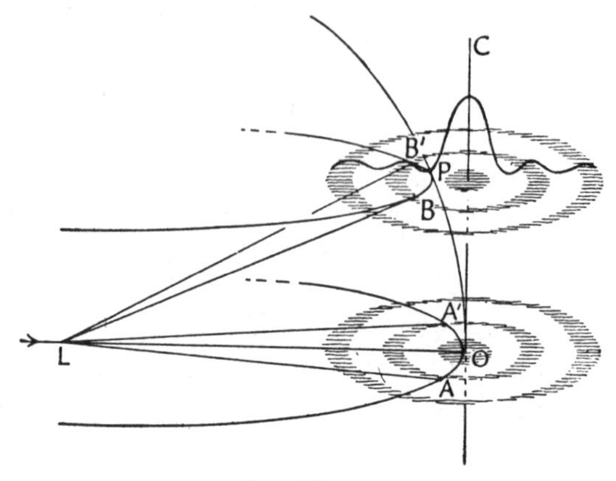

FIG. 209

rings represent the rather diffuse maxima of the function, $1 + J_0(2\pi ar)$, r being the radial distance from the axis OC. The sphere of reflection of centre L and radius $1/\lambda$, is shown for radiation incident in a direction perpendicular to c. The arcs AOA′ and BPB′ are the intersections of the sphere with the planes $\zeta = 0$ and $\zeta = 1$, respectively. If A and A′ are the points where one of the circles of maximum $\bar{I}_0$ meets the sphere, LA and LA′ are the directions of the corresponding diffraction maxima, which in this case lie upon the equatorial layer line if the radiation is allowed to fall on a photographic plate with its plane perpendicular to LO. LB and LB′ give the directions of the corresponding maxima on the first layer line, $\zeta = 1$. The maxima will be very broad and diffuse along the layer lines, but narrow transverse to them.

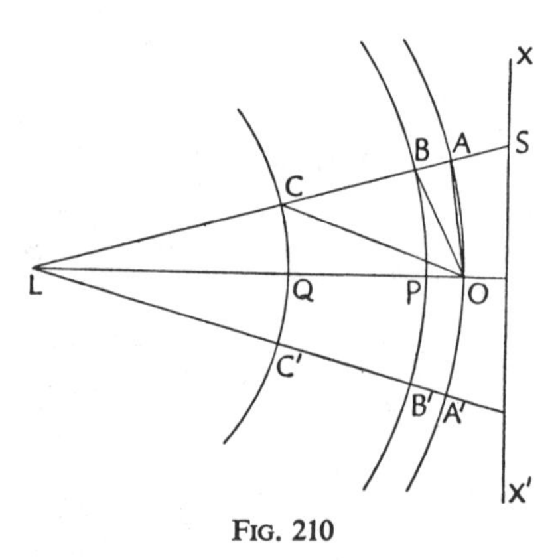

FIG. 210

Fig. 210 is a projection of fig. 209 on the plane $\zeta = 0$. The circle of intersection of the sphere of reflection with the plane $\zeta = 0$ is AOA′, which passes through the origin. BPB′, CQC′, are the projections of the corresponding circles of intersection with the planes $\zeta = 1$, $\zeta = 2$, and so on. LA, prolonged, meets XX′, the trace of the plate, at S, which is the position of the maximum on the equatorial layer line. LS cuts the circles BPB′, CQC′, at B and C, at different radii, OB and

OC, and the values of $2\{1+J_0(2\pi ar)\}$ for these radii give the values of the intensity on the first and second layer lines at points on the line through S perpendicular to the equatorial layer line. These points will not, in general, coincide with the maxima on the higher layer lines, but it is easy from the figure to map out the distribution for any particular case.

For this simple example, the characteristics are (i) the diffuseness of the spots along the layer lines, corresponding to the small number of rows in directions perpendicular to the fibre axis, (ii) the narrowness of the maxima across the layer lines, corresponding to the large number of repetitions of the scattering unit along the fibre axes.

(c) *Effects of lack of parallelism of the fibres:* In no actual specimen can we expect exact parallelism of the component fibres. It is not difficult to see the effect of a distribution of the direction of the fibre axes about a mean direction coinciding with the axis OC of fig. 209. We must suppose OC to assume a variety of directions, O remaining fixed. In fig. 211, the horizontal lines represent diagrammatically

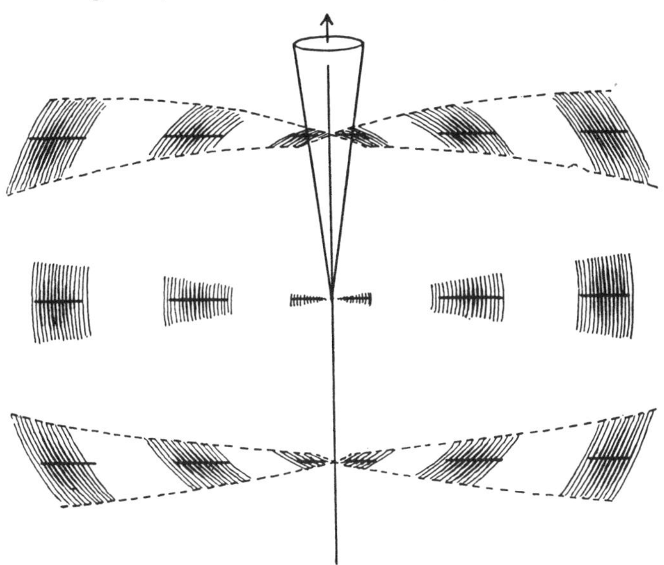

FIG. 211. Effect on fibre diagram of disorientation of the irradiated fibres

sections through the rings of maximum intensity by a plane containing the axis OC. For a fixed direction of OC, the regions of appreciable density are indefinitely thin, but if OC can assume a variety of directions lying within the cone indicated in the diagram, the average density distribution will be enlarged in the manner shown. Any particular portion of the density distribution will be smeared out into a region lying on a sphere having O as centre, and the corresponding spot on the photographic plate will be elongated along the arc of a circle having the point of impact of the primary beam as centre, in fact, along a Debye-Scherrer ring. The general result, illustrated in fig. 211, is to widen the equatorial spots in directions perpendicular to the equator, and to broaden the spots near the meridian along the layer lines, but to alter the breadth across them very little. All the main features discussed above are well shown on a photograph obtained by Astbury from natural silk, reproduced in fig. 212 (Pl. XVI). Cu $K\alpha$ radiation was used, and the plate was about 4 cm. from the specimen. The extension along the Debye-Scherrer circles, and its increasing amount as the distance from the centre of the plate increases are well shown, as well as the broadness of the maxima in a direction transverse to the equatorial layer lines, and the narrowness in the corresponding direction of the spots near the meridian.

Although there is a general resemblance in type between the pattern for silk and that predicted from the simple structure consisting of two rows, it is clear that the spots on the actual photograph are much too well defined to be produced by crystallites having such a small number of repetitions perpendicular to the fibre axis. We shall therefore consider in the next paragraph the effect of increasing the number of lateral rows.

(*d*) *The elongated two-dimensional structure with a number of rows:* We now consider the effect of increasing the number of points in rows parallel to a, while keeping N_2 equal to 1, so that the crystallites now consist of a number of parallel rows, side by side, in one plane.

The general form of the interference function in the plane $\zeta = \text{integer}$ is now

$$I_0(\xi) = N^2 \frac{\sin^2(N_1 \pi \xi)}{\sin^2(\pi \xi)} = N^2 \left\{ N_1 + 2 \sum_{n=1}^{n=N_1-1} (N_1 - n) \cos 2\pi n \xi \right\}, \quad (10.134)$$

by (*10*.12), and it is clear that by applying the process of averaging that led to (*10*.132) we obtain for the average value of the interference function at a distance r from the axis in a plane $\zeta = \text{integer}$,

$$\bar{I}_0(r) = N^2 \left\{ N_1 + 2 \sum_{n=1}^{n=N_1-1} (N_1 - n) J_0(2\pi n a r) \right\}. \quad (10.135)$$

If $N_1 = 4$ for example,

$$\bar{I}_0(r) = N^2 \{4 + 6J_0(2\pi a r) + 4J_0(4\pi a r) + 2J_0(6\pi a r)\}, \quad (10.136)$$

and each additional row gives an extra 'harmonic' in the series. The effect of the extra terms is to sharpen the maxima, and to displace them. This is illustrated by fig. 213, which shows the distribution of intensity along the zero layer line for $N_1 = 4$, plotted as a function of $4\pi a (\sin \theta)/\lambda$, or μa, in the notation used earlier in this chapter.

For the zero layer line $r = 2(\sin \theta)/\lambda$, so that $\bar{I}_0(r)$ can in this case be

PLATE XVI

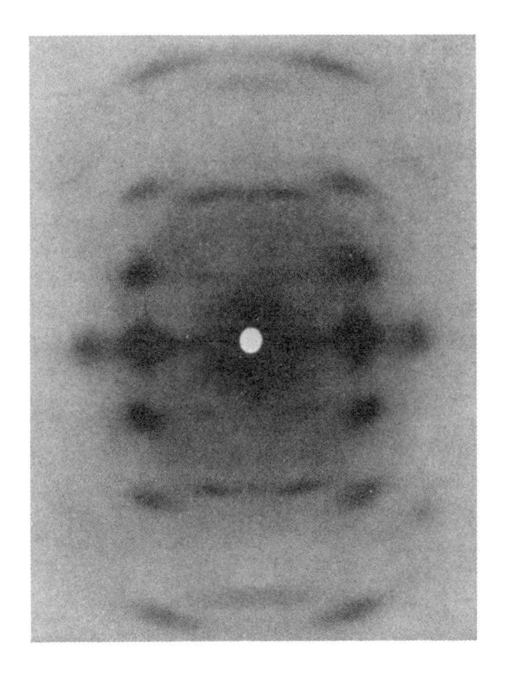

FIG. 212. Fibre photograph from natural (domestic) silk, Cu Kα: fibre axis vertical (Astbury)
(*Textile fibres under the X-rays*, Imp. Chem. Industries)

(P. 578)

Plate XVI — See other side

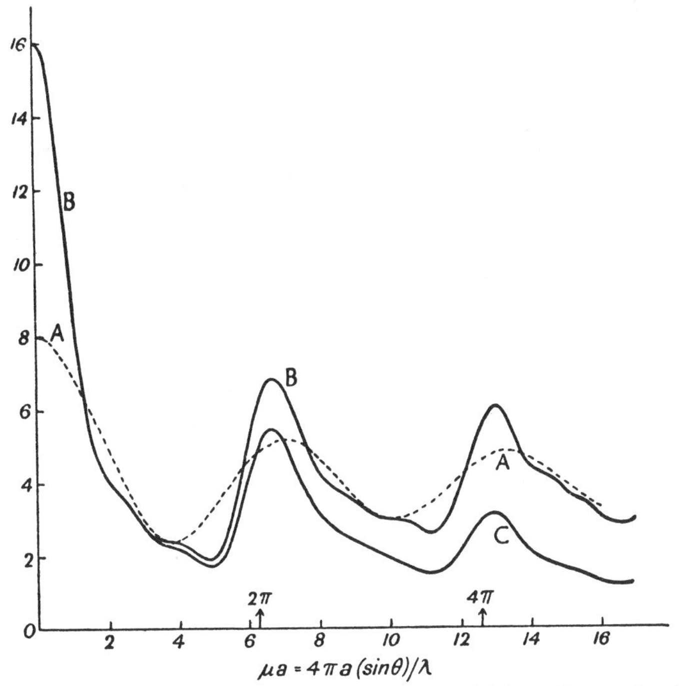

FIG. 213. Distribution of intensity on the equatorial layer line produced by grids of two (A) and four (B) rows, with random orientation about the direction of the rows. Curves A and B are drawn for point atoms. Curve C illustrates the effect of the atomic scattering-factor

written $\bar{I}_0(\mu a)$. This will not, of course, hold good for the higher layer lines.

The dotted curve A is that for the grid of two rows with a spacing a between them (equation *10*.132), and the continuous line B is that for a grid of four rows with the same spacing. The two curves are plotted on a comparable scale. It is assumed that in each case there are equal numbers of rows of N points. In case B, they form crystallites consisting of four rows, each giving an intensity proportional to 16 at small angles; in the other case, A, they form two independent crystallites of two rows each, each giving an intensity proportional to 4, and so a total intensity proportional to 8 at small angles. Zero-angle scattering is neglected (cf. Chapter IX, § 2(a)).

For a crystal of spacing a, the maximum 100 on the equatorial layer line would occur at an angle given by

$$(\sin\theta)/\lambda = 1/2a, \quad \text{or} \quad \mu a = 4\pi a(\sin\theta)/\lambda = 2\pi.$$

For the grids of both 2 and 4 rows, the maxima occur at angles greater than this, although the approximation to it is nearer for the greater number of rows. The first maximum of $J_0(x)$ occurs for $x = 7{\cdot}016$, or when $r=7{\cdot}016/2\pi a$. On the equatorial layer line, $r=2(\sin\theta)/\lambda$, so that the maximum 100 for two rows occurs when

$$(\sin\theta)/\lambda=7{\cdot}016/4\pi a=1{\cdot}115/2a,$$

or when $\mu a=7{\cdot}016$. This is illustrated in fig. 213, where the arrows at $\mu a=2\pi$ and 4π show the positions of the spectra 100 and 200 for a crystal of spacing a. It will be noted, too, that the background is considerable for the two-dimensional grids.

If the rows consist of actual atoms, instead of scattering points, the maxima will be less well marked, for the interference function then has to be multiplied by the square of the atomic scattering factor, which decreases rapidly with increasing $(\sin\theta)/\lambda$. The effect is illustrated by curve C, fig. 213. Its magnitude depends on the value of the spacing a relative to the atomic dimensions. In plotting curve C, it was assumed that $a=3{\cdot}98$A, corresponding to $(\sin\theta)/\lambda=\mu a/50$, and the values of f for the oxygen atom, as given in the tables of James and Brindley, were used. It will be seen that the maxima are still quite prominent, particularly the first. For larger values of a, and the same type of atom, the difference between the curves B and C would be less. Physically, this means that any given maximum, such as 100, occurs at a smaller value of $(\sin\theta)/\lambda$ the larger the a spacing, while the rate at which f^2 falls off with increasing $(\sin\theta)/\lambda$ depends only on the type of atom of which the structure is composed. For small spacings, of the order of 2A, the effect of the falling away in f^2 in reducing the maxima would be much greater, and we have already considered an example of this in dealing with scattering by diatomic molecules such as oxygen and nitrogen, (Chapter IX, § 3(c)), in which maxima, quite well marked so long as the atoms scatter as points, virtually disappear when the atomic scattering factor is taken into account.

(e) *The elongated three-dimensional crystallite:* If both N_1 and N_2 are greater than unity, the infinite lines of appreciable density parallel to b^* on the planes $\zeta=$integer characteristic of the grid-like crystallite break up into isolated regions surrounding the reciprocal-lattice points. The interference function in any such plane is now a function both of ξ and η, and can be written

$$I_0(\xi,\eta)=N^2\frac{\sin^2(\pi N_1\xi)}{\sin^2(\pi\xi)}\cdot\frac{\sin^2(\pi N_2\eta)}{\sin^2(\pi\eta)}. \qquad (10.137)$$

We saw in § 1(f) of this chapter that (*10*.137) may be expressed as a Fourier series, the sum of a constant term, and a series of cosine terms. For any one cosine term, the lines of constant phase in the reciprocal-lattice distribution are parallel to some row of reciprocal-lattice points in the plane $\zeta=$integer, and the wave-length is the spacing of the corresponding rows, and so the reciprocal of the distance between two points in the plane ab of the actual lattice. For example, if the base of the crystal unit consists of four points at the corners of a parallelogram ABCD, there will be one cosine term in the Fourier series for (*10*.137) for each of the distances AB, AD, AC, and DB. The coefficient of any term is equal to the number of times the corresponding distance occurs in the figure ABCD, counted once in each direction, thus 4 for AB and AD, and 2 for AC and BD; and the wave-lengths are the reciprocals of the corresponding distances. Each of these cosine terms can be treated on averaging for rotation about the c axis exactly in the same way as in § 4(*b*), and each gives rise to a term containing a Bessel function.

If the base ABCD of the crystal is square, so that AB = AD, and AC = BD, the average value of the interference function in the planes ζ = integer for a crystallite consisting of 4 rows is

$$\bar{I}_0(r) = N^2\{4 + 8J_0(2\pi ar) + 4J_0(2\pi\sqrt{2}ar)\}, \qquad (10.138)$$

and if the base consists of 9 points arranged in a square net so that the crystallite has 9 rows,

$$\bar{I}_0(r) = N^2\{9 + 24J_0(x) + 16J_0(\sqrt{2}x) + 12J_0(2x) + 16J_0(\sqrt{5}x) + 4J_0(2\sqrt{2}x)\}, \qquad (10.139)$$

where $x = 2\pi ar$, corresponding to the distances between points a, $\sqrt{2}a$, $2a$, $\sqrt{5}a$, and $2\sqrt{2}a$, which occur, when counted once in both directions, 24, 16, 12, 16, and 4 times respectively, in the basal plane of the crystallite.

The distributions in the equatorial layer lines corresponding to (*10*.138) and (*10*.139) are shown in fig. 214, curves A and B respectively. In curve B, corresponding to 9 rows, the peaks 100 and 110 are well resolved, but 200 and 120 are not. The positions of the corresponding spectra, as given by Bragg's law, are indicated in the diagram. It is clear that such crystallites would give rise to quite well-marked diffraction patterns. Even for the crystallite of 4 rows, the maxima are fairly well defined, although 110 and 100 are not resolved. A characteristic feature of these patterns produced by three-dimensional arrangements—by prisms, as distinct from grids—is the smaller amount of background, and in particular, the nearly zero value between the central maximum and 100. As we have seen, the repetition of the pattern in two directions perpendicular to c breaks up the distribution corresponding to the interference function into isolated regions surrounding the reciprocal-lattice points, and when the distribution is rotated about an axis through one of the points, it is clear that the average density will be very small in a region surrounding the point about which rotation occurs. For the same reason, the first maxima will remain well marked even when the falling off of the atomic scattering factor is taken into account. In the case of the prism of 9 rows, the subsidiary maxima between the main maxima give rise to a very small peak lying within this region, which shows up clearly in fig. 214.

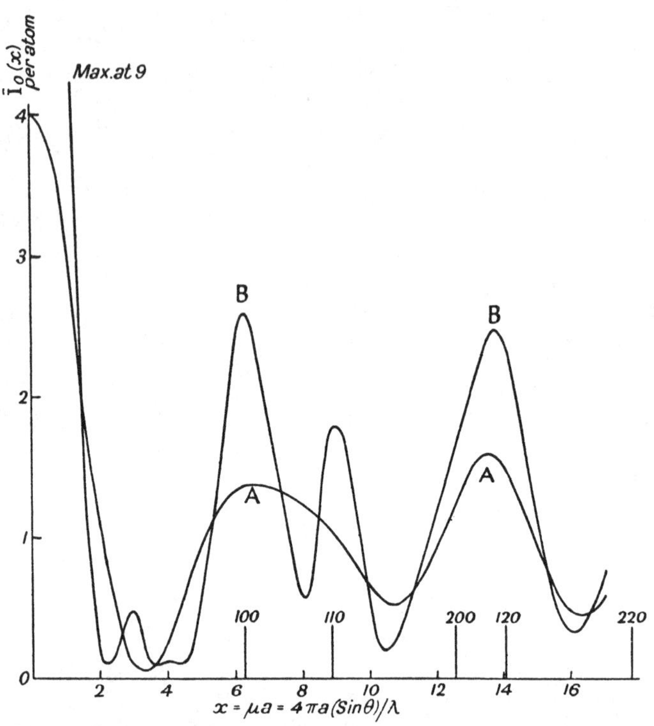

FIG. 214. Distribution of intensity on the equatorial layer line for parallel elongated crystallites of square cross-section, A, with four rows, B with nine, and random orientation about the direction of the rows

(*f*) *The external interference terms :* We have hitherto assumed the different crystallites to be optically independent, and have made no allowance for that part of the interference pattern due to their lateral approach. It would be exceedingly difficult to discuss this problem in detail, but we may get some idea of the general nature of the effects to be expected by considering a simple case.

Consider first two infinitely long, straight, parallel chains of identical atoms, each with an interatomic spacing c, the lateral distance between them being always a; but let all relative positions of the chains parallel to their own lengths be equally probable. Suppose the diffracting specimen to be made up of a very large number of such pairs, all with their c axes parallel to a common direction, the distance between the individual pairs being large enough for them to be considered as optically independent. If we now suppose that all orientations of each pair about the direction of its c axis are equally probable, it is possible to determine the distribution of the average interference function for the assemblage.

Let fig. 215(a) represent one configuration of such a pair of chains, (1) and (2), that in which the points form a rectangular grid with a

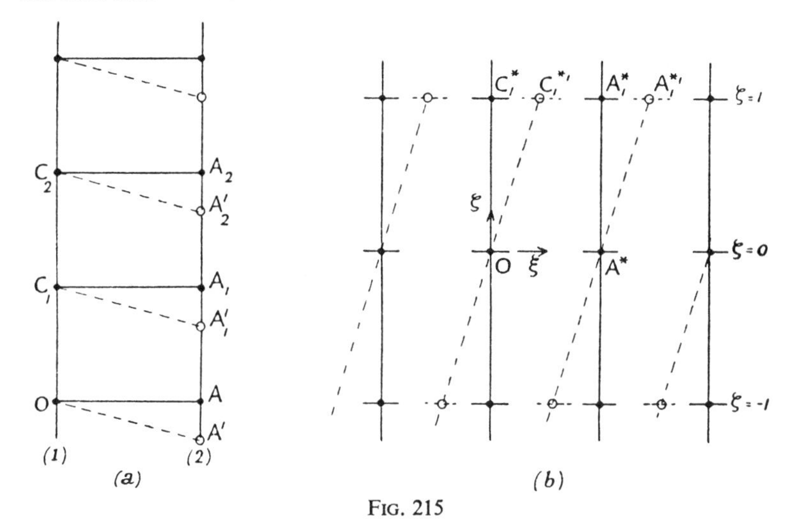

FIG. 215

mesh such as OAA_1C_1. The corresponding interference function has a distribution of the same type as that shown in fig. 207, and is founded on the rectangular reciprocal-lattice net with translations OA*, OC_1^* (fig. 215(*b*)). The short lines parallel to ξ in the planes $\zeta=0$ represent the extensions of the reciprocal-lattice points parallel to a^*, corresponding to two points in rows such as OA, parallel to a. The distribution is independent of η, and must be thought of as extending to infinity in directions perpendicular to the plane of the diagram.

Let the points A′, A_1' ... , fig. 215(*a*), represent another position of chain (2) relative to chain (1). The corresponding distribution of the interference function is founded on the reciprocal net with translations OA*, $OC_1'^*$, where $C_1'^*$ lies in the line $C_1^*A_1^*$. As chain (2) takes up all possible positions relative to chain (1), $C_1'^*$ assumes all possible positions between C_1^* and A_1^* with equal probability, the c^* axis of the reciprocal net pivoting about the origin O. The distribution of the interference function in the plane $\zeta=0$ is not changed by the displacement, but in all the other planes $\zeta=$integer it becomes a uniform sheet distribution. If all orientations about the direction of the c axis are equally probable, we must allow the distribution to rotate about the axis OC* in order to get the average distribution, which will be uniform in all the planes $\zeta=$integer, except in the plane $\zeta=0$, in which it will be similar to that represented in fig. 209. The average value at a radial distance r from the axis will be proportional to $4\{1+J_0(2\pi ar)\}$.

Let radiation be incident in a direction perpendicular to the axis c; then the interference pattern in a plane perpendicular to the direction of incidence will show a distribution of diffuse spots along the equatorial layer line, but a continuous distribution of intensity along all the other layer lines.

If the chains are not all exactly parallel, but have a distribution about the c direction, we must suppose the whole intensity distribution to oscillate about O through the corresponding cone of angles. The section of the distribution by any plane containing OC* will be of the type indicated in fig. 216, which is a rotation diagram, in the sense in which this term is used by Bernal.

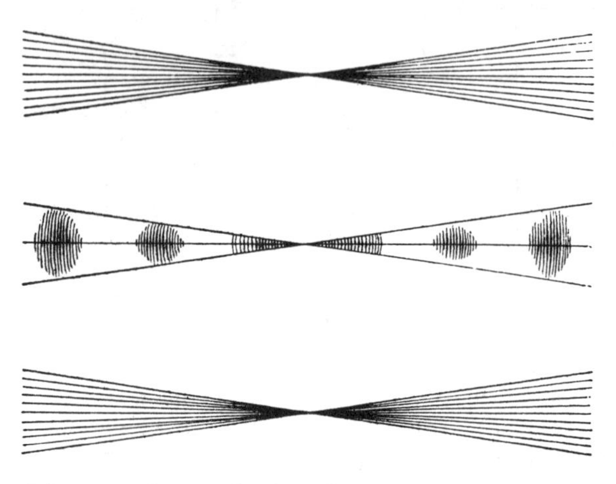

FIG. 216. Diagrammatic representation of pattern due to external interference

The problem considered above corresponds, of course, to conditions much simpler than any occurring in nature. As a next step, we might assume the specimen to consist of a number of long chains, all parallel, and with any relative positions parallel to their lengths, but so arranged that in any section of the assemblage perpendicular to the lengths of the chain, the positions of the fibre axes correspond to irregular close-packing. An extension of the analysis given above for the simple case indicates that again there will be concentric maxima in the equatorial plane of the intensity distribution, not unlike those corresponding to diffraction by a monatomic liquid, but a uniform distribution in all the other planes $\zeta = \text{integer}$. Interference spots will occur only on the equatorial layer lines, and while the first maximum may perhaps be expected to be rather more distinct than that given by the pairs of chains considered above, it will still be diffuse.

It may be stated generally that if definite interference spots occur on layer lines other than the equatorial one, it is very strong evidence that the individual crystallites consist of more than one row of points, and that the rows are in a lattice-like array. Single rows of points give diffuse interference effects on the zero layer lines in virtue of the external interference, but can give no interference spots on the other layer lines unless the relative positions of the different rows parallel to their lengths are no longer random. It is true that if the units of which the individual chains are built up are complex, it is not to be expected that the relative positions of the chains in the direction of their lengths will be entirely random, and this would in principle produce interference effects on the other layer lines. One would, however, expect the spots produced in this way to be very diffuse. It is quite clear that some of the fibre photographs figured by Astbury, for example those given by silk and hemp, correspond to quite a high degree of crystalline organisation; while others, such, for example, as some of those from mammalian hair, suggest a much simpler arrangement. The effect of the structure factor of the individual units of which the chains are built up, when they are complex, will be considered in the next paragraph, but we shall find that this produces no serious modification of the conclusions already reached.

When the individual crystallites are more complex than simple rows of points it is not possible to take into account, with any certainty, the effects of external interference. The greater the number of chains in the individual crystallites, the less important the external effects will become, and the less definite will be the closest distance of lateral approach ; but the same general principles must apply. Any interposition of material amongst the crystallites with a less definite structure, or with a lack of definite orientation, will tend to produce diffuse haloes superposed on the more regular patterns.

(g) *The structure factor and the atomic scattering factor :* When the units of which the chains are composed are not points, but themselves have a structure, the interference function corresponding to any crystallite has to be multiplied by the square of the structure factor of a single unit before the averaging process is carried out. As we have already seen in § 1(*b*) of this chapter, when the maxima of the interference function are diffuse we must consider the structure factor as a continuous function of ξ, η, and ζ. We may take, as an example, a single chain of units each of which is a regular hexagonal ring of atoms, the points of adjacent rings in the chain being separated by a distance equal to one side of a ring. If the chains are long, the interference function I_0 has a finite, uniform density in sheets coinciding with the planes $\zeta = $ integer, and is zero elsewhere; but in the distribution corresponding to the total intensity function these uniform sheets will be broken up by the variation of the square of the structure factor. In this particular case, indeed, the variation of $|F|^2$ in the planes $\zeta = $ integer constitutes the intensity distribution.

The values of $|F|^2$ for the case under consideration are easily calculated from equation (*10*.4), and are as follows:

$$\text{For } \zeta = 0, \quad F^2(\xi, \eta, 0) = 4f^2(3 + 4\cos u + 2\cos 2u),$$
$$\text{for } \zeta = 1, \quad F^2(\xi, \eta, 1) = f^2(3 - 4\cos u + 2\cos 2u),$$
$$\text{for } \zeta = 2, \quad F^2(\xi, \eta, 2) = f^2(3 + 4\cos u + 2\cos 2u),$$

and so on, where $u=\pi\xi/\sqrt{3}$, and f is the atomic scattering-factor of one of the atoms of the ring. It was assumed in working out the expressions in this form that the chain was a degenerate case, with N_1 and N_2 equal to unity, of a lattice with an a spacing equal to the c spacing, which was that of the units of the chain. Thus, the parameter ξ corresponds to an actual distance $x=\xi/c$ measured in the a^* direction from the c^* axis; while the parameter ζ corresponds to a distance $z=\zeta/c$ measured parallel to the c^* axis, from the a^* axis. $|F|^2$ is independent of η.

If the chains can have any orientation about the c axis, the radial distribution of the average intensity function in the plane $\zeta=0$, at a distance r from the c^* axis, will be given by

$$\overline{I(u)}=4f^2\{3+4J_0(u)+2J_0(2u)\},$$

where $u=\pi cr/\sqrt{3}$; and there are corresponding expressions for the other planes $\zeta=$integer.

The values of $\overline{I(u)}$ for $\zeta=0$ and $\zeta=1$ are shown in curves A and B respectively of fig. 217, in which it has been assumed that each atom scatters as a point. It will be seen that the maxima and minima of the intensity function in the plane $\zeta=0$ are very well marked. The corresponding maxima on the zero layer line when a photograph is taken with the direction of incidence perpendicular to the fibre axes will not, however, be so definite, for the atomic scattering factor f is itself a function of ξ, η, and ζ, spherically symmetrical about the origin, but diminishing rapidly as the distance from it increases. To investigate this effect, let us assume that the atoms of which the chain we have just considered are composed scatter in the same way as O^{--} in the tables of James and Brindley. When the radiation is incident in a direction perpendicular to the fibre axes, we may put for the equatorial layer line $r=|S|/\lambda=2(\sin\theta)/\lambda$, so that $u=2\pi c(\sin\theta)/\sqrt{3}\lambda$. If we then assume $c=5{\cdot}52$A, we find $(\sin\theta)/\lambda=u/20$.

In fig. 217, a scale of $(\sin\theta)/\lambda$ corresponding to this case is shown below the scale of u, and in curve C the values of the ordinates of curve A have been multiplied by the appropriate values of f^2. It will be seen that little remains of the maxima that are so prominent in curve A.

It must be emphasised that the effect of the atomic scattering factor on the distribution of intensity in the diffraction pattern is a matter to be considered for each individual case, and that no very definite rules can be given. The diffraction maxima in fig. 217, which are due to the structure of the individual units of the chain, occur at values of $(\sin\theta)/\lambda$ greater than those corresponding to the maxima in fig. 213, which have their origin in the lateral spacing of different chains in the structure; and this must necessarily be so, since the structure of the units themselves must be on a smaller scale than that of the repetition of the units. As a consequence, the maxima of fig. 217, with the spacing chosen, occur at values of $(\sin\theta)/\lambda$ at which the falling off in value of the atomic scattering factor f^2 with increasing $(\sin\theta)/\lambda$ has a very large effect, while the principal maxima of fig. 213 occur at values of $(\sin\theta)/\lambda$ for which it is not yet very appreciable. It is possible to make the general statement that if the lateral spacing of the chains is of the order of 5A

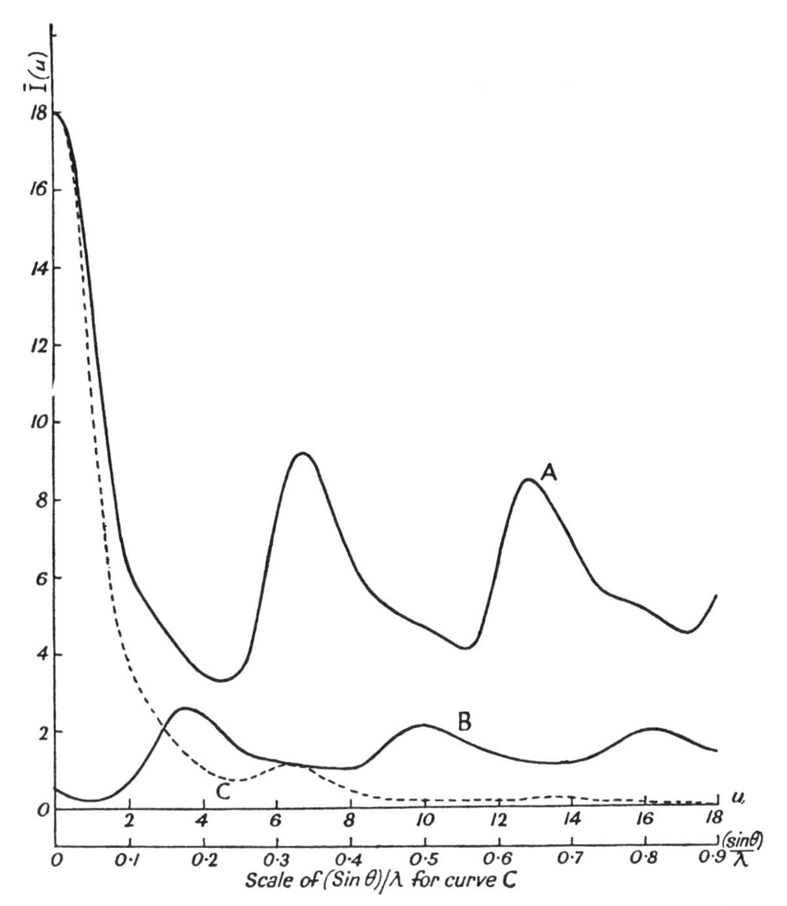

FIG. 217. Illustrating the maxima produced by the structure factor of the units of a chain

or more, the effect of the atomic scattering factor in reducing the maxima due to these spacings will be relatively small; but it will be much greater where the maxima due to the structure of the individual units are concerned, because these must necessarily be on a scale smaller than that of the spacing of the units, and so will correspond to larger values of $(\sin\theta)/\lambda$, and therefore to a much greater reduction of intensity on account of the atomic scattering factor.

(*h*) *Summary of conclusions:* It is entirely beyond the scope of this volume to discuss in any critical manner the work that has been done on the structure of fibres and the detailed interpretation of the diffraction pattern. All that we have tried to do is to indicate briefly the main optical principles that must govern the quantitative calculation of the distribution of intensity when X-rays are diffracted by assemblages of fibres. The results reached are necessarily incomplete, because of the great complexity of the problem in any actual case, but nevertheless certain broad conclusions can be stated.

(i) A parallel assemblage of single chains composed of equally spaced units, each of which is a scattering point, or a single atom, will give rise to layer lines, but to no maxima and minima of intensity along those layer lines, so long as the individual chains are so far apart that external interference effects may be neglected.

(ii) If the chains are still single, and still so far apart that external interference effects can be neglected, it may yet be possible for the diffraction pattern to show maxima and minima if the individual units of the chain themselves have a structure. The effect of the atomic scattering factors of the atoms of which the chains are composed will, however, in general tend to make the maxima so produced weak and ill-defined.

(iii) If the assemblage consists of single chains of scattering points, which are parallel and now no longer so far apart that external interference may be neglected, maxima due to the external effects will develop *on the equatorial layer line only*. These maxima may be expected to be of a degree of diffuseness not less than that of the haloes in diffraction patterns due to liquids. It is necessary to consider rather carefully some of the simpler fibre photographs, such, for example, as those from mammalian hair, in the light of this result.

(iv) When well-defined maxima occur not only on the equatorial layer line, but also on other layer lines, the structure must consist of a lateral repetition of chains in a definite array. Astbury's photographs of natural silk, and of native cellulose, to take two examples, are clearly produced by structures having this type of crystalline array, although the diffuseness of the maxima along the direction of the layer lines shows that the number of repetitions is not very large. The photograph of stretched rubber, reproduced in fig. 127, Plate XVII, of Volume I, therefore shows clearly that in the stretched substance a definitely crystalline type of array is reached, and that in this case the stretching does not merely render parallel a set of single and laterally unrelated chains.

(v) The maxima will be better defined, and the background weaker, if the chains form a three-dimensional structure, rather than a two-dimensional grid.

(vi) If the individual fibres remain straight, but are not strictly parallel, the interference maxima spread along the Debye-Scherrer rings, in the manner illustrated in fig. 211. A simultaneous bending or crumpling of the chains will cause a radial spreading of the spots, and at the same time extension along the rings, and ultimately the development of diffuse haloes. Any admixture of amorphous and unorientated material will cause the superposition of diffuse haloes on the pattern.

REFERENCES

1. P. Scherrer, *Nachr. Göttinger Gesell.*, 98 (1918): Zsigmondy's *Kolloid-chemie*, 3rd Edn., p. 394.
2. M. v. Laue, *Zeit. f. Krist.*, **64**, 115 (1926).
3. N. Seljakov, *Zeit. f. Physik*, **31**, 439, **33**, 648 (1925).
4. B. E. Warren, *Zeit. f. Krist.*, **99**, 448 (1938).
5. A. L. Patterson, *Phys. Rev.*, **56**, 972 (1939).
6. A. R. Stokes and A. J. C. Wilson, *Proc. Camb. Phil. Soc.*, **38**, 313 (1942).
7. R. Brill, *Zeit. f. Krist.*, **68**, 387 (1928); **75**, 217 (1930).
8. R. Brill and H. Pelzer, *Zeit. f. Krist.*, **72**, 398 (1929); **74**, 147 (1930).
9. G. H. Cameron and A. L. Patterson, American Society for Testing Materials, *Symposium on Radiography and X-ray Diffraction*, 1937.
10. F. W. Jones, *Proc. Roy. Soc.*, A, **166**, 16 (1938).
11. A. L. Patterson, *Zeit. f. Krist.*, **66**, 637 (1928).
12. W. L. Bragg, *Nature*, London, **127**, 738 (1931).
13. M. v. Laue, *Ann. d. Physik*, **26**, 55 (1936).
14. M. v. Laue and H. Riewe, *Zeit. f. Krist.*, (A), **95**, 408 (1936).
15. H. Lassen, *Physikal. Zeit.*, **35**, 172 (1934).
16. F. Kirchner and H. Lassen, *Ann. d. Physik*, **24**, 113 (1935).
17. M. v. Laue, *Berliner Sitzungsberichte*, p. 89 (1936).
18. R. Straubel, *Dissert.*, Jena, 1888; *Wied. Ann.*, **56**, 746 (1895).
19. F. W. Jones and C. Sykes, (i) *Proc. Roy. Soc.*, A **157**, 213 (1936); (ii) A **166**, 376 (1938).
20. O. S. Edwards and H. Lipson, *Proc. Roy. Soc.*, A **180**, 268 (1942).
21. A. J. C. Wilson, *Proc. Roy. Soc.*, A **180**, 277 (1942).
22. A. R. Stokes and A. J. C. Wilson, *Proc. Phys. Soc.*, **56**, 174 (1944).
23. A. Kochendörfer, *Zeit f. Krist.*, **101**, 149 (1939).
24. V. Daniel and H. Lipson, *Proc. Roy. Soc.*, A **181**, 368 (1943).
25. S. H. Chao and W. H. Taylor, *Proc. Roy. Soc.*, A **176**, 76 (1940).
26. V. Daniel and H. Lipson, *Proc. Roy. Soc.*, A **182**, 378 (1944).
27. D. H. Saunder, *Proc. Roy. Soc.*, A **188**, 31 (1946); A **190**, 508 (1947).
28. H. Mark and K. Andress, *Zeit. f. physikal. Chemie.*, **4**, 431 (1929).
29. G. Herzog, W. Jancke and M. Polanyi, *Zeit. f. Physik*, **3**, 196, 343 (1920).
30. J. R. Katz, *Naturwiss.*, **13**, 411 (1925).
31. W. T. Astbury and A. Street, *Phil. Trans. Roy. Soc.*, A **230**, 75 (1931); *Trans. Faraday Soc.*, **29**, 193 (1933); *Science Progress*, No. **133** (1939).

APPENDICES

APPENDIX I

SUMMARY OF VECTOR FORMULAE

For the convenience of those readers who are unfamiliar with vector methods, a brief summary of the principal results of vector algebra used in this book is given here.

(*a*) *Addition of vectors:* A vector is a quantity of which both the direction and magnitude must be given in order to specify it completely. To add two vectors **A** and **B** we represent them in both magnitude and direction by lines. Each vector is thus represented by a certain displacement, and their sum is the resultant displacement, *i.e.* the single displacement **C** that is equivalent to the displacements **A** and **B** applied successively. In fig. i, the line **C** represents the vector that is the sum of the vectors represented by the lines **A** and **B**. We may write

$$\mathbf{C} = \mathbf{A} + \mathbf{B}, \tag{1}$$

$$\text{and} \quad \mathbf{B} = \mathbf{C} - \mathbf{A}. \tag{2}$$

The figure illustrates the significance of the operators + and − in equations (1) and (2). Any number of vectors may be added by the repetition of this same process. The order in which they are taken is immaterial.

Fig. i

It is plain that the reversal of the sign of a vector is equivalent to a reversal of its direction.

A vector quantity will be represented by heavy type. Its magnitude alone, without the idea of direction, will be represented either by ordinary type A, or by the symbol $|\mathbf{A}|$.

(*b*) *Unit vectors:* A unit vector in any direction is a vector of unit magnitude in that direction. Thus if **i** is a vector of unit magnitude in the direction of the vector **A**,

$$\mathbf{A} = \mathbf{i}|\mathbf{A}| = \mathbf{i}\mathrm{A}. \tag{3}$$

The use of unit vectors enables us to separate the ideas of direction and magnitude.

Let **i**, **j**, **k** be unit vectors in the directions of the three co-ordinate axes *x*, *y*, *z*, not necessarily rectangular. Then if A_x, A_y, A_z are the components of the vector **A** referred to the three axes,

$$\mathbf{A} = \mathbf{i}\mathrm{A}_x + \mathbf{j}\mathrm{A}_y + \mathbf{k}\mathrm{A}_z, \tag{4}$$

a result illustrated by fig. ii.

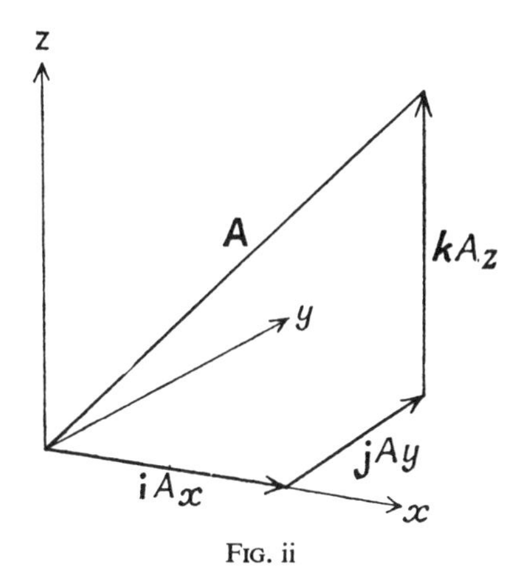

FIG. ii

(*c*) *The scalar product:* The scalar product of two vectors **A** and **B** is a scalar quantity equal to the product of the magnitudes of **A** and **B** and the cosine of the angle between their directions. It is denoted by **A·B**. Thus

$$\mathbf{A}\cdot\mathbf{B} = \mathrm{AB}\cos\theta, \tag{5}$$

θ being the angle between the directions of the two vectors. The scalar product is equal to the projection of **A** on the direction of **B**, multiplied by the magnitude of **B**, or to the projection of **B** on the direction of **A** multiplied by the magnitude of **A**. If **n** is a unit vector, **A·n** is the projection of **A** on the direction of **n**. It is clear from (5) that the scalar multiplication of vectors is commutative: the order of multiplication does not matter, and **A·B** = **B·A**.

If θ is the angle between the directions of the vectors **A** and **B**,

$$\cos\theta = (\mathbf{A}\cdot\mathbf{B})/|\mathbf{A}|\,|\mathbf{B}|. \tag{6}$$

Thus, if two vectors are mutually perpendicular, their scalar product vanishes, and conversely.

Let A_x, A_y, A_z, B_x, B_y, B_z be the components of the vectors **A** and **B** referred to rectangular axes, and let **i**, **j**, **k** be unit vectors in the directions of these axes. Then

$$\mathbf{A}\cdot\mathbf{B} = (\mathbf{i}\mathrm{A}_x + \mathbf{j}\mathrm{A}_y + \mathbf{k}\mathrm{A}_z)\cdot(\mathbf{i}\mathrm{B}_x + \mathbf{j}\mathrm{B}_y + \mathbf{k}\mathrm{B}_z). \tag{7}$$

Scalar multiplication of vectors is distributive, and since **i**, **j**, **k** are unit vectors, while **i** and **j**, **j** and **k**, **k** and **i** are mutually perpendicular,

$$\mathbf{i}\cdot\mathbf{i} = \mathbf{j}\cdot\mathbf{j} = \mathbf{k}\cdot\mathbf{k} = 1; \quad \mathbf{i}\cdot\mathbf{j} = \mathbf{j}\cdot\mathbf{k} = \mathbf{k}\cdot\mathbf{i} = 0. \tag{8}$$

Equation (7) therefore becomes

$$\mathbf{A}\cdot\mathbf{B} = \mathrm{A}_x\mathrm{B}_x + \mathrm{A}_y\mathrm{B}_y + \mathrm{A}_z\mathrm{B}_z. \tag{9}$$

We notice also that the scalar product of a vector with itself is equal to the square of its magnitude, so that

$$|\mathbf{A}|^2 = A_x^2 + A_y^2 + A_z^2, \tag{10}$$

a result that is, of course, otherwise obvious.

By (6), (9), and (10),

$$\cos\theta = (A_xB_x + A_yB_y + A_zB_z)/\sqrt{(A_x^2 + A_y^2 + A_z^2)(B_x^2 + B_y^2 + B_z^2)}. \tag{11}$$

If $\mathbf{n}$ is a unit vector, its projections on the axes are equal to its direction cosines λ, μ, ν, or

$$\lambda = \mathbf{n}\cdot\mathbf{i}, \quad \mu = \mathbf{n}\cdot\mathbf{j}, \quad \nu = \mathbf{n}\cdot\mathbf{k},$$

and the angle θ between two vectors whose direction cosines are (λ, μ, ν), (λ', μ', ν'), is thus, by (11), given by

$$\cos\theta = \lambda\lambda' + \mu\mu' + \nu\nu'. \tag{12}$$

(*d*) *The equation to a plane wave:* Let $\mathbf{r}$ be the vector from the origin to any point on a plane the perpendicular distance of which from the origin is p, and let $\mathbf{n}$ be the unit vector in the direction of the normal to the plane. Then p is the projection of $\mathbf{r}$ on the direction of $\mathbf{n}$, or $\mathbf{n}\cdot\mathbf{r}$, and so

$$p = \mathbf{n}\cdot\mathbf{r} \tag{13}$$

is the equation to the plane in vector notation. If (x, y, z) are the co-ordinates of a point on the plane referred to rectangular axes, these are the components of $\mathbf{r}$ relative to the axes, and since λ, μ, ν are the corresponding components of $\mathbf{n}$, by (13) and (9), we obtain the usual form of the equation to a plane in terms of the direction cosines of its normal,

$$\lambda x + \mu y + \nu z = p.$$

The displacement at time t at any point in a plane wave travelling in the direction of $\mathbf{n}$ may be written

$$u(\mathbf{r}, t) = a \sin\frac{2\pi}{\lambda}(vt - \mathbf{n}\cdot\mathbf{r}), \tag{14}$$

since the plane of constant phase travels with a speed v. If $\boldsymbol{\kappa}$ is a vector of magnitude $2\pi/\lambda$ in the direction of the wave-normal, (14) can be written

$$u(\mathbf{r}, t) = a \sin(\omega t - \boldsymbol{\kappa}\cdot\mathbf{r}). \tag{15}$$

These forms of the equation to a plane wave are frequently employed in the text. The amplitude factor a may itself be either a scalar or a vector quantity, depending on the nature of the wave. In the examples met with in this book it is usually a vector.

(*e*) *The vector product:* The vector product of two vectors is itself a vector whose direction is perpendicular to those of both $\mathbf{A}$ and $\mathbf{B}$, and whose magnitude is equal to the products of the magnitudes of $\mathbf{A}$

and **B** and the sine of the angle between their directions. It is written $\mathbf{A}\times\mathbf{B}$. (Alternative notations are $[\mathbf{A}, \mathbf{B}]$ and $\mathbf{A}\wedge\mathbf{B}$.)

Thus
$$|\mathbf{A}\times\mathbf{B}| = AB\sin\theta. \tag{16}$$

The positive direction of $\mathbf{A}\times\mathbf{B}$ is given by the following rule. Let **A** and **B** be represented in direction and magnitude by lines drawn from a point O, as in fig. iii, and let the parallelogram defined by **A** and **B** be completed. Then the magnitude of $\mathbf{A}\times\mathbf{B}$ is equal to the area of this parallelogram, and its direction is normal to the plane of the parallelogram. The positive direction of $\mathbf{A}\times\mathbf{B}$ is from O towards that side of the plane of the parallelogram from which the rotation through an angle less than 180° that would bring the direction of **A** into the direction of **B** would appear positive, or anti-clockwise. It is clear from this that

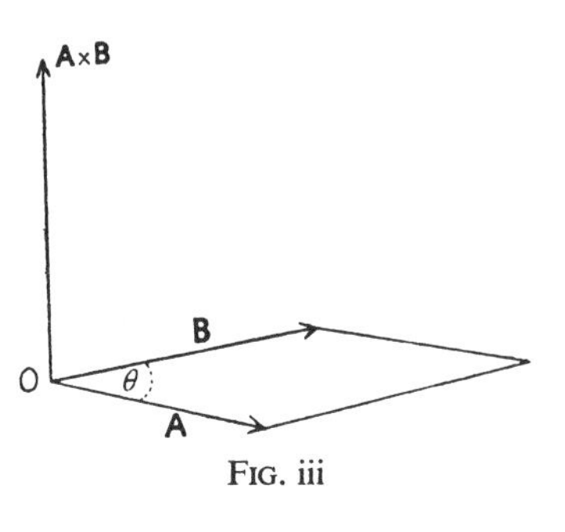

FIG. iii

$$\mathbf{A}\times\mathbf{B} = -(\mathbf{B}\times\mathbf{A}), \tag{17}$$

so that the vector product is non-commutative.

If the two vectors are parallel, their vector product vanishes, while if they are mutually perpendicular, the magnitude of the vector product is the product of their magnitudes.

If, as before, **i**, **j**, **k** are unit vectors in the direction of the rectangular axes x, y, z, the results just quoted show that

$$\mathbf{i}\times\mathbf{i} = \mathbf{j}\times\mathbf{j} = \mathbf{k}\times\mathbf{k} = 0$$

and

$$\mathbf{i}\times\mathbf{j} = \mathbf{k}, \quad \mathbf{j}\times\mathbf{i} = -\mathbf{k} \tag{18}$$

$$\mathbf{j}\times\mathbf{k} = \mathbf{i}, \quad \mathbf{k}\times\mathbf{j} = -\mathbf{i}$$

$$\mathbf{k}\times\mathbf{i} = \mathbf{j}, \quad \mathbf{i}\times\mathbf{k} = -\mathbf{j}. \tag{19}$$

Then

$$\begin{aligned}\mathbf{A}\times\mathbf{B} &= (\mathbf{i}A_x + \mathbf{j}A_y + \mathbf{k}A_z)\times(\mathbf{i}B_x + \mathbf{j}B_y + \mathbf{k}B_z)\\ &= \mathbf{i}(A_yB_z - A_zB_y) + \mathbf{j}(A_zB_x - A_xB_z) + \mathbf{k}(A_xB_y - A_yB_x),\end{aligned} \tag{20}$$

by the application of (18) and (19). The three quantities multiplying **i**, **j**, and **k** in (20) are the x, y, and z components of the vector product. It is easily verified that reversal of the order of multiplication reverses the sign of the product.

The vector product can be written in determinant notation,

$$\mathbf{A}\times\mathbf{B} = \begin{vmatrix} \mathbf{i} & \mathbf{j} & \mathbf{k} \\ A_x & A_y & A_z \\ B_x & B_y & B_z \end{vmatrix}. \tag{21}$$

It will be seen that (20) is the expansion of the determinant (21).

The double vector product $\mathbf{A}\times(\mathbf{B}\times\mathbf{C})$ is given by

$$\mathbf{A}\times(\mathbf{B}\times\mathbf{C})=\mathbf{B}(\mathbf{A}\cdot\mathbf{C})-(\mathbf{A}\cdot\mathbf{B})\mathbf{C}. \tag{22}$$

It is a vector lying in the plane containing **B** and **C**. The result follows immediately from a second application of the method used to obtain (20).

(*f*) *Vectors involving differential coefficients with respect to the co-ordinates :* (i) The *gradient* of a *scalar* function of the co-ordinates, $\phi(x, y, z)$, written grad ϕ, is defined by

$$\text{grad}\,\phi=\mathbf{i}\frac{\partial\phi}{\partial x}+\mathbf{j}\frac{\partial\phi}{\partial y}+\mathbf{k}\frac{\partial\phi}{\partial z}. \tag{23}$$

It is a vector quantity, and may be considered as the vector operator $\left(\mathbf{i}\frac{\partial}{\partial x}+\mathbf{j}\frac{\partial}{\partial y}+\mathbf{k}\frac{\partial}{\partial z}\right)$ applied to the scalar ϕ. We may write

$$\mathbf{i}\frac{\partial}{\partial x}+\mathbf{j}\frac{\partial}{\partial y}+\mathbf{k}\frac{\partial}{\partial z}=\nabla. \tag{24}$$

If, for example, ϕ represents the electrostatic potential, a scalar quantity, $-$grad ϕ gives the electric field at the point (x, y, z), a vector quantity. The direction of the gradient is that direction in which ϕ changes most rapidly with the co-ordinates.

(ii) The *divergence* of a *vector* quantity, written div **A**, is a scalar. If **A** is a vector whose value depends on the co-ordinates x, y, z,

$$\text{div}\,\mathbf{A}=\frac{\partial A_x}{\partial x}+\frac{\partial A_y}{\partial y}+\frac{\partial A_z}{\partial z}. \tag{25}$$

By (9) and (24), the divergence may be considered as the *scalar* product of the vector operator ∇ and the vector **A**, and may be written

$$\text{div}\,\mathbf{A}=\nabla\cdot\mathbf{A}. \tag{26}$$

By (23),

$$\text{div grad}\,\phi=\nabla\cdot\nabla\phi=\nabla^2\phi=\frac{\partial^2\phi}{\partial x^2}+\frac{\partial^2\phi}{\partial y^2}+\frac{\partial^2\phi}{\partial z^2}. \tag{27}$$

(iii) The *curl* or *rotation* of a vector **A**, written curl **A**, or sometimes rot **A**, is a vector whose components C_x, C_y, C_z referred to rectangular axes are given by

$$C_x=\frac{\partial A_z}{\partial y}-\frac{\partial A_y}{\partial z},\quad C_y=\frac{\partial A_x}{\partial z}-\frac{\partial A_z}{\partial x},\quad C_z=\frac{\partial A_y}{\partial x}-\frac{\partial A_x}{\partial y},$$

or

$$\text{curl}\,\mathbf{A}=\mathbf{i}\left(\frac{\partial A_z}{\partial y}-\frac{\partial A_y}{\partial z}\right)+\mathbf{j}\left(\frac{\partial A_x}{\partial z}-\frac{\partial A_z}{\partial x}\right)+\mathbf{k}\left(\frac{\partial A_y}{\partial x}-\frac{\partial A_x}{\partial y}\right). \tag{28}$$

It will be seen by comparing (28) and (20) that curl **A** may be considered as the *vector* product of the operator ∇ and the vector **A**. Thus, in this notation,

$$\text{curl}\,\mathbf{A}=\nabla\times\mathbf{A}. \tag{29}$$

The vector curl curl **A** is sometimes required. From (29), and (22),

$$\begin{aligned}\text{curl curl } \mathbf{A} &= \nabla \times \nabla \times \mathbf{A}\\ &= \nabla(\nabla\cdot\mathbf{A}) - (\nabla\cdot\nabla)\mathbf{A}\\ &= \text{grad div } \mathbf{A} - \nabla^2\mathbf{A}.\end{aligned} \tag{30}$$

The curl of a curl is, of course, itself a vector, and the x component is, by (30),

$$(\text{curl curl } \mathbf{A})_x = \frac{\partial}{\partial x}\left(\frac{\partial A_x}{\partial x} + \frac{\partial A_y}{\partial y} + \frac{\partial A_z}{\partial z}\right) - \left(\frac{\partial^2 A_x}{\partial x^2} + \frac{\partial^2 A_x}{\partial y^2} + \frac{\partial^2 A_x}{\partial z^2}\right).$$

(*g*) *Some applications to a plane vector wave :* Let the displacement in the wave be represented at the point defined by the vector **r** from the origin of co-ordinates by

$$\mathbf{u}(\mathbf{r}, t) = \mathbf{a} \sin \kappa(vt - \mathbf{n}\cdot\mathbf{r}). \tag{32}$$

The displacement is a vector quantity having a constant magnitude and direction at any time t anywhere on the plane $\mathbf{n}\cdot\mathbf{r} = \text{const.}$ The vector **a** gives the direction of the displacement and its maximum magnitude, and is supposed constant everywhere.

If λ, μ, ν are the direction cosines of the vector **n**, (32) may be written

$$\mathbf{u} = \mathbf{a} \sin \kappa\{vt - (\lambda x + \mu y + \nu z)\}.$$

Then

$$\begin{aligned}\text{div } \mathbf{u} &= \frac{\partial u_x}{\partial x} + \frac{\partial u_y}{\partial y} + \frac{\partial u_z}{\partial z}\\ &= -\kappa(\lambda a_x + \mu a_y + \nu a_z)\cos \kappa(vt - \mathbf{n}\cdot\mathbf{r})\\ &= -\kappa(\mathbf{n}\cdot\mathbf{a})\cos \kappa(vt - \mathbf{n}\cdot\mathbf{r}).\end{aligned} \tag{33}$$

If the wave is transverse, **a** is perpendicular to **n**, $\mathbf{n}\cdot\mathbf{a} = 0$, and so div **u** vanishes.

The curl of **u** may be obtained in a similar manner, using (24) and (22).

$$\begin{aligned}\text{curl } \mathbf{u} &= \nabla \times \mathbf{u}\\ &= -\kappa(\mathbf{a} \times \mathbf{n})\cos \kappa(vt - \mathbf{n}\cdot\mathbf{r}).\end{aligned} \tag{34}$$

It is a vector perpendicular both to **n** and to **a**, and for a transverse wave lies in the wave-front.

APPENDIX II

THE RECIPROCAL LATTICE

(*a*) *Definition of the reciprocal lattice :* Let the primitive translations of a space-lattice L be represented by the vectors **a**, **b**, **c**, which define a unit cell containing no lattice-points other than those lying at its corners. The vector $\mathbf{r}(u, v, w)$ to any lattice-point is then given by the vector equation

$$\mathbf{r} = u\mathbf{a} + v\mathbf{b} + w\mathbf{c}, \tag{1}$$

where u, v, w are any integers, positive, negative, or zero. We may call the lattice-point defined by (1) the point (u, v, w). A second lattice R, known as the reciprocal lattice, may be defined in terms of the lattice L in the following way. Denote the primitive translations of R by $\mathbf{a}^*$, $\mathbf{b}^*$, $\mathbf{c}^*$, which lie in the directions of the reciprocal axes. Then the reciprocal axis $\mathbf{a}^*$ is perpendicular to the axes **b** and **c**, that is to say, to the *a* plane of L, the axis $\mathbf{b}^*$ is perpendicular to **c** and **a**, and $\mathbf{c}^*$ to **a** and **b**. These relations are summed up in the vector equations

$$\mathbf{a}^*\cdot\mathbf{b} = \mathbf{a}^*\cdot\mathbf{c} = \mathbf{b}^*\cdot\mathbf{a} = \mathbf{b}^*\cdot\mathbf{c} = \mathbf{c}^*\cdot\mathbf{a} = \mathbf{c}^*\cdot\mathbf{b} = 0, \tag{2}$$

where the dot denotes the scalar product.

The magnitudes of the reciprocal vectors are fixed by the relations

$$\mathbf{a}^*\cdot\mathbf{a} = \mathbf{b}^*\cdot\mathbf{b} = \mathbf{c}^*\cdot\mathbf{c} = k^2, \tag{3}$$

k being a constant.

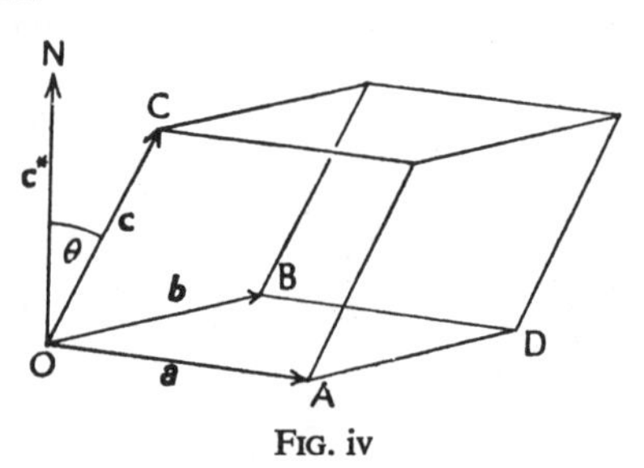

FIG. iv

Equation (3) has a simple geometrical significance. In fig. iv, OA, OB, and OC are the *a*, *b*, and *c* translations of the lattice L. ON, normal to the plane AOB, is then in the direction of the reciprocal vector $\mathbf{c}^*$. Now $\mathbf{c}\cdot\mathbf{c}^* = |\mathbf{c}|\,|\mathbf{c}^*| \cos\theta$, where θ is the angle between ON and OC, or

$$|\mathbf{c}^*| = k^2/(|\mathbf{c}| \cos\theta) = k^2/(c \text{ spacing of L}). \tag{4}$$

Thus the magnitude of the vector $\mathbf{c}^*$ is proportional to the reciprocal of the *c* spacing of the lattice L; and similarly for $\mathbf{b}^*$ and $\mathbf{a}^*$. In crystal geometry, it is customary to take $k^2 = 1$, a practice followed everywhere in this book unless otherwise stated, so that the magnitudes of the vectors $\mathbf{a}^*$, $\mathbf{b}^*$, and $\mathbf{c}^*$ are respectively equal numerically to the reciprocals of the a, b, and c spacings of the lattice L, and equations (3) take the form

$$\mathbf{a}\cdot\mathbf{a}^* = \mathbf{b}\cdot\mathbf{b}^* = \mathbf{c}\cdot\mathbf{c}^* = 1. \tag{5}$$

(*b*) *The lengths of the reciprocal axes:* The reciprocal vector $\mathbf{a}^*$ is perpendicular to $\mathbf{b}$ and $\mathbf{c}$. It is therefore in the direction of their vector product, and so can be written

$$\mathbf{a}^* = \beta(\mathbf{b} \times \mathbf{c}), \tag{6}$$

β being constant; and there are of course corresponding relations for $\mathbf{b}^*$ and $\mathbf{c}^*$. The value of β may be determined as follows. Let V be the volume of the unit cell of L, defined by the vectors $\mathbf{a}$, $\mathbf{b}$, $\mathbf{c}$. Then

$$V = \mathbf{a}\cdot(\mathbf{b} \times \mathbf{c}) = \mathbf{b}\cdot(\mathbf{c} \times \mathbf{a}) = \mathbf{c}\cdot(\mathbf{a} \times \mathbf{b}). \tag{7}$$

For if γ is the angle between $\mathbf{a}$ and $\mathbf{b}$, $|\mathbf{a} \times \mathbf{b}| = |\mathbf{a}|\ |\mathbf{b}| \sin\gamma$, and is the area of the face OADB of the cell in fig. iv, so that $\mathbf{c}\cdot(\mathbf{a} \times \mathbf{b})$ is $|\mathbf{c}| \cos\theta \times$ area of the base of the cell, and so is equal to its volume V. Thus, from (6) and (7),

$$V = (\mathbf{a}\cdot\mathbf{a}^*)/\beta = 1/\beta \tag{8}$$

by (5), and

$$\mathbf{a}^* = (\mathbf{b} \times \mathbf{c})/V, \quad \mathbf{b}^* = (\mathbf{c} \times \mathbf{a})/V, \quad \mathbf{c}^* = (\mathbf{a} \times \mathbf{b})/V, \tag{9}$$

or

$$|\mathbf{a}^*| = bc \sin\alpha/V, \quad |\mathbf{b}^*| = ca \sin\beta/V, \quad |\mathbf{c}^*| = ab \sin\gamma/V, \tag{10}$$

where α, β, and γ are respectively the angles between $\mathbf{b}$ and $\mathbf{c}$, $\mathbf{c}$ and $\mathbf{a}$, and $\mathbf{a}$ and $\mathbf{b}$.

We can write $V = Nabc$, where N is a function of α, β, and γ given by

$$N^2 = 1 + 2\cos\alpha\cos\beta\cos\gamma - \cos^2\alpha - \cos^2\beta - \cos^2\gamma, \tag{11}$$

and equations (10) then become

$$|\mathbf{a}^*| = \sin\alpha/Na, \quad |\mathbf{b}^*| = \sin\beta/Nb, \quad |\mathbf{c}^*| = \sin\gamma/Nc. \tag{12}$$

It is clear that for an orthogonal lattice the lengths of the reciprocal axes are just the reciprocals of the lengths of the lattice axes, and that the two sets of axes coincide in direction. From the way in which the reciprocal lattice is defined, and from the symmetry of equations (2) and (3), it is plain that if R is the reciprocal lattice of L, then L is the reciprocal lattice of R. It is also plain that for the orthogonal lattice $V^* = 1/V$, where V^* is the volume of the unit cell of R, and it is not difficult to show that this is quite generally true for any axes. (See §(*d*) below).

(*c*) *The angles of the reciprocal cell:* The angles α^*, β^*, γ^* between the pairs of reciprocal axes $\mathbf{b}^*$ and $\mathbf{c}^*$, $\mathbf{c}^*$ and $\mathbf{a}^*$, $\mathbf{a}^*$ and $\mathbf{b}^*$, respectively, can be calculated most easily in terms of α, β, and γ by means of a spherical triangle.

Let A, B, C, fig. v, be the intersections of the axes of the lattice L with a sphere of centre O. The spherical triangle ABC has sides α, β, and γ. OA*, normal to the plane of the great-circle through B and C, OB*, normal to the plane of the great-circle through C and A, and OC*, normal to the plane of the great-circle through A and B, are in

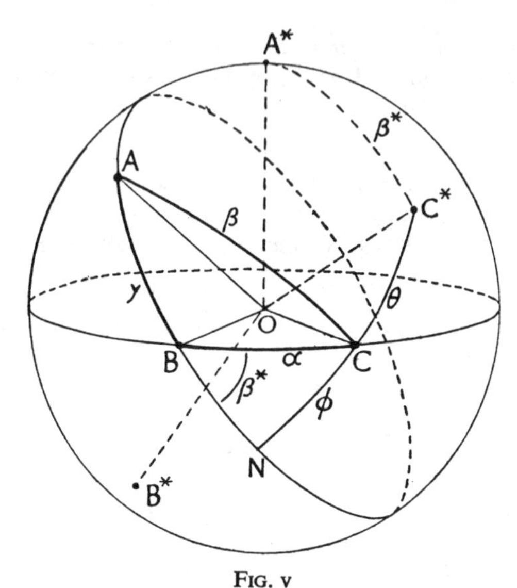

FIG. v

the directions of the axes **a***, **b***, and **c***, respectively, and the points A*, B*, C*, form a spherical triangle A*B*C* polar to the triangle ABC. The angle β^*, between OA* and OC*, is equal to the angle between the planes of the great-circles through BC and AB, and so to $\pi - \mathrm{B}$, where B is the angle of the spherical triangle ABC. Then, from the ordinary formulae for spherical triangles,

$$\cos \beta^* = -\cos \mathrm{B} = (\cos \gamma \cos \alpha - \cos \beta)/\sin \gamma \sin \alpha$$

and similarly

$$\cos \gamma^* = (\cos \alpha \cos \beta - \cos \gamma)/\sin \alpha \sin \beta$$

$$\cos \alpha^* = (\cos \beta \cos \gamma - \cos \alpha)/\sin \beta \sin \gamma. \tag{13}$$

From these formulae, the angles of the reciprocal cell can be calculated if those of the primary cell are known. In general, the geometrical formulae for the space-lattice take much simpler forms if the reciprocal quantities are used instead of the primary quantities.

(*d*) *The volumes of the unit cells of the primary and reciprocal lattices:* The volume of the unit cell of L is given by

$$\mathrm{V} = abc \sin \gamma \cos \theta,$$

where θ is the angle between $\mathbf{c}^*$ and $\mathbf{c}$ (see fig. iv). In fig. v, let C*CN, the great-circle through C* and C, cut the great-circle through A and B in N. Then the arc C*N is $\pi/2$, while $CC^* = \theta$, so that $CN = \pi/2 - \theta = \phi$ say, so that

$$\cos\theta = \sin\phi = \sin\alpha \sin\beta^*, \tag{14}$$

from the right-angled spherical triangle BCN.

$$\therefore\ V = abc \sin\alpha \sin\beta^* \sin\gamma. \tag{15}$$

By symmetry, equation (15) can be put in the alternative forms

$$V = abc \sin\alpha^* \sin\beta \sin\gamma = abc \sin\alpha \sin\beta \sin\gamma^*.$$

Thus N of equation (11) is given by

$$N = \sin\alpha^* \sin\beta \sin\gamma = \sin\alpha \sin\beta^* \sin\gamma = \sin\alpha \sin\beta \sin\gamma^*, \tag{16}$$

forms which are much more convenient for calculation than that of (11), which, however, follows at once from (16) and (13). Equation (16) leads to the relations

$$\sin\alpha^*/\sin\alpha = \sin\beta^*/\sin\beta = \sin\gamma^*/\sin\gamma. \tag{17}$$

The reciprocal relationship of V and V* now follows readily. By (9),

$$\mathbf{c}^* = (\mathbf{a} \times \mathbf{b})/V, \quad \text{and conversely} \quad \mathbf{c} = (\mathbf{a}^* \times \mathbf{b}^*)/V^*.$$

Since $\mathbf{c}\cdot\mathbf{c}^* = 1$, the reciprocal relation of V and V* follows if it can be shown that $(\mathbf{a} \times \mathbf{b})\cdot(\mathbf{a}^* \times \mathbf{b}^*) = 1$.

Now $(\mathbf{a} \times \mathbf{b})$ is in the direction of $\mathbf{c}^*$, and $(\mathbf{a}^* \times \mathbf{b}^*)$ is in the direction of $\mathbf{c}$. Their scalar product can therefore be written

$$\begin{aligned}(\mathbf{a} \times \mathbf{b})\cdot(\mathbf{a}^* \times \mathbf{b}^*) &= ab \sin\gamma a^*b^* \sin\gamma^* \cos\theta \\ &= ab \sin\gamma a^*b^* \sin\gamma^* \sin\alpha \sin\beta^* \text{ by (14).}\end{aligned}$$

Moreover, if ψ and χ are respectively the angles between $\mathbf{a}$ and $\mathbf{a}^*$, and $\mathbf{b}$ and $\mathbf{b}^*$, it follows by symmetry from (14) that

$$\cos\psi = \sin\beta \sin\gamma^* = \sin\beta^* \sin\gamma \text{ (from (17))}$$

$$\cos\chi = \sin\gamma \sin\alpha^* = \sin\gamma^* \sin\alpha.$$

Equation (18) can now be written

$$(\mathbf{a} \times \mathbf{b})\cdot(\mathbf{a}^* \times \mathbf{b}^*) = aa^* \cos\psi\, bb^* \cos\chi = (\mathbf{a}\cdot\mathbf{a}^*)(\mathbf{b}\cdot\mathbf{b}^*) = 1 \quad \text{by (5).}$$

Therefore $VV^* = 1$.

(*e*) *Some properties of the reciprocal lattice:* We shall now prove some properties of the reciprocal lattice that make it of great use in crystal structure calculations.

(i) The vector $\mathbf{r}^*(hkl)$, defined by

$$\mathbf{r}^* = h\mathbf{a}^* + k\mathbf{b}^* + l\mathbf{c}^*, \tag{19}$$

where h, k, and l are integers, to the point (h, k, l) in the reciprocal lattice R, is normal to the planes whose Miller indices are (hkl) in the primary lattice L, and (ii), the magnitude $|\mathbf{r}^*(hkl)|$ of this vector is the reciprocal of the spacing $d(hkl)$ of the planes (hkl) of the lattice L, the term *spacing* being here used in the somewhat special sense explained below.

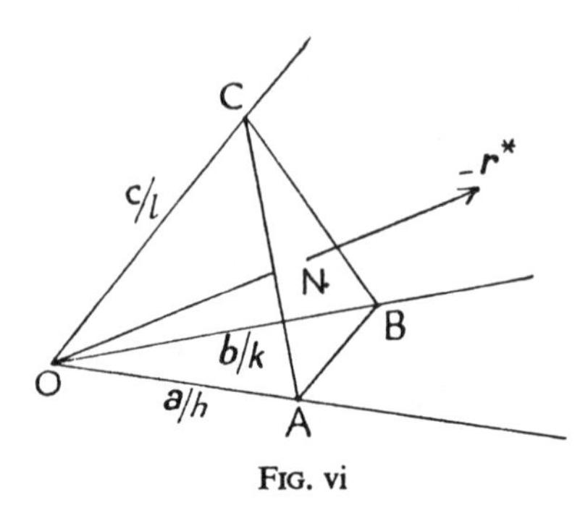

FIG. vi

Let ABC, fig. vi, be the plane of the set (hkl) that passes nearest to the origin O. Then the lengths OA, OB, OC are respectively a/h, b/k, c/l. We have to show that the vector $\mathbf{r}^*(hkl)$ is perpendicular to the plane ABC. To do this it is sufficient to show that it is perpendicular to any two of the vectors representing the sides of the triangle ABC. Expressed as a vector, $\overrightarrow{AB}$ is equal to $\mathbf{b}/k - \mathbf{a}/h$. We take the scalar product of this vector with $\mathbf{r}^*(hkl)$. This gives

$$(h\mathbf{a}^* + k\mathbf{b}^* + l\mathbf{c}^*)\cdot(\mathbf{b}/k - \mathbf{a}/h) = \mathbf{a}\cdot\mathbf{a}^* - \mathbf{b}\cdot\mathbf{b}^* = 0,$$

by (2) and (5). Since the scalar product of the two vectors vanishes, they must be mutually perpendicular; and in an exactly similar way we can show that $\mathbf{r}^*(hkl)$ is perpendicular to the vectors $\overrightarrow{BC}$ and $\overrightarrow{CA}$. Thus the first part of the proposition is proved.

To prove the second part we proceed as follows. The spacing $d(hkl)$ of planes (hkl) of the lattice L is equal to the length of the normal ON onto the plane ABC from the origin O. ON is the projection of OA onto the direction of the normal, so that if $\mathbf{n}$ is a unit vector in the direction of the normal, the length of ON is the scalar product of the vector $\overrightarrow{OA}$ with $\mathbf{n}$. Since $\mathbf{r}^*$ is in the direction of ON, $\mathbf{n} = \mathbf{r}^*/|\mathbf{r}^*|$. Thus

$$d(hkl) = \mathbf{n}\cdot\mathbf{a}/h = (\mathbf{r}^*\cdot\mathbf{a})/h|\mathbf{r}^*| = 1/|\mathbf{r}^*|, \tag{20}$$

by (2) and (3). Thus the magnitude of the vector $\mathbf{r}^*(hkl)$ is the reciprocal of the spacing of the planes (hkl) of the primary lattice L.

In this proof it has been assumed that h, k, and l, being the Miller indices of the plane (hkl), have no common factor. If θ_m is the glancing angle of reflection for the mth order spectrum from the planes (hkl) with wave-length λ, then

$$2d(hkl) \sin \theta_m = m\lambda$$

or

$$(2 \sin \theta_m)/\lambda = m|\mathbf{r}^*(hkl)|$$
$$= |\mathbf{r}^*(mh, mk, ml)|. \tag{21}$$

Every point in the reciprocal lattice can therefore be considered as representing a spectrum. The mth point from the origin in a given row of reciprocal-lattice points passing through the origin represents the mth order spectrum from the planes (hkl). If then we remove all restrictions as to common factors in the indices, and define the *spacing* $d(hkl)$ corresponding to the three indices h, k, l by the equation

$$2d(hkl) \sin\theta = \lambda, \tag{22}$$

we can write quite generally

$$d(hkl) = 1/|\mathbf{r}^*(hkl)|, \tag{23}$$

where h, k, and l are any integers whatsoever, $d(hkl)$ being in fact that spacing which would produce, as a first order, a spectrum at the glancing angle θ with the wave-length λ.

It is the property of the reciprocal lattice discussed above that makes it of particular use in problems of crystal geometry. The consideration of sets of planes is replaced by the geometrically much simpler consideration of sets of points. Examples of the use of the reciprocal lattice in diffraction problems are given in Chapter I, and at many other points in this book. This appendix provides the necessary mathematical background.

(*f*) *The use of reciprocal lattices in crystal geometry:* Quite apart from the great value of the reciprocal lattice in considering diffraction problems, its use simplifies greatly many of the problems that arise in the geometry of space-lattices. We shall give a few examples.

(1) Calculation of lattice spacings. By equation (23),

$$d(hkl) = 1/|\mathbf{r}^*(hkl)|.$$

Now

$$\begin{aligned}|\mathbf{r}^*(hkl)|^2 &= |h\mathbf{a}^* + k\mathbf{b}^* + l\mathbf{c}^*|^2 \\ &= h^2\mathbf{a}^{*2} + k^2\mathbf{b}^{*2} + l^2\mathbf{c}^{*2} + 2hk\mathbf{a}^*\cdot\mathbf{b}^* + 2kl\mathbf{b}^*\cdot\mathbf{c}^* + 2lh\mathbf{c}^*\cdot\mathbf{a}^* \\ &= h^2a^{*2} + k^2b^{*2} + l^2c^{*2} + 2hka^*b^*\cos\gamma^* + 2klb^*c^*\cos\alpha^* \\ &\qquad + 2lhc^*a^*\cos\beta^*,\end{aligned} \tag{24}$$

whence $d(hkl)$ can be calculated.

Except for crystals of high symmetry, it is generally advantageous to calculate a^*, b^*, c^*, and α^*, β^*, γ^*, from equations (10) and (13), and to do all subsequent calculations in terms of these quantities. Equation (24), for example, is much simpler than the corresponding one in terms of the lattice quantities a, b, c, and α, β, γ.

If θ is the glancing angle for the spectrum (hkl),

$$\sin^2\theta = \frac{\lambda^2}{4}\{h^2a^{*2} + k^2b^{*2} + l^2c^{*2} + 2hka^*b^*\cos\gamma^* + 2klb^*c^*\cos\alpha^* + 2lhc^*a^*\cos\beta^*\}. \tag{25}$$

This is the general form for a triclinic crystal. For a monoclinic crystal, $\alpha^* = \gamma^* = 90°$, and

$$\sin\theta = \tfrac{1}{2}\lambda\sqrt{h^2a^{*2} + k^2b^{*2} + l^2c^{*2} + 2hlc^*a^*\cos\beta^*}. \tag{26}$$

For a rhombohedral crystal, $\alpha^* = \beta^* = \gamma^*$, $a^* = b^* = c^*$, and

$$\sin\theta = \tfrac{1}{2}\lambda a^*\sqrt{h^2 + k^2 + l^2 + 2(hk + kl + lh)\cos\alpha^*}. \tag{27}$$

(2) The angle between two sets of planes $(h_1k_1l_1)$, $(h_2k_2l_2)$. This is equal to the angle between the normals to the planes, and so to that between the two reciprocal vectors $\mathbf{r}^*(h_1k_1l_1)$, $\mathbf{r}^*(h_2k_2l_2)$, or $\mathbf{r}_1^*$, $\mathbf{r}_2^*$. If ϕ is the angle,

$$\mathbf{r}_1^*\cdot\mathbf{r}_2^* = r_1^* r_2^* \cos\phi,$$

$$\cos\phi = (h_1\mathbf{a}^* + k_1\mathbf{b}^* + l_1\mathbf{c}^*)\cdot(h_2\mathbf{a}^* + k_2\mathbf{b}^* + l_2\mathbf{c}^*)\,d_1d_2,$$

where d_1 and d_2 are the corresponding spacings. Or, multiplying out,

$$\cos\phi = \{h_1h_2a^{*2} + k_1k_2b^{*2} + l_1l_2c^{*2} + (h_1k_2 + h_2k_1)a^*b^*\cos\gamma^*$$
$$+ (k_1l_2 + k_2l_1)b^*c^*\cos\alpha^* + (l_1h_2 + l_2h_1)c^*a^*\cos\beta^*\}\,d_1d_2. \tag{27}$$

(3) The vector product of two space-lattice vectors: Let $\mathbf{r}_1$ and $\mathbf{r}_2$ be the vectors to the points $(u_1, v_1, w_1), (u_2, v_2, w_2)$ of the space-lattice, so that

$$\mathbf{r}_1 = u_1\mathbf{a} + v_1\mathbf{b} + w_1\mathbf{c}$$

$$\mathbf{r}_2 = u_2\mathbf{a} + v_2\mathbf{b} + w_2\mathbf{c}.$$

Then

$$\mathbf{r}_1 \times \mathbf{r}_2 = (v_1w_2 - v_2w_1)\,\mathbf{b}\times\mathbf{c} + (w_1u_2 - w_2u_1)\,\mathbf{c}\times\mathbf{a}$$
$$+ (u_1v_2 - u_2v_1)\,\mathbf{a}\times\mathbf{b}, \tag{29}$$

since $\mathbf{a}\times\mathbf{a} = \mathbf{b}\times\mathbf{b} = \mathbf{c}\times\mathbf{c} = 0$, and $(\mathbf{a}\times\mathbf{b}) = -(\mathbf{b}\times\mathbf{a})$, and so on. Using equation (9), we can now write (29) in the form

$$\mathbf{r}_1 \times \mathbf{r}_2 = \mathrm{V}_p\,\{(v_1w_2 - v_2w_1)\,\mathbf{a}^* + (w_1u_2 - w_2u_1)\,\mathbf{b}^* + (u_1v_2 - u_2v_1)\,\mathbf{c}^*\} \tag{30}$$

$$= \mathrm{V}_p \begin{vmatrix} \mathbf{a}^* & \mathbf{b}^* & \mathbf{c}^* \\ u_1 & v_1 & w_1 \\ u_2 & v_2 & w_2 \end{vmatrix}. \tag{31}$$

V_p is here the volume of a primitive cell of the lattice.

(4) The scalar product of a lattice vector and a vector in the reciprocal lattice:

$$\mathbf{r}\cdot\mathbf{r}^* = (u\mathbf{a} + v\mathbf{b} + w\mathbf{c})\cdot(h\mathbf{a}^* + k\mathbf{b}^* + l\mathbf{c}^*)$$
$$= hu + kv + lw, \tag{32}$$

by (2) and (5). This scalar product must always be an integer.

(5) Condition that the cell defined by three lattice vectors shall be primitive: Let **a**, **b**, **c** be the primitive translations of a lattice. There are an infinite number of other sets of three vectors which can be used

as primitive translations. The necessary condition is that the volume of the cell defined by the three vectors should be the same as that of the cell defined by the vectors **a**, **b**, **c**. Consider the three vectors $\mathbf{r}_1$, $\mathbf{r}_2$, $\mathbf{r}_3$, from the origin to the points (u_1, v_1, w_1), (u_2, v_2, w_2), (u_3, v_3, w_3) of the lattice. V, the volume of the cell defined by the vectors, is given by

$$\begin{aligned} \mathrm{V} &= \mathbf{r}_1 \cdot (\mathbf{r}_2 \times \mathbf{r}_3) \\ &= \mathrm{V}_p (u_1\mathbf{a} + v_1\mathbf{b} + w_1\mathbf{c}) \cdot \{(v_2w_3 - v_3w_2)\mathbf{a}^* + (w_2u_3 - w_3u_2)\mathbf{b}^* \\ &\qquad + (u_2v_3 - u_3v_2)\mathbf{c}^*\}, \quad \text{by (30)}, \\ &= \mathrm{V}_p\{u_1(v_2w_3 - v_3w_2) + v_1(w_2u_3 - w_3u_2) + w_1(u_2v_3 - u_3v_2)\} \\ &= \mathrm{V}_p \begin{vmatrix} u_1 & v_1 & w_1 \\ u_2 & v_2 & w_2 \\ u_3 & v_3 & w_3 \end{vmatrix}. \end{aligned} \tag{33}$$

If the cell defined by the three vectors is to be primitive, $\mathrm{V} = \mathrm{V}_p$ numerically, so that the condition may be written

$$\begin{vmatrix} u_1 & v_1 & w_1 \\ u_2 & v_2 & w_2 \\ u_3 & v_3 & w_3 \end{vmatrix} = \pm 1. \tag{34}$$

The negative sign must be included, since the determinant may have a negative value. If the determinant has the value $\pm n$, the corresponding cell contains n lattice-points.

(6) Zone rules: (i) Any row of points in a space-lattice is common to an indefinite number of lattice-planes. All such planes, containing a common direction, are said to belong to the same zone. Any vector $\mathbf{r} = u\mathbf{a} + v\mathbf{b} + w\mathbf{c}$ in the space-lattice defines a row of points, and so a zone. The symbol of this zone is $[uvw]$, and in writing it, it is customary to use values of u, v, and w containing no common factor. It is required to find the condition that a plane (hkl) shall belong to the zone $[uvw]$.

Since the plane, if it belongs to the zone, either contains or is parallel to the vector **r**, any normal to the plane is perpendicular to **r**. But the reciprocal vector $\mathbf{r}^*$ to the point (h, k, l) of the reciprocal lattice is normal to the plane (hkl), and so the required condition is

$$\mathbf{r} \cdot \mathbf{r}^* = 0, \quad \text{or} \quad hu + kv + lw = 0, \quad \text{by (32)}. \tag{35}$$

(ii) Any two planes $(h_1k_1l_1)$, $(h_2k_2l_2)$ define a zone. To find its symbol. Let it be $[uvw]$. Then, by (35),

$$h_1u + k_1v + l_1w = 0,$$
$$h_2u + k_2v + l_2w = 0,$$

whence $$u : v : w = (k_1l_2 - k_2l_1) : (l_1h_2 - l_2h_1) : (h_1k_2 - h_2k_1). \tag{36}$$

(iii) The condition that three planes $(h_1k_1l_1)$, $(h_2k_2l_2)$, $(h_3k_3l_3)$ shall belong to the same zone is evidently that the three corresponding reciprocal vectors shall lie in one plane through the origin, and shall therefore define a cell in the reciprocal lattice whose volume is zero. By equation (33), the required condition is therefore

$$\begin{vmatrix} h_1 & k_1 & l_1 \\ h_2 & k_2 & l_2 \\ h_3 & k_3 & l_3 \end{vmatrix} = 0.$$

(iv) The condition that a given lattice-point lies in a given plane: Let the plane be one of the set (hkl), and the point (u, v, w). The normal to the plane is in the direction of $\mathbf{r}^*(hkl)$, or $\mathbf{r}^*$, for brevity. Now if (u, v, w) lies in any plane the projection of $\mathbf{r}(uvw)$ onto the direction of $\mathbf{r}^*$ is equal to p, the perpendicular distance of the plane (hkl) from the origin. If the plane is the nth of the set from the origin,

$$p = nd(hkl) = n/|\mathbf{r}^*|.$$

Thus

$$\mathbf{r}\cdot\mathbf{r}^*/|\mathbf{r}^*| = p = n/|\mathbf{r}^*|,$$

or

$$hu + kv + lw = n, \quad \text{by (32).} \tag{38}$$

(v) Layer-line conditions: The last result has a useful reciprocal interpretation. The zone axis $[uvw]$ of the space-lattice L, (*i.e.* the vector $\mathbf{r}$) is perpendicular to the *plane* (uvw) of the reciprocal lattice R. If (h, k, l) are the co-ordinates of a point lying in the nth reciprocal-lattice plane of the set (uvw) from the origin,

$$hu + kv + lw = n, \tag{39}$$

as above. The points fulfilling this condition are those that give rise to the nth layer line in a rotation photograph when the crystal is rotated about the axis $[uvw]$. For example, for rotation about [100], the condition is simply $h = n$. All points on the nth layer line have $h = n$, and there is no other restriction. For rotation about [111], $h + k + l = n$, and so on.

(vi) Change of indices corresponding to change of axes: Let hkl be the indices of a spectrum referred to axes with translations $\mathbf{a}$, $\mathbf{b}$, $\mathbf{c}$. Suppose a new set of axes with translations $\mathbf{A}$, $\mathbf{B}$, $\mathbf{C}$, given by

$$\begin{aligned} \mathbf{A} &= u_1\mathbf{a} + v_1\mathbf{b} + w_1\mathbf{c} \\ \mathbf{B} &= u_2\mathbf{a} + v_2\mathbf{b} + w_2\mathbf{c} \\ \mathbf{C} &= u_3\mathbf{a} + v_3\mathbf{b} + w_3\mathbf{c} \end{aligned} \tag{40}$$

to be chosen. It is required to find HKL, the indices of the same spectrum referred to the new axes.

Let $\mathbf{A}^*$, $\mathbf{B}^*$, $\mathbf{C}^*$ be the vectors reciprocal to the new translations, so that

$$\mathbf{A}\cdot\mathbf{A}^* = \mathbf{B}\cdot\mathbf{B}^* = \mathbf{C}\cdot\mathbf{C}^* = 1, \quad \mathbf{A}\cdot\mathbf{B}^* = \mathbf{A}\cdot\mathbf{C}^* = \ldots = 0. \tag{41}$$

The vector to a given reciprocal-lattice point, corresponding to a given spectrum, is the same to whatever axes the co-ordinates of the point may be referred, and so

$$\mathrm{H}\mathbf{A}^* + \mathrm{K}\mathbf{B}^* + \mathrm{L}\mathbf{C}^* = h\mathbf{a}^* + k\mathbf{b}^* + l\mathbf{c}^*. \tag{42}$$

On multiplying each side of (42) by **A**, we obtain, using (41),

$$\mathrm{H}\mathbf{A}\cdot\mathbf{A}^* = (u_1\mathbf{a} + v_1\mathbf{b} + w_1\mathbf{c})\cdot(h\mathbf{a}^* + k\mathbf{b}^* + l\mathbf{c}^*),$$

or

$$\mathrm{H} = hu_1 + kv_1 + lw_1,$$

and similarly

$$\mathrm{K} = hu_2 + kv_2 + lw_2 \qquad (43)$$

$$\mathrm{L} = hu_3 + kv_3 + lw_3,$$

which are the required transformations.

TABLES FOR ESTIMATING THE CORRECTION TO BE APPLIED TO THE SCATTERING FACTOR ON ACCOUNT OF DISPERSION BY THE K ELECTRONS

Tables I and II of this appendix give respectively the values of $\Delta f'_K$ and $\Delta f''_K$, the real and imaginary parts of the correction to be applied to the scattering factor on account of dispersion by the K electrons, calculated by means of Hönl's formula, equation (*4*.61) of Chapter IV, p. 160. The tables show the values of the corrections as functions of λ/λ_K and δ_K, λ being the wave-length of the radiation scattered, λ_K the wave-length of the K absorption-edge of the scattering element, and δ_K a parameter characteristic of the scattering element the significance of which is explained in Chapter IV, § 1 (*n*). The values of δ_K for a number of elements will be found in Table IV. 1, p. 159, calculated for an element of atomic number Z from the formula

$$\delta_K = (A - 911/\lambda_K)/A, \tag{1}$$

where

$$A = (Z - 0{\cdot}3)^2 + 1{\cdot}33 \times 10^{-5}(Z - 0{\cdot}3)^4 + 3{\cdot}55 \times 10^{-10}\,(Z - 0{\cdot}3)^6 + 11{\cdot}7 \times 10^{-15}\,(Z - 0{\cdot}3)^8 + \ldots \tag{2}$$

TABLE I

Δf as a function of λ/λ_K and δ_K

λ/λ_K \ δ_K	0·12	0·14	0·16	0·18	0·20	0·22	0·24	0·26	0·28	0·30
0·05	0·02	0·02	0·02	0·02	0·02	0·02	0·03	0·03	0·03	0·03
0·10	0·05	0·06	0·06	0·06	0·07	0·07	0·07	0·08	0·08	0·08
0·15	0·10	0·10	0·11	0·11	0·12	0·12	0·13	0·13	0·14	0·15
0·20	0·14	0·15	0·15	0·16	0·17	0·18	0·18	0·19	0·20	0·21
0·25	0·18	0·19	0·20	0·21	0·22	0·23	0·24	0·25	0·26	0·27
0·30	0·22	0·23	0·24	0·25	0·26	0·27	0·28	0·29	0·30	0·32
0·35	0·25	0·25	0·26	0·28	0·29	0·30	0·31	0·32	0·34	0·35
0·40	0·26	0·27	0·28	0·29	0·30	0·31	0·32	0·33	0·35	0·36
0·45	0·25	0·26	0·27	0·28	0·29	0·30	0·32	0·33	0·34	0·35
0·50	0·23	0·24	0·24	0·25	0·26	0·27	0·28	0·29	0·29	0·30
0·55	0·19	0·19	0·20	0·20	0·21	0·22	0·22	0·23	0·23	0·24
0·60	0·12	0·12	0·12	0·12	0·12	0·12	0·12	0·12	0·12	0·12
0·65	0·02	0·02	0·01	0·01	0·00	−0·00	−0·01	−0·02	−0·03	−0·04
0·70	−0·12	−0·13	−0·14	−0·15	−0·16	−0·18	−0·19	−0·21	−0·23	−0·25
0·75	−0·30	−0·32	−0·34	−0·35	−0·38	−0·40	−0·43	−0·46	−0·49	−0·53
0·80	−0·52	−0·54	−0·57	−0·60	−0·63	−0·66	−0·70	−0·74	−0·78	−0·83
0·85	−0·89	−0·93	−0·97	−1·02	−1·07	−1·12	−1·18	−1·24	−1·31	−1·39
0·90	−1·38	−1·44	−1·50	−1·57	−1·65	−1·72	−1·81	−1·90	−2·00	−2·10
0·95	−2·22	−2·31	−2·41	−2·51	−2·62	−2·74	−2·87	−3·00	−3·14	−3·30
0·975	−2·93	−3·04	−3·17	−3·30	−3·44	−3·59	−3·75	−3·92	−4·10	−4·29
0·980	−3·17	−3·30	−3·43	−3·57	−3·72	−3·88	−4·06	−4·23	−4·43	−4·54
0·985	−3·46	−3·60	−3·75	−3·90	−4·06	−4·23	−4·42	−4·61	−4·82	−5·04
0·990	−3·91	−4·06	−4·23	−4·40	−4·57	−4·76	−4·99	−5·21	−5·44	−5·69

TABLE I—*Continued*

λ/λ_K \ δ_K	0·12	0·14	0·16	0·18	0·20	0·22	0·24	0·26	0·28	0·30
1·005	−4·72	−4·89	−5·20	−5·30	−5·52	−5·75	−5·99	−6·26	−6·53	−6·83
1·010	−4·09	−4·25	−4·42	−4·59	−4·79	−4·99	−5·20	−5·43	−5·67	−5·93
1·015	−3·73	−3·88	−4·03	−4·19	−4·37	−4·55	−4·75	−4·96	−5·18	−5·45
1·020	−3·49	−3·62	−3·77	−3·92	−4·08	−4·26	−4·54	−4·64	−4·85	−5·07
1·025	−3·28	−3·41	−3·55	−3·69	−3·85	−4·01	−4·19	−4·37	−4·57	−4·78
1·030	−3·13	−3·25	−3·39	−3·52	−3·67	−3·83	−3·99	−4·17	−4·36	−4·56
1·035	−3·00	−3·12	−3·25	−3·38	−3·52	−3·67	−3·83	−4·00	−4·18	−4·38
1·040	−2·90	−3·02	−3·14	−3·27	−3·40	−3·55	−3·71	−3·87	−4·04	−4·23
1·045	−2·81	−2·92	−3·03	−3·16	−3·29	−3·43	−3·58	−3·74	−3·91	−4·10
1·050	−2·72	−2·83	−2·95	−3·07	−3·20	−3·33	−3·48	−3·63	−3·80	−3·98
1·055	−2·65	−2·76	−2·87	−2·99	−3·12	−3·25	−3·39	−3·53	−3·70	−3·88
1·060	−2·59	−2·69	−2·80	−2·92	−3·04	−3·17	−3·31	−3·46	−3·62	−3·79
1·065	−2·53	−2·63	−2·74	−2·85	−2·97	−3·10	−3·23	−3·38	−3·53	−3·70
1·070	−2·47	−2·57	−2·68	−2·79	−2·91	−3·03	−3·17	−3·31	−3·46	−3·62
1·08	−2·38	−2·47	−2·57	−2·68	−2·80	−2·91	−3·04	−3·18	−3·33	−3·48
1·09	−2·30	−2·39	−2·50	−2·59	−2·70	−2·82	−2·94	−3·08	−3·22	−3·37
1·10	−2·23	−2·32	−2·41	−2·51	−2·62	−2·73	−2·86	−2·98	−3·12	−3·27
1·15	−1·98	−2·05	−2·14	−2·23	−2·32	−2·42	−2·53	−2·65	−2·77	−2·92
1·20	−1·82	−1·89	−1·97	−2·05	−2·14	−2·23	−2·33	−2·44	−2·55	−2·69
1·25	−1·70	−1·77	−1·85	−1·92	−2·01	−2·09	−2·19	−2·29	−2·40	−2·51
1·30	−1·64	−1·71	−1·78	−1·86	−1·94	−2·02	−2·11	−2·21	−2·32	−2·43
1·40	−1·50	−1·56	−1·63	−1·69	−1·77	−1·85	−1·93	−2·02	−2·11	−2·22
1·50	−1·42	−1·48	−1·54	−1·61	−1·68	−1·75	−1·83	−1·92	−2·01	−2·11
1·60	−1·37	−1·43	−1·49	−1·55	−1·62	−1·69	−1·76	−1·85	−1·93	−2·03
1·80	−1·29	−1·35	−1·40	−1·46	−1·53	−1·59	−1·67	−1·75	−1·83	−1·92
2·00	−1·25	−1·30	−1·35	−1·41	−1·47	−1·54	−1·61	−1·68	−1·76	−1·85
3·0	−1·15	−1·20	−1·25	−1·30	−1·36	−1·42	−1·49	−1·56	−1·63	−1·71
∞ $(\Delta f'_K)_\infty = g_K$	−1·09	−1·14	−1·18	−1·23	−1·29	−1·35	−1·41	−1·47	−1·55	−1·62

TABLE II

$\Delta f''_K$ as a function of λ/λ_K and δ_K

λ/λ_K \ δ_K	0·12	0·14	0·16	0·18	0·20	0·22	0·24	0·26	0·28	0·30
0·0	0·00	0·00	0·00	0·00	0·00	0·00	0·00	0·00	0·00	0·00
0·1	0·04	0·04	0·05	0·05	0·05	0·05	0·05	0·06	0·06	0·06
0·2	0·16	0·17	0·17	0·18	0·19	0·20	0·21	0·22	0·23	0·25
0·3	0·35	0·36	0·38	0·40	0·42	0·44	0·46	0·48	0·51	0·54
0·4	0·60	0·63	0·65	0·68	0·72	0·75	0·79	0·83	0·87	0·92
0·5	0·91	0·95	0·99	1·03	1·08	1·13	1·18	1·24	1·31	1·37
0·6	1·26	1·32	1·37	1·43	1·50	1·56	1·64	1·72	1·80	1·89
0·7	1·66	1·73	1·80	1·88	1·96	2·04	2·14	2·24	2·34	2·46
0·8	2·09	2·17	2·26	2·35	2·45	2·56	2·67	2·79	2·92	3·05
0·9	2·55	2·65	2·75	2·86	2·98	3·10	3·23	3·37	3·52	3·67
1·0	3·03	3·14	3·26	3·38	3·52	3·65	3·80	3·96	4·12	4·30

$\Delta f''_K = 0$ for $\lambda/\lambda_K > 1$

If f_0 is the scattering factor for waves short in comparison with the wave-length of the absorption edge of the scattering element, that is to say, the scattering factor as usually tabulated, the scattering factor f corrected for dispersion by the K electrons is given by

$$f = f_0 + \Delta f'_K + i\,\Delta f''_K, \tag{3}$$

so that

$$|f| = \sqrt{(f_0 + \Delta f'_K)^2 + (\Delta f''_K)^2}. \tag{4}$$

Equation (4) can be written quite nearly enough

$$|f| = f_0 + \Delta f'_K + \tfrac{1}{2}\frac{(\Delta f''_K)^2}{f_0 + \Delta f'_K}, \tag{5}$$

whence the value of $|f|$ can be calculated by means of Tables I and II, and tables of scattering factors, such as Tables XIV and XV of Vol. I.

Examples : (1) Suppose the value of f for the scattering of Cu Kα radiation by Ni is required at an angle at which $(\sin\theta)/\lambda = 0{\cdot}5$. In this case, $\lambda = 1{\cdot}539$A, $\lambda_K = 1{\cdot}484$A, the K absorption-edge of Ni, so that $\lambda/\lambda_K = 1{\cdot}037$.

From equations (1) and (2), δ_K for Ni (Z = 28) is 0·208, and by simple interpolation in Table I we find $\Delta f'_K = -3{\cdot}53$.

$$\text{Since } \lambda/\lambda_K > 1, \quad \Delta f''_K = 0.$$

The value of f_0 for Ni for $(\sin\theta)/\lambda = 0{\cdot}5$ is 12·8 according to Table XV, Vol. I.

Thus,

$$|f| = f_0 + \Delta f'_K = 12{\cdot}8 - 3{\cdot}5 = 9{\cdot}3.$$

(2) As a second example, we consider the scattering of Mo Kα radiation by Ni. Here, $\lambda = 0{\cdot}710$, and $\lambda/\lambda_K = 0{\cdot}478$. From Table I, $\Delta f'_K = +0{\cdot}28$, and from Table II, $\Delta f''_K = +1{\cdot}02$. Then

$$\delta f = |f| - f_0 = 0{\cdot}28 + \tfrac{1}{2}\frac{(1{\cdot}02)^2}{12{\cdot}8 + 0{\cdot}3} = 0{\cdot}32$$

and

$$|f| = 12{\cdot}8 + 0{\cdot}3$$
$$= 13{\cdot}1.$$

APPENDIX IV

DERIVATION OF THE FOURIER INTEGRAL

We take as an example the derivation of equation (7.96), the integral actually used in deducing the radial distribution.

Let
$$f(x) = \int_0^{\infty} A(p) \sin px \, dp. \tag{1}$$

To determine the value of $A(p)$ for some particular value, q, of p we multiply both sides of (1) by $\sin qx \, dx$, and integrate with respect to x between the limits $\pm\infty$. Then

$$\int_{-\infty}^{+\infty} f(x) \sin qx \, dx = \int_{-\infty}^{+\infty}\int_0^{\infty} A(p) \sin px \sin qx \, dp \, dx$$
$$= \int_{-\infty}^{+\infty}\int_0^{\infty} \tfrac{1}{2}A(p)\{\cos(p-q)x - \cos(p+q)x\} \, dp \, dx. \tag{2}$$

$A(p)$ is a fairly slowly varying function of p, but the cosine term oscillates rapidly unless p is nearly equal to q. We therefore take $A(p)$ outside the integral, giving p the value q. Thus

$$\int_{-\infty}^{+\infty} f(x) \sin qx \, dx = \tfrac{1}{2}A(q)\int_{-\infty}^{+\infty}\int_0^{\infty} \{\cos(p-q)x - \cos(p+q)x\} \, dp \, dx. \tag{3}$$

Consider first the integral

$$I_1 = \int_{-\infty}^{+\infty}\int_0^{\infty} \cos(p-q)x \, dp \, dx. \tag{4}$$

We integrate first with respect to x, but take the limits as X_1 and X_0, X_1 being a positive number that may be made afterwards to tend to $+\infty$, and X_0 a negative number that may be made to tend to $-\infty$. We also put $M = p - q$, so that $dM = dp$, and the limits of integration for M become $-q$ and $+\infty$. Therefore,

$$I_1 = \int_{-q}^{+\infty} \{(\sin MX_1)/M - (\sin MX_0)/M\} \, dM. \tag{5}$$

Now
$$\int_{-\infty}^{+\infty} \frac{\sin XM}{M} dM = \pi \text{ if } X>0, \quad \text{and} \quad -\pi \text{ if } X<0.$$

In fig. vii, $(\sin XM)/M$ is shown plotted as a function of M. The height of the central maximum is equal to X, and the half breadth to π/X. As X tends to infinity, the curve becomes narrower and narrower, the area remaining the same. For $X = \infty$, therefore, only the regions in the immediate neighbourhood of $M = 0$ $(p = q)$ make any contribution

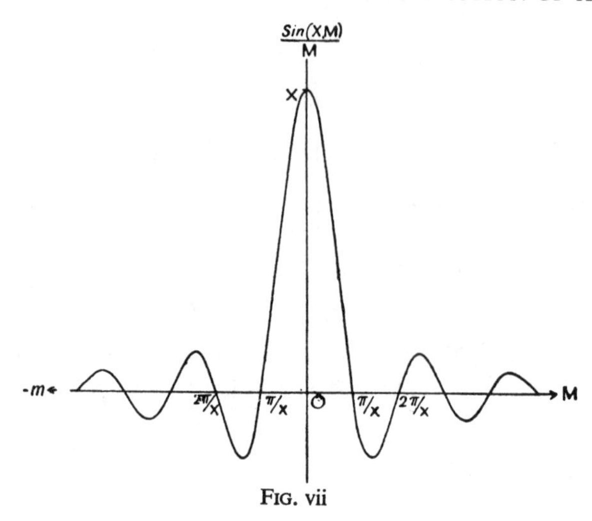

FIG. vii

to the integral. $M=0$ is included in the range of integration $-q$ to $+\infty$, which may therefore be taken equally well as $-\infty$ to $+\infty$. The value of the integral (5) is therefore 2π.

By a similar argument it may be shown that the second term in the integrand of (3) contributes nothing to the integral. This time, we should put $M=p+q$, and the range of integration would be from $+q$ to $+\infty$, and would not include $M=0$.

We may note that the behaviour of the integral (5) justifies our taking $A(q)$ outside the integration in (3). The final value of the integral on the right-hand side of (3) is $A(q)$, and so

$$A(q) \quad \frac{1}{\pi}\int_{-\infty}^{+\infty} f(x)\sin(qx)\,dx, \tag{6}$$

which was the result to be proved. Formula (7.92) can be proved by a slight modification of the above method.

APPENDIX V

THE APPLICATION OF THE FOURIER TRANSFORM TO DIFFRACTION PROBLEMS

(*a*) *Introductory:* The idea that the diffraction patterns obtainable from a given distribution of scattering matter may be derived from a corresponding distribution in reciprocal space, which has been used throughout this volume, finds its neatest and most elegant expression in terms of Fourier-transform theory,[1] a very brief account of the application of which to Fraunhofer diffraction will be given in this Appendix.

Let $\rho(\mathbf{r})$ be the density of scattering matter, expressed in electrons per unit volume, at a point defined relative to the origin O by a vector $\mathbf{r}$. When the directions of the incident and scattered waves are defined by the unit vectors $\mathbf{s}_0$ and $\mathbf{s}$ respectively, the scattered amplitude, which we shall call $\mathrm{T}(\mathbf{R})$, is given by

$$\mathrm{T}(\mathbf{R}) = \int \rho(\mathbf{r}) e^{2\pi i \mathbf{R}\cdot\mathbf{r}}\, d\mathbf{r}, \tag{1}$$

in which $\mathbf{R}$ is the vector $(\mathbf{s}-\mathbf{s}_0)/\lambda$, which we have previously denoted by $\mathbf{S}/\lambda$, and is determined by the directions of incidence and scattering, and the wave-length. The amplitude is expressed in terms of that scattered under the same conditions by a single classical electron situated at the origin. The integration is to be taken throughout the region occupied by the scattering matter, supposed for the moment to be finite, of which $d\mathbf{r}$ is an element of volume. Absorption of the radiation, and rescattering of the scattered radiation are neglected.

$\mathbf{R}$ has the dimensions of the reciprocal of a length, but we may conveniently represent $\mathrm{T}(\mathbf{R})$ geometrically as a distribution in the space defined by a vector having the direction and magnitude of $\mathbf{R}$, drawn from the origin O. This space we call *reciprocal space*. The distribution $\mathrm{T}(\mathbf{R})$ sums up the possible Fraunhofer diffraction patterns that can be produced under any conditions of incidence and scattering, with any wave-length, by the distribution $\rho(\mathbf{r})$. To determine the pattern in any given case, we draw in the distribution $\mathrm{T}(\mathbf{R})$ the so-called sphere of reflection, more appropriately perhaps called the sphere of diffraction. This sphere has radius $1/\lambda$, passes through the origin, and has its centre at P in the direction $-\mathbf{s}_0$ from the origin. The value of $\mathrm{T}(\mathbf{R})$ at any point Q on the sphere, gives the amplitude scattered in the direction PQ. (See fig. 3, p. 7.)

From eqn. (1) it will be seen that $\mathrm{T}(\mathbf{R})$, which is the structure factor, or scattering factor, of the distribution $\rho(\mathbf{r})$, is, mathematically, the Fourier transform of this distribution. To every such distribution there corresponds the appropriate transform, which, plotted in reciprocal space, allows us with the help of the sphere of reflection to determine the nature of the diffraction pattern given by the distribution under any conditions.

The integration in (1) may be supposed to extend to infinity, even if in practice $\rho(\mathbf{r})$ will differ appreciably from zero only over a finite region. Applying Fourier's integral theorem, we may write

$$\rho(\mathbf{r}) = \int \mathrm{T}(\mathbf{R}) e^{-2\pi i \mathbf{r} \cdot \mathbf{R}}\, d\mathbf{R}, \tag{2}$$

where the integration now extends to infinity in reciprocal space, of which $d\mathbf{R}$ is an element of volume. Equation (2) is the fundamental relation upon which the determination of a distribution of scattering matter from its Fraunhofer diffraction must depend. If we know the transform completely for the whole of reciprocal space we can determine $\rho(\mathbf{r})$ by carrying out the integration in (2), but only if we do know $\mathrm{T}(\mathbf{R})$ completely is it in principle possible to determine $\rho(\mathbf{r})$ correctly.

(*b*) *Restrictions imposed by finite wave-length :* For a given wave-length, all regions of the transform accessible to observation lie inside the limiting sphere (p. 15) of radius $2/\lambda$, obtained by rotating the sphere of reflection into all possible positions about the origin. The radius of this sphere increases as λ decreases, but there will in practice always be a region of $\mathbf{R}$ space, lying outside it, for which $\mathrm{T}(\mathbf{R})$ can never be determined by direct optical observation. The distribution $\rho'(\mathbf{r})$ determined from that part of $\mathrm{T}(\mathbf{R})$ lying within the limiting sphere will be one having this portion of the actual transform as its complete transform, and this distribution will differ from $\rho(\mathbf{r})$ unless the part of the transform lying outside the limiting sphere is in practice negligible. It is only when this is so that a faithful optical representation of a distribution of scattering matter can be obtained.

Consider a very simple example, that of a spherical shell of diffracting matter of uniform density unity, and radius a. The transform of this distribution is very easily shown to be

$$\mathrm{T}(|\mathbf{R}|) = 4\pi a^2 (\sin 2\pi a\,|\mathbf{R}|)/2\pi a\,|\mathbf{R}|. \tag{3}$$

It is spherically symmetrical in $\mathbf{R}$ space, with its principal maximum at the origin. It has nodal spheres of zero value at radii $n/2a$, and subsidiary, alternately positive and negative, loops between the nodes. The central maximum is much stronger than the others, which die away rapidly in amplitude with increasing $|\mathbf{R}|$, and, if the limiting sphere includes two or three of the nodal spheres, enough information about $\mathrm{T}(\mathbf{R})$ is available to give a fairly accurate idea of the distribution. But if a/λ is so small that the sphere of radius $1/2a$, containing the central maximum of the transform, extends beyond the limiting sphere, the sphere of reflection will always lie inside the central maximum, and little can be inferred from the observed scattering about the nature of the scatterer. The apparent dimensions of the scattering region are roughly determined by the radius of the first nodal sphere, and when a is much smaller than λ this will in effect be the radius of the limiting sphere. The observed scattering from the shell will then be very much like that given by a sphere of radius $\lambda/4$. If a is less than this, nothing at all can be inferred about its true magnitude. This is a well-known result considered in an unfamiliar way.

(*c*) *The phase of the transform:* The limitations considered in the last paragraph were those imposed by the finite wave-lengths used in optical observations. There is, however, a second limit imposed on the knowledge we can obtain of $T(\mathbf{R})$. We shall assume $\rho(\mathbf{r})$ to be real, which means that there is no change in phase produced by the act of scattering, although our argument will still apply if there is a phase change that is independent of $\mathbf{r}$. Even when $\rho(\mathbf{r})$ is real, $T(\mathbf{R})$ may be complex, which means physically that the phase of the resultant scattered radiation, relative to the phase scattered by an electron at the origin, is a function of $\mathbf{R}$, and so of the angle of scattering.

In evaluating the integral (2) it is necessary to take this phase variation into account. In ordinary vision, with or without optical aid, the lenses do this automatically, but when X-rays are scattered by crystals or other matter, we have nothing corresponding to a lens to do this automatic integration. Nor is there any direct and generally applicable way of observing the relative phases. What is measured is always the scattered *intensity*, from which $|T(\mathbf{R})|$, and not $T(\mathbf{R})$, is determined. We can consider $|T(\mathbf{R})|^2$, plotted as a distribution in reciprocal space, as representing all that can be determined by direct optical means from the scattering of X-rays. We shall call $|T(\mathbf{R})|^2$ the *intensity transform*, and the type of information that can be derived from it has already been discussed in Chapter VII, § 1 (*p*). We shall return to it later in this Appendix. It can never give the actual distribution $\rho(\mathbf{r})$, although it may give useful information about it. The idea of the intensity transform has also been used extensively in Chapter X, although not under that name. The question of phase determination has also been discussed briefly in Chapter VII, but will be dealt with in greater detail by the authors of Volume III.

(*d*) *The symmetry of the transform:* When $\rho(\mathbf{r})$ is real, we see from (1) that $T(\mathbf{R})$ and $T(-\mathbf{R})$ are conjugate complex quantities, so that $|T(\mathbf{R})|^2=|T(-\mathbf{R})|^2$. The intensity transform of a real scattering distribution is thus always centro-symmetrical about the origin, however unsymmetrical the distribution itself may be. The well-known fact that the Laue symmetry of a crystal, which is that of the distribution of the intensities of its diffraction spectra in the reciprocal lattice, may be higher than its true symmetry by the possession of a centre of symmetry is an example of this. (See also Chapter X, p. 551).

If $\rho(\mathbf{r})$ is complex, so that there is a phase change on scattering depending on $\mathbf{r}$, it is no longer true that the intensity transform is necessarily centro-symmetrical, for $T(\mathbf{R})$ and $T(-\mathbf{R})$ are then no longer conjugate complex quantities. An example of this, due to the effects of dispersion, has been considered on p. 32.

(*e*) *The transform of a periodic distribution of scattering matter :* The possibility of analysing a crystal by means of X-rays depends on the fact that the transform of a periodically repeating distribution containing a large number of repetitions of the pattern has appreciable values only at or nearly at certain discrete points, so that the experimental work of determining the intensity-transform is limited to measuring it at these points.

Let a density-distribution occupying a limited volume, and constituting the unit of pattern, be repeated by successive translations through a vector distance $\mathbf{a}$ until a row of N_1 similar units has been built up. This row is multiplied into a set of N_2 similar, parallel, rows by successive translations through a vector distance $\mathbf{b}$, thus forming a set of similar units lying on a two-dimensional net with primitive vectors $\mathbf{a}$ and $\mathbf{b}$. Finally, by successive translations through a vector distance $\mathbf{c}$, a set of N_3 parallel nets is built up, the whole forming a three-dimensional pattern based on a space-lattice with primitive translations $\mathbf{a}$, $\mathbf{b}$, $\mathbf{c}$, and constituting a parallelepipedal crystal with $N_1N_2N_3$ units. It is a simple matter to calculate the transform of this crystal.

Let $F(\mathbf{R})$ be the transform of the original unit of structure. When it is moved a vector distance $\mathbf{a}$ its transform becomes $F(\mathbf{R})\exp(2\pi i\mathbf{a}\cdot\mathbf{R})$, and the transform of the row of N_1 units in the direction $\mathbf{a}$ is therefore

$$F(\mathbf{R})\sum_{n=0}^{n=N_1-1}\exp(2\pi in\mathbf{a}\cdot\mathbf{R})=F(\mathbf{R})G_a(\mathbf{R}), \tag{4}$$

where $$G_a(\mathbf{R})=\exp\{\pi i(N_1-1)\mathbf{a}\cdot\mathbf{R}\}\sin(\pi N_1\mathbf{a}\cdot\mathbf{R})/\sin(\pi\mathbf{a}\cdot\mathbf{R}), \tag{4a}$$

and is the transform of a row of N_1 points, each of unit scattering power, equally spaced, at a distance $\mathbf{a}$ apart, along a line. It is a function of $\mathbf{a}\cdot\mathbf{R}$, and so is constant over any plane perpendicular to the direction of $\mathbf{a}$. If N_1 is large, $G_a(\mathbf{R})$ will have appreciable values only over planes at a distance $1/a$ apart, one of which passes through the origin. The magnitude of $G_a(\mathbf{R})$ over these planes is N_1.

In the second operation, the row whose transform is (4) is turned into N_2 similar rows by successive translations $\mathbf{b}$. The transform of the resulting net is evidently obtained by multiplying the transform of the single row, (4), by $G_b(\mathbf{R})$, in which $\mathbf{b}$ replaces $\mathbf{a}$ and N_2 replaces N_1 in (4a). The new transform is therefore $F(\mathbf{R})G_a(\mathbf{R})G_b(\mathbf{R})$. If N_2 is also large, $G_b(\mathbf{R})$ has appreciable values only over a set of planes perpendicular to $\mathbf{b}$ with spacing $1/b$, and the transform of the net has appreciable values only over the lines of intersection of the two sets of planes, which are of course perpendicular to both $\mathbf{a}$ and $\mathbf{b}$. The value of

$G_a(\mathbf{R})G_b(\mathbf{R})$ everywhere on these lines is N_1N_2, and they are the transform of an infinite network of unit scattering points.

Clearly, the transform of the parallelepipedal crystal is now obtained by multiplying the transform of the net by $G_c(\mathbf{R})$, which has appreciable values only over a set of planes perpendicular to **c** with a spacing $1/c$. The final transform is therefore

$$T(\mathbf{R}) = F(\mathbf{R})G_a(\mathbf{R})G_b(\mathbf{R})G_c(\mathbf{R}) = F(\mathbf{R})G(\mathbf{R}). \quad (5)$$

$G(\mathbf{R})$ is the transform of a lattice of point scatterers with primitive translations **a**, **b**, **c**, and N_1, N_2, N_3 points parallel to these directions respectively. In the form in which $G(\mathbf{R})$ has been given, the origin lies at the corner of the parallelepipedal block.

If the number of points in all three directions is large, the transform has appreciable values only at the points of a lattice formed by the intersections of the three sets of planes discussed above, one set perpendicular to **a** with a spacing $1/a$, one perpendicular to **b**, with a spacing $1/b$, and the third perpendicular to **c**, with a spacing $1/c$. These points of intersection form a space-lattice, which is clearly the lattice reciprocal to that upon which the crystal is based, as defined in Appendix II, (*a*). The Fourier transform of an infinite space-lattice is thus the lattice reciprocal to it. If the primary lattice is not of infinite extent, the transform $G(\mathbf{R})$ has maxima at the reciprocal-lattice points, and an appreciable density about these points through regions with dimensions of the order of the reciprocals of those of the finite lattice. The nature of this distribution has been discussed in Chapter II, § 1 (*c*).

If X, Y, Z, are the components of **R** parallel to the reciprocal axes $\mathbf{a}^*$, $\mathbf{b}^*$, $\mathbf{c}^*$, respectively, we can conveniently express them, as we have done elsewhere, in terms of these translations, writing

$$X = \xi a^*, \ Y = \eta b^*, \ Z = \zeta c^*.$$

Then
$$\mathbf{a}\cdot\mathbf{R} = \mathbf{a}\cdot\mathbf{a}^*\xi = \xi, \quad \mathbf{b}\cdot\mathbf{R} = \eta, \quad \mathbf{c}\cdot\mathbf{R} = \zeta.$$

Plainly then, $G(\mathbf{R})$ has appreciable values only when ξ, η, ζ are integers, say *h*, *k*, *l*. The intensity-transform of the lattice is $|G(\mathbf{R})|^2$, and is the interference function, $I_0(\xi, \eta, \zeta)$ of Chapters II, V and X. It is interesting to see how naturally the idea of the reciprocal lattice emerges from that of the transform.

The Fourier transform of a finite piece of crystal is thus the transform of the unit cell multiplied by $G(\mathbf{R})$, or $G(\xi, \eta, \zeta)$, the transform of the finite point-lattice on which the crystal is based. If the crystal has a very large number of units, $G(\mathbf{R})$ has appreciable values only at the reciprocal-lattice points (*h*, *k*, *l*), so that the transform is represented for all practical purposes by the values of $F(hkl)$, the structure factors of the unit cell for the different spectra given by the crystal. We need therefore only sample the transform of the unit cell at certain discrete points in order to get the information needed to determine the structure.

The transform G(**R**) of the point lattice is truly periodic, and extends to infinity. F(**R**), on the other hand, is not periodic. The scattering factors of the atoms of which the unit cell is composed are themselves moreover functions of **R**, which decrease rapidly as | **R** | increases, and when | **R** | is large enough F(**R**) becomes inappreciable. We can think of the transform of the crystal as the reciprocal lattice weighted at each point with the appropriate value of F(*hkl*); and to get a true representation of the structure we must be sure that all appreciable values of F(*hkl*) are included in the domain of T(**R**) open to observation. This can rarely be completely attained in practice, and we must therefore be on our guard against the occurrence of false detail in the final representation of $\rho(\mathbf{r})$.

(*f*) *The crystal structure expressed as a Fourier integral:* A real crystal must have a finite volume, and the representation of $\rho(\mathbf{r})$, the actual density distribution in this fragment, involves the evaluation of the integral (2). If (*u*, *v* *w*) are the fractional coordinates of a point in the crystal, so that $x = ua$, $y = vb$, $z = wc$, the transform of the whole crystal may be written

$$\mathrm{T}(\xi, \eta, \zeta) = \mathrm{V}\iiint_{-\infty}^{\infty} \rho(u, v, w)\, e^{2\pi i(\xi u + \eta v + \zeta w)}\, du\, dv\, dw, \tag{6}$$

$\rho(u, v, w)$ being the density at the point (*u*, *v*, *w*), and V the volume of the unit cell. But, as we have just seen,

$$\mathrm{T}(\xi, \eta, \zeta) = \mathrm{F}(\xi, \eta, \zeta)\mathrm{G}(\xi, \eta, \zeta), \tag{7}$$

$\mathrm{F}(\xi, \eta, \zeta)$ being the transform of a single cell, and $\mathrm{G}(\xi, \eta, \zeta)$ that of the basic point lattice. By Fourier's theorem, using (2), we obtain

$$\mathrm{V}\rho(u, v, w) = \iiint_{-\infty}^{\infty} \mathrm{F}(\xi, \eta, \zeta)\mathrm{G}(\xi, \eta, \zeta)\, e^{-2\pi i(u\xi + v\eta + w\zeta)}\, d\xi\, d\eta\, d\zeta. \tag{8}$$

So far there have been no approximations. The density $\rho(u, v, w)$ is given by a Fourier integral, taken throughout reciprocal space. We now assume the number of units to be large, so that $\mathrm{G}(\xi, \eta, \zeta)$ has appreciable values only when ξ, η, ζ are very nearly integral. We shall then make no appreciable error by treating the integral (8) as the sum of a series of integrals, one over each cell of the reciprocal lattice.

Moreover, because the maxima of $\mathrm{G}(\xi, \eta, \zeta)$ are very sharp when the lattice contains many points, and can be made as sharp as we please by increasing the number of points indefinitely, while $\mathrm{F}(\xi, \eta, \zeta)$ varies relatively slowly, we can replace $\mathrm{F}(\xi, \eta, \zeta)$ in the integration over each cell by its value at the corresponding reciprocal-lattice point; and the same is true of the factor $e^{-2\pi i \mathbf{R}\cdot\mathbf{r}}$, for although this oscillates rapidly when | **R** | is large, by increasing the number of cells we may make G(**R**) in the neighbourhood of the reciprocal-lattice points vary even more rapidly. Thus, in the limit, when the number of points is very large, we may write without sensible error

$$V\rho(u, v, w) = \sum_h \sum_k \sum_l F(hkl) e^{-2\pi i(hu+kv+lw)} \iiint_{Cell} G(\xi, \eta, \zeta)\, d\xi\, d\eta\, d\zeta. \quad (9)$$

The integral over the cell approaches unity as the number of lattice cells increases. The proof of this follows the lines of that given on p. 43 for $I_0(\xi, \eta, \zeta)$, and it is true not only for a parallelepipedal crystal, but can be generalised to one of any form. In the limit therefore, for a truly periodic crystal,

$$V\rho(u, v, w) = \sum_h \sum_{k=-\infty}^{\infty} \sum_l F(hkl) \exp\{-2\pi i(hu + kv + lw)\}. \quad (10)$$

The Fourier series discussed in Chapter VII has thus been obtained as the limit, when the number of crystal units tends to infinity, of a Fourier integral applicable to a finite crystal.

(*g*) *Sections of Fourier transforms, and two-dimensional projections:* Let (x, y, z), (X, Y, Z) be the rectangular coordinates of points in real and reciprocal space respectively, so that $\mathbf{R}\cdot\mathbf{r} = xX + yY + zZ$. Then the transform $T(X, Y, Z)$ of a distribution $\rho(x, y, z)$ is

$$T(X, Y, Z) = \iiint \rho(x, y, z) \exp\{2\pi i(Xx + Yy + Zz)\}\, dx\, dy\, dz. \quad (11)$$

The section of this transform by the plane $Z = Z_1$ is

$$T(X, Y, Z_1) = \iint \sigma_1(x, y) \exp\{2\pi i(Xx + Yy)\}\, dx\, dy, \quad (12)$$

where $$\sigma_1(x, y) = \int \rho(x, y, z) \exp(2\pi i Z_1 z)\, dz, \quad (13)$$

which is the density of the projection of the scattering distribution $\rho(x, y, z)$ on the plane $z = 0$, weighted in such a way that the projection of any element of the distribution is multiplied by the factor $\exp(2\pi i Z_1 z)$, z being the distance of the element from the plane $z = 0$. Transforming (12), we obtain

$$\sigma_1(x, y) = \iint T(X, Y, Z_1) \exp\{-2\pi i(xX + yY)\}\, dX\, dY, \quad (14)$$

that is to say the section $T(X, Y, Z_1)$ of the transform of $\rho(x, y, z)$ is the two-dimensional transform of the weighted projection of $\rho(x, y, z)$ on the plane $z = 0$. If $Z_1 = 0$, the weighted projection becomes the actual projection, which can thus be calculated by integrating (14) with $Z_1 = 0$. It is easy to generalise these results to axes that are not rectangular, and to apply them, by an argument exactly analogous to that used in §(*f*), to a periodic distribution containing a large number of units. If the projection is made on the *ab* plane we obtain

$$A\sigma_L(u, v) = \sum_h \sum_k F(hkL) \exp\{-2\pi i(hu + kv)\}, \tag{15}$$

$$\sigma_L(u, v) = c\int_0^1 \rho(u, v, w) \exp(2\pi i L w) dw, \tag{16}$$

A being the area of the *ab* cell-face.

The coefficients $F(hkL)$ are now the structure-factors associated with the layer L of the reciprocal lattice, *i.e.*, with the spectra on the Lth layer-line of a rotation photograph about the *z* axis

If $L = 0$, $\sigma_L(u, v)$ becomes $\sigma(u, v)$, the ordinary unweighted projection of the matter in the unit cell on the *ab* plane, and equation (15) becomes equation (7.21) of p. 353.

If C_L and S_L are the real and imaginary parts of σ_L, we may write, in the special case in which the crystal has a centre of symmetry at the origin, so that $F(hkL)$ is real,

$$AC_L(u, v) = \sum_h \sum_k F(hkL) \cos 2\pi x(hu + kv), \tag{17}$$

$$AS_L(u, v) = -\sum_h \sum_k F(hkL) \sin 2\pi(hu + kv), \tag{18}$$

with

$$C_L(u, v) = c\int_0^1 \rho(u, v, w) \cos(2\pi L w)\, dw, \tag{19}$$

$$S_L(u, v) = c\int_0^1 \rho(u, v, w) \sin(2\pi L w) dw, \tag{19a}$$

giving two different weighted projections, which may be calculated from the Fourier series (17) and (18). If $F(hkL) = F(\bar{h}\bar{k}L)$, $S_L(u, v)$ vanishes, and $\sigma_L(u, v) = C_L(u, v)$.

As Cochran and Dyer[2] have shown, very useful information can be obtained from weighted projections in which $L \neq 0$. For example, if $L = 1$, $\cos 2\pi L w = 0$ when $w = \frac{1}{4}$. Atoms in the cell which have this *w* coordinate therefore make little or no contribution to the projection $C_1(u, v)$. If $w = \frac{1}{2}$, $\cos 2\pi L w = -1$, and atoms with $w = \frac{1}{2}$ appear in C_1 as negative hollows, instead of as peaks as they do in the $\sigma(u, v)$, or $L = 0$, projection. By comparing $\sigma(u, v)$ and $C_1(u, v)$ valuable information may be obtained about the distribution of atoms in the unit cell. Cochran and Dyer discuss a very interesting example in which two long molecules at different levels, crossing each other and therefore not separable in the $L = 0$ projection, are clearly separated by comparing the $L = 0$ and $L = 1$ projections.

(*h*) *Projection of spherically symmetrical atoms:* The scattering factor, $f(\mathbf{R})$, of an atom is the transform of the atomic density distribution. $|\mathbf{R}| = 2(\sin\theta)/\lambda$, and $f(\mathbf{R})$ is usually assumed to be spherically symmetrical. The projection of the atom on the plane $z = 0$ is, by (14),

$$\rho_p(x, y) = \iint f(X, Y, 0) \exp\{-2\pi i(xX + yY)\}\, dX\, dY. \tag{20}$$

If the atom is spherically symmetrical, we may write this

$$\rho_p(r)=2\int_0^{\pi}\int_0^{\infty} \mathrm{R}f(\mathrm{R})\cos(2\pi \mathrm{R}r\cos\alpha)\,d\mathrm{R}\,d\alpha, \tag{21}$$

R and r being now radii in the plane of projection, and α the angle between them. Integration with respect to α gives

$$\rho_p(r)=2\pi\int_0^{\infty} \mathrm{R}f(\mathrm{R})\mathrm{J}_0(2\pi \mathrm{R}r)\,d\mathrm{R}, \tag{22}$$

which can be integrated numerically for a series of values of r if $f(\mathbf{R})$ is given. It is also possible to determine the projection approximately by supposing the atom to form a unit of a crystal with a spacing much larger than the effective atomic radius. Equation (20) for this periodic structure becomes a Fourier series, the coefficients of which are the samples of $f(\mathbf{R})$ at the points of the reciprocal lattice of the assumed crystal.

If $f(\mathbf{R})$ is constant, and equal to $f(0)$, the scattering factor is that of a point. Assuming the limits of integration in (22) to be 0 and R_1, we obtain, as in equation (7.83), p. 399,

$$\rho_p(r)=2\pi f(0)\,\mathrm{R}_1^2 J_1(2\pi r\mathrm{R}_1)/2\pi r\mathrm{R}_1. \tag{23}$$

The function $2\mathrm{J}_1(2\pi r\mathrm{R}_1)/2\pi r\mathrm{R}_1$ is unity when $r\mathrm{R}_1=0$, and has its first zero value when $r=0{\cdot}61/\mathrm{R}_1$. As R_1 becomes greater, the maximum of $\rho_p(r)$ becomes narrower, and in the limit, when R_1 becomes infinite, the projection tends to a region of negligible dimensions and infinite density, that is to say, a point. The integrated value of $\rho_p(r)$ over the plane of projection is $f(0)$, and remains so in the limit. If a region of the transform of finite radius R_1 is used, the projection shows false detail in the form of rings of alternate positive and negative density, the subsidiary maxima and minima of the function $\mathrm{J}_1(m)/m$.

If in (22) we interchange R and r, we may consider $\rho_p(\mathrm{R})$ as the transform of a uniform distribution of scattering matter over a circle of radius R_1, and can apply it directly to diffraction by a circular aperture, the scatterers being now the virtual Huyghens sources of optical theory. The function $\rho_p(\mathrm{R})$ is actually the section of the transform of the aperture by a plane perpendicular to its axis, and is the same for all such sections.

If $\rho(\mathbf{r})$ in the atom has a Gaussian distribution it is easy to show by direct integration of (11) and then of (20) that both the transform and the projection of the atom on a plane are also Gaussian. If $f(\mathrm{R})=\mathrm{Z}\exp(-\pi\mathrm{R}^2a^2)$, the projection of the corresponding atom is $\rho_p(r)=(\mathrm{Z}/a^2)\exp(-\pi r^2/a^2)$. The scattering factor and the projection fall to $e^{-\pi}$ of their maximum values at $\mathrm{R}=1/a$ and $r=a$ respectively, a result that is sometimes useful in approximate calculations.

If $\rho(x, y, z)=\rho_0\exp\{-\pi(x^2/a^2+y^2/b^2+z^2/c^2)\}$, in which the surfaces of constant density are ellipsoids, the corresponding transform is, by (11),

$$f(\mathrm{X}, \mathrm{Y}, \mathrm{Z})=abc\exp\{-\pi(a^2\mathrm{X}^2+b^2\mathrm{Y}^2+c^2\mathrm{Z}^2)\} \tag{24}$$

The projections on the axial planes are easily determined from (20). These results may be of use in considering the effect on the transform of anisotropic thermal motion in crystals.

(*i*) *The transform of a sinusoidal scattering distribution of infinite extent :* Let the density of the distribution be independent of y and z, and be given as a function of x by

$$\rho(x) = \mathrm{A} \cos 2\pi p x, \tag{25}$$

so that there are p cycles of density variation in unit length parallel to x. The transform is given by

$$\mathrm{T(X, Y, Z)} = \mathrm{A}\iiint \cos 2\pi p x \exp 2\pi i(\mathrm{X}x + \mathrm{Y}y + \mathrm{Z}z)\, dx\, dy\, dz, \tag{26}$$

which can be written as the product of three independent integrals, taken over x, y, and z respectively. To evaluate the integral with respect to x we take limits x_1 and $-x_1$, which can later be made to tend to infinity. With these limits the integral is equal to

$$x_1\mathrm{A}\left\{\frac{\sin 2\pi(p+\mathrm{X})x_1}{2\pi(p+\mathrm{X})x_1} + \frac{\sin 2\pi(p-\mathrm{X})x_1}{2\pi(p-\mathrm{X})x_1}\right\}, \tag{27}$$

which has two main maxima, when $\mathrm{X} = \pm p$. The first lateral zeros accompanying these maxima occur when $\mathrm{X} = \pm p \pm 1/2x_1$. As x_1 increases these maxima become narrower and higher until, in the limit, when the wave train is infinite, the expression (27) is zero except over the planes $\mathrm{X} = \pm p$. The integrals with respect to y and z are of the form $(\sin u)/u$, and vanish when the limits become infinite unless $\mathrm{Y}=0$ and $\mathrm{Z}=0$, so that the transform T(X, Y, Z) of an infinite plane sinusoidal distribution of linear frequency p parallel to x is zero except at the points $(\pm p, 0, 0)$ of reciprocal space, a result we have used in Chapter VII, §1 (*d*), p. 345. It is perhaps more easily seen in the converse sense, for it is clear from §(*e*) of this appendix that the transform of two points is an infinite plane sinusoidal distribution.

The transform of a *linear* distribution of scattering matter with a density proportional to $\cos 2\pi p x$ is zero except over the planes $\mathrm{X} = \pm p$.

(*j*) *The convolution theorem* : We consider now the transform of the product of two functions $f(x)$ and $s(x)$. Let $\mathrm{F}(\xi)$ and $\mathrm{S}(\xi)$ be the transforms of these functions, so that

$$\begin{aligned} \mathrm{F}(\xi) &= \int f(x) e^{2\pi i \xi x}\, dx, & \mathrm{S}(\xi) &= \int s(x) e^{2\pi i \xi x}\, dx, \\ f(x) &= \int \mathrm{F}(\xi) e^{-2\pi i \xi x} d\xi, & s(x) &= \int \mathrm{S}(\xi) e^{-2\pi i \xi x} d\xi. \end{aligned} \tag{28}$$

The required transform is

$$\begin{aligned} \int f(x) s(x) e^{2\pi i x \mathrm{X}}\, dx &= \iint \mathrm{F}(\xi) s(x) e^{2\pi i x(\mathrm{X}-\xi)}\, dx\, d\xi \\ &= \int \mathrm{F}(\xi) \mathrm{S}(\mathrm{X}-\xi) d\xi. \end{aligned} \tag{29}$$

The integral on the right-hand side of (29) is known as the *convolution* or *folding* (German *Faltung*) of the functions $F(\xi)$ and $S(\xi)$, and the above theorem states that the transform of the product of two functions is the convolution of their transforms, a result that has a number of applications in diffraction theory.

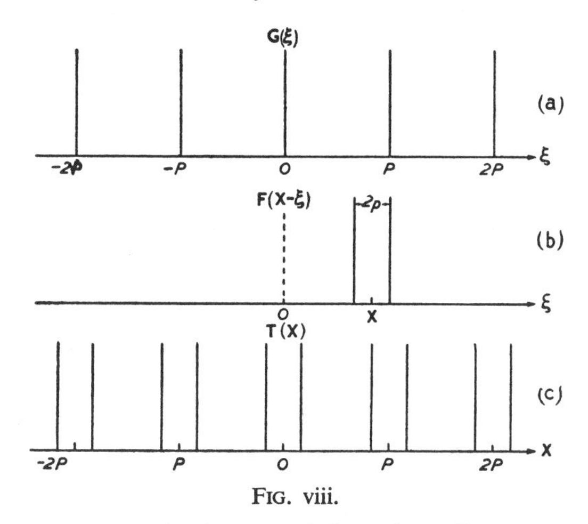

FIG. viii.

Equation (29) can easily be extended to three dimensions. We may write

$$\int f(\mathbf{u})s(\mathbf{u})e^{2\pi i\mathbf{R}\cdot\mathbf{u}}\,d\mathbf{u} = \int F(\mathbf{t})S(\mathbf{R}-\mathbf{t})\,d\mathbf{t}, \tag{30}$$

and also

$$\int F(\mathbf{R})S(\mathbf{R})e^{-2\pi i\mathbf{R}\cdot\mathbf{u}}\,d\mathbf{R} = \int f(\mathbf{r})s(\mathbf{u}-\mathbf{r})\,d\mathbf{r}, \tag{31}$$

where **u**, **R**, **r**, and **t** are now vectors.

As a simple example of the application of this theorem we consider the transform of an infinite row of scattering points, P in unit distance in the x direction, and such that the scattering power of a point distant x from the origin is $f(x)=a\cos 2\pi px$. There are therefore p cycles of scattering power in unit length. For simplicity, we consider the problem in one dimension only. The distribution may be considered as the product of the function $f(x)$ and a second function $g(x)$ that is zero except when $x=n/P$. Let $F(\xi)$ and $G(\xi)$ be the transforms of $f(x)$ and $g(x)$. Then, by the convolution theorem, the required transform is the convolution of F and G, or $\int G(\xi)F(X-\xi)\,d\xi$. The integral vanishes unless $G(\xi)$ and $F(X-\xi)$ are simultaneously different from zero. As we have seen in § (*i*), $F(X-\xi)$ will vanish unless $X-\xi=\pm p$, while $G(\xi)$, the transform of a row of points, will vanish unless $\xi=mP$. The transform is thus zero unless $X=mP\pm p$. The coincidences of the nonzero values of F and G may be understood graphically from fig. (viii), by supposing one transform to be displaced relative to the other through different distances X.

It is easy to extend this to three dimensions. The transform of a row of points varying sinusoidally in scattering power plainly consists of pairs of planes $\pm p$ distant from the planes $X = mP$ that are the transform of a row of points of equal scattering power. It is illustrated by fig. viii *c*. Had the scattering power been $f_0 + a \cos 2\pi px$, the planes $X = mP$ would also have occurred in the transform, flanked by the ' ghost ' planes at $X = mP \pm p$.

(*k*) *The transform of a helix :* An interesting example of the use of the convolution theorem to calculate the transform of a set of points on a helix, which is of importance in the X-ray analysis of certain biological substances, has been given by Cochran, Crick, and Vand[3]. Consider first a continuous linear helix of pitch P, the axis of which is the z axis of coordinates. Any point on the helix is fixed relative to an origin 0 by an axial vector $\mathbf{z}$ and a radial vector $\mathbf{r}$. Let $\mathbf{Z}$ and $\mathbf{R}$ be corresponding axial and radial vectors fixing a point in reciprocal space.

The transform of the helix is then

$$T(\mathbf{R}, \mathbf{Z}) = \int \exp \{2\pi i(\mathbf{R}\cdot\mathbf{r} + \mathbf{Z}\cdot\mathbf{z})\}\, dv, \tag{32}$$

dv being an element of volume of the helix.

Let $\mathbf{r}$ make an angle ϕ with a fixed axial plane containing the radius of the helix that passes through the origin, and let $\mathbf{R}$ make an angle ψ with the same plane. Then $\phi = 2\pi z/P$ and $\mathbf{R}\cdot\mathbf{r} = Rr \cos(\phi - \psi)$, r being constant. Apart from a constant factor relating dz and dv, the transform of *one complete turn* of the helix may now be written

$$F(\mathbf{R}, \mathbf{Z}) = \int_0^P \exp [2\pi i\{Rr \cos(\phi - \psi) + Zz\}]\, dz. \tag{33}$$

The whole helix is produced by repetition of this unit at intervals P parallel to $\mathbf{z}$, and, as in § (*e*) of this appendix, the transform is obtained by multiplying $F(\mathbf{R}, \mathbf{Z})$ by $G_P(\mathbf{Z})$, the transform of a row of points of spacing P parallel to $\mathbf{z}$. If the number of turns of the spiral is large, $G_P(\mathbf{Z})$ has appreciable values only over the planes $Z = n/P$, and the required transform is thus given by the values of $F(\mathbf{R}, \mathbf{Z})$ when $Z = n/P$. Thus

$$T(R, \psi, n/P) = \int_0^P \exp [2\pi i\{Rr \cos(2\pi z/P - \psi) + nz/P\}]\, dz. \tag{34}$$

By making use of the identity

$$2\pi i^n J_n(x) = \int_0^{2\pi} \exp(ix \cos\theta) \exp(in\theta)\, d\theta,$$

$J_n(x)$ being a Bessel function of integral order n, we may write T, again apart from a constant factor,

$$T(R, \psi, n/P) = J_n(2\pi Rr) \exp \{in(\psi + \tfrac{1}{2}\pi)\}. \tag{35}$$

The amplitude of T, which is a function of R only, is given by the Bessel function. $|T|$ has a maximum value at $R=0$ when $n=0$, but for all the other planes $Z=n/P$ it is zero when $R=0$, and its values for small R diminish rapidly as n increases. $|T|$ has maxima for certain values of R, the radii of corresponding circles of maximum amplitude increasing, and the amplitudes themselves diminishing, as n increases. For the behaviour of $J_n(x)$ for small values of x reference may be made to fig. 205, p. 567.

T is not real, for a helix has no symmetry centre. The phase factor is unity for $n=0$. For other values of n it varies with the azimuth ψ, but is independent of R.

(*l*) *The discontinuous helix*: We now consider the transform of a set of points lying on the helix at a constant axial interval p. Such a set of points may be represented as a product of two functions, one of which, H, is zero everywhere except on the helix, and the other, G, zero everywhere except over a set of planes of spacing p, upon which its value is unity. The transform of the discontinuous helix is the transform of the product HG, and this, by the theorem proved in § (*j*) above, is the convolution of the transforms of H and G. The transform of H we have just calculated: it is the function T, and has values different from zero only on the planes $Z=n/P$. The transform of G, g, is a row of points of spacing $1/p$, with $Z=m/p$. The convolution of the transforms may be written

$$C(X, Y, Z) = \int T(X-\xi, Y-\eta, Z-\zeta) g(\xi, \eta, \zeta)\, d\xi\, d\eta\, d\zeta. \tag{36}$$

$T(X-\xi, Y-\eta, Z-\zeta)$ vanishes unless $Z-\zeta=n/P$, but with this restriction $X-\xi$ and $Y-\eta$ may have any values. The function $g(\xi, \eta, \zeta)$ vanishes unless $\xi=0$, $\eta=0$, $\zeta=m/p$. C(X, Y, Z) will have a value different from zero only if both factors of the integrand are simultaneously different from zero, or when

$$Z = \zeta + n/P = m/p + n/P. \tag{37}$$

If Z has one of these values, X and Y may have any value. The required transform is therefore zero except over the set of planes given by (37), and its value over any such plane is proportional to $J_n(2\pi Rr)$. For $m=0$ there is a set of planes identical with that giving the transform of the continuous helix. For $m=1$, there is an exactly similar set of planes based on $Z=1/p$ as origin; and similarly, every one of the points $Z=\pm m/p$ serves in its turn as the origin for a similar set of planes of spacing $1/P$. The whole transform is the totality of these planes.

If P/p cannot be expressed as a ratio of whole numbers, the totality of planes will in principle form a continuous distribution; but in fact it is only for small values of n that the amplitude $|T|$ is appreciable for small R, and only a few planes of each set n/P will make much contribution. If P/p can be expressed as a ratio of whole numbers there will be coincidences between certain planes of the different sets, and the total number of planes in a finite region will be finite.

For an example of the application of the theorems outlined above to an actual problem reference may be made to the paper of Cochran, Crick, and Vand.

The formal similarity between the results just considered and those discussed in Chap. X, § 3 (*c*), p. 563, in connection with periodic errors in lattices, should be noted. The discontinuous helix is in fact a row of points subjected to a certain type of periodic distortion, and the groups of planes n/P associated with each plane m/p are the corresponding optical ghosts accompanying the spectra of the undistorted row.

(*m*) *The auto-correlation coefficient and the Patterson series:* We now put $f(\mathbf{r}) = \rho(\mathbf{r})$, and $s(\mathbf{r}) = \rho^*(-\mathbf{r})$. Then, if $T(\mathbf{R})$ is the transform of $\rho(\mathbf{r})$, $F(\mathbf{R}) = T(\mathbf{R})$ and $S(\mathbf{R}) = T^*(\mathbf{R})$. Equation (31) now becomes

$$P(\mathbf{u}) = \int |T(\mathbf{R})|^2 e^{-2\pi i \mathbf{R}\cdot\mathbf{u}} d\mathbf{R} = \int \rho(\mathbf{r})\rho^*(\mathbf{r}-\mathbf{u})\, d\mathbf{r}$$
$$= \int \rho^*(\mathbf{r})\rho(\mathbf{r}+\mathbf{u})\, d\mathbf{r}. \qquad (38)$$

If we apply this result to a periodic distribution, as on p. 617, putting $T(\mathbf{R}) = F(\mathbf{R})G(\mathbf{R})$, we get at once Patterson's series, equation (7.48), p. 373.

In studying the statistical characteristics of fluctuating quantities use is made of the so-called auto-correlation function, which, for a function $f(x)$, is defined as $p(\xi) = \int f^*(x)f(x+\xi)\,dx \Big/ \int f^*(x)f(x)\,dx$. It will be seen that the function $P(\mathbf{u})$ of equation (38) is proportional to the auto-correlation function of the distribution of scattering matter, and is the transform of the intensity transform of that distribution. The Patterson series gives the auto-correlation function of the scattering distribution in the crystal unit, and contains all the information about that distribution that can be obtained directly from diffraction experiments when the phases of the diffracted waves are unknown. Booker, Ratcliffe, and Shinn[4] have applied the auto-correlation function to the study of the scattering of electric waves by the ionosphere, and reference may be made to their paper for a further account of its use.

(*n*) *The use of molecular transforms in structure determination:* As we have already seen, the transform of a crystal structure is the transform of its unit cell sampled at the points of the lattice reciprocal to that upon which the crystal is built. Suppose the unit ceil to contain a group of points the configuration of which is known, which may for convenience be called a molecule, and that its orientation in the unit cell is to be determined from the X-ray spectra. Let $M(\mathbf{R})$ be the transform of the molecule, plotted as a distribution in reciprocal space. The operation of obtaining the best agreement between observed and calculated structure factors may be thought of as that of rotating the reciprocal lattice relative to the transform $M(\mathbf{R})$ into such a position that the closest agreement between the observed amplitudes and the values of the transform at the positions of the corresponding reciprocal-lattice points is obtained.[5] If the molecule has a symmetry centre, which is used as origin, the transform $M(\mathbf{R})$ is real, but, if not, it will be necessary to plot $|M(\mathbf{R})|^2$, and to obtain a fit between observed and calculated intensities. Once the best fit has been obtained, the phases of the spectra corresponding to this orientation of the molecule in the cell can be calculated from the real and imaginary parts of the transform.

This method is most easily applied when the molecule itself is flat, and of fairly simple known configuration, a benzene ring, for example. Axes X and Y would then be chosen in the plane of the molecule, and Z perpendicular to it. The transform is independent of Z, and its sections by all planes $Z = \text{constant}$ are the same. If the projection of the molecule on the *ab* plane is being considered, only the points of the plane $Z = 0$ of the reciprocal lattice are relevant. The rotation of this plane in the transform may be accomplished in effect by a suitable distortion of the reciprocal net in the plane $Z = 0$ of the transform.

The unit cell will usually contain not one molecule, but several, related by the symmetry operations of the space-group, and the resultant transform for any relative positions and orientations of the molecules could be built up from those of the constituent molecules, due regard being taken of their phase differences for each value of R, but this would as a rule be a complicated process. It may, however, happen that the projections of the molecules on a plane must, by symmetry, be identical and parallel, and we shall for the moment suppose this to be so.

If there are two molecules, distant $\mathbf{d}$ apart in the projection, their two-dimensional transform will be $T(\mathbf{R}) = M(\mathbf{R})\{1 + \exp(2\pi i\mathbf{R}\cdot\mathbf{d})\}$, with the intensity transform $I(\mathbf{R}) = 4|M(\mathbf{R})|^2 \cos^2(\pi\mathbf{R}\cdot\mathbf{d})$. The intensity transform of the pair of molecules will be that of a single molecule modulated by a set of interference fringes, the zero lines of which form lines of zero intensity, or nodal lines, in the transform, perpendicular to the direction $\mathbf{d}$ of the relative displacement of the molecules, and at a distance $1/d$ apart. Although certain spectra, which would have been strong had the unit cell contained only one molecule, may now be weak or absent, we can still be sure that strong spectra correspond to strong regions of the transform of the single molecule. If the transform of this molecule is characteristic and known it may still be possible to recognise it when the array of spectra on the net $Z = 0$ of the reciprocal lattice is considered as a whole, and perhaps also to identify the interference fringes, or nodal lines, and so to get an idea of the relative displacement and orientation of the molecules.

If the unit cell contains four molecules, related in such a way by the symmetry elements that they appear in the projection as four identical and parallel molecules, lying at the corners of a parallelogram of sides **u** and **v**, their intensity transform in two dimensions is $16 \mid M(\mathbf{R}) \mid^2 \cos^2 (\pi\mathbf{R}\cdot\mathbf{u}) \cos^2 (\pi\mathbf{R}\cdot\mathbf{v})$, which is that of a single molecule multiplied by the transform of a net of four points. The latter has broad maxima at the points of the net reciprocal to itself, which must have a wider spacing than that of the reciprocal net of the crystal. The array of spectra on the reciprocal net will thus tend to show groups of stronger spectra in the regions of these broad maxima, separated by nodal lines of weaker or zero intensity. It may be possible in this way to obtain information about this pattern within a pattern. Even if the elements of the subpattern are not related by symmetry, it may happen that the unit cell contains a regular repetition of certain simple units or groups of atoms, which will give rise to a corresponding set of reciprocal maxima in the transform. An interesting example of the application of this idea has been given by Vand.[6]

(*o*) *Nodal surfaces in transforms of symmetrical distributions:* The transform of a molecule with no centre of symmetry, referred to any origin, may be written in the form

$$M(\mathbf{R}) = A(\mathbf{R}) + iB(\mathbf{R}). \tag{39}$$

If the molecule is inverted about the origin as centre, and if its scattering distribution is everywhere real, the transform of the inverted molecule is $A(\mathbf{R}) - iB(\mathbf{R})$, and that of the original molecule and its inversion together is $2A(\mathbf{R})$, which is real and centro-symmetrical. Now suppose the molecule and its inversion to be moved apart through distances $\mathbf{d}/2$ and $-\mathbf{d}/2$ respectively from their initially coincident origins. They will still form a centro-symmetrical distribution, with a transform

$$\begin{aligned} T(\mathbf{R}) &= M(\mathbf{R}) \exp (i\pi\mathbf{R}\cdot\mathbf{d}) + M^*(\mathbf{R}) \exp (-i\pi\mathbf{R}\cdot\mathbf{d}) \\ &= 2\{A(\mathbf{R}) \cos (\pi\mathbf{R}\cdot\mathbf{d}) - B(\mathbf{R}) \sin (\pi\mathbf{R}\cdot\mathbf{d})\} \end{aligned} \tag{40}$$

This is a general expression for the transform of a centro-symmetrical distribution, which may always be thought of as consisting of two halves, themselves either symmetrical or unsymmetrical, related by inversion across a centre.

It will be seen that $T(\mathbf{R})$ is real, and has certain nodal surfaces, given by the equation

$$\tan (\pi\mathbf{R}\cdot\mathbf{d}) = A(\mathbf{R})/B(\mathbf{R}), \tag{41}$$

upon which it must vanish. These surfaces are not usually of any very simple form. The choice of **d** may at first sight seem arbitrary, but it should be noticed that both $A(\mathbf{R})$ and $B(\mathbf{R})$ depend on this choice, which is a matter of convenience and does not affect the surfaces given by (41).

As a simple example, consider four scattering points at the corners of a rectangle of sides **a** and **b**. The points at the ends of one diagonal have equal scattering powers f_1, those at the ends of the other diagonal scattering power f_2. Taking X and Y parallel to **a** and **b**, we may write for the transform of one pair of points f_1 and f_2, referred to an origin at $x=a/2$, $y=0$ midway between them,

$$\begin{aligned} M(\mathbf{R}) &= f_1 \exp(i\pi bY) + f_2 \exp(-i\pi bY) \\ &= (f_1+f_2)\cos(\pi bY) + i(f_1-f_2)\sin(\pi bY). \end{aligned} \tag{42}$$

By (41), the nodal surfaces are given by

$$\tan(\pi Xa) = (f_1+f_2)\cos(\pi bY)/(f_1-f_2)\sin(\pi bY)$$

or

$$\tan\pi\xi\tan\pi\eta = (f_1+f_2)/(f_1-f_2), \tag{43}$$

if $X=\xi/a, Y=\eta/b$. The points $\xi=h, \eta=k$ are those of a net reciprocal to one with primitive translations **a** and **b**.

If $f_2=0$, so that there is only one pair of points, the transform is $2f_1\cos\pi(\xi+\eta)$, which is zero when $\xi+\eta=n+\frac{1}{2}$. The nodal lines are

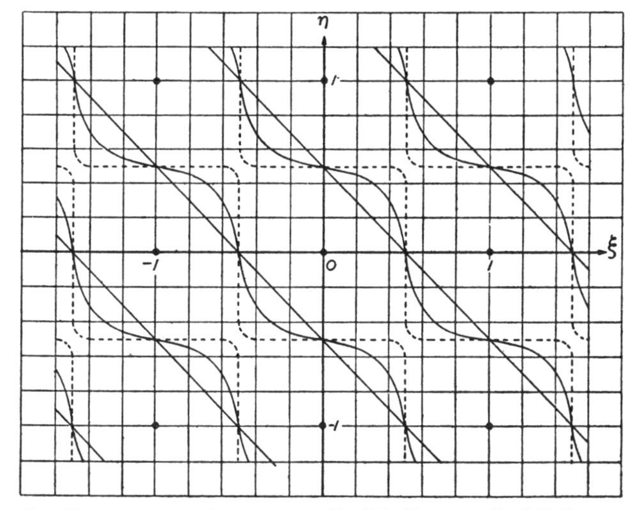

FIG. ix. The curves $\tan\pi\xi=p\cot\pi\eta$. Straight lines, $p=1$; full-line curves, $p=4$; dotted-line curves, $p=100$. The dots indicate points of the reciprocal net. Each curve separates a positive from a negative region of the corresponding transform.

parallel to that set of diagonals of the reciprocal net perpendicular to the line joining the two scatterers f_1, and passing through points lying midway between adjacent reciprocal-net points. They are shown by the diagonal straight lines in fig. ix. In this case, equation (43) becomes $\tan\pi\xi=\cot\pi\eta$, which is satisfied when $\xi+\eta=n+\frac{1}{2}$, and so gives the same result. If now $f_1\neq 0$, the equation becomes $\tan\pi\xi=p\cot\pi\eta$, with

$p > 1$, which is satisfied by $\xi = h$, $\eta = k + \frac{1}{2}$, or $\xi = h + \frac{1}{2}$, $\eta = k$. These are the points on the sets of diagonals discussed above that lie midway between adjacent reciprocal-net points. The nodal lines still pass through these points, but are no longer straight. They have a wave-like form lying alternately on one side or other of the diagonals. As f_1 approaches f_2, the waves approximate to rectangular zig-zags, and when $f_1 = f_2$, so that p is infinite, fuse together to form the two sets of nodal lines, intersecting at right-angles and passing through the mid-points of the reciprocal-net cells, which are characteristic of the transform of four equal points at the corners of a rectangle. The nodal lines for $p = 4$ and $p = 100$ are shown in fig. ix.

(*p*) *Comparison of the transforms of symmetrical and unsymmetrical distributions :* If the distribution is not centro-symmetrical, the transform is not real, but $I(\mathbf{R})$, the intensity transform, is always real for a real distribution, and it is always of the form $A(\mathbf{R})^2 + B(\mathbf{R})^2$, which need not be zero except by chance, if $A(\mathbf{R})$ and $B(\mathbf{R})$ should happen to vanish for the same value of $\mathbf{R}$. We may look at this rather differently. The phase of the transform of a centro-symmetrical distribution, relative to that of a wave scattered by a point at the origin, must be either 0 or π. The nodal surfaces separate regions in which the phases have these values, and correspond to a change in sign of the transform from positive to negative, so that the transform consists of a series of positive and negative loops, separated by nodal surfaces.

If the distribution is not centro-symmetrical the phase will not in general vary discontinuously, but may have any value, depending on the nature of the distribution and the way in which $A(\mathbf{R})$ and $B(\mathbf{R})$ vary with $\mathbf{R}$; and there is no necessity for a system of surfaces of zero value in the intensity transform.

This essential difference in the nature of the two types of transform is the optical basis of Wilson's[7] method of determining statistically, from the distribution of the intensities of its X-ray spectra, whether the unit cell of a crystal has or has not a centre of symmetry. Since there are necessarily nodal surfaces, or surfaces of zero intensity, in the intensity transform of the unit cell of a centro-symmetrical crystal, the relative probability that a spectrum from such a crystal has a low intensity is rather greater than it would have been if the crystal had had no symmetry centre; for in this case there would be no such necessary surfaces of zero intensity. The distribution curve showing the fraction of the total number of spectra which have intensities between any given limits is thus of a different shape for the two types of crystal, and tends to be higher for the lower intensities when the crystal is centro-symmetrical than when it is not.

(*q*) *Optical representations of Fourier transforms :* The calculation of the transforms of even relatively simple atomic distributions is a very tedious operation and methods have been devised to produce the intensity transform of a given trial distribution of atoms optically, for comparison with the observed spectra. One of these, the method of the ' fly's eye ' due originally to W. L. Bragg,[8] produces by means of a multiple lens a repetition of negative images on a photographic plate of a set of black discs, arranged on a white sheet of paper to form the assumed projection of the unit cell. The photographic negative so produced is now used as a grating to form the Fraunhofer diffraction pattern of a point source, which is the required intensity transform of the projection.

Lipson[9] and his fellow-workers have modified and simplified the method. Even if the grating consists of quite a small number of repetitions, for example four, it will give spectra sharp enough to allow the distribution of intensities in the transform to be compared with observation. By means of a special drilling machine, fitted with a pantagraph arrangement, very fine holes are drilled in a metal plate to represent the atoms, and in this way a grating is constructed, consisting of perhaps two repetitions of the pattern in each axial direction. This grating is now used to produce the Fraunhofer pattern of a bright point-source of light, which can be viewed in an eyepiece, or photographed; and by varying the pattern of the holes the transform may be made to approximate to that actually given by the crystal. Many most interesting optical effects can be studied with this beautiful device, but the consideration of the method belongs properly to the volume on techniques.

References

1. P. P. Ewald, *Proc. Phys. Soc. Lond.*, **52,** 167 (1940).
2. W. Cochran and H. B. Dyer, *Acta Cryst.*, **5**, 634 (1952).
3. W. Cochran, F. H. C. Crick, and V. Vand. *Acta Cryst.*, **5**, 581 (1952).
4. H. G. Booker, J. A. Ratcliffe, and D. H. Shinn, *Phil. Trans. Roy. Soc.*, **A. 242,** 579 (1950).
5. G. Knott, *Proc. Phys. Soc. Lond.*, **52**, 229 (1940), A. Klug, *Acta Cryst.*, **3**, 176 (1950).
6. V. Vand, *Acta Cryst.*, **4**, 104 (1951).
7. A. J. C. Wilson, *Acta Cryst.*, **2**, 318 (1949), E. R. Howells, D. C. Phillips and D. Rogers, *Acta Cryst.*, **3**, 210 (1950).
8. W. L. Bragg, *Nature Lond.*, **154**, 69 (1944), A. R. Stokes, *Proc. Phys. Soc. Lond.*, **58**, 306 (1946); P. J. G. de Vos, *Acta Cryst.*, **1**, 118 (1948).
9. C. A. Taylor and H. Lipson, *Nature Lond.*, **167**, 809 (1951); C. A. Taylor, R. M. Hinde and H. Lipson, *Acta Cryst*, **4**, 261 (1951); H. Lipson and C. A. Taylor, *Acta Cryst.*, **4**, 458 (1951); A. W. Hanson and H. Lipson, *Acta Cryst.*, **5**, 145 (1952); A. W. Hanson, *Acta Cryst.*, **6**, 35 (1953).

APPENDIX VI

THE DYNAMICAL THEORY OF ABSORPTION AND PROPAGATION OF ENERGY IN PERFECT CRYSTALS

(*a*) *Introductory:* In this appendix the treatment of the dynamical theory given in Chapter VIII is extended to crystals that absorb X-rays. As early as 1941 Borrmann[1] discovered that the absorption coefficient of a perfect crystal was abnormally small for X-rays incident upon it so as to fall on the atomic planes at or near the Bragg angle. The intensities of Laue reflections in perfect crystals were discussed by Zachariasen in 1945,[2] and more fully by von Laue[3] himself in 1949, and the theory was extended to Bragg reflection by Wagner[9] in 1956. The experimental verification of the theoretical results by Borrmann,[4,5,6] and others has done much to confirm the general correctness of the dynamical point of view, and to show that it is essential in considering the propagation of waves in perfect crystals. The treatment given here is based for the most part on the work of von Laue and Wagner. References to a number of relevant papers are given in the bibliography at the end of the appendix.

It is easiest to consider absorption by supposing formally that the dielectric constant of the crystal is a complex quantity, and this in its turn implies that the quantity ϕ, defined in equation (*8*.4), and its Fourier coefficients, defined in (*8*.5), must also be complex. By (*8*.8)

$$\phi = -(e^2/\pi m v^2)\rho = -(e^2/mc^2)(\lambda^2/\pi)\rho, \tag{1}$$

ρ being the electron density at any point in the crystal, now of course also complex, which can be taken as equal to $|\psi|^2$, where ψ is the electronic wave-function, suitably normalised, at the point considered. The quantity ϕ is a lattice-function, that is to say it is a function of the coordinates repeating itself at corresponding points in each unit cell of the crystal, and its Fourier coefficients are given by

$$\phi_m = \frac{1}{V}\int_{\text{Cell}} \phi e^{j(\mathbf{r}\cdot\mathbf{r}^*_m)}\, dv = -q\mathrm{F}(m), \tag{2}$$

where $\mathrm{F}(m)$ is the structure factor of the unit cell for the spectrum m, and

$$q = (e^2/mc^2)(\lambda^2/\pi \mathrm{V}), \tag{3}$$

a constant for a given crystal and radiation. If therefore the structure factor is complex, so are ϕ and ϕ_m.

A structure factor may be complex simply because the origin of coordinates of the unit cell is not a centre of symmetry, but this does not of course lead to absorption, which occurs only if the scattering factors, f, of the atoms themselves are complex. So long as we may consider the atomic electrons as free Thomson electrons the f-factors are real and positive, and the coherent radiation is scattered with a lag of phase π, but a complex scattering factor $f' + if''$ corresponds to scattering with a phase in advance of that characteristic of free electrons, and this, as we have seen in Chap. IV § (c), p. 137, leads to absorption of the primary wave. The phase advance is not large unless the frequency of the radiation is near an atomic absorption frequency, but it is never entirely negligible, for a non-absorbing crystal is an ideal body that does not exist in nature. In what follows, to avoid complications, we shall suppose the frequency of the radiation not to lie close to an absorption edge.

(*b*) *The complex Fourier coefficients:* We now suppose the quantities ϕ, ϕ_m and $\phi_m\phi_{\bar{m}}$, which occur throughout the expressions of Chap. VIII,* to be complex, putting

$$\phi = \phi_r + i\phi_i, \tag{4}$$

in which ϕ_r and ϕ_i are by definition real. We also write

$$\phi_m = \frac{1}{V}\int_{\text{Cell}} (\phi_r + i\phi_i)\, e^{j(\mathbf{r}\cdot\mathbf{r}^*_m)}\, dv = \phi_{mr} + i\phi_{mi}. \tag{5}$$

The quantities ϕ_{mr} and ϕ_{mi} are given by the two terms of the integral, and are real only if the origin of the cell is a centre of symmetry; but ϕ_{mr}, $\phi_{\bar{m}r}$ and ϕ_{mi}, $\phi_{\bar{m}i}$, if complex, are pairs of conjugate quantities, which follows at once from (5), since ϕ_r and ϕ_i are real, and $\mathbf{r}^*_{\bar{m}} = -\mathbf{r}^*_m$. Generally therefore,

$$\begin{aligned} \phi_{mr} &= |\,\phi_{mr}\,|\, e^{i\eta_m}, & \phi_{\bar{m}r} &= |\,\phi_{mr}\,|\, e^{-i\eta_m}, \\ \phi_{mi} &= |\,\phi_{mi}\,|\, e^{i\omega_m}, & \phi_{\bar{m}i} &= |\,\phi_{mi}\,|\, e^{-i\omega_m}, \end{aligned} \tag{6}$$

where η_m and ω_m represent phases. By (5) and (6) then,

$$\phi_m\phi_{\bar{m}} = P_{mr} + iP_{mi}, \tag{7}$$

where

$$P_{mr} = |\,\phi_{mr}|^2 - |\,\phi_{mi}\,|^2; \quad P_{mi} = 2|\,\phi_{mr}\,|\,|\,\phi_{mi}\,| \cos \nu_m, \tag{8}$$

and

$$\nu_m = \eta_m - \omega_m. \tag{9}$$

P_{mr} and P_{mi} are both real, and if the cell has a symmetry centre, as will be assumed in what follows, $\cos \nu_m = \pm 1$, since if ϕ_{mr} and ϕ_{mi} are real η_m and ω_m are either 0 or π.

(*c*) *The absorption coefficient and the complex wave-vector:* Let a progressive wave, travelling in the direction of x, be represented by

$$U(x, t) = a \exp\{2\pi i(\nu t - Kx)\}. \tag{10}$$

Now suppose the wave-vector K to be complex and equal to $K_r + iK_i$. Equation (10) may then be written

* We here, for compactness, write $\phi_{\bar{m}}$ instead of ϕ_{-m}.

$$\mathrm{U}(x, t) = a \exp(2\pi \mathrm{K}_i x) \exp\{2\pi i(\nu t - \mathrm{K}_r x)\}, \tag{11}$$

which represents a progressive wave of wave-vector K_r propagated in the direction of x with an amplitude varying exponentially with x. If K_i is negative, the amplitude decreases with increasing x, and there is in effect a linear absorption coefficient μ for *intensity* given by

$$\mu = -4\pi \mathrm{K}_i. \tag{12}$$

Consider now radiation passing through a crystal in a direction such that no appreciable diffracted waves other than the direct beam are set up. The field can then be represented by a single wave-train, and by (*8*.36) its wave-vector may be written

$$\mathbf{K}_0 = k(1 + \tfrac{1}{2}\phi_0)\mathbf{s}_0, \tag{13}$$

$\mathbf{s}_0$ being a unit vector in the direction of the wave-normal. When ϕ_0 is complex, $\mathbf{K}_{0r}$ and $\mathbf{K}_{0i}$, the real and imaginary parts of the wave-vector, are

$$\begin{aligned} \mathbf{K}_{0r} &= k(1 + \tfrac{1}{2}\phi_{0r})\mathbf{s}_0 \\ \mathbf{K}_{0i} &= \tfrac{1}{2}k\phi_{0i}\mathbf{s}_0. \end{aligned} \tag{14}$$

In all actual examples ϕ_{0r} and ϕ_{0i} are negative, and so therefore is $\mathbf{K}_{0i}$, and, using (12), we may write for the linear absorption coefficient of the wave-train

$$\mu = 2\pi k \mid \phi_{0i} \mid. \tag{15}$$

This will be the ordinary linear absorption coefficient of the crystal for the radiation, when no diffracted wave-trains are produced.

The measured value of μ may now be used to determine the magnitude of $\mid \phi_{0i} \mid$; and if the frequency of the radiation is not too near an atomic absorption frequency the value of $\mid \phi_{0r} \mid$ may be calculated with fair accuracy from the formula applicable to the non-absorbing crystal, certainly so far as order of magnitude is concerned. By (*8*.37) therefore, we write approximately

$$\mid \phi_{0r} \mid = (e^2/mc^2)(\lambda^2/\pi)\,\mathrm{F}(0)/\mathrm{V}. \tag{16}$$

The values of the linear absorption coefficient for rock salt for Mo $\mathrm{K}\alpha$ ($\lambda = 0{\cdot}71$A), and Cu $\mathrm{K}\alpha$ (1·54A) are respectively 16·4 cm^{-1} and 160 cm^{-1}, giving values $1{\cdot}85 \times 10^{-8}$ and $3{\cdot}92 \times 10^{-7}$ for $\mid \phi_{0i} \mid$. The corresponding values of $\mid \phi_{0r} \mid$, calculated from (16), are $2{\cdot}84 \times 10^{-6}$ and $1{\cdot}33 \times 10^{-5}$, so that the ratio $\mid \phi_{0i} \mid / \mid \phi_{0r} \mid$ has the values $6{\cdot}5 \times 10^{-3}$ and $2{\cdot}92 \times 10^{-2}$ for rock salt for Mo and Cu $\mathrm{K}\alpha$ radiation respectively. It is less easy to estimate accurately the values of $\mid \phi_{mi} \mid / \mid \phi_{mr} \mid$ when other waves are excited, but we may assume it to be of the same order of magnitude, and that little error will be made in neglecting the squares of such quantities, so long as dispersion effects are not of importance, and in all that follows it will be assumed that we may do so. To this degree of approximation therefore, by (8),

$$\mathrm{P}_{mr} > 0\ , \quad \mathrm{P}_{mi} \ll \mathrm{P}_{mr}\ , \quad \mathrm{P}_{mr} = \mid \phi_{mr} \mid^2. \tag{17}$$

(*d*) *Absorption when diffracted wave-trains occur:* We shall consider the case discussed in Chapters II and VIII, when only two reciprocal-lattice points lie near the Ewald sphere, and two wave-trains are associated with each wave-point on the hyperboloidal dispersion surface. For Case I (Transmission) fig. 38, p. 85, applies, and it should be considered in conjunction with fig. 148, p. 424, which, although drawn for Case II (Reflection), is relevant generally to the discussion in the present paragraph.

In fig. 38, the straight line LPM represents within the limits of the figure a circle of radius $k(=1/\lambda)$ and centre O passing through L, the centre of the Ewald sphere. We may conveniently call it the circle of incidence. The vector $\overrightarrow{PO}$ is the incident wave-vector $\mathbf{k}_0$. The distance PA to a wave-point is equal to kg, g being the quantity we have called the accommodation (see p. 424). We consider two states of polarisation, (*a*) in which $\mathbf{D}$ lies perpendicular to the plane containing the vectors $\mathbf{K}_0$ and $\mathbf{K}_m$, and (*b*) in which $\mathbf{D}$ lies in that plane. For each state of polarisation there are two possible values of g for any angle of incidence, each of which gives a wave-field that is an independent solution of the wave-equation holding in the crystal. The two fields for a given state of polarisation are, however, coherent, and the relation between them is determined by the boundary conditions.

By fig. 148,

$$\begin{aligned} \mathbf{K}_0 &= \mathbf{k}_0 - gk\mathbf{z} \\ \mathbf{K}_m &= \mathbf{k}_0 + \mathbf{r}_m^* - gk\mathbf{z}, \end{aligned} \tag{18}$$

where $\mathbf{z}$ is a unit vector in the direction of the normal to the surface. The accommodation g is a function of ϕ_0 and $\phi_m\phi_{\bar{m}}$, and will now be complex, and so consequently will be the wave-vectors $\mathbf{K}_0$ and $\mathbf{K}_m$. Since, however, their difference is the real vector $\mathbf{r}_m^*$, the complex components of the two vectors are equal, and each lies in the direction of $\mathbf{z}$. Putting

$$g = g_r + ig_i, \tag{19}$$

we have therefore, using an obvious notation,

$$\mathbf{K}_{0i} = \mathbf{K}_{mi} = -kg_i\mathbf{z}, \tag{20}$$

so that, by (12), each wave-train suffers the same absorption in depth and σ, the relevant absorption coefficient, is given by

$$\sigma = -4\pi |\mathbf{K}_{0i}| = 4\pi k g_i. \tag{21}$$

The determination of the relevant absorption coefficient thus reduces to the determination of the imaginary part of the accommodation g_i. Since g differs for the two wave-points that correspond to any state of polarisation we may expect the absorption coefficients for the two fields to differ. The physical reason for such an effect is not difficult to understand. Within the crystal, the two wave-trains corresponding to a single wave-point give rise to a pattern of stationary maxima and minima of intensity, not related to the wave-length, but repeating themselves in each unit cell. If an atom lies at a maximum or a minimum of intensity in one cell, the corresponding atom will do so in all cells. The maxima of intensity will be stronger and the minima weaker than the average uniform intensity in the progressive wave that would pass through the crystal in the absence of diffraction, and since the absorption is proportional to the average intensity to which the atoms are exposed, it may be much greater or much less than the ordinary absorption, according to the positions of the atoms relative to the maxima and minima. Since these positions differ for the two fields, the absorption coefficients may well differ considerably. Marked maxima and minima will appear only in that range of angles of incidence in which diffracted beams are excited, and the most marked anomalies of absorption may therefore be expected to be confined to the same narrow range, an expectation confirmed by experiment.

(*e*) *The complex incidence-parameters:* By (*8*.71), the accommodation is equal to

$$g = -(1/2\gamma_0)(\phi_0 + x\phi_{\bar{m}}), \tag{22}$$

and on substituting the value of x from (*8*.47) we obtain

$$g = -\frac{\phi_0}{2\gamma_0} + \frac{1}{4\gamma_m}\left\{\eta \pm \sqrt{\eta^2 + 4|\mathrm{C}|^2\phi_m\phi_{\bar{m}}\frac{\gamma_m}{\gamma_0}}\right\}, \tag{23}$$

where

$$\eta = 2\alpha_m - \phi_0\{1 - (\gamma_m/\gamma_0)\}. \tag{24}$$

Equation (*8*.47) refers to state (*a*) of polarisation alone, in which **D** lies perpendicular to the plane of incidence, and equation (23) has been made more general by the insertion of the factor $|\mathrm{C}|$, which is equal to unity for state (*a*) and to $|\cos 2\theta_0|$ for state (*b*). The equations apply to both Case I and Case II if we remember that in transmission $\gamma_m = |\gamma_m|$, and in reflection, $\gamma_m = -|\gamma_m|$.

The quantity η is an incidence parameter. Its value is discussed in Chap. VIII § (*g*), but its geometrical meaning may be more easily understood from fig. 38. As P moves along the circle of incidence LM, the angle of incidence alters, and the parameter η becomes zero when P is in such a position that the perpendicular drawn through it to the crystal surface passes through Q, the point of intersection of the asymptotes of the hyperbola. We denote this position of P by $\mathrm{P_Q}$. The angle of incidence referred to that corresponding to $\mathrm{P_Q}$ as zero we denote by $\Delta\psi_{\mathrm{Q}}$. Then, as in (*8*.51), in which, however, $\Delta\psi_1 = -\Delta\psi_{\mathrm{Q}}$, since it denotes a change in the glancing angle of incidence,

$$\Delta\psi_{\mathrm{Q}} = \frac{\eta}{2\sin 2\theta_0}, \tag{25}$$

and is positive if P lies to the right of $\mathrm{P_Q}$. The quantity $2\alpha_m$ in (24) is the corresponding incidence parameter referred to the Bragg angle as zero. For most purposes, η is the more convenient parameter.

It is, however, often better to use a parameter p, proportional to η, and defined by an equation analogous to (8.64) as

$$\eta = b(\phi_m\phi_{\bar{m}})^{\frac{1}{2}}p, \tag{26}$$

where
$$b = 2|\,C\,|\sqrt{|\,\gamma_m\,|/\gamma_0}. \tag{27}$$

Both η and p are now complex quantities. Using (24), we write $\eta = \eta_r + i\eta_i$, with

$$\eta_r = 2\alpha_m + |\,\phi_{0r}\,|\left(1 - \frac{\gamma_m}{\gamma_0}\right) \tag{28}$$

$$\eta_i = |\,\phi_{0i}\,|\left(1 - \frac{\gamma_m}{\gamma_0}\right), \tag{29}$$

since ϕ_{0i} and ϕ_{0r} are always negative quantities. By (7) and (26),

$$p = p_r + ip_i = \frac{1}{b\sqrt{P_{mr}}}(\eta_r + i\eta_i)(1 + iQ)^{-\frac{1}{2}}, \tag{30}$$

where
$$Q = P_{mi}/P_{mr}, \tag{31}$$

and is always small in comparison with unity, as also is η_i. On expanding (30), neglecting terms of second order in Q and η_i, and equating real and imaginary parts, we obtain, using (17),

$$p_r = \frac{\eta_r}{b\,|\,\phi_{mr}\,|}\,,\quad p_i = \frac{\eta_i - \eta_r Q/2}{b\,|\,\phi_{mr}\,|}. \tag{32}$$

To this degree of accuracy therefore p_r is still a linear function of α_m, and can be used as an incidence parameter in place of the parameter p used in the theory of the non-absorbing crystal.

(*f*) *The calculation of the absorption coefficients:* Equation (23) may be written

$$g = -\frac{\phi_0}{2\gamma_0} + \frac{1}{4\gamma_m}\{\eta \pm W\}, \tag{33}$$

where
$$W = \sqrt{\eta^2 \pm b^2\phi_m\phi_{\bar{m}}}. \tag{34}$$

In equation (33) the positive sign gives the value of g_1 and the negative sign the value of g_2, corresponding respectively to fields (1) and (2), associated in Case I with wave-points on the lower and upper branches of the hyperbola. In equation (34), the positive sign inside the square-root refers to Case I, in which $\gamma_m = |\,\gamma_m\,|$, and the negative sign to Case II, in which $\gamma_m = -|\,\gamma_m\,|$.

Using (31), (32), (7) and (29), we write

$$\begin{aligned} W = W_r + iW_i &= \sqrt{(\eta_r + i\eta_i)^2 \pm b^2 P_{mr}(1 + iQ)} \\ &= b\,|\,\phi_{mr}\,|\sqrt{(p_r + iB)^2 \pm (1 + iQ)}, \end{aligned} \tag{35}$$

if
$$B = \frac{\eta_i}{b\sqrt{P_{mr}}} = \frac{1}{b}\left|\frac{\phi_{0i}}{\phi_{mr}}\right|\left(1 - \frac{\gamma_m}{\gamma_0}\right). \tag{36}$$

B is a small quantity whose value can be estimated numerically in any given case. It becomes zero in transmission if the relevant crystal planes are perpendicular to the surface, so that $\gamma_m = \gamma_0$, and $|\phi_{0i}/\phi_{mr}|/|C|$ in reflection if the planes are parallel to the surface.

By (31), (8), and (17),

$$Q = \frac{P_{mi}}{P_{mr}} = 2\left|\frac{\phi_{mi}}{\phi_{mr}}\right| \cos \nu_m = 2\left|\frac{\phi_{0i}}{\phi_{mr}}\right| A, \tag{37}$$

where

$$A = \left|\frac{\phi_{mi}}{\phi_{0i}}\right| \cos \nu_m , \tag{38}$$

and is always numerically less than unity, although not as a rule much less, since the imaginary part of the structure factor does not depend greatly on the order of the spectrum.

In (35) we now neglect B^2 in comparison with unity, and obtain

Case I: $$W = b\,|\phi_{mr}|\,\{p_r^2 + 1 + i(2Bp_r + Q)\}^{\frac{1}{2}}, \tag{39}$$

Case II: $$W = b\,|\phi_{mr}|\,\{p_r^2 - 1 + i(2Bp_r - Q)\}^{\frac{1}{2}}, \tag{40}$$

where B differs in the two cases, since $\gamma_m = |\gamma_m|$ in Case I, and $-|\gamma_m|$ in Case II.

By (29) and (33), the real and imaginary parts of g may now be written

$$g_r = \frac{|\phi_{0r}|}{2\gamma_0} + \frac{1}{4\gamma_m}\{\eta_r \pm W_r\} \tag{41}$$

$$g_i = \frac{|\phi_{0i}|}{4}\left(\frac{1}{\gamma_0} + \frac{1}{\gamma_m}\right) \pm \frac{1}{4\gamma_m} W_i. \tag{42}$$

The positive sign applies to field (1) and the negative to field (2). By (15) and (21), if σ is the absorption coefficient in depth,

$$\sigma = 4\pi k g_i = 2\mu g_i/|\phi_{0i}| ,$$

and so, by (42),

$$\sigma = \frac{\mu}{2}\left\{\frac{1}{\gamma_0} + \frac{1}{\gamma_m} \pm \frac{W_i}{\gamma_m |\phi_{0i}|}\right\}. \tag{43}$$

If σ_1 and σ_2 are the absorption coefficients for fields (1) and (2) respectively,

$$\tfrac{1}{2}(\sigma_1 + \sigma_2) = \frac{\mu}{2}\left(\frac{1}{\gamma_0} + \frac{1}{\gamma_m}\right) = \bar{\sigma} \tag{44}$$

$$\tfrac{1}{2}(\sigma_1 - \sigma_2) = \frac{\mu}{2\gamma_m} \frac{W_i}{|\phi_{0i}|} = \kappa , \tag{45}$$

and the absorption coefficients may be written

$$\sigma_{1,2} = \bar{\sigma} \pm \kappa, \tag{46}$$

a convenient form, since $\bar{\sigma}$ does not depend on p_r and is the same for both fields.

(*g*) *The absorption coefficients in Case I* (*Transmission*): In (39), the term $(2\mathrm{B}p_r + \mathrm{Q})$ is always small in comparison with p_r^2+1, for both B and Q are proportional to $|\phi_{0i}/\phi_{mr}|$. We therefore expand the square-root, and neglecting terms of the second order obtain

$$\mathrm{W}_r = b\,|\phi_{mr}|\sqrt{p_r^2+1} \tag{47}$$

$$\mathrm{W}_i = \frac{b\,|\phi_{mr}|(2\mathrm{B}p_r+\mathrm{Q})}{2\sqrt{p_r^2+1}}, \tag{48}$$

and hence, by (36), (37), and (45),

$$\kappa = \tfrac{1}{2}(\sigma_1 - \sigma_2) = \frac{\mu}{2\sqrt{p_r^2+1}}\left\{\left(\frac{1}{\gamma_m} - \frac{1}{\gamma_0}\right)p_r + \frac{2\,|\mathrm{C}|\,\mathrm{A}}{\sqrt{\gamma_0|\,\gamma_m\,|}}\right\}, \tag{49}$$

$$\sigma_{1,2} = \frac{\mu}{2}\left[\frac{1}{\gamma_0} + \frac{1}{\gamma_m} \pm \frac{(1/\gamma_m - 1/\gamma_0)\,p_r + 2\,|\,\mathrm{C}\,|\mathrm{A}/\sqrt{\gamma_0|\,\gamma_m\,|}}{\sqrt{p_r^2+1}}\right], \tag{50}$$

which gives the absorption coefficients for the two fields as functions of the angle of incidence, and is valid over the whole angular range. If the reflecting planes are perpendicular to the crystal surface, so that $\gamma_m = \gamma_0$, equation (50) takes the simpler form

$$\sigma_{1,2} = \frac{\mu}{\gamma_0}\left\{1 \pm \frac{|\,\mathrm{C}\,|\mathrm{A}}{\sqrt{p_r^2+1}}\right\}, \tag{51}$$

and since A is less than but not very different from unity, while $|\,\mathrm{C}\,|$ is equal to unity in state (*a*) of polarisation and $|\cos 2\theta_0|$ in state (*b*), it is clear that both σ_1 and σ_2 are positive, which can also be shown to be true in the more general case. It is clear too that when $p_r = 0$ one coefficient may be only a small fraction of the normal absorption coefficient, while the other may have nearly double its value. Which of the fields has the larger coefficient of absorption depends on the sign of A, and so on the phase of the scattering.

It will be seen that equation (50) is not symmetrical in p_r. We may write

$$\sigma_{1,2} = \tfrac{1}{2}(\sigma_1 + \sigma_2) \pm (\sigma' + \sigma''), \tag{52}$$

where

$$\sigma' = \frac{\mu}{2}\left(\frac{1}{\gamma_m} - \frac{1}{\gamma_0}\right)\frac{p_r}{\sqrt{p_r^2+1}} \tag{53}$$

$$\sigma'' = \frac{\mu\mathrm{A}|\,\mathrm{C}\,|}{\sqrt{\gamma_0\,|\,\gamma_m\,|}}\,\frac{1}{\sqrt{p_r^2+1}}. \tag{54}$$

The term σ'' is symmetrical in p_r, but σ', which vanishes for symmetrical transmission, changes sign with p_r when the crystal planes are inclined to the surface. It is zero for $p_r = 0$, and tends to $\pm(\mu/2)(1/\gamma_m - 1/\gamma_0)$ when p_r has large positive or negative values.

In fig. x, σ_1 and σ_2 are shown plotted as functions of p_r for rock salt

200 when the surface of the crystal is parallel to (111). The planes (200) then make an angle 52° 42′ with the surface. The figure is drawn for Cu Kα radiation ($\lambda = 1{\cdot}54$A, $\mu = 160$ cm^{-1}), and the value of $|\phi_{mi}/\phi_{mr}|$ has been taken as 0·95. The numerical values are based on those given in Chap. VI § (*j*), p. 325. In this case, $\cos \nu_m = 1$, and so $A = 0{\cdot}95$, and is also positive. We thus find

$$\bar{\sigma} = 213 \text{ cm}^{-1}, \quad \sigma' = \frac{42{\cdot}7\, p_r}{\sqrt{p_r^2+1}} \text{ cm}^{-1}, \quad \sigma'' = \frac{198}{\sqrt{p_r^2+1}} \text{ cm}^{-1}.$$

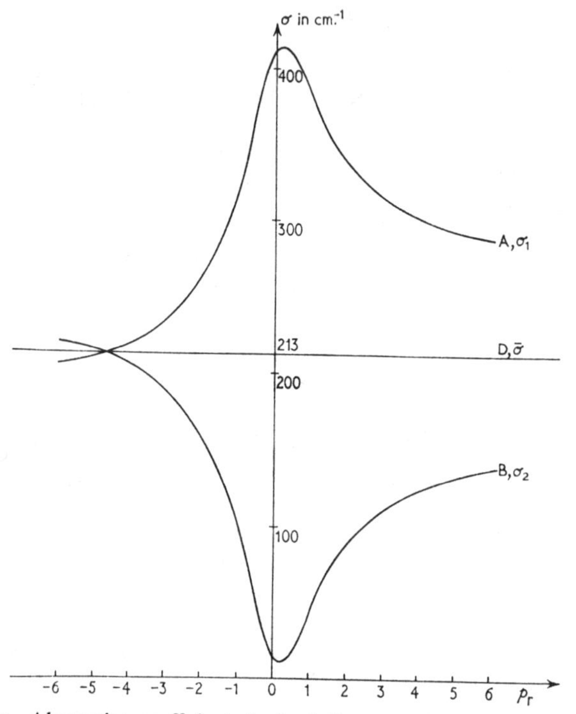

FIG. x Absorption coefficients in depth for transmission for NaCl 200. Cu Kα radiation, $\mu = 160$ cm^{-1}. (111) parallel to surface.

The figure shows clearly the asymmetry due to the inclination of the reflecting planes to the surface. The curves A and B for σ_1 and σ_2 are mirror images of each other in the line D, which gives the constant value of $\bar{\sigma}$. It should be remembered that the whole range of p_r shown in this figure corresponds to an angle of 12 or 15 seconds of arc on either side of $p_r = 0$, so that the angular range of the abnormal absorption is very small.

It is interesting to consider a few actual figures. The intensity of Cu Kα radiation passing through a slice of crystal 0·1 cm in thickness would be reduced by the average absorption $\bar{\sigma} = 213$ cm^{-1} to $e^{-21{\cdot}3} = 5{\cdot}6 \times 10^{-10}$ of its surface value. The maximum value of σ_1 is

415 cm^{-1}, which corresponds to a reduction factor of about 10^{-18}; but the minimum value of σ_2 is about 12 cm^{-1}, for which the reduction factor is only 0·30. Over a very small range of angles therefore, the slice, while virtually opaque for field (1), is relatively transparent for field (2). The abnormal transmission is less marked for state (*b*) of polarisation. In the example we are considering $|C|A$ is about 0·81 for state (*b*), and the numerical coefficient in σ'' is 168 instead of 198, as in state (*a*). This leads to a minimum value of σ_2 equal to about 40 cm^{-1}, which while still much smaller than the normal coefficient is yet considerably greater than the minimum value of σ_2 for state (*a*) If the crystal slice is more than a millimetre or two in thickness the transmitted radiation in the range of abnormal transmission will consist of field (2) almost entirely in state (*a*) of polarisation.

The existence of this abnormal transmission has been conclusively demonstrated by the beautiful photographs obtained by Borrmann[4] and his fellow workers in Berlin, which show the sharp traces of the Kossel cones of reflection and transmission when radiation from a copper anticathode passes through crystal slices of calcite and germanium, in some cases more than 3 mm in thickness. The photographs show with beautiful clarity not only the existence of the phenomenon, but the smallness of the angular range over which it occurs. Very perfect crystals are of course necessary to show this effect satisfactorily. For experimental details, the reader must be referred to the original papers.

(*h*) *The absorption coefficients in Case II* (*Reflection*): The relevant expression for W is now that of equation (40), and an expansion of the type used in Case I is clearly no longer valid if p_r lies in the neighbourhood of ± 1. We must distinguish moreover between range (i) $p_r < -1$, range (ii) $-1 < p_r < +1$, and range (iii) $p_r > +1$.

In ranges (i) and (iii), in both of which $p_r^2 > 1$, we put

$$x = p_r^2 - 1, \quad y = 2Bp_r - Q = \frac{2}{|C|}\left|\frac{\phi_{0i}}{\phi_{mr}}\right|\left\{\frac{\gamma_0 + |\gamma_m|}{2\sqrt{\gamma_0 |\gamma_m|}}\, p_r - |C|A\right\}, \tag{55}$$

so that

$$W = b\,|\phi_{mr}|(x + iy)^{\frac{1}{2}} \tag{56}$$

$$W_i = b|\phi_{mr}|\{\tfrac{1}{2}(\sqrt{x^2 + y^2} - x)\}^{\frac{1}{2}}. \tag{57}$$

By (44) and (45), since now $\gamma_m = -|\gamma_m|$,

$$\bar{\sigma} = \tfrac{1}{2}(\sigma_1 + \sigma_2) = (\mu/2)(1/\gamma_0 - 1/|\gamma_m|), \tag{58}$$

and

$$\kappa = \tfrac{1}{2}(\sigma_1 - \sigma_2) = -\frac{\mu|C|}{\sqrt{\gamma_0 |\gamma_m|}}\left|\frac{\phi_{mr}}{\phi_{0i}}\right|\{\tfrac{1}{2}(\sqrt{x^2 + y^2} - x)\}^{\frac{1}{2}}, \tag{59}$$

an expression that is valid over the whole of ranges (i) and (iii), and is easy to evaluate as a function of p_r if the numerical value of y is known.

When $p_r = \pm 1$, $x = 0$ and $W = b|\phi_{mr}|\sqrt{iy}$. By (55), since $|C|$ and $|A|$ are both less than unity, while $\gamma_0 + |\gamma_m| > 2\sqrt{\gamma_0 |\gamma_m|}$, y is positive when $p_r \geqslant 1$ and negative when $p_r \leqslant -1$. Using the result that the imaginary part of $\sqrt{\pm i|y|}$ is equal to $\pm\sqrt{\tfrac{1}{2}|y|}$, we therefore find for the value of κ when $p_r = \pm 1$

$$\kappa(\pm 1) = \mp \frac{\mu|C|}{\sqrt{\gamma_0|\gamma_m|}} \left|\frac{\phi_{mr}}{\phi_{0i}}\right| \sqrt{\tfrac{1}{2}|y|}, \tag{60}$$

where
$$|y| = \frac{2}{|C|}\left|\frac{\phi_{0i}}{\phi_{mr}}\right| \left\{\frac{\gamma_0 + |\gamma_m|}{2\sqrt{\gamma_0|\gamma_m|}} \mp A|C|\right\}. \tag{61}$$

In those parts of ranges (i) and (iii) in which $y^2 \ll x^2$, (59) may be expanded in the form

$$\kappa = -\frac{\mu|C|}{2\sqrt{\gamma_0|\gamma_m|}} \left|\frac{\phi_{mr}}{\phi_{0i}}\right| \frac{y}{\sqrt{x}} \tag{62}$$

$$= -\frac{\mu}{2\sqrt{p_r^2 - 1}} \left\{\left(\frac{1}{\gamma_0} + \frac{1}{|\gamma_m|}\right)p_r - \frac{2|C|A}{\sqrt{\gamma_0|\gamma_m|}}\right\}. \tag{63}$$

In range (ii), $p_r^2 - 1 < 0$, and we write

$$W = ib|\phi_{mr}|(x' - iy)^{\frac{1}{2}} \tag{64}$$

where $x' = 1 - p_r^2$ and y has the same value as before. Then

$$\kappa = -\frac{\mu|C|}{\sqrt{\gamma_0|\gamma_m|}} \left|\frac{\phi_{mr}}{\phi_{0i}}\right| \{\tfrac{1}{2}(\sqrt{x'^2 + y^2} + x')\}^{\frac{1}{2}}, \tag{65}$$

which is valid over the whole range. In that part of the range in which $y^2 \ll x'^2$, and this in practice may be a considerable fraction of it,

$$\kappa = -\frac{\mu|C|}{\sqrt{\gamma_0|\gamma_m|}} \left|\frac{\phi_{mr}}{\phi_{0i}}\right| \sqrt{1 - p_r^2}\left(1 + \frac{y^2}{8x'^2}\right) \tag{66}$$

$$= -\frac{2\pi k|C|\phi_{mr}|}{\sqrt{\gamma_0|\gamma_m|}} \sqrt{1 - p_r^2}\left(1 + \frac{y^2}{8x'^2}\right), \tag{67}$$

by (15). The numerical value of κ is here virtually that of the primary extinction coefficient in the range of total reflection for the perfect non-absorbing crystal. (See eqn. (*8*.74).)

(*i*) *Symmetrical reflection:* The detailed numerical discussion will be confined to the case in which the reflecting planes are parallel to the crystal surface, so that $\gamma_0 = |\gamma_m| = \sin\theta_0$, and $\bar{\sigma} = 0$, by (58). Then

$$\sigma_{1,2} = \pm\kappa, \tag{68}$$

and the absorption coefficients for the two fields are equal in magnitude but opposite in sign. By (55),

$$y = \frac{2}{|C|}\left|\frac{\phi_{0i}}{\phi_{mr}}\right|(p_r - |C|A). \tag{69}$$

In ranges (i) and (iii)

$$\sigma_{1,2} = \pm\kappa = \mp\frac{\mu\,|\,C\,|}{\sin\theta_0}\left|\frac{\phi_{mr}}{\phi_{0i}}\right|\{\tfrac{1}{2}(\sqrt{x^2+y^2}-x)\}^{\frac{1}{2}}, \tag{70}$$

or, if the expanded form is valid,

$$\sigma_{1,2} = \mp\frac{\mu}{\sin\theta_0}\,\frac{p_r - |\,C\,|A}{\sqrt{p_r{}^2-1}}. \tag{71}$$

Since, as we have seen, y is negative in range (i) and positive in range (iii) κ must be positive in range (i) and negative in range (iii), and so, by (68), σ_1 is positive and σ_2 negative in range (i) and σ_1 negative and σ_2 positive in range (iii). If the crystal is indefinitely thick, since only a positive absorption coefficient is then physically possible, only field (1) can exist in range (i) and field (2) in range (iii); and since, by (60), the same results are true when $p_r = \pm 1$ they will clearly apply throughout the ranges (i) and (iii). A similar argument applied to (67) shows that in range (ii) only field (2) is possible in a thick crystal.

When the slice has a finite thickness we may not draw these conclusions, for there may then be reflected wave-trains travelling from the lower to the upper surface, diminishing in intensity as they do so, and for these a negative absorption coefficient is appropriate.[9,17] They must therefore belong to field (2) in range (i) and to field (1) in range (iii).

If we put
$$\mu/\sin\theta_0 = \sigma_n, \tag{72}$$
the absorption coefficient in depth if only the normal linear absorption were active, and consider only the numerical value $|\,\sigma\,|$ of the coefficient σ, we may write

$$\frac{|\,\sigma\,|}{\sigma_n} = \frac{|\,p_r - |\,C\,|A\,|}{\sqrt{p_r{}^2-1}}. \tag{73}$$

If A is positive, as it is for example for rock salt, this is greater than unity in range (i) and less than unity in range (iii), but tends to unity for large $|\,p_r\,|$ in both ranges.

When $p_r = \pm 1$, we may use (60) and (61) and write for symmetrical reflection

$$\frac{|\,\sigma(\pm 1)\,|}{\sigma_n} = \sqrt{|\,C\,|}\left\{\left|\frac{\phi_{mr}}{\phi_{0i}}\right|\left|1 \mp |\,C\,|\,A\right|\right\}^{\frac{1}{2}}. \tag{74}$$

The smaller value will apply to $p_r = +1$ when A is positive.

We apply the results to rock salt 200, for which we may take $|\,\phi_{mr}/\phi_{0i}\,| = 25$ and $A = +0{\cdot}95$. In state (*a*) of polarisation $|\,C\,| = 1$. Equation (74) then gives $|\,\sigma\,|/\sigma_n = 6{\cdot}98$ and $1{\cdot}12$ for $p_r = -1$ and $+1$ respectively. When p_r becomes a little greater than $+1$, $|\,\sigma\,|/\sigma_n$ falls away rapidly to about $0{\cdot}31$ at $p_r = 1{\cdot}05$, and thereafter rises, tending to unity from below as p_r becomes large. Below $p_r = -1$, $|\,\sigma\,|/\sigma_n$ falls steadily from a value about seven times that for normal absorption at $p_r = -1$ as the negative value of p_r increases, tending again to unity for large negative values, this time from above. In the example considered, equation (73) may be applied in range (i) when p_r is less than about $-1{\cdot}75$, and in range (iii) when p_r is greater than about $1{\cdot}04$.

In range (ii), which corresponds closely to Darwin's range of total reflection, the values of $|\sigma|/\sigma_n$ can be calculated for the example considered with enough accuracy from the expression

$$\frac{|\sigma|}{\sigma_n} = \left|\frac{\phi_{mr}}{\phi_{0i}}\right| \sqrt{1 - p_r^2}, \tag{75}$$

which follows at once from (66), if p_r lies between $-0{\cdot}7$ and $+0{\cdot}95$. Over the rest of the range (65) must be used. When $p_r = \text{A}$ the two formulae agree. The curve of fig. xi was calculated from (65).

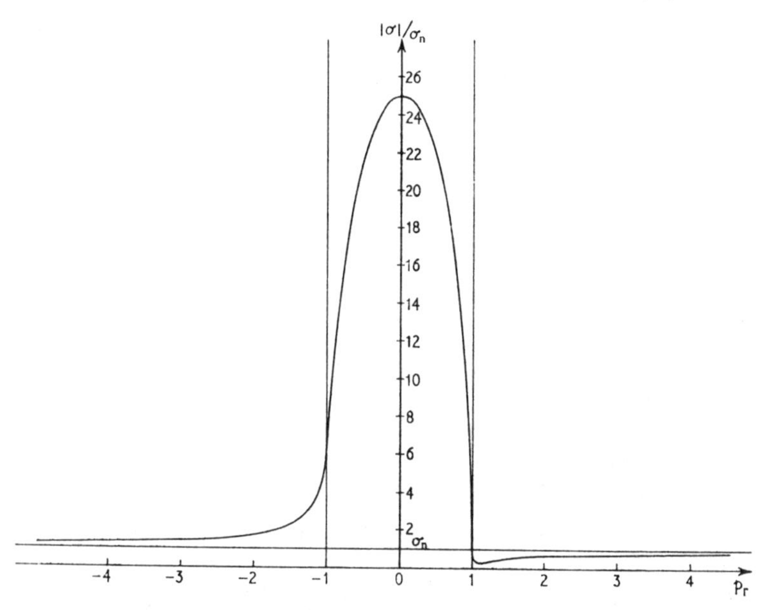

FIG. xi $|\sigma|/\sigma_n$ for symmetrical Bragg reflection from rock salt, 200. Cu Kα radiation, $\mu = 160\ \text{cm}^{-1}$, $\sigma_n = 586\ \text{cm}^{-1}$.

In that part of range (ii) in which (75) is valid the absorption differs very little from the primary extinction; but extinction becomes zero at both ends of the range, and here there is a definite absorption effect, much greater when p_r is near -1 than when it is near $+1$. The reason for this may perhaps be understood from the discussion in Chap. VIII § (*l*), in which the nature of the stationary distribution of intensity in a non-absorbing crystal was considered. It was there shown that the intensity of the field on the atomic planes was a minimum for $p = 1$, and that it rose as p passed through the range of total reflection, reaching a maximum at $p = -1$. (See fig. 150, p. 433.) True absorption can occur only if the atoms that produce it lie in an appreciable field, so that we should expect the greatest absorption to occur near $p_r = -1$. The unsymmetrical shape of Prins's reflection curves for absorbing crystals, illustrated in fig. 27, p. 65, is due to the same cause. It should be noticed that in the curves there illustrated, due to Renninger, $\eta = +1$ corresponds to our $p_r = -1$.

As p_r passes through the range of large extinction the field penetrates so slightly into the crystal that the atoms have little opportunity to cause absorption, and the reduction of the field-strength with depth depends virtually on extinction alone, whether or no the crystal is absorbing. At the edges of the range the extinction becomes small, and the radiation penetrates to depths that are governed by true absorption.

(*j*) *The propagation of energy in perfect crystals:* The complete group of wave-trains associated with a wave-point is a solution of the equation governing wave propagation in the crystal, but its constituent trains, taken alone, are not. The direction and magnitude of the energy current-density due to such a group of waves may be calculated from Poynting's vector

$$\mathbf{S} = (c/4\pi)\,\mathbf{E} \times \mathbf{H}, \tag{76}$$

in which **E** and **H** are the resultant electric and magnetic fields at the point considered due to the whole set of waves. When radiation in a given state of polarisation is incident on the crystal the fields associated with the wave-points on different branches of the dispersion surface, the positions of which are determined by the boundary conditions on incidence, are coherent. For extended wave-trains, the fields to be used in (76) are then the resultant fields due to all the wave-points, and we shall first consider this case, but it must be remembered that the energy current associated with any one wave-point can, under suitable conditions, be propagated independently.

In calculating the direction and magnitude of the energy current, we may without appreciable error replace **E** by **D** in (76). The time average of **S**, which is the physically measurable quantity, is equal to

$$\overline{\mathbf{S}}^{t} = (c/8\pi)\,\mathrm{R}(\mathbf{D} \times \mathbf{H}^*), \tag{77}$$

where R denotes the real part of the expression that follows it.

Let a and b refer to wave-points and m and n to reciprocal-lattice points that correspond to wave-trains in the field. Then, by (*8*.12) and (*8*.12*a*), we may put

$$\begin{aligned} \mathbf{D} &= \sum_a \exp\{j(\nu t - \mathbf{K}_{0a}\cdot\mathbf{r})\} \sum_m \mathbf{D}_{ma} \exp\{-j(\mathbf{r}_m^*\cdot\mathbf{r})\} \\ \mathbf{H}^* &= \sum_b \exp\{-j(\nu t - \mathbf{K}_{0b}^*\cdot\mathbf{r})\} \sum_n \mathbf{H}_{nb}^* \exp\{j(\mathbf{r}_n^*\cdot\mathbf{r})\}, \end{aligned} \tag{78}$$

whence

$$\mathbf{D}\times\mathbf{H}^* = \sum_a\sum_b \exp\{j(\mathbf{K}^*_{0b}-\mathbf{K}_{0a})\cdot\mathbf{r}\}\sum_m\sum_n \mathbf{D}_{ma}\times\mathbf{H}^*_{nb}\exp\{j(\mathbf{r}^*_n-\mathbf{r}^*_m)\cdot\mathbf{r}\}. \quad (79)$$

We take absorption into account by using equations (18), (19) and (21), from which

$$\mathbf{K}^*_{0b}-\mathbf{K}_{0a} = -(g^*_b - g_a)\,k\mathbf{z} = \{(g_{ar}-g_{br})+i(g_{ai}+g_{bi})\}\,k\mathbf{z}$$

$$= (g_{ar}-g_{br})\,k\mathbf{z} + (i/2\pi)\,\bar{\sigma}_{ab}\mathbf{z}, \quad (80)$$

where

$$\bar{\sigma}_{ab} = \tfrac{1}{2}(\sigma_a+\sigma_b), \quad (81)$$

σ_a and σ_b being the absorption coefficients in depth associated with fields (a) and (b).

Substituting (80) in (79), we obtain

$$\mathbf{D}\times\mathbf{H}^* = \sum_a\sum_b \exp\{jk(g_{ar}-g_{br})z\}$$

$$\times \exp(-\bar{\sigma}_{ab}z)\sum_m\sum_n \mathbf{D}_{ma}\times\mathbf{H}^*_{nb}\exp\{j(\mathbf{r}^*_n-\mathbf{r}^*_m)\cdot\mathbf{r}\}, \quad (82)$$

where $z=\mathbf{z}\cdot\mathbf{r}$, and is the actual depth below the surface of the crystal.

The first periodic factor in (82), and the absorption factor, vary very little in the depth of a single cell, and the last double sum varies in the same way in all unit cells. To obtain the mean value of the energy current-density in any unit cell we therefore treat the first two factors as constant, and take the average of the last exponential factor over the cell, which is equal to unity if $m=n$, but otherwise vanishes. Taking the real part of (82) and using (77), we obtain for the time average of the energy current-density in a unit cell at depth z in the crystal

$$\bar{\mathbf{S}} = (c/8\pi)\sum_a\sum_b \cos\{2\pi k(g_{ar}-g_{br})z\}\exp(-\bar{\sigma}_{ab}z)\sum_n \mathbf{D}_{na}\times\mathbf{H}^*_{nb}. \quad (83)$$

Let $\mathbf{s}_{nb}$ be a unit vector in the direction of the wave-vector $\mathbf{K}_{nb}$. Each constituent plane wave-train is transverse, so that $\mathbf{s}_{nb}$, $\mathbf{D}_{nb}$ and $\mathbf{H}_{nb}$ are mutually perpendicular; and with the units used here the numerical values of **D** and **H** associated with any wave-train may be taken with sufficient accuracy to be equal. We may therefore write

$$\mathbf{H}_{nb} = \mathbf{s}_{nb}\times\mathbf{D}_{nb},$$

and so

$$\mathbf{D}_{na}\times\mathbf{H}_{nb} = \mathbf{D}_{na}\times(\mathbf{s}_{nb}\times\mathbf{D}_{nb}) = \mathbf{s}_{nb}(\mathbf{D}_{na}\cdot\mathbf{D}_{nb}) - (\mathbf{s}_{nb}\cdot\mathbf{D}_{na})\mathbf{D}_{nb}.$$

In state (a) of polarisation the scalar product $\mathbf{s}_{nb}\cdot\mathbf{D}_{na}$ vanishes, and in state (b) it is always very small, so that (83) may be written

$$\bar{\mathbf{S}} = (c/8\pi)\sum_a\sum_b \cos\{2\pi k(g_{ar}-g_{br})z\}\exp(-\bar{\sigma}_{ab}z)\sum_n (\mathbf{D}_{na}\cdot\mathbf{D}^*_{nb})\mathbf{s}_{nb}. \quad (84)$$

The typical term in this summation is a vector in the direction of the wave-vector $\mathbf{K}_{nb}$, from the wave-point (b) to the reciprocal-lattice point

n, and the resultant energy current-density is the vector sum of all such terms over all wave-points and all reciprocal-lattice points. The corresponding formula for a non-absorbing crystal was given by von Laue.[10]

(k) *The energy current-density when there are two relevant reciprocal-lattice points:* We shall confine such detailed discussion as space permits to this case, and shall further restrict it to state (a) of polarisation, when the electric vector is perpendicular to the plane of incidence.

We now have to consider two reciprocal-lattice points, o and m, and two wave points, (1) and (2), on the lower and upper branches of the hyperbola respectively. It will be accurate enough to replace both $\mathbf{s}_{n1}$ and $\mathbf{s}_{n2}$ by the unit vector $\mathbf{s}_n$ in the direction of the radius of the Ewald sphere to the reciprocal-lattice point n. The directions of $\mathbf{s}_0$ and $\mathbf{s}_m$ are those of incidence and reflection for a Bragg reflection at the relevant lattice planes. In its expanded form, (84) for this case becomes

$$\mathbf{S} = (c/8\pi)[e^{-\sigma_1 z}(|\mathbf{D}_{01}|^2\,\mathbf{s}_0 + |\mathbf{D}_{m1}|^2\,\mathbf{s}_m) + e^{-\sigma_2 z}(|\mathbf{D}_{02}|^2\,\mathbf{s}_0 + |\mathbf{D}_{m2}|^2\,\mathbf{s}_m)$$
$$+ e^{-\bar{\sigma} z}\cos(2\pi z/\Delta)\{(\mathbf{D}_{01}\mathbf{D}^*_{02} + \mathbf{D}^*_{01}\mathbf{D}_{02})\,\mathbf{s}_0 + (\mathbf{D}_{m1}\mathbf{D}^*_{m2} + \mathbf{D}^*_{m1}\mathbf{D}_{m2})\mathbf{s}_m\}], \quad (85)$$

where
$$\Delta = \frac{1}{k(g_{1r} - g_{2r})} = \frac{\lambda\sqrt{\gamma_0|\gamma_m|}}{|\mathrm{C}||\phi_{mr}|\sqrt{p_r^2+1}} \quad (86)$$

by (41), (47) and (27). In state (a) of polarisation $|\mathrm{C}| = 1$.

The first two terms on the right-hand side of (85) represent the energy currents due to the fields associated with wave-points (1) and (2) respectively, each multiplied by its appropriate absorption factor. Either of these energy currents could be propagated independently in the crystal, but because the two fields are coherent their interaction must be considered when extended wave-fronts are dealt with, and this is given by the third term, the interference term, which gives rise to a beat phenomenon, periodic in depth Δ, which will be discussed later. The interference term has an absorption coefficient $\bar{\sigma}$, the mean of σ_1 and σ_2.

(l) *The energy current-density associated with a single wave-point when absorption is negligible:* The values of the induction vectors in (85) are fixed by the boundary conditions on incidence, given by (8.83), p. 435, from which, if $\mathbf{D}_0^i$ is now the induction vector in the incident wave, and

$$x_1 = \mathbf{D}_{m1}/\mathbf{D}_{01}, \quad x_2 = \mathbf{D}_{m2}/\mathbf{D}_{02},$$

we deduce

$$\mathbf{D}_{01} = \frac{x_2}{x_2 - x_1}\mathbf{D}_0^i, \quad \mathbf{D}_{02} = \frac{-x_1}{x_2 - x_1}\mathbf{D}_0^i. \quad (87)$$

Using (8.47), (8.62) and (8.27) it is easy to show that

$$x = -\mathrm{G}(p \pm \sqrt{p^2+1}), \quad (88)$$

where $\mathrm{G} = \sqrt{\gamma_0/|\gamma_m|}\,e^{i\theta m}$, as in Chap. VIII § ($j$). The positive sign of the square-root applies to field (1) and the negative to field (2). Then, from (87) and (88),

$$\mathbf{D}_{01}=\tfrac{1}{2}\{1-(p/\sqrt{p^2+1})\}\,\mathbf{D}_0^i,\quad \mathbf{D}_{02}=\tfrac{1}{2}\{1+(p/\sqrt{p^2+1})\}\,\mathbf{D}_0^i,\qquad (89)$$

$$\mathbf{D}_{m1}=-\tfrac{1}{2}(\mathrm{G}/\sqrt{p^2+1})\mathbf{D}_0^i\quad,\quad \mathbf{D}_{m2}=\tfrac{1}{2}(\mathrm{G}/\sqrt{p^2+1})\,\mathbf{D}_0^i,\qquad (90)$$

The energy current $\overline{\mathbf{S}}_2$, due to field (2) alone, is given by the second term on the right-hand side of (85), the absorption factor being now unity. Its two components lie in the directions $\mathbf{s}_0$ and $\mathbf{s}_m$, as indicated in fig. xii. Equations (89) and (90) show that when p is large and positive only $\mathbf{D}_{02}$ of the four Fourier coefficients has appreciable value. Virtually the whole energy current is carried by field (2) and its magnitude is $|\,\overline{\mathbf{S}}_0\,|$, given by

$$|\,\overline{\mathbf{S}}_0\,|=(c/8\pi)|\,\mathbf{D}_0^i\,|^2,\qquad (91)$$

the average energy current-density in the incident radiation.

As p becomes smaller $|\,\mathbf{D}_{02}\,|$ diminishes, and the diffracted wave $|\,\mathbf{D}_{m2}\,|$ grows. The two components of $\overline{\mathbf{S}}_2$ become equal when $p=0$, if the atomic planes concerned are perpendicular to the surface, and the resultant energy current is in the direction $\mathbf{j}$ in fig. xii, perpendicular to the surface and parallel to the atomic planes. The magnitude of $\overline{\mathbf{S}}_2$ is then $\tfrac{1}{2}|\,\overline{\mathbf{S}}_0\,|\cos\theta_0$, half the normal energy current-density in the incident beam. The other half is of course now carried by field (1).

As p becomes negative both $|\,\mathbf{D}_{02}\,|$ and $|\,\mathbf{D}_{m2}\,|$ diminish, $|\,\mathbf{D}_{02}\,|$ much the more rapidly, and, for large negative values of p, $\overline{\mathbf{S}}_2$ is in the

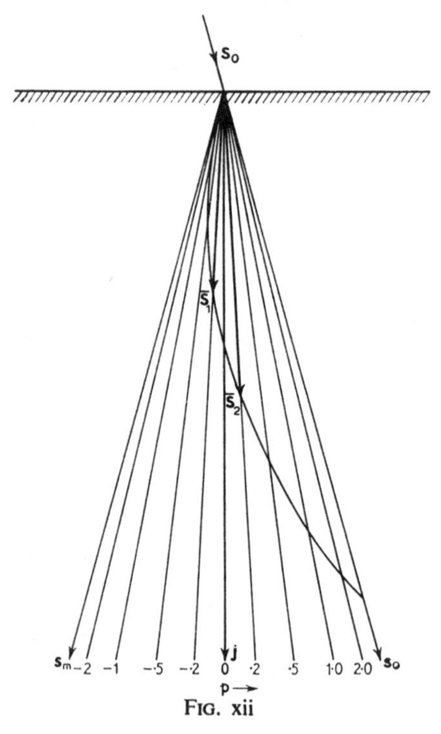

FIG. xii

direction $\mathbf{s}_m$, but is very small. The range of angles of incidence involved is very small, and $|p|=10$, corresponding to a change of angle of incidence of perhaps 30″, is in this context a large value. Within a range of angles of this order on either side of that corresponding to the diffraction maximum the energy current due to field (2) swings in direction from $\mathbf{s}_0$ to $\mathbf{s}_m$. The curve in fig. xii shows the locus of the extremity of the vector $\bar{\mathbf{S}}_2$ as p varies, and some of the corresponding values of p are indicated. The range of variation of angles of incidence is too small to show in the figure.

From (85), we see that when absorption is neglected $\bar{\mathbf{S}}_2(-p)=\bar{\mathbf{S}}_1(p)$, so that the extremity of $\bar{\mathbf{S}}_1$ has the same locus as that of $\bar{\mathbf{S}}_2$. The energy current for field (1) is, however, in the direction $\mathbf{s}_m$ for large positive p, and is very small. It swings across to $\mathbf{s}_0$ for large negative p, when it carries virtually the whole energy current, since $\bar{\mathbf{S}}_2$ has here become very small. The directions of both $\bar{\mathbf{S}}_1$ and $\bar{\mathbf{S}}_2$ are shown in fig. xii for NaCl 200 when $p=+0{\cdot}2$ and the atomic planes are normal to the surface. The two directions coincide when $p=0$, but a change of angle of incidence of about 0·8 seconds of arc is enough to make them diverge by about $6\frac{1}{2}°$. The two energy currents differ in magnitude, and their resultant lies between $\mathbf{j}$ and $\mathbf{s}_0$ for all values of p.

Ewald[12] and Kato[11] have shown generally that the direction of the energy current associated with any wave-point is normal to the dispersion surface at that point, and it is easy to see in this way how the directions of $\bar{\mathbf{S}}_1$ and $\bar{\mathbf{S}}_2$ vary as the wave-points move along the branches of the hyperbola of fig. 38.

(*m*) *The interference term and Ewald's pendulum solution* (*Pendellösung*)*:* We now substitute the values of the induction vectors given in (89) and (90) in (85), and using (91) obtain for the resultant energy current in a crystal with negligible absorption

$$\bar{\mathbf{S}}=|\bar{\mathbf{S}}_0|\left[\mathbf{s}_0+\frac{\sin^2(\pi z/\Delta)}{p^2+1}\{|G|^2\mathbf{s}_m-\mathbf{s}_0\}\right]. \tag{92}$$

The last term of (92), which is periodic in depth Δ, is a vector parallel to the crystal surface. This can be seen at once from fig. xiii, in which $\overrightarrow{AB}$ is the unit vector $\mathbf{s}_0$ and BC is parallel to the surface. Since $\gamma_0=\cos\psi_0$ and $\gamma_m=\cos\psi_m$, $\overrightarrow{AC}$ is the vector $(\gamma_0/\gamma_m)\mathbf{s}_m=|G|^2\mathbf{s}_m$, and BC is therefore the vector $|G|^2\mathbf{s}_m-\mathbf{s}_0$.

Whenever $z=n\Delta$, n being an integer, the periodic term in (92) vanishes, and

$$\bar{\mathbf{S}}=|\mathbf{S}_0|\mathbf{s}_0.$$

The energy current at these depths is entirely in the direction of the incident beam, whatever the direction of incidence. When $z=(n+\frac{1}{2})\Delta$

the periodic term has its greatest value, and the direction of the energy current is that of the vector $\overrightarrow{AT}$ in fig. xiii, where

$$\overrightarrow{AT}=\overrightarrow{AB}+\overrightarrow{BT}, \quad \text{and} \quad \overrightarrow{BT}=\overrightarrow{BC}/(p^2+1).$$

As the depth below the surface of the crystal increases the resultant energy current swings backwards and forwards between the direction of the incident beam $\mathbf{s}_0$, when $z=n\Delta$, and $\overrightarrow{AT}$, when $z=(n+\frac{1}{2})\Delta$. The variation of direction is greatest when $p=0$, for then $BT=BC$, and

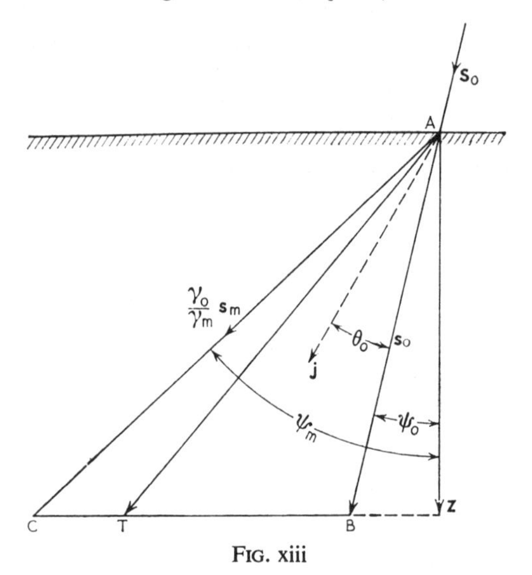

FIG. xiii

the energy current oscillates between the direction of incidence $\mathbf{s}_0$ and the direction of reflection $\mathbf{s}_m$. A closer investigation shows that the actual directions are those of the vectors $\frac{1}{2}(\mathbf{K}_{01}+\mathbf{K}_{02})$ and $\frac{1}{2}(\mathbf{K}_{m1}+\mathbf{K}_{m2})$, the difference being due to the approximations made in deducing (85). For our purpose it is negligible.

When $p=0$ the beam emerging from the lower surface of a slice of crystal will lie entirely in the direction of incidence, $\mathbf{s}_0$, when the thickness is $n\Delta$ and entirely in the direction of reflection, $\mathbf{s}_m$, when it is $(n+\frac{1}{2})\Delta$. For intermediate thicknesses there are emergent beams in both directions. This is Ewald's well known pendulum solution, (*Pendellösung*), considered from the point of view of energy transmission.

We may call the depth Δ the *beat period* in depth in the crystal. By (86) it is greatest when $p=0$, and diminishes quite rapidly as $|p|$ increases. At the same time, the range of variation in the direction of $\bar{\mathbf{S}}$ decreases. $\bar{\mathbf{S}}$ is always in the direction $\mathbf{s}_0$ when $z=n\Delta$, and deviates less and less from that direction with changing depth as $|p|$ increases. The actual value of the beat thickness is small, about $3{\cdot}2\times10^{-3}$ cm.

for NaCl 200, with Mo Kα radiation, but this is some tens of thousands of times the depth of the unit cell.

It is difficult to check these results experimentally. A very small angular spread in the direction of the incident beam, such as is almost inevitable in practice, will cause the direction of energy transmission to vary over a fan of directions between $\mathbf{s}_0$ and $\mathbf{s}_m$, and since at the same time the beat period itself varies the oscillatory effect will be largely smeared out. Reference should, however, be made to recent work by Kato and Lang.[13]

(*n*) *Incident beams of limited width:* Suppose now a beam of limited width and sharply defined direction to be incident on the crystal, and for simplicity let the atomic planes concerned be normal to the surface. Then, when $p=0$, the two energy currents $\bar{\mathbf{S}}_1$ and $\bar{\mathbf{S}}_2$ follow the same path, also normal to the crystal surface. For any other value of p, however, their directions will differ, and since each can be propagated independently, it will be only in the region of their overlap, near the crystal surface, that the type of interference field we have just considered will be set up. The two beams will be propagated in directions lying one on each side of the normal, and in neither will there be any separa-

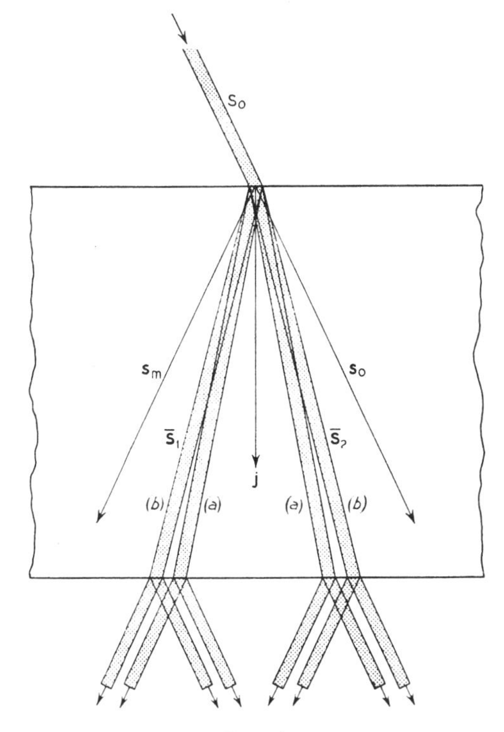

FIG. xiv

tion into primary and reflected wave-trains. For both beams such separation takes place only on emergence from the lower surface of the crystal. Here there will be four emergent beams, one parallel to $\mathbf{s}_0$ and one to $\mathbf{s}_m$, from each area of exit. Since by (25), (26) and (27) the value of p for a given angle of incidence depends on the state of polarisation, and is larger for state (*b*) than for state (*a*), there will be two close pairs of beams, $\bar{\mathbf{S}}_{1a}$, $\bar{\mathbf{S}}_{1b}$ and $\bar{\mathbf{S}}_{2a}$, $\bar{\mathbf{S}}_{2b}$, in the crystal, and eight beams in all on emergence if the incident beam contains radiation in both states of polarisation, and this is represented diagrammatically in fig. xiv.

On the geometrical theory, we should expect the incident beam to travel through the crystal in the direction $\mathbf{s}_0$, giving rise along its whole path to diffracted radiation in the direction $\mathbf{s}_m$; and this does in fact occur with crystals of ordinary imperfection.

(*o*) *Limited beams in absorbing crystals:* In a perfect crystal with negligible absorption the effects discussed in the last paragraph would be largely masked by the spread in direction of the beams $\bar{\mathbf{S}}_1$ and $\bar{\mathbf{S}}_2$ due to the unavoidable small divergence of any actual incident beam. Some of the results can, however, be confirmed by making use of the absorption effects that occur in all real crystals, which were discussed in the earlier part of this appendix. The absorption coefficients for fields (1) and (2) may differ considerably, and over a small range of angles in the neighbourhood of the diffraction maximum the crystal may be abnormally transparent for one of them. Suppose that field (2) has the smaller absorption coefficient. Then it will be clear from fig. x that if the crystal is thick enough only the second term of (85) need be considered, and that, too, over only a very small range of angles. The abnormal transparency in this range will in effect select a slightly divergent beam for transmission, belonging to a single energy current, in this case $\bar{\mathbf{S}}_{2a}$. In this way, Borrmann[6] was able to show that the divergence of primary and reflected beams does occur at the lower surface of the crystal.

The principle of the experiment may be understood from fig. xv, in which the atomic planes are assumed to be perpendicular to the crystal surface. The incident beam $\mathbf{S}_0$ gives rise to an energy current $\bar{\mathbf{S}}$ within the crystal, propagated in this case normal to the surface and parallel to the crystal planes. On emergence, this gives rise to the transmitted and reflected beams T and R, which fall upon a photographic film. Because the incident beam is not entirely monochromatic, radiation that takes no part in the interference phenomena is transmitted directly through the crystal and gives rise to a spot marked O on the film, which serves as a reference point. The variation in the separation of the points O, T and R when the distance of the film from the crystal is altered enables one to determine the point of divergence of the rays T and R, which is found to lie in the lower surface of the crystal. Very good crystal material is necessary for success in experiments of this kind. In some experiments with a calcite crystal 3·2 mm in thickness Borrmann and Hildebrandt[17] found that a temperature difference of 0·6° C. between the two sides of the crystal produced enough distortion of the lattice to destroy the effect.

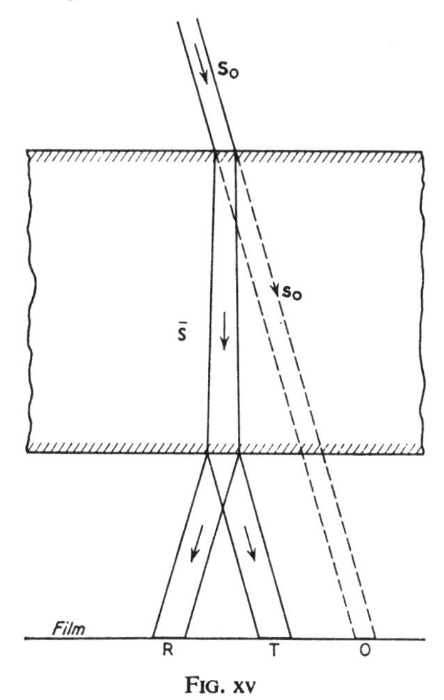

FIG. XV

REFERENCES

1. G. Borrmann, *Physikal. Zeit.*, **42**, 157 (1941).
2. W. H. Zachariasen, *Theory of X-Ray Diffraction in Crystals*, New York Wiley (1945).
3. M. von Laue, *Acta Cryst.*, **2**, 106 (1949).
4. G. Borrmann and W. Gerisch, *Zeit. f. Krist.*, **106**, 99 (1954).
5. G. Borrmann, *Zeit. f. Krist.*, **106**, 109 (1954).
6. G. Borrmann, *Zeit. f. Physik*, **127**, 297 (1950).
7. P. B. Hirsch, *Acta Cryst.*, **5**, 176 (1952).
8. G. Borrmann, G. Hildebrandt and H. Wagner, *Zeit. f. Physik*, **142**, 406 (1955).
9. H. Wagner, *Zeit. f. Physik*, **146**, 127 (1956).
10. M. von Laue, *Acta Cryst.*, **5**, 619 (1952).
11. N. Kato, *Acta Cryst.*, **11**, 885 (1958).
12. P. P. Ewald, *Acta Cryst.*, **11**, 887 (1958).
13. N. Kato and A. R. Lang, *Acta Cryst.*, **12**, 787 (1959).
14. G. Hildebrandt, *Zeit. f. Krist.*, **112**, 312 and 340 (1959).
15. G. Borrmann and G. Hildebrandt, *Zeit. f. Physik*, **156**, 189 (1959).
16. N. Kato, *Acta Cryst.*, **13**, 349 (1960).
17. M. von Laue, *Röntgenwellenfelder in Kristallen, S.B. dtsch. Akad. Wiss. Nr* **1** (1959).

INDEX OF SUBJECTS

ABBE's theory of image formation, 390
Absorption,
  allowance for in powder method, 48, 333, 337
  atomic, variation of with frequency, 147
  calculation of dispersion terms from, 148
  effect of on measurements of critical angle, 168, 173
  effect of on reflection by perfect crystals, 64, 429
  effect of on wave-field in perfect crystals, 428, 632
  elimination of by method of mixed powders, 185, 337
  in medium containing dipole oscillators, 137, 139
  line, width of in relation to damping, 138
  relation of to complex refractive index, 62, 138, 168
  relation of to photo-electric effect, 140
  treatment of by means of complex amplitudes, 62, 138
Acoustic waves in lattices, 203, 226
Aluminium,
  atomic scattering factor of, 303
  integrated reflection from, 281, 286, 303
  mean-square amplitude in, 238
  state of ionisation in, 303
  zero-point energy in, 303
Anomalous dispersion of X-rays (*see* Dispersion)
Anomalous scattering, 180, 181, 182, 184, 187, 188
  in Heusler alloys, 185
  of tungsten for L edge, 187
Argon, scattering by, 472, 473
Atomic scattering factor (*see* Scattering factor, atomic)

BASIS, of lattice structure, 27
Benzil, diffuse scattering from, 241, 243
Boundary conditions,
  in dynamical theory of reflection, 83, 427
  for vibrations of a finite lattice, 197
Boundary waves, 68
  in semi-infinite crystal, 80, 84
Bragg's reflection law,
  deviations from 52, 54, 168, 426, 436
  deviations from, measurement of refractive index from, 168
Breit-Dirac correction factor, 260, 463

CALCITE,
  refractive index of for X-rays, 170, 179
  test of Prins's formula for, 330
  width of reflection curves from, 306, 332

Carbon tetrachloride,
  amplitude of thermal motion in, 492
  scattering by, 486
  temperature correction for, 491
Cerussite, refractive index of, 171
Characteristic temperature, 218
  as a function of the crystal temperature, 225, 235
  calculated from the elastic constants, 222
  determined from specific heats, 220
  relation of to compressibility, 223
  table of, 221
  value appropriate to temperature factor, 219, 220, 222
Charge-density,
  in crystals, determination of by Fourier series, 342
  Schrödinger, 102
  Schrödinger, in many-electron atom, 113
Charge-distribution, atomic,
  contribution of to different parts of f-curve, 125
  Hartree's calculation of, 117
  Pauling-Sherman, 123
  Thomas-Fermi, 123
Close packing, irregular, in liquids, 476
Coherent scattering,
  by assemblages of atoms, 463
  Compton-Raman theory of, 98, 100
  Wentzel-Waller theory of, 103
Composite lattice, diffraction by, 27
  in perfect crystals, 65, 325
Compton-Raman theory of scattering, 99 100
Critical angle (*see* Total reflection)
Crossed spectra, application of method of to X-rays, 174, 176
Crystal-form factor, 548
  as a line integral, 552
  as a surface integral, 550
  symmetry properties of, 551
Cybotaxis, 460, 481

DAMPING due to radiation, 136
  influence of on scattering factor, 154
Darwin-Prins formula,
  derivation of, 55, 62, 428
  test of for calcite, 330
  test of for rock salt, 322, 325
Debye-Scherrer rings, widths of,
  in terms of reciprocal lattice, 529
  integrated line breadth of (q.v.)
  Laue's treatment of, 529, 537
  Stokes-Wilson formula for, 531, 534
Debye temperature (*see* Characteristic temperature)

Debye-Waller temperature factor, 24, 210
effect of expansion of crystal on, 225, 234
experimental tests of,
for aluminium, 232, 233
for diamond, 232
for rock salt and sylvine, 232, 233
for crystals with non-cubic symmetry, 227
formal representation of for complex crystals, 226
influence of variation of characteristic temperature on, 234
limitations of, 225
measurement of at low temperatures, 233
numerical evaluation of for cubic crystals, 215
relation of to elastic constants, 238
Deficiency lines in divergent beam photography, 453
Diamond,
as perfect crystal, 318
characteristic temperature of, 224, 232
extra reflections from, 240, 252
forbidden reflections from, 253, 321
integrated reflection from, 320
scattering factor of carbon in, 322
temperature factor for, 232
types of, 252
widths of reflections from, 305, 318
Diffraction effects in Fourier projections, 396
Diffraction of electrons,
by gas molecules, 493
by small crystals, 545, 547
Diffraction of X-rays,
by distorted lattices, 560
by fibrous materials, 459, 571
by linear crystallites, 518, 573
by parallel chains of atoms, 572
by parallel elongated crystallites, 578, 580
by parallelepipedal lattices, 1, 4
by periodically distorted lattices, 563
by regular lattices with non-identical cells, 555
by space-lattices,
dynamical theory of, 66, 413
geometrical theory of, 1
when source lies inside crystal, 438
Diffuse maxima,
bridges between, 244, 256
contribution of different elastic waves to, 209
distribution of intensity in, 207, 210
dependence of shape of on crystal structure, 248
intensities of, 257
Diffuse reflections,
considered as optical ghosts, 205
dependence of on elastic type of crystal, 214, 250
effect of temperature on, 247
experimental study of, 239
Faxén-Waller theory of, 195, 201, 207
from aluminium, 241, benzil, 241, 243, rock salt, 246, sodium, 250, 251, sorbic acid, 247, 250, sylvine, 242, tungsten, 251, urea nitrate, 247, 249
radial streaks in, 239, 244
relation of to Bragg and Laue spectra, 246
relation of to elastic waves in crystal, 208
shape of, for chain structures, 250
for layer lattices, 248
study of on Laue photographs, 241
Diffuse scattering,
dependence of on temperature, 257
from sylvine, 261, sodium fluoride, 263, zinc, 264
investigation of with ionisation spectrometer, 253
Jauncey's theory of, 258
Diffusion, surfaces of equal, 210
for cubic crystals, 210, 213
dependence of form of on elastic type, 213
dependence of form of on order of spectrum, 214
Dipole oscillator,
scattering by, 135
scattering factor for, 137
virtual, in dispersion theory, 140, 143
Dipole terms in scattering factor, 163
Dipole-waves in dynamical theory, 66
amplitude of accompanying wave, 69
description of in terms of reciprocal lattice, 70
electromagnetic field due to, 67
in semi-infinite crystal, 80
relation of to interference wave-field, 72
resonance error of, 70
Dispersion equation, 75
Dispersion formula,
experimental verification of, 167
from refractive indices, 167
from scattering factors, 180
for many-electron atom, 143
Kramers-Heisenberg, 143
oscillator strengths in (q.v.)
Dispersion of X-rays, 135
calculated from absorption coefficients, 148
calculated from atomic wave-functions, 157
Dispersion surface in dynamical theory, 75
for two waves, 76, 419
relation between phases on different sheets of, 422
significance of, 421
transition from one to two waves, 79
Dispersion terms in scattering formulae,
for bound electrons, 112, 140
for many-electron atom, 143
Displacements, root-mean-square,
of atoms in crystals, 236, 238
Distortion of crystals,
effect of on scattering, 560

Divergent-beam photography, 452
  with gamma rays, 455
Double-crystal spectrometer,
  effect of polarisation in, 315
  integrated reflection from, 316
  parallel setting for, 308
    absence of dispersion in, 312
    reflection curve for, 310, 312
  (1, 1) setting for, 314
  principle of, 306
  reflection curves from diamond, 318
  test of Darwin-Prins formula with, 322
  use of in measuring refractive indices, 180
  widths of double and single reflection curves, 317
Dynamical theory of reflection, 66, 413
  boundary conditions in, 83, 427
  dipole wave in, 66
  dispersion surface in, 75
  interchange solution in, 86
  in terms of continuous charge-distribution, 413
  Laue's form of, 413
  solution of for two waves in finite crystal, 422
  symmetrical reflection in, 86
  unsymmetrical reflection in, 88
  wave-field in (q.v.)

ELASTIC spectrum of lattice,
  contribution of different frequency ranges to, 223
  dependence of on temperature, 224
  discontinuities in for real crystals, 226
Electrons,
  diffraction of, by gas molecules, 493, by small crystals, 545, 547
  scattering by, on classical theory, 29, on quantum theory, 101
Electron density, radial, in atoms, 97
  false detail in, 408
  Fourier series for, 408
  from scattering factors, 404
  numerical determination of, 117, 123, 125
Electron spin,
  effect of on exchange terms, 116
Energy propagation in perfect crystals, 632, 645
Ewald's construction, 8, 9
  use of in dynamical theory, 72
Ewald's dynamical theory, 66
Exchange,
  effect of on *f*-factor for Cl^-, 122
  in incoherent scattering, 116
Exclusion principle, 114, 117
Extinction, primary, 60, 268
  calculation of in range of total reflection, 430
  correction for, 270
  dependence of on wave-length, 330
  effect of in divergent beam photography, 454
  effect of on measurement of temperature factor, 232
  estimate of for rock salt and diamond, 61
Extinction, primary and secondary,
  simultaneous existence of, 294
Extinction, secondary, 49, 268, 274
  coefficient of, 292
  correction for in mosaic crystal, 276
  correction of absorption for, 282
  Darwin's theory of, 270, 276
  in crystal powders, 332
  methods of estimating, 292
  value of for rock salt, 290
Extra reflections,
  primary & secondary, from diamond, 252
f-CURVES (*see* scattering factor, atomic)
  comparison of for atomic models, 129
  correction of for dispersion, 154, App. III
  limitations of applicability of, 132
False detail,
  in Fourier projections, 385, 395, 396, 401
  in radial electron distributions, 409
Faults in lattices,
  effect of on diffraction spectra, 555
  periodic, 563
  Wilson's theory of, 557
Fibrous materials,
  diffraction by, 459, 571
  diffraction by, general remarks on, 588
Finite lattice,
  normal modes in, 197
Fourier integrals,
  application of to radial density in atoms, 403
  application of to radial distribution in powdered crystals, 503, 504
  application of to scattering by liquids, 477, 496
  derivation of, App. IV
  limitations of, 406
Fourier projections,
  considered as optical images, 385
  diffraction effects in, 385, 396
  equivalent numerical aperture of, 397
  false detail in, 396, 401
  of unit cell, 351, 360
  of phthalocyanines, 363
  of slice of structure, 355
  relation of to two-dimensional grating, 386, 388, 392
Fourier series,
  application of to crystal analysis, 342
  double, application of to centrosymmetrical structures, 360, 361
  double, application of to polar structures, 365
  double, method of handling, 356
  form of for crystal with symmetry centre, 347
  for radial distribution in atoms, 408
  interference function expressed as, 522
  in terms of reciprocal lattice, 348, 356
  one-dimensional, 369

Fourier series,
Patterson's application of (*see* Patterson series)
phases of terms in, 346, 365
physical interpretation of terms of, 345
stability of solution by, 366
triple, derivation of, 343
triple, evaluation of coefficients of, 345
triple, use of, 349, 355, 384
Fourier transforms, application of, App. V
Frequency limit of lattice vibrations, 197
Fresnel's reflection law, application to X-rays, 172, 177
Friedel's law, 33

GASES, scattering of X-rays by, 458
Geranylamine hydrochloride, analysis of, 384
Ghosts, optical,
due to thermal movements, 205
due to periodic distortions, 563

HARTREE's method of self-consistent field, 117
Helium, scattering, by, 472
Heusler alloys, structure of, 185

IMAGINARY part of *f*-factor,
effect of on scattered phase, 151
Imperfections of crystals, 284
evidence of from intensities of spectra, 285
Incoherent scattering,
allowance for in scattering experiments, 478
Breit-Dirac correction for, 463
calculation of from Thomas-Fermi distribution, 462
Compton-Raman formula for, 99, 462
exchange terms in, 116, 461
numerical estimation of, 462
Wentzel-Waller theory of, 103, 110
Integral line-breadth (*see also* Line-breadth)
calculation of in special cases, 536
definition of, 529
from strained lattices, 562
from structures with faults, 558
Laue's theory of, 529
Stokes and Wilson's calculation of, 531
Integrated reflection, 38, 41
comparison of experiment and theory, 285
correction of for primary extinction, 270
from aluminium crystals, 303
from crystal slab without absorption, 37
from diamond, 320
from mosaic crystal face, 44, 269
from mosaic crystal slice, 45, 287
from mosaic of small blocks, 281
from perfect crystal, 55, 59, 62, 269
from powdered crystals, 46, 48, 332
from rock salt, 287, 299, 302
from small crystal of any form, 39
from sylvine, 302
measurement of by double-crystal spectrometer, 316
measurement of by ionisation spectrometer, 268
summary of formulae for, 50
treatment of in terms of reciprocal lattice, 41
Intensity,
integrated (*see* integrated reflection)
of reflection of X-rays by crystals, 27
effect of thermal movements on, 20, 193
maxima, subsidiary, from parallelepipedal lattice, 10
Interchange solution in dynamical theory, 86
Interference,
conditions, in terms of reciprocal lattice, 6
external, 98, 459, 481, 582
internal, 459, 481
Interference function,
definition of, 8, 9, 201, 514
effect of crystal form on, 548
expressed as Fourier series, 522
for crystal with faults, 557
for lattice in thermal motion, 203
optical analogy to, 555
Iron pyrites, refractive index of, 170
Iso-diffusion surfaces (*see* Diffusion, surfaces of equal)

JAUNCEY's scattering formula, 258
experimental tests of, 261
relation of to Compton and Debye formulae, 259
Woo's modification of, 260

KIKUCHI lines, 455
Laue's theory of, 455
Kossel lines, 439
complementary cones for, 444
composite cones, 445
diffraction cones in, details of, 441
excited by X-rays, 447
experimental production of, 446
geometry of, 448
intersection of cones, 448
spacing measurements from, 451
Kramers-Heisenberg dispersion formula, 143

LADENBURG's absorption formula, 139
Lattice,
basis of, 27
irregularities of, effect on diffraction, 20, 555, 560, 563
normal modes in finite, 197
parallelepipedal, diffraction by, 1
phase, 198
Laue,
form of dynamical theory, 413
method, 19, 244

Laue,
  point, 78, 421
  theory of diffraction by lattices, 1
  theory of Kikuchi lines, 455
  theory of line-breadth, 529
Layer lines, 16
Limiting sphere, 15
Line-breadth,
  for colloidal gold, 544
  influence of structure factor on, 531, 545
  integral (*see also* Integral line-breadth)
    experimental determination of, 540
    generality of expression for, 535
    Stokes-Wilson formula for, 534
    Scherrer constant for, 536
    use of approximation functions in calculating, 537
  relation of to crystal size, 8, 12, 528
Linear crystallite,
  interference function for, 518
  diatomic molecule as special case of, 520
Liquid crystals, 460

MATRIX elements of perturbing field, calculation of, 106
Maximum frequency of lattice vibrations,
  calculation of by Debye's method, 217
  calculation of from elastic constants, 222
  for rock salt and diamond, 223, 224
Mercury, scattering by liquid, 477, 478
Mixed powders, method of, 185, 187, 337
  precautions necessary in, 337
Momentum in electromagnetic problems, 104
Mosaic crystals, 21, 43
  reflection from (*see* Integrated reflection)
Mosaic formulae,
  derivation of, 44
  quantitative tests of, 299, 302
  summary of, 50

NICKEL,
  interference of X-rays in film of, 175
  total reflection of X-rays from, 174
Normal co-ordinates,
  of lattice vibrations, 195
  root-mean-square displacement in terms of, 198
Normal modes,
  Debye's approximation to for cubic crystal, 216
  in a finite lattice, 197
  in silicon-tetrachloride molecule, 489, 492

OCTOPOLE terms in scattering factor, 163
Optical frequencies in lattice vibrations, 226
Optical images, relation of to Fourier projections, 385
Oscillating-crystal method, 15
Oscillator-density,
  determination of from photoelectric absorption, 146
  for continuum of positive energy states, 144
  of K continuum for hydrogen-like electron, 157, 159
Oscillator strength,
  calculation of from atomic wave-functions, 157
  effect of forbidden transitions on, 143
  expression of in terms of matrix elements, 142
  for hydrogen-like electron, 144, 157
  from photoelectric absorption, 146
  table of for K and L continua, 159, 160
  Thomas-Reiche-Kuhn rule for, 143
  total, of K continuum, 159
  values of for K electrons of iron, 146, 174

PARALLELEPIPEDAL lattice,
  intensity scattered by, 4
  interference function for, 9
  subsidiary scattering maxima from, 10
Path difference,
  vector expression for in scattering, 2
Patterson series, 371
  analysis of pentaerythritol by, 380
  Harker's application of, 377
  properties of, 373
  resolution of, 376
  two-dimensional, 374
  use of in preliminary analysis, 383
Pentaerythritol, structure of, 380
Perfect crystals, reflection by (*see* Dynamical theory and Integrated reflection)
Perfection of crystals,
  artificial alteration of, 295, 297, 298
  classification of, 296
  effect of thermal strain on, 298
  effect of piezo-electric vibrations on, 298
Periodically distorted lattices,
  diffraction by, 563
  experimental evidence of, 568
    in Cu_4FeNi_3, 568
    in felspars, 568
    in organic molecular compounds, 569
Perturbed wave-functions, calculation of for electromagnetic waves, 105
Phase,
  difference of between points in vibrating lattice, 196
  difference of for sheets of dispersion surface, 422
  of terms in Fourier series, 346
  lattice, 198
Phase-change on scattering,
  effect of on structure factor, 32, 151
  expressed as a complex amplitude, 32 151
Photoelectric absorption,
  determination of oscillator strengths from, 146
Phthalocyanines, Fourier projection of, 364

Plane waves,
  in lattices, 196
  velocities of in cubic crystals, 211
Polarisation factor, 31
  in perfect crystal formula, 59, 65
  with double-crystal spectrometer, 315, 327
Polarity of crystal,
  effect of on structure factor, 33, 65
Potassium hydrogen phosphate,
  application of Fourier series to, 361
Powdered-crystal method, 18
  focusing condition for, 335
  integrated reflection for, 46
  intensity measurements by, 332
    mixed powder method, 185, 337
    reflection method, 334
    substitution method, 337, 338
    transmission method, 48, 333
  (*see also* Debye-Scherrer rings)
Primary extinction (*see* Extinction, primary)
Probability function for intermolecular distances, 460, 469, 474
  determination of for mercury, 477
  for irregular close-packing, 476
  in liquids with polyatomic molecules, 495

QUADRUPOLE terms in scattering factor, 161
  for iron, 164
  for Zn $K\alpha$ scattered by copper, 166

RADIAL distribution, average,
  of atoms in powdered crystals, 501, 502, 504, 508
  of charge in atoms,
    determination of by Fourier integrals, 404, 408
Random distribution, impossibility of for real molecules, 458
Reciprocal lattice,
  analogy between, and grating spectra, 11, 386
  definition of, 6
  interference conditions in terms of, 7
  Laue pattern in terms of, 19, 244
  properties of, 6, App. II
  scattering function in terms of, 202
  thermal scattering in terms of, 203
  use of in describing dipole waves, 71
Reciprocity theorem, 439
  application of to Kossel lines, 440
Reflecting plane, 2
Reflection,
  condition in diffraction, 5
  integrated (*see* Integrated reflection)
  multiple, 25, 339
  sphere of, 8, 15
Reflection of X-rays,
  angular range of for real crystals, 282
    examples of, 283
    relation of to crystal perfection, 284
  by crystal mosaics, 44
  by infinite slab, 36
  by perfect crystals, 52, 452
    with absorption, 62, 428
  by sheet of atoms, 35
  dynamical theory of (*see* Dynamical theory)
  total, at critical angle (*see* Total reflection)
  total, from perfect crystal, 58, 87, 425
  unsymmetrical, from crystal mosaic, 278
    in dynamical theory, 88, 425, 428
Refractive index,
  of calcite, 179
  of crystals for X-rays, 53, 74, 79, 167
  of medium containing dipole oscillators, 137
  measurement of,
    from critical angle, 171
    from deviations from Bragg's law, 168
    from deviations produced by prisms, 177
    from unsymmetrical reflection, 169
  relation of to scattering, 137
Resonance error, 70
Reststrahlen frequencies,
  relation of to acoustic frequencies, 226
  relation of to characteristic temperature, 220, 226
Rock salt,
  as a perfect crystal, 325
  determination of scattering factors from, 287
  diffuse reflections from, 246
  integrated reflection from, 299, 325
  maximum frequency for, 223
  primary extinction in, 61, 273, 328
  secondary extinction in, 287
  surfaces of equal diffusion for, 213, 215
  temperature factor for, 232, 236
  tests of mosaic formula from, 299
  thermal motions in, 237
Rotating crystal method, 15
Rotations of groups of atoms in crystals, 229
  criterion of possibility of, 229
  effect of on apparent symmetry, 230
  effect of on structure factors, 230
  in ammonium nitrate, 230
  in long-chain compounds, 231
Root-mean-square displacements,
  calculation of in lattice, 198
  measurement of in crystals, 236
  relation of to temperature factor, 23, 193, 219
Row-lines, 18
Rubidium bromide, anomalous scattering by, 181

SCATTERING factor, atomic,
  calculation of,
    theoretical, classical, 94, 95, 101
      quantum-mechanical, 101
    numerical,
      by Hartree's method, 117, 121
      by Lenz and Jensen's method, 131

Scattering factor, atomic (*contd.*)
calculation of, numerical,
by Pauling-Sherman method, 123
by Thomas-Fermi method, 123
comparison of for different atomic models, 129
contribution of atomic electron groups to, 125
correction of for dispersion, 154, App. III
definition of, 30, 93
dispersion terms in, 112, 140, 145
effect of quadrupole terms on, 161
effect of on scattering by diatomic molecules, 484
experimental determination of,
from monatomic gases, 472
from powdered crystals, 332
from rock salt, 287, 299
for dipole oscillator, 137
for many-electron atom, 113
imaginary part of, 148
effect of on amplitude and phase, 151
influence of damping on, 154
limitations of tabulated values of, 132
methods of calculation from charge-distributions, 125
test of dispersion formula by measurement, of, 180
total, 149, 164
variation of with frequency, 149
Scattering moment, for coherent radiation, 108
Scattering of X-rays,
anomalous, 33, 135
by assemblages of atoms of finite size, 469
by bound electrons,
Compton-Raman theory of, 99
quantum theory treatment of, 101, 107
by classical electron, 29
by dipole oscillator, 135
by fibrous materials, 459, 571
by gases, 458
diatomic, 483
effect of temperature on, 488
monatomic, 469
polyatomic, 480
with point atoms, 465
by liquids, 458
monatomic, 475
polyatomic, 494
by single sheet of atoms, 35
by vitreous solids, 508
diffuse (*see* Diffuse, maxima, reflections, scattering)
incoherent (*see* Incoherent scattering)
in terms of fictitious dielectric constant, 415
total, Hartree-Waller formula for, 116
Scherrer constant, 536
Schrödinger charge-density, 102
for many-electron atom, 113
Schrödinger current-density,
calculation of scattering from, 103, 108
Self-consistent field, method of, 117
with exchange, 121
Silicon tetrachloride,
normal modes in, 489
scattering by, 486, 492
thermal vibrations in molecule of, 492
Size of crystals, effect of on widths of spectra, 8, 13, 528 (*see also* Integral line-breadth)
Sodium, scattering by liquid, 479, 507
Space-lattice, diffraction by simple, 1
diffraction by composite, 27
Spacing, generalised interpretation of, 5, 6, 603
Specific heats, Debye theory of, 193, 216, 218
Spectrometer, double-crystal (*see* Double-crystal spectrometer)
Sphere, limiting, 15
Sphere of reflection, 8, 15
use of in dynamical theory, 71, 72
Structure amplitude, 28
Structure factor, 31
as Fourier amplitude, 345
effect of polarity of crystal on, 33
geometrical, 32
in diffraction by fibres, 585
influence of on intensity function, 514
influence of on line-breadth, 545
in terms of fictitious dielectric constant, 415
Sulphur, rhombic, radial distribution in, 506
Summation rule, Thomas-Reiche-Kuhn, 143
Superlattice structure,
diffuse lines due to, 556, 560
in alloys, 556
Sylvine,
Debye factor for, 233
diffuse reflections from, 242, 261
integrated reflection from, 302
zero-point energy in, 302

TEMPERATURE,
characteristic (*see* Characteristic temperature)
relation of root-mean-square displacement to, 199
Temperature factor (*see* Debye-Waller temperature factor)
Thermal vibrations, amplitude of, 236
Thomas-Fermi charge-distribution, 123
calculation of incoherent scattering from, 462
Thomson scattering formula, 29
Total reflection,
at critical angle, 171
effect of absorption on sharpness of, 172
from nickel, 174
measurement of refractive index by, 171
in dynamical theory, 58, 87, 425, 430

Transmission coefficient for perfect crystal, 435

UMWEGANREGUNG, 26, 339
Unsymmetrical reflection,
  dynamical theory of, 88, 428
  measurement of refractive index by, 169

VECTOR formulae, summary of, App. I
Vitreous silica, structure of, 508
Vitreous solids,
  radial distribution in, 509
  scattering by, 501, 508

WATER, structure of, 497
Wave-equation,
  derivation of for scattering problems, 103
  numerical solution of, 117
Wave-field in dynamical theory, 73, 416
  complementary, in transmission problems, 437
  details of for crystal with small absorption, 429
  equations of in crystal, 416
  for two waves, 76, 427, 431
  in absorbing crystal, 428
  intensity of in reflection problem. 431
  relation of to dispersion equation, 419
Wave-function, atomic,
  approximate form of for many-electron atom, 114
  calculation of by self-consistent fields, 117, 121
  calculation of oscillator strengths from, 157
  calculation of perturbed, 105
  interpretation of, 102
Wave-point, 73, 75
  determination of from boundary conditions, 83, 422, 425
Wentzel-Waller theory of scattering, 103, 110
Widths of diffraction maxima,
  dependence of on crystal size (*see* Line-breadth)
  from calcite crystals, 306, 332
  from diamond, 305
  in dynamical theory, 58, 64, 426

X-RAY analysis, parallel with microscopic vision, 385, 390, 396
X-rays,
  anomalous dispersion of (*see* Anomalous dispersion)
  interference of in thin films, 175
  refractive index of matter for, 53, 75, 79, 167, 171, 177
  total reflection of (*see* Total reflection)

ZERO-POINT energy,
  allowance for in temperature factor, 219
  contribution of to Debye factor, 223
  for aluminium, 303
  for diamond, 225
  for rock salt, 236, 300, 302
  for sylvine, 236, 302
Zincblende, effect of polarity on reflection from, 33, 182, 429

INDEX OF AUTHORS

ABBE, E., 390, 391, 555
Allison, S. K., 64, 306, 307, 322, 330
Amaldi, E., 497
Andrade, E. N. da C., 455
Andress, K., 572
Armstrong, A. H., 182
Astbury, W. T., 572, 578, 585

BACKHURST, I., 232
Baltzer, O. J., 240, 246, 258
Barrett, C. S., 405, 472
Baxter, A., 128, 187, 188, 190, 339
Bearden, J. A., 49, 160, 177, 178, 180, 304, 333, 409
Beevers, C. A., 356, 357, 371
Berg, O., 25
Bernal, J. D., 231, 500, 584
Bethe, H., 143, 456
Bewilogua, L., 125, 462, 486, 487
Bilinsky, S., 220, 225, 234, 235
Black, M., 161
Blackman, M., 225
Bloch, F., 491
Boersch, H., 457
Bohlin, H., 335
Booth, A. D., 356
Born, M., 193, 194, 195, 225, 226, 240, 257, 258
Borrmann, G., 439, 446, 447, 448, 632, 641, 652, 653
Bothe, W., 146
Bosanquet, C. H., 34, 50, 101, 270, 287, 295, 299, 306, 327, 332, 333, 409
Bradley, A. J., 48, 184, 189, 339, 569
Bragg, W. H., 50, 93, 232, 241, 270, 280, 299, 305, 318, 321, 335, 343, 370
Bragg, W. L., 6, 19, 34, 50, 54, 91, 93, 101, 130, 270, 287, 294, 295, 299, 305, 306, 327, 332, 333, 343, 352, 355, 371, 386, 401, 403, 409, 547, 631
Breit, G., 462
Brentano, J. C. M., 185, 187, 188, 190, 335, 336, 337, 338, 339
Brill, R., 302, 322, 541
Brillouin, L., 194
Brindley, G. W., 61, 130, 184, 185, 186, 187, 190, 228, 233, 238, 257, 262, 263, 281, 286, 302, 303, 320, 334, 338, 485
Bruce, W. A., 264
Burwell, J. T., 507
Bush, V., 124

CALDWELL, S. H., 124
Cameron, G. H., 541
Chao, S. H., 568
Charlesby, A., 516
Claasen, A., 339
Clark, G. L., 239, 438
Claus, W. D., 264
Cochran, W., 620, 624
Collins, E. H., 232
Compton, A. H., 34, 93, 99, 101, 171, 259, 287, 305, 306, 332, 343, 370, 405, 408, 409, 462, 473
Cooper, E. R., 26
Cork, J. M., 370
Coster, D., 33, 154, 182
Cox, E. G., 380, 382
Crick, F. H. C., 624
Crowther, J. A., 261

DANIEL, V., 566, 568, 569
Darbyshire, J. A., 26
Darwin, C. G., 34, 36, 50, 52, 55, 57, 59, 60, 62, 66, 93, 168, 270, 272, 275, 282, 304, 305, 323, 328, 332
Davis, B., 90, 169, 170, 180, 306, 322
Debye, P., 21, 22, 47, 93, 99, 193, 215, 217, 218, 219, 231, 259, 262, 339, 343, 458, 459, 460, 470, 477, 478, 480, 481, 485, 487, 494
de Smedt, J., 460
Dirac, P. A. M., 102, 261, 463
Duane, W., 168, 239, 343, 344, 438
Dyer, H. B., 620

EDWARDS, O., 556, 558
Ehrenberg, W., 232, 305, 306, 318, 319
Ehrenfest, P. S., 343, 344, 459
Ehrhardt, F., 487
Epstein, P., 343, 344
Ewald, P. P., 8, 9, 14, 44, 52, 53, 55, 66, **67**, 68, 70, 73, 74, 76, 78, 83, 84, 87, 89, **90**, 168, 169, 170, 232, 304, 305, 318, **319**, 321, 343, 348, 413, 424, 516

FAXÉN, H., 25, 193, 195, 239
Fermi, E., 102, 123, 129, 130, 462
Field, J. E., 171
Finch, G. I., 516
Firth, E. M., 61, 233, 236, 274, 293, 299, 326
Fock, V., 121, 129
Forster, R., 173
Fowler, R. H., 500
Fox, J. J., 240, 252
Freeman, N. L., 332
Friedel, G., 33
Friedrich, W., 14, 239, 459
Fukushima, E., 299

GAJEWSKI, H., 485
Gibbs, W., 8
Gingrich, N. S., 240, 343, 372, 478, **504**, **506**
Glocker, R., 155, 185, 189
Goodwin, T. H., 380, 382
Gottfried, C., 371
Gregg, R. Q., 240

Grimm, H. G., 302, 322

Hall, J. H., 240
Harker, D., 378, 380
Hartree, D. R., 101, 116, 118, 119, 121, 122, 125, 129, 161, 184, 185, 263, 300, 302, 326, 401, 405, 461
Hartree, W., 121, 122, 185
Harvey, G. G., 239, 259, 261, 262, 263
Hatley, C. C., 170, 171
Havighurst, R. J., 182, 238, 336, 337, 343, 350, 409
Heisenberg, W., 103, 125, 143, 462
Heitler, W., 104
Hendricks, S. B., 230, 231, 361
Hermann, C., 302, 322
Herzog, G., 405, 472, 572
Hildebrandt, G., 653
Hönl, H., 135, 147, 157, 158, 161, 163, 171, 179, 186, 189, 321, 326
Hope, R. A. H., 184, 189
Horrocks, H., 359
Houston, R. A., 139, 174
Hull, A. W., 459
Hultgren, R., 478
Hylleraas, E. A., 158

Jahn, H., 210, 213, 251, 257
James, R. W., 34, 50, 61, 95, 99, 101, 130, 184, 232, 233, 236, 238, 257, 262, 263, 270, 274, 281, 286, 287, 293, 295, 297, 299, 300, 303, 305, 326, 332, 334, 359, 401, 409, 485, 486, 489, 493, 571
Jancke, W., 572
Jauncey, G. E. M., 101, 239, 240, 246, 258, 259, 260, 261, 263, 264
Jeffrey, C. A., 384
Jensen, H., 131
Jentzsch, F., 173
Jones, F. W., 541, 543, 544, 556, 559
Jönsson, E., 160

Kallmann, H., 146
Kármán, T. v., 193, 195, 225, 226
Kato, N., 651, 653
Katz, J. R., 572
Katzoff, S., 497
Keesom, W. H., 460
Kiessig, H., 174, 177
Kikuchi, S., 455
King, G., 359
Kirchner, F., 555
Kirkpatrick, P., 240
Klein, O., 463
Knaggs, I., 241
Knipping, P., 14
Knol, K. S., 33, 154, 182
Knott, G., 516
Kochendörfer, A., 565
Kohler, M., 87, 413, 427, 428, 429
Kossel, W., 26, 439, 446, 449, 451, 454
Kracek, F. C., 230
Kramers, H. A., 103, 143, 147, 158, 174
Kronig, R. de L., 146, 147, 158, 174
Krutter, H., 504, 508
Kuhn, W., 143
Kulenkampff .H., 295, 306

Ladenburg, R., 139
Lameris, A. J., 176
Larsson, A., 177, 179
Lassen, H., 555
Laue, M. v., 1, 8, 9, 14, 21, 83, 93, 306, 312, 413, 414, 429, 437, 439, 442, 455, 456, 528, 535, 537, 540, 550, 555, 632, 647, 653
Laval, J., 195, 239, 243, 244, 247, 253, 254, 255, 256, 257, 258, 264
Lenz, W., 131
Lindsay, G. A., 171
Lipson, H., 356, 357, 556, 558, 566, 568, 569, 631
Llewellyn, F. J., 380, 382
Loeck, V., 439, 446
Lonsdale, K., 240, 241, 242, 243, 247, 249, 250, 251, 252, 257, 258, 264, 439, 452, 454
Lorentz, H. A., 136

McNatt, E. M., 264
Mark, H., 146, 181, 232, 305, 306, 318, 319, 493, 572
Martin, A. E., 240, 252
May, H. L., 261
Menke, H., 343, 460, 477, 478, 494, 496, 497
Meyer, H. H., 497
Morgan, J., 497, 500
Morningstar, O., 504, 508
Moseley, H. G. J., 270
Müller, A., 231, 249, 452

Nagy, B. v. Sz., 132
Nilakantan, P., 240
Nishikawa, S., 298
Nishina, Y., 463

Ott, H., 22, 194

Parker, H. M., 403
Parratt, L. G., 322, 330, 331, 332
Patterson, A. L., 343, 371, 376, 380, 529, 541
Patterson, R. A., 168
Pauli, W., 115, 117
Pauling, L., 123, 130, 229, 405
Peierls, R., 205
Pelzer, H., 541
Pennell, F., 263
Peters, C., 302, 322
Petrashen, M., 121
Pisharoty, P. R., 253
Planck, M., 219
Polanyi, M., 572
Porter, A. B., 392
Posnjak, E., 230
Preston, G. D., 240, 241, 243, 247, 258
Prins, J. A., 33, 55, 57, 62, 146, 152, 154, 155, 172, 173, 174, 176, 182, 328, 343, 460, 477, 478
Purks, H., 306

RAMAN, C. V., 99, 240, 252, 259, 460, 462
Ramanathan, K. R., 460
Randall, J. T., 95, 461, 572
Reiche, F., 143
Renninger, M., 26, 64, 320, 322, 325, 326, 329, 339, 446
Richter, H., 485
Ridley, P., 187, 302, 339
Riewe, H., 553
Robertson, J. M., 250, 343, 355, 356, 363, 364, 366, 385
Robertson, R., 240, 252
Rodgers, J. W., 184, 185
Rusterholz, A., 166, 185, 186, 187, 190
Rutherford, E., 455

SAKISAKA, Y., 295, 296, 298
Sarginson, K., 194
Saunder, D. H., 569, 571
Schaefer, C., 104, 168
Schäfer, K., 155, 185, 189, 337,
Scherrer, W., 47, 93, 459, 472, 528, 529, 539
Schrödinger, E., 102, 103, 143, 463
Schwarzschild, M., 306
Seemann, H., 335, 452
Selenyi, P., 439
Seljakov, N., 529
Shaw, C. H., 180
Sherman, J., 123, 130, 405
Shonka, J. J., 238, 263
Siegbahn, M., 55, 169, 177, 180
Siegel, S., 239
Simon, F., 220
Slack, C. M., 180
Slater, J. C., 121
Smith, H., 240, 241, 242, 243, 247, 249, 251, 252
Smith, L. P., 306
Sommerfeld, A., 66
Spencer, R. C., 306
Spiers, F. W., 185, 186, 190, 338
Stäger, A., 472
Stauss, H. E., 177
Stempel, W., 306, 322
Stenström, W., 55, 90, 168
Stewart, G. W., 460, 481, 497
Stokes, A. R., 529, 531, 535, 537, 557, 560, 561
Straubel, R., 555
Süsich, G. v., 306
Sugiura, Y., 144, 157, 158
Sumoto, I., 298
Sykes, C., 556, 559
Szilard, L., 181

TARASOV, L. P., 479
Taylor, W. H., 568
Teague, D. S., 240
Terrill, H. M., 169
Thomas, W., 123, 124, 143, 405, 462
Thomson, J. J., 29
Trumpy, B., 491

VAND, V., 624, 628
Van der Grinten, W., 486, 488
van Niekerk, J. N., 366
Voigt, W., 213, 238
Voges, H., 439, 445, 446, 447, 449, 457
von Nardroff, R., 170

WADLUND, S. P. R., 239
Wagner, E., 295, 306
Wagner, H., 632, 653
Waller, I., 22, 25, 94, 103, 110, 116, 143, 177, 193, 195, 225, 236, 238, 262, 263, 299, 300, 302, 326, 401, 463
Warren, B. E., 343, 372, 461, 478, 479, 500, 504, 506, 507, 508, 509, 510
Wentzel, G., 94, 103, 110, 111
West, J., 130, 294, 361, 371, 401, 403
Wheeler, J. A., 160
Whitehouse, W. J., 403
Wierl, R., 493
Williams, E. J., 101, 146, 151, 159, 160
Williams, J. H., 307
Williams, P. S., 263
Wilman, H., 516
Wilson, A. J. C., 529, 531, 535, 537, 557, 558, 559, 560, 561, 630
Wollan, E. O., 405, 408, 472
Woo, Y. H., 99, 239, 260
Wood, R. G., 184, 233, 238, 281, 286, 303, 334
Woodward, I., 250
Wyckoff, R. G. W., 182, 184, 189

ZACHARIASEN, W., 239, 240, 497, 510, 632
Zener, C., 219, 220, 225, 228, 234, 235
Zernicke, F., 343, 460,477, 478